A Companion to Environmental Geography

Wiley Blackwell Companions to Geography

Wiley Blackwell Companions to Geography is a blue-chip, comprehensive series covering each major subdiscipline of human geography in detail. Edited and contributed by the disciplines' leading authorities each book provides the most up to date and authoritative syntheses available in its field. The overviews provided in each *Companion* will be an indispensable introduction to the field for students of all levels, while the cutting-edge, critical direction will engage students, teachers, and practitioners alike.

Published

A Companion to Feminist Geography
Edited by Lise Nelson and Joni Seager

A Companion to Environmental Geography
Edited by Noel Castree, David Demeritt, Diana Liverman, and Bruce Rhoads

A Companion to Health and Medical Geography
Edited by Tim Brown, Sara McLafferty, and Graham Moon

A Companion to Social Geography
Edited by Vincent J. Del Casino Jr., Mary Thomas, Ruth Panelli, and Paul Cloke

The Wiley-Blackwell Companion to Human Geography
Edited by John A. Agnew and James S. Duncan

The Wiley-Blackwell Companion to Economic Geography
Edited by Trevor J. Barnes, Jamie Peck, and Eric Sheppard

The Wiley-Blackwell Companion to Cultural Geography
Edited by Nuala C. Johnson, Richard H. Schein, and Jamie Winders

The Wiley Blackwell Companion to Tourism
Edited by Alan A. Lew, C. Michael Hall, and Allan M. Williams

The Wiley Blackwell Companion to Political Geography
Edited by John Agnew, Virginie Mamadouh, Anna J. Secor, and Joanne Sharp

The New Blackwell Companion to the City
Edited by Gary Bridge and Sophie Watson

The Blackwell Companion to Globalization
Edited by George Ritzer

The Handbook of Geographic Information Science
Edited by John Wilson and Stewart Fotheringham

A Companion to Environmental Geography

Edited by

Noel Castree, David Demeritt,
Diana Liverman & Bruce Rhoads

WILEY Blackwell

This edition first published 2016
© 2009 John Wiley & Sons, Ltd.

Edition history: Blackwell Publishing Ltd (hardback, 2009)

Registered Office
John Wiley & Sons, Ltd, The Atrium, Southern Gate, Chichester, West Sussex, PO19 8SQ, UK

Editorial Offices
350 Main Street, Malden, MA 02148-5020, USA
9600 Garsington Road, Oxford, OX4 2DQ, UK
The Atrium, Southern Gate, Chichester, West Sussex, PO19 8SQ, UK

For details of our global editorial offices, for customer services, and for information about how
to apply for permission to reuse the copyright material in this book please see our website at
www.wiley.com/wiley-blackwell.

The right of Noel Castree, David Demeritt, Diana Liverman and Bruce Rhoads to be identified as the
authors of the editorial material in this work has been asserted in accordance with the UK Copyright,
Designs and Patents Act 1988.

Library of Congress Cataloging-in-Publication Data

A companion to environmental geography / edited by Noel Castree … [et al.].
    p. cm. – (Blackwell companions to geography)
  Includes bibliographical references and index.
  ISBN 978-1-4051-5622-6 (hardcover : alk. paper) ISBN 978-1-119-25062-3 (papercover)
1. Environmental geography. I. Castree, Noel, 1968–
  G143.C66 2009
  333.709–dc22

                                        2008040245

A catalogue record for this book is available from the British Library.

Cover image: Antony Gormley, *Time Horizon*, 2006, cast iron, 100 elements, 189 × 53 × 29 cm.
Parco Archeologico di Scolacium, Roccelletta di Borgia, Italy. Photo © Remo Casilli/ Reuters.
Reproduced courtesy of the artist.

Set in 10/12pt Sabon by SPi Global, Pondicherry, India

Contents

Acknowledgements

We would like to thank the contributors for writing and revising their chapters in such a timely fashion. At Wiley-Blackwell, Justin Vaughn secured the contract and Ben Thatcher was a terrific point-person throughout. We extend our thanks to both of them. We also thank Maria Escobar for creating the index and Peter Howard for helping to convert a number of figures from colour to black and white.

Every effort has been made to trace copyright holders and to obtain their permission for use of copyright material. The publisher apologises for any errors or omissions and would be grateful if notified of any corrections that should be incorporated in future reprints or editions of the book.

Finally, NC would like to dedicate this big book to a very small person: Felicity Marie Brasset. DL would like to dedicate this book to her parents, Peggy and John Liverman.

Contributors

Neil Adger is Professor of Environmental Economics in the School of Environmental Sciences at the University of East Anglia, Norwich, UK. He has led the research programme on adaptation in the Tyndall Centre for Climate Change Research since its inception in 2000 and is a member of the Resilience Alliance. His latest book is *Governing Sustainability* (Cambridge University Press).

Karen Bakker is an Associate Professor in the Department of Geography, University of British Columbia, Vancouver, Canada. Her research interests span political economy, political ecology and environmental policy. She has published in the *Annals of the Association of American Geographers, Antipode, Economic Geography, Political Geography, Transactions of the Institute of British Geographers,* and *World Development.* Her books include *An Uncooperative Commodity: Privatizing Water in England and Wales* (2004, Oxford University Press) and the forthcoming *Beyond Privatization: Environment, Development, and the Commons* (Cornell University Press).

Heiko Balzter is Professor in Physical Geography at the University of Leicester (UK) and Director of the Centre for Environmental Research. He obtained his PhD in 1998 from Justus-Liebig-University in Giessen, Germany. His current research interests include the fire regimes in Africa and Siberia, carbon cycle research and biosphere/climate interactions, SAR and LiDAR remote sensing of forest structure and dynamics, land surface and climate impacts modelling. His key publications so far have been published in *IEEE Transactions in Geoscience and Remote Sensing; Journal of Climate; Remote Sensing of Environment; International Journal of Remote Sensing; Water, Air and Soil Pollution; Geophysical Research Letters; Canadian Journal of Remote Sensing; Forest Ecology; and Management and Ecological Modelling.*

Bruce Braun teaches environmental and political geography at the University of Minnesota. His work focuses on the conceptualisation of nature–society relationships in social and political thought, technicity and human life, and the intersections of biopolitics, biocapital and biosecurity.

Gavin Bridge is a Reader in Economic Geography in the School of Environment and Development at the University of Manchester. His research centres on the political economy of extractive industries, and examines the new geographies of resource access and environmental governance associated with the oil, gas and mining sectors. Publications include essays in the *Annual Review of Environment and Resources, Journal of Economic Geography, Antipode, Geoforum, Environment and Planning A, Progress in Human Geography* and *Professional Geographer.*

James D. Brown works for the National Oceanic and Atmospheric Administration, USA. His work is broadly concerned with locating, quantifying and reducing uncertainty in hydrologic models. He has held university positions in the UK, the Netherlands and USA.

Katrina Brown is Professor in the School of Development Studies at the University of East Anglia, Norwich, UK. She is an environmental social scientist and Deputy Director of the Tyndall Centre for Climate Change Research. Since graduating with a PhD examining women's collective action and coping strategies in semi-arid Kenya, she has been involved in research on environmental change including land use change, climate change and biodiversity conservation. She is author of *Making Waves: Integrating Coastal Conservation and Development* (Earthscan).

Noel Castree is Professor of Geography at Manchester University, England, and the University of Wollongong, Australia. Editor of *Social Nature* (2001) and author of *Making Sense of Nature* (2013), his current research focuses on how people and Earth are represented by expert communities cross the disciplines.

Jason Chilvers is a Lecturer in Environment and Society at the School of Geography, Earth and Environmental Sciences, University of Birmingham. His research interests centre on relations between environment, science, policy and society – spanning the governance of, public understanding of, and public engagement in environmental risk, science and technology. He has published widely in peer-reviewed international journals, including *Environment and Planning A, Science, Technology, and Human Values* and *Journal of Risk Research*, and book chapters in edited volumes such as *Science and Citizens: Globalization and the Challenge of Engagement* (edited by Melissa Leach, Ian Scoones and Brian Wynne).

Sarah L. Damery is a Lecturer in Geography at the University of Birmingham, UK. Her work is concerned with environment–society relations in three main areas: the management and communication of risk, particularly flood risk; public participation in environmental policy making; and scientific inputs to environmental policy making.

David Demeritt is a Professor in the Department of Geography at King's College London. He is especially interested in the politics of environmental science and the role of scientific knowledge in the governance of society–environment relations.

Georgina Endfield is a Reader in Environmental History in the School of Geography at the University of Nottingham. Her research focuses on climate and environmental history, with particular interests in colonial Mexico and nineteenth-century Africa.

Robert A. Francis is a Lecturer in Ecology at the Department of Geography, King's College London. His interdisciplinary research focuses mainly on plant ecology, landscape ecology, biodiversity and invasive species within freshwater and urban ecosystems. He has published in a range of ecological and geographical journals such as *Forest Ecology and Management, Earth Surface Processes and Landforms, Area, Restoration Ecology* and *Wetlands Ecology and Management.*

George Henderson teaches and writes about the political economy of American capitalism, the relationship between capitalism and agriculture, and geographic thought. He is the author of *California and the Fictions of Capital* (Temple University Press, 2003) and editor (with Mary Waterstone) of *Geographic Thought: A Praxis Perspective* (Routledge, 2008). He is Associate Professor of Geography at the University of Minnesota.

Scott Jiusto of Worcester Polytechnic Institute has research interests in environmental policy and philosophy, particularly energy policy and the pursuit of sustainability at sub-national scales. Research themes include carbon emissions analysis and the politics of energy across geographic scales, electrical power sector restructuring, sustainable community development in South Africa, and experiential learning and sustainability in higher education. Recent publications appear in *Energy Policy* and the *Journal of Engineering Education.*

Owain Jones is a Research Fellow at the Countryside & Community Research Institute, England. He has conducted research on a range of 'nature-society' topics, including trees and place; children, nature and the city; and linkages between biodiversity and food chains. His publications include *Tree Cultures: The Place of Trees, and Trees in their Place* (2002, with Paul Cloke).

Hilda Kurtz is an Associate Professor of Geography at the University of Georgia. Her previous research explores the geography of social movements, both in the social construction of movement grievances and the geographic constitution of movement strategies and outcomes. Her current research considers the role of women and the construction of citizenship claims within the movement for environmental justice.

Richard Le Heron is Professor of Geography, School of Geography, Geology and Environmental Science, University of Auckland, Auckland, New Zealand. His research interests include agri-food developments in the globalising economy, nature-society relations, geographies of accumulation, new economic spaces under neo-liberalism, trans-disciplinarity and knowledge production and conditions for co-learning. He has published in *Geoforum* and *Journal of Rural Studies* and has recently co-edited *Agri-food Commodity Chains and Globalising Networks* (Ashgate, 2008).

Diana Liverman is Co-Director of the Institute of the Environment and Regents Professor of Geography and Development at the University of Arizona. She has published widely on environmental change and policy.

Becky Mansfield is an Associate Professor of Geography at The Ohio State University. Her research interests are in nature–society relations, political economy, agro-food studies, and health geographies. She has published on these topics in a variety of journals, including the *Annals of the Association of American Geographers, Progress in Human Geography* and *Antipode*, and she is the editor of *Privatization: Property and the Re-Making of Nature-Society Relations* (2008, Blackwell).

Steven M. Manson is in the Department of Geography at the University of Minnesota. He combines environmental research, social science approaches and geographic information science to understand global environmental change, environmental decision making and complex human–environment systems. His work has appeared in the *Annals of the Association of American Geographers*, *Geoforum, Environment and Planning, International Journal of Geographic Information Science* and *Proceedings of the National Academy of Sciences*.

James McCarthy is an Associate Professor in the Department of Geography at the Pennsylvania State University. His research centres on political ecology and political economy, with particular interests in neoliberalism, environmental politics and social movements. His research has appeared in the *Annals of the Association of the Association of American Geographers, Antipode, Environment and Planning A, Geoforum, Progress in Human Geography* and other interdisciplinary journals.

Tom Mels is a Lecturer in Human Geography at the University of Kalmar, Sweden. His main research interest is in historical geographies of social power, with a focus on the politics of landscape and nature. He has recently published two books, as well as papers in *Cultural Geographies, Geografiska Annaler B, Journal of Historical Geography* and *Society and Space*.

Arthur P. J. Mol is Professor and Chair in Environmental Policy at Wageningen University, the Netherlands. His research fields are in social theory, environmental governance, globalisation and environment, information and environment, and marine policy. His latest books are *Environmental Reform in the Information Age: The Contours of Informational Governance* (2008, Cambridge UP) and *Environmental Governance in China* (edited with Neil Carter, 2007, Routledge).

Daanish Mustafa is in the Department of Geography, Environment Politics and Development Group, King's College, London. His research interests include development geography, water resources, hazards geography and geography of violence and terror. He has published in journals such as *Annals of the Association of American Geographers, Antipode, Professional Geographer, Environment and Planning D: Society and Space* and *World Development*.

Professor Kenneth R. Olwig is in the Department of Landscape Architecture, Planning and Heritage at the Swedish University of Agricultural Sciences, SLU-Alnarp. His major research interests are in geographical theory, landscape philology, æsthetics, law and the concept of nature. He is author of *Landscape, Nature and the Body Politic* (Madison: University of Wisconsin Press, 2002) and editor (with Don Mitchell) of *Justice, Power and the Political Landscape* (London: Routledge, 2008–9) and (with Michael Jones) of *Nordic Landscapes: Region and Belonging on the Northern Edge of Europe* (Minneapolis: University of Minnesota Press, 2008).

Marianna Pavlovskaya is an Associate Professor at Hunter College, City University of New York. Her research focuses on the constitution of class and gender in post-socialist Moscow and New York City, rethinking neo-liberal transition in Russia, and critically re-reading geo-spatial technologies. Her work has been published in *Environment and Planning A, Annals of the Association of American Geographers*, and *Gender, Place, and Culture*. For a complete list of publications and author preprints go to http://www.geo.hunter.cuny.edu.

Tom Perreault is Associate Professor of Geography, Syracuse University. His research focuses on the political ecologies of resource use, rural development and indigenous peoples' social movements in Bolivia and Ecuador. His work has appeared in the *Annals of the Association of American Geographers, Antipode, Environment and Planning A, Environment and Planning D: Society and Space* and *Political Geography*, and has been published (in Spanish) by Abya Yala (Quito), Plural (La Paz) and the Instituto de Estudios Peruanos (Lima).

George Perry is a Biogeographer and Senior Lecturer in the School of Geography, Geology and Environmental Science at The University of Auckland, New Zealand. His research and teaching interests span plant ecology, spatial analysis and environmental modelling, combining empirical fieldwork with various forms of statistical and simulation modeling.

Scott Prudham is an Associate Professor in the Department of Geography and the Centre for Environment at the University of Toronto. He is the author of *Knock on Wood: Nature as Commodity in Douglas-fir Country*, co-editor of *Neoliberal Environments: False Promises and Unnatural Consequences*, and an editor of the journal *Geoforum*. His research is broadly concerned with the intersection of environmental politics, environmental change and the political economy of capitalism.

Bruce Rhoads is a Professor and Chair of the Department of Geography at the University of Illinois, Urbana-Champaign. His research interests include the fluvial dynamics of stream confluences and river meanders; stream naturalisation and restoration; the connection between geomorphological conditions and physical habitat for fish communities; human impacts on river systems; and philosophical and conceptual issues in geomorphology and physical geography.

Nathan Sayre is Assistant Professor of Geography at the University of California-Berkeley. He studies the political, ecological and economic interactions of people and landscapes in semi-arid rangelands, with a focus on the borderlands of north-western Mexico and the southwestern US. His publications include *Ranching, Endangered Species, and Urbanization in the Southwest* (University of Arizona Press), *Working Wilderness* (Rio Nuevo Publishers) and *The New Ranch Handbook* (QuiviraCoalition).

Kevin St. Martin is an Assistant Professor at Rutgers, The State University of New Jersey. His research focuses on the discourse and practice of fisheries science and its implications for both resource management and community-based economic development. His work has been published in *Antipode, Environment and Planning A, Annals of the Association of American Geographers*, as well as other journals and edited volumes. Author preprints of his articles can be found at http://geography.rutgers.edu.

Karen E. Smoyer-Tomic is a Senior Research Analyst for HealthCore, as well as Research Fellow and Adjunct Associate Professor in Geography at the University of Delaware and Adjunct Associate Professor in Public Health Sciences at the University of Alberta. Her expertise is in the physical, built and social environmental determinants of health, focusing on climate and air pollution impacts on human health, spatial accessibility to health resources and the role of urban sprawl on environment and population health.

Gert Spaargaren is Professor of Environmental Policy for Sustainable Lifestyles and Consumption at the Environmental Policy Group of Wageningen University, The Netherlands. His main research interests and publications are in the field of environmental sociology, sustainable consumption and behavior, and the globalisation of environmental reform.

James Tansey is an Associate Professor in the Sauder School of Business at the University of British Columbia and an Associate Fellow at the James Martin Institute, University of Oxford. His major research interested are in integrated assessment, science and technology studies, social innovation. His work has appeared in the journals *Global Environmental Change, Climatic Change* and *Public Understanding of Science*, among others.

B. L. Turner II is the Gilbert F. White Professor of Environment and Society, School of Geographical Sciences, and School of Sustainability, Arizona State University. His research interests are human-environment relationships, ranging from sustainability and land change science to political and cultural ecology. He is co-author of *Cultivated Landscapes of Native Middle America* (Oxford: Oxford University Press 2001) and *Global Change and the Earth System* (Berlin: Springer-Verlag, 2004), and co-editor of *The Earth as Transformed by Human Action* (Cambridge: Cambridge University Press 1990) and *Land Change Science* (New York: Kluwer, 2004), among others.

Matthew D. Turner is a Professor of Geography at the University of Wisconsin-Madison. He works in West Africa with research interests including political ecology, science studies, resource-related conflict, savanna biogeography, property theory and the politics of conservation.

John Wainwright is in the Department of Geography at the University of Sheffield, where he is also a member of the Sheffield Centre for International Drylands Research and the Catchment Science Centre. His research focuses on human–environment interactions in drylands, especially from land-degradation and geo-archaeological perspectives, processes of runoff and erosion, landscape evolution and theoretical geomorphology; he may (or may not) be an 'earth-system scientist'. He co-wrote *Environmental Issues in the Mediterranean* and co-edited *Environmental Modelling: Finding Simplicity in Complexity, Landscape Evolution: Temporal and Spatial Scales of Denudation, Climate and Tectonics* and *Le Laouret et la Montagne d'Alaric à la Fin de l'Âge du Bronze*, as well as authoring a number of articles in journals such as *Earth Surface Processes and Landforms, Water Resources Research, Ecohydrology, Geomorphology* and *Catena*.

Jamie Woodward holds a Chair in Physical Geography at The University of Manchester where he leads the Quaternary Environments and Geoarchaeology research group in the School of Environment and Development. He is a geomorphologist with particular interests in Quaternary environmental change, landscape dynamics and human-environment interactions in the Mediterranean region and in the Nile Valley. A good deal of his research takes place in collaboration with archaeologists and he is the Co-Editor of *Geoarchaeology: An International Journal* as well as being the Quaternary Science and Geomorphology Editor for the *Journal of the Geological Society of London*. He has just completed a major project editing *The Physical Geography of the Mediterranean* for Oxford University Press (2009).

Karl S. Zimmerer is Professor in Human-Environment Geography and Head of the Geography Department at the Pennsylvania State University where he leads the research group on biodiversity, resource management, and land use in environmental and development change. His work is situated in the environmental sciences and political and human-cultural ecology, with particular interests in agrobiodiversity and agricultural systems, water and soil resources, environmental conservation, climate change, and the development and use of geographic models and analysis in planning and policy. He has written numerous journal articles and several books and monographs, most recently *Globalization and New Geographies of Conservation* (2006, University of Chicago Press). He is appointed as the Editor of the Nature-Society section of the *Annals of the Association of American Geographers* and is active in various organisations involved with sustainability, social-environmental change, and food security.

Chapter 1

Introduction: Making Sense of Environmental Geography

Noel Castree, David Demeritt and Diana Liverman

On the evening of Monday, 31 January 1887, Halford Mackinder delivered a now famous address to London's Royal Geographical Society. In his lecture – entitled 'On the scope and methods of geography' – he explained how and why geography should take its place alongside other disciplines within the academic division of labour. His strategy, at once simple and audacious, was to call that division of labour into question. Geography, Mackinder (1887) argued, can 'bridge one of the greatest of all gaps': namely, that separating 'the natural sciences and the study of humanity' (p. 145). He was not alone in defining geography as 'the science whose main function is to trace the interaction of man [*sic*.] in society and so much of his environment as varies locally'. At points east and west, others were doing much the same, such as William Morris Davis in America and Friedrich Ratzel in Germany. The three men soon occupied important university positions and were followed by similarly vigorous prosleytisers who quickly built on the foundations their forebears had laid.

So began geography's career as a university subject and what historian of geographical thought David Livingstone (1992, p. 177) called 'the geographical experiment'. A century on that experiment continues. Although space and region have since joined human-environment relations as central organising concepts for the discipline, many still see geography as the 'original integrated environmental science' (Marston, 2006). Geography remains one of the few disciplines committed to bridging the divide between the natural and physical sciences, on the one side, and the social sciences and humanities on the other. Quite how successful that bridging has been is a matter of some debate (see, for example, Matthews and Herbert's [2004] book *Unifying geography*). Despite the hopes invested by Turner (2002) and others (e.g., Marston, 2006; Zimmerer, 2007) in human–environment relations as the unifying link holding the discipline together, many geographers prefer to study other things. There is no shortage of 'pure' human and physical geographers. Even so, the scale and richness of geographers' attempts to understand the entanglements of people and the non-human world are highly impressive. These many geographers, their findings and their ideas are what we are calling here 'environmental geography'

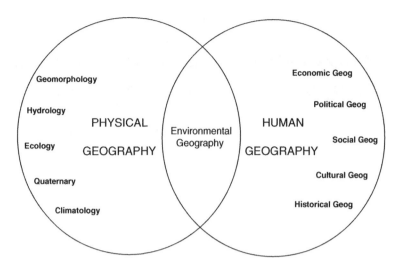

Figure 1.1 Environmental geography as disciplinary 'middle ground'.

(or what has sometimes also been called the 'human-environment' or 'man-land' traditions of geography'). By whatever name, environmental geography occupies the fertile 'borderlands' where geography's various traditions of scholarship – not only human and physical, but also regional and GIS – come together and connect with each other and with cognate traditions of environmental work outside geography (figure 1.1).

Though the term is perhaps less familiar than are 'human' and 'physical' geography, environmental geography deserves greater recognition both within and beyond the discipline. As this *Companion* is designed to show, environmental geography is much more than simply the residual intersection of geography's two halves. Environmental geography is a large, diverse and vibrant field of knowledge with few, if any, equivalents elsewhere in the conventional academic division of labour. The 32 chapters of this book will, we hope, offer readers both an incisive and accessible introduction to this field and set the agenda for its future development.

What makes this book distinctive is its catholic vision for environmental geography. There are now myriad texts focusing on human or physical geography respectively or some subfield thereof, including several previous *Companions* (see, for instance, Agnew et al., 2001). There are also now numerous volumes focusing on some specific approach to, or branch of, the study of human–environment relations, such as 'political ecology' (see, for instance, Robbins, 2004) or 'hazards geography' (see, for instance, Pelling, 2003). What is long overdue is a book that demonstrates the size, breadth and multiplicity of geographical work at the people–environment interface. In short, the *Companion* casts its net far wider than most recent texts about one or other subfield of geography has been prepared to do. As a result, the book is not beholden to the now conventional view – among many geographers at least – that geography comprises two 'halves' and only a vanishing centre.

The volume has four parts: 'Concepts', 'Approaches', 'Practices' and 'Topics'. They comprise epistemic 'cuts' into the body of environmental geography, four ways of organising a wide-ranging set of contributions. In each case, authors were asked to address some specific issue or aspect of this broader terrain. Consequently, each chapter can be read alone and in no particular order since their authors were not instructed to formally situate their 'part' within a wider 'whole'. As even a quick glance at the chapter titles reveals, these parts together cover an enormous range of material and perspectives. We trust that this will make the *Companion* a lively, interesting and synoptic account of the field. Depending on your background and predilections, there will be material in this book that is (variously) familiar, surprising, challenging and even unsettling. Specialists will find insightful discussions of the 'state of the art' in specific conceptual, methodological and topical areas. Teachers should find the chapters to be useful pedagogical resources, while for students of geography and related fields, it offers accessible introductions to a wide range of key ideas, methods and debates. In all cases, the *Companion* aims to be as intelligible to readers with no geographic education as to those who have studied or practised geography for years. Indeed, a key claim of the book is that the field and discourse of environmental geography *exceed* the discipline of geography. At the same time, it is important to note that although the field of environmental geography is increasingly international in its scope and membership, our contributors hail largely, but by no means exclusively from the UK and North America. In part, this is a function of our own personal and professional histories of living, studying and working on both sides of the Atlantic. (The anglophone focus of this *Companion* partly reflects the barriers which need to be overcome to create a truly international environmental geography, although some contributions certainly acknowledge the considerable influence of non-English-speaking theorists and analysts of environment [in environmental discourses or development theory for example] and cite important international collaborative work [in land science for example].)

Rather than trying to summarise the contents of each and every chapter, we want instead to provide an overview of the wider landscape of research, practice and knowledge to which they contribute. As a result, the next three sections of this introduction are devoted to making sense of the complicated intellectual landscape that is environmental geography. There are a number of important and interesting issues to consider here, starting with definitional ones.

Defining Environmental Geography

The term 'environmental geography' is not one that most geographers to whom it could reasonably apply usually use to identify themselves or their work. Instead, geographers more typically imagine their discipline as one of two halves – human and physical. Within those two broad churches, there are numerous subfields, like economic geography or geomorphology, with which specialists identify. Although activity and interaction between human and physical geography (e.g., by geographers of 'natural hazards' and 'natural resources') is being increasingly acknowledged, through, for example, various conference sessions designed to speak across 'the divide' (e.g., Harrison et al., 2004), this dualism still dominates the organisation of the discipline in which *Progress in Physical Geography* is imagined as something separate from *Progress in Human Geography* (these names, for readers unfamiliar with them, refer to two leading geography journals).

This view of things may surprise non-geographers or pre-university geography students. After all, geography's public image is partly that of an 'integrative' discipline, while much of the subject's popularity in schools is due precisely to its focus on human-environment interactions. Yet the reality is that for most academic geographers 'environmental geography' is a small and often pretty elusive thing compared with the dominant human and physical wings of the discipline. (It may also be less familiar to North American readers where environmental geography has maintained more of a central role in some departments and topics, following for example, the traditions of human-environment geographers such as Carl Sauer or Gilbert White.)

One impetus for this book is to raise the profile of environmental geography both within and beyond the discipline. The environment is now widely touted as one important reason for 'Rediscovering Geography', to quote the title of a US National Academy of Sciences (1997) report on the future of geography. Echoing such calls, Billie Lee Turner (2002; cf. Zimmerer, 2007) is just one of a number of prominent figures urging geographers to embrace their long-ignored human-environment tradition so as to revitalise the discipline and secure its historically precarious place in the academy. Environmental geography, according to this way of thinking, provides a unifying link holding the two parts of the discipline together. It promises to make good on the integrative vision of geography celebrated by Mackinder, Davis and Ratzel but foiled as the discipline has become progressively more segmented and specialised since the Second World War.

While we certainly support those aspirations, they will only be achieved by overcoming three misconceptions about environmental geography. The first is about its place in the discipline of geography. Though environmental geography is often understood as a sort of middle ground between human and physical geography, this greatly oversimplifies the shape of the discipline and thus the problems we face in forging closer bonds of collective connection, collaboration and solidarity among its various parts and branches. Rather than thinking about geography divided horizontally between human and physical geography, we also need to recognise that the heterogeneity within those very broad divisions means they are also stretched out in the vertical dimension (figure 1.2), as indeed in a third temporal dimension of

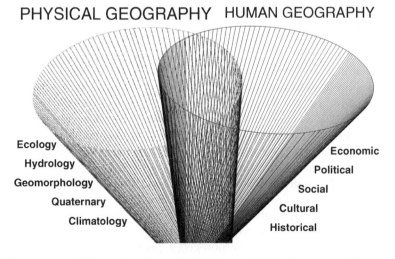

PHYSICAL GEOGRAPHY HUMAN GEOGRAPHY

Ecology
Hydrology
Geomorphology
Quaternary
Climatology

Economic
Political
Social
Cultural
Historical

Figure 1.2 The multidimensionality of disciplinary divides in geography.

time. The implications of this verticality are several. First, the vertical gaps within human geography between, say, modellers of land-use change and various post-natural theorists of the environment can be even more yawning than the putative human-physical divide. But second, acknowledging this verticality also implies that there ought to be many more potential points of contact than is suggested by the simplistic ideas of environmental geography as some kind of halfway house between human and physical geography. (Third, it indicates the multiple points of possible connections with other disciplines and communities.)

The second misconception stems from this first one. Seeing environmental geography as the mid-point of a one-dimensional divide between human and physical geography leads to a very narrow definition of what environmental geography is and ought to be. Implicit in many geographers' thinking today – so implicit that it is now arguably part of geographical lore – is the idea that only a fully 'symmetrical' approach to human–environment relations counts as 'real' environmental geography. By symmetrical we mean an approach that pays *equally* detailed attention to both people and non-humans as they interact. For instance, a symmetrical approach to the study of a new urban greenspace would need to account for how this patch of country in the city sustains migratory and local wildlife, reduces surface rainfall runoff, moderates solar radiation and so on, but it would also need to examine how people perceive and use this greenspace, taking care to differentiate age, gender, ethnic groups and so forth, while also considering issues of leisure as well as crime.

Historically, this kind of symmetrical understanding of human–environment relations was achieved and embodied by the individual geographer. Indeed, Mackinder made little distinction between individual geographers and the wider discipline they comprised. For him the integrative role as bridge between the natural and social sciences applied equally to both. But specialisation within the sciences, along with the exponential increase in the stock of scientific knowledge, has meant that even at the smallest geographical scale, this kind of all-encompassing and fully symmetrical account of human–environment relations is very difficult, if not impossible for any one individual to achieve: it requires broad expertise and a great deal of time if it is to be done well. Furthermore, the sorts of integrative and symmetrical understandings that individual geographers could provide also run the risk of being dismissed by specialists as trivial for failing to advance knowledge in more narrowly defined areas of research. For all these reasons, few geographers even try to achieve fully symmetrical understandings ideal typically associated with environmental geography.

One response to this dilemma is to relocate the sites for symmetrical environmental explanation to the level of discipline or research programme. When Marston (2006) refers to geography as the 'original integrative environmental science', the claim is not about the knowledge of individual geographers but about the potential of the discipline as a whole to bridge the divides between the various kinds of specialist expertise germane to understanding human–environmental relations. Similarly, many science-funding agencies are now looking to support large, multi-component research programmes that bring together the different sorts of specialist expertise to address the pressing problems of our times. Because the discipline of geography combines specialists from both sides of the divide who ideally have had some undergraduate-level training in both human and physical geography, geographers ought to be well placed to respond to environmental initiatives like the ongoing Rural Economy and Land Use (RELU) of the UK Research Councils (www.relu.ac.uk) (or the calls for integrative environmental research initiatives within the EU or US National Science Foundation). However, as the development of Earth System Science (see Wainwright, this volume) shows, the discipline of geography has not always profited from such initiatives.

This is at least partly because the lingering hold of Mackinder's normative vision of geographical knowledge as fully symmetrical has been so great that we have not always recognised the valuable contributions to be made by the profusion of 'asymmetrical' environmental research evident within geography today. By this term, we mean research and teaching that stitches together separately fashioned pieces of the human-environment jigsaw. People and the non-human world are connected in a multiplicity of ways; there are varying degrees and kinds of interactions, associations, couplings, feedbacks, interferences, transformations and accommodations going on. It is perfectly possible – and for a variety of reasons defensible, even necessary – to examine human-environment connectivities in 'asymmetrical' ways. For instance, physical geographers who are expert in river restoration may go about their work without having to know why certain social groups like restored rivers or why government planning regulations prohibit more restoration projects from occurring. Likewise, the 'Third World political ecologist' can say important things about how and why peasant farmers use their land in the ways they do, without having to know all the biological intricacies of crop rotation, soil fertility and plant germination.

This book is mostly about environmental geography in this asymmetrical sense – which is to say, the form in which it predominantly exists today. This does not, as we are suggesting, make the research reported in its many chapters an ersatz version of 'symmetrical' environmental geography. The latter has become a hard-to-achieve and highly normative ideal that many geographers have, understandably, found of little use to describe their own and others' work. In our view, the expanded definition of environmental geography that we are working with here – namely, *any form of geographical inquiry which considers formally some element of society or nature relative to each other* – is usefully open-ended. It opens up a much broader landscape of shared knowledge and practice, whose richness and potential only becomes apparent once we shake off the older vision of environmental geography as necessarily symmetrical.

This more expansive sense of environmental geography highlights a third misconception about environmental geography, namely that it is confined to the *discipline* of geography. Environmental geography bleeds into other disciplines and fields that share its interest in 'the geographical experiment' (and human environment interactions). As noted above, we can formalise both points by drawing a distinction between the 'discipline' of environmental geography and a wider *discourse* that goes beyond it (cf. Gregory, 1995). This includes specialised fields like environmental sociology and environmental economics, as well as relatively young, purposefully cross-disciplinary fields like environmental science, 'science studies', 'environmental studies' and the already mentioned Earth Systems Science. Unsurprisingly, little of the work done in these and cognate fields uses the term 'environmental geography'. But it does share the same commitment to investigating the social and non-human worlds in relation to one another (albeit 'asymmetrically' in many cases). On the social sciences side of all this, something of the scale and diversity of the discourse of environmental geography is captured well in Pretty et al.'s (2008) recent *Hand-*

book of Environment and Society. (And on the science side, a series of reports by the US National Research Council on sustainability, human dimensions of global change and common property resources acknowledge the value of engaging the social sciences [www.nrc.edu].)

While fairly definite, the borders that demarcate geography from these various other fields in the wider discourse of environmental geography are sufficiently porous that two-way traffic occurs quite readily, as many of our chapters bear out. In some cases, environmental geographers feel as much *part* of these other fields as their own. In other cases, they either draw upon the other fields to make their own distinctive contributions or else seek to shape them by 'exporting' their particular skills, perspectives and insights. Whatever the 'terms of engagement', an important common denominator applies here: most environmental geographers happily see themselves as part of a *wider* project, which they can learn from and shape. Today, 'the geographical experiment' is far, far more extensive than Mackinder could have possibly envisaged. Indeed, one might argue that there has never been more interest in the study of human–environment relations – from students, publics, states, firms and a range of other stakeholders – than there is today.

Geography, it is fair to say, does not occupy centre stage in the wider discourse of environmental geography. No one subject does. This fact might well have disappointed Mackinder, but if he were alive today, we would suggest to him that centrality is not the issue. Far more important is that environmental geographers are able to contribute distinctive and significant things to researchers, teachers, students and other stakeholders involved in the wider discourse.

Environmental Geography: Unity and Difference

Having loosely defined environmental geography, some further questions arise. What, it may be asked, is to be gained by abandoning the narrow, normative 'symmetrical' definition of the field and embracing a broader, more inclusive one? The answer to this question depends upon us answering another: namely, what do environmental geographers – ecumenically defined – have in common? Some obvious answers come immediately to mind.

First, as per our enlarged definition of environmental geography, they all study some aspect of society or nature in relation to one another rather than alone. They all take as axiomatic David Harvey's (1996) observation that 'all social . . . projects are . . . projects about environment, and vice versa' (p. 189). Second, they are all engaged in discussion about the character, purpose, meaning and proper management of these socio-natural relations (in peer review journals, edited books like this one, monographs, textbooks, lectures, seminars, policy briefs, etc.). These discussions involve various semantically rich terms, metaphors and analogies – such as 'dependent' and 'independent' variables, cause and consequence, condition and outcome, feedback and perturbation, hybrid actants, dialectical contradiction, force and resistance, co-constitution, and so on. Third, the specific knowledge claims in question are produced largely by professionals who regard it as their job – an occupational objective – to produce them. In other words, the discourse of environmental geography is not generated by accident or happenstance but intentionally and as a formal, full-time pursuit. Fourth, and relatedly, this knowledge has the specific qualities of all academic discourse: namely, it is derived from disciplined thought and inquiry, is somewhat (or very) esoteric, and commands a certain authority from students and others dependent on academic expertise. Put differently, the discourse of environmental geography is not colloquial, tacit or everyday. In the fifth place, whether couched in 'realist' or more 'constructivist' language, the claims advanced by environmental geographers are intended to tell us something about the actualities (today, yesterday or tomorrow) of human–environment relations. It is – at least usually – the opposite of science fiction, ungrounded speculation or metaphysics, a feature very much in keeping with geography's long-standing reputation as a 'practical' discipline that has its feet firmly on the ground. Finally, as all the chapters of this book make clear and as we have had already noted, environmental geographers of all stripes are intellectually outward-looking. They draw upon (and seek to contribute to) debates in cognate fields in both the social and the biophysical sciences, as well as in the humanities (see, e.g., the chapters by Zimmerer, Mels, Olwig, Turner, and Jones).

These various commonalities are real enough, but they may – understandably – strike many readers as being far too generic to define a real, as opposed to a contrived, field of research, teaching and practice. Indeed, the final commonality mentioned above may appear to render questionable the very idea of 'environmental geography' since the field routinely blurs into so many others as to lack any defining features of its own. Not surprisingly, we beg to differ with this rather dim assessment. True, environmental geography is diverse and lacks coherence philosophically, theoretically, methodologically and in terms of its practical applications. Its exponents produce an array of cognitive, evaluative, expressive, methodological and applied knowledges; and they vary greatly in the spatio-temporal scale and topical foci of their concern. Whatever unity environmental geography possesses is, *pace* the six commonalities listed above, certainly quite general. However, the field's diversity is nonetheless a structured one and we regard the heterodoxy of environmental geography as a strength not a weakness. Let us explain.

Even though environmental geography – like the wider discipline of which it is a major part – does not posses the sort of 'hard' external boundaries one finds in, say, the discipline of economics, it nonetheless has a very real identity – a 'structure of feeling' in Raymond Williams' evocative but nonetheless definite sense of the term. Over a century on, the legacy of Mackinder, Davis, Ratzel and like-minded pioneers is tangible: Geography remains one of the few places where it is possible to find social science, humanities and physical science perspectives on the environment rubbing shoulders. In other words, academic geography is constituted so as to permit something that one still finds rarely elsewhere: namely, a 'full spectrum' approach to understanding human–environment relations, albeit in the form of separate, asymmetrical contributions. For this reason, geography is 'recognized as possessing unusual strength in integrated, human-environment science' (Turner, 2002, p. 63). Compare this with, say, earth science (which excludes the human factor) or sociology (which has 'rural' and 'environmental' branches but both of these bracket biophysical issues for the most part).

This internal permissiveness – this encouragement and toleration of widely divergent research, teaching and policy work on human–environment relations – can be regarded as a virtue. This may seem counter-intuitive. Typically, the ongoing debates about the (dis)unity of geography as a whole depicts intellectual diversity as synonymous with fragmentation, and thus, intellectual weakness. This much is obvious in the book *Unifying Geography*, whose normative, aspirational title speaks to the editors' desire to reconnect the discipline's many (in their eyes) amputated limbs.

However, underlying such a negative judgement about disunity are some question-able presumptions that are not always made manifest. One is that there is a single reality 'out there' that demands an intellectual and practical approach able to respect its integrity. Another, relatedly, is that otherwise different perspectives on the world can ultimately be commensurated and synthesised (perhaps via a meta-language like 'complexity theory'). The idea that there might be multiple realities and/or a range of legitimately different perspectives on them is barely entertained. As sociologist of knowledge Tim Dant (1991) once noted, 'We tend to live as if knowledge could be settled, as if there is only *one* true knowledge we are striving for' (p. 1, emphasis added).

This belief reflects the enduring power of the idea of 'science' in the 21st century. In William Whewell's (1794–1866) original sense, 'science' simply meant any form of systematic inquiry undertaken according to a procedure that suitably qualified others that could replicate or validate. However, over time, the term has become polysemic, signifying (among other things) a form of 'objective inquiry' into a world that exists independently of the inquirer and whose 'real' properties can be correctly understood given time and adequate resources. Geography's enchantment with science in this specific sense was most intense between the mid-1950s and mid-1970s. Somewhat diminished, it nonetheless continues to this day, notably in most branches of physical geography, some parts of human geography and in elements of environmental geography too. The commitment to science conceived thus has a 'strong' and a 'weaker' form. The former (which few environmental geographers or, indeed, any geographers would publicly defend) supposes that there is only *one* 'true method' for interrogating reality: namely, 'the scientific method', which would today be understood practically as a form of hypothesis testing (or problem-solving) using melange of inductivism, deduction, inference, retroduction, verification and falsifica-tion depending on the case. The latter ('weak scientism') is a modern version of Auguste Comte's (1798–1857) Enlightenment conception of human knowledge as a giant jigsaw puzzle, the pieces of which can be identified by different disciplines and sub-disciplines and ultimately pieced together. It supposes that there may be different ways of deriving true knowledge, but that these knowledges (once derived) can be married together on the grounds that reality is continuous not partitioned into the mental boxes we typically use to comprehend it.

The commitment to science in either of these forms cannot be dismissed, even after several decades of questioning the whole idea that science = truth (or at least the quest for truth). However, our own view – and that of environmental geogra-phers as a whole, if this book is anything to go by – is that 'science' is in fact plural and, thus, best seen as one approach to, and form of knowledge, among many – rather than a privileged or Archimedean one. To argue otherwise entails suggesting that 'non-scientific' forms of knowledge are less valid and that reality is, ontologi-cally speaking, singular and consistent rather than discontinuous, differentiated and stratified. There is also the questionable implication that science is value-free.[1]

In this light, we might look favourably upon the 'multi-paradigm' condition of environmental geography (and note too that many other fields of knowledge in the humanities, social sciences and humanities are today similarly heterodox). The field's astonishing intellectual diversity can, perhaps, be seen to reflect a very impor-tant fact: namely, that a topic as broad as 'human–environment' relations simply cannot be understood through one – let alone one putatively 'objective' – approach, worldview or method. You do not have to be an epistemic 'conventionalist' or 'nominalist' to acknowledge this fact, let alone a 'relativist'. One can happily insist that there is a 'real world' out there, while still conceding that it is sufficiently complex and differentiated such that no one mode of knowing it will suffice for all our wishes and purposes. (Even traditions of environmental modelling can approach the same question using very different assumptions about human behaviour and societal dynamics, and reflect different approaches to explaining atmospheric or ecosystem dynamics.) In short, environmental geography's diversity should not be sacrificed on the altar of 'unity' – or least not the sort of 'strong' unity that presumes epistemic variety to be symptomatic of intellectual confusion about the 'true' nature of human–environment relations.

This said, our reluctance to define environmental geography in terms of the narrow and highly normative standard of symmetry does not mean that we are agnostic about its current condition. On the contrary, we believe some positive change is required. There is one obvious problem with a 'let many flowers bloom' stance towards the field. It is not so much a problem of epistemic relativism – as we have explained, there is no consensus about whether we can know reality independently of our various mental and physical engagements with it as researchers. Instead, it is more a problem of mutual ignorance and indifference. This risk was identified many years ago for geography as a whole by John Pickles and Michael Watts. As they put it, the '. . . unwillingness to debate the merits of competing frameworks encourages reliance on values: assertion, training and faith become sufficient conditions for selection. A new [plural] dogmatism is asserted . . .' (Pickles and Watts, 1992, p. 303). What they were calling for was the development of a critical culture within the discipline. Nominally at least, environmental geographers share a common object of analysis and concern: 'the environment'. While there will always be real limits to communication to do with the sheer inability of one group of environmental geographers to understand what other equally specialised groups are 'up to', there is nonetheless room for greater cross-group dialogue and critique.

What would be the virtues of this and how might it be engendered? We can answer the first part of this question by analogising environmental geography to a nation state composed of highly diverse populations – think the USA, Britain or Australia, for example. A monocultural polity environmental geography is not. So is it, in analogical terms, a multicultural or a republican one? In our view, it is currently multicultural when it ought to be far more republican. What does this mean? We are using the term multicultural here (contentiously, we admit) to denote different ways of life that are spatially juxtaposed but which ignore or talk past one another. Some might call this 'communitarianism'. 'True' republicanism, by contrast, corresponds to what philosopher of science Karl Popper (1945) famously called 'the open society'. In Popper's view, all knowledge claims – along with their practical consequences – are only robust once they have withstood, been modified by, or enriched through an encounter with criticisms issuing from various quarters. Republicanism in knowledge (as in politics) ought to involve a genuine engagement between rival perspectives on the basis of common sensibilities – not so much to reduce epistemic differences in the name of 'one truth' but, instead, to ensure the socio-practical robustness of otherwise divergent knowledge claims.

The sort of open, critical culture being described here is difficult to engineer. It is underpinned by an ethic of responsibility rather than (*pace* Fuller, Pickles and Watts) an ethic of conviction, one that many or most members of any given academic discipline would need to share. It entails both mutual recognition and respect between the parties who might stand to gain through an epistemic encounter. Though environmental geographers, like geographers writ large, would find it far easier to continue with business as usual, it would nonetheless be far more possible (and desirable) to create an 'epistemic republicanism' within a generation than it would be to create the sort of 'strong' intellectual unity and 'symmetrical' environmental geography we have already discussed. Quite how one does this practically speaking remains uncertain. It would doubtlessly require a small number of respected intellectual leaders to set an example, along with a strong steer from professional associations like the Association of American Geographers and from academic journal editors too. It would also likely occur most readily by otherwise different researchers communicating about shared and specific topical concerns or problem-sets, such as water management, animal conservation and climate change.

Fortunately, we are not entirely bereft of precedents and current examples of critical engagements between various strands of environmental geography. The sheer diversity of environmental geography has presented researchers and teachers with the possibility (if not the obligation) of becoming critical and creative synthesisers. Contrast this with a discipline like economics, where intellectual plurality is not tolerated nearly so much. In other words, the plain lack of orthodoxy in environmental geography as a whole has arguably made it easier for certain individuals to avoid encampment in one of other of its subfields. Think of Third World political ecology, which is a critical synthesis and application of a plethora of otherwise different concepts, methods and approaches. Think of 'new resource geography', which often combines neo-Marxist, institutionalist and Foucauldian concepts to make sense of modern mining or forestry.

Environmental Geography in the 'Knowledge Society'

Most environmental geography, as this book's contents attest, is produced in universities by professional academics. While the discipline and discourse of environmental geography are not entirely academic – (researchers and) non-academics in the environmental movement, for example, contribute richly to the discourse (see Porritt [2005], for instance) – they are largely so. Though a seemingly banal observation, it actually strikes us as being quite important. To understand why, we need to consider the meaning of the now-familiar term 'the knowledge society'.

As Fuller (2002) wryly notes, '. . . saying that we live in a 'knowledge society' would seem to be no more informative than saying that we live in a 'power society' or a 'money society' . . .' (p. 2). However, the term has a more precise meaning that is associated variously with commentators like Peter Drucker, Daniel Bell and Manuel Castells. In this more specific sense, the term denotes two distinct but related shifts in knowledge that were initially characteristic of the advanced capitalist economies but which are now more widespread. The first is a deliberate move to increase the range and volume of *formal* (as opposed to tacit) knowledge, something coincident with its intensified modularisation (as in the proliferation of software systems that can perform specific functions; as in the profusion of different databases, and so on). Second, 'the knowledge society' refers to an equally deliberate move to *put this knowledge to work* in a variety of ways as a means, an end or both – not the least of which is to make money ('commodified knowledge', such as patented gene codes). In this second sense, knowledge is not a goal in itself but, instead, a medium for realising particular ends and an instrument for action.

If, in even only a general sense, the idea of a knowledge society holds good, then it obliges us to look again at the functions of the university as well as the wider context in which it now operates. Historically, as Bjorn Wittrock (1985) has argued, there are three models of the university operative in the West (archetypes if you will), and in all cases the university held a virtual social monopoly on the creation and dissemination of canonical as well as new formalised knowledge. In the British model, the post-medieval university aimed to create the 'well-rounded' or 'whole' person; in the French model, higher education was, as per Napoleon I's intentions, geared to the national interest; finally, in the German, Humboldtian model, universities are geared to the pursuit of pure understanding. In the late 20th century, there is plenty of evidence to suggest that Western universities have, en masse, moved closer to the archetypal French model. They have, according to one line of criticism, become 'corporatised' and very mindful of their contributions to 'national competitiveness' and 'the public interest'. At the same time, it is clear that the near monopoly that universities once held on the creation and dissemination of canonical, as well as new, formalised knowledge has been challenged. Today, research and teaching at a high level goes on, variously, in think tanks, foundations, non-governmental organisations, charitable bodies, colleges funded by benefactors, large firms and so on.

What has all this got to do with environmental geography? A good deal. Because of its intellectual breadth, environmental geography – like its parent discipline – has, historically, been able to meet the demands of *all three* models of the university. Importantly, its inability to be disciplined by the demands of any one of these models explains why, along with some other university subjects, it has been able to resist current pressures to make universities 'relevant' in a fairly instrumental sense. The knowledge that geographers produce, teach and disseminate outside the university remains sufficiently diverse that, while the latter pressures can be accommodated, they do not 'skew' the discipline unduly.

Skewing presents real dangers to any field. If, through financial or other levers, a discipline is steered heavily by outside interests, then there is the strong possibility exists for a reduction in epistemic diversity and the rise of new paradigms in Kuhn's original, subject-wide sense. The possibilities are already evident in so-called 'big science', where huge resources are being channelled into certain lines of inquiry but not others courtesy of biotechnology, biomedical, energy and pharmaceutical firms – sometimes aided by national governments. But similar pressures are also on the horizon (perhaps already here) for those disciplines that study human–environment relations. The sort of 'land change science' discussed in Billie Lee Turner's chapter is exciting, as are the closely related fields of 'earth system science' and 'sustainability science'. (Similarly, the growing focus on payments for environmental services, which engages many physical geographers in the measurement of such services, can too easily become the servant of a naïve market environmentalism.) But they could, in time, become the focus of enormous intellectual and fiscal inputs as societies become increasingly alarmed about global environmental change. In the USA, we have already seen the Global Change Research Program (created in 1990) become one of the largest ever foci of public research funds in American history. As currently constituted, environmental geography's plurality can make it a player in such grand endeavours yet without sacrificing its capacity to offer multiple insights and perspectives on human–environment relations. Indeed, environmental geographers were key players in the creation of the current 'global environmental change' research agenda going back 20-plus years. This bespeaks an admirable capacity to set their sites on big agenda issues, while refusing to be corralled into intellectual orthodoxies of a theoretical, methodological or policy-political kind.

It is no accident that environmental geography's diversity and vitality is coincident with its basis in the university system. Despite being subject to varying degrees of 'corporatisation', Anglophone universities remain, for the most part, publicly funded and public in their identities. Though managerialism has, to some extent, eroded its potency, 'academic freedom' remains a critical ideal and reality for researchers, teachers and consultants based in university geography departments – so too for all those other academics whose work constitutes the 'discourse of environmental geography'. A reflection of the relative autonomy of academics from outside interests and their historical claim to self-government, such freedom is precisely what – even today – allows environmental geographers and those working in cognate fields to determine how and why they will do the work that they do. Contrast this with knowledge producers and disseminators working in the 'knowledge society's' many other institutions, like think tanks, privately funded foundations (and even NGOs). In these institutions, the sort of environmental knowledge created is very much determined by the specific agendas of patrons, benefactors, shareholders and owners. This does not render it illegitimate of course. But it does circumscribe its likely interest and relevance to the enormous array of people and groups who have some stake in the drama – as well as the quotidian course – of human–environment relations.

This raises some critical questions about who is authorised to produce and validate particular sorts of environment-society knowledge today. In relation to the so-called 'expert' knowledge, the days of ivory-tower elitism are thankfully behind us. Universities are no longer recognised as being dispensaries of indisputable truth and wisdom. But they still play a vitally important role in our 'knowledge societies'. There is much debate about the nature of this role and how it might be sustained or altered. One well-known view is that academic experts 'enter the fray' as part of a new epistemic condition that Michael Gibbons and colleagues (1994) termed 'mode 2 knowledge'. 'Mode 1' knowledge has, historically, been produced by those (like academics) inhabiting a few 'authorised' institutions. By contrast, a mode 2 society (in Gibbons et al.'s view) is one where many knowledge workers in a range of sites come together to create robust knowledge about issues and problems of common concern (like climate change). This mode 2 way of operating is not beholden to old expert-lay distinctions and nor is it interested in the preservation of academic disciplines – unless the members of those disciplines can contribute meaningfully to the many, changing epistemic collectives that produce mode 2 knowledge.

In contrast to this vision of where universities sit within a wider knowledge society, others suggest that we update older ideas of academic expertise and non-partisanship. For instance, in his book *The Governance of Science*, Steve Fuller (2000) suggests that universities are becoming 'clearing houses' for the airing, testing and encounter between diverse knowledges. In his view, basic and applied research should in future be undertaken *outside* universities in all those other institutions mentioned earlier in this section. The role of university experts is then, in his view, to scrutinise these knowledges according to an array of criteria (cognitive, moral, aesthetic, etc.). These experts will not seek to eliminate knowledges on the grounds of their 'falsity'. Instead, they will undertake both 'translation work' (making apparently incommensurable knowledges speak to one another) and check for the 'robustness' of knowledge (i.e., can it be made meaningful to a wide array of stakeholders or not?).

These and other views on the future of the university and its disciplines matter greatly for environmental geography and cognate fields. 'The environment' and the way humans use it is of such widespread and fundamental social importance that the creation, validation, disputation and circulation of human-environment knowledge will become ever more important for ourselves and the future of the biophysical world. To date, practitioners of environmental geography have gone about their research largely unmindful of the big debates on the university and the knowledge society. Looking to the future, this ought to change for the simple reason that the institutional and social context of knowledge production profoundly affects its content and aims. There is no 'context-free' knowledge and the precise role that environmental geographers play in wider epistemic debates on human–environment relations in academia and society will depend almost entirely upon how the university (re)defines itself as an institution.

Conclusion

This book is by no means an exhaustive introduction to environmental geography. For various reasons, certain things were left out (e.g., the Approaches section would have benefited from chapters on 'urban political ecology' and 'environmental restoration'). So this could have been a much larger, more comprehensive volume. Even so, it offers a fairly complete sense of what environmental geography currently is. In so doing, this book – and our attempt in this introduction to explain its aims – will, we hope, remind professional geographers that the 'middle ground' is not nearly as small as many often think it to be, while showing other readers outside geography that the discipline offers a virtually unique suite of theories, approaches, investigative methods and substantive insights into human–environment relations. As we have explained above, environmental geography does not 'represent itself': rather, it needs actively to be made sense of given the apparent dominance of geography's two halves. We hope very much that this book helps environmental geography to be seen by readers as what many of our contributors already regard it as being: that is, a major area of activity, at least equal in size and significance to human and physical geography, respectively.

This book, with its expansive sense of environmental geography, clearly says much about how 'the geographical experiment' is currently being conducted, and we in this introduction have suggested how it might be altered in years to come. It almost goes without saying that this experiment needs to continue on into the future and to have a proper institutional home in universities and other research, teaching and policy environments. Geography remains one important place for investigations of human–environment relations to be undertaken and communicated, though not the only one. It ultimately matters not where and under what banner such investigations occur. What is far more important is that societies continue to properly fund and resource them. After all, even in our supposedly digital, post-industrial, knowledge-intensive, 'weightless', information technology era, all of us draw upon the non-human world ineluctably as fleshy, emotional, thinking and acting beings. Current worries about the nature and impacts of 'global environmental change' are only the most dramatic reminder of this fact.

We will never not need cognitive, moral, aesthetic and applied knowledge about how we currently (and ought in the future to) interact with the non-human world. Such knowledge covers a wide spectrum of functions and uses, such as problem solving (how can we reduce soil erosion?), moral guidance (what shared values might underpin global environmental accords?), the satisfaction of curiosity (how do wild animals adapt to urban life?) and much more besides. In humanity's various attempts to engage with the biophysical world materially and imaginatively, the sort of diverse, high-level inquiries reported here will be vital tools. In our capacity as citizens, workers, family members, tourists, activists, local residents and any number of other roles, we surely need the sort of research, teaching and policy knowledge that environmental geography offers alone and as part of a wider, societal discourse.

NOTE

1. These arguments and the counter-arguments to them were aired not altogether productively in the so-called 'science wars' of the late 1990s in the USA. See Ashman and Baringer (2001) for a post-mortem.

BIBLIOGRAPHY

Agnew, J., Mitchell, K. and Toal, G. (eds) (2001) *A Companion to Political Geography*. Oxford: Blackwell.

Ashman, K. and Baringer, P. (eds) (2001) *After the Science Wars*. New York: Routledge.

Dant, T. (1991) *Knowledge, Ideology and Discourse*. London: Routledge.

Fuller, S. (2000) *The Governance of Science*. Buckingham: Open University Press.

Fuller, S. (2002) *Knowledge Management Foundations*. Woburn: Butterworth-Heinemann.

Fuller, S. (2003) *Popper vs Kuhn*. Cambridge: Icon Books.

Gibbons, M. (1994) *The New Production of Knowledge*. London: Sage.

Gregory, D. (1995) *Geographical Imaginations*. Oxford: Blackwell.

Harvey, D. (1996) *Justice, Nature and the Geography of Difference*. Oxford: Blackwell.

Harrison, S., Massey, D., Richards, K., Magilligan, F. J., Thrift, N. and Bender, B. (2004) Thinking across the divide: perspectives on the conversations between physical and human geography. *Area*, 36, 435–42.

Livingstone, D. (1992) *The Geographical Tradition*. Oxford: Blackwell.

Mackinder, H. (1887) On the scope and methods of geography. *Proceedings of the Royal Geographical Society*, 9, 141–60.

Marston, R. (2006) Geography: the original integrated environmental Science. Presidential Plenary address to the Association of American Geographers, 8 March, Chicago, IL.

Matthews, J. and Herbert, D. (eds) (2004) *Unifying Geography*. London: Routledge.

Pelling, M. (ed.) (2003) *Natural Disasters and Development in a Globalizing World*. London: Routledge.

Pickles, J. and Watts, M. (1992) Paradigms for inquiry? In R. Abler, M. Marcus and J. Olson (eds), *Geography's Inner Worlds*. New Brunswick: Rutgers University Press, pp. 301–26.

Popper, K. (1945) *The Open Society and Its Enemies* London: Routledge & Kegan Paul.

Porritt, J. (2005) *Capitalism as if the World Matters*. London: Earthscan.

Pretty, J., Ball, A.S., Benton, T., Guivant, J., Lee, D.R., Orr, D., Pfeffer, M. and Ward, H. (eds) (2008) *The Sage Handbook of Environment and Society*. London: Sage.

Robbins, P. (2004) *Political Ecology*. Oxford: Blackwell.

Wittrock, B. (1985) Dinosaurs or dolphins? In B. Wittrock and A. Elzinga (eds), *The University Research System*. Buckingham: Open University Press, pp. 13–38.

Part I   Concepts

Chapter 2

Nature

Bruce Braun

Introduction

In 2001, researchers from University College London documented a massive imbalance in the sex ratio of Blue Moon butterflies on the Samoan Islands of Savali and Upolu. Males, they discovered, accounted for only 1 percent of the population. Biologists now believe that the imbalance was caused by the parasitic Wolbachia bacteria, which is passed down from mothers and kills male embryos before they hatch. When they surveyed the islands five years later, however, they were surprised to find that males accounted for about 40 percent of the population. What explains the dramatic recovery? Scientists postulate that the comeback was due to 'suppressor' genes that controlled the bacteria and that this was, in the words of one member of the research team, 'the fastest evolutionary change that has ever been observed' (quoted by *BBC News*, 2007). The same researcher went on to suggest that the example further strengthened the view that parasites may be one of the major drivers in evolution.

Shift to a somewhat different context in Birmingham, England. Here conservationists have noted some peculiar changes in the behaviour of water voles. In this urbanised setting, the voles have apparently learned to live with the brown rat, usually considered to be a predator (Hinchliffe, 2008). Urban ecologies, it appears, can give rise to new capacities in animals, scrambling a system of classification that presupposes that all members of a species of vole are the same, and thus, interchangeable. Yet, urban and rural voles, it seems, do not exhibit the same behaviours.

Finally, consider our endless battle with infectious diseases. In the years after the Second World War, public health officials, at least in the 'developed' world, imagined that through quarantine and immunisation as well as the wide use of antibiotics and vaccines, such diseases would be eliminated, and we would experience an 'epidemiological transition' where infectious diseases would increasingly be dispatched to the dustbin of history. Today, we seem threatened with new and emerging infectious diseases like never before. More than this, though, new work in microbiology suggests that far from microbes being that which most threatens humans, they may have been the most significant historical actors in the genetic composition of humans. In the words of Melinda Cooper (2006, p. 115), 'We are literally born of ancient alliances between bacteria and our own cells; microbes are inside us, in our history, but are also implicated in the continuing evolution of all the forms of life on earth'. Microbes, apparently, have made us what we are, right down to our DNA.

In each of these examples, scientists – both physical and social – are saying something about nature. On some matters they agree. All, for instance, bear witness to a world of fabulous transformations, where bodies and their capacities are actualised in surprising and unpredictable ways. There is nothing static about living beings in these accounts. Regardless of whether the focus is on genetic mutation and the evolution of species, or the developmental trajectory of individual organisms, nature is presented as a realm of dynamic change in which bodies have no fixed or eternal form. Admittedly, these accounts are heterodox – the majority of scientists continue to point to processes that remain relatively static or predictable, and in everyday life, 'nature' is often taken to name things that are eternal and immutable. But even as these examples share certain assumptions about nature, they disagree about others. For the butterfly scientists, nature names a realm *external* to humans, reduced to 'predator–prey' relationships that help shape the direction of evolutionary change. For these scientists, nature is something 'out there' to be studied and science tells us what is going on. For the virologist, on the other hand, the human body is an emergent effect of its interaction with the non-human world: much like the butterfly that adapts to parasitic bacteria, bacteria and viruses have made humans what they are. But can we say that these exchanges are *natural* processes? And if they were in the past, can we still say so today?

We may have always lived in a viral ocean, but as SARS, HIV and avian flu suggest, it is difficult to imagine today's viral economies apart from the sociotechnical networks – airplanes, food chains, virtual research communities, immigration law and antiviral medications – that stretch these viral geographies across immense distances or seek to regulate their form. Hence, if we agree that human bodies are part of nature, must we also say that technology is too? Where does nature end and society begin? This question is equally evident in the work of the geographer Steve Hinchliffe and his fellow conservationists, for whom apparently 'natural' beings like voles cannot be separated from the urban environments in which they have come to embody unique characteristics not shared by other voles. In a world of ongoing differentiation, it is not just the case that natural kinds (voles) differ from others (rats), but that they *differ from themselves*, resulting in quite a problem for conservationists, who have long imagined that conservation takes the 'species' as its concern, and that one water vole is the same as another. Nor is arriving at knowledge about these shape-shifting creatures a straightforward process: voles afford themselves to observation only in certain ways – through their traces, for instance, rather than direct observation – and so the observer of voles must engage in certain disciplined bodily practices by which he or she can be 'affected' by the voles, and thus 'make present' the voles within orders of knowledge. What counts as 'nature' is not separate from its representation; but representation, in turn, is irrevocably tied to the embodied actions of the observer. In these cases familiar culture-nature and representation-matter binaries fail us.

If anything comes clear from these examples it is that nature is an immensely difficult word to define. For some, it names the essence of things, such as when we say that it is the 'nature' of something to be a certain way. We might even say, for instance, that the nature of nature is to perpetually change! For others, nature names that which exists separate from humanity – the 'natural world' studied by physical scientists, for instance, or the 'nature' that some environmentalists feel needs to be saved from humans, who, in turn, are imagined as unnatural. For others still, nature includes humans as part of the ongoing processes by which the physical world is constituted, including the physical nature of humans themselves. For those who hold this view, the boundaries between nature and society, or the ecological and the technological, are indistinct. And then there are those for whom any statement about nature must necessarily be provisional, since nature, like any sign, is meaningful only within a larger semiotic system, or because each and any knowledge of nature is situated and partial. What is of interest to 'constructivists', as they are often called, is how nature comes to be known and represented in certain ways, and not others, or how certain things come to be gathered under the sign 'nature' at particular historical moments while others are excluded.

It would be impossible to cover the diverse meanings that this word carries today in the space of a short essay. Lengthy treatises have been dedicated to the topic, many of them rich in historical detail and philosophical insight (see Collingwood, 1945; Glacken, 1967; Williams, 1980). My concern here is to focus more narrowly on a number of debates within contemporary geography about the nature of nature. The first concerns how we understand the relation between society and nature. Do these terms name two separate ontological domains, or are non-dualist ontologies better suited for thinking about the world in which we dwell? The second concerns what we might call the *temporality* of nature. Does nature name that which is eternal and immutable or is it chaotic and 'eventful'? The third has to do with our ability to make the sorts of claims found in the previous questions. Does knowledge about nature result from detached observation? Is it mediated by culture, language and images, all of which precede our encounter with things? Or does it result from our practical activities in the world? I will end the essay by suggesting that how we answer each of these questions leads to a fourth set of questions about ethics and politics, or about how we are to live in a world of human and non-human others.

The Matter of Nature

Does nature name a realm external to humanity or is the boundary between humanity and nature indistinct? At least within Western thought this question has been answered in many different ways. For Aristotle man was an animal with the capacity for politics, a definition that created an internal division *within* man between animality and humanity (see Agamben, 2004). Aristotle also distinguished between nature and artifice: 'natural' things were governed by a final cause (the oak tree the final cause of the acorn, for instance), while things made by humans were not (a table from the oak tree, which could just as well have been made into a chair). The Roman poet and physicist Lucretius, influenced by Epicurus, rejected final causes, and was far less certain about human uniqueness, instead situating humans fully within the flux and flow of a tumultuous atomic world. Human life, he suggested, was characterised by just as much contingency and chance as nonhuman life – in the 'swerve' of atoms emerged new and wondrous forms, both human and nonhuman. Christian theologians, on the other hand, imagined a created order, in which humans had been granted their own special place. Nature existed apart from humans, who were conceived either as having dominion over it, or as called to care for it. Natural theology, in turn, took as its task understanding God's design. For Enlightenment thinkers like René Descartes, the human subject was conceived to be different in kind from animals, since animals did not have the capacity for reason. Human reason, in turn, was exercised upon nature as if from a position outside of nature – a 'brain in a vat', as Bruno Latour (1999) has famously put it. And, of course, in the eyes of Romantic poets, nature existed as an external realm in which humans could glimpse an eternal, transcendental order, should they bother to immerse themselves in it. This sublime nature, seen to be at risk from the depredations of industrial modernity, still beckons to us today in the form of national parks and wilderness preserves, where the last vestiges of pristine nature are imagined to exist, and the promise is held out of a return to a more original and more authentic existence. Romanticism teaches us that we are no longer natural, although perhaps we once were – that at some point in the past we managed to extract ourselves from nature into an entirely different realm called society, which now places nature at risk.

Clearly, the relation between 'nature' and 'humanity' has had a tumultuous history. In geography today, however, it is virtually a truism that the separation of the world into two distinct ontological domains – nature and society – is a habit of thought that demands to be challenged, both on conceptual and ethical-political grounds. Hence, any inquiry into the status of 'nature' in geographical thought today must necessarily take up the question of dualism and attempts to overcome it. Before proceeding further, however, I should note that the question of nature and its relation to humanity has been a more pressing one among human geographers than physical geographers. At first blush this may seem counterintuitive, for is it not physical geographers who study nature? And have not human geographers been accused of too often *ignoring* nature, labouring under the false impression that society followed its own rules and logics, entirely separate from nonhuman nature? While both statements are certainly true, there are a number of intellectual and historical reasons why human geography has been the side of the discipline more preoccupied with the question.

One very simple reason is that physical geographers, and others in the environmental sciences, rarely work with such grand abstractions as 'nature'. The concern of field scientists and lab workers alike is to understand specific physical processes. How is fluvial gravel entrained, transported and deposited in different kinds of rivers? And how is this different in humid and arid environments? To answer questions such as these, a geomorphologist like Marwan Hassan has no need of such baggy concepts as 'nature' or 'society' (Hassan et al., 2006). On the other hand, many physical geographers *do* work with an implicit and largely unquestioned nature/society dualism. As Urban and Rhoads (2003) explain, most physical geographers understand their task to be to ascertain the physical processes or events that have shaped the earth's biotic, geomorphological and climatological systems, and have conceived humans to be separate from and external to these 'natural' systems, which are assumed to be independent from, prior to, or unaffected by humans (see Gregory, 2000). At most, humans enter physical geographers' accounts in one of two ways: either as scientific practitioners (with all the attendant questions about method), or, as an external force that 'disturbs' or exerts an 'impact' on physical processes. While more recent work by some environmental geographers has begun to study human activities among the processes shaping physical landscapes, the underlying terms of the human–nature dualism are held intact within physical geography so long as human activity is understood as something 'unnatural'. Indeed, the language of 'modified' landscape is telling in this regard, since it places *human* modification in a class of its own (after all, every landscape is modified by the organisms that live in it, although the scale of effects varies dramatically).

One place where physical geography's dualist ontology has begun to erode is in the work of hydrologists and geomorphologists. In part this has resulted from a growing understanding that these sciences actively *order* the world, such that knowledge of physical landscapes is invariably bound up with the world of the observer (see Church, 1996; Beven, 2002). But it has also followed from a growing recognition that at least today *human* processes are in many respects the most important ones to understand in order to grasp the development and evolution of specific physical systems. Likewise, the growing focus on urban environments by climatologists, biogeographers and hydrologists has led to more integrative work, where urban ecologies are studied as complex systems in their own right, without the implicit dualism inherent in the language of 'human impact'. Such studies, however, are still a minor strand within physical geography. Indeed, within an otherwise excellent discussion of key philosophical questions in physical geography, including a number of epistemological questions first raised by human geographers, Rob Inkpen (2005, p. 144) devotes only one paragraph on the last pages of his volume to the possibility of a post-dualist ontology, noting that 'the interpenetration of the physical and human means that it is difficult to justify that processes of environmental change are purely physical or that social structures rely solely upon human processes'.

Human geographers, on the other hand, have for some time debated a set of explicitly ontological questions about the relation between humans and nature, and over the past three decades this has given rise to a diverse literature. We might suggest several reasons for this. On the one hand, the flourishing of such work can be seen as a reaction to the fact that the discipline was surprisingly unprepared to respond to, and analyse, the environmental effects of industrial society as these effects were articulated in public discourse in the 1970s. The 'spatial science' approach, for instance, with its isotropic planes and rational economic actors, had for the most part dispensed with nonhuman nature entirely, and offered very little in the way of a conceptual framework through which to understand human–environment relations. Society might be reduced to law-like behaviour, even modelled after physics, but in no way was the actual *physical* world to be part of this!

Even with the emergence of radical theoretical alternatives, the physical world was often ignored, as Margaret Fitzsimmons pointed out in a key 1989 essay. On the other hand, those human geographers who did attend to questions of the environment tended to focus most of their attention on rural landscapes, or, in the case of many cultural ecologists, 'pre-modern' cultures. This resulted in theories of cultural adaptation to environmental conditions that were not well suited to the complexity of modern technological societies. When the question of the environment exploded in the 1970s and 1980s, human geographers found themselves trying to cover a lot of ground quickly, with various attempts made to place the question of society and nature on a firm analytical footing (e.g., see Harvey, 1974; Hewitt, 1983; Smith, 1984; Turner et al., 1990). This renewed emphasis on the question of nature was given further impetus by two additional developments. The first was the strong neo-Malthusian flavour of 1970s environmentalism, which was received with considerable skepticism by those who worried over the misanthropic and often racist conclusions drawn by proponents. Against calls for 'lifeboat ethics' or the 'culling' of human populations in the face of a looming 'environmental crisis', radical geographers found themselves compelled to explore different ways of conceptualising human–environment relations, and the social and political causes of environmental change and so-called 'natural disasters'. The second was a growing critique in the 1970s and 1980s of dualist thought in general, which was taken by some to lie at the core of many of modernity's pathologies, including its instrumental relation to the nonhuman world (see, for instance, Merchant, 1990). The problem for geographers, then, was to move beyond dualist conceptions of nature and society, a concern that they eagerly took up over the next two decades.

Beyond Dualism? Marxist Geography and the Production of Nature

One of the most influential efforts by human geographers to conceptualise the matter of nature has been that of Marxist geographers who sought to develop an understanding of nature consistent with the tenets of historical materialism.[2] Key to these efforts were a number of close readings of Marx's scattered reflections on the topic, the first by the Frankfurt School author, Alfred Schmidt (1971), and the second by the geographer Neil Smith (1984). As Smith explained in his book *Uneven Development*, although Marx's writings on nature were far from systematic, it was possible to identify within them a strong challenge to ontological dualism, since he consistently situated humans *within* nature, as one of its constituent parts. As Marx famously put it in his *Economic and Philosophical Manuscripts*:

> Nature is man's inorganic body. . . . Man lives from nature, i.e. nature is his body, and he must maintain a continuing dialogue with it if he is not to die. To say that man's physical and mental life is linked to nature simply means that nature is linked to itself. (1975[1844], p. 328)

Elsewhere Marx would emphasise labour as that which mediated the relation between society and nature:

> Labour is, in the first place, a process in which both man and Nature participate, and in which man of his own accord starts, regulates, and controls the material reactions between himself and Nature. He opposes himself to Nature as one of her own forces, setting in motion arms and legs, head and hands, the natural forces of his body, in order to appropriate Nature's productions in a form adapted to his own wants. By thus acting on the external world and changing it, he at the same time changes his own nature. (1967[1887], p. 173)

It is not difficult to see why such a view was immensely attractive and yet at the same time jarring to readers fed a steady diet of ontological dualism. On the one hand, it pulled the rug out from beneath those who claimed that nature named an external realm separate from humans, governed by immutable laws to which humans must conform. Against the Malthusian discourse of 'natural limits', and the biological reductionism of socio-biologists, for instance, the return to Marx presented both an analytical provocation and a political intervention, for if society and nature were presented as an internal relation, it was no longer possible to invoke external nature as a source of authority or legitimation for specific social arrangements (see Smith, 1984; Castree, 2000). On what basis could it be said that nature existed as an immutable external force, to which humans must submit, if through their labour humans transformed both external nature and their own internal nature? For writers like Smith (1984) nature was best understood as something 'produced', rather than something timeless and eternal. Nature did not stand outside history; its history was still to be written.

An equally important contribution of historical materialist approaches to nature came in the form of an explicit challenge to the 'deep green' or 'preservationist' impulse found in so much of the North American environmental movement of the 1970s and 1980s. For many deep green environmentalists, nature was taken to be a realm entirely separate from, and threatened by, humans. According to this view nature was that place where humans were not, and thus the presence of humans, considered by some as a cancer on a preexisting natural world, was taken to signal the imminent *destruction* of nature (see, for instance, McKibben 1989). As numerous commentators pointed out, this introduced a contradiction into ecological thought, for if humans signaled the 'end' of nature, then the only way to save nature would be to remove humans entirely. In short, such a perspective provided no basis on which to determine *how* to live in the world (Cronon, 1995; White, 1995). From Smith's perspective, nature did not need to be 'saved' from humans, since humans were part of nature. It is here where we can begin to see the importance of the production of nature thesis, for the insistence that humanity and nature stood in an internal relation, rather than an external one, pointed to an important analytical project: if nature is something produced, then the question becomes *how* and *why* it is that human and nonhuman natures are produced in the forms they are at any particular historical moment. Likewise, the thesis provided radical environmental geographers and environmental activists with a political project, for as Smith (1996, p. 50) put it, eco-politics could no longer be about saving nature from humans, but instead must find answers to the question: 'how, and by what social means and through what social institution is the production of nature to be organized?'

Others in this volume have provided a thorough discussion of attempts by Marxist geographers to account for specifically *capitalist* productions of nature, and how they answered Smith's questions about social means and social institutions (see George Henderson's chapter). Here I merely wish to note that not everyone was convinced that historical materialists overcame the nature-society dualism as successfully as they imagined. Critiques proceeded along several lines. On the one hand, critics argued that Marxists conceived of the production of nature in much too narrow a way, tending towards an economic reductionism that underplayed other social and cultural processes that shaped nature's material transformation (see Haraway, 1997), and paid inadequate attention to the connections between the production of nature and relations of race, gender and sexuality. One of the strongest challenges came from cultural and political geographers – many influenced by post-structuralist writers such as Jacques Derrida and Michel Foucault – who suggested that Marxist geographers had underplayed the role of ideas and images in shaping how environments were valued and transformed. The argument here was that the discursive construction of nature was *generative* in its own right, and not simply epiphenomenal to the economy. How the nonhuman world was *framed* as an object of knowledge or aesthetic appreciation was taken to be an integral part of how nature was made an object of economic and political calculation, and the sorts of cultural politics in play in specific environmental practices (see Braun, 2002). This applied equally to science as it did to art, each of which carried a force that could not be reduced to the dictates of capital.

As we will see later, the emphasis on nature's cultural construction was not inimical to the notion that nature was materially produced. Where such accounts differed was on how and why nature was produced in the form it was in any given context, and how scholars and critics could claim to know this nature in any direct or unmediated way. Others took exception to what they considered the anthropocentrism of *both* positions. We can understand this in two ways. On the one hand, constructionist accounts were said to be anthropocentric in an ethical-political sense, since they privileged the needs and desires of humans, and tended to treat nonhuman nature as mere means for human ends. On the other hand, these accounts were also said to be anthropocentric in an analytical sense, since they tended to place *human action* at the heart of their accounts of nature's production, rendering nature a static and inert realm. In other words, it was not clear that the *matter* of nonhuman nature mattered. Hence, while Marxist accounts sought to overcome the nature-society dualism, they tended to retain a subject-object dichotomy, and by doing so collapsed nature into society (see Castree, 1995).

As is discussed elsewhere in this volume, recent work by Marxist scholars has responded in a robust fashion to this charge, with various degrees of success. David Harvey (1996), for instance, expanded his dialectical approach to include the environment as a constitutive moment within a larger 'relational' ontology. How this dialectic unfolded, then, depended upon the specific elements of the economy and environment in question in any given occasion. Likewise, James O'Connor (1996) proposed that Marxist theory should be augmented by noting a 'second contradiction' to capitalism, in which the degradation of what Marx called the 'conditions of production' created a specific form of economic crisis. The material properties and processes of nonhuman nature, then, had some influence on how economic crises occurred, and on the social forms that emerged in attempts to overcome them. A great many Marxist geographers have explicitly taken up this question. In his work on forestry in the northwest United States, for instance, Scott Prudham (2005) gives full weight to the specific biological features of Douglas Fir forests and the mountain topography of Oregon, both of which presented immense challenges to capital, and shaped the technologies, work regimes, politics and labour relations that emerged in the region. A similar argument has been made by Karen Bakker (2004) who has shown how the physical properties of water repel attempts at commodification. Others, like Noel Castree, James McCarthy, Gavin Bridge, Becky Mansfield, Matthew Gandy, and Eric Swyngedouw have all registered the ways in which nonhuman nature is both a *problem* and an *opportunity* for capital, at once interrupting circuits of capital, and providing new spaces for commodification, a point which is perhaps made best by George Henderson (1999), in his classic work *California and the Fictions of Capital*.

A final point of contention with historical materialist accounts of nature has turned on the adequacy of *dialectics* for overcoming dualist conceptions of nature and society. The problem, in the eyes of critics, is that in important respects dialectical approaches still presume the existence of the initial categories (nature and society) even as they seek to multiply the connections between them. For the sociologist Bruno Latour (1993) dialectics remains too crude an analytical device that at best renders the nature-society divide more permeable, and at worst deepens the original error. Likewise, for Sarah Whatmore (1999, p. 25) the problem with dialectics is that instead of challenging the *a priori* categorisation of things into 'nature' and 'society' it raises this binary logic 'to the level of a contradiction and engine of history'. The question that Latour and Whatmore ask, then, is on what basis a dialectical relation can be said to exist between 'nature' and 'society' if the categories do not hold in the first place?

Univocity and Ontogenesis: Non-essentialist Ontologies and the Matter of Nature

If dialectics provided one influential approach to the matter of nature, a substantially different solution to the problem of Enlightenment antinomies has been put forward by a number of scholars who have attempted to sidestep the categories 'nature' and 'society' altogether. For these 'new materialists', a term that I will say more about below, nature and society are categories that we have imposed upon a world that can never be so neatly divided.

In some respects this position has become easier to grasp with the rapid growth of biotechnology and the proliferation of so-called hybrid entities like genetically modified organisms or through the use of pharmaceuticals to transform the habits and capacities of bodies. With each innovation it becomes ever more difficult to divide the world into separate domains. But the argument put forward here is an ontological one, rather than an historical one, and, as we will see, it places emphasis on the univocity of being. For new materialists, the point is *not* that we have recently entered a world in which nature and society are increasingly indistinct, but that these categories *have never been distinct to begin with*, since the world has never been divisible into separate planes. In a sense, such arguments take Enlightenment dualisms and stand them on their heads. Whereas dualist thought begins from an original separation, and then worries over how these separate domains might be related, the new materialists posit a single ontological plane – philosopher Baruch Spinoza's 'Substance', for instance – from which emerges the differentiated and differentiating worlds that we inhabit. Hence there is not a 'social' realm in one location and a separate 'natural' realm elsewhere, nor a dialectical relation between them; rather the things that we consider to be 'natural' or 'social' can be considered so only through practices of purification by which objects are assigned to either pole (see Latour, 1993; Haraway, 1997).[1] Indeed, from such a perspective, the figure of the human is not something that magically appears on the scene from elsewhere, but emerges from its involvements in the world – from its entanglements with tools, animals, minerals and viruses.

We can see the influence of these decidedly non-essentialist ontologies in one of the most creative and sustained efforts in the past two decades to rethink the matter of nature, led in large part by a group of geographers in the United Kingdom, including Sarah Whatmore, Nick Bingham, Steve Hinchliffe and Gail Davies, whose originally unorthodox views on the matter have rapidly gained traction. For our purposes we can identify four significant contributions that they have made to how we think about the nature of nature, including some that return us to the work and insights of a number of physical geographers. The *first* is a methodological emphasis on immanence. Each writer refuses to take recourse to any supplemental dimension or transcendental cause – whether this be God, Capital, Spirit or History – which lies above, beyond or behind worldly phenomena, and which determines their form. In the words of Gilles Deleuze (1988, p. 122), there is only one 'common plane of immanence on which all bodies, all minds, and all individuals are situated'. Whatever 'is' must therefore be understood to have emerged from the flux of bodies and matter in the practices of everyday life. This is related closely to a *second* emphasis on individuation rather than identity, which holds that the world is not characterised by discrete classes of being, or by the eternal repetition of the same, but by ongoing differentiation and individuation, both in human and nonhuman nature. We can make this concrete if we think back to Steve Hinchliffe's example of water voles in the urbanised ecologies of Birmingham, England. Hinchliffe argues that the capacities of the urban voles were not simply the innate qualities of a species in general; they were emergent effects of an assemblage within which *these voles in particular* came to take on unique qualities and capacities. This flies in the face of identity-thinking, which assumes that the world can be unproblematically divided into different classes of being, as well as conservation policies like habitat trading that are based on the same assumption, and thus fail to recognise that organisms are constituted not simply through genetic selection, but through their activities in specific environments. Voles do not just differ from rats and mice, they differ from themselves. It is equally important to stress that Hinchliffe's point is not that this sort of ontogenesis is a uniquely urban, and hence 'unnatural' phenomenon – although the particular environment makes a great difference to the emergent capacities of the organisms composed in its spaces – but that the contingent composition of the organism is an underlying *ontological* truth equally valid in all contexts, urban or rural.

Here we might pause to note that this emphasis on contingency and self-organisation is not far removed from similar discussions in the physical sciences, a point that I will return to below. Before I do, let me note that if taken to their logical conclusion, new materialist approaches present a sharp challenge to the subject-object dualism that has long characterised Western thought, as well as how we think about and locate 'agency' in the world. Within these heterogenous networks entities are simultaneously subject *and* object, or, in the words of Michel Serres, 'quasi-subjects, quasi-objects', since all entities, human and nonhuman alike, have the capacity for *affect* (Serres and Latour, 1995). This is to say that they can *receive* affections from other entities (consider the way that a cyclist gains bodily knowledge of the resistance of hills, the ratios of gears, or the pressure required to activate brakes), and in turn can *cause* affects in others (such as in the training of a dog, although the trainer is just as often the trainee).

The *third* contribution of new materialists, then, is not only to have shed new light on the age-old matter of nature's agency, but to have given geographers some radically new ways to think about the what is *meant* by agency. For writers like Latour, Callon, Whatmore and Hinchliffe, agency is not an innate property that belongs to things, but an emergent effect of the ways in which entities enter into combination with others (Callon and Law, 1995; Whatmore, 1999). Gilles Deleuze and Felix Guattari may have captured this best in their concept *agencement*, a concept that lies behind Bruno Latour's more widely referenced figure of the actant. *Agencement* relates and combines two different ideas in a clever wordplay in which the idea of a layout or a coming together of disparate elements contains within it also the idea of agency or the capacity to cause affects. The effect is to neatly relate the coming together of things with the capacity to act, where the latter is seen to be an effect of the former (see Callon and Law, 1995; Hardie and MacKenzie, 2006; Phillips, 2006; Palmas, 2007). Hence, as writers like Sarah Whatmore and Steve Hinchliffe both emphasise, the capacity for any particular thing to cause affects – whether this thing be human or nonhuman – must be seen to belong not to the individual thing, as humanism teaches us, but to the larger collectivity out of which any actor is composed. Indeed, the term 'actant' has long been confusing in this regard, since its singular emphasis – *an* actant – obscures the distributed notion of agency, which Latour borrows from Deleuze and Guattari, and thus loses some of the force of *agencement*. Some have suggested that all of this gives us a 'flat' ontology that cannot account for the differential power relations that we see in our everyday lives. In response, advocates of this position have suggested that there is nothing in the above which suggests that agency is ever evenly distributed, only that power – or the capacity to cause effects – does not exist apart from the arrangements that constitute entities with more or less power. So, while it is true that humans wield far more power than grizzly bears, to draw upon an example from the interior mountains of British Columbia, if you strip the human of her car, binoculars and rifle, the tables are quickly turned. In other words, power is itself an emergent effect of heterogeneous networks, not an innate quality of autonomous bodies.

All of this has crucial implications for a *fourth* area: epistemology. Earlier I noted that one of the challenges to Marxist theories of nature came from critics, such as this author, who felt that its economism led it to underplay the role of nature's cultural construction. Not only were ideas about nature provisional and power-laden, we argued, they also had very material effects, and were part and parcel of *how* the production of nature occurred. It is precisely this constructivist emphasis, however, that new materialists have vigorously questioned. Or, more to the point, they have questioned the assumption that knowledge about human and nonhuman nature can or should be understood primarily in *representational* terms. There are several reasons for this. The first is because for the most part constructivist accounts remain wedded to a subject-object dichotomy, even if the subject is itself constituted in and through ideology or in relation to particular disciplinary practices. In other words, constructivist accounts of nonhuman world leave no room for the nonhuman world! It is presumed that knowledge is acquired through a detached contemplation, or through an arbitrary and differential system of signs, where signs obtain their meaning through their relation to other signs. What this elides, new materialist argue, is the possibility that we know the world through our practical engagements with it, rather than through a passive and detached observation. By this view, science is not just about 'seeing', or about the application of a disembodied reason, but about a set of embodied practices through which nonhuman entities are encountered and subsequently translated into matters of fact. Scientists are not merely detached observers and nature is never a passive or inert field (Latour, 2004b). Put in more philosophical terms, science is located on the same plane of immanence as the things it purports to study. Hence the 'matters of fact' produced by science are seen to emerge from the conjoined capacity of scientists and nonhuman nature to affect and be affected by each other.

Equally as important, critics of constructivism have argued that there are myriad non-cognitive ways of knowing that cannot be reduced to representation, such as through touch or smell, or in relation to movement and rhythm (Harrison, 2000; McCormack, 2004; Lorimer, 2005; McCormack, 2005; see also Ingold, 2000). These non-cognitive knowledges can be traced in activities as diverse as gardening, mountain climbing or driving a car: in each we encounter the world not first and foremost as a set of visual images, but as a set of physical affects, by which our bodies register the feel of soil, the grade of a climb, or the torque on a steering wheel. In short, nonhuman entities are not merely vessels that humans fill with meaning; through their 'performances' they add something of their own to the story.

Neo-Vitalism, Cosmopolitics and Ethics

Recently, Hayden Lorimer (2005, pp. 84–85) has pushed this one step further. The problem with representationalism, Lorimer suggests, is 'that it framed, fixed and rendered inert all that *ought* to be most lively' (emphasis added). Here we find an increasingly common theme in many writings on nature by new materialist scholars: the *vitality* or *liveliness* of nonhuman nature. This vitalism is implicit in Hinchliffe's water voles, which made a mess of conservationists' systems of classification. But we can find it stated explicitly in the work of numerous contemporary geographers, to the point where this position is rapidly becoming as orthodox as were earlier approaches that assumed nature to be an inert realm of timeless essences. Rose and Wylie (2006, p. 476), for instance, tell us that life is characterised by a 'burgeoning, proliferating, even wondrous topology'. Likewise, Matthew Kearnes (2006, p. 67), drawing upon Gilles Deleuze, suggests that 'the singularity of matter is alive with the creative potential of endless evolutions and innovations'. Nature, it seems, has a sense of humor, as do the socio-technical networks out of which new entities are continuously born (Davies, 2007). The earth is 'volatile' (Clark, 2007). Everywhere life is 'feral' (Clark, 2003), 'being summoned' (Thrift, 2004), or simply being 'added to' (Bingham, 2006). Even technological objects are now seen as 'ontologically unstable', putting in question our dreams of mastery (Kearnes, 2006; Thrift, 2006).

It merits comment that these vitalist tendencies in human geography mirror, and to some extent draw upon, the growing influence of complexity theory, non-linear dynamics and notions of self-organisation within the physical sciences, including physical geography. In many fields, notions of equilibrium are decidedly out of fashion, for, as physical geographer Barbara Kennedy (1994, p. 703) argues: 'If there *is* any non-transient part of our planet's surface in something we might term 'equilibrium' it is surely a real oddity and what, if anything, would it tell us about the rest of the globe?'. Whether equilibrium theories are entirely outmoded, or, indeed, whether complexity theory can be said to be entirely opposed to them, is not entirely clear. What *is* emphasised today is that any sort of equilibrium is best understood as an achieved state, rather than an eternal essence, or, in the language of complexity theory, an *emergent order* that is the property of the whole, rather than something that can be reduced to, or predicted by, the component parts of a system. In other words, emergence cannot be predicted in advance, but can only be known in its effects. As a number of commentators have noted, drawing upon writers such as Gilles Deleuze, Henri Bergson, Gilbert Simondon, Keith Ansell Pearson and Brian Massumi, this places a premium on the 'inventiveness' of the earth.

It is not entirely clear how far this emphasis on 'emergence' can or should be pushed. Physical scientists, for instance, have tended to be less concerned with pointing to novelty for its own sake, and have placed equal emphasis on processes that *sustain* certain material forms, or the ways that apparently chaotic phenomena at one scale resolve into forms of meta-stability at another, or the significance of specific thresholds (singularities) for shifts from one steady state to another. Nor is it clear whether vitalism should be read literally – as giving us a true description of the world – or taken as a cautionary tale against modern(ist) dreams of mastery. Drawing upon the arguments of Canguilhem (1994), for instance, Fraser et al. (2005, p. 2) suggest that 'Vitalism remains vital partly because of its epistemological role within the history of the life sciences. . . . [It] functions in part as an ongoing form of resistance to reductionism and to the temptation of premature satisfaction, closure, denial or ignorance' (see also Greco, 2005). Even philosopher Henri Bergson, whose work from the first decades of the 20th century has inspired many of today's new vitalists, emphasized the *ethical* force of the position. If nothing else, he explained, 'the "vital principle" . . . is at least a sort of label affixed to our ignorance'. Because it gives us a world filled with contingency, and calls attention to that which is permanently suspended between being and non-being, it perhaps best names a discipline of thought (Greco, 2005; see also Stengers, 1997), that in turn informs an ethical relation to life and a political orientation.

What this helpfully illuminates are the close connections between how one answers a set of ontological questions about the 'nature' of the material world, and what one holds as a set of ethical and political commitments. There is no hard and fast rule that a particular ontology leads necessarily to a particular politics, but neither can any ontology be said to be neutral. If we imagine that nature names an immutable realm, for instance, and see humans to be part of it, we have ample justification in support of existing social relations, since these can be passed off as 'natural'. Likewise, if we imagine nature to name a realm entirely external to humans, it may be possible, as some have suggested, to treat it merely as so many objects of utility, or, as others have suggested, to imagine that it has 'inherent value'.

If this is true, then it follows that the accounts of nature given by new materialists are no more innocent than any other accounts; they too can underwrite a particular orientation to the world. Indeed, it is possible to argue that the non-essentialist materialisms of these writers leads to a politics of nature that must invariably be a kind of active experimentation, since 'we do not know in advance which way a line is going to turn' (Deleuze and Parnet, 2002, p. 134; see also Braun, 2006b). From this perspective, the discipline of geography – as earth-writing – does not stand outside this experimentation, but participates in it. They may also teach us that dreams of mastery, or reductionist accounts of such things as nanotechnology, which presume that we can build things 'atom by atom' without any surprises, are the height of hubris, and harbour the possibility of catastrophe. For, as Kearnes (2006, p. 59) puts it, 'in the application of force and control we can also see the radical possibility for creativity and escape'.

It is precisely this radical uncertainty that has informed the ethical and political positions of post-dualist geographers such as Sarah Whatmore, Steve Hinchliffe and Nick Bingham. For each of them humans exist in the midst of things. Thus, as Bingham explains, being is always already being-with-one-another, not in terms of a pluralism that imagines a world of diverse yet discrete things, but in terms of a 'community of singularities' in which different forms of life are constituted through what circulates between them. If we add to this the vitalist intuition that the world is not a fixed and eternal order, but is instead continuously 'added to' through the performances of people and things, then the most pressing task we face today may be to develop institutional spaces and procedures that allow us to work through, in an agonistic manner, how this composition of common worlds should proceed (see Stengers, 2000; Latour, 2004; Latour and Weibel, 2005). Who or what must we take into account in our biotechnological innovations? What does it mean to 'add' something like GM crops to a world where human and nonhuman lives coexist, and where nonhumans also have the capacity to affect wider collectivities? Whose lives should flourish, and whose should be abandoned or excluded from our collectives? Given what post-dualist and non-essentialist ontologies suggest about the interwoven nature of human and nonhuman lives, how might we slow down the process of assembly, in order to properly weigh the propositions that continuously confront the collectives in which we dwell with new and often strange matters of concern?

These concerns have increasingly come to the forefront in the work of environmental geographers, and perhaps suggest a common ground shared by human and physical geographers alike. This does not mean that all contemporary geographers pose ethical and political questions in these exact terms. The environmental justice movement, for instance, has tended to place attention on questions of social inequality within these assemblages of people and matter, taking up Neil Smith's (1996) appeal for a 'political theory' of nature that attends to its social production (see Di Chiro, 1995). Others, influenced by the cultural turn, have suggested the need for a 'deconstructive responsibility' that never loses sight of the violence inherent in any closure around being, ethics and politics, even as it acknowledges the necessity of making provisional claims about all three (see Braun, 2002). Still others have asked why it is that we draw limits around whom or what is allowed 'representation' in our political arenas. If animals are part of our 'communities of singularities', if their forms of life are constituted in relations to ours, why should they not be taken into account when we design new biotechnologies, burn fossil fuels or clear forests?

Once dualism is abandoned, it seems, nature becomes political, and politics finds itself filled to the brim with nature, which it never really had left behind. It is this attention to the making of common worlds – what Isabelle Stengers (2000; 2003) rightly calls cosmopolitics – that is the task left to us. In this task, vitalism may offer a valuable ethical and practical orientation, one that recognizes the ontological instability of matter, and thus takes precaution as its central principle. For if we live in a world in which 'intersection, transfer, emergence and paradox are central to life' (Thrift, 2004, p. 83), then we face a situation that is equally terrifying and hopeful, in which 'anything is possible – the worst disasters or the most flexible evolutions' (Guattari, 2000, p. 66).

NOTES

1. It is important to recognise that the term 'hybridity' is not a term that fits well in the lexicon of the new materialists since it presupposes the existence of the two separate domains. It is better seen as a 'middle term' that names an impasse in dualist thought. The new materialists discussed in this section *begin* with the middle term and drop the two poles.
2. The following sections draw in part on arguments developed in Braun 2007, 2008.

BIBLIOGRAPHY

Agamben, G. (2004) *The Open: Man and Animal.* Translation by Kevin Attell. Stanford: Stanford University Press.

Bakker, K. (2004) *An Uncooperative Commodity: Privatizing Water in England and Wales.* Oxford: Oxford University Press.

Bakker, K. and Bridge, G. (2006) Material worlds? Resource geographies and the 'matter of nature'. Progress in Human Geography, 30(1), 1–23.

Barry, A. (2001) *Political Machines: Governing a Technological Society.* London: Athlone.

Barry, A. (2002) The anti-political economy. *Economy and Society,* 31(2), 268–84.

BBC News (2007) Butterfly shows evolution at work. http://news.bbc.co.uk/go/pr/fr/-2/hi/science/nature/6896753.stm (accessed 12 July).

Bergson, H. (1988) [1911] *Matter and Memory.* Translation by Nancy Margaret Paul and W. Scott Palmer. New York: Zone Books

Beven, K. (2002) Towards a coherent philosophy for modeling the environment. *Proceedings of the Royal Society of London A,* 458, 2465–84.

Bingham, N. (2006) Bees, butterflies and bacteria: biotechnology and the politics of nonhuman friendship. *Environment and Planning A,* 38, 483–98.

Bingham, N. (2008) Slowing things down: lessons from the GM controversy, *Geoforum,* 39(1), 111–22.

Braun, B. (2006a) Global natures in the space of assemblage. *Progress in Human Geography,* 30(5), 644–54.

Braun, B (2006b) Toward a new earth and a new humanity: nature, ontology, politics. In N. Castree and D. Gregory (eds), *David Harvey: A Critical Reader.* Oxford: Blackwell, pp. 191–222.

Braun, B. (2007) Theorizing the nature-culture divide. In K. Cox, M. Low and J. Robinson (eds), *Handbook of Political Geography.* London: Sage.

Braun, B. (2008) Inventive life. *Progress in Human Geography,* Prepublished May, DOI: 10.1177/0309132507088030.

Bridge, G. (2000) The social regulation of resource access and environmental impact: production, nature and contradiction in the US Copper Industry. *Geoforum,* 31, 237–56.

Callon, M. (1998) *The Laws of the Markets.* Oxford: Blackwell.

Callon, M. (2002) Technology, politics and the market. Interview with Andrew Barry and Don Slater. *Economy and Society,* 31(2), 285–307.

Callon, M. and Law, J. (1995) Agency and the hybrid collectif. *South Atlantic Quarterly,* 94(2), 481–507.

Callon, M. and Caliskan, K. (2005) New and old directions in the anthropology of markets. Paper presented to Wenner-Gren Foundation for Anthropological Research, New York. April 9.

Canguilhem, G. (1994) *A Vital Rationalist.* New York: Zone Books.

Castree, N. (1995) The nature of produced nature: materiality and knowledge construction in Marxism *Antipode,* 27(1), 12–48.

Castree, N. (2003) Environmental issues: relational ontologies and hybrid politics. *Progress in Human Geography,* 27, 203–11.

Castree, N. (2006) A congress of the world. *Science as Culture,* 15(2), 159–70.

Church, M. (1996) Space, time and the mountain – how do we order what we see? In B. L. Rhoads and C. E. Thorn (eds), *The Scientific Nature of Geomorphology.* Proceedings of the 27th Binghamton Symposium in Geomorphology. New York: John Wiley, pp. 147–70.

Collingwood, R. G. (1960) *The Idea of Nature.* Oxford: Oxford University Press.

Cooper, M. (2006) Pre-empting emergence: the biological turn in the war on terror. *Theory, Culture and Society,* 23(4), 113–35.

Cosgrove, D. (1985) Prospect, perspective and the evolution of the landscape idea *Transactions of the Institute of British Geographers,* 10(1), 45–62.

Crouch, D. (2001) Spatialities and the feeling of doing. *Social and Cultural Geography,* 2(1), 61–75.

Crouch, D. (2003) Performances and constitutions of nature: a consideration of the performance of lay geographies. In B. Szerszynski, W. Heim and C. Waterton (eds), *Nature Performed: Environment, Culture and Performance*. Oxford: Blackwell, pp. 17–30.

Clark, N. (2003) Feral ecologies: performing life on the colonial periphery. In B. Szerszynski, W. Heim and C. Waterton (eds), *Nature Performed: Environment, Culture and Performance*. Oxford: Blackwell, pp. 203–18.

Clark, N. (2007) Living through the tsunami: vulnerability and generosity on a volatile earth. *Geoforum*, 38(6), 1127–39.

Collingwood, R. (1945) *The Idea of Nature*. Oxford: Clarendon Press.

Daniels, S. (1993) *Fields of Vision : Landscape Imagery and National Identity in England and the United States*. Cambridge: Polity Press.

Davie, G. (2007) The funny business of biotechnology: better living through chemistry comedy. *Geoforum*, 38(2), 221–3.

Deleuze, G. (1988) *Spinoza: Practical Philosophy*. Translated by R. Hurley. San Francisco: City Lights Books.

Deleuze, G. (1994) *Difference and Repetition*. Translated by Paul Patton, London: Athlone Press.

Deleuze, G. and Guattari, F. (1987) *A Thousand Plateaus: Capitalism and Schizophrenia*. Minneapolis: University of Minnesota Press.

Deleuze, G. and Parnet, C. (2002) *Dialogues*. New York: Columbia University Press.

Derrida, J. (1978) *Writing and Difference*. London: Routledge.

Di Chiro, G. (1995) Nature as community: the convergence of environment and social justice. In William Cronon (ed.), *Uncommon Ground: Toward Reinventing Nature*. New York: W.W. Norton & Company, pp. 298–320.

Fiser, A. (2000) Theory as event: the ontogenesis of virtual-structuralism in Deleuze and Guattari. *Space and Culture*, 7/8/9, 59–69.

Foucault, M. (1989) Introduction. In G. Canguilhem (ed.), *The Normal and the Pathalogical*. New York: Zone Books.

Fraser, M., Kember, S. and Lury, C. (2005) Inventive life: approaches to the new vitalism. *Theory, Culture and Society*, 22(1), 1–14.

Glacken, C. (1967) *Traces on the Rhodian Shore: nature and culture in Western thought from ancient times to the end of the eighteenth century*. Berkeley: University of California Press.

Greco, M. (2005) On the vitality of vitalism. *Theory, Culture and Society*, 22(1), 15–27.

Greenhough, B. and Roe, E. (2006) Toward a geography of bodily technologies. *Environment and Planning A*, 38, 416–22.

Gregory, K. J. (2000) *The Changing Nature of Physical Geography*. London: Arnold Publications.

Haraway, D. (1997) *Modest_Witness@Second_Millennium: FemaleMan_Meets_Oncomouse*. London: Routledge.

Hardie, I. and MacKenzie, D. (2006) Assembling an economic actor: the *agencement* of a hedge fund. Paper presented at workshop 'New actors in a financialised economy and implications for varieties of capitalism', Institute of Commonwealth Studies, London, 11–12 May.

Harrison, P. (2000) Making sense: embodiment and the sensibilities of the everyday. *Environment and Planning D: Society and Space*, 18, 497–517.

Harvey, D. (1974) Population, resources and the ideology of science. *Economic Geogrpahy*, 50(2), 256–77.

Harvey, D. (1982) *Limits to Capital*. Oxford: Blackwell.

Harvey, D. (1996) *Justice, Nature and the Geography of Difference*. Oxford: Blackwell.

Hassan, M. A., Egozi, R. and Parker, G. (2006) Effect of hydrograph characteristics on vertical sorting in gravel-bed rivers: humid versus arid environments. *Water Resources Research*, 42, W09408, doi:10.1029/2005WR004707.

Henderson, G. (1999) *California and the Fictions of Capital*. Oxford: Oxford University Press.

Hewitt, K. (ed.) (1983) *Interpretations of Calamity*. Boston: Allen & Unwin.

Hinchliffe, S. (2008) Reconstituting nature conservation: towards a careful political ecology. *Geoforum*, 39, 88–97.

Hitchings, R. (2003) People, plants and performance: on actor-network theory and the material pleasures of the private garden. *Social and Cultural Geography*, 4(1), 99–113.

Ingold, T. (2000) *The Perception of the Environment: Essays on Livelihood, Dwelling and Skill*. London: Routledge.

Inkpen, R. (2005) *Science, Philosophy and Physical Geography*. London: Routledge.

Kearnes, M. (2006) Chaos and control: nanotechnology and the politics of emergence. *Paragraph*, 29(2), 57–80.

Kennedy, B. (1994) Requiem for a dead concept. *Annals of the Association of American Geographers*, 84(4), 702–5.

Kosek, J. (2006) *Understories: The Political Life of Forests in Northern New Mexico*. Durham: Duke University Press.

Latour, B. (1993) *We Have Never Been Modern*. Cambridge, MA: Harvard University Press.

Latour, B. (1999) *Pandora's Hope: Essays in the Reality of Science Studies*. Cambridge, MA: Harvard University Press.

Latour, B. (2004a) How to talk about the body? The normative dimension of science studies. *Body and Society*, 10, 205–29.

Latour, B. (2004b) *Politics of Nature: How to Bring the Sciences into Democracy*. Cambridge, MA: Harvard University Press.

Latour, B. (2005) *Reassembling the Social*. Oxford: Oxford University Press.

Lewis, N. (2000) The climbing body: nature and the experience of modernity. *Body and Society*, 6(3–4), 58–80.

Lorimer, H. (2005) Cultural geography: the busyness of being 'more-than-representational'. *Progress in Human Geography*, 29(1), 83–94.

Lorimer, H. (2006) Herding memories of humans and animals. *Environment and Planning D: Society and Space*, 24, 497–518.

Lorimer, J. (2006) What about the nematodes? Taxonomic partialities in the scope of UK biodiversity conservation. *Social and Cultural Geography*, 7(4), 539–58.

Lorimer, J. (2007) Nonhuman charisma. *Environment and Planning D: Society and Space*, 25, 911–32.

Mansfield, B. (2003) Spatializing globalization: a 'geography of quality' in the seafood industry. *Economic Geography*, 79(1), 1–16.

McCormack, D. (2005) Diagramming practice and performance. *Environment and Planning D: Society and Space*, 23, (119–47).

McKibbon, B. (1889) *The End of Nature*. New York: Random House.

Merchant, C. (1990) *The Death of Nature: Women, Ecology and the Scientific Revolution*. New York: Harper and Row.

Merleau-Ponty, M. (2002) [1945] *Phenomenology of Perception*. Translated by Colin Smith. London: Routledge.

Michaels, M. (2000) These books are made for walking . . . : mundane technology, the body and human-environment relations. *Body and Society*, 6, 107–26.

Mitchell, T. (2008) Rethinking economy. *Geoforum*, 39(3), 1116–21.

Nash, C. (2000) Performativity in practice: some recent work in cultural geography. *Progress in Human Geography*, 24, 653–64.

O'Connor, J. (1996) The second contradiction of capitalism. In Ted Benton (ed.), *The Greening of Marxism*. New York: Guilford Press, pp. 197–221.

Palmas, K. (2007) Deleuze and DeLanda: a new ontology, a new political economy? Paper presented at the Economic Sociology Seminar Series, Department of Sociology, London School of Economics and Political Science, 29 January 2007.

Pels, D., Hetherington, K. and Vandenberghe, F. (2002) The status of the object: performances, mediations and techniques. *Theory, Culture and Society*, 19(5–6), 1–21.

Perkins, H. (2007) Ecologies of actor-networks and (non)social labour within urban political economies of nature. *Geoforum*, 38(6), 1152–62.

Phillips, J. (2006) *Agencement*/Assemblage. *Theory, Culture and Society*, 23(2–3), 108–9.

Prudham, S. (2005) *Knock on Wood: Nature as Commodity in Douglas-Fir Country*. London: Routledge.

Robbins, P. (2007) *Lawn People: How Grasses, Weeds and Chemicals Make Us Who We Are*. Philadelphia: Temple University Press.

Rose, M. and Wylie, J. (2006) Animating landscape. *Environment and Planning D: Society and Space*, 24(4), 475–80.

Serres, M. and Latour, B. (1995) *Conversations on Science, Culture and Time*. Ann Arbor: University of Michigan Press.

Smith, N. (1984) *Uneven Development: Nature, Capital and the Production of Space*. New York: Blackwell.

Spinoza, B. (1994) *A Spinoza Reader : The Ethics and Other Works*, edited and translated by Edwin Curley. Princeton: Princeton University Press.

Stengers, I. (1997) *Power and Invention*. Minneapolis: University of Minnesota Press.

Stengers, I. (2000) *The Invention of Modern Science*, translated by D. W. Smith. Minneapolis: University of Minnesota Press.

Stengers, I. (2003) *Cosmopolitiques I*. Paris: La Découverte.

Stiegler, B. (1998) *Technics and Time, 1: The Fault of Epimetheus*, translated by Richard Beardsworth and George Collins. Stanford: Stanford University Press.

Swyngedouw, E. (2006) Metabolic urbanization: the making of cyborg cities. In N. Heynen, M. Kaika, and E. Swyngedouw (eds.), *In the Nature of Cities – Urban Political Ecology and the Politics of Urban Metabolism*. London: Routledge, pp. 21–40.

Szerszynski, B., Heim, W. and Waterton, C. (eds) (2003) *Nature Performed: Environment, Culture and Performance*. Oxford: Blackwell.

Thrift, N. (2000) Still life in nearly present time. *Body and Society*, 6(3–4), 34–57.

Thrift, N. (2004) Summoning Life. In P. Cloke, P. Crang and M. Goodwin (eds), *Envisioning Human Geographies*. London: Arnold, pp. 81–103.

Thrift, N. (2005) From born to made: technology, biology and space. *Transactions of the Institute of British Geographers*, 30, 463–76.

Turner, B. L., Clark, W. C., Kates, R. W., Richards, J. F., Matthews, J. T. and Meyer, W. B. (1990) *The Earth as Transformed by Human Action*. Cambridge: Cambridge University Press.

Urban, Michael and Bruce Rhoads. 2003. Conceptions of nature: implications for an integrated geography. In S. Trudgill and A. Roy (eds), *Contemporary Meanings in Physical Geography: From What to Why*. London: Arnold, pp. 211–32.

Wainwright, J. (2005) Politics of nature: a review of three recent works by Bruno Latour. *Capitalism, Nature, Socialism*, 16(1), 115–27.

Whatmore, S. (2002) *Hybrid Geographies: Natures, Cultures, Spaces*. London: Sage.

Whitehead, A. (1979) [1929]. *Process and Reality: An Essay in Cosmology*. Edited by David Ray Griffin and Donald W. Sherburne. New York: Free Press.

Williams, R. (1980) *Problems in Materialism and Culture*. London: New Left Books.

Wolf-Meyer, M. (2007) *Nocturnes: Sleep, Medicine, Governmentality and the Production of American 'Everyday Life'*. Dissertation, Department of Anthropology, University of Minnesota.

Wylie, J. (2005) A single day's walking: narrating self and landscape on the South West Coast Path. *Transactions of the Institute for British Geographers, New Series*, 30, 234–47.

Zimmerer, K. (1994) Human geography and the 'new ecology': the prospect and promise of integration. *Annals of the Association of American Geographers*, 84(1), 108–25.

Zimmerer, K. (2000) The reworking of conservations geographies: non-equilibrium landscapes and nature-society hybrids. *Annals of the Association of American Geographers*, 90(2), 356–69.

Chapter 3

Sustainability

Becky Mansfield

Introduction

'Sustainability' is wildly popular as a way of thinking about how to simultaneously meet the needs of people and the environment by enhancing human well-being without undermining ecological integrity. Since it came into prominence in the 1980s, debate about sustainability has underscored the political nature of conservation, economic development, human well-being, and links among them. Sustainability also highlights the political nature of socio-ecological processes that produce environmental degradation, poverty, and injustice – in short, the political nature of *un*sustainability. At the same time, it is striking the extent to which politics – relations of power – have been written out of the vast majority of discussions about sustainability. While most will recognise that discussion about sustainability is itself contentious and therefore political, the orthodox view is that achieving sustainability is a technical issue. According to this orthodox perspective, all that is needed is better knowledge, incentives, and technology. This orthodoxy, however, ignores relations of power that create problems and impede solutions, and ignores ways 'sustainability', in its attempt to solve problems while avoiding politics, is itself a political project.

This chapter identifies several ways in which sustainability is political. First, in the shallowest sense, sustainability is political because it is the outcome of heated debate, much of it in the formal policy arena. Second, sustainability research and policy addresses itself to real-world processes that are always political in that they are shaped, at least in part, by relations of power. The political nature of these processes must be understood and addressed. Third, the concept of sustainability is inherently political because it is normative; it fundamentally involves value-laden choices. Finally, the current usage of sustainability in many academic and policy circles hides the latter two forms of politics by making sustainability appear to be technical. This chapter argues that this retreat from politics is a form of hidden politics.

The first section provides an overview of the trajectory of the global politics of sustainability, focusing on convergence between sustainability and neoliberalism in the international policy arena. The second section provides an overview of academic responses to this global politics. The strength of sustainability is that it bridges the social and the ecological both materially (sustainability as the search for 'win-win' solutions) and conceptually (sustainability as a way of thinking about how nature and society are interconnected). The central weakness is that much of the sustainability literature undermines this promise by making sustainability a technical issue. Subsequent sections demonstrate these strengths and weaknesses in three fields to which geographers contribute: conservation biology, sustainability science, and geography more generally.

Sustainability in Global Environmental Politics

Sustainability as it is used today usually references the term 'sustainable development', an idea that became enshrined in global policy discussion in the 1980s. The concept has much deeper roots in Twentieth Century resource management, which used calculations of 'maximum sustained yield' to regulate use of renewable resources such as fish and trees (Larkin, 1977). Sustainability is the level of use that matches the long-term rate of regeneration; using less is wasteful because resources go unused, while using more depletes the resource. The concept has been criticised from many angles (Larkin, 1977, in geography, see, e.g., Demeritt, 2001; Prudham, 2005), and explicit use of this approach was waning just as the term sustainability was coming into prominence in the context of sustainable development.

It was the 1987 UN-commissioned report *Our common future* (the 'Brundtland report') that launched sustainability into everyday use, defining sustainable development as 'development that meets the needs of the present without compromising the ability of future generations to meet their own needs'. The aim of the 1992 UN Conference on Environment and Development (the 'Earth Summit', held in Rio de Janeiro) was to implement sustainable development as it had been defined in the Brundtland report (see Adams, 2001; Mansfield, 2008). Sustainable development represented a shift regarding issues of environment and development, which until then had been considered to be largely separate. This shift represents a major victory for governments of the global South who had argued for decades that environmental concerns could not be considered separately from concerns about economic growth and equity. For them, the causes of environmental degradation are the same as those of Third World poverty: exploitative behaviour of governments and corporations from the North in the past and present. Further, attempts to get countries of the South to forgo development in the name of conservation were seen largely as neo-colonial efforts to control resources of the South for the benefit of the North. The concept of sustainable development, then, reflected North-South politics in policy discussions, and it reflected the realities of power relations between the North and South.

What is fascinating, however, is how the term 'sustainable development' managed to subvert politics at the very moment politics seemed to erupt most explicitly. It does so by entrenching the idea that economic growth is good for people and the environment. In the 1970s, conservationists considered the major causes of environmental problems to be economic growth through industrialisation (largely in the North) and population growth (largely in the South). In the Brundtland report and at the Earth Summit, policymakers maintained their focus on population but reversed their stance on economic growth. They argued that population is still a problem, but it is the outcome of poverty, rather than its cause. Because poverty is the problem, economic development becomes the solution to both socio-economic and environmental problems; industrialisation in the South will create economic growth that will decrease poverty, reduce population growth, relieve direct pressure on resources, and provide economic resources for conservation. No longer seen as an environmental threat or cause of global inequality, development became the route to sustainability. Governments around the world could embrace the broad outlines of this sustainable development agenda because they could sidestep discussion of politically difficult changes necessary to reduce poverty, increase equity, and create more environmentally friendly ways of living. Critics responded by claiming that the notion of sustainable development promotes the status quo, i.e. global economic activity that exploits the environment and dispossess the poor of access to resources (The Ecologist, 1993; Chatterjee and Finger, 1994).

Ten years later, the 2002 World Summit on Sustainable Development (the WSSD, held in Johannesburg), further entrenched the idea that sustainable development should be linked to capitalist development and neoliberal globalisation (Luke, 2005; Sneddon et al., 2006; Mansfield, 2008). The approach institutionalised at the WSSD not only subordinates sustainable development to neoliberalism but promotes neo-liberalism as the central means to achieve sustainable development. WSSD agree-ments are emblematic of a private, market-based approach to environmental protection and poverty alleviation. They promote free trade and investment in general, encourage developing countries to increase their level of participation in global trade, and explicitly state that it is necessary to implement agreements of the World Trade Organization to achieve sustainability. They also promote 'voluntary partnerships' in which governments work with the private sector to achieve particu-lar goals. Thus, the WSSD represents the triumph of neoliberalism as a framework for sustainable development. By using the term sustainability, proponents can cast neoliberal, market-based approaches as a form of egalitarianism, justice and ecologi-cal economics (Okereke, 2006; Krueger and Gibbs, 2007; Mansfield, 2008).

Sustainability as a Bridging Concept: Promises and Pitfalls

Academic commentators have responded to the troubling trajectory of sustainable development within global politics in different ways. Whereas some argue that the entire concept of sustainability should be abandoned because of its problematic political commitments (e.g., Luke, 2005), others argue that sustainability should be 'resuscitated and rescued from those proponents of sustainable development who use it to advance a development agenda that is demonstrably unsustainable' (Sneddon et al., 2006, p. 264, see also Krueger and Gibbs, 2007). For Sneddon, sustainability is precisely a way of bringing politics back into the debate, asking key questions about what is meant by sustainability and who will benefit from it. Because sustain-ability is a malleable concept, it has the potential to create bridges among very different people. Discussion about sustainability can be a way in which people recognise their differences and work through the politics of human-environment interactions (Sneddon, 2000; Padoch and Sears, 2005).

A resuscitated sustainability also creates bridges between the human and the natural, and between the social and physical sciences (e.g., Costanza et al., 2007).

Sustainability is generally represented as being at the nexus of environmental, economic and social concerns, such that sustainability can only exist if all three are addressed together (e.g., Sneddon, 2000; Whitehead, 2007). This offers a different way to think about the major problems of our time, and holds out the promise that something can be done to address these problems. In so doing, sustainability also recognises the integration of humans and nature as an inescapable reality. The question is *not* how do we re-integrate humans and nature in order to have a sustainable existence (suggesting that humans are currently an external disturbance to nature), but why do we have socio-ecological systems that are unsustainable and what do we need to create more ecologically friendly and socially just human-environment relations? It is in this sense that the bridging capacity of sustainability as a concept raises key political issues.

However, just as policy debates manage to subvert the political potential of sustainability at the very moment politics seemed to erupt, the same is true of academic debates. As Redclift (1994; 2005) has long emphasised, it is not always clear to what sustainability refers, or what is being sustained. Sustainability can refer to maintaining ecological processes, sustained resource production, or sustained profitability. There is 'strong' sustainability that focuses on ecosystem services in the broadest sense and 'weak' sustainability that focuses on protecting only those parts of nature for which people cannot develop substitutes (Neumayer, 2003). Sustainability can refer to fostering the well-being of all people, now and in the future (both intra- and inter-generational equity). Or it can refer to any set of practices that can be maintained over the long-term, regardless of their effect on particular people or environments! These differences make the neat triangle of sustainability – environment, economy, society – a little less neat. If people mean different things by these terms, and tend to prioritise one over the others, then reference to sustainability becomes a means to avoid hard discussions.

This suggests that using the term sustainability in any seriousness requires having some answer to the question 'sustainability of what?' Further, answers to this question cannot be found through scientific analysis. While research can certainly answer questions about the social and ecological effects of certain actions, it can only tell us if those outcomes are 'sustainable' *if we have already defined* sustainability. In other words, the process of defining sustainability is an inherently normative, political process. Yet many academic researchers fail to address these political issues, trying instead to use supposedly objective research about sustainability to answer questions about what sustainability should mean. In other words, researchers often *try to turn sustainability into a technical, rather than political, issue.*

The outcome is that there is tension between the promise of sustainability as a bridging concept and the pitfalls of sustainability as a retreat into the technical. In his review of contributions from ecology, ecological economics, and livelihoods, Sneddon (2000) argues that these fields all push the sustainability framework away from that offered by mainstream sustainable development, and do so by creating bridges between social and ecological processes (see also Sneddon et al., 2006). But he also argues that these fields 'tend to side step the power discrepancies embedded within social relations . . . which lie at the heart of many environment and development dilemmas' (2000, p. 538). In other words, they tend to avoid and ignore politics, thus blunting their effectiveness. The following sections build from and illustrate these insights regarding the potential and pitfalls of sustainability by examining several fields to which geographers have contributed most centrally.

Conservation biology

Because the field of conservation biology is centrally concerned with maintaining biological diversity (Society for Conservation Biology, 2007), it necessarily examines interconnections among social and ecological processes: what actions degrade the environment, and which will contribute to conserving it? Yet some conservation biologists criticise the notion of sustainability precisely because it embraces social questions about economics and equity. They worry that sustainability 'poses the particular risk that ecological and biodiversity concerns will be cast aside in favor of more pressing human wants' (Newton and Freyfogle, 2005, p. 23). Writing in direct response, Padoch and Sears (2005) point out that this view is part of the long history of global conservation politics, in which 'poor rural people around the planet have repeatedly received and rejected already too-simplified versions of urban and developed-country conservation priorities' (p. 40). In contrast, they see sustainability as an opportunity for those concerned about the environment to work with, rather than against, poor people of the world to address interlocked 'problems that affect the health and well-being of our own and other communities and of the environments in which we live. We need to know what our roles are in creating those problems and be engaged collectively in solving them' (p. 41).

Geographers are pushing discussion about importance of social issues within conservation in important directions. Campbell (2002) examines debates about sustainable use of the environment (in this case, endangered sea turtles and their eggs), finding that managers have a hard time addressing social concerns; biological science 'remains the privileged language' of the experts she interviewed (p. 1243). McSweeney (2005) engages debates about effects of population growth among indigenous peoples on tropical forests. She finds that in place of strategies such as fertility reduction, conservationists should use social science to address broader social dynamics regarding women's conservation activities and enforcement of indigenous territorial rights. Further, these social dynamics are fundamentally political, in that they are about power relations among various different groups of people.

Sustainability is particularly useful in the context of debates such as this about the necessity of addressing social dynamics. Because it explicitly forges a bridge between social and ecological concerns, reference to sustainability prevents withdrawal from politics into the technical. It does so by highlighting ways that politics are a key part of human-environment interactions, and by showing that a retreat into seemingly objective concerns about the environment is a political tactic. Such a retreat makes a political statement not only about what is important, but about what gets to count as relevant knowledge that can contribute to forging more sustainable human-environment relations.

Sustainability science

Another field to which geographers have made major contributions is sustainability science, which provides information regarding socio-economic and environmental patterns, causes of problems, and potential solutions (Kates et al., 2001; Clark and Dickson, 2003; Clark, 2007). Sustainability science is founded on the premise of bridging and integrating. As one of its founders put it, its 'core focus' is 'coupled human-environment systems' (Clark, 2007, p. 1737). The field is also explicitly interdisciplinary, and some of the major figures in the field, such as R. Kates, R. Kasperson and B. L. Turner, are geographers. Although young, sustainability science has been recognised by some of the top scientific journals, including *Science* and *Proceedings of the National Academy of Sciences*, which in 2007 started a 'sustainability science' section. This new prominence – which clearly not all fields have been able to achieve – means that both the interdisciplinary, human-environment approach and questions to which it addresses itself are being recognised as legitimate and important. This sort of prominence also gives the field the imprimatur of science (as its name, too, claims), such that the field is seen as the best way to produce rigorous and useful knowledge regarding coupled human-environment systems.

While rising visibility and legitimacy for this kind of integrative approach is to be applauded, one concern is that integration is fairly superficial; the field looks at both social and environmental issues, but does so in ways that do not carefully link them. One example is a pair of synthetic articles by Kates and Parris. The first lists and briefly describes 26 trends related to sustainability (e.g., 'slowing and differential population growth' and 'modification of grasslands and pasturelands') (Kates and Parris, 2003). These are based on trends identified in the NRC report on sustainability, for which Kates was co-chair of the board (National Research Council 1999). The second focuses on the status of four goals (reducing hunger, promoting literacy, stabilising greenhouse gas concentrations, and maintaining freshwater availability) (Parris and Kates, 2003). The trends and goals they address do include those that are social and those that are environmental, yet there is no effort to link them; little in the discussion of each target or goal is actually integrative. Most telling, they themselves say 'two of the goals . . . are selected from the consensus on meeting human needs, and the other two . . . are selected from the consensus on preserving life-support systems' (Parris and Kates, 2003, p. 8068). 'Human needs' and 'life-support systems' may both be important, but they are not treated as interconnected, either materially or analytically.

A troubling outcome of superficial integration is that researchers rarely attend to complexity of the socio-environmental processes they claim to be examining. Much of the research in sustainability science fails, in particular, to properly identify key social factors, such that not only the analyses but the problems themselves are treated as fairly technical. Parris and Kates fail to address key structural issues that lead to chronic hunger; as a result they advocate kinds of international aid policies that others suggest contribute to the problem in the first place (cf. Lappe et al., 1998). In a project quantifying water needs associated with adequately feeding everyone in the world, the researchers treat the challenge as the need to grow more food, and hence use more water (Rockstrom et al., 2007). They never address how water needs might change if developing countries stopped producing luxury foods (such as coffee) for elite consumers (cf. Lappe et al., 1998). This is a perfect example of the need to ask what it is we are trying to sustain! In a project on socio-environmental tradeoffs related to agroforestry in Indonesia, the researchers claim that in addition to examining local market forces they also address 'rarely considered cultural factors' (Steffan-Dewenter et al., 2007, p. 4973). Instead they treat in-migration to the study region as an apolitical process of cultural exchange (i.e. learning how to be market-oriented from these outsiders). This fails to analyse changes due to migration as complex political ecologies in which issues of ethnicity, access to resources, control of markets, access to government officials, and the like may be important (cf. Peluso, 1992; 2005). Reference to 'culture' allows the authors to avoid addressing relations of power in their study site. Against this trend, a more carefully integrative approach is offered by Turner et al. (2003a; 2003b), in their development of a framework for analysis of vulnerability to environmental change (i.e. the likelihood of experiencing harm). As they present it, vulnerability analysis aims not just to understand effects of environmental change on people, but also how those effects are shaped by ongoing coupled human-environment interactions at multiple scales. In other words, it is not enough to note there is a connection between humans and the environment, but one must carefully identify links among multiple, intersecting human-environment interactions.

Another key term in sustainability science (related to vulnerability) is *resilience*, which refers to the ability of systems to bounce back from (or at least not change state completely after) a stress or perturbation. Systems of humans and nature are 'interlinked in never-ending adaptive cycles of growth, accumulation, restructuring, and renewal' that occur at multiple temporal and spatial scales (Holling, 2001, p. 392). According to this framework, resilience is a function of the ability of the system to restructure and renew, rather than grow and accumulate. A wealth of research has refined the model, including making it more precise and useful for empirical measurement of systems and their sustainability (e.g., Cumming et al., 2005). Resilience theory improves on the literature cited above in that very little distinction is made between human and non-human aspects of systems. The problem, however, is that social dynamics are not well understood or addressed. There is little effort to understand why people do what they do; the resilience model is not explanatory. As a result, scholarship on resilience has very little to say about some of the supposed pillars of sustainability, such as equity or social justice. Instead, resilience is mainly about maintaining a given system and its ability to accumulate resources, with no discussion about who or what benefits from it. Resilience may be about 'understanding complexity' but that understanding is seen as objective and technical, rather than normative and political.

This reflects a larger problem with sustainability science, which is that scholars in this field tend to downplay political aspects of their work. Researchers do recognise that they are participating in a political process. Because their work is problem-oriented, sustainability scientists actively and openly 'promote a sustainability transition' (Clark, 2007, p. 1737); this requires engaging in political debates. But they claim to do so only on the basis of their research findings. That is, sustainability scientists see their science as a way of avoiding, and even trumping, the politics of sustainability. Thus Kates and Parris, cited above, imply they are engaged in an apolitical action of characterising goals and trends that already exist, rather than in a political action of choosing which goals and trends are important (cf. Morse, 2004). Similarly, researchers use the resilience model to characterise complex systems and identify key times and places for intervention, and they do so without seeming to engage in subjective and political discussions about which systems and interventions are good for whom and in what ways. Indeed, it is partly the ability to seem apolitical that gives sustainability science its legitimacy, and proponents themselves claim that they are trying to move away from overt politics. As Kates et al. (2001) state, 'during the late '80s and early '90s . . . much of the science and technology community became increasingly estranged from the preponderantly societal and political processes that were shaping the sustainable development agenda. This is now changing as efforts to promote the sustainability transition emerge from international scientific programs, the world's scientific academies, and independent networks of scientists' (p. 641).

Thus sustainability scientists argue that their scientific approach, because it is objective, can replace the overly subjective and politicised approach prominent in ongoing debates regarding what counts as sustainability and how to achieve it. In other words, science can define for us what should count as sustainable and what processes contribute to a sustainability transition. What should be obvious is that this completely ignores the key questions raised earlier about the normative – rather than objective – nature of decisions about what counts as sustainability. As Redclift points out, the idea that sustainability 'speak[s] to objective scientific method, without the complication of human judgement' has been present in debates about sustainability since at least the 1992 Earth Summit (2005, p. 17). This idea therefore precedes (and even suggests the need for) sustainability science as a new field. Sustainability science aims to bypass politics by making sustainability a technical question, yet in so doing scholars in this area ignore the extent to which they are actually participating in the politics of sustainability. They do so by claiming that their approach to sustainability is objectively better than others, which is also a claim about what kinds of knowledge get to count. Not only does this leave little room for non-academic forms of knowledge, it also denigrates other forms of academic research that are not seen as appropriately scientific. This, fundamentally, is the politics of sustainability.

Sustainability in geography

A very different understanding of sustainability is presented in the more general geographical literature in both human and nature-society geography. (Physical geographers have largely been absent from explicit discussion of what is meant by sustainability, yet a large proportion of the work that physical geographers are engaged in is related to sustainability, in that it is about understanding environmental change, especially as related to human action). Within this geographical literature, there is no unifying approach to the study of sustainability, yet there are some overarching contributions. The first is that geographic work, especially on nature-society relations, on the whole does a better job at integrating social and ecological concerns and processes, giving special attention to the complexity of these processes. The second is that geographers treat sustainability itself as diverse, rather than singular. It is context dependent, influenced by space, place, and scale, and – above all – is the *outcome* of diverse and complex socio-ecological relations. Although certainly not alone in addressing these issues, geographers contribute to sustainability discussions especially on the basis of their unique, long-standing spatial and human-environment traditions.

Turning to the first contribution, geographical literature presents a much different, more textured sense of what human-environment integration means. One way it does this is by challenging the notion that nature and society exist as two separate realms that interact. Instead, geographers demonstrate that the idea that they are separate is itself historical, and is based on a complicated politics of knowledge that is tied-up with the history of science, colonialism, capitalism, and the exploitation of both people and nature (Castree 2005). Further, views of nature, and of a human-nature split, influence actions. Dualistic views of nature not only justify actions that degrade the environment, but they also influence conservation strategies, which are often based on the idea of protecting external nature (Braun, 2002). This suggests that, because it fundamentally refers to notions such as economy, society, and environment, any discussion of 'sustainability' is always caught up in this politics of knowledge. It also suggests that strategies to achieve sustainability may be based on faulty foundations, and hence may contribute to problems rather than solutions. For example, Benjaminsen et al. (2006) show that deeply held visions of ideal landscapes and human-environment relations influence seemingly objective scientific notions (such as 'carrying capacity') and serve to obscure socio-ecological relations that do not fit these models, thus leading scientists and policymakers to privilege environmental sustainability over its social and economic dimensions.

Another way geographers present a very different sense of human-environment integration is by working to *explain* (rather than simply describe elements of) particular socio-ecological systems. In political ecology (broadly defined) researchers reject both the idea that people are only agents of destruction (i.e. humans are outside of nature) and simplistic explanations of environmental degradation and poverty (e.g., overpopulation or backwardness of local people) (Robbins, 2004; Castree, 2005). Researchers document various ways that people – in multiple times and places – have managed to create healthy (sustainable!) socio-ecological relations, and they document the breakdown of these healthy relations as a result of struggles over control of resources. Asking why people do what they do, researchers have found that environmental degradation often results from extensive political and economic processes, including state intervention and integration into capitalist markets (e.g., Prudham, 2005). Problems in one place may be caused at least in part by practices that are quite distant. By offering alternative explanations of both environmental degradation and poverty, this research provides the basis for a critique of orthodox approaches to both development and conservation, such as those offered by the World Bank and major conservation organisations (Robbins, 2004; Goldman, 2005).

This research also provides the basis for a critique of sustainability as a dimension of both global environmental politics and academic discussion. For one, most literature on sustainability fails to address these relations of power that shape what people do. As outlined above, even research that claims to be 'integrative' avoids addressing the politics of socio-ecological relations. Additionally, and partly because of this failure, sustainability is itself part of the politics of control over resources. Reference to the idea of sustainability is a way of making claims about who should have access to resources, on what basis, and for what purpose. In this vein, Adams' (2001) influential work on the history of sustainability gives attention to the deep roots of the idea of sustainability in colonial conservation practices, and shows how orthodox approaches to sustainability reproduce faulty explanations of environmental problems and their solutions. Other recent research argues that the prominence of sustainable development in international debate reflects that it is a form of geo-politics and extension of state power (e.g., Luke, 2005), and shows that efforts to create sustainable livelihoods must attend to gender dynamics, which are key to understanding how people organise access to resources and use of the environment (Hovorka, 2005). These examples show that sustainability participates in and must take into account power dynamics of multiple types and also at multiple scales, including households, states, and international relations.

Attention to sustainability in multiple contexts and scalar configurations brings us to the second major contribution that geographers make to the study of sustainability, which is to treat it as a geographical outcome rather than a transcendent reality. The political ecological literature discussed above clearly demonstrates this. What we consider to be 'local' is produced by processes that cut across scales, yet these 'contexts' do not erase the uniqueness of particular situations, which are the outcome of the intersection of multiple processes. Recent scholarship in human geography also emphasises that sustainability is an inherently geographical project, and space and scale should be central to any attempt to define, plan for, or implement sustainability (Whitehead, 2007). For example, noting that policymakers now emphasise local action as the best means to implement sustainability, a recent set of articles examines local capacities for sustainable development (Gibbs and Krueger, 2005). One finding is that people in specific contexts interpret sustainability on their own terms – such that it is impossible for local people to implement global policy – and these local interpretations are influenced by social relations of power within the locality (Houghton, 2005). Cowell (2003) argues that the scale at which people frame environmental 'assets' profoundly influences what other issues (such as equity) are visible or invisible, and therefore choice of scale influences what is meant by sustainability and who will benefit from it. In one of their contributions to literature on measuring sustainability, Morse and Fraser (2005) contend that focusing on national-scale indicators is particularly misleading because these indicators overgeneralise across the nation-state, which then 'reinforces the prevailing view that the West is better than the developing world' (p. 638). What these articles demonstrate quite clearly is that the production of scale also 'restructures the objects of sustainability' (Cowell, 2003, p. 343). Sustainability is not a universal concept that is scale and space neutral, but instead the choice of scale shapes what we think we know about particular places and how they relate to each other; these ideas subsequently shape actions, which, of course, have material outcomes. Not only is sustainability inherently political, but sustainability politics is a geographical practice.

Conclusions

This chapter has outlined some of the complexities of sustainability as an organising concept. Sustainability has become the dominant way of framing issues of environment and development at the global scale. In this global politics, sustainability has merged with neoliberalism, such that the capitalist market is offered as the only solution to environmental degradation, poverty, and injustice. In this sense, sustainability is clearly not an apolitical concept, but instead serves to legitimise the status quo. Academic responses to this politics of sustainability have varied. Some suggest that we reject the idea completely; others embrace the term despite its shortcomings. Of those who embrace it, some – particularly in the field of sustainability science – try to overcome the political problems of sustainability debates by claiming to reject politics. These scholars try to turn sustainability into a set of technical questions about the right way to live on earth, questions that can be answered through careful science.

Others recognise that making sustainability into a technical question is impossible – these questions are inherently political. Those who treat sustainability as a technical problem engage in this politics implicitly and without examining the political commitments they are making. Many of those who do recognise the political nature of sustainability embrace the term precisely because it is political, and is so on many levels. Its greatest strength is that it challenges the dominant tendency to prioritise either social or environmental issues at the expense of the other. At its best, sustainability offers a vision of socio-ecological integration that breaks down the categories 'humans' and 'nature' and instead focuses on intersections of multiple and complex processes that do not obey our efforts to neatly categorise them. In so doing, sustainability can also open the door to deep understanding of the causes of environmental degradation and social injustice, and how these are interconnected, a project that requires attending to relations of power.

The malleability of sustainability as a concept should be seen in light of this politics. One outcome of the fact that people can use the term to refer to very different things is that people can avoid difficult discussions about what they really mean simply by reference to 'sustainability'; the malleability of sustainability masks relations of power by subverting political discussion regarding the causes of global inequity, injustice, and environmental problems. Yet malleability is also a reflection of the fact that sustainability is not a closed concept, but is constantly open to revision. Anyone who engages the idea – whether as a scholar, policymaker, lay person, or some combination – is actively shaping what sustainability means. As scholars, and especially as geographers, we can participate in 'writing the story of sustainability' in a way that makes it into 'a progressive project that ameliorates the negative externalities of economic activity for everyone' (Krueger and Gibbs, 2007). In other words, it is impossible to categorically decide whether sustainability is a progressive idea or not; to pretend to do so is, once again, to treat sustainability as an externally given idea that we can know objectively. Instead, we must recognise that sustainability is the outcome of power-laden discussions regarding what is right, what should be done and by whom, and to whose benefit.

BIBLIOGRAPHY

Adams, W. M. (2001) *Green Development: Environment and Sustainability in the Third World*, 2nd edn. London: Routledge.

Benjaminsen, T. A., Rohde, R., Sjaastad, E., Wisborg, P. and Lebert, T. (2006) Land reform, range ecology, and carrying capacities in Namaqualand, South Africa. *Annals of the Association of American Geographers*, 96(3), 524–40.

Braun, B. (2002) *The Intemperate Rainforest: Nature, Culture, and Power on Canada's West Coast*. Minneapolis: University of Minnesota Press.

Campbell, L. M. (2002) Science and sustainable use: views of marine turtle conservation experts. *Ecological Applications*, 12(4), 1229–46.

Castree, N. (2005) *Nature*. London: Routledge.

Chatterjee, P. and Finger, M. (1994) *The Earth Brokers: Power, Politics, and World Development*. London: Routledge.

Clark, W. (2007) Sustainability science: a room of its own. *Proceedings of the National Academy of Sciences*, 104(6), 1737–38.

Clark, W. C. and Dickson, N. M. (2003) Sustainability science: the emerging research program. *Proceedings of the National Academy of Sciences*, 100(14), 8059–61.

Costanza, R., Graumlich, L. J. and Steffen, W. (eds) (2007) *Sustainability or Collapse? An Integrated History and Future of People on Earth*. Cambridge, MA: MIT Press.

Cowell, R. (2003) Substitution and scale politics: negotiating environmental compensation in Cardiff Bay. *Geoforum*, 34, 343–58.

Cumming, G., Barnes, G., Perz, S., Schmink, M., Sieving, K., Southworth, J., Binford, M., Holt, R., Stickler, C. and Van Holt, T. (2005) An exploratory framework for the empirical measurement of resilience. *Ecosystems*, 8, 975–87.

Demeritt, D. (2001) Scientific forest conservation and the statistical picturing of nature's limits in the Progressive-era United States. *Environment and Planning D: Society and Space*, 19, 431–59.

Gibbs, D. and Krueger, R. (2005) Exploring local capacities for sustainable development. *Geoforum*, 36, 407–9.

Goldman, M. (2005) *Imperial Nature: The World Bank and Struggles for Social Justice in the Age of Globalization*. New Haven: Yale University Press.

Holling, C. (2001) Understanding the complexity of economic, ecological, and social systems. *Ecosystems*, 4, 390–405.

Houghton, J. (2005) Place and the implications of 'the local' for sustainability: an investigation of the Ugu District Municipality in South Africa. *Geoforum*, 36, 418–28.

Hovorka, A. J. (2005) The (re) production of gendered positionality in Botswana's commercial urban agriculture sector. *Annals of the Association of American Geographers*, 95(2), 294–313.

Kates, R., Clark, W., Corell, R., Hall, J. M., Jaeger, C., Lowe, I., et al. (2001) Sustainability science. *Science*, 292(5517), 641–2.

Kates, R. W. and Parris, T. M. (2003) Long-term trends and a sustainability transition. *Proceedings of the National Academy of Sciences*, 100(14), 8062–67.

Krueger, R. and Gibbs, D. (2007) Introduction: problematizing the politics of sustainability. In R. Krueger and D. Gibbs (eds), *The Sustainable Development Paradox: Urban Political Economy in the United States and Europe*. New York: Guilford.

Lappe, F. M., Collins, J. and Rosset, P. (1998) *World Hunger: 12 Myths*, 2nd edn. New York: Grove Press.

Larkin, P. A. (1977) An epitaph for the concept of maximum sustained yield. *Transactions of the American Fisheries Society*, 106(1), 1–11.

Luke, T. W. (2005) Neither sustainable nor development: reconsidering sustainability in development. *Sustainable Development*, 13, 228–38.

Mansfield, B. (2008) Global environmental politics. In K. Cox, M. Low and J. Robinson (eds), *Handbook of Political Geography*. London: Sage, pp. 235–346.

McSweeney, K. (2005) Indigenous population growth in the lowland neotropics: social science insights for biodiversity conservation. *Conservation Biology*, 19(5), 1375–84.

Morse, S. (2004) *Indices and Indicators in Development: An Unhealthy Obsession with Numbers*. London: Earthscan.

Morse, S. and Fraser, E. D. (2005) Making 'dirty' nations look clean? The nation state and the problem of selecting and weighting indices as tools for measuring progress towards sustainability. *Geoforum*, 36, 625–40.

National Research Council (1999) *Our Common Journey: A Transition toward Sustainability*. Washington, DC: National Academy Press.

Neumayer, E. (2003) *Weak versus Strong Sustainability: Exploring the Limits of Two Opposing Paradigms*. Cheltenham, UK: Edward Elgar.

Newton, J. L. and Freyfogle, E. T. (2005) Sustainability: a dissent. *Conservation Biology*, 19(1), 23–32.

Okereke, C. (2006) Global environmental sustainability: intragenerational equity and conceptions of justice in multilateral environmental regimes. *Geoforum*, 37, 725–38.

Padoch, C. and Sears, R. R. (2005) Conserving concepts: in praise of sustainability. *Conservation Biology*, 19(1), 39–41.

Parris, T. M. and Kates, R. W. (2003) Characterizing a sustainability transition: goals, targets, trends, and driving forces. *Proceedings of the National Academy of Sciences*, 100(14), 8068–73.

Peluso, N. L. (1992) *Rich Forests, Poor People: Resource Control and Resistance in Java*. Berkeley: University of California Press.

Peluso, N. L. (2005) Seeing property in land use: local territorializations in West Kalimantan, Indonesia. *Geografisk Tidsskrift*, 105(1), 1–15.

Prudham, S. (2005) *Knock on Wood: Nature as Commodity in Douglas-fir Country*. New York: Routledge.

Redclift, M. (1994) Reflections in the 'sustainable development' debate. *International Journal of Sustainable Development and World Ecology*, 1, 3–21.

Redclift, M. R. (2005) General Introduction. In M. R. Redclift (ed.), *Sustainability: Critical Concepts in the Social Sciences*, Vol. I. London: Routledge, pp. 1–22.

Robbins, P. (2004) *Political Ecology: A Critical Introduction*. Malden, MA: Blackwell.

Rockstrom, J., Lannerstad, M. and Falkenmark, M. (2007) Assessing the water challenge of a new green revolution in developing countries. *Proceedings of the National Academy of Sciences*, 104(15), 6253–60.

Sneddon, C. S. (2000) 'Sustainability' in ecological economics, ecology and livelihoods: a review. *Progress in Human Geography*, 24(4), 521–49.

Sneddon, C., Howarth, R. B. and Norgaard, R. B. (2006) Sustainable development in a post-Brundtland world. *Ecological Economics*, 57: 253–68.

Society for Conservation Biology. (2007) About us. http://www.conbio.org/AboutUs/ (accessed 12 June 2007).

Steffan-Dewenter, I., Kessler, M., Barkmann, J., Bos, M. M., Buchori, D., Erasmi, S., Faust, H., Gerold, G., Glenk, K., Gradstein, S. R., Guhardja, E., Harteveld, M., Hertal, D., Hohn, P., Kappas, M., Kohler, S., Leuschner, C., Maertens, M., Marggraf, R., Migge-Kleian, S., Mogea, J., Pitopang, R., Schaefer, M., Schwarze, S., Sporn, S. G., Steingrebe, A., Tjitrosoedirdjo, S. S., Tjitrosoemito, S., Twele, A., Weber, R., Woltmann, L., Zeller, M. and Tscharntke, T. (2007) Tradeoffs between income, biodiversity, and ecosystem functioning during tropical rainforest conversion and agroforestry intensification. *Proceedings of the National Academy of Sciences*, 104(12), 4973–8.

The Ecologist (1993) *Whose Common Future? Reclaiming the Commons*. Philadelphia: New Society.

Turner, B. L., II, Kasperson, R. E., Matson, P. A., McCarthy, J. J., Corell, R. W., Christensen, L., Eckley, N., Kasperson, J. X., Luers, A., Martello, M. L., Polsky, C., Pulsipher, A. and Schiller, A. (2003a) A framework for vulnerability analysis in sustainability science. *Proceedings of the National Academy of Sciences*, 100(14), 8074–9.

Turner, B. L., II, Matson, P. A., McCarthy, J. J., Corell, R., Christensen, L., Eckley, N., Hovelsrd-Brodha, G. K., Kasperson, J. X., Kasperson, R. E., Luers, A., Martello, M. L., Mathiesen, S., Naylor, R., Polsky, C., Pulsipher, A., Schiller, A., Selin, H. and Tyler, N. (2003b) Illustrating the coupled human-environment system for vulnerability analysis: three case studies. *Proceedings of the National Academy of Sciences*, 100(14), 8080–85.

Whitehead, M. (2007) *Spaces of Sustainability: Geographical Perspectives on the Sustainable Society*. London: Routledge.

Chapter 4

Biodiversity

Karl S. Zimmerer

Introducing Biodiversity

Biodiversity is one of the most central and versatile themes of environmental geography. It is defined as 'the variety and variability among organisms and the ecological complexes in which they occur' (OTA, 1987). This definition has served as a mainstay for the establishment of biodiversity as a major theme in both environmental geography and in large interdisciplinary currents across the environmental sciences and environmental studies (Lubchenco, 1998; Botkin, 2000). In addition to interdisciplinary exchanges, the interest in biodiversity has expanded via exchanges between environmental geography and diverse disciplines in the natural sciences, social sciences, and humanities. As a result, the analysis of biodiversity in this chapter requires both the outward looking view to broader currents and, at the same time, close examination within environmental geography per se.

My analysis begins with biodiversity concepts and concerns of policy and management that are relevant, but not restricted, to the realm of environmental geography ('Perspectives on Biodiversity'). It then constructs a brief overview of the historical and geographical parameters of biodiversity science and related themes within the social sciences and humanities ('Biodiversity: Concepts and Concerns: an Overview'). The main part of the analysis is centred on the understandings of biodiversity concepts that are developed within the subfield approaches of biogeography and physical geography along with ecology and the geosciences ('Biodiversity: Biogeography, Ecology, Geosciences, and Genetics'), nature-society geography ('Biodiversity: Nature-Society and Human-Environment'), and human geography and its related fields in the social sciences and humanities ('Biodiversity: Human Geography and Related Fields'). These subfields overlap and coalesce into the 'borderlands' of environmental geography (Zimmerer, 2007). Indeed, the multi-faceted qualities of biodiversity are entwined intricately, as shown throughout this entire chapter, with the approach of environmental geography. Biodiversity's intricate interweaving, emblematic of environmental geography, is centred on the complex interactions, agency, and embedding of biodiversity as biogeophysical nature within the lives and livelihoods of humans as created through social, economic, and cultural practices (for an earlier discussion see Zimmerer, 1996, pp. 15–25).

Perspectives on biodiversity

Several interdisciplinary perspectives delimit the contemporary status of biodiversity as a far-reaching concept in contemporary environmental studies and sciences. Primary perspectives include: (i) biological and ecological sciences; (ii) environmentalism and conservation; (iii) economics and ethics; and (iv) public environmental science.

The biological sciences, associated particularly with ecology and evolution, provide a predominant perspective on biodiversity as the 'the variety and variability among organisms and the ecological complexes in which they occur' (the familiar definition mentioned above). Taxonomy frequently functions as a scientific *lingua franca*. It sees biodiversity as objects of nature that are classified according to systematic categories ('things' is the term chosen in environmental geography using a humanities inflection; see Section see pages 60–61 and Bakker and Bridge [2006] *inter alia*). The species is the most common taxonomic unit of biodiversity. Approximately 1.4 million species have been identified, but the actual number is likely between 10 and 100 million. Subspecific units (e.g., genetic- and population-levels) and multi-specific ecological groupings (e.g., guild-, habitat- and ecosystem-levels) are also integral to biodiversity. Taxonomic treatments of biodiversity are increasingly dependent upon genetic analysis and genome-based assessments, albeit not without sharp debate (Greene, 2005). The genetic-level emphasis has spawned the growth of bioinformatics. This young field, which is the fusion of computational structures and organised biological information, has become an integral part of the taxonomic advances applied to biodiversity. One example is the new model linking the genomics and taxonomy of the plant family Solanaceae (the 'nightshade' family) with the support of the Planetary Biodiversity Inventories initiative of the US National Science Foundation (NSF) (Knapp et al., 2004; see also Soberon, 1999; Graham et al., 2004; Blakey et al., 2007). In general, bioinformatics draws upon a new geographical and spatial emphasis as discussed below (see pages 54–56).

Ecology and evolution offer a second and equally prominent view within the biological sciences that are being applied to biodiversity. The ecological and evolutionary sciences are concerned with the processes, functions, and spatial patterns that support the evolution and maintenance of biodiversity across a variety of scales from local and regional to global (Wilson, 1988; Nabhan, 1995; Reid, 1997; Ehrlich and Levin, 1998). Geographical and spatial analysis has gained a growing centrality in the application of these sciences to biodiversity. In the field of evolution and bioinformatics, for example, one recent editorial is entitled 'putting the geography into phylogeography' (Kidd and Ritchie, 2006; see also Moritz, 2002). Similar to the taxonomic approaches, ecology now relies more heavily on genetic and genomic-level analysis in the treatment of biodiversity. Ecology also involves an increased geographical and spatial emphasis. This shift is evident in the approaches of landscape genetics and conservation genetics, discussed further below (see pages 54–56), which are new pillars of biodiversity science. Pioneering contributions, ranging from biogeography and landscape ecology to conservation biology, are driving this shift that now marks more than one decade of advances (e.g., Jelinski, 1997; Manel et al., 2003; Parker and Jorgensen, 2003; Rigg, 2003)

Environmentalism, broadly conceived, provides a second perspective on biodiversity and one that is inextricably entwined with the biological, ecological, and evolutionary sciences. Escalated environmentalist concerns have been fueled by the widespread occurrence and worsening of human-induced biological extinctions. Current human-induced extinctions are estimated at 1,000 to 10,000 times higher than existed in pre-human times, although this estimate must be treated as a coarse approximation (see pages 54–60). The biologist and taxonomist E. O. Wilson – one of the earliest and probably the most influential proponent of biodiversity as both a scientific concept and environmentalist concern – has regularly drawn attention to the worsening threats of extinction, combined with newly available data on deforestation and advances in tropical biology, as one of the main forces behind the explosion of interest in biodiversity (Wilson, 1988).

Wilson and other advocates of biodiversity conservation have frequently traced scientific and environmentalist concerns for biodiversity to the 1986 founding of the Society for Conservation Biology (SCB). The SCB commitment to biodiversity conservation (parks, reserves, protected areas) has been applied globally in tens of thousands of management units during the course of the past couple decades (Zimmerer et al., 2004). The proliferation of policies and management must be seen as a global geographic phenomenon that has arisen in response to acceleration of the anthropogenic extinctions of biodiversity. Still the relations of biodiversity conservation to environmentalism, policy, and management, while dynamic and undeniable, are complex ones, which have spurred dynamic nodes of geographic interest and understanding (see pages 56–61).

Economics along with non-economic frameworks of human valuation furnish yet another perspective on biodiversity (NRC, 1999). Indeed economic value is invariably one of the reasons highlighted in accounts of the nature and importance of biodiversity. Conventional economic approaches attribute 'raw material'-type value to biodiversity as the growth stock of new sources of foods, pharmaceuticals, fibers, petroleum substitutes, and other products. 'Biofuels', 'bioenergy' and a mushrooming array of 'bioproducts', generated through applications of biotechnology, are widely recognised as derived from and dependent upon biodiversity. This expansive arena of economic growth hinges on the contributions of biotechnology (see Section IV below). Explicit environmental accounting has grown via the sophisticated subfields of ecological and environmental economics, along with application in various neoliberal policies. These approaches assign economic weight according to the valuation of various ecosystem goods and services, such as ecosystem resilience and carbon sequestration, that occur through the ecological functioning of biodiversity (see Ehrlich and Levin, 1998; Costanza et al., 2007).

Non-economic human values are also widely assigned to biodiversity in relation to human societies and cultures (NRC, 1999). Such values of biodiversity may accrue through livelihood, ethical, and humanistic beliefs and practices (see Sections IV and V below). These non-economic values have been widely documented in diverse social and cultural settings. Such values are advocated as important counterpoints needed to balance the potentially reductionist and strictly economistic or utilitarian valuations of biodiversity (Nabhan, 1995). The 'biophilia hypothesis' also belongs within the broad umbrella of non-economic valuations of biodiversity. It attributes the value of biodiversity to human co-evolution with biodiversity-rich nature and within the context of biodiversity-rich environments (Kellert and Wilson, 1993; Martin-Lopez et al., 2007).

Public environmental science, institutions, and governance approaches offer a fourth perspective on biodiversity. Diverse organisations have acquired unprecedented importance as key institutional contexts for the management of biodiversity. Indeed the theme gained much initial visibility and influence through interest and influence through the 'National Forum on BioDiversity', convened in 1986, that was funded and organised through the National Resource Council and National Academy of Sciences of the United States, with additional support from the Smithsonian Institution, the World Wildlife Fund, and other prominent public and Non-Governmental Organizations. This event coincided with a report entitled 'Technologies to Maintain Biodiversity' that was issued in 1987 through the US Office of Technology Assessment (OTA, 1987). Subsequent influence on biodiversity initiative is widely demonstrated through both many individual countries, which have adopted pioneering approaches, and the global-scale and international organisations, many headquartered in Europe and the United States (such as the IUCN and UNEP, see pages 53–54 and 56–61). These influences – which are relatively tractable and well-documented – reveal how the prevailing idea of biodiversity came about though the activities and ideas of specific institutions and individuals (i.e., its 'constructedness'), who have held influential positions in science, policy, and management (see pages 56–61; see also Takacs, 1996; Farnham, 2007).

Biodiversity concepts and concerns: overview

The concepts and concerns of biodiversity are rooted in a complex scientific and social web that is historically and geographically extensive. Biodiversity, as a term, has become imbued with multiple and sometimes contested meanings and interpretations that stem from these highly varied strands. This realisation is not meant to detract from the validity or worthiness of the concepts and concerns of biodiversity. Rather, quite the opposite, my analysis urges engagement with the fuller range of meanings of biodiversity. Future advances depend on fuller engagement across the gamut of scientific analysis to activist interpretation in ways that are both constructive and critically aware.

Multiple geographic scales distinguish the formative phase of contemporary biodiversity interests that began in the mid- and late-1980s. Concurrent with US national-level undertakings – principally the report and workshops organised by the US Office of Technology Assessment and the National Forum on BioDiversity in 1987 and 1988 that are described above – there co-existed global-scale framings of the idea. The global scale was prioritised, for example, in the Interagency Task Force on Biological Diversity, formed by the US Congress in 1985, which was the outcome of an amendment (Section 119) to the Foreign Assistance Act that authorised the US AID (Agency for International Development) to assist developing countries in conservation programs, with an emphasis at that time on protecting wildlife habitat and endangered species. By the late 1980s US AID was supporting the Biodiversity Support Program, with substantial involvement and assistance from global environmental organisations, such as the World Wildlife Fund, The Nature Conservancy, and World Resources Institute (Oldfield and Alcorn, 1991). Well-publicised scientific analysis of the biodiversity crisis, along with coordinated institutional and political efforts aimed at conservation, has thus relied on the global and international scales as key frames of reference.

The global framing of biodiversity became still more explicit and predominant in the Convention on Biological Diversity (CBD) that was adopted by more than 100 countries following the 1992 United Nations Conference on Environment and Development held in Río de Janeiro, Brazil (UNCED that has become well known also as the 'Earth Summit'). Article 1 of the CBD asserts that the main objective of the global suite of signatory countries includes 'the conservation of biological diversity; the sustainable use of its components and the fair and equitable sharing of the benefits arising out of the utilisation of genetic resources'. Extensive international negotiations and support, along with key issues of protracted disagreement, marked the continued evolution of the CBD as a framework-style agreement. (One main source of disagreement has been the position of the United States that has led efforts to block or alter the provisions on intellectual property rights proposed and supported by tropical biodiversity-rich countries such as Brazil and Indonesia.) Two global organisations, the International Union for the Conservation of Nature (IUCN, based in Switzerland) and the United Nations Environment Program (UNEP), were central to the processes and preparation that had led to formalisation of the CBD at the 'Earth Summit.'

Economic and political issues are one persistent source of uncertainty stemming from deeper disagreement about how to finance the 'global approach to the conservation of biological diversity' that is called for in the CBD (McNeely, 1988). To date, the UNEP, the World Bank, and the latter's Global Environmental Facility (GEF), along with international development organisations, have provided noticeable financing for biodiversity assessments and conservation, although not without controversy (see Sections IV and V below). This funding has often targeted national and regional or local counterparts throughout the world (in such projects as the Global Biodiversity Assessment and the Global Biodiversity Strategy). The country-level agencies and 'on-the-ground' organisations have served as crucial institutions – albeit sometimes overlooked – in the consolidation of biodiversity-related interests as a global phenomenon (Bassett and Zuéli, 2003; Zimmerer, 2006a,b).

Biodiversity: Biogeography, Ecology, Geosciences, and Genetics

Global, country-level, and regional biogeographic scales analysis serve as principal frames of reference for biodiversity science. The global scale consistently provides a vital outermost framing. It is evidenced, for example, in the Planetary Biodiversity Initiative of the US National Science Foundation. The global scale of biodiversity science is also featured in many environmental and conservation organisations, such as Conservation International (CI), World Wildlife Fund (WWF), and The Nature Conservancy (TNC) in the United States, and the World Conservation Monitoring Centre (WCMC) in Great Britain. General references to sub-global biogeographic units have been similarly central – 'the tropics' has been highlighted throughout the recent wave of interest. As the same time, the biogeographic scales of region-, landscape- and local-level have also become core concepts within biodiversity thinking (MacDonald, 1995; Reid, 1997). Similarly foundational are country-level framings, such as the so-called 'megadiversity countries' (e.g., applied frequently to Indonesia, Madagascar, and Brazil).

Biogeography is the root of the productive growth of biodiversity science, especially through the theoretical and applied usefulness of the Theory of Island Biogeography (pioneered by biologists Robert MacArthur and E. O. Wilson), with subsequent revisions and continued widespread use (Lomolino, 2000; Whittaker, 2000). Environmental complexity, in addition to spatial area *per se*, has become a principal theme in evaluating and estimating biodiversity-supporting habitats. Changes such as forest fragmentation typically require the analysis of both human drivers and biophysical factors. Patterning of these changes can be modelled using algorithms to evaluate and select the design of reserves or protected areas (PAs) for biodiversity conservation. This type of analysis is often associated with the widespread and still expanding approaches of 'conservation biology' and 'conservation ecology', along with several highly cited scientific journals, including *Biodiversity and Conservation, Conservation Ecology, Conservation Biology,* and *Diversity and Distributions.*

Species richness (i.e., the number of species) estimated per unit area is an approximation of the extent of biodiversity that is highly useful. Species-level estimation is often used alongside ones of other taxonomic levels at either the same or different scales (e.g., within-species diversity, ecosystem-level diversity). Combined with biogeographic analysis, studies are able to identify areal concentrations and spatially underrepresented areas (e.g., various forms of 'gap analysis'), while it also demarcates the areas of concentrated biodiversity (e.g., the concept of biodiversity 'hot spots'). Noteworthy too are the occasional pro-conservation arguments that oppose the logic of biodiversity 'hot spots' as the grounds for baseline conservation priorities. This latter logic, while scientifically sound, widely accepted, and persuasive for policy purposes, may overlook methodological differences as well as create philosophical and political concerns about too narrow or one-dimensional a view of biodiversity.

Environmental processes underpin the patterning of biodiversity that is of interest to biogeographers. The most well-known factors behind spatial patterning are those of the geo-environment, such as landforms, soils, and climate, which operate at multiple spatio-temporal scales (e.g., Rosenzweig 2003). Environmental variation thus contributes a primary dimension to the differentiation of biodiversity at a range of taxonomic levels (e.g., species, intra-specific populations, multi-specific guilds). Modelling approaches, such as ecologic niche modelling (ENM), can relate the spatial patterning of biodiversity occurrences (typically species-level) across landscapes to raster GIS coverages. Biodiversity-differentiating factors also are often distinguishable as historical events at the time scale of geo-environmental time spans. Innumerable such events have that led to both the increase of biodiversity (e.g., through the geographic differentiation of species or intra-specific populations) or the decrease (e.g., through extinctions generated through processes that are either human-influenced or entirely unrelated to humans) (Young et al., 2002). The latter distinction draws the contrast between 'natural' or autogenic disturbances, as the creation of tree fall gaps within forests as a result of such factors as windthrow or pests), on the one hand, and anthropogenic disturbances, on the other hand. It is the properties of scale, magnitude, and frequency that are used to determine the resemblance of these disturbance regimes (Zimmerer and Young, 1998; Botkin, 2000).

Biodiversity is also influenced through myriad ecological interactions within and among groups of organisms ranging from communities to ecosystems; these interactions are highly spatially dependent. Particular species play key roles in the biodiversity-support functions of various communities and ecosystems. The roles of the so-called keystone species are documented in an expanding number of case studies as well as modelling and theoretical treatments of biodiversity. One well-known example of a keystone species is the California sea otter, which preys on sea urchins and thus, indirectly, on the diverse kelp forests that are grazed by the sea urchins; another example is nitrogen-fixing bacteria in many soils environments (Ehrlich and Levin, 1998). Geographic scale and spatial analysis are important to the ecological perspective on biodiversity. For example, geographic scale influences the ecological interactions of keystone species and thus the regulation of biodiversity-related processes (e.g., Foster, 2002). Ecological disturbances represent another major form of interactions that influence biodiversity. Spatial and geoenvironmental analysis serves as a main avenue for understanding these ecological interactions involving disturbance (Parker et al., 2001).

Increasingly biodiversity is understood through the approach of genetic analysis, including molecular-level genomics. This approach is resulting in vast quantities of data on genetic variation in diverse organisms. Spatial and geographical frameworks have emerged as one of the principal means of organising, modelling, and analysing the previously unimagined quantities and types of information on biodiversity at the genetic level. These include the use of spatial autocorrelation techniques including correlograms (Smouse and Peakall, 1999); spatial distance measures (e.g., Epperson, 1995); spatial classification estimators such as regionalisation methods (Monmonier, 1973); polynomic models of geographic distributions; and spatial-statistical models, such as wombling, of the patterning of gradients ('clines') and patchiness (Sokal and Thomson, 1998). The use of spatial statistics in genetic analysis is increasingly associated with biodiversity conservation. It includes the development of bioinformatics with applications centred on biodiversity and conservation issues (e.g., the new journal *Biodiversity and Bioinformatics* and contributions to the Convention on Biological Diversity (CBD); see Silva, 2004). Landscape genetics and conservation genetics, two growing approaches, are potentially integral to environmental geography. Significant contributions are demonstrated, for example, in the contribution to understanding landscape and geographical factors in the partitioning of within- and between-population genetic diversity (e.g., Jelinski, 1997; Zimmerer and Douches, 1991; Manel et al., 2003; Parker and Jorgensen, 2003; Rigg, 2003). In sum, the theme of spatial and geographic structuring has clearly emerged as one of the primary means of organising the vast quantities of genetic-level information on biodiversity that is fast becoming available.

Geo-environmental change across spatial and temporal scales, such as in global climate, is essential to understanding biodiversity in the context of evolutionary and ecological processes. This view enables both basic scientific understandings and management-policy information about the threat of potentially irreversible losses. In the case of global climate change, biodiversity science has identified several crucial themes, which include range shifts (i.e., changes in biogeographic distributions), taxon-specific abundance changes (numbers within the group of interest), phenological alterations (pertaining to timing of seasonal and interannual behaviours), and general identification of species (and groups of species) that will become more or less important as a consequence of global warming (Lovejoy and Hannah, 2005). It also evaluates the responses of biodiversity to other forms of environmental change – examples of the latter include land degradation, atmospheric acidification, and the general accumulation of toxic substances. Alteration of the spatial patterning of habitats is also a major theme; habitat fragmentation receives ever more sophisticated analysis. Such change is subject to interactions with other kinds of human-driven changes (such as changes in land use, see below). In general, evolutionary ecology highlights the multiple spatial and temporal scales of biodiversity processes, while it can also be used to draw attention to potential irreversible losses.

Biodiversity: Nature-Society And Human-Environment

Perspectives centred on nature-society and human-environment interactions (such as political ecology, cultural ecology, human dimensions of global change, and resource management) offer several cornerstone contributions towards the understanding of biodiversity within environmental geography. Human activities and management determine the status of biodiversity (including biodiversity-supporting processes) in a wide range of environments. The perspectives in this section centre on humanised landscapes (anthropogenic habitats) that vary from near-wilderness-type settings to ones that are extremely modified. As a result, there is a gradient of impacts on biodiversity that begins, on one end, with such activities as relatively low-impact land use, exemplified through the gathering of non-timber forest products (NTFPs) and rotational shifting cultivation. Forest regrowth and regeneration as a result of land use abandonment is another example of low-impact, indeed generally positive, effects on biodiversity (similar to low-impact autogenic disturbances described on pages 54–56). At the other end of this gradient are activities with high-level impacts on biodiversity, such as permanent forest clearing, agricultural land use that varies from conventional systems to expansion of biotechnology-based agriculture along with urban and industrial development.

Biodiversity-impacting activities are related to socio- and political economic processes at scales ranging from local to regional and global. The latter scale is especially salient, since biodiversity impacts are a major form of global human-environmental change. The 'Global Change' and 'Global Human Environmental Change' networks of researchers, scientific institutions, and policy specialists have singled out biodiversity loss, along with climate change, desertification, and water resources, as key issues of planetary biogeophysical systems involving human-environment interactions. 'Scaling up' the estimates and understanding of biodiversity impacts, from local and regional studies to the global scale, is an important and continued challenge. Many human-environment interactions involving biodiversity do not lend themselves to straightforward spatial extrapolation – they are uneven as a result of underlying spatial variation in both the human-social dynamics as well as the environment-biodiversity interactions. Nonetheless, considerable progress in understanding in biodiversity-scale relationships have been made recently thanks to new or expanding techniques, many including Geographic Information Systems (GIS) and remote sensing (RS), as well as other innovative forms of research design and analysis, such as multiple case studies, cross-regional comparisons, and meta-analysis.

Research into Land Use/Cover Change, or LUCC, involves several of the above techniques that are frequently applied to understanding the impacts of human activities on biodiversity in forest ecosystems (Velazquez et al., 2003). Frequently it evaluates the changes in the spatial parameters of forest cover (e.g., the extent and patterning of forest edge, overall shape, and other geometric and distance-related features) in comparison to non-forest areas. Typically cast as diachronic comparisons involving two or more time periods this approach offers a means of estimating cover-related impacts with inferences about biodiversity. Also, LUCC is increasingly linked to intensive studies of human-social and ecological-change processes that are georeferenced and coded into the frameworks of spatial analysis. The emphasis of LUCC on forest-and-other-land-use-areas, while well suited to remote sensing and other land-cover analysis, has thus far precluded the analysis using this approach of other forms of biodiversity impacts, such as changes *within* agricultural and urban land use. These latter changes include the increasingly important impact of biotechnology-based agriculture. Here the impact on biodiversity is concentrated within agricultural systems (agrobiodiversity), which is of interest in the global-change research and policy networks as well as those related to food security and rights.

Prospects for biodiversity and environmental conservation can be analysed by identifying the space- and time-based parameters of change processes. Analytical approaches include quantitative spatial-environmental methodologies, along with quantitative and/or qualitative assessments of economic, political, and historical factors (e.g., regression-tree statistics and rule-based, expert knowledge analysis) and 'threat analysis' in conservation-centred approaches. Core techniques include Geographic Information Science (GIScience), cross-regional comparisons, and remote sensing analysis. Potential 'win-win' scenarios offer combinations of equitable socio-economic development, on the one hand, and favorable environmental and biodiversity outcomes, on the other hand (Adams et al., 2004; Naughton-Treves et al., 2005; Zimmerer, 2006a; 2006b). The potential existence of this combination is frequently complex, yet it is often of primary interest. Identification of potentially favorable combinations of conditions suited to the design and establishment of protected areas (PAs), for example, is a high priority for biodiversity conservation.

Successful expansion of initiatives for biodiversity conservation is linked, in several cases, to well-developed concerns for human rights and environmental justice. Such concerns are centred on biodiversity conservation measures that have led to the loss of resource access and livelihood among local inhabitants (Peluso, 1993; Neumann, 2004). The latter include poorer, less socially powerful, and, in many cases, indigenous people. These people reside and practice land use in many of the tropical and less-accessible areas that are prioritised for biodiversity conservation. The often long-term and still unfolding relations of these people to biodiversity and biodiversity conservation have become a major subject of geographic research (e.g., through the approaches of political ecology and cultural ecology). Indeed if much biodiversity-related environmental geography perceives people as a threat to biodiversity, then the perspective of human rights and environmental justice can be seen as inverting the focal point. Here the question of how biodiversity initiatives may pose a threat to people becomes the primary focus. Biodiversity conservation initiatives, often framed as global and integral to sustainability policies, have become a main avenue for development programmes at national, regional, and local scales in many places across the world – this elevation of biodiversity distinguishes the present historical moment.

Cultural activities often do support certain types of biodiversity and, more generally, are interwoven with various biodiversity-influencing processes. These relations have led to interest in biodiversity that exists in close relation to the activities and habits of people (e.g., utilised and known-about biota) in relation to cultural diversity (e.g., livelihood practices, food customs and cuisine, ethnic and language group differences) and sociocultural and development change processes (e.g., increased influence of commodification, market relations, and labour migration). Geographic contexts range widely for these intensive interactions between biodiversity and humans (Naughton-Treves et al., 2006). Analysis of human-environment interactions in such contexts include local- and region-scale differences in land use activities that range from utilitarian and 'backyard'-type to deeply cultural and religious practices (Hecht, 2004, Zimmerer, 2004). Accelerating change tends to typify these interactions. The biodiversity of agricultural plants and ecosystems ('agrobiodiversity') in Africa, for example, is on the verge of becoming subject to technology-based advances in 'bio-fortification' – the process of creating, either through conventional breeding or genetic modification, and subsequently disseminating genetically improved food crops with enhanced levels of bio-available micronutrients.

Historical factors strongly contour the relations of biodiversity to humans. Historical analysis abounds in nature-society and human-environment approaches, with varying degrees of similarity to the closely related approaches of historical geography, environmental history, and ecological history. These historical perspectives offer important insights into biodiversity. Such insights include the following: (i) much biodiversity exists in environments, often geographically extensive, that have undergone long histories of interaction with human activities (*contra* the so-called Pristine Myth) (Denevan, 1992; Balée 2006); (ii) it is the type, magnitude, and scale of human impacts that determines relations of human-modified environments to biodiversity (Zimmerer and Young, 1998); and (iii) biodiversity, like other environmental concerns, is appreciated and understood among many audiences, including both specialists and non-specialists, through the kinds of stories, or narratives, that are used to present and describe such issues. These perspectives also have cast much new light on the roles of indigenous people and other non-Western groups as neither 'Noble Savages' nor 'Ignorant Natives' in their relations to biodiversity and biodiversity-supporting landscapes (Oldfield and Alcorn, 1991).

Relations of biodiversity to humans are also deeply rooted in the nature of nature itself. Although the latter might be thought of as the domain of the natural sciences, it is also a vital theme for understand biodiversity through the lens of human-environment and nature-society interactions (e.g., fields such as cultural ecology, political ecology and resource management). For example, many dynamics of human relations to biodiversity are dependent upon change-prone processes that are triggered by so-called disturbance events and that do not tend towards a well-defined or easily identifiable 'balance of nature'. Examples include the biodiversity of such economically important landscapes as range ecosystems as well as such icons of more pristine-type conservation as renowned wildlife populations (Zimmerer and Young, 1998). The dependence of biodiversity on disturbances is resonant with the interpretive perspective of humans-in-nature, as opposed to the conventional dichotomy of humans and nature.

Human-environment interactions and biodiversity are increasingly paired with pathbreaking progress in genetic analysis genomics. Such advances are opening new vistas for the future analysis of biodiversity and human-environment interactions within environmental geography. These developments include a focus on the geographic dimension of major human migrations and such correlates as the spatial distribution of languages (e.g., the spread of European languages; see Barbujani and Sokal, 1991; Piazza et al., 1995). Recent molecular-level genetic analysis also opens new vistas on geographic dimensions of the formative plant and animal domestications and dispersals, including the consequences of these human-environment interactions (agriculture, livestock-raising) on the genetic systems of domestication organisms (e.g., Hanotte et al., 2002; Doebley, 2006; Doebley et al., 2006; Parker et al., 2007). The new developments also include a wave of Green Revolution-style questions that is driven through advances and debates in genomics (e.g., on food production and quality), including whether the use of biodiversity can improve food security and other benefits among resource-poor land users (e.g., Dawson and Powell, 1999; Reece, 2007). The role of biodiversity in genomics and biotechnology-based advances is central to the development of possible plant and land use adaptations to climate change and other key agroenvironmental factors. A new volume, *Darwin's Harvest* (Motley et al., 2006), along with papers in the 2003 issue of *Physical Geography*, bring together many of the advances in genetics that are relevant to human-environment interactions and environmental geography (e.g., Rigg, 2003). The latter addresses the spatial dynamics of genetic introgression, for example, which is a major topic in the treatment of invasive species (Blumler, 2003).

Biodiversity: Human Geography and Related Fields

The economic valuation of biodiversity is owed in substantial part to the rise of its usefulness and potential promise as genetic raw material to the biotechnology industry. Chronologies have coincided closely in the growth of these two spheres of interest since the 1980s. Legal frameworks, such as property laws and intellectual property rights, as well as technological innovations, such as DNA banking, have continued to offer new facets to the biodiversity-biotechnology relation. The economic significance of biodiversity is also incorporated into the valuation of ecosystem goods and services. These modes of valuation are best understood as not merely environmental and economic but also broadly social and political as well. Contemporary economic geography has generated numerous insights into the powerful human social dynamics surrounding and infusing biodiversity issues.

The predominant modes of biodiversity valuation belong to the current economic philosophy, policy frameworks, and politics of neoliberalism. Neoliberalism espouses market-based rationales for the protection and conservation of nature and, more specifically, for the valuation of biodiversity. The extensive capacity for valuation of biodiversity through marketing under neoliberal policies has led to critiques within economic geography that demonstrate the scenario of 'saving nature to sell it' (McAfee, 1999). Markets have gained the status as possible saviours, in addition to still serving as threats, to biodiversity. In biodiversity-rich places worldwide, particularly in the Global South, economic valuation is also distinguished by costs that are incurred among local residents who may lose access to land and other resources as a result of Protected Areas (PAs) designated for the purpose of biodiversity protection and conservation (Adams et al., 2004; Neumann, 2004).

Potentially the economic value of biodiversity can be used to provide local and national benefits through such agreements as bioprospecting and the commercialisation of biodiverse genetic material. Indeed this 'geography of hope' has stimulated diverse works on market-based conservation through the lens of cultural and political ecology, with focus especially on the people and resources of the 'Global South', (e.g., Coomes et al., 2004). Recent global and international agreements on biodiversity such as the Cartagena Protocol on Biosafety, which was adopted in 2000 in order to regulate agricultural biotechnologies internationally, also offer potentially hopeful developments. The Cartagena Protocol illustrates the increased role and influence of regulations involving the global valuation of biodiversity along with the place of the countries and citizens of the Global South in this regulation. A distinct yet generally related example, set in the Global North, is the new framework of the new Agri-Environmental Policies (AEP) of the European Union. The treatment of biodiversity issues within this EU framework reveal that new regulatory approaches can be vital, and that market-based mechanisms do not represent all-encompassing avenues for environmental management.

Politics of biodiversity issues range from national resource concerns and identity, on the one hand, to international treaties and relations, including the processes of globalisation, on the other hand. The politics of nations is central to many biodiversity issues; for example, biodiversity is commonly viewed as a feature of national heritage. For example, while international and global organisations may finance much of the drive for biodiversity conservation, these efforts typically are enacted through institutions and agencies at the national, regional, and local scales. The latter group of institutions influences 'what gets understood as and comes to be understood as biodiversity in a national context' (Lorimer, 2006, p. 540). The largest of the World Bank-funded initiatives for biodiversity conservation, for example, has required National Environmental Action Plans (NEAPs), while another line of Bank-funded projects has established National Biodiversity Conservation Areas (NBCAs) in many countries (Bassett and Zuéli, 2003). The international and global biodiversity initiatives show tendencies that express an underlying politics of nature (e.g., the environmentally and geographically skewed emphasis on tropical rain forest conservation; Zimmerer, 2006). These global environmental politics may run starkly counter to national politics and identity practices, which often emphasise the utilitarian-type landscapes of agrarian ideals.

Ethics and moral geographies infuse the understanding of biodiversity in myriad ways. The 'cultural valuation' *per se* of biodiversity (and, more commonly, biodiversity-incorporating attributes of nature) is embedded in a host of belief systems. But ethics and morals may also be thought of in a broader sense, and thus acquire still more wide-ranging relevance to biodiversity issues along with those of biotechnology (e.g., Greenhough, 2007). For example, ethics offers an appropriate framework for understanding the beliefs and values associated with biodiversity-containing landscapes that are also of vital cultural importance (e.g., ideas of ancestral domain related to biodiversity issues in the Philippines; Bryant, 2000). Ethics also inform beliefs concerning moral order, which is seen as a positive force in several relations of humans to nature and biodiversity – such as in ideas of stewardship and the place of people-in-nature. At the same time, however, the valence of moral order is not inherently positive, and its influence may be manipulated in many ways. Rationalisation of the abuse of human rights of local residents and the justification of deadly violence against wildlife poachers in African national parks is the result of the 'radical [discursive] re-ordering of moral standing' (Neumann, 2004, p. 234). It lowers these local people to a sub-human level of ethical status. Most recently, the ethical dimension of biodiversity has mushroomed in the question of 'who owns the human species'. Addressing this epochal question is sure to spawn a new phase of biodiversity analysis within environmental geography.

Conclusion: Biodiversity and Environmental Geography

Entwining of the human and non-human in biodiversity is increasingly relevant not only to biodiversity but to various aspects of human conditions and social dynamics. Various powerful new developments in the economics, policy, and management of biodiversity (e.g., biotechnology, conservation) have intensified this entwining. In response, perspectives in the social sciences and humanities use the ideas of 'hybrid' and 'socionature hybridity' in order to describe those elements of nature, from landscapes to organisms, which are deeply entwined with the human social world. Biodiversity offers many illustrations of the inseparability of the non-human and the human. One example is the biotechnology industry's coining of the 'small molecules' throughout nature as a so-called lexicon of biodiversity to be marshaled for genetic engineering. The perspective of this hybridity is useful also since it reveals the powerful tendency of modern belief systems, including in academic disciplines, to pry apart the 'natural' and the 'cultural' as separate spheres of activity and analysis (Castree, 2005). Human social dynamics, which are pervasive in everyday discourses as well as disciplinary divisions within the academy, have tended to separate 'nature' and 'culture', contrary to the realities that are represented in most types of biodiversity issues.

Ultimately the rapid and successful rise of biodiversity – both as an influential multi-faceted concept crossing the sciences and humanities and, at the same time, as one of the most pressing and urgent present-day environmental issues – must be seen as rooted in still larger historical and geographic scales. One persistently powerful force is the centuries-old and still vigorous legacy of natural history, which offers an important deep-time precursor to present-day biodiversity science and ideas. Natural history has typically combined scientific and emotive interests in the variety of the natural world, similar to some of the main threads of contemporary biodiversity-centred activities. Moreover, natural history is drawn from field studies and international milieus at the global scale, which are similarly a signature of biodiversity science and conservation. Persistence of the deep cultural premium placed on the value of natural diversity may be traced to Enlightenment and Romantic views of nature, as evidenced for example in the works of Alexander von Humboldt. Indeed, Humboldt's scientific and human-environmental legacy has offered specific precursors to current interests in biodiversity, along with its general influences on contemporary environmental geography (Zimmerer, 2006).

Social studies of science and technology can be used to reflect also on current trends and interest in biodiversity. These studies suggest that the power of scientific ideas and concerns typically emanate from immediate circumstance as well as wider social, ideological, and environmental contexts and currents. Rapidly expanding interest in biodiversity must be evaluated, therefore, within the matrix of the enormous scientific and economic growth of biotechnology and the emergence of a 'bioeconomy' (including 'biofuels' and 'bioenergy'). Genomics and bioinformatics, for example, are fueling a new wave of advances that bring into play an unprecedented emphasis on the role of the spatial and geographical dimension of biodiversity dynamics.

ACKNOWLEDGEMENTS

I am grateful for the expert editorial feedback and comments of David Demeritt. This research is carried out in conjunction with the Geo-Genetics research group and the colloquium series (2008–2010) of the Huck Institutes of the Biological Sciences at Pennsylvania State University. It has benefitted from interactions with Geo-Genetics colleagues from several departments. Funding is provided through Peter Hudson and the Huck Institutes.

BIBLIOGRAPHY

Adams, W. M., Aveling, R. and D. Brockington (2004). Biodiversity conservation and the eradication of poverty. *Science*, 306(5699), 1146–49.
Bakker, K. and Bridge, G. (2006) Material worlds? Resource geographies and the 'matter of nature'. *Progress in Human Geography*, 30, 5–27.
Balée, W. (2006) The research program of historical ecology. *Annual Review of Anthropology*, 35, 75–98.

Barbujani, G. and Sokal, R. R. (1991) Genetic population structure of Italy. I. Geographic patterns of gene frequencies. *Human Biology*, 63(3), 253–72.

Bassett, T. and Zuéli, K. B. (2003) The Ivorian savanna: global narratives and local knowledge of environmental change. In K. S. Zimmerer and T. J. Bassett (eds), *Political ecology: An Integrative Approach to Geography and Environment-Development Studies*. New York: Guilford Publications, pp. 115–136.

Blakey, C. A, Costich, D., Sokolov, V. and Island-Faridi, M. N. (2007) *Tripsacum* genetics: From observations along a river to molecular genomics. *Maydica*, 52, 81–99.

Blumler, M. A. (2003) Introgression as a spatial phenomenon. *Physical Geography*, 24, 414–23.

Botkin, D. B. (2000) *Environmental Science: Earth as a Living Planet*, 3rd edn. New York: John Wiley.

Bryant, R. L. (2000) Politicized moral geographies – Debating biodiversity conservation and ancestral domain in the Philippines. *Political Geography*, 19(6), 673–705.

Castree, N. (2005) *Nature*. London: Routledge.

Coomes, O. T., Barham, B. L. and Takasaki, Y. (2004) Targeting conservation-development initiatives in tropical forests: insights from analyses of rain forest use and economic reliance among Amazonian peasants. *Ecological Economics*, 51(1–2): 47–64.

Costanza, R., Fisher, B., Mulder K, Liu, S. and Christo, A. (2007) Biodiversity and ecosystem services: a multi-scale empirical study of the relationship between species richness and net primary production. *Ecological Economics*, 61(2–3), 478–91.

Dawson, I. K. and Powell, W. (1999) Genetic variation in the Afromontane tree *Prunus africana*, an endangered medicinal species. *Molecular Ecology*, 8, 151–56.

Denevan, W. M. (1992) The pristine myth – the landscape of the Americas in 1492. *Annals of the Association of American Geographers*, 82(3), 369–85.

Doebley, J. (2006) Unfallen grains: how ancient farmers turned weeds into crops. *Science*, 312, 1318–19.

Doebley, J., Gaut, B. S. and Smith, B. D. (2006) The molecular genetics of crop domestication. *Cell*, 127, 1309–21.

Ehrlich, P. R. and Levin, S. A. (1998) Biodiversity: What is it and why we need it. In L. Koebner (ed.), *Scientists on Biodiversity*. New York: American Museum of Natural History, pp. 20–23.

Epperson, B. K. (1995) Spatial distribution of genotypes under isolation by distance. *Genetics*, 140, 1431–40.

Farnham, T. J. (2007) *Saving Nature's Legacy: Origins of the Idea of Biological Diversity*. New Haven: Yale University Press.

Foster, D. R. (2002) Insights from historical geography to ecology and conservation: lessons from the New England landscape. *Journal of Biogeography*, 29(10–11), 1269–75.

Graham, C. H., Ferrier, S., Huettman, F., Moritz, C., and Townsend Peterson, A. (2004) New developments in museum-based informatics and applications in biodiversity analysis. *TRENDS in Ecology and Evolution*, 19(9), 497–503.

Greene, H. W. (2005) Organisms in nature as a central focus for biology. *TRENDS in Ecology and Evolution*, 20(1), 23–27.

Greenhough, B. (2007) Situated knowledges and the spaces of consent. *Geoforum*, 38, 1140–51.

Hanotte, O., Bradley, D. G., Ochleng, J. W., Verjee, Y., Hill, E. W. and Rege, J. E. O. (2002) African pastoralism: genetic imprints of origins and migrations. *Science*, 296, 336–39.

Hecht, S. B. (2004) Invisible forests: the political ecology of forest resurgence in El Salvador. In R. Peet and M. Watts (ed.), *Liberation Ecologies: Environment, Development, Social Movements*, 2nd edn. London: Routledge, pp. 64–103.

Jelinski, D. E. (1997) On genes and geography: a landscape perspective on genetic variation in natural plant populations. *Landscape and Urban Planning*, 39, 11–23.

Kellert, S. R. and Wilson, E. O. (eds) (1993) *The Biophilia Hypothesis*. Washington, DC: Island Press.

Kidd, D. M., and Ritchie, M. G. (2006) Phylogeographic information systems: putting the geography into phylogeography. *Journal of Biogeography*, 33, 1851–65.

Knapp, S., Bohs, L., Nee, M., and Spooner, D. M. (2004) Solanacease – a model for linking genomics with biodiversity. *Comparative and Functional Genomics*, 5, 285–91.

Lomolino, M. V. (2000) A call for a new paradigm of island biogeography. *Global Ecology and Biogeography*, 9(1), 1–6.

Lorimer, J. (2006) What about the nematodes? Taxonomic partialities in the scope of UK biodiversity conservation. *Social and Cultural Geography*, 7(4), 539–58.

Lovejoy, T. E. and Hannah, L. (2005) *Climate Change and Biodiversity*. New Haven: Yale University Press.

Lubchenco, J. (1998) Entering the century of the environment: a new social contract for science. *Science*, 279(5350), 491–97.

MacDonald, G. B. (1995) The case for boreal mixed-wood management: an Ontario perspective. *Forestry Chronicle*, 71(6), 725–34.

Manel, S., Schwartz, M. K., Luikart, G., and Taberlet, P. (2003) Landscape genetics: combining landscape ecology and population genetics. *TRENDS in Ecology and Evolution*, 18(4), 189–97.

Martin-Lopez, B., Montes, C. and Benayas, J. (2007) The non-economic motives behind the willingness to pay for biodiversity conservation. *Biological Conservation*, 139(1–2), 67–82.

McAfee, K. (1999) Selling nature to save it? Biodiversity and green developmentalism. *Environment and Planning D – Society and Space*, 17(2), 133–54.

McNeely, J. A. (1988) *Economics and Biological Diversity: Developing and Using Incentives to Conserve Biological Resources*. Gland, Switzerland: International Union for Conservation of Nature (IUCN).

Monmonier, M. S. (1973) Maximum-difference barriers: an alternative numerical regionalization method. *Geographical Analysis*, 5(3), 245–61.

Moritz, C. (2002) Strategies to protect biological diversity and the evolutionary processes that sustain it. *Systematic Biology*, 51(2), 238–54.

Motley, T. J., Zerega, N. and Cross, H. (eds) (2006) *Darwin's Harvest: New Approaches to the Origins, Evolution, and Conservation of Crops*. New York: Columbia University Press.

Nabhan, G. P. (1995) The dangers of reductionism in biodiversity conservation. *Conservation Biology*, 9(3), 479–81.

Naughton-Treves L., Holland, M. B. and Brandon, K. (2005). The role of protected areas in conserving biodiversity and sustaining local livelihoods. *Annual Review of Environment and Resources*, 30, 219–52.

NRC (National Research Council) (1999) *Perspectives on Biodiversity: Valuing Its Role in an Ever-changing World*. Washington, DC: Committee on Noneconomic and Economic Value of Biodiversity.

Neumann, R. P. (2004) Moral and discursive geographies in the war for biodiversity in Africa. *Political Geography*, 23(7), 813–37.

Oldfield, M. L. and Alcorn, J. B. (1991) *Biodiversity: Culture, Conservation, and Ecodevelopment*. Boulder: Westview Press.

OTA (Office of Technology Assessment) (1987) *Technologies to Maintain Biodiversity*. Washington, DC: Congress of the U.S. and Office of Technology Assessment.

Parker, K. C. and Jorgensen, S. M. (2003) Examples of the use of molecular markers in biogeographic research. *Physical Geography*, 24(5), 378–98.

Parker, K. C., Hamrick, J. L. and Parker, A. J. (2001) Fine-scale genetic structure in *Pinus clausa* (Pinaceae) populations: effects of disturbance history. *Heredity*, 87, 99–113.

Parker, K. C., Hamrick, J.L. and Hodgson, W. C. (2007) Genetic consequences of pre-columbian cultivation for Agave murpheyi and A-delamateri (Agavaceae). *American Journal of Botany*, 94(9), 1479–90.

Peluso, N. L. (1993) Coercing conservation – the politics of state resource control. *Global Environmental Change-Human and Policy Dimensions*, 3(2), 199–217.

Piazza, A., Rendine, S., Minch, E., Menozzi, P., Mountain J., and Cavalli-Sporza, L. L. (1995) Genetics and the origin of European languages. *Proceedings of the National Academy of Sciences*, 92, 5836–40.

Reece, J. D. (2007) Does genomics empower resource-poor farmers? Some critical questions and experiences. *Agricultural Systems*, 94, 553–65.

Reid, W. V. (1997) Strategies for conserving biodiversity. *Environment*, 39(7), 16.

Rigg, L. S. (2003) Genetic applications in biogeography. *Physical Geography*, 24(5), 355–57.

Rosenzweig, M. L., Turner, W. R. and Cox, J. G. (2003) Estimating diversity in unsampled habitats of a biogeographical province. *Conservation Biology*, 17(3), 864–74.

Silva, M. (2004) Bioinformatics, the clearing-house mechanism, and the Convention on Biological Diversity. *Biodiversity Informatics*, 1, 23–29.

Smouse, P. E., and Peakall, R. (1999) Spatial autocorrelation analysis of individual multiallele and multilocus genetic structure. *Heredity*, 82, 561–73.

Soberon, J. (1999) Linking biodiversity information sources. *TRENDS in Ecology and Evolution*, 14(7), 291.

Sokal, R. R. and Thomson, B. A. (1998) Spatial genetic structure of human populations in Japan. *Human Biology*, 70(1), 1–22.

Takacs, D. (1996) The *Idea of Biodiversity: Philosophies of Paradise*. Baltimore: Johns Hopkins University Press.

Velazquez, A., Duran, E., Ramirez, I., Masa, J. F., Bocco, G., Ramíreza, G. and Palacio, J. L. (2003) Land use-cover change processes in highly biodiverse areas: the case of Oaxaca, Mexico. *Global Environmental Change-Human and Policy Dimensions*, 13(3), 175–84.

Whittaker, R. J. (2000) Scale, succession and complexity in island biogeography: are we asking the right questions? *Global Ecology and Biogeography*, 9(1), 75–85.

Wilson, E. O. (1988) *Biodiversity*. Washington, DC: National Academy Press.

Young, K. R., Ulloa, C. U., Luteyn, J. L. and Knapp, S. (2002) Plant evolution and endemism in Andean South America: an introduction. *Botanical Review*, 68(1), 4–21.

Zimmerer, K. S. (1996) *Changing Fortunes: Biodiversity and Peasant Livelihood in the Peruvian Andes*. Los Angeles and Berkeley: University of California Press.

Zimmerer, K. S. (2004) Cultural ecology: placing households in human-environment studies – the cases of tropical forest transitions and agrobiodiversity change. *Progress in Human Geography*, 28(6), 795–806.

Zimmerer, K. S. (2006a) Cultural ecology: at the interface with political ecology – the new geographies of environmental conservation and globalization. *Progress in Human Geography*, 30, 63–78.

Zimmerer, K. S. (2006b) Conclusion: rethinking the compatibility, consequences, and geographic strategies of conservation and development. In K. S. Zimmerer (ed.), *Globalization and New Geographies of Conservation*. Chicago: University of Chicago Press, pp. 315–346.

Zimmerer, K. S. and Douches, D. S. (1991) Geographical approaches to crop conservation – the partitioning of genetic diversity in Andean potatoes. *Economic Botany*, 45(2), 176–89.

Zimmerer, K. S. and Young, K. R. (eds) (1998) *Nature's Geography: New Lessons for Conservation in Developing Countries*. Madison: The University of Wisconsin Press.

Zimmerer, K. S., Galt, R. E., and Buck, M. V. (2004) Globalization and multi-spatial trends in the coverage of protected-area conservation (1980–2000). *Ambio: A Journal of the Human Environment*, 33, 520–29.

Zimmerer, K. S. (2007) Cultural ecology (and politcal ecology) in the 'environmental borderlands': Exploring the expanded connectivities within geography. *Progress in Human Geography*, 31(2), 227–44.

Chapter 5

Complexity, Chaos and Emergence

Steven M. Manson

Introduction

Geographers use concepts of complexity, chaos, and emergence in their research, whether focused on society and space, human-environment systems, geographic information science, or ecological and biophysical systems. At the same time, non-geographers increasingly find that complexity research – the general term applied to work on complexity, chaos, and emergence – leads them to concepts of space and place that undergird the geographical enterprise (Byrne, 1998; Cilliers, 1998; Lissack, 2001; Manson, 2001; Reitsma, 2002; Urry, 2003). The synergy between research in geography and complexity is supported by some shared characteristics. Geography and complexity both span a broad array of substantive areas, synthesise across multiple disciplines, and focus on an array of human and environmental systems that encompass multiple spatial, temporal and organisational scales. More broadly, complexity research is found in a variety of fields that have varying levels of engagement with geography, ranging from policy (McKelvey, 1999; Gatrell, 2005) to the natural sciences (Rind, 1999; Phillips, 2003; Brose et al., 2004), social sciences (Arthur, 1999; Batten, 2001; Sampson et al., 2002), and the humanities (Nowotny 2005; Portugali, 2006).

While the combination of complexity theory with geography in general and environmental geography in particular has excellent prospects for continued growth, it also confronts a series of methodological and conceptual challenges. Perhaps the greatest issue in complexity research is that there is no single or widely shared definition of complexity. Clear definitions are also lacking for more specific topics such as chaos and self-organisation, which have been used in various ways in complexity research across disciplines. In many respects, complexity follows the old adage that 'geography is what geographers do' because complexity researchers are often self-identified. Complexity is therefore usefully seen as an interdisciplinary endeavor in which individual disciplines and practitioners borrow techniques and approaches from other fields.

Thus, the terms complexity theory or complexity sciences serve as placeholders for a wide array of research. It is possible to identify three distinct, but highly interrelated approaches to complexity: algorithmic complexity, deterministic complexity, and aggregate complexity. Within each of these approaches and the larger field of complexity research, we can also identify and critically evaluate several areas that host the latest debates and larger challenges. Among the most pressing are questions about the novelty of complexity, reconciling simplicity with complexity, understanding the balance between equilibrium and change, bridging various disciplinary divides, and understanding how complexity affects our assessment and use of spatial, temporal and organisational scale.

Approaches to Complex Systems

Complexity research examines systems. A system is a set of entities connected in a way that gives the system an overall identity and behaviour. Systems can be of almost any scale, from atoms bound together in a molecule, to households in an economy, or the planets of our solar system. Complexity research centers on identifying the most important system elements and describing relationships among them. Systems are defined in part by these internal elements as well as by their larger environments. An ecosystem is self-contained in terms of much of its structure and function, for example, but also has many connections to the larger climatic, geophysical, and biotic environment.

Complexity research tends to fall into three broad areas of theory and practice (Manson, 2001), although many categorisations and definitions exist (cf. Byrne, 1998; Cilliers, 1998; Lissack, 2001; Reitsma, 2002). The first kind of complexity research can be termed algorithmic because it measures the structure of a system in terms of the computational processes needed to replicate the system. The second form is deterministic complexity, which explores systems via mathematical approaches that have become known as non-linear dynamics and chaos theory. The third form of complexity research examines aggregate complexity, or the manner in which systems such as ecosystems emerge from the local interactions of individual elements such as animals or plants.

Algorithmic complexity

Algorithmic complexity encompasses mathematical and computational approaches that attempt to calculate or characterise how difficult it is to represent or model a system in mathematical or algorithmic terms. The field of computational complexity theory measures the difficulty of solving mathematical or computational problems, particularly with respect to how changes in the size of a system affect the difficulty of representing a system. One common problem in environmental geography is determining the time or computational resources required to calculate all permutations in a resource allocation situation, such as choosing a set of conservation areas designed to maximise biodiversity in a given region (Aerts et al., 2003). For problems of moderate size, say involving the allocation of 100 areas of interest (e.g., represented as a 10×10 raster grid or 100 discrete regions) there are billions of different ways of ordering the permutations of suitable areas. Solving this problem in a Geographic Information System or spatial model is very difficult without recourse to approximation or heuristics. A related subfield of mathematics, information theory, quantifies the 'complexity' of a system as the shortest algorithm that can reliably describe the system and reproduce its behaviour (Chaitin, 1992). In essence, simple algorithms are used to describe simple systems while longer and more sophisticated algorithms are necessary for complicated systems. These measures also often focus on entropy, or the amount of order versus randomness in a system.

Computational complexity and information theory provide measures of how complicated a system is, but not necessarily how 'complex' the system is in the way meant by most researchers in complexity science interested in deterministic or aggregate complexity because it does not distinguish between systems that are merely complicated and those that exhibit processual elements such as feedback or emergence (Gilbert, 1995; Reitsma, 2003; Perry, this volume). Algorithmic complexity is useful to complexity researchers, however, because it provides straightforward measures of how complicated a system will be to represent or how difficult it is to solve a problem. In particular, these measures identify problems that cannot be solved analytically, but instead must be approximated . Information-theoretic measures such as entropy are also useful because they assess the degree of order in a system; as discussed below, complexity research is very interested in systems that move between randomness and order (Phillips, 2003). More broadly, however, the use of algorithmic complexity by geographers employing complexity approaches has been limited (Manson, 2001; O'Sullivan, 2004).

Deterministic complexity

Deterministic complexity is comprised of approaches that describe the underlying dynamics of a system that determine its state and trajectory of evolution. Systems can have both negative feedback, whereby changes in the state of the system tend to diminish over time, and positive feedback, where system dynamics make changes self-reinforcing. For example, in the case of climate change, warming of the tropical oceans will generate more cloud cover that will reflect incoming solar radiation and thereby dampen the warming effects of anthropogenic greenhouse gas emissions, while melting of the polar ice caps is a positive feedback that will accelerate global warming by increasing the amount of solar radiation absorbed at high latitudes (Rind, 1999; Schneider, 2004). Deterministic complexity provides a framework for understanding and predicting the dynamics of the climate and other systems by deriving equations to describe the behaviour of and relationships among their component parts and examining how feedback among these equations (and thereby the system components they describe) affects the system overall. This area of complexity is also concerned with understanding how feedback can make the system sensitive to small perturbations, as detailed below (Malanson et al., 1990).

Deterministic complexity takes its name from the idea that a few key variables in a small set of equations can describe a system. The deterministic aspect stems from the way in which system behaviour is 'determined' by the equations and their initial values. To capture animal population dynamics, for example, we can look to a population growth model developed in 1838 by Pierre François Verhult:

$$X_{t+1} = \alpha X_t (1 - X_t) \tag{5.1}$$

This equation predicts the future size of a population X_{t+1} as a function of the present population size X_t (measured on a scale of zero to one) and a rate of growth α that represents natural factors such as the rate at which births and deaths occur, availability of food, or threat of predation. This equation is used iteratively or, in other words, the answer from one time step (t) becomes the input for the next step (t + 1). This population model is simple because it uses very straightforward mathematics, but it can capture complex population dynamics that are highly sensitive to the value of α representing the rate of growth. Figure 5.1 illustrates these complex dynamics by showing a single end-point or 'attractor' for thousands of different iterations. The value of each system end point, given by the y-axis, varies widely with changes in α along the x-axis.

Systems often have feedback. In the population model, the use of iteration creates feedback between the present population X_t and future population X_{t+1}. When α is between one and three, for example, negative feedback causes the population to settle over time to a single value of 1 - 1/α. This value is an example of a stable attractor, or the value within a mathematical system towards which a variable inevitably tends or reaches. When α = 3, for example, the population value settles down to become 1 - $\frac{1}{3}$ or $X = \frac{2}{3}$. Another stable attractor occurs when α ranges between zero and one. In this case, the population dies out over time due to negative feedback. The population can also expand endlessly via positive feedback when α is greater than four.

Deterministic systems can be both sensitive, in that changes in their overall behaviour may occur as a result of small changes or perturbations in one of their

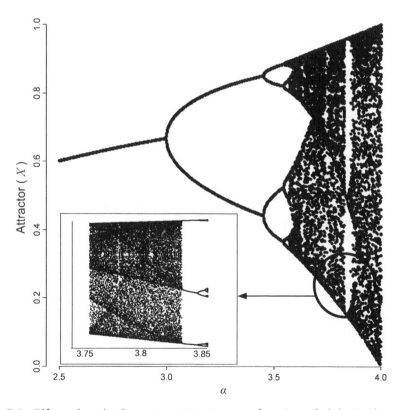

Figure 5.1 Bifurcations in the system attractor as a function of alpha in the population growth model.

parts, and non-linear because such small changes can lead to large changes in other parts (Phillips, 2003). The population model is a particularly interesting example, because it is only sensitive to changes in α over a small range of values. So long as its value remains between one and three, a small change in α has a correspondingly minor impact on X. Similarly, any change in α, when it is less than one or greater than four, has no impact in the sense that X drops to zero or becomes infinite regardless of the size of change in α. In contrast to those ranges, the system is very sensitive to changes in α when it takes a value between three and four. Any small shift in the value of α in this range results in large shifts in X among multiple attractors, which are equivalent to population boom-bust cycles in real-world populations (figure 5.1). The sudden shifts that occur due to this sensitivity are termed bifurcations or catastrophic changes (May, 1976; Feigenbaum, 1980; Brown, 1995). The term butterfly effect, which metaphorically suggests that the flapping of a butterfly's wings may cause severe weather elsewhere (Lorenz, 1973), also describes sensitivity, particularly in the initial values of a model. As discussed below, the potential for sensitivity to initial conditions raises fundamental questions about our ability to model complex systems and predict their behaviour, as well as, more broadly, about the nature of equilibrium and change.

Many mathematical systems have stable attractors, but those describing deterministically complex systems can also have strange attractors, or sets of values towards which the system tends, but never quite reaches. In our population model, the population size X seemingly becomes completely random when $\alpha = 3.8$ (figure 5.1, inset). In terms of deterministic complexity, however, this system is not truly chaotic or unknowable because we can model it with a single equation and know exactly which value of α generates the seemingly random values of X. Moreover, the values of X will generally cluster around a certain set of values that define the strange attractor. A system that exhibits these two characteristics – being modelled with equations and having attractors – is termed deterministically chaotic as opposed to truly chaotic (Leiber, 1999).

One kind of strange attractor that has garnered much attention is the fractal. This term refers to a pattern that remains unchanged over the spatial or temporal scale of observation. Trees and river systems, for example, are fractal in the sense that they appear to have a branching structure at scales ranging from the very small, such as veins in leaves or the smallest stream branches, to the very large, such as the branching structure of the entire tree or river system (Pecknold et al., 1997). So too is the general branching structure of the population system in figure 5.1, which is mirrored at the small scale in the figure inset. Systems that exhibit fractal patterns are interesting because the appearance of similar patterns at different scales implies that similar underlying processes may exist. Thus, understanding the processes operating at one scale may lead to understanding of the processes operating at others. As examined below, however, actually using the discovery of fractal patterns at one scale to understand or predict the behaviour of a system at other scale is fraught with difficulty.

Aggregate complexity

Aggregate complexity examines many of the same features of systems as algorithmic and deterministic complexity, such as feedback or non-linearity. However, aggregate complexity is more concerned with how systems are created by the simple and local interactions of system components, and less with the measures or variables that define systems. Aggregate complexity places particular emphasis on the role of individual entities and the relationships among them in defining system structure and behaviour within its larger environment. The role of adaptation, learning, and change in both system components and the system as a whole is critically important to research in aggregate complexity and is especially important to research related to coupled human-environment systems.

Systems like an ecosystem or an economy are driven to a great extent by individual components and their relationships. Within ecology, for example, biotic entities such as plants and animals have relationships defined by exchanges of matter, energy, and information with other entities in larger ecological systems. Importantly, most entities in the system have multiple relationships and play multiple roles. A tree, for example, cannot survive without relationships with entities and systems ranging from bacteria to other trees to weather systems. Of course, some relationships are more important than others to any given component. Especially tight links between entities will join them into larger collective groups that act as entities in and of themselves (Allen and Holling, 2002). For example, the odds of a single tree thriving are very dependent on whether a sufficient number of other trees exist to form a stand to protect individual trees from wind damage, while the existence of many stands in close proximity is important to defining a larger forest that in turn creates its own self-sustaining microclimate and habitat to which arboreal vegetation is better adapted than competing grassland species.

One particularly important kind of interaction is self-organisation, in which entities within a system change their relationships in a manner that enables the system as a whole to adapt its structure and behaviour to better suit its environment (Easterling and Kok, 2002). Sometimes these changes to internal relationships are slow and gradual. At other times, outside forces or internal perturbations may encourage the system to make sudden, large changes similar to the bifurcations of deterministic complexity. Even small disturbances such as fires have the capacity to reorder entities and relationships throughout an ecosystem, causing it to move through cycles of destruction and rejuvenation (Holling, 2001). Self-organisation can lead to self-organised criticality, where the system hovers on the edge of collapse and, as a result, can quickly shift resources and internal relationships to respond to internal or external changes (Bak, 1996). The evolution of a forested landscape in the face of both environmental and human perturbations, for example, can be understood as a system governed in part by self-organised criticality in which there is a balance between disturbances (human ones such as building roads and environmental ones such as fires) and orderly succession of land use and cover (Bolliger et al., 2003; Crawford, 2005).

Self-organisation is closely tied to the concept of emergence. Systems that are treated as 'complex' by aggregate complexity can possess emergent qualities that do not result from superposition (i.e., additive effects of system components), but instead from synergistic interactions among components. In other words, the behaviour of a system can be greater than the sum of the behaviour of system's constituent parts. Individual cells within an organ such as the brain or liver, for example, band together to allow it to act in ways not easily surmised from examining the characteristics and behaviour of individual cells. Some authors go so far as to claim that a system is only complex if it displays emergent properties that cannot be fully explained by analysing its components in isolation (Holland, 1998). Emergence can be difficult to define beyond the general focus on synergy. This has led some researchers to the more grounded concept of supervenience, which asserts that changes in a system at one level are tied to changes at another level, and even small changes in one level can lead to large changes in another (Sawyer, 2002). We examine emergence below in greater detail.

Is Complexity New?

Complexity research is often promoted as a fundamentally 'New Kind of Science' (after Wolfram, 2002), but it has deep conceptual roots. Such research reflects long-standing philosophical ideas, among them Aristotle's metaphysical work on synergy and Whitehead's philosophy of organism, which contends that nature is not merely a set of fixed laws or circumstances, but instead is a continually evolving process (Whitehead, 1925). Complexity also shares features with cybernetics, the study of how feedback in systems relates to communication and control in entities ranging from organisms to machines to social institutions (Wiener, 1961). Complexity can also be traced to specific computational and analytical approaches like neural networks, computer or mechanical programs that mimic biological brain functioning (McCulloch and Pitts, 1943), and cellular automata, simple computer programs that interact with one another (von Neumann, 1966). It also shares attributes with general systems theory, which posits that many human and natural systems can be understood by holistically treating them as stocks and flows of energy, matter, or information (von Bertalanffy, 1968).

Complexity theory differs from earlier movements in general systems theory, computer science, or philosophy in its treatment of relationships among entities (Phelan, 1999). These earlier efforts typically concentrated on fixed entities and stocks, such as animal populations linked by linear flows of energy or matter. Complexity focuses more on how systems evolve or emerge from simple, local interactions among individual system components. Systems theory, and much of current systems dynamics modelling, focuses on parameterising flows and stocks of energy or matter existing in equilibrium. Conversely, much complexity research contends that systems often exist in disequilibrium or near the edge of chaos. Riverbanks, for example, can be modelled as systems where bank erosion balances deposition of silt and other matter. The balance between these two forces, however, is not always a gradual to-and-fro, but instead a system that shuttles between them, which results in periods of stability marked by sudden riverbank failures (Fonstad and Marcus, 2003). Similarly, complex systems as envisioned by aggregate complexity may have emergent characteristics that cannot be explicitly specified in advance of running a model as a series of entities and their interrelationships. While the notion of a system being more than the sum of its parts has long been central to systems thinking, complexity research is interested specifically in how systems evolve over time as a function of relationships among the entities that comprise them.

Simplicity and Complexity

A perennial tension in using complexity concepts is reconciling the theory and ethos of complex systems with the complicated nature of the real world. In particular, it is difficult to computationally represent environmental phenomena like ecosystems or global climate with a few simple rules or equations as called for by many complexity approaches. An important drawback of algorithmic complexity, for example, is the conflation of data with knowledge or meaning; some systems simply may not be amenable to representation with bits and bytes in an algorithm. Rates of change in environmental systems can be represented using mathematically well-understood non-linear differential equations, for example, but these are difficult to solve. As a result, 'modellers have had to find various ways to approximate them using methods such as numerical iteration or finite difference calculation that provide analytically tractable solutions' (Mulligan and Wainwright, 2003; Demeritt and Wainwright, 2005, p. 215). A related problem with implementing concepts of deterministic complexity is finding appropriate and measurable variables for use in mathematically tractable equations. Choosing and defining these variables is a difficult task that can result in the omission of important factors. This said, while fewer systems than hoped for are deterministically complex (Zimmer, 1999), strong examples from systems ranging from river systems to earthquakes do exist in environmental geography and cognate fields such as ecology, biogeography, and geomorphology (Phillips, 2003).

Aggregate complexity also poses challenges to encoding real systems into data and models because it posits that system characteristics emerge 'bottom up' from interactions among entities at small scales. Translating these straightforward principles into a model that can use empirical data or test existing theories is a difficult task because both quantitative models and qualitative theories can quickly become more complicated (but not more complex in the sense meant in complexity research) in order to describe real systems (Torrens and O'Sullivan, 2001). For example, a host of different complexity-based methods are used to examine land change, including neural networks, cellular automata, and agent-based modelling. As these approaches have become more common and more sophisticated, however, modellers are increasingly tempted to capture a large number of features in human-environment systems. In doing so, they run the risk of moving away from a basic tenet of complexity science, namely that seemingly complex systems or dynamics can be generated by a small set of rules, such as transition rules for cellular automata or simple decision-making strategies of agents (Parker et al., 2003).

Complexity researchers often risk focusing on patterns that they believe signal the presence of complex processes instead of the complex processes as such. Algorithmic, deterministic, and aggregate complexity all search for hallmark patterns of complexity such as information-theory measures or fractals. This is because these patterns can indicate the existence of processes including deterministic chaos, emergence, and self-organised criticality (Goodchild and Mark, 1987; Bak, 1996; Barabási and Bonabeau, 2003). Patterns associated with complexity do not necessarily indicate the existence of complex processes, however, because many processes may create a single pattern and a single process may create many patterns. This is the case for deforestation in many parts of the world, for example, where a single pattern such as runaway tree felling can be driven by a broad range of social and economic processes, and a single process, such as infrastructure development, can result in a range of different deforestation patterns (Geist and Lambin, 2002). It is also possible to create complex patterns using complex processes that have no correspondence to real-world processes. A growing body of work questions the validity of using generic complex processes such as self-organised criticality to model systems ranging from species composition to biogeographical regions to social systems (Malanson, 1999; Plotnick and Sepkoski, 2001; Stewart, 2001).

Equilibrium and Change

Science in general has had great success in analysing systems such as economies or ecologies by assuming they are in equilibrium. Complexity moves away from the common definition of equilibrium, in which opposing forces are in balance, and towards more dynamic forms of system stability and resilience. On a surficial level, sensitivity and non-linearity in a system appear antithetical to equilibrium because small perturbations in one part of a system can lead to large shifts in system behaviour elsewhere. However, sensitivity and non-linearity are typically found only at particular thresholds, with the result that sudden shifts in system behaviour or structure are fairly limited and occur as shifts among multiple varying attractors. A host of physical phenomena, ranging from vegetation-soil dynamics to stream systems, demonstrate the ability of sensitive systems to reach two or three stable states (Sivakumar, 2000; Phillips, 2006). The question becomes whether to focus on the large shifts among attractors due to sensitivity or on the fact that the system is insensitive in the sense that it ends up being defined by attractors regardless of initial values. These subtle, yet important, differences in the meaning of sensitivity are evident in two related definitions: sensitivity to initial conditions (emphasising the effect of small changes) and independence of initial conditions (emphasising attractors) (Phillips, 2003).

The interplay among sensitivity, non-linearity, and equilibrium forces researchers and policymakers to question the extent to which models can help project the future of human or natural systems, especially if a small change in one location results in large changes elsewhere (Ortegon-Monroy, 2003). At the same time, we can understand the general characteristics of a system even when its precise state may be beyond prediction. These general characteristics are gleaned through simple rules under aggregate complexity, equations with deterministic complexity, and measures under algorithmic complexity (Byrne, 2005). A complex system can also be path-dependent; that is, its present state can be contingent on past states. In the extreme, a complex system such as an ecosystem can lock into a fixed state due to positive or negative feedback (Hendry and McGlade, 1995). Wildfires can be path-dependent, for example, because the ability of a fire to spread is largely a function of its size; large fires have greater capacity than small ones to expand until they run out of fuel or encounter adverse weather conditions (Moritz et al., 2005).

Two characteristics of complex systems help offset the destabilising effects of sensitivity and non-linearity. The first is resilience, which is the ability of the system to change without drastically affecting the relationships among its components. The second is transformability, or the capacity of a system to move to new configurations (Walker et al., 2004). The combination of resilience and transformability can give a complex system a form of stability and equilibrium in the larger sense that its internal components remain intact even if some of their relationships shift. Deterministic complexity focuses on the manner in which attractors and sensitivity capture this dynamic, while aggregate complexity places greater emphasis on how systems hover between randomness and stasis through self-organisation and self-organised criticality.

Scales of Space, Time and Organisation

Complexity research also complicates accepted notions of geographical scale. Spatial scale is most commonly treated as a nested hierarchy in which areal extents act as containers for those at a smaller scale level and are themselves encompassed by a single container at a larger scale level (Haggett, 1965). A single watershed may contain multiple reaches and may itself be contained by a larger watershed hemmed in by continental divides and oceans. Nested hierarchies assume that all the components at one level, such as river reaches, fit completely within a single component at a larger scale, here, a watershed. In this spatial hierarchy, the effects of an event at the local level work their way up to larger levels, but their impact is vanishingly small, as when a single drop of water works its way from a single subwatershed through its encompassing regional and continental watersheds.

Scale becomes less straightforward in the face of deterministic complexity. Sensitivity and non-linearity ensure that local actions can have disproportionate effects at larger scales. Instead of being dampened out, small changes may become amplified through the non-linear interactions among components across scales. This can occur in purely physical systems, as when we see the butterfly effect noted above in climate systems, or in human-environment systems, as when environmentalists use iconic imagery, such as polar bears facing extinction, to skip over regional and national political arenas to discuss climate change on the world stage (Slocum, 2004).

Emergence and supervenience result in dynamic system structures. Some ecologists argue that scale levels in an environmental system should not be seen as fixed, but instead, should be defined by interactions and relationships among entities (O'Neill, 1988). Complexity adds the notion of emergence to the mix by positing that scale levels are emergent phenomena that arise from interactions among entities as when institutions and organisations emerge from the interactions among individuals (Ostrom, 2005). For example, ecological landscapes are usefully treated as complex systems composed of interactions among human and natural actors that generate multiple scales of analysis (Easterling and Kok, 2002; Bousquet and Le Page, 2004).

Complexity research also contributes the concept of scale invariance, defined as a single process or pattern that is identical across spatial or temporal scales. Fractals, for example, are scale invariant because their appearance does not vary with the scale at which they are observed. As noted above, invariant patterns such as fractals may indicate that processes giving rise to their existence may also be similar across scales (White and Engelen, 1993; Marquet, 2000). These processes in turn are often of interest to complexity scientists, such as self-organisation, self-organised criticality and emergence (Prigogine and Allen, 1982; Bak, 1996; Lee, 2004; Crawford, 2005).

Given the emphasis on how systems emerge or grow from the bottom up, there is less research on how systems evolve when the components are conscious of their own part in the wider system. Research on how social norms emerge from interactions among people, for instance, is often emphasised at the expense of understanding how these emergent norms affect the people themselves in turn (Ostrom, 2005). Some definitions of emergence, particularly in the natural sciences, posit that constituent elements are unaware of the role they play in creating emergence in a system (Forrest, 1990). This approach to emergence may not adequately reflect the importance of human decision making to coupled human-environment systems given the role of reflexivity in decision making. As noted by Nigel Gilbert (1995), 'people are routinely capable of detecting, reasoning about and acting on the macro-level properties (the emergent features) of the societies of which they form part' (p. 71).

Complexity and Geographical Divides

The conceptual breadth and multidisciplinarity of complexity research offer the potential for greater integration and reconciliation among human geography, physical geography and geographic information science (O'Sullivan, 2004).

Researchers in the humanities and social sciences using interpretivist approaches have long used complexity concepts such as chaos and catastrophe (Hayles 1991). This view of complexity often relies on social constructionism, which contends that our understanding of reality is molded through societal features such as language and power, focusing on 'understanding the plurality of constructions, how various assertions are made, how these are related to various interests of stakeholder groups and how outcomes are affected by power relations' (Jones, 2002, p. 248). Postmodern, post-structural and other interpretivist perspectives explore systems through the rubric of knowledge, language and power. Features such as sensitivity and nonlinearity are powerful as metaphors because they capture the importance of nuance, context and contingency, all bywords of an interpretivist and post-modern understanding of the world (Portugali, 2006). The importance of interactions among entities, particularly to aggregate complexity, also maps well onto various flavors of research that examines networks defined by relationships among individuals and communities that form and contest the larger social, cultural and human-environmental systems of which they are part (Cilliers, 1998; Thrift, 1999; Urry, 2003; Byrne, 2005; Nowotny, 2005; Braun, this volume).

Complexity is also a wellspring for quantitative research. This is especially true for computer simulations that act as virtual laboratories for exploring 'would-be' worlds as they unfold (Casti, 1997). Geographers use simulation and modelling for research, policy and education. Computer simulations allow examination of how systems appear and of their many possible futures or pasts. They additionally allow researchers to determine what we do and do not know about the system in question. Growth in complexity science relies to a great extent on advances in approaches such as computational intelligence, neural networks, cellular modelling and agent-based modelling. These methods or their antecedents have been available for decades, but their use has exploded with complexity research as such and the better availability of computer processing power and tools (Manson and O'Sullivan, 2006).

Complexity offers a way to bridge divides – including quantitative/qualitative or among subdisciplines like human and physical geography – because it accommodates a range of ontological perspectives and highlights that understanding complex systems requires triangulation among approaches and viewpoints. O'Sullivan (2004) argues that complexity points to the potential for greater engagement between groups ordinarily having little engagement, such as post-structuralist human geographers and modelers, because both approaches allow for competing explanations of many systems. More generally, this work courts the notion of complexity as seeing the world through a lens of 'imaginable surprise' that treats seemingly unexpected system outcomes as explainable when taking into account characteristics of complexity such as emergence, non-linearity, sensitivity and self-organisation (Schneider et al., 1998). In examining the potential for sudden tipping points or shifts in the global climate system, for example, researchers increasingly recognise the potential for sensitivity and non-linearity while also inviting greater involvement from various publics to answer questions that are beyond science, such as the political, cultural and ethical ramifications of climate change (Rind, 1999; Schneider, 2004).

Conclusion

Complexity and environmental geography have much to offer each other. Algorithmic, deterministic and aggregate complexity offer a range of methods and concepts to the study of environmental and human-environment systems. Environmental geography, in turn, offers a host of real-world systems and theories with which to test and expand complexity science. We can identify several areas of future research that link these two fields. The contest between simplicity and complexity may be perennial, but studies within environmental geography highlight the need to join generalised hallmarks of complexity to field-based observations. Experiments in hydrology, geomorphology and land-cover change, for example, are leading the way in establishing real-world examples of complexity science that go beyond use of complexity as a metaphor or analog (Sivakumar, 2000; Crawford, 2005; Phillips, 2006). The same holds true for competing views on equilibrium and change in that we can tie general complexity concepts to specific geographical examples, such as the tug of war between ecological zones along ecotones (Malanson et al., 2006). We can also triangulate among a range of quantitative and qualitative methods as mixed method research becomes more popular (Phillips, 2004; Moss and Edmonds, 2005). The field also offers a long history of research on scale as such and a deep understanding of, and expertise in, many systems that span spatial and temporal scales (Sheppard and McMaster, 2004). Thus, this is an exciting time for research at the interface of complexity science and environmental geography.

BIBLIOGRAPHY

Aerts, J. C. J. H., Eisinger, E., Heuvelink, G. B. M. complexity and Stewart, T. J. (2003) Using linear integer programming for multi-site land-use allocation. *Geographical Analysis*, 35(2), 148–69.

Allen, C. R. and Holling, C. S. (2002) Cross-scale structure and scale breaks in ecosystems and other complex systems. *Ecosystems*, 5(4), 315–18.

Arthur, W. B. (1999) Complexity and the economy. *Science*, 284(5411), 107–9.

Bak, P. (1996) *How Nature Works: The Science of Self-Organized Criticality*. New York: Copernicus Books.

Barabási, A.-L. and Bonabeau, E. (2003) Scale-free networks. *Scientific American*, 288(5), 50–59.

Batten, D. F. (2001) Complex landscapes of spatial interaction. *The Annals of Regional Science*, 35(1), 81–111.

Bolliger, J., Sprott, J. C. and Mladenoff, D. J. (2003) Self-organization and complexity in historical landscape patterns. *Oikos*, 100, 541–53.

Bousquet, F. and Le Page, C. (2004) Multi-agent simulations and ecosystem management: a review. *Ecological Modelling*, 176(3–4), 313–32.

Brose, U., Ostling, A., Harrison, K. and Martinez, N. D. (2004) Unified spatial scaling of species and their trophic interactions. *Nature*, 428(6979), 167–71.

Brown, C. (1995) *Chaos and Catastrophe Theories*. Thousand Oaks, CA: Sage.

Byrne, D. (1998) *Complexity Theory and the Social Sciences*. London: Routledge.

Byrne, D. (2005) Complexity, configurations and cases. *Theory Culture Society*, 22(5), 95–111.

Casti, J. (1997) *Would-Be Worlds: How Simulation Is Changing the Frontiers of Science*. New York: John Wiley and Sons.

Chaitin, G. J. (1992) *Information, Randomness, and Incompleteness*. Singapore: World Scientific.

Cilliers, P. (1998) *Complexity and Postmodernism: Understanding Complex Systems*. New York: Routledge.

Crawford, T. W. (2005) Spatial fluctuations as signatures of self-organization: a complex systems approach to landscape dynamics in Rondônia, Brazil. *Environment and Planning B: Planning and Design*, 32(6), 857–75.

Demeritt D. and Wainwright J. (2005) Models, modeling, and geography. In N. Castree, A. Rogers and D. Sherman (eds), *Questioning Geography*. Oxford: Blackwell, pp. 206–25.

Easterling, W. E. and Kok, K. (2002) Emergent properties of scale in global environmental modeling – are there any? *Integrated Assessment*, 3(2–3), 233–46.

Feigenbaum, M. J. (1980) Universal behavior in nonlinear systems. *Los Alamos Science*, 1, 4–27.

Fonstad, M. and Marcus, W. A. (2003) Self-organized criticality in riverbank systems. *Annals of the Association of American Geographers*, 93(2), 281–96.

Forrest, S. (1990) Emergent computation: self-organization, collective, and cooperative phenomena in natural and artificial computing networks. *Physica D*, 42, 1–11.

Gatrell, A. C. (2005) Complexity theory and geographies of health: a critical assessment. *Social Science and Medicine*, 60(12), 2661–71.

Gilbert, N. (1995) Emergence in social simulation. In N. Gilbert and R. Conte (eds), *Artificial Societies: The Computer Simulation of Social Life*. London: UCL Press, pp. 144–56.

Goodchild, M. F. and Mark, D. M. (1987) The fractal nature of geographic phenomena. *Annals of the Association of American Geographers*, 77(2), 265–78.

Haggett, P. (1965) *Locational Analysis in Human Geography*. London: Edward Arnold.

Hayles, N. K. (1991) *Chaos and Order: Complex Dynamics in Literature and Science*. Chicago: University of Chicago Press.

Hendry, R. J. and McGlade, J. M. (1995) The role of memory in ecological systems. *Proceedings of the Royal Society of London*, 259, 153–59.

Holland, J. H. (1998) *Emergence: From Chaos to Order*. Philadelphia: Perseus Press.

Holling, C. S. (2001) Understanding the complexity of economic, ecological and social systems. *Ecosystems*, 4(5), 390–405.

Jones, S. (2002) Social constructionism and the environment: through the quagmire. *Global Environmental Change*, 12, 247–51.

Lee, C. (2004) Emergence and universal computation. *Metroeconomica*, 55(2–3), 219–38.

Leiber, T. (1999) Deterministic chaos and computational complexity: the case of methodological complexity reductions. *Journal for General Philosophy of Science*, 30(1), 87–100.

Lissack, M. (2001) Special issue: what is complexity science? *Emergence*, 3(1), 3–188.

Lorenz, E. N. (1973) *Predictability: Does the Flap of a Butterfly's Wings in Brazil Set off a Tornado in Texas?* AAAS Convention of the Global Atmospheric Research Program, Cambridge, MA, MIT Press.

Malanson, G. (1999) Considering complexity. *Annals of the Association of American Geographers*, 89(4), 746–53.

Malanson, G. P., Butler, D. R. and Walsh, S. J. (1990) Chaos theory in physical geography. *Physical Geography*, 11(4), 293–304.

Malanson, G. P., Zeng, Y. and Walsh, S. J. (2006) Complexity at advancing ecotones and frontiers. *Environment and Planning A*, 38(4), 619–32.

Manson, S. M. (2001) Simplifying complexity: a review of complexity theory. *Geoforum*, 32(3), 405–14.

Manson, S. M. and O'Sullivan, D. (2006) Complexity theory in the study of space and place. *Environment and Planning A*, 38(4), 677.

Marquet, P. A. (2000) Ecology: invariants, scaling laws, and ecological complexity. *Science*, 289(5484), 1487–91.

May, R. (1976) Simple mathematical models with very complicated dynamics. *Nature*, 261, 459–67.

McCulloch, W. S. and Pitts, W. (1943) A logical calculus of the ideas imminent in nervous activity. *Bulletin of Mathematical Biophysics*, 5, 115–37.

McKelvey, B. (1999) Complexity theory in organization science: seizing the promise or becoming a fad? *Emergence*, 1(1), 5–32.

Moritz, M. A., Morais, M. E., Summerell, L. A., Carlson, J. M. and Doyle, J. (2005) Wildfires, complexity, and highly optimized tolerance. *Proceedings of the National Academy of Sciences*, 102(50), 17912–17.

Moss, S. and Edmonds, B. (2005) Sociology and simulation: statistical and qualitative cross-validation. *American Journal of Sociology*, 100(4), 1095–131.

Mulligan, M. and Wainwright, J. (2003) Modelling and model building. In J. Wainwright and M. Mulligan (eds), *Environmental Modelling: Finding Simplicity in Complexity*. Chinchester: John Wiley and Sons, pp. 7–73.

Nowotny, H. (2005) The increase of complexity and its reduction: emergent interfaces between the natural sciences, humanities and social sciences. *Theory Culture Society*, 22(5), 15.

O'Neill, R. V. (1988) Hierarchy theory and global change. In T. Rosswall, R. G. Woodmansee and P. G. Risser (eds), *Scales and Global Change*. New York: John Wiley and Sons, pp. 29–45.

Ortegon-Monroy, M. C. (2003) Chaos and complexity theory in management: an exploration from a critical systems thinking perspective. *Systems Research and Behavioral Science*, 20(5), 387–403.

Ostrom, E. (2005) *Understanding Institutional Diversity*. Princeton, NJ: Princeton University Press.

O'Sullivan, D. (2004) Complexity science and human geography. *Transactions of the Institute of British Geographers*, 29(3), 282–95.

Parker, D. C., Manson, S. M., Janssen, M., Hoffmann, M. J. and Deadman, P. J. (2003) Multi-agent systems for the simulation of land use and land cover change: a review. *Annals of the Association of American Geographers*, 93(2), 316–40.

Pecknold, S., Lovejoy, S., Schertzer, D. and Hooge, C. (1997) Multifractals and resolution dependence of remotely sensed data: GSI to GIS. In D. A. Quattrochi and M. F. Goodchild (eds), *Scale in Remote Sensing and GIS*. New York: Lewis Publishers, pp. 361–94.

Phelan, S. E. (1999) Note on the correspondence between complexity and systems theory. *Systemic Practice and Action Research*, 12(3), 237–38.

Phillips, J. D. (2003) Sources of nonlinearity and complexity in geomorphic systems. *Progress in Physical Geography*, 27(1), 1–23.

Phillips, J. D. (2004) Laws, contingencies, irreversible divergence, and physical geography. *The Professional Geographer*, 56(1), 37–43.

Phillips, J. D. (2006) Deterministic chaos and historical geomorphology: a review and look forward. *Geomorphology*, 76(1–2), 109.

Plotnick, R. E. and Sepkoski, J. J., Jr. (2001) A multiplicative multifractal model for originations and extinctions. *Paleobiology*, 27(1), 126–39.

Portugali, J. (2006) Complexity theory as a link between space and place. *Environment and Planning A*, 38(4), 647–64.

Prigogine, I. and Allen, P. M. (1982) The challenge of complexity. In W. C. Schieve and P. M. Allen (eds), *Self-Organization and Dissipative Structures: Applications in the Physical and Social Sciences*. Austin: University of Texas Press, pp. 3–39.

Reitsma, F. (2002) A response to 'simplifying complexity'. *Geoforum*, 34(1), 13–16.

Rind, D. (1999) Complexity and climate. *Science*, 284(5411), 105–7.

Sampson, R. J., Morenoff, J. D. and Gannon-Rowley, T. (2002) Assessing 'neighborhood effects': social processes and new directions in research. *Annual Review of Sociology*, 28, 443.

Sawyer, R. K. (2002) Nonreductive individualism, part I: supervenience and wild disjunction. *Philosophy of the Social Sciences*, 32(4), 537–59.

Schneider, S. H. (2004) Abrupt non-linear climate change, irreversibility and surprise. *Global Environmental Change Part A*, 14(3), 245–58.

Schneider, S., Turner, B. L., II and Garriga-Morehouse, H. (1998) Imaginable surprise in global change science. *Journal of Risk Research*, 1(2), 165–85.

Sheppard, E. and McMaster, R. (2004) *Scale and Geographic Inquiry: Nature, Society and Method*. Oxford: Blackwell.

Sivakumar, B. (2000) Chaos theory in hydrology: important issues and interpretations. *Journal of Hydrology*, 227(1–4), 1–20.

Slocum, R. (2004) Polar bears and energy efficient light bulbs: strategies to bring climate change home. *Environment and Planning D: Society and Space*, 22(3), 413–43.

Stewart, P. (2001) Complexity theories, social theory, and the question of social complexity. *Philosophy of the Social Sciences*, 31(3), 323–60.

Thrift, N. (1999) The place of complexity. *Theory, Culture and Society*, 16(3), 31–69.

Torrens, P. M. and O'Sullivan, D. (2001) Cellular automata and urban simulation: where do we go from here? *Environment and Planning B*, 28(2), 163–68.

Urry, J. (2003) *Global Complexity*. Cambridge, UK: Polity.

von Bertalanffy, L. (1968) *General Systems Theory: Foundation, Development, Applications*. London: Allen Lane.

von Neumann, J. (1966) *Theory of Self-Reproducing Automata*. Champaign-Urbana: University of Illinois Press.

Walker, B., Holling, C. S., Carpenter, S. R. and Kinzig, A. (2004) Resilience, adaptability and transformability in social–ecological systems. *Ecology and Society*, 9(2). http //www.ecologyandsociety.org/vol9/iss2/art5/ (accessed 08-01-08).

White, R. and Engelen, G. (1993) Cellular automata and fractal urban form: a cellular modelling approach to the evolution of urban land-use patterns. *Environment and Planning D: Society and Space*, 25(8), 1175–99.

Whitehead, A. N. (1925) *Science and the Modern World*. New York: Macmillan.

Wiener, N. (1961) *Cybernetics: Or, Control and Communication in the Animal and the Machine*. Cambridge, MA: MIT Press.

Wolfram, S. (2002) *A New Kind of Science*. Champaign, IL: Wolfram Media.

Zimmer, C. (1999) Life after chaos. *Science*, 284(5411), 83–86.

Chapter 6

Uncertainty and Risk

James D. Brown and Sarah L. Damery

Introduction

Uncertainties and risks pervade all aspects of scientific research and decision making. They are apparent both in the *processes* through which knowledge is gained and in its *outcomes*. Uncertainty refers to a lack of confidence about our knowledge (Brown, 2004). Risk involves *deciding* with a lack of confidence, where the precise outcome is unknown, but one or more possible outcomes may cause harm. While these concepts are not new, they have received increasing attention from scientists, policy-makers, and social theorists, and are linked to claims about paradigm shifts within science and in its relationship with society (Gibbons, 1999). While traditional modes of enquiry emphasised the primacy of scientific knowledge and its ability to resolve, and ultimately control, the 'true' state of the world, scientific determinism has since been criticised for its inability to tackle the worst problems facing our modern societies (Beck, 1992; Jasanoff, 1996). Problems such as climate change, environmental degradation, and 'natural disasters' are characterised by paralysing uncertainties (Handmer et al., 2001), multiple vested interests (Winstanley et al., 1998), and extensive inequality (Parry et al. 2007).

Numerous typologies and techniques have been developed to conceptualise, assess, control, and communicate uncertainty. These include probability theory (Bernardo and Smith, 2000), possibility theory (Zadeh, 1978) and game theory (von Neumann and Morgenstern, 1944). Such treatments have varied substantially, both within and between disciplines, and between the social and physical sciences. For example, many social science researchers emphasise the social and psychological origins of uncertainty and risk (e.g., Adams, 1995). Conversely, many physical scientists have ignored these aspects or attempted to control them artificially through model inputs and outputs (Shackley and Wynne, 1995). Indeed, an important argument of this chapter is that current treatments of uncertainty, far from increasing transparency and accountability in geographical research, often provide little more than a probabilistic façade on traditional, deterministic practices.

In environmental decision making, uncertainties may be sufficiently large to generate persistent conflicts and indecision. Uncertainties in decision outcomes are typically formulated as risks, where the consequences of a given outcome are evaluated alongside their probability of occurring. This may involve a prescriptive model of the decision process, such as Cost Benefit Analysis (Nas, 1996) or Multi-criteria Decision Analysis (Herath and Prato, 2006), in which risks are evaluated, and possibly aggregated, systematically. Risk assessments typically ignore wider processes of decision making, where problem definition and choice of decision framework are crucial. Consequently, they are often criticised for their cursory treatment of people as producers of uncertainty and risk. Particularly when risk assessments are undertaken for policy and decision-making purposes (often the case in society-nature issues), risk is now recognised as having multiple sources, multiple means of conceptualisation, and multiple influences on decision making.

This chapter provides an overview of the origins, nature and implications of uncertainty and risk for environmental research and decision making. First, it establishes a theoretical framework and consistent terminology for discussing uncertainty and risk (second section). The third section focuses on causes of uncertainty and risk in geographical enquiry, which are separated into psychological, social, and situational factors. Approaches for assessing and communicating uncertainty and risk are considered in the fourth section. Here, quantitative approaches to assessing and controlling uncertainty are compared with more recent, deliberative understandings (see Chilvers, this volume).

Alongside strategies for assessing uncertainty and risk, there are several prescriptive strategies for managing them. These include the Precautionary Principle (Harremoës et al., 2002), Life Cycle Analysis (Ciambrone, 1997), adaptive environmental management (Holling, 1978), and ecological modernisation (Young, 2000). This chapter focuses on the conceptual aspects of uncertainty and risk. (Further details on these applied aspects can be found in Chapters 26 and 28 of this volume.)

The Nature of Uncertainty and Risk

Discussions about uncertainty and risk are complicated by the varying ways in which these concepts are defined and applied, both within and between disciplines. Numerous taxonomies of imperfect knowledge have been proposed in recent years. These include taxonomies for general types or levels of imperfect knowledge, such as error, indeterminacy, uncertainty, and ignorance (Suter et al., 1987; Smithson, 1989), schemes that focus on particular sources of imperfect knowledge (Wätzold, 2000; Regan et al., 2002), and schemes that employ some combination of the two (Walker et al., 2003). As the major sources of uncertainty vary between cases, it is common for detailed studies to employ different terminologies.

Here, *uncertainty* is regarded as a lack of confidence about our knowledge. Our confidence may vary from being certain that something is correct, incorrect (i.e., in error) or irrelevant, to accepting that we have no useful knowledge for some practical application, such as decision making. Uncertainty occurs at varying levels in between. It may be viewed as temporary, where some aspects of the environment cannot be resolved in practice (e.g., direct observations of groundwater flow), or a permanent condition, where some aspects cannot be resolved in principle (e.g., the evolutionary state of humans 10^6 years into the future). The latter is known as *indeterminacy*. No distinction is made here between 'scientific uncertainty' and the numerous 'other' uncertainties that affect environmental decision making, such as political, social, and economic uncertainties. Such a distinction wrongly implies that the outcomes of science are independent of the processes that generated them, which are necessarily subject to social, political, and economic uncertainties.

In common usage, *risk* focuses on the potential negative impacts of being exposed to harm and is therefore synonymous with loss. It extends the concept of uncertainty to decision making, where the potential for loss is known (e.g., in terms of time, money, property, environmental quality or human life) but the precise nature of the loss, whether it will occur, or even how probable it is, are unclear. In this context, various technical definitions of risk have been proposed in the social and physical sciences, including the treatment of risk as the probability of an undesirable event or the probability of an accident multiplied by the expected loss (Bernstein, 1998). However, risk is often embraced in the pursuit of some gain, and others associate risk with opportunity and the entrepreneurial balancing of uncertain costs and benefits (Baker and Simon, 2002). While the latter implies choice, the extent to which risks are chosen, and the ways in which they are managed, will depend on a range of individual, social and situational factors (Section 4). Similarly, while risk implies uncertainty about the costs and benefits of a decision, this uncertainty may be unevenly distributed among those involved in, and affected by, a decision. Indeed, multiple perspectives can originate from disagreement about what a decision *should* achieve, including what represents a good or bad outcome (e.g., more housing versus more greenbelt land), as much as any uncertainty about the precise consequences of an action.

As an expression of confidence, uncertainty will vary between individuals and groups of scientists and is, therefore, subjective (Brown, 2004). Similarly, perceptions of risk vary with personality and culture. This will in turn influence decision making: 'risk perceived is risk acted upon' (Adams, 1995). Uncertainty and risk will depend on our level of *awareness* or recognition of a 'problem', its perceived importance, and our apparent ability to resolve it. All of these factors are psychologically and socially motivated (Section 3). While uncertainty and risk imply that we are aware of a problem or potential loss, our precise level of awareness may vary, and it is the interaction of awareness and confidence that leads to the various expressions of imperfect knowledge considered here.

Any decision that involves uncertain costs and benefits, i.e., risk, entails the possibility of surprise. An important result of considering awareness and confidence jointly is that our capacity to be surprised is endemic, since it may originate from a lack of awareness, misplaced confidence, or some combination of the two. For example, most 'natural disasters' can be attributed to a lack of awareness about where and when an event will occur and misplaced confidence about existing levels of protection. Related to this is the *act of ignoring*, where some information is dismissed for reasons of efficiency, simplicity, or self-interest. For example, a modelling problem may be simplified by using a limited space-time domain, a finite resolution or a reduced process description. However, in non-linear systems, the impacts of these assumptions can only be predicted in general terms (and often not at all), such that the accumulation of uncertainties may ultimately lead to surprise.

A basic result of uncertainty, both in scientific research and environmental decision making, is the presence of multiple possible outcomes or explanations (referred to as non-uniqueness). For example, a single causal pathway can lead to different outcomes in different geographical contexts (Von Engelhardt and Zimmerman, 1988). Similarly, multiple causal pathways can lead to the same outcome in any given context, for which the term 'equifinality' has been used (e.g., Beven, 2002). Here, the distinction between possible *outcomes* and *probabilities* of particular outcomes is useful, as it points to different types and levels of uncertainty, which should be addressed with different methods (Brown, 2004). For example, uncertainties associated with climate scenarios are of a different type and magnitude to those associated with numerical parameters in a climate model, and are addressed with correspondingly different methods (Houghten et al., 2001). Indeed, for some types of outcomes, commonly assessed in the physical sciences, the distinction between possibilities and probabilities is largely a question of methodology. Here, formal methods are available at varying levels of detail to describe possible outcomes (scenario analysis, possibility theory) and probabilities of outcomes based on precise definitions of 'events' (subjective and frequency-based probability methods). Common to these formalisms is an assumption that *only one* outcome exists in principle. In contrast, other types of outcomes, commonly addressed in the social sciences, are non-unique in principle, such that discussions of probability are irrelevant (e.g., perceptions on the fair distribution of wealth). This can lead to tensions between social and physical scientists on issues of uncertainty.

Causes of Uncertainty and Risk

Uncertainties and risks may be psychological, social or situational in origin. Psychological factors include the propensity for risk aversion and fear of the unknown. Social factors include language, and the development of scientific networks, which are built on trust and consensus. Situational factors concern the types of problem addressed, including their transparency, scale, variability, and complexity. Clearly, these factors are closely related in practice; for example, trust (a social factor) is built on personality (an individual factor) and depends on the complexity of the problem in question (a situational factor). Establishing the relationships among these sources of uncertainty is a key research challenge, as the accumulated uncertainties in decision making will be sensitive to these relationships. For example, the overall uncertainty in a flood inundation model will depend on the type (e.g., linearity versus non-linearity) and degree of association between physical parameters in adjacent locations. Similarly, the levels of uncertainty associated with a flood warning will depend on the modes of message construction and dissemination (e.g., expert-driven versus community-driven) between planning authorities, the emergency services, and the general public, who develop, issue and respond to those warnings (Handmer, 2001). It follows that the sources of uncertainty are manifest both in the outcomes of research and decision making, i.e., 'what we know', and in the processes through which those outcomes are produced, i.e., 'how we came to know'.

Research in the cognitive sciences has shown that an individual's perceptions of uncertainty and risk are determined partly by the structure of the human brain and partly by their experiences and personality. Knowledge is embodied in cognitive structures that are commonly referred to as 'mental models' (Morgan et al., 2001). These models are implicit, intuitive, and frequently wrong. In particular, they are sensitive to 'framing effects', which originate from the presentation of a single problem in different ways (e.g., glass half full versus half empty). Other biases include the positive weighting of events that are easily remembered ('availability heuristic'); the tendency to rate two events more probable than a single event ('conjunction fallacy'); the selective processing of information that confirms an expected result ('confirmation bias'); and the tendency to over- or underestimate some uncertainties (over-/under-confidence: see Kahneman et al., 1982 for an overview).

Much research in the 1970s and 1980s focused on the psychological aspects of uncertainty and risk. This research aimed to: (i) improve methods of eliciting opinions about uncertainty and risk; (ii) provide a basis for understanding and anticipating public responses to various hazards; and (iii) improve the communication of risk information between lay people, technical experts and policymakers (Slovic et al., 1981). The so-called 'psychometric' approach, built on advances in cognitive and social psychology, became popular in the 1970s (Slovic et al., 1974). Data were typically gathered by questionnaire, and statistical relationships were developed between various aspects of personality and risk perception. Perhaps the best-known outcome of this work is the scatter diagram created by Paul Slovic and colleagues whereby lay attitudes towards a variety of hazards are plotted on two axes, labelled 'unknown' risk, and 'dread' risk. The resulting scatter often indicates a preference for stricter controls of less familiar, more frightening risks. Public perceptions were thus conceived as exaggerating the unknown when accident statistics reveal that many familiar risks are incrementally more costly (Adams, 1995).

The psychometric approach clarified some important psychological controls on risk perception, such as voluntariness, familiarity, the nature of the hazard, and the types of people exposed to or benefiting from risk (e.g., children versus adults). However, it was also premised on an assumption that 'gaps' between expert and public understandings of uncertainty and risk are caused by erroneous public perceptions of 'true' risks. More recent social and political analyses have questioned this long-standing dichotomy, suggesting instead that all perceptions of risk, whether expert or lay, represent partial and selective views of events (Wynne, 1996).

Attitudes towards uncertainty and risk are socially as well as psychologically constructed. Language is crucial in translating general concepts, such as 'eutrophication' (nutrient enrichment of water bodies) into particular entities, such as 'algal blooms', and then into measurable quantities, such as 'chlorophyll-a' (Richards et al., 1997). Theories may perform badly against observations if this translation is ambiguous *or* if the theories or observations are inadequate *or* if the criteria used to compare them (the 'demarcation criteria') are inadequate (Brown, 2004). Trust in the sources of information and the processes through which knowledge is gained also affect confidence in research outcomes. In terms of the former, MacKenzie (1990) describes a 'certainty trough' whereby those nearest to and most alienated from scientific research will harbour the greatest uncertainties about its outcomes. MacKenzie argues that scientific output will appear most certain at an 'intermediate distance', where it is sufficiently close to be valued by its users but sufficiently remote to avoid detailed criticism of its methods. Building on this concept, Shackley and Wynne (1995) suggest that climate impacts modellers are frequently overconfident in integrated assessment models because they are unfamiliar with some of the disciplinary assumptions made. This may be reinforced by the use of naïve coupling procedures, whereby disciplinary models are connected through their numerical inputs and outputs (cf. climate-forcing scenarios) without considering the possibility of structural feedbacks, such as human responses to climate change (Shackley and Wynne, 1995).

Several authors have sought to emphasise the cultural aspects of uncertainty and risk. 'Cultural Theory' was developed by Mary Douglas and colleagues in the 1970s and 1980s. This came from dissatisfaction with psychometric approaches and their failure to account for cultural influences on risk perception: '. . . psychometricians [have] isolated the cultural factors and treated them as another variable in an experimentally derived technical framework . . . [rather than] explore the cultural underpinnings of risk perception' (Plough and Krimsky, 1987, p. 8). Crucially, Douglas (1986) argued that the psychometric approach simply studied *what* people perceive as risky rather than *why* they hold such views, and emphasised that risks are actually built on a network of social and institutional relations defining acceptable behaviour. Risk perceptions were argued to vary systematically according to four cultural 'biases': individualist, fatalist, hierarchist, and egalitarian. Each of these was seen as defending a particular way of life and a corresponding set of institutional arrangements (Pidgeon and Beattie, 1997). However, Cultural Theory omits the ambiguity of interpretation that is central to the social construction of risk. It also ignores the possibility that an individual could have more than one cultural bias in any given context (Horlick-Jones, 1998). Several other risk frameworks have emerged since the 1980s. One such framework was the Social Amplification of Risk Framework (SARF) attributed to Kasperson et al. (1988). SARF was based on communications theory, in which the media was seen as an important source of risk information, along with the symbols and imagery with which events and hazards were portrayed to the public (Petts et al., 2001).

In human geography, technical, economic, and psychological approaches to studying risk have increasingly given way to cultural and sociological approaches (Adams, 1995; Lash et al., 1996). These approaches have been influenced heavily by Ulrich Beck's claim that we now live in a 'risk society' (Beck, 1992; 1995). This has a number of implications. First, contemporary risks, such as avian flu, genetically modified (GM) crops, and climate change are larger, more complex, and more uncertain than those experienced in the past. Second, scientific knowledge is 'both the medium through which risks are defined and the source of their solution' (p. 155). According to Beck, science is the primary *cause* of many environmental problems, such that 'science becomes more and more necessary, but at the same time less and less sufficient for the socially binding definition of truth' (p. 156). Finally, it becomes increasingly difficult to identify and solve problems: 'The boundaries of the problem are diffuse, so it can hardly be separated from other problems [. . .] Conflicting values and facts are interwoven, and many actors become involved in the policy process' (Hisschemoller and Hoppe, 1996, p. 43).

Of course, Beck's ideas have been criticised. Goldblatt (1996) considers Beck's writings to be '. . . not so much rigorous analytical accounts of modernity as surveys of the institutional bases of the fears and paradoxes of modern societies – societies that no longer correspond to the classical sociological descriptions or possess cultural resources that allow them to live comfortably with the world' (p. 154). In this way, Beck's theory can be seen as narrowly focused on the hazards generated by industrial society. A key example cited is the risk of environmental disaster caused by nuclear reactor accidents like Chernobyl.

Alongside the psychological and social aspects of uncertainty and risk, the focus of research will have an important influence on uncertainty and risk. 'Situational factors' include the definition of a problem, its complexity, scale, spatial and temporal variability, and transparency to investigation (Brown, 2004). Traditionally, these factors are treated as methodological issues, controlled by computing power, model resolution, field methodology, or sample size, rather than inherent properties of environmental systems. In practice, however, deeper investigation often reveals greater complexity and non-linearity in social and environmental systems than expected. This is particularly apparent when addressing large, interdisciplinary, problems, such as climate change, where the range of responses, and capacity of the system to adapt (e.g., through structural changes in the land-surface and ocean currents), will continue to generate large uncertainties.

The concepts of 'complexity', 'scale' and 'variability' are referred to as situational because they allude to external structures, like the environment or society. However, their meaning is derived through our *representations* of these structures. Thus, people may be uncertain about the environment because it appears more complex than our abstraction and simplifications imply, because it is too variable for us to capture, too large to observe everything at once or too small to observe in sufficient detail (Brown, 2004). This is evidenced by the close relationship between environmental scales and scales of measurement, modelling and presentation (Van Asselt and Rotmans, 1996). There are also close connections between environmental variability and the variance (un)explained by statistical modelling, as variable processes are more difficult to model than stationary ones (Wainwright and Mulligan, 2004). These factors have been widely examined in geographical research. For example, place (time, space and location) is recognised as an important control on the operation and observed outcomes of geographical processes (Richards et al., 1997).

Assessing and Managing Uncertainty and Risk

In many respects, the literatures on assessing and managing uncertainty have followed a similar trajectory to those on risk. Both are dominated by attempts to quantify, minimise, and control uncertainty. In recent years, there has been a proliferation of studies in which theories of uncertainty have been devised and applied to geographical problems. Most of these studies have focused on the quantification of uncertainties in geographic data (Cressie, 1993) and the propagation of uncertainties through geographic models (Heuvelink, 1998). In terms of the latter, probability distributions may be developed for the uncertain inputs and parameters of a model, sampled randomly to create different input and parameter combinations and then propagated through the model by repeat simulation (Hammersley and Handscomb, 1979). The propagated uncertainties can then feed directly into quantitative studies of risk, where probabilities of outcomes are combined with their expected costs and benefits (Ayyub, 2001). Nevertheless, a distinct spectrum of methods, not all statistical, has emerged for characterising uncertainty. For example, the Numeral, Unit, Spread, Assessment and Pedigree (NUSAP) scheme of Funtowicz and Ravetz (1990) employs a combination of numerical scoring and qualitative assessment to address a range of uncertainties in scientific information.

An uncertainty analysis is typically limited to a few sources of uncertainty, which may be selected by expert judgement or sensitivity testing (Saltelli et al., 2004). In practice, the uncertainties associated with model inputs and parameters have received much greater attention than those associated with model structure (Refsgaard et al., 2006). The latter refers to uncertainty about social and environmental processes (e.g., what are the dominant process controls?) and how they are manifest in observations (e.g., is a linear regression appropriate?). Typically, structural uncertainty will lead to several methods providing reasonable accounts of the observed data or plausible explanations of system behaviour. For example, in a study of groundwater vulnerability to nitrate pollution, Refsgaard et al. (2006) report how six groups of engineering consultants developed six different accounts of groundwater vulnerability. Each consultant worked from a common database, with major differences related to choice of method and to the assumptions made in assessing vulnerability. Similarly, in a review of multi-criteria decision models, Myšiak (2006) found that model selection was often based on prejudiced views about the strengths and weaknesses of the candidate methods, rather than a careful analysis of the decision problem. Unsurprisingly, most scientific studies show a strong partiality for whichever method conforms best to the worldview of the policy advisor. Indeed, when consensus is lacking, other factors often influence the selection of methodology, such as institutional arrangements (Fisher et al., 2002) or historical precedent (Shackley and Wynne, 1995).

Despite the success of mathematical approaches, there are still many situations in which a technical assessment of uncertainty cannot establish the reliability of data and models, or may itself lack credibility. For example, probabilities of extreme events may be highly unreliable, as extreme events are rare by definition. Also, their probabilities will vary with the trajectory of the system (e.g., with climate change), and their process controls may be qualitatively different from those operating during smaller events (e.g., Powell et al., 2003). In order to evaluate these probabilities, observations must be pooled into groups of similarly behaving or 'stationary' samples, yet the concept of stationarity may be difficult to justify for extreme occurrences. In principle, therefore, the types and levels of uncertainty should be reflected in the methodologies chosen to assess and propagate them. In practice, however, this link between types and levels of uncertainty and methods of assessment is frequently missed (Brown, 2004), leading to spurious notions of precision, unreliable uncertainties, or the omission of key sources of uncertainty, such as those associated with social and political processes.

Early approaches to assessing risk also focused on quantifying, minimising, and controlling uncertainty. They typically distinguish between expert and lay understandings or 'real' versus 'perceived' risk (e.g., Irwin and Wynne, 1996a,b; Wynne, 1992a,b), with most research devoted to expert understandings of 'real' risk (Owens et al., 2004). These views can be seen in successive reports on risk published in the 1980s and early 1990s. For example, the Royal Society (1983) clearly distinguishes between objective risks, identified by science, and subjective perceptions of those risks, which are considered poor approximations of the former. A later report (Royal Society, 1985) lamented the public 'misunderstanding' of risk and called for wider education on its scientific basis, while Royal Society (1992) proposed a series of remedial approaches to better inform ignorant publics of the 'real' risks they faced (Owens, 2000).

These ideas, often referred to as the information deficit model (IDM), are based on a number of contentious assumptions about the primacy of scientific knowledge. First, they view the environment as a physical phenomenon, separate from society, and measurable through objective, scientific, procedures. Many commentators (e.g., Wynne, 1996) have argued that this distinction is artificial because scientific practices are necessarily complicated by social and political processes.

Secondly, technical approaches assume that risk can be measured objectively. The modern image of science and technology has been that '. . .given enough information and powerful enough computers, it could predict with certainty in a quantitative form, which would make it possible to control natural systems' (Tognetti, 1999,

p. 690). However, others (e.g., Wynne, 1992a,b) suggest that the full implications of uncertainty remain under explored in environmental management, where many problems are essentially indeterminate. This is particularly evident in policy-related research where non-uniqueness originates from a combination of uncertainty and multiple perspectives on what policy *should* achieve. In such cases, decisions may depend largely on the values of the experts involved (Rowe et al., 2005). Furthermore, contemporary environmental risks may be 'trans-scientific' (Weinberg, 1972) in that the scientific inputs to 'hard' policy decisions are *irredeemably* 'soft', uncertain, contested, and extremely complex (Funtowicz and Ravetz, 1993).

Another assumption behind the IDM is that science and scientific experts are inherently trusted by the public, and that a wider exposure to scientific thinking will encourage its acceptance (Szerszynski et al., 1996). Here, a lack of public dissent to scientific information is equated with public acceptance. However, this 'acceptance' may be explained by a failure of science to address public concerns and to the social conditions of its consumption and negotiation, including feelings of resignation and a lack of power to effect change (Irwin, 1995). In practice, public attitudes may be intimately connected with attitudes to institutions and political control. Eden (1998) notes that lay people are not passive in the face of scientific knowledge, but actively construct their own knowledge (and their own ignorance). Therefore, the very notions of 'expert' and 'public' are flexible and contingent, contrary to their representation in general frameworks of risk understanding and perception.

More recent research has recast the notion of public understandings of risk in several important ways. The first concerns lay assessments of risk, which are traditionally viewed as ignorant and irrational. Commentators such as Wynne (1996) have argued that public risk assessments, far from being ignorant, have their own rationality, which may differ from the 'expert', but is not always inferior. As such, the public can play an important role in generating new understandings of risk (e.g., O'Connor, 1999). Indeed, in certain cases (such as through the expression of smaller scale, locally embedded contextual knowledges), public understandings of risk can be *at least as* robust and well informed as expert understandings, despite differences in status and power between the two groups. Of course, citizen knowledge is not necessarily better than expert knowledge. Rather, in accepting the possibility of a rational public, it follows that no unique understanding of risk is available in any given context, not least because tolerance to risk varies between individuals and groups of people (Irwin and Wynne, 1996a,b; Lash et al., 1996).

There is evidence that diverse understandings of uncertainty and risk are being increasingly accepted by policymakers. Indeed, in a recent report on environmental standards, the Royal Commission on Environmental Pollution (RCEP) asserted that 'better ways need to be developed for articulating people's values and taking them into account from the earliest stage in what have been hitherto relatively technocratic procedures' (RCEP, 1998, para. 8.37). Taking account of alternative and complementary knowledges in policy and decision-making processes has been termed the 'deliberative' model by social scientists. This model stresses the *interaction* between scientists and the public (Burgess, 2005). It is argued that such interactions will support publicly defensible decisions in the face of seemingly irreducible uncertainties and risks. Lay understandings are argued to usefully complement more traditional scientific input into the policy process, especially at the local scale. In this way, the traditional 'top-down' hierarchy of knowledge can be recast in favour of a balanced diversity of knowledges, without simply perpetuating the expert-lay divide.

Numerous methods have been used in deliberative geographical research, including focus groups, citizens' juries, consensus conferences, and multi-criteria deliberation. These methods seek to couple risk assessments, including fundamental decisions about what and whose risks are being considered, with strategies for managing risk. Traditionally, approaches to assessing and managing risk have evolved separately, with the latter developed in response to the former. For example, in the early 1980s, there was concern in the United States that risk assessments were being diluted by 'irrelevant' social policy issues (Gerrard and Petts, 1998). The deliberative model aims to incorporate the social, political and economic dimensions of risk (to which science is equally exposed) in an explicit and transparent way *without* compromising scientific methods. Here, risk has both scientific and social dimensions. Thus, while scientific studies can provide valuable insights into hazardous events, whether a risk is tolerable or requires a particular action will ultimately depend on individual and social judgement.

However, the deliberative model is still relatively new, and has been more widely used in some areas than others. For example, in recent years, public deliberation has been a key element of dialogue regarding the potential risks of GM foods (Horlick-Jones et al., 2004); in outlining UK energy policy (DTI, 2003); in discussions regarding radioactive waste disposal methods (CoRWM, 2005), and in debates surrounding the restoration and rehabilitation of degraded rivers (e.g., McDonald et al., 2004).

Conclusions

Uncertainty affects both the processes and outcomes of environmental research and decision making. It has been a critical factor in debates on climate change, acid rain, desertification, GM crops, and other contemporary environmental problems, where it has led to persistent conflict and indecision. It poses philosophical challenges regarding the nature, origins, and value of knowledge, ethical challenges regarding acceptable levels of risk, and political challenges concerning how to act and who has the mandate to decide. In particular, it is increasingly acknowledged that quantitative analyses, while important for addressing some types and sources of uncertainty, are often too narrowly defined and esoteric for public decision making. Technical approaches to assessing uncertainty and risk can also exacerbate an expert-lay divide, thereby complicating management efforts. More recent work has focused on deliberative and participatory approaches to understanding, in which technical analyses are only one input to environmental decision making (and not necessarily the most important).

In many areas of interdisplinary work, such as climate change, the social and political aspects of uncertainty and risk have been vigorously debated. However, the significance of these debates is easily overstated. First, they are largely confined to interdisciplinary work involving both social and physical scientists. Yet many of the crucial inputs to environmental decision making come from the technical disciplines. For example, numerical weather prediction has improved dramatically in recent years, partly due to improvements in the underlying physics and partly to improved numerical methods. In this context, Shackley and Wynne (1995) allude to an important problem with some interdisciplinary work, namely that much of the communication (both between scientists and between scientists and the public) is facilitated by 'gatekeepers'. Typically, these practitioners do not have a detailed understanding of the disciplinary methods employed, which can result in misinformation and overconfidence about the integrated results. Claims that a 'paradigm shift' is occurring *within* science would, therefore, appear overstated.

Secondly, in terms of the relationship between science and society, there is little evidence that new understandings of uncertainty and risk are having a profound impact on environmental management. For example, reaction to the GM Nation (Rowe et al., 2005) highlighted several problems with the deliberative approach to evaluating risk, including not only the specific methodology employed, but also the extent to which public understandings really *are* that informed. Further work transcending the public-expert dichotomy is clearly needed (Rowe and Frewer, 2004).

Finally, it is easy to underestimate inertia towards change among the various government and other agencies that use scientific research in environmental management. Many structures and procedures employed by these agencies are built on determinism; for example, databases store only deterministic data, models are built for deterministic predictions, and modes of communication are often top-down rather than deliberative. Changing these practices will require more than financial investment. In particular, it will require changes in the institutional cultures, politics and legal mechanisms through which uncertainty and risk are conceived.

BIBLIOGRAPHY

Adams, J., (1995) *Risk*. London: UCL Press.

Ayyub, B. M. (2001) *Elicitation of Expert Opinions for Uncertainty and Risks*. Florida: CRS Press.

Baker, T. and Simon, J. (eds) (2002) *Embracing Risk: The Changing Culture of Insurance and Responsibility*. Chicago: University of Chicago Press.

Beck, U. (1992). *Risk Society: Towards a New Modernity*. London: Sage.

Beck, U. (1995) *Ecological Politics in an Age of Risk*. Cambridge: Polity.

Bernardo, J. M. and Smith A. F. M. (2000). *Bayesian Theory*. Chichester: John Wiley & Sons, Ltd.

Bernstein, P. L. (1998) *Against the Gods: The Remarkable Story of Risk*. Chichester: John Wiley & Sons, Ltd.

Beven, K. J. (2002) Towards a coherent philosophy for modelling the environment. *Proceedings of the Royal Society of London A*, 458, 2465–84.

Brown, J. D. (2004) Knowledge, uncertainty and physical geography: towards the development of methodologies for questioning belief. *Transactions of the Institute of British Geographers*, 29(3), 367–81.

Burgess, J. (2005) Follow the argument where it leads: some personal reflections on 'policy-relevant' research. *Transactions of the Institute of British Geographers*, 30(2), 273–81.

Ciambrone, F. D. (1997) *Environmental Life Cycle Analysis*. New York: Lewis.

Committee on Radioactive Waste Management (CoRWM) (2005) *Learning from the Past – Listening for the Future: How Should the UK Manage Radioactive Waste?* London: Committee on Radioactive Waste Management.

Cressie N. A. C. (1993) *Statistics for Spatial Data*, 2nd edn. New York: John Wiley & Sons, Ltd.

Department of Trade and Industry (DTI) (2003) *Our Energy Future: Creating a Low Carbon Economy*. London: Department for Trade and Industry.

Douglas, M. (1986) *Risk Acceptability According to the Social Sciences*. London: Routledge and Keegan Paul.

Eden, S. (1998) Environmental issues: knowledge, uncertainty and the environment. *Progress in Human Geography*, 22(3), 425–32.

Fisher, P. F., Comber, A. J. and Wadsworth, R. A. (2002) The production of uncertainty in spatial information: the case of land cover mapping. In G. Hunter and K. Lowell (eds), *Accuracy 2002: Proceedings of the 5th International Symposium on Spatial Accuracy Assessment in Natural Resources and Environmental Sciences, Melbourne, Australia, 10–12 July 2002*.

Funtowicz, S. O. and Ravetz, J. R. (1990) *Uncertainty and Quality in Science for Policy*. Dordrecht: Kluwer.

Funtowicz, S. O. and Ravetz, J. R. (1993) Science for the post-normal age. *Futures*, 25, 739–55.

Gerrard, S. and Petts, J. (1998) Isolation or integration: The relationship between risk assessment and risk management. In R. E. Hester and R. M. Harmson (eds), *Risk Assessment and Risk Management. Issues in Environmental Science and Technology 9*, Royal Society of Chemistry, Great Britain, pp. 1–19.

Gibbons, M. (1999) Science's new social contract with society. *Nature*, 402, C81–84.

Goldblatt, D. (1996) *Social Theory and the Environment*. Cambridge: Polity.

Hammersley, J. M. and Handscomb, D. C. (1979) *Montecarlo Methods*. London: Chapman and Hall.

Handmer, J. (2001) Improving flood warnings in Europe: a research and policy agenda. *Global Environmental Change, B: Environmental Hazards*, 3(1), 19–28.

Handmer, J., Norton, T. and Dovers, S. (eds) (2001) *Uncertainty, Ecology and Policy: Managing Ecosystems for Sustainability*. Harlow: Prentice-Hall.

Harremoës, P., Gee, D., MacGarvin, M., Stirling, A., Keys, J., Wynne, B. and Guedes Vaz, S. (eds) (2002) *The Precautionary Principle in the 20th Century: Late Lessons from Early Warnings*. London: Earthscan, pp. 155–65.

Herath, G. and Prato, T. (eds) (2006). *Multicriteria Approaches in Natural Resource Management*. Hampshire: Ashgate Publishing.

Heuvelink, G. B. M. (1998) *Error Propagation in Environmental Modelling with GIS*. London: Taylor & Francis.

Hisschemoller, M., and Hoppe, R. (1996) Coping with intractable controversies: the case for problem structuring in policy design and analysis. *Knowledge and Policy*, 8, 40–60.

Holling, C. S. (ed.) (1978) *Adaptive Environmental Assessment and Management*. International Series on Applied Systems Analysis. Chichester: John Wiley & Sons, Ltd.

Horlick-Jones, T. (1998). Meaning and Contextualisation in Risk Assessment. *Reliability Engineering and System Safety*, 59, 79–89.

Horlick-Jones, T., Walls, J., Rowe, G., Pidgeon, N., Poortinga, W., and O'Riordan, T. (2004) *A Deliberative Future?* Understanding Risk Working Paper 04-02. Cardiff: Cardiff University.

Irwin, A. (1995) *Citizen Science: A Study of People, Expertise and Sustainable Development*. London: Routledge.

Irwin, A., and Wynne, B. (eds) (1996a) *Misunderstanding Science? The Public Reconstruction of Science and Technology*. Cambridge: Cambridge University Press.

Irwin, A., and Wynne, B. (1996b) Conclusions. In A. Irwin, A. and B. Wynne (eds), *Misunderstanding Science? The Public Reconstruction of Science and Technology*. Cambridge: Cambridge University Press, pp. 213–21.

Jasanoff, S. (1996) Is science socially constructed – and can it still inform public policy? *Science and Engineering Ethics*, 2(3), 263–76.

Kahneman, D., Slovic, P. and Tversky, A. (1982) *Judgment Under Uncertainty: Heuristics and Biases*. Cambridge: Cambridge University Press.

Kasperson, R. E., Renn, O. and Slovic, P. (1988) Social amplification of risk: a conceptual framework. *Risk Analysis*, 8, 177–87.

Lash, S., Szerszynski, B. and Wynne, B. (eds) (1996) *Risk, Environment and Modernity: Towards a New Ecology*. London: Sage.

MacKenzie, D. (1990) *Inventing Accuracy: A Historical Sociology of Nuclear Missile Guidance*. Cambridge, MA: MIT Press.

Macnaghten, P. and Urry, J. (1998) *Contested Natures*. London: Sage.

McDonald, A., Lane, S. N., Haycock, N. E. and Chalk, E. A. (2004) Rivers of dreams: on the gulf between theoretical and practical aspects of an upland river restoration. *Transactions of the Institute of British Geographers*, 29, 257–81.

Morgan, M.G., Fischhoff, B., Bostrom, A. and Atman, C. (2001) *Risk Communication: The Mental Models Approach*. New York: Cambridge University Press.

Myšiak, J. (2006) Consistency of the results of different MCA methods: a critical review. *Environment and Planning C: Government and Policy*, 24(2), 257–77.

Nas, T. F. (1996) *Cost-Benefit Analysis: Theory and Application*. Thousand Oaks, CA: Sage.

O'Connor, M. (1999) Dialog and debate in a post-normal practice of science: a reflection. *Futures*, 31, 671–87.

Owens, S. E., Rayner, T. and Bina, O. (2004) New agendas for appraisal: reflections on theory, practice and research. *Environment and Planning A*, 36, 1943–59.

Owens, S., 2000. Engaging the public: information and deliberation in environmental policy. *Environment and Planning A*, 32, 1141–48.

Parry, M., Canziarii, O., Palutikof, J., van der Linden, P. and Hansen, C. (eds) (2007) *Climate Change 2007*: impacts adaptation and vulnerability. Contribution of Working Group II to the forth assessment report of the intergovernmental panel on climate change. Cambridge, UK: Cambridge University Press.

Petts, J., Horlick-Jones, T. and Murdock, G. (2001) *Social Amplification of Risk: The Media and the Public*. HSE Research Report, CR 329/2001. London: HMSO.

Pidgeon, N. and Beattie, J. (1997). The psychology of risk and uncertainty. In P. Calow (ed.), *Handbook of Environmental Risk Assessment and Management*. Oxford: Blackwell Science, pp. 289–318.

Plough, A. and Krimsky, S. (1987) The emergence of risk communication studies: social and political context. *Science, Technology and Human Values*, 12(3–4), 4–10.

Powell, M. D., Vickery, P. J. and Reinhold, T. A. (2003) Reduced drag coefficient for high wind speeds in tropical cyclones. *Nature*, 422, 279–83.

Refsgaard, J. C., van der Sluijs, J. P., Brown, J. D. and van der Keur, P. (2006) A framework for dealing with uncertainty due to model structure error. *Advances in Water Resources*, 29(11), 1586–97.

Regan, H. M., Colyvan, M. and Burgman, M. A. (2002) A taxonomy and treatment of uncertainty for ecology and conservation biology. *Ecological Applications*, 12(2), 618–28.

Richards, K. S., Brooks, S. M., Clifford, N. J., Harris, T. R. M. and Lane, S. N. (1997) Real geomorphology in physical geography: theory, observations and testing. In D. R Stoddard (ed.), *Process and Form in Geomorphology*. London: Routledge, pp. 269–92.

Rowe, G. and Frewer, L. J. 2004. Evaluating public-participation exercises: a research agenda. *Science Technology Human Values*, 29(4), 512–56.

Rowe, G., Horlick-Jones, T., Walls, J., and Pidgeon, N., 2005. difficulties in evaluating public engagement initiatives: reflections on an evaluation of the UK *GM Nation?* Public Debate About Transgenic Crops. *Public Understanding of Science*, 14, 331–52.

Royal Commission on Environmental Pollution (RCEP) (1998) *Setting Environmental Standards: 21st Report of the Royal Commission on Environmental Pollution*. London: HMSO.

Royal Society (1983) *Risk Assessment: A Study Group Report*. London: Royal Society.

Royal Society (1985) *The Public Understanding of Science*. London: Royal Society.

Royal Society (1992) *Risk: Analysis, Perception and Management: Report of a Royal Society Study Group.* London: Royal Society.

Saltelli, A., Tarantola, S., Campolongo, F. and Ratto, M. (2004) *Sensitivity Analysis in Practice, a Guide to Assessing Scientific Models.* New York: John Wiley & Sons, Ltd.

Shackley, S. and Wynne, B. (1995) Integrating knowledges for climate change: pyramids, nets and uncertainties. *Global Environmental Change,* 5(2), 113–26.

Slovic, P., Fischhoff, B. and Lichtenstein, S. (1981) *The Assessment and Perception of Risk.* London: Royal Society.

Slovic, P., Kunreuther, H. C., and White, G. (1974) Decision processes, rationality and adjustment to natural hazards. In G. F. White (ed.), *Natural Hazards, Local, National, Global.* Oxford University Press, Oxford. pp. 187–205.

Smithson, M. (1989) *Ignorance and Uncertainty: Emerging Paradigms.* New York: Springer-Verlag.

Suter, G. W., Barnthouse, L. W. and O'Neill, R. V. (1987) Treatment of risk in environmental impact assessment. *Environmental Management,* 11, 295–303.

Szerszynski, B., Lash, S. and Wynne, B. (1996) Ecology, realism and the social sciences. Introduction in Lash, S., Szerszynski, B., and Wynne, B. (eds), *Risk, Environment and Modernity: Towards a New Ecology.* London: Sage, pp. 1–26.

Tognetti, S. (1999) Science in a double bind: Gregory Bateson and the origins of post-normal science. *Futures,* 31, 689–703.

Van Asselt, M. B. A. and Rotmans, J. (1996) Uncertainty in perspective. *Global Environmental Change,* 6, 121–57.

Von Engelhardt, W. and Zimmerman, J. (1988) *Theory of Earth Science.* Cambridge: Cambridge University Press.

von Neumann, J. and Morgenstern, O. (1944) *Theory of Games and Economic Behavior.* Princeton: Princeton University Press.

Wainwright, J. and Mulligan, M. (eds) (2004) *Environmental Modelling: Finding Simplicity in Complexity.* Chichester: John Wiley & Sons, Ltd.

Walker, W. E., Harremoës, P., Rotmans, J., van der Sluijs, J. P., van Asselt, M. B. A., Janssen, P. and Krayer von Krauss, M. P. (2003) Defining uncertainty: a conceptual basis for uncertainty management in model-based decision support. *Integrated Assessment,* 4(1), 5–18.

Wätzold, F. (2000) Efficiency and applicability of economic concepts dealing with environmental risk and ignorance. *Ecological Economics,* 2, 299–311.

Weinberg, A. (1972) Science and trans-science. *Minerva,* 10, 209–22.

Winstanley, D., Lackey, R. T., Warnick, W. L. and Malanchuck, J. (1998). Acid rain: science and policy making. *Environmental Science and Policy,* 1(1), 51–57.

Wynne, B. (1991) Knowledges in context. *Science, Technology and Human Values,* 15, 111–21.

Wynne, B. (1992a) Uncertainty and environmental learning: reconceiving science and policy in the preventive paradigm. *Global Environmental Change,* 2, 111–27.

Wynne, B. (1992b) Misunderstood misunderstanding: social identities and the public uptake of science. *Public Understanding of Science,* 1(2), 281–304.

Wynne, B. (1996) May the sheep safely graze? A reflexive view of the expert-lay knowledge divide. In S. Lash, B. Szerszynski and B. Wynne (eds), *Risk, Environment and Modernity: Towards a New Ecology.* London: Sage, pp. 44–83.

Young, S. C. (2000) *The Emergence of Ecological Modernisation: Integrating the Environment and the Economy?* London: Routledge.

Zadeh, L. A. (1978) Fuzzy sets as a basis for a theory of possibility. *Fuzzy Sets and Systems,* 1, 3–28.

Chapter 7

Scale

Nathan F. Sayre

Introduction: The Many Meanings of Scale

In his Robert H. MacArthur Award lecture in 1989, Princeton ecologist Simon Levin declared: 'The problem of relating phenomena across scales is the central problem in biology and in all of science' (Levin, 1992, p. 1961). Levin is not alone: inside and outside the academy, there is an effective consensus that scale is of the utmost importance to matters of humans and the environment. Consider these assertions: 'The history of human cultural evolution has been the story of cross-scale subsidies', from a paper on the resilience of social-ecological systems (Carpenter et al., 2001, p. 767); and 'Scale is a nonreductionist unifying concept in ecology', by two other prominent theorists (Peterson and Parker, 1998, p. 521). Scale is discussed with comparable gravity and still greater rhetorical flourish in more popular venues. Science journalist Elizabeth Kolbert, for example, opens her book on climate change, *Field Notes from a Catastrophe*, with the claim that: 'For better or (mostly) for worse, global warming is all about scale' (Kolbert, 2006, p. 3). Pulitzer prize-winning columnist and neoliberal enthusiast Thomas Friedman puts it this way: 'Hey, the more energy-saving bulbs Wal-Mart sells, the more innovation it triggers, the more prices go down. That's how you get scale. And scale is everything if you want to change the world' (*New York Times*, 22 December 2006, p. A31). For many people, scale is the fundamental conceptual challenge in the human and natural sciences, critical to progress in understanding and ameliorating human-environment interactions.

It remains remarkably unclear exactly what scale means and how to use it, however, and within geography the confusion is particularly acute. Biophysical geographers understand and employ scale much as ecologists do (where it is also much debated), but cartographers, Geographic Information Scientists, and especially human geographers have various other ideas of scale and its theoretical and methodological implications. The editors of a recent volume on the subject conclude that 'conceptions of geographic scale range across a spectrum of almost intimidating diversity' (Sheppard and McMaster, 2004, p. 256). Marston et al. (2005, p. 416), after noting that 'scholarly positions on scale are divergent in the extreme', conclude that the concept is fundamentally flawed and should be banished from human geography altogether. A review paper in the journal *Ecological Economics* confirms the diagnosis but prescribes the opposite cure: 'Now, scale issues are found at the center of methodological discussions in both physical and human geography', the authors observe, but 'common definitions do not exist for scale – even within disciplines – and especially in the social sciences' (Gibson et al., 2000, pp. 226, 236). Nonetheless, they issue an unequivocal call: 'The challenge of global environmental change requires that both the physical and social sciences be included in its study. If researchers are to generate accurate analyses of environmental change, the first step, we believe, is to push beyond the present cacophony and construct a common understanding of issues related to scale' (p. 237).

The problem with scale derives in large part from a surfeit of meanings and uses. The word occupies nearly four pages of the original *Oxford English Dictionary*, and a search of the BIOSIS database finds the term in more than 85,000 abstracts since 1990. Richard Howitt (1998; 2003) discerns three 'aspects' or 'facets' of scale: as size, level, and relation. The first two are relatively well understood, he argues; they predominate in non-technical, quotidian contexts, and even in academic writings scale is usually a simple descriptor, not a concept. But it is only as relation that scale assumes the importance ascribed to it in recent decades, and the apparent clarity of the first two meanings has made understanding the third much more difficult. Conflating scale and level may be convenient and non-problematic when neither term is the focus of inquiry, but collapsing the two risks evacuating scale of conceptual importance altogether. In short, distinguishing scale as relation from its more casual or colloquial meanings is necessary if its significance for environmental geography is to be clarified, let alone realised.

In what follows, I first review the various uses and meanings of scale in geography, including 'the scale question' in human geography. Scale as size, level, and relation are not mutually exclusive – indeed, they build on and presuppose one another – but they are analytically distinct; many, if not all, of the debates about scale in recent human geography can be traced to conflation among these meanings. I then examine the emergence of scale in ecology, in order to clarify why it is considered of such overriding importance to our understanding of ecosystems and environmental problems. For ecologists, scale is intrinsically spatio-temporal, playing a key role in the critique of equilibrium models and assumptions that has gathered momentum over the past three decades. The conclusion develops a framework for theorising scale to advance research in environmental geography.

The edited volume *Scale and Geographic Inquiry* provides a valuable overview of geographical scale. In their introductory essay, the editors note that 'different concepts of scale are employed in geography's various subdisciplines', but that 'there has been little attempt to integrate across these subdisciplinary perspectives' (Sheppard and McMaster, 2004, p. 2f.). A brief summary of scale's various meanings in geography is therefore warranted. It is useful to organise them into the three facets of size, level and relation.

Scale as Size

In this first and simplest sense, scale refers to measurements expressed in terms of standardised units. 'Space and time are not scales until they are divided into segments that can be used for measurement' (Rykiel, 1998, p. 488). A scale is used to ascertain some attribute of an object or phenomenon – such as length, mass, volume, velocity, and so forth; in geographical contexts, scale in this sense generally refers to *size*. *Cartographic scale* is the oldest kind of geographical scale, having emerged with the science of cartography during the eighteenth century. It refers to the mathematical relationship between a map and what it represents: the 'representative fraction' or ratio of a unit space on a map to space in the world, such as 1:62,500 for maps in which one inch represents one mile. Expressed in this way, smaller scale maps depict larger areas than do maps of larger scales, resulting in the peculiarity that cartographers employ 'large scale' and 'small scale' in the opposite way from scholars in other fields. Choice of scale has obvious implications for cartographic generalisation: Smaller scale maps (depicting larger areas) necessarily sacrifice details that can readily be included on maps of larger scales (depicting smaller areas). (Hereinafter, I will use 'small' and 'large' scale the way non-cartographers do, to avoid confusion.) Scale is a central conceptual and representational issue in cartography because it strongly determines selection, simplification, classification, and symbolisation. Different tasks – depicting a neighborhood, a city, a region or a continent, for instance – call for the use of different cartographic scales.

Developments in Geographical Information Science (GISc) raise the possibility of overcoming constraints of cartographic scale, at least in theory. Digitised data can be assembled and analysed across multiple scales, such that details visible at small scales are not lost (to the computer, at least) when one 'zooms out' to much larger scales. As Sheppard and McMaster (2004, p. 4) note, however, this does not mean that 'there is no scale' in GISc, because the underlying data are themselves typically derived from scaled sources. (Think of what happens when one zooms in on Google Earth, for example: the image becomes blurry at certain scales, then regains focus when the programme shifts to an image taken at another scale.) The technical details and particularities of GISc cannot be adequately reviewed here, but the issues of scale discussed below are nonetheless relevant to that field.

Cartographic scale is principally a representational issue, but in the second half of the twentieth century other fields in geography identified empirical corollaries: situations in which spatial analysis resulted in different (or even opposite) conclusions depending on the scale employed. The distribution and intensity of poverty, for example, might look very different if the smallest unit of analysis were city blocks rather than census tracts, cities, or entire states. Openshaw (1977; 1984) famously demonstrated that the boundaries and size of units for spatial aggregation could determine whether two variables correlated positively, negatively, or not at all: a form of ecological fallacy known as the modifiable areal unit problem. *Observational scale* refers to this methodological issue, which at face value resembles cartographic scale: At what spatial dimensions can one best perceive and analyse particular phenomena? Even when the question is not posed as such, scientists cannot avoid this issue: 'Because science is about the search for and explanation of patterns, all scientific inquiry explicitly or implicitly incorporates scale into the process of identifying research objects: the very act of identifying a particular pattern means that scale, extent, and resolution have been employed' (Gibson et al., 2000, p. 221).

Observational scale has two components. *Resolution*, or *grain*, is the smallest unit of measurement: it determines the precision or detail captured by a certain method. *Extent* is the overall dimensions of a study: the area (and time period) over which measurements are made. 'Small scale', in this context, typically denotes a finer resolution, while 'large scale' indicates a large extent; practical constraints generally dictate a small extent for fine-grained studies and a coarse grain for studies that have a large extent. Combining a fine grain with a large extent is difficult because a fine grain captures greater variability, which in turn necessitates larger sample sizes, even at a small or medium extent. Often, grain and extent are constrained by the technical capacities of available instruments for measurement (and of computers for analysing the resulting data): If one has only a meter stick, for example, the grain can be no smaller than a millimeter, and extents of greater than, say, fifty meters are likely to be impractical. Likewise with temporal scale: annual rainfall, for example, is too coarse a resolution to understand vegetation patterns where seasonal variability is high. Choosing one's grain and extent carefully is important precisely because 'patterns that appear at one level of resolution or extent may be lost at lower or higher levels' (Gibson et al., 2000, p. 221). Conversely, the advent of new observational tools and technologies can strongly affect the kinds of questions that scholars pose and the theories they construct. As Church (1996, p. 153) puts it: 'The space and time scales of observation constrain the structure and physical content of functionalist theories [in geomorphology] through their control of the resolution of information in the theory. Our theoretical construction of order in nature is bound by the tyranny of the scales'.

Observational scale is principally an epistemological issue, but subsequent work in ecology and biophysical geography indicates that scale may have ontological implications as well. *Operational scale* refers to the idea that phenomena occur at determinate spatial (and temporal) scales in the real world: that scale is an actual, material property of processes, not simply a matter of how they are observed. The Coriolis force, for example, determines patterns of winds and weather systems at very large scales: It is why low-pressure systems rotate counterclockwise in the northern hemisphere and clockwise in the southern hemisphere. However, contrary to popular belief, it does not affect which way water spins down the drain, a process at much smaller spatial and temporal scales. Similarly, tectonic drift occurs over very long time periods and very large areas, but at smaller scales it is, practically speaking, not only invisible but generally irrelevant. For both ecologists and biophysical geographers, operational scale is ontologically real.

A key point of agreement among geographers and ecologists is that no single 'correct' scale exists for either field: different processes operate at different scales and must be studied accordingly. Identifying the operational scales of processes and reconciling them with observational scales are therefore central challenges of research. The former may be termed the *ontological moment* of scale, the latter its *epistemological moment* (Sayre, 2005). One must work back and forth between the two moments (dialectically or, at the least, hermeneutically), incrementally reducing epistemological obstacles and thereby strengthening ontological insights. Over time, the observational scales utilised by scientists should more closely reflect the operational scales of material processes.

Scale as Level

That different processes have different operational scales raises difficult questions about their interactions. If the Coriolis force can give direction to something as big and powerful as a hurricane, shouldn't it also affect water going down the drain? Can one 'scale down' from a large or slow process to a smaller or faster scale in a simple, linear fashion, or not? How can a relationship identified at a small scale be extrapolated 'up' to larger scales? Multiscale analyses, and the study of cross-scale linkages, aim to address questions such as these. In order to do this, scientists classify phenomena into various levels based on the scales at which they can be observed or measured: the organism level and the community level in ecology, for example. The spatial and temporal units of measurement appropriate for each level tend to coalesce in a pattern: larger areas with longer time-periods, smaller areas with shorter time-periods (figure 7.1). Whether such levels are onto-logically real or merely artifacts of observation can only be determined by empiri-cal research.

It is easy to see how level and scale might become confused, since they are inter-changeable *in this sense of scale*. In common usage, for example, one can generally refer to 'the urban level' as 'the urban scale' without loss of meaning (even though the extent of this level may vary depending on historical and geographical context). Epistemologically, scale as level involves choices of what will and will not be observed and analysed: A study conducted at 'the community scale' focuses on phenomena of certain (more or less determinate) spatial and temporal dimensions, and it may choose to ignore (or hold constant) processes at other levels for the purposes at hand.

Of course, phenomena that scientists classify at different levels do interact with one another in the real world, and studies of such interactions require some kind of ordering principle among levels. Various metaphors have been used: a pyramid, ladder, scaffold, or the famous 'Russian doll' of nested, recursive systems (Herod

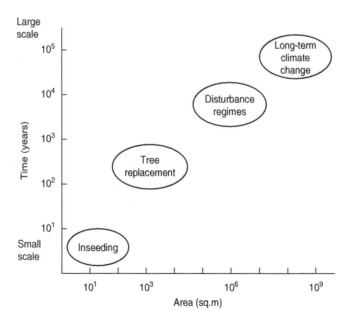

Figure 7.1 The hierarchy of space-time scales. The space/area and time/rate of processes tend to co-vary, lending support to the notion of hierarchically ordered levels in space-time. *Source:* Sheppard and McMaster (2004, p. 12), reproduced with permission.

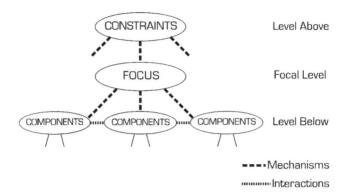

Figure 7.2 Relationships between levels in a system, as conceived in hierarchy theory. Processes at the focal level are constrained by the level above. They are driven by interactions among components at the level below. Figure by Darin Jensen.

and Wright, 2002). The prevailing approach in biophysical geography, as in ecology, is hierarchy theory (Allen and Starr, 1982), in which phenomena are classified based on functional relations or operational scales. Wu and Loucks (1995, p. 451) argue that ecological studies should examine (at least) three levels: the level of the process at issue, plus the levels above and below it (figure 7.2). 'The higher level provides a context and imposes top-down constraints on the focal level, and the lower level provides mechanisms and imposes bottom-up constraints'. Note that causality here is not unidirectional (*contra* Leitner and Miller, 2007): The outcome at a given level is determined both at that level and by the interaction of processes that link it 'upwards' and 'downwards' to adjacent levels.

Most biophysical systems are theorised as constitutive hierarchies. This means that relations are not simply bureaucratic, in which 'higher' levels dictate what happens at 'lower' ones (known to political scientists as an exclusive hierarchy). Nor are they inclusive, as in taxonomy, in which each level simply encompasses those below it. In a constitutive hierarchy, units at one level, when combined at the next level up, may display patterns of self-organisation and 'emergent properties' that cannot be discerned in, or deduced from, their behaviour at the focal level (Gibson et al., 2000). The idea is often expressed as 'the whole is greater than the sum of its parts'. Landscape ecology descends in part from this insight, sometimes glossed as 'holism' or the study of 'holons' (Naveh and Lieberman, 1984). Similarly, 'complex adaptive systems' are defined by heterogeneity and unpredictability as 'pattern emerges from the interplay between processes that generate novelty and those that winnow that novelty' (Chave and Levin, 2004, p. 31). Chaos theory and panarchy (Gunderson and Holling, 2002) are other recent attempts to make sense of such phenomena, which Church (1996, p. 167) locates 'in the zone between mechanistic and contingent explanation'.

Scale as Relation

It is here that scale as relation emerges. Not only is there no single 'correct' scale for understanding social or ecological systems, but neither can one assume linearity across scales. As Chave and Levin (2004, p. 32) note, scaling relationships – between metabolic rate and body size, for example, or between area and species richness – 'are among the most robust empirical generalisations found' in ecological systems – but they are not linear. They cite financial crashes and traffic jams as 'typical of the dynamics found in complex adaptive systems': as the component parts interact and adapt, positive feedback loops can trigger abrupt, extreme, unpredictable change. Economies of scale are another example: how the division of labour and the expansion of production result in non-linear increases in output and qualitatively new social phenomena can only be understood relationally.

Scale as relation requires a strong conceptual distinction from level. It is, so to speak, an order removed from scale as level, defined by the spatial and temporal relations among (processes at different) levels. To address scale as relation, then, one must eschew the conventional synonymy of scale and level.

Among ecologists, scale as relation is part of a larger critique of equilibrium models and assumptions. In a famous 1977 article, Robert May presented mathematical models of systems with 'a multiplicity of stable states', inspired by empirical cases of grazing ecosystems, fisheries, insect outbreaks in forests, and host-parasite systems. He likened ecosystem dynamics to a marble in a cup. If the cup formed 'a single valley', then the system would always return to a single stable state following disturbance, and historical effects would be unimportant. But if the cup were a 'dynamical landscape pockmarked with many different valleys, separated by hills and watersheds', then 'the state into which the system settles depends on the initial conditions: the system may return to this state following small perturbations, but large disturbances are likely to carry it into some new region of the dynamical landscape'. Scale is thus not only a spatial issue but also a temporal one. Any equilibrium presupposes some period of time over which stability persists; it might turn out to be unstable if evaluated at a different temporal scale. Moreover, 'if there are many alternative locally stable states, historical accidents can be of overriding significance' (May, 1977, p. 471).

Understood in this way, scale is central to current notions of sustainability and resilience in complex adaptive systems involving humans and the environment. Once one admits the possibility of multiple stable states, one cannot avoid the issue of thresholds or 'breakpoints' between them. May (1977, p. 477) emphasised that 'continuous variation in a control variable can produce discontinuous effects' and that 'increasingly severe nonlinearities can make the dynamical behaviour range from a stable point, through a bifurcating hierarchy of stable cycles, into a regime which is in many ways indistinguishable from random noise'. In the three decades since, ecologists have struggled to model complex systems and quantify thresholds of non-linear change. Predictive knowledge of thresholds has remained elusive, but theory and conceptual models have advanced considerably and empirical observations are accumulating (Crumley, 1994; Westoby et al., 1989). There is also growing interest in the hypothesis that unsustainable resource use results from 'mismatches of scale' between human and natural processes (Lee, 1993; Cumming et al., 2006). Determining the relevant processes involved, and their operational scales, thus becomes a necessary prerequisite for advancing both research and management (for an example involving fisheries, see Perry and Ommer, 2003).

While ecologists turn to ever more sophisticated mathematics and models to understand scale as relation, human geographers explore the matter through metaphors and theory. Howitt (1998) examines musical scales, pointing out that the value of each note is determined not simply by its individual qualities but also by the other notes and its position among them. Any change in one note affects the scale as a whole, and vice-versa; the scale is more than the sum of its parts, and this 'gestalt' can be perceived in the way certain scales provoke spontaneous cultural associations. Howitt's metaphor is suggestive, even if exactly how musical scales might elucidate human and environmental processes remains unclear. He argues that 'scale is better understood dialectically than hierarchically' (Howitt, 1998, p. 52). Ecologists rarely employ such terminology, but the underlying point strongly resembles the idea of emergent properties or panarchy: shifting scales results in qualitative, rather than merely quantitative, change.

Scale in the Discipline of Human Geography

Scale has a rather different genealogy in human geography, although the underlying methodological and theoretical issues converge with those elsewhere in the discipline and in ecology. As in the physical sciences, social science disciplines have divided and defined themselves – intentionally or unwittingly – by scale (as size, both operational and observational): psychology studies individuals; anthropology villages, clans or tribes; sociology neighborhoods or cities; political science governments and states, etc. Each discipline could thus take its own scale more or less for granted. (The separation of micro- from macroeconomics is the exception that proves the rule.) As Gibson et al. (2000, p. 221) observe: 'Overt choices of particular scales to identify specific patterns are generally taken more consciously in the natural sciences than in the social sciences'. Human geography, with its diversity of subdisciplines and methods, could not so easily avoid the issue, but many topics had operational and thus observational scales that seemed obvious and could therefore remain implicit. In recent decades, however, the economic, political and cultural dynamics of globalisation have called into question the scales of previous human geographic research.

A typical classification of human geographical scales includes the body; the household; the neighborhood; the city; the metropolitan area; the province or state; the nation-state; the continent; and the earth as a whole (Sheppard and McMaster, 2004, p. 4). (The region is another oft-employed geographical scale, albeit one whose position in this classification is variable. . . .) By the preceding analysis, this is simply a list of levels; the implied nested hierarchy resembles the way ecologists conventionally imagined organisms, populations, communities, ecosystems and biomes. If one questions the stability of these categories, however – how they are produced, reproduced or transformed – or if one asks how multiple levels interact, then the issue of scale as relation is raised. This is how 'the scale question' in human geography has emerged.

For political ecology in particular, and environmental geography more generally, one might trace recent debates about scale to Piers Blaikie and Harold Brookfield's landmark book, *Land Degradation and Society*, which addressed problems of aligning observational and operational scales and working across scales (1987, pp. 64–74). '[I]t is very evident that we must take care to define the scale at which we are working if the social causes and consequences of degradation are to be described adequately'.

But the scale question in critical human geography also has its roots in political economy: an article by Peter Taylor (1982) that defined the local, national and global as the scales of experience, ideology and capital accumulation, respectively. Taylor characterised the global scale as the most 'real', reflecting the Marxian-materialist priority given to production and simultaneously reinforcing a top-down, hierarchical notion of scale.

Building on the work of Henri Lefebvre and David Harvey, Neil Smith (1993, p. 96f.) criticised geography for taking its scales – 'localities, regions, nations and so forth' – for granted, and for trivialising geographical scale 'as merely a question of methodological preference'. Focusing on the ontological rather than the epistemological moment, he stressed the importance of scale in spatial differentiation. '[S]cale is produced in and through societal activity which, in turn, produces and is produced by geographical structures of social interaction . . . [T]he production of geographical scale is the site of potentially intense political struggle'. Smith proceeded to offer a typology of geographical scales similar to the list given above, but he treated them as operational rather observational. He specified the processes that produced each scale materially: for example, daily commuting for the urban scale, and capital circulation and uneven development for the global. Insofar as Smith considered how each scale is determined by interactions with the others, he pointed beyond scale as level towards scale as relation.

Human geographers have proceeded to explore the production and politics of scale further, particularly in regard to the city, the nation-state, and the global economy. Scale as level provides the framework for these studies, insofar as the nation-state is construed as 'above' the city and 'below' the global in a socio-spatial, hierarchical order. But the point usually is to understand the historical-geographical constitution and reconfiguration of levels in relation to one another – such that scale is construed, at least implicitly, as relational. Erik Swyngedouw (1997) introduced the term 'glocalisation', for example, to capture the combination of upward and downward shifts in the scale of accumulation and regulation with the advent of globalisation. Neil Brenner (1998, p. 464) argued that 'scales are not merely the platforms within which spatial fixes are secured, but one of their most fundamental geographical dimensions, actively and directly implicated in the historical constitution, reconfiguration, and transformation of each successive configuration of capitalist territorial organization'. Viewed as a process of *rescaling*, globalisation 'entails less an obliteration of the national spatial scale than its rearticulation with the subnational and supranational spatial configurations on which it is superimposed' (Brenner, 1997, p. 299).

In different ways, both Swyngedouw and Brenner shift attention away from scale *per se* and towards the processes that produce (patterns that have) scales. Like Smith, they are concerned with operational scale. Swyngedouw (1997, p. 141) is explicit: 'The theoretical and political priority . . . never resides in a particular geographical scale, but rather in the process through which particular scales become (re)constituted . . . . In short, scale . . . is not and can never be the starting point for sociospatial theory . . . the kernel of the problem is theorising and understanding "process"'. Swyngedouw's (2004; 2007) empirical research reflects this approach and is widely credited for bringing ecological processes (such as hydrologic cycling) into cogent relation with political-economic processes such as capital accumulation and governance. Brenner (1998, p. 466) emphasises 'the relational, mutually interdependent character of geographical scales under capitalism', and he develops a thesis that clearly transcends scale as size or level:

> the forms of territorialization for capital are always *scaled* within historically specific, multitiered territorial-organizational arrangements. The resultant scale-configurations, or 'scalar fixes,' simultaneously *circumscribe* the social relations of capitalism within determinate, if intensely contested, geographical boundaries and *hierarchize* them within relatively structured, if highly uneven and asymmetrical, patterns of sociospatial interdependence (Brenner, 1998, p 464, emphases in original).

Terms such as 'scaling', 'rescaling', 'scale effects' and 'jumping scales' all draw attention not only to the ongoing production of scale (and therefore its historical contingency and malleability) but also to the non-linear, complex outcomes that are hallmarks of scale-as-relation.

Research along these lines has more recently opened into vociferous debates about the conceptual status of scale throughout human geography. In an oft-cited article, Marston (2000) reviewed the literature and argued persuasively that geographical scale is socially constructed. Bodies, neighborhoods, cities and so forth are not given *a priori* but produced through social processes; geographers have gone astray, she argued, by taking their scales for granted and by privileging certain scales – such as the nation-state or the global economy – over others such as the household. Marston's article provoked a response by Brenner (2001), followed by several further contributions (Marston and Smith, 2001; Purcell, 2003; Sayre, 2005). Subsequently, Marston et al. (2005) changed course and expanded the controversy by making a case 'to expurgate scale from the geographic vocabulary' altogether; a flurry of responses ensued, almost all of them critical of this position (e.g., Collinge, 2006; Jonas, 2006; Leitner and Miller, 2007). There is neither need nor space to review these exchanges in detail here. Two points suffice to defuse much of the controversy.

First, the debate has suffered from a confounding of scale's epistemological and ontological moments. The critique of conventional geographical scales stemmed initially from epistemological considerations: Taking the local, the national and the global as *a priori* givens may obscure the interactions among various scales; a crude hierarchy theory risks overlooking actors and processes at 'smaller' or 'lower' scales by privileging 'larger' or 'bigger' ones. These are important points, but in choosing a scale for observation one is not necessarily making any ontological commitments or claims. Most of the substantive issues raised in the debate, however, concern the ways that the operational scales of governance, reproduction, regulation and accumulation have shifted in recent decades and how people contest and transform the scales of actual processes in the world. This is not to say that the two moments are separate or unrelated – on the contrary, their dialectical relation is of the utmost importance. But confounding the two moments collapses the dialectic (Sayre, 2005).

Second, the acrimony and confusion reflects a persistent failure to distinguish between scale as size, level and relation. Almost all contributors employ scale both in its second sense (where scale and level are interchangeable) and in its third sense (where they are not) without recognising the problems this entails. Marston et al. (2005, p. 420) argue that scale may 'be simply and effectively collapsed into' level; they proceed to use the terms interchangeably or together, as in the phrase 'levels of scale' (p. 422). But they do not even acknowledge the existence of scale as relation (despite citing Howitt's papers on the subject), and collapsing scale into level compels them to make hierarchy into an inherent attribute of scale. Since their real animus is hierarchy, they indict scale *tout court*. It is true that Brenner, among others, understands scale as inherently hierarchical, but this reflects his own failure to distinguish scale as level from scale as relation. The former does entail hierarchy (or some such principle of ordering); the latter does not. Furthermore, there is nothing inherently hierarchical (or 'vertical') about emergent properties, complex interactions, or thresholds of nonlinear change. Other frameworks that have been proposed in recent years, such as networks (Leitner, 2004; Taylor, 2004) and heterarchy (Crumley, 2005), confirm that one can critique hierarchy theory yet retain a strong emphasis on scale. Leitner (2004, p. 246) notes that 'networks are themselves scaled', and that '[n]etwork scales are emergent properties of sociospatial processes operating inside and beyond networks'. It is precisely by rescaling processes that networks have the potential to bypass or subvert conventional hierarchies of power.

Conclusion: Towards an Integrated Conceptual Framework

A remarkable and apparently unwitting convergence has occurred in ecological and geographical conceptions of scale in the past two or three decades. From very different starting points, drawing on ideas and insights from across the social and natural sciences, scholars in both fields have moved from scale as size and level to scale as relation. The common interests and ideas include emergent properties, hierarchies and networks, non-equilibrium, thresholds of change, spatio-temporality, path dependence and self-organisation. The challenges and opportunities for integrative work and collaboration are growing in number and importance.

How to integrate ecological and geographical scale for purposes of environmental geography? The following six principles can be derived from the preceding analysis of geographical and ecological scale:

1. Scale is *relational*. Its scientific value lies not in absolute or discrete measurements of a phenomenon in terms of size, duration, or magnitude, but rather in exploring relations among phenomena so measured.
2. The focus of theorising about scale must therefore fall on *processes* rather than on scale *per se*, because it is through processes that relations among phenomena are manifest.
3. Processes are simultaneously spatial and temporal; while many uses of scale are implicitly spatial, the concept as developed here is intrinsically spatio-temporal.
4. There is no single 'correct' scale for studying or understanding societies, ecosystems, or their interactions; any given process may, however, have an appropriate or best scale for research.
5. Scales are *produced*, whether by human-social, geophysical or biological processes. They have an ontological moment, insofar as they are integral to the constitution of material processes; they have an epistemological moment, insofar as one's scale of observation determines the patterns (or lack thereof) that one observes. The two moments are dialectically related.
6. A major topic for further research and theorising on scale concerns *thresholds* of non-linear or qualitative change across scales (for any given process) and between processes of different scales. It is at these points that *scaling effects, mismatches of scale* or *rescaling* are manifest, and where critical issues of social-ecological change and sustainability may be engaged most fruitfully.

It remains to be seen whether and how collaboration and integration can be achieved, both practically and theoretically. There are growing numbers of interdisciplinary research projects and funding opportunities aimed at the social-ecological interface, such as the National Science Foundation's Coupled Natural and Human Systems programme, in which scale figures prominently. Vogt et al. (2002, p. 168) point out a more theoretical challenge:

> To assist in the integration of social and natural sciences for natural resource management, researchers will need to explicitly recognize and address issues of scale differently from their traditional, disciplinary approaches. Instead of emphasizing the need for scale-dependent information that may be associated with their respective disciplines, it may be more important to determine what is the most appropriate scale(s) to address various natural resource issues. Integrating the social and natural sciences will require improving our understanding of how space is currently perceived by each discipline.

Beyond this, of course, lie still deeper philosophical questions. Bruce Rhoads (2006, p. 14) has argued convincingly that geomorphology should embrace a process-philosophical metaphysics, in which 'the nature of reality, including geomorphological phenomena, is fundamentally processual'. This is also where Erik Swyngedouw (1997, p. 140) starts: 'I insist that social life is process-based, that is, in a state of perpetual change, transformation, and reconfiguration'. Obviously, the geomorphological and the social processes in question are likely to unfold on temporal scales that differ by several orders of magnitude – such is the challenge and the potential of the problem of scale. It will also require, as Church (1996, p. 166f.) has argued, a general recognition that 'the scales of enquiry determine the most appropriate mode of explanation', and that some process-scale combinations may not yield to mechanistic, quantitative, or predictive methods.

BIBLIOGRAPHY

Allen, T. F. H. and Starr, T. B. (1982) *Hierarchy: Perspectives for Ecological Complexity.* Chicago and London: University of Chicago Press.

Brenner, N. (1997) State territorial restructuring and the production of spatial scale: urban and regional planning in the Federal Republic of Germany, 1960–1990. *Political Geography*, 16, 273–306.

Brenner, N. (1998) Between fixity and motion: accumulation, territorial organisation and the historical geography of spatial scales. *Environment and Planning D: Society and Space*, 16, 459–81.

Carpenter, S., Walker, B., Anderies, J. M. and Abel, N. (2001) From metaphor to measurement: resilience of what to what? *Ecosystems*, 4, 765–81.

Chave, J. and Levin, S. (2004) Scale and scaling in ecological and economic systems. In P. Dasgupta and K.-G. Muller (eds), *The Economics of Non-convex Ecosystems.* Dordrecht, Boston and London: Kluwer Academic Publishers, pp. 29–59.

Church, M. (1996) Space, time and the mountain – How do we order what we see? In B. L. Rhoads and C. E. Thorn (eds), *The Scientific Nature of Geomorphology: Proceedings of the 27th Binghamton Symposium in Geomorphology Held 27–29 September 1996.* Chichester, NY: John Wiley and Sons, pp. 147–70.

Collinge, C. (2006) Flat ontology and the deconstruction of scale: a response to Marston, Jones and Woodward. *Transactions of the Institute of British Geographers*, 31, 244–51.

Crumley, C. (ed.) (1994) *Historical Ecology: Cultural Knowledge and Changing Landscapes.* Santa Fe, NM: School of American Research Press.

Crumley, C. (2005) Remember how to organize: heterarchy across disciplines. In C. S. Beekman and W. S. Baden (eds), *Nonlinear Models for Archaeology and Anthropology.* Aldershot (Hampshire), UK: Ashgate Press, pp. 35–50.

Cumming, G. S., Cumming, D. H. M. and Redman, C. L. (2006) Scale mismatches in social-ecological systems: causes, consequences, and solutions. *Ecology and Society,* 11(1), 14.

Gibson, C. C., Ostrom, E. and Ahn, T. K. (2000) The concept of scale and the human dimensions of global change: a survey. *Ecological Economics,* 32, 217–39.

Gunderson, L. H. and Holling, C. S. (2002) *Panarchy: Understanding Transformations in Human and Natural Systems.* Washington, DC: Island Press.

Herod, A. and Wright, M. W. (eds) (2002) *Geographies of Power: Placing Scale.* Oxford: Blackwell.

Howitt, R. (1998) Scale as relation: musical metaphors of geographical scale. *Area,* 30, 49–58.

Howitt, R. (2003) Scale. In J. Agnew, K. Mitchell and G. O'Tuathail (eds), *A Companion to Political Geography.* Oxford: Blackwell, pp. 138–57).

Jonas, A. E. G. (2006) Pro scale: further reflections on the 'scale debate' in human geography. *Transactions of the Institute of British Geographers,* 31, 399–406.

Kolbert, E. (2006) Field notes from a catastrophe: man, nature, and climate change. New York: Bloomsbury.

Lee, K. N. (1993) Greed, scale mismatch, and learning. *Ecological Applications,* 3, 560–64.

Leitner, H. (2004) The Politics of scale and networks of spatial connectivity: transnational interurban networks and the rescaling of political governance in Europe. In E. Sheppard and R.B. McMaster (eds). *Scale and Geographic Inquiry: Nature, Society, and Method.* Malden, MA: Blackwell Publishing, pp. 236–55).

Leitner, H. and Miller, B. (2007) Scale and the limitations of ontological debate: a commentary on Marston, Jones and Woodward. *Transactions of the Institute of British Geographers,* 32, 116–25.

Levin, S. A. (1992) The Problem of pattern and scale in ecology. *Ecology,* 73, 1943–1967.

Marston, S. A. (2000) The social construction of scale. *Progress in Human Geography,* 24, 219–42.

Marston, S. A. and Smith, N. (2001) States, scales and households: limits to scale thinking? A response to Brenner. *Progress in Human Geography,* 25, 615–19.

Marston, S. A., Jones, J. P. III and Woodward, K. (2005) Human geography without scale. *Transactions of the Institute of British Geographers,* 30, 416–32.

May, R. (1977) Thresholds and breakpoints in ecosystems with a multiplicity of stable states. *Nature,* 269, 471–77.

Naveh, Z. and Lieberman, A. S. (1984) *Landscape Ecology: Theory and Application.* New York: Springer.

Openshaw, S. (1977) A geographical study of scale and aggregation problems in region-building, partitioning and spatial modeling. *Transactions, Institute of British Geographers* (new series), 2, 459–72.

Openshaw, S. (1984) *The Modifiable Unit Area Problem.* Concepts and techniques in modern geography 38. Norwich: Geo Books.

Perry, R. I. and Ommer, R. E. (2003) Scale issues in marine ecosystems and human interactions. *Fisheries Oceanography,* 12, 513–22.

Peterson, D. L. and Parker, V. T. (1998) Dimensions of scale in ecology, resource management, and society. In D.L. Peterson and V.T. Parker (eds), *Ecological Scale: Theory and Application.* New York: Columbia University Press, pp. 499–522.

Purcell, M. (2003) Islands of practice and the Marston/Brenner debate: toward a more synthetic critical human geography. *Progress in Human Geography*, 27, 317–32.

Rhoads, B. L. (2006) The dynamic basis of geomorphology reenvisioned. *Annals of the Association of American Geographers*, 96, 14–30.

Rykiel, E. J. J. (1998) Relationships of scale to policy and decision making. In D. L. Peterson and V. T. Parker (eds), *Ecological Scale: Theory and Application*. New York: Columbia University Press, pp. 485–98.

Sayre, N. F. (2005) Ecological and geographical scale: parallels and potential for integration. *Progress in Human Geography*, 29, 276–90.

Sheppard, E. and McMaster, R.B. (eds) (2004) *Scale and Geographic Inquiry: Nature, Society, and Method*. Malden, MA: Blackwell Publishing.

Smith, N. (1993) Homeless/global: Scaling places. In J. Bird, B. Curtis, T Putnam, G. Robertson and L. Tickner (eds), *Mapping the Futures: Local Cultures, Global Change*. London: Routledge, pp. 87–119.

Swyngedouw, E. (1997) Neither global nor local: 'glocalization' and the politics of scale. In K. R. Cox (ed.), *Spaces of Globalization: Reasserting the Power of the Local*. New York and London: Guilford Press, pp. 137–66.

Swyngedouw, E. (2004) *Social Power and the Urbanization of Water: Flows of Power*. Oxford and New York: Oxford University Press.

Swyngedouw, E. (2007) Technonatural revolutions: the scalar politics of Franco's hydro-social dream for Spain, 1939–1975. *Transactions of the Institute of British Geographers*, 32, 9–28.

Taylor, P. (1982) A materialist framework for political geography. *Transactions of the Institute of British Geographers (new series)*, 7: 15–34.

Taylor, P. J. (2004) Is there a Europe of cities? World cities and the limitations of geographical scale analysis. In E. Sheppard and R. B. McMaster (eds). *Scale and Geographic Inquiry: Nature, Society, and Method*. Malden, MA: Blackwell Publishing, pp. 213–35.

Vogt, K. A., Grove, M., Asbjornsen, H., Maxwell, K. B., Vogt, D. J., Sigurdardottir, R., Larson, B. C., Schibli, L. and Dove, M. (2002) Linking ecological and social scales for natural resource management. In J. Liu and W. W. Taylor (eds), *Integrating Landscape Ecology into Natural Resource Management*. Cambridge and New York: Cambridge University Press, pp. 143–75.

Wu, J. and Loucks, O. L. (1995) From balance of nature to hierarchical patch dynamics: a paradigm shift in ecology. *Quarterly Review of Biology*, 70: 439–66.

Westoby, M. B., Walker, B. and Noy-Meir, I. (1989) Opportunistic management for rangelands not at equilibrium. *Journal of Range Management*, 42, 266–74.

Chapter 8

Vulnerability and Resilience to Environmental Change: Ecological and Social Perspectives

W. Neil Adger and Katrina Brown

The Vulnerability and Resilience of Society and Environments

Vulnerability and resilience are attractive concepts for geographers. Vulnerability captures the idea that there are inherent risks that are experienced by people and communities living in particular places. Resilience captures the ability of people and ecosystems together to adapt to changing risks and opportunities. Physical and biological phenomena that we describe as hazards are pervasive. Hence, vulnerability is often measured as the extent to which a threshold to some undesirable state has been crossed while resilience focuses on the capacity to tolerate disturbance without collapsing into a qualitatively different state that is controlled by a different set of processes.

Vulnerability in this context is thus about the susceptibility of groups or individuals to harm from social or environmental change. Vulnerability is an important characteristic of individuals, communities and larger social groups. The vulnerability of a group or individual depends on its capacity to respond to external stresses that may come from environmental variability or from change imposed by economic or social forces outside of the local domain. Thus, vulnerability does not exist in isolation from the wider political economy but rather is related to inadvertent or deliberate action that reinforces self-interest and the distribution of power. Vulnerability is made up of a number of components including exposure and sensitivity to hazard or external stresses and the capacity to adapt. The definition of key terms is outlined in table 8.1.

The definitions and elements of vulnerability in table 8.1 represent a convergence of perspectives derived from different underlying paradigms in geography. Burton et al. (1993) developed the integrative notion of vulnerability as a characteristic of interacting forces that create environmental hazards as well as opportunities. A critique of this approach within human geography effectively pointed to the underlying structural factors and power relations that create and maintain social vulnerabilities. Hewitt (1983), for example, attempted to explain why the poor and marginalised have been most at risk from natural hazards: what he termed the human ecology

Table 8.1 Attributes of vulnerability to environmental and social change and perturbations

Element of vulnerability	Definition
Exposure	The nature and degree to which a system experiences environmental or socio-political stress.
Sensitivity	The extent to which a human or natural system can absorb the impacts without suffering long-term harm or some significant state change. This concept of sensitivity, closely related to resilience, can be observed in physical systems with impact-response models, but requires greater interpretation in ecological and social systems, where harm and state change are more contested.
Adaptive capacity	The ability of a system to evolve in order to accommodate environmental perturbations or to expand the range of variability with which it can cope.

of endangerment. He concluded, for example, that poorer households tend to live in riskier areas in urban settlements, making them more exposed to flooding, disease and other chronic stresses. A number of other geographers have also highlighted the distinction between outcomes and processes of vulnerability in its analysis and measurement (e.g. Liverman, 1990; Watts and Bohle, 1993; Blaikie et al., 1994; Cutter et al., 2003; Turner et al., 2003; Leichenko and O'Brien, 2008; on methods to measure vulnerability, see Eakin and Luers, 2006).

Vulnerability is socially differentiated: virtually all natural hazards and human causes of vulnerability impact differently on different groups in society. Many comparative studies have noted that the poor and marginalised have historically been most at risk from natural hazards. Poorer households are forced to live in higher-risk areas and so are more likely to be affected, and to a greater extent, by earthquakes, landslides, flooding, tsunamis, and poor air and water quality, particularly in the increasingly urbanised world (Mitchell, 1999; Pelling, 2003). Women are differentially at risk from many environmental hazards, including, for example, the burden of work in recovery of home and livelihood after an event (Fordham, 2003). In many studies of the impact of earthquakes, including analysis of the Asian tsunami of 2004, women and other household dependants suffered much greater mortality than adult males.

Flooding in low-lying coastal areas associated with monsoon climates or hurricane impacts, for example, is seasonal and usually short-lived, yet can have significant unexpected impacts for vulnerable sections of society. But of course one person's flood is another person's irrigation water. Periodic flooding is an integral part of many farming systems as it provides nutrients in fertile floodplain areas. Hence, natural hazards are often a disadvantageous aspect of a phenomenon at one point in time that is predominantly beneficial.

The concept of resilience has its roots in ecology and, when applied to interactions between society and nature, provides a powerful framework for analysing the integrated, or coupled, nature of such interacting systems. Ecology has promoted notions of resilience, both to explain how ecosystems can radically change from one state to another very different one and also as a guiding principle for ecosystem management to avoid rigidities and so-called social traps (Folke, 2006; see Francis and Turner, both this volume). At its core, ecological resilience is measured by the magnitude of the perturbations that can be absorbed before the system flips to another state (Gunderson and Holling, 2002). No ecological state is unambiguously 'better' than another. For example, grassland-dominated ecosystems can flip into scrubby vegetation. Or coral reef ecosystems can shift from being dominated by live coral to being covered by algae so that the corals become less productive and the diversity of fisheries dependent on them declines. All these states are natural, but those that provide services to humans are, from our perspective, more valuable.

Resilience of an ecological system relates to the functioning of the system as a whole, rather than the stability of its component populations, or even the ability to maintain a steady ecological state. Ecosystems have diverse properties, which ecologists have sought to measure. These form the basis of normative statements about sustainability and sustainable utilisation of ecosystems. Many tropical terrestrial ecosystems have stable and diverse populations but are relatively low in resilience. For example, in tropical rainforests, most of the nitrogen (which plants need to grow) is cycled and stored biologically in the biomass itself, rather than in the soil, which is often very poor as a result. Consequently, when tropical forests are logged and cleared, the nitrogen needed for plant growth is removed too, and the land is unable to support more than scrub grassland with much lower biological productivity. By contrast, many temperate forest ecosystems in temperate regions with apparently low diversity can exhibit greater resilience in the face of disturbance.

From declining fish stocks in the Pacific, through to land-use change in the Sahel, ecosystems have been shown to be subject to periodic shifts into states which are often less desirable for, but often triggered by, human use (Scheffer et al., 2001). These shifts are often triggered by single events such as a tropical storm impacting on coral reefs or through fires and their impact on forest ecosystems. Sometimes they are caused by longer-term events such as the removal of one predator from an ecological system (Folke et al., 2005).

The resilience of a social-ecological system is made up of a number of elements: the amount of perturbation a system can handle and still retain the same characteristics and controls on function and structure; the degree to which a system is capable of self-organisation; and the ability to build and increase the capacity for learning and adaptation (Carpenter et al., 2001; Berkes et al., 2003). Resilient systems can, in other words, cope; they adapt and reorganise in the face of change without losing their ability to provide valuable ecosystem services. A loss of resilience in social-ecological systems is often associated with irreversible change, the creation of vulnerabilities for marginalised elements of society, and the reduction of flows of ecosystem services.

Interactions between Ecological Systems and Society

There are three primary sets of interactions between ecological changes described above and society. First, human action drives ecological change. In ecological resilience analysis, ecosystems are characterised as having multiple possible equilibria that are regulated by fast and slow variables, ranging from physical disturbance, natural response to nutrient availability cycles, through to accumulation of persistent pollutants. Some of these are driven by human action. Indeed, the impacts of altering global carbon and nitrogen cycles are classic 'slow variable' impacts on many of the world's ecosystems.

The second interaction comes from the impact of ecosystem state changes on the availability of ecosystem services to society. The Millennium Ecosystem Assessment (2005) demonstrated that all ecosystems contribute to the well-being of humanity in providing or regulating services that provide the basic needs for everyone on the planet, what they termed, a 'good quality of life'. Clearly, this step into ecosystem services involves values that are socially contingent and change over time and space. It also raises the issue of whether ecosystem services have intrinsic value above and beyond any human use or appreciation of them. Clearly, when ecosystems undergo regime shifts, the flow of ecosystem services is altered. Folke (2006) and others argue that the majority of such changes observed indeed reduce the flow of ecosystem services in aggregate, a threat to human well-being.

The third interaction between ecological resilience and society is reflected in the question of whether whole systems, incorporating ecological and social elements, are themselves resilient. In other words, do the characteristics that make ecosystems resilient also make social-ecological systems (such as ocean ecosystems, fisheries and fishing communities taken together) resilient to change? Interdisciplinary research spanning the social and ecological sciences in these areas increasingly argues that environments co-evolve with the institutions and rules that mediate human use of resources. Rapid changes in either can create vulnerabilities as well as opportunities for both ecosystems and humans alike (Folke et al., 2005). The social elements of resilience are bound up with the ability of groups or communities to adapt in the face of external social, political or environmental stresses and disturbances (Adger, 2000). If formal and informal institutions (such as local-level watershed management committees, fisheries collectives and the like) themselves are resilient, they can promote wider resilience.

Institutions can be persistent, sustainable and resilient, but clearly not always for the benefit of everyone. Anderies et al. (2004) and Walker et al. (2004) suggest that there are inherent trade-offs involved in making resource use more efficient at providing goods and services to human users, which can often make them less resilient or able to adapt to changing circumstances. In northeast Brazil, for example, interventions to reduce the risk of periodic drought on the farming community have been carried to such an extreme that the principal government adaptation to drought is now humanitarian aid (Nelson et al., 2007). Efforts to reduce the level of vulnerability or to increase resilience are overshadowed by the levels of resources dedicated to maintaining the food and water supply during droughts.

In beginning to analyse the social implications of changes in ecosystems and in their resilience, there is no escaping the social construction of demands for environmental services, and increasingly, the construction of markets designed to promote the conservation or enhancement of such services. Advocates of creating markets for ecosystem services argue that they make use of natural resources more efficient by making explicit the linkages between ecosystem services and human development (Millennium Ecosystem Assessment, 2005). The benefits provided by ecosystem services are, in most cases, public goods: in other words, the benefits do not accrue exclusively to those people managing the resources.

There have been increasing numbers of markets created associated with forest services, particularly watershed regulation, biodiversity conservation, and especially the carbon sink function of forests (Pagiola et al., 2002). In the case of carbon sinks, the argument is not only that forests are potentially conserved by such markets, but that mitigation of net emissions of greenhouse gas emissions to the atmosphere can be undertaken more cost-effectively this way than by reducing fossil fuel use. But emerging critiques of these markets question whether payments for ecosystem services go to those actually providing the services. If they do not, then ultimately the sustainability of these markets is in question. Corbera et al. (2007), for example, examined the impact on two forest communities involved in a project for carbon sequestration services of forests in the state of Chiapas in Mexico and found that most of the benefits follow political affiliation, while the poorest farmers and women have been excluded from project design and implementation. They argue that these pitfalls reinforce existing uneven power structures, inequities and vulnerabilities: such markets are in fact highly limited in delivering more legitimate forms of decision making or a fair distribution of benefits.

In summary, the state of knowledge on how social-ecological systems interact is focused primarily on how ecosystem services are produced and maintained. But a further important normative set of knowledge relates to how to provide a stable environment for human use of these services. Economic growth involving unsustainable resource use or chronic stress on ecosystems creates vulnerabilities and makes society more sensitive to shocks. Discontinuous changes in ecosystem functions are associated with a loss of productivity and of ecosystem services. In addition, losing resilience reduces what economists have termed positive option values of the environment. Arrow and colleagues (1995) argue that the loss of ecosystem resilience and shifts to more unfamiliar states increase the uncertainties associated with environmental interactions. In other words, dealing with unfamiliar and undesirable states involves added (and often unacknowledged) costs. The nature of resilience and vulnerability is manifest in specific places and resource systems. Hence, these principles and issues can be examined using various techniques of environmental geographers. Resilience and vulnerability are also manifest across a range of spatial and political scales, as discussed below.

What Is a Resilient Community?

The causes of vulnerability are linked across space and time. We have highlighted how the resilience of social-ecological systems involves multiple facets and changing parameters. How can we recognise and identify the interlinked stimuli that influence resilience and vulnerability in a given location? This is a complex issue given the context- and place-specific dynamics of resilience and vulnerability within diverse societies. There are also issues about how the concepts of resilience and vulnerability are applied and understood within different disciplinary traditions. Many of the problems of their application are revealed if we examine how resilience has been approached and studied, and how it is manifest at a community level in different contexts. This section therefore presents two examples which relate and amplify different components and understandings of resilience, as applied to communities and how they respond to change.

The first example examines how rural households and communities were able to respond to the external shock of an 'economic crisis' and the associated impacts on livelihoods and resource use in the humid forests of southern Cameroon. In southern Cameroon, a range of events and changes had profoundly affected rural livelihoods within the past generation. During the mid-1980s and 1990s, Cameroon faced an economic crisis and currency devaluation that led to a significant macroeconomic reform programme implemented as a so-called Structural Adjustment Programme. Many public sector agencies were forced to lay off their staff causing widespread unemployment and reverse migrations from the cities back to rural areas. These rural areas also faced reductions in real producer prices for agricultural commodities, shifts in cropping patterns, as well as more intensive exploitation of natural resources, such as forests, through increased commercial licences to domestic and international companies.

Brown and Lapuyade (2001) examined the effects of these broad socio-economic and environmental changes on rural households and showed how the resilience of different sections of the population was differentiated. Men and women are able to adapt to changes in quite different ways. For instance, men moved into the production of food crops for cash, previously an activity primarily done by women, while women were found to rely increasingly on food processing as a means of livelihood, and at the same time, because of greater exploitation of forests, lost traditional rights of access to non-timber forest products. Women almost unanimously described the changes as negative, expressing it as 'maybe this is the end of the world', whereas men recognised that although the changes were tough, 'our standard of living is improving constantly'. Hence, the multiple stressors of social change and economic crisis interact to cause particular impacts in time and space.

In rural Cameroon, social and environmental change is experienced very differently by individuals even within the same household; vulnerability and resilience are not simply system characteristics, but are also differentiated individually. Social status and gender both matter. Understanding the political economy of resilience requires addressing the question of whose resilience counts (Lebel et al., 2006). In southern Cameroon, men's and women's adaptive capacity was acutely differentiated, enabling men to diversify their livelihoods in the face of multiple stressors, whereas women fell (or rather, were pushed) into poverty traps. Key factors were their rights and access to resources and markets, which critically affected individuals' adaptive capacity.

A second case highlights the role of perceptions of resilience and vulnerability and their potential to act as barriers to adaptation. In research on the management of coastal resources in the light of climate change in the Orkney Islands north of Scotland, resilience of the social-ecological system was expressed as a culturally dependent phenomenon, representing the ways in which island life can be sustained and the communities remain distinct and independent (Brown et al., 2005). It appears that Orkney Islands have a high degree of adaptive capacity articulated, for example, by participants in focus groups, highlighted by the voices of Orcadians in Box 8.1.

Despite the island's dependence on grant aid and subsidies, residents had positive perceptions of its autonomy and potential for self-organised local development. The threat of climate change and the possible impacts and changes already experienced – greater storm intensities, windier conditions and warmer winters – were seen as providing an opportunity to enhance independence and sustainability, by encouraging local production on islands instead of relying on inter-island transport and imports from the mainland which could be at risk from climate change. The exploration of possible responses to climate change reveals that Orkney society has many attributes associated with resilience. These include, in particular, a continued reference to shared history and a manifest social memory; sensitivity to environmental

---

Box 8.1   Positive perceptions of resilience in the context of future climate change among residents of the Orkney Islands

'Generally people here don't see weather or the movement of the sea as a problem. It is something to accommodate, accept and work around'.

'Orkney may need to become more self-sufficient in many food products to reduce dependence on the importation of stocks'.

'Orcadians should think ahead to alleviate prospective problems and need to start planning now'.

Source: Brown et al. (2005) based on field notes taken in 2004–5.

stimuli and an apparently adaptive approach; and anticipatory as well as reactionary responses.

These examples show that processes of decision making and perception are important in determining both individual and collective vulnerability and resilience. Indeed, social psychologists have long made a link between perceived vulnerability and marginalisation and the actual ability to take positive adaptive action (Satterfield et al., 2004). In general, we can say that resilient communities are promoted through integrating features of social organisation such as trust, norms and networks. These cultural contexts and local knowledge tend to be overlooked in many policy interventions that focus simply on economic efficiency of sustainable use of natural resources. The emphasis in this section on communities and social interaction may, at first glance, appear difficult to reconcile with the systems-based analysis of resilience in the ecological literature (Nelson et al., 2007). But they are indeed compatible. A systems approach to communities does not simply focus on the economic relations between agents, but is fundamentally concerned with factors such as inclusivity, degrees of trust and the mental models that individuals hold of the world and the decisions they face.

Vulnerability and Resilience Across Scales

These examples about the nature of resilience in particular places also show the multiple scales of analysis required to understand resilience and vulnerability. Often external forces, such as international development assistance, risks of climate change, or the vagaries of world commodity markets, are as important as local-scale responses to change. Vulnerability and resilience are not static phenomena: they can be accelerated and amplified by processes of global, as well as local, change. The integration of the world economy, for example, not only creates new challenges and opportunities; it also exacerbates trends in vulnerability and contributes to the production and mitigation of vulnerability in distant places. In one sense, economic integration and liberalisation have contributed to reduced poverty levels for many millions of people in the past 30 years, particularly in Asia. As we highlighted above, however, markets are not a panacea for environmental sustainability: the development of new markets for ecosystem services challenges existing property rights and institutions for forests and other resources. Trade liberalisation, while creating opportunities for economic growth in some parts of the world, drives ecological resource exploitation in others.

There are, according to Adger et al. (2009), potentially three mechanisms of interdependence linking vulnerabilities and resilience of socio-environmental systems around the world. First, there are the linked physical, biological and social processes that constitute global environmental change. Due to the accelerating and increasingly global nature of environmental change processes, the impacts of environmental change in one locality are connected to regional and global systems through human action and response. Some environmental changes involve changes to global systems such as the carbon and nitrogen cycles involving oceans, atmospheres and land. Other issues become global concerns, according to Turner et al. (1990), due to the local effects of trends observed everywhere on the planet, such as local water scarcity, local habitat fragmentation or degradation or local air pollutants. Of course these physical and biological processes are themselves interrelated at various scales, many with crucial thresholds (Scheffer et al., 2001; Steffen et al., 2004). Hence, global environmental change is a collection of processes that are manifest in localities, but with causes and consequences at multiple spatial, temporal and socio-political scales.

Second, economic market linkages are not only tied up with global environmental change, but can also themselves be a driver of interdependent vulnerabilities. The processes of global environmental change are indeed amplified by the social, political and economic trends of globalisation. Economic policies such as trade liberalisation and the integration of economies into world markets can make the incomes of the poor insecure, open to vagaries and price fluctuations, and ultimately more vulnerable when other shocks and stresses come along. Such places are 'doubly exposed' to social and environmental change (Leichenko and O'Brien, 2008). In India, for example, both climate change and market liberalisation for agricultural commodities are changing the context for agricultural production. Some farmers may be able to adapt to these changing conditions, including discrete events such as drought and rapid changes in commodity prices, while other farmers may experience predominately negative outcomes. O'Brien and colleagues (2004) argue that a combination of biophysical, socio-economic and technological conditions influence the resilience of places and populations. These factors range from groundwater availability to literacy, gender equity and the distribution of the proceeds of farming to landowners and waged labour. Together, these factors suggest which districts are most and least able to adapt to drier conditions and variability in the Indian monsoons and to import competition resulting from liberalised agricultural trade. Inland areas exposed to high-potential temperature increases and water stress and where there is an increasing dependency on internationally traded agricultural commodities are relatively more vulnerable than those where diversity of agricultural production is higher. The reduction of landscape scale diversity in crop variety in India also reduces the resilience of rural communities.

A further important trend is the observed widening disparity in income and access to resources in many regions of the world including China and the former Soviet republics. The reasons why inequality is important in terms of environmental degradation and management have been examined by Boyce (2002) who demonstrates theoretically that in resource-allocation decisions, the unequal power relationships that are inherent in unequal distributions of wealth lead to undesirable outcomes. If it is, in general, the powerful who gain most from environmentally damaging

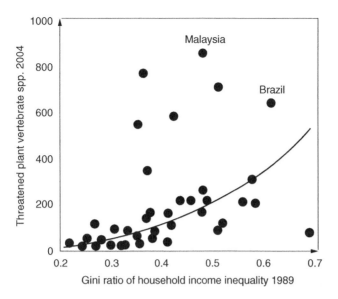

Figure 8.1 Relationships between the Gini ratio of income inequality and early indicators of biodiversity loss. (Source: Mikkelson et al., 2007)

activities, then the bargained solution between these winners and the less well-off losers (sufferers of the impacts of the environmentally damaging activity) will be skewed towards the benefits of the powerful.

The direct consequences of inequality are difficult to discern, but global patterns of inequality and ecological vulnerabilities are striking. Figure 8.1 reports the analysis of Mikkelson et al. (2007) showing that societies with more unequal distributions of income experience greater loss of biodiversity. Figure 8.1 shows the country-level Gini coefficient of household income inequality (a standard measure of inequality) in 1989 has a significant power relationship with the number of threatened plant and vertebrate species in 2004. A 1 percent increase in the Gini ratio for the data in figure 8.1 is associated with an almost 2 percent rise in the number of threatened species. Vulnerabilities are transmitted through the mechanisms of skewed land ownership and lack of accountability. Countries such as Brazil and Malaysia are prominent in figure 8.1 because where land ownership is also highly skewed, there are high rates of ecological threat (Mikkelson et al., 2007). Similarly, recent research on corruption and environmental degradation show similar patterns of loss (Smith et al., 2003). Many countries have experienced increases in inequality in the past two decades, despite contested evidence of overall convergence of world income levels.

The third mechanism of interdependence of social-ecological systems across space and time is the closer connection between places in the world through movements of people and resources around the world. This mechanism has several consequences, both positive and negative in terms of vulnerability. Demographic changes and migration flows produce new forms of sensitivity to risk, while providing some populations with new opportunities or access to resources that enable them to mitigate vulnerability. Population movements in Asia, for example, from lowland to uplands in Vietnam, and rapid urbanisation in China, Thailand and Malaysia over the past decades bring new opportunities and challenges for environmental sustainability. Increasing proportions of very old or very young people in a population, for example, change the nature of susceptibility to emerging diseases and pathogens. Informal settlements of migrant populations are often the most vulnerable to hurricanes, landslides and earthquakes (Mitchell, 1999), and, because of the complexity of cultural integration, addressing these vulnerabilities can themselves produce challenging policy problems.

The actual movement of resources for energy, food and primary production has both direct and indirect consequences. The food eaten at dinner tables across the world, for example, has increasing environmental impact due to energy and fertilizer inputs, food miles travelled to the table and land use changes associated with new production because of the flows of commodities and materials. Agricultural and economic policies in one part of the world have direct consequences on producers in another part of the world, and the globalisation of consumer tastes is now driving commodity production and economic decisions in local places. The consequences of the movement of materials round the world also are increasingly apparent in bio-invasive species (Perrings et al., 2005), demand for land that leads to habitat conversion and over-exploitation of species, and even the emergence of new diseases.

The susceptibility of populations and ecosystems to changes that affect their resilience in particular places is not only comparable but are actually linked to vulnerabilities elsewhere. This is apparent in the realm of human health. Certain sections of all populations are more vulnerable to emerging diseases than others, but global interdependence connects these vulnerabilities in new and surprising ways. Over 30 infectious diseases new to medicine emerged between the mid-1970s and 2000 according to the World Health Organization (see Epstein, 2002). These include HIV/AIDS, Ebola fever, Lyme disease, a new strain of cholera and toxic *E. coli*. In addition, there has been a resurgence and redistribution on a global scale of well-known diseases such as malaria and dengue fever, both transmitted by mosquitoes.

The factors influencing the observed emergence of new diseases include urbanisation, increased human mobility, changing land use patterns and the decline of public health infrastructure in parts of the world (McMichael, 2001). The emergence in 2003 of SARS (a virus recognised in several animal species that has crossed into human populations) in South East Asia illustrates the mechanisms for teleconnections of nested vulnerabilities outlined above. First, the interdependence of 'globalised flows' in this case of people increases the global scope of human transmission of emerging diseases such as SARS. Second, the underlying environmental drivers are common to the rise of emerging diseases (Ebola fever, SARS and HIV), the global biodiversity crisis and significant global environmental change associated with land use.

Infectious diseases such as SARS are transmitted around the world through movements of people. In early 2003, the SARS virus was recorded in Guangdong Province in southern China. Within a month, it had spread to Vietnam, Hong Kong, Singapore and Canada with over 8,000 cases and almost 700 fatalities worldwide. The SARS case also highlights another aspect of nested vulnerabilities: the links between environmental changes and emerging diseases. The cases of SARS were traced back to individuals who handled animals sold live in food markets in Guangdong. The SARS virus jumped the species barrier to humans, probably from masked palm civet cats and possibly raccoon dogs. Bell et al. (2004) suggest that it is the trade in wild animals, wrecking havoc with local biodiversity in South East Asia, that causes the risk and vulnerability in the first place.

Vulnerabilities to SARS are therefore connected with other vulnerabilities through markets and demographic changes and through biological feedbacks and linkages. Wildlife trade networks spread not only the risk, but also cause localised biodiversity loss, as new species are exploited and others become scarce. In this way SARS illustrates the mechanisms that communicate human exposure to disease as well as the nested nature of global environmental change. Thus, the economic changes associated with increasing incomes and changing consumption patterns combine with land use and environmental change to create the conditions for populations to be vulnerable to emerging diseases (Adger et al., 2009). Globalisation of travel and economic linkages in this case spread vulnerability of susceptible populations across the globe and created a global public health crisis.

In summary, the resilience of social-ecological systems is challenged by several trends in the modern world including rising connectedness of places, declining diversity of function and even of species in natural and managed landscapes (Young et al., 2006). It is also challenged by the so-called spatial stretching of systems of governance to deal with ever more complex issues such as ocean acidification, fisheries exploitation and climate change.

Conclusions

This chapter has outlined the concepts of vulnerability and resilience, pointed to their origins in the social and natural sciences, and showed how they are influenced by geographical factors and observed at various scales. Vulnerability and resilience have evolved from different disciplines and research traditions. Vulnerability, from its beginnings, in geography, risk and hazards research, has had a strong focus on economic and political structures as causes of social vulnerability. Resilience, derived from ecological sciences, is based on complex systems studies with a focus on adaptive capacity and maintaining the ability to deal with future, uncertain change. A resilience framework provides a dynamic perspective on processes of change within social and natural systems and the effects of these processes at different spatial and temporal scales.

Observations of how societies cope with hazards and with underlying risks show that some elements of society are inherently vulnerable and others are inherently resilient. This chapter highlights two important geographical aspects to this story. First, the scale at which vulnerability and resilience are observed matters. Global interdependencies and movement of people, resources and capital mean that vulnerabilities to change in one place are often linked to unforeseen consequences elsewhere. Second, the elements of where people reside and what they are vulnerable to are intimately bound up with the places that are valuable to people.

Of course, resilience and vulnerability to environmental change are neither static nor passive states. People and biological organisms adapt to changing conditions in order to make themselves less vulnerable to unforeseen or uncontrollable perturbations or changes. Adaptation by people is categorically different to adaptation in biological systems in that it can involve significant foresight, and hence, people adapt in anticipation or in expectation of change.

From recent research into how adaptation takes place in general, it would appear that vulnerability can be reduced through adaptation. People and systems are not passive to the risks they face and adaptation is indeed the norm. If we look at risks associated with climate change, such as flood risk, property and livelihoods due to coastal loss of land, planned adaptation is initiated because the benefits generally outweigh the costs (Adger et al., 2007). But, as we have illustrated in this chapter, adaptation often does not occur because of the unevenness of adaptive capacity and the persistence of various barriers to action. For risks such as exposure of elderly people to increasing heatwaves and extreme heat, which caused more than 30,000 excess deaths in Europe in 2003, vulnerabilities persist despite clear knowledge of the risks and recognition of the cognitive and economic barriers to addressing them.

The key message of this chapter is that vulnerability and resilience are important characteristics of places, people and combined social-ecological systems. Vulnerabilities are usually defined in terms of perturbations and changes outside the control of localities, and hence, usually portrayed as a negative state and something to be avoided. Resilience, deriving from the ecological sciences, involves the ability to retain system function and essential character. In some ways, it is the flip side or antonym to vulnerability. These concepts are embedded in distinct research traditions, but they are converging over time towards a common agenda that recognises the place-specific nature of resilient communities, the range of scales that vulnerability and resilience can be assessed, and the need to understand the winners and losers from interventions and adaptations that seek to promote resilience and the capacity to adapt.

BIBLIOGRAPHY

Adger, W. N. (2000) Social and ecological resilience: are they related? *Progress in Human Geography*, 24, 347–64.

Adger, W. N., Agrawala, S., Mirza, M. et al. (2007) Assessment of adaptation practices, options, constraints and capacity. In M. L. Parry, O. F. Canziani, J. P. Palutikof, C. E. Hanson and P. J. van der Linden (eds), *Climate Change 2007: Impacts, Adaptation and Vulnerability. Contribution of Working Group II to the Fourth Assessment Report of the Intergovernmental Panel on Climate Change.* Cambridge: Cambridge University Press, pp. 719–43.

Adger, W. N., Eakin, H. and Winkels, A. (2009) Nested and teleconnected vulnerabilities to environmental change. *Frontiers in Ecology and the Environment* 7, DOI: 10.1890/07014.

Anderies J. M., Janssen, M. A. and Ostrom E. (2004) A framework to analyze the robustness of social-ecological systems from an institutional perspective. *Ecology and Society*, 9, 18, http://www.ecologyandsociety.org/vol9/iss1/art18 (accessed 02-09-08).

Arrow, K., Bolin, B., Costanza, R. et al. (1995) Economic growth, carrying capacity and the environment. *Science*, 268, 520–21.

Bell, D., Robertson, S. and Hunter, P. R. (2004) Animal origins of SARS coronavirus: possible links with the international trade in small carnivores. *Philosophical Transactions of the Royal Society B*, 359, 1107–114.

Berkes, F., Colding, J. and Folke, C. (eds) (2003) *Navigating Social-ecological Systems: Building Resilience for Complexity and Change.* Cambridge: Cambridge University Press.

Blaikie, P., Cannon, T., Davis, I. and Wisner, B. (1994) *At Risk: Natural Hazards, People's Vulnerability and Disasters.* London: Routledge.

Brown, K., Few, R., Tompkins, E., Tsimplis, M. and Sortti, T. (2005) *Responding to Climate Change: Inclusive and Integrated Coastal Analysis*. Technical Report 24, Tyndall Centre Final Report, Tyndall centre for Climate Change Research, University of East Anglia, Norwich.

Brown, K. and Lapuyade, S. (2001) A livelihood from the forest: gendered visions of social, economic and environmental change in southern Cameroon. *Journal of International Development*, 13, 1131–49.

Boyce, J. K. (2002) *The Political Economy of the Environment*. Cheltenham: Edward Elgar.

Burton, I., Kates, R. W. and White, G. F. (1993) *The Environment as Hazard*, 2nd edn. New York: Guilford Press.

Carpenter, S., Walker, B., Anderies, J. M. and Abel, N. (2001) From metaphor to measurement: resilience of what to what? *Ecosystems*, 4, 765–81.

Corbera, E., Brown, K. and Adger, W. N. (2007) The equity and legitimacy of markets for ecosystem services. *Development and Change*, 38, 587–613.

Cutter, S. L., Boruff, B. J. and Shirley, W. L. (2003) Social vulnerability to environmental hazards. *Social Science Quarterly*, 84, 242–61.

Eakin, H. and Luers, A. L. (2006) Assessing the vulnerability of social-environmental systems. *Annual Review of Environment and Resources*, 31, 365–94

Epstein, P. R. (2002) Detecting the infectious disease consequences of climate change and extreme weather events. In P. Martens and A. J. McMichael (eds), *Environmental Change, Climate and Health*. Cambridge: Cambridge University Press, pp. 172–96.

Folke, C. (2006) Resilience: the emergence of a perspective for social-ecological systems analyses. *Global Environmental Change*, 16, 253–67.

Folke, C., Carpenter, S., Walker, B. et al. (2005) Regime shifts, resilience and biodiversity in ecosystem management. *Annual Review of Ecology Evolution and Systematics*, 35, 557–81.

Fordham, M. (2003) Gender, disaster and development: the necessity for integration. In M. Pelling (ed.), *Natural Disasters and Development in a Globalizing World*. London: Routledge, pp. 57–74.

Gunderson, L. H. and Holling, C. S. (eds) (2002) *Panarchy: Understanding Transformations in Human and Natural Systems*. Washington, DC: Island Press.

Hewitt, K. (ed.) (1983) *Interpretations of Calamity from the Viewpoint of Human Ecology*. Boston: Allen and Unwin.

Lebel, L., Anderies, J. M., Campbell, B. et al. (2006) Governance and the capacity to manage resilience in regional social-ecological systems. *Ecology and Society*, 11(1), 19, http://www. ecologyandsociety.org/vol11/iss1/art19/ (accessed 02-09-08).

Leichenko, R. and O'Brien, K. L. (2008) *Environmental Change and Globalization: Double Exposures*. Oxford: Oxford University Press.

Liverman, D. M. (1990) Drought impacts in Mexico: climate, agriculture, technology, and land tenure in Sonora and Puebla. *Annals of the Association of American Geographers*, 80, 49–72.

Millennium Ecosystem Assessment (2005) *Ecosystems and Human Well-being: Synthesis*. Washington, DC: Island Press.

McMichael, A. J. (2001) *Human Frontiers, Environments and Disease: Past Patterns, Uncertain Futures*. Cambridge: Cambridge University Press.

Mikkelson, G. M., Gonzalez, A. and Peterson, G. D. (2007) Economic inequality predicts biodiversity loss. *PLoS ONE* 2(5), e444. doi:10.1371/journal.pone.0000444.

Mitchell, J. K. (ed.) (1999) *Crucibles of Hazard: Mega-Cities and Disasters in Transition*. Tokyo: UNU Press.

Nelson, D. R., Adger, W. N. and Brown, K. (2007) Adaptation to environmental change: contributions of a resilience framework. *Annual Review of Environment and Resources*, 32, 395–419.

O'Brien, K. L., Leichenko, R., Kelkarc, U., et al. (2004) Mapping vulnerability to multiple stressors: Climate change and globalization in India. *Global Environmental Change*, 14, 303–13.

Pagiola, S., Landell-Mills, N. and Bishop, J. (eds) (2002) *Selling Forest Environmental Services*. London: Earthscan.

Pelling, M. (2003) *The Vulnerability of Cities: Natural Disasters and Social Resilience*. London: Earthscan.

Perrings, C., Dehnen-Schmutz, K., Touza, J. and Williamson, M. (2005) How to manage biological invasions under globalization. *Trends in Ecology and Evolution*, 20, 212–15.

Satterfield, T. A., Mertz, C. K. and Slovic, P. (2004) Discrimination, vulnerability, and justice in the face of risk. *Risk Analysis*, 24, 115–29.

Scheffer, M., Carpenter, S., Foley, J. A., Folke, C. and Walker, B. (2001) Catastrophic shifts in ecosystems. *Nature*, 413, 591–96.

Smith, R. J., Muir, R. D. J., Walpole, M. J., Balmford, A. and Leader-Williams, N. (2003) Governance and the loss of biodiversity. *Nature*, 426, 67–70.

Steffen, W., Sanderson, A., Tyson, P. D. et al. (eds) (2004) *Global Change and the Earth System: A Planet under Pressure*. Berlin: Springer.

Turner, B. L. II, Kasperson, R. E., Meyer, W. B. et al. (1990) Two types of global environmental change: definitional and spatial-scale issues in their human dimensions. *Global Environmental Change*, 1, 14–22.

Turner, B. L. II, Kasperson, R. E., Matson, P. A. et al. (2003) A framework for vulnerability analysis in sustainability science. *Proceedings of the National Academy of Sciences US*, 100, 8074–79.

Walker, B., Holling, C. S., Carpenter, S. and Kinzig, A. (2004) Resilience, adaptability and transformability in social-ecological systems. *Ecology and Society*, 9(2), 5. http://www.ecologyandsociety.org/vol9/iss2/art5 (accessed 02-09-08).

Watts, M. J. and Bohle, H. G. (1993) The space of vulnerability: the causal structure of hunger and famine. *Progress in Human Geography*, 17, 43–67.

Young, O. R., Berkhout, F., Gallopin, G. C., Janssen, M. A., Ostrom, E. and van der Leeuw, S. (2006) The globalization of socio-ecological systems: an agenda for scientific research. *Global Environmental Change*, 16, 304–16.

Chapter 9

Commodification

Scott Prudham

Introduction

The nexus of commodification with environmental change and environmental politics is of immense and growing interest to geographers and activists alike. There are several good reasons for this. First, the global criss-cross of commodities via far-flung networks of production, investment, coordination, distribution, and exchange leaves behind traces of myriad kinds with important and intertwined social and environmental implications. This includes by-products such as persistent organic pollutants, gaseous emissions from combustion and other chemical processes, and an assortment of organic and inorganic wastes. It also includes ecosystems transformed by and for production, for example, forests converted to plantations for fibre or other products, and land devoted to agricultural production. Even the city itself, emerging from dense intersecting networks of commodity production and exchange, is sustained in part by complex metabolic transformations of biophysical nature in the production of urban spaces (Cronon, 1991; Gandy, 2002; 2005; Swyngedouw and Heynen, 2003).

Second, direct forms of the commodification of what we understand as nature (both non-human *and* human, it must be said) seem to have proliferated in recent years. This includes new or reinvigorated commercialisation of discrete resources from water to fish to seeds to genes (see, e.g., Bakker, 2003; McAfee, 2003; Mansfield, 2004a; McCarthy, 2004; Swyngedouw, 2005), propelled in substantial measure by private firms seeking new avenues for the circulation of capital in and through discrete biophysical processes (Kloppenburg, 2004). Yet, it bears noting, no small amount of the impetus for this recent acceleration in nature's commodification comes from explicit policy prescriptions advocating privatisation and market exchange as means to better conserve and rationally manage natural resources and the environment (McAfee, 1999; Liverman, 2004). A proliferation of so-called 'market-based' mechanisms in environmental governance has deepened the commodification of particular biophysical processes and entities under the influence of a broad 'neoliberalisation' of nature (McCarthy and Prudham, 2004; Heynen et al.,

2007), including the emergence of carbon offset markets as well as biodiversity conservation programmes and wetland banking systems (see, e.g., respectively Mac-Donald, 2005; Robertson, 2006).

Finally, it is not only in the strictly material sense that nature is increasingly commodified. Rather, what we come to know as nature seems ever more tied to commodity circuits. From representations of pristine and wild spaces circulated to sell travel and adventure tourism, to the invocation of pastoral mythologies in the sale of everything from cheese to wine, and even to scientific representations that help render biophysical entities alienable and commensurable (Bridge and Wood, 2005; Robertson, 2006), 'nature' in the semiotic sense of the term is also subject to processes of commodification.

In this context, a growing and diverse range of scholarship and activism has tackled in various ways the commodification of nature, the nature of commodification, and the social and environmental implications of commodification. Though I cannot do justice to this full range, I would argue most of this literature is animated by various forms of three key questions: (i) What does commodification entail, in general terms and specifically with respect to nature? (ii) How exactly are discrete elements of nature (non-human *and* human, material and symbolic) made to circulate in the commodity-form? (iii) What are the interlinked social and environmental implications of commodifying nature, and of commodification more generally?

Definitions

Despite the ubiquity of commodities and a rich and growing literature on commodities and commodification, there are in fact longstanding, enduring and important differences in the ways that these terms are conceptualised and deployed. For instance, some have invoked more generic notions of commodity as *anything* that is exchanged or is exchangeable (e.g., Appadurai, 1986). This expansive sense of the term implicitly recognises the diverse historical, geographical, and cultural circumstances under which peoples have met their needs and desires by means of exchange. It also suggests (again, somewhat implicitly) that things become commodities through exchange; thus, 'commodity' or commodity-form is an acquired trait (Castree, 2001) representing but one phase in the 'complex social life of things'.

Yet, reference exclusively and simply to exchange as the defining feature of a commodity misses some potentially important distinctions, particularly in a contemporary world of seemingly rampant commodification (Sayer, 2003). For some, then, a crucial role in increasingly far-flung contemporary commodity circuits is played by money, not least in providing a common metric of value and thus allowing production and exchange to be separated by great gulfs of time and space. Castree, for instance, defines commodification as '. . . a process where qualitatively distinct things are rendered equivalent and *saleable through the medium of money*' (Castree, 2003, p. 278, emphasis added). Similarly, Ben Page (2005) states that '. . . a commodity is an object that is bought and sold *with money*' and that commodification is '. . . the process during which a thing that previously circulated outside monetary exchange is brought into the nexus of a market . . .' (p. 295, emphasis added). And Peter Jackson (1999, p. 96) argues that 'commodification' refers '. . . literally, to the extension of the commodity form to goods and services that were not previously commodified'. He goes on to point to the 19th century as a period of exploding commodification (first and most particularly evident in Britain) as the commodity-form became the dominant vehicle by which economic value was expressed (and made to travel), and the predominant way in which human needs and wants (i.e., use values) were secured.

An important connotation of Jackson's notion of commodification is that it points to interlinked transformations, including in the realms of *both* production and exchange. That is, as consumption or demand is increasingly met via exchange, production becomes increasingly oriented towards exchange. This is perhaps why economic historian Karl Polanyi, in considering the significance of what he viewed as modern, market-centred economies, defined commodities '. . . as objects produced for sale on the market' (Polanyi, 1944, p. 75). This is a simple yet subtle statement that does two things. First it links the dynamics of production and consumption in commodification, seemingly important (as we will see) if we want to know not only how and why commodities are exchanged, but also something of where they come from, and how they travel through various stages from inputs of raw materials and labour, through transportation, storage and distribution, and ultimately to markets (and waste disposal!). This is important to note not least because overly singular focus on the realms of either production or exchange has been a consistent source of tension in the commodification of nature literature, and in the literature on commodities more generally.[1] Second, however, Polanyi's deceptively simple framing implicates a shift towards economic production increasingly motivated *by or for exchange*. This shift has profound implications. The significance of production motivated increasingly by exchange has long been noted, including in the writing of Aristotle, in the work of numerous classical political economists (including Adam Smith and Karl Marx), and of course by Polanyi. This lineage of thought views with suspicion economic production driven primarily or even exclusively by the pursuit of profit and money as ends in and of themselves (rather than, for instance, commodity exchange purely as an outlet for surplus production), and this is a concern evident in more popular and pejorative invocations of the term 'commodification' (see, e.g., Booth, 1994; Sayer, 2003).[2] Whether one shares this normative concern or not, historically, the notion of commodification '. . . as a change from producing what previously or otherwise might have been simply use values to producing things for their exchange value' (Sayer, 2003, p. 343) points to a sociological transformation particularly apparent in and an important feature of capitalist political economies.

Synthesising these observations, and recognising the need to consider what might be distinct about the complex socio-spatial and institutional networks of contemporary commodity circuits in an increasingly integrated global economy, we might usefully define commodification as interlinked processes whereby: production for use is systematically displaced by production for exchange; social consumption and reproduction increasingly relies on purchased commodities; new classes of goods and services are made available in the commodity-form[3]; and money plays an increasing role in mediating exchange as a common currency of value. And given this, it might be useful to consider two distinct moments in commodification. The first of these is the development of relations of exchange spanning across greater distances of space and time (market expansion) or *stretching*. The second is the systemic provisioning of more and more types of things (goods and services) in the commodity-form, or *deepening* (see figure 9.1).[4]

Note here in particular that an emphasis on commodification suggests dynamism, change, and process, pointing to transformations always more (or less) in a state of

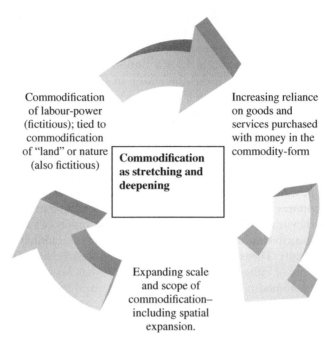

Figure 9.1 Commodification as integrated processes of stretching and deepening, including the increasing commodification of biophysical nature (i.e., the circulation of discrete socio-natures in the commodity-form).

flux and transition (Castree, 2001). Despite evident tendencies, there is a diversity of ways in which discrete goods and services come to be produced, circulated and exchanged in the commodity-form, shaped in part by the material and discursive character of what is being commodified, as well as the geographical and historical context in which these processes occur. In no way does any of this imply that there is a single path to commodity status (this is a particularly important theme in the commodification of nature literature). Moreover, and as I return to below, the process-oriented valence of commodification suggests the possibility of reversal, and thus of (de)commodification (Page, 2005; Sayer, 2003).

Capitalism and Commodification

No one has proposed – not even Karl Marx – that commodities and processes of commodification are in and of themselves unique features of capitalist political economies. Nor is it true that all of the commodities circulating in our (more than) capitalist world are produced and exchanged under the auspices of the private sector, profit driven economy. States, for instance, clearly produce commodities (given the definitions above), not least via state-owned companies, utilities, etc. (e.g., electricity, water, public transportation services). One can even trace complex histories of energy and water service delivery which ebb and flow between state and private provisioning, and yet which remain commodified in important respects throughout (Bakker, 2005; Page, 2005). And it is quite clear that the historical origins of far flung commodity regimes – e.g. the sugar trade (Mintz, 1985) – are just as much tied to the emergence of capitalism as they are products of this emergence (Wolf, 1982).

Even still, conceptualising commodification serves as an invitation to consider what differences, if any, characterise the development of a system of generalised commodity production and circulation in a capitalist political economy. Many scholars *have* chosen to make this distinction, though not all for the same reason. Marx, for instance, while recognising that commodities predate capitalism, also theorised commodification under capitalism as a switch from the mercantilist sale of commodities to secure money to buy commodities (represented in the abstract by C-M-C) towards the outlay of money as capital to produce commodities in order to sell for more money-capital (M-C-M'). Marx argues that this represents an important transition towards a more generalised system of commodity production and exchange, one whose culmination is in many ways signified by the commodification of labour, or what he called labour-power.

Why mark this transition and the emergence of ostensibly commodified labour, particularly if our primary interest is in environmental geographies of commodification? At one level, the commodification of labour-power, that is, the development of markets in labour and the emergence of large numbers of people (indeed the majority in capitalist societies) who work for wages in order to secure their own social reproduction (as well as to satisfy all manner of aspirations necessary and otherwise) is pivotal to the *deepening* of commodification mentioned above. This is because the availability of people to work in a wider and wider range of commodity producing sectors is tied in turn to the economic demand created by these same people who buy what they need (and want) to live. From this perspective, it is hard to imagine the generalised character of commodification, including the commodification of nature in various respects, without considering the character of wage labour and the labourers themselves who comprise a primary, though by no means sole market for commodities. Food provides an excellent example, since it is only the existence of large numbers of people who cannot or do not produce their own food that allows food to be produced primarily in the commodity-form. Moreover, as numerous scholars in the agrarian and food literatures have observed, the shifting dynamics of labour markets over time (e.g., the entry of large numbers of women into the labour force in industrialised countries since about the middle of the 20th century) are tied directly to the commodification of food (e.g., the increasing sale of pre-cooked and pre-prepared meals) (Guthman, 2002).[5] This is in one sense a specific example of a more fundamental connection between the commodification of labour-power through the emergence of wage labour, and the commodification of land in so much as the latter entails separation of labour from 'land' broadly understood (Polanyi, 1944; Marx, 1977). However, these should not be understood as stages in the prehistory of capitalism but rather as systemic tendencies that continue to be manifest in a variety of guises (Kloppenburg, 2004; Glassman, 2006).

A second reason to mark the commodification of labour-power and the historically and sociologically distinct character of M-C-M' – again particularly emphasised by Marx and many Marxist scholars – is that it is integral to an account of the uniquely dynamic and growth oriented character of capitalist production and capital accumulation on an ever-expanding scale. The extraction and reinvestment of surplus (signified by a positive difference between M' and M) fuels a restless drive to reproduce and expand the scale and scope of commodification via stretching and deepening in order (i) to provide outlets for the productive capacity of this expanded capital; and (ii) to renew conditions of profitability eroded by capitalist competition. Commodification under capitalism thus entails the proliferation of circuits (including biophysical ones) through which this capital as value-in-motion may flow. This in part propels the restless, growth driven logic of capitalist political economies, with important geographical implications, including a tendency to expand and rework the space economy (Harvey, 1982; 1985), and with it, to make and remake, transform and 'produce' nature (Smith, 1984; 1996). These tendencies are manifest in demands for greater and greater amounts but also more and more different kinds of raw material inputs while at the same time generating waste products (typically) on an expanding scale and in frequently novel forms.[6] All of this gives capitalism its own specific form of socio-natural *metabolism* (Foster, 2000), distinct ways in which biophysical nature is appropriated, made and remade.

Commodification and/of Nature

All that said, considerable recent scholarship in geography and related fields has examined the commodification of specific natures as a sort of collective 'special case' based in part on the 'difference' that biophysical processes make in shaping and conditioning trajectories of commodification (e.g., Bridge, 2000; Sayre, 2002; Bakker, 2003; Prudham, 2005). At a basic level, the idea here is that the commodification of any particular 'nature' relies on ecological production processes whose subordination to the realm of market-coordination can only ever be partial. One might say further that this includes both non-human and human nature, in as much as the reproduction of labour-power by market coordination alone is a project in the commodification of human nature (as bodies, as identities, etc.) and is, similarly, a dubious if not impossible project.

These seemingly basic observations underpin Polanyi's (1944) argument that labour and nature can only ever be *fictitious* commodities. According to Polanyi, nature and labour are special categories of commodity in that they are not literally produced exclusively or even primarily for sale. For instance, if we consider non-human, biophysical nature, ecological functions of myriad kinds remain clearly important in the provision of all manner of environmental inputs and services, and these are only incompletely coordinated by social decision making, including market coordination (see discussion and elaboration in Prudham, 2005). Recognising this basic fictitiousness points to all manner of problems with calls to privatise nature and to extend markets in order to meet environmental objectives. If nature is only a fictitious commodity, then market coordination in the allocation of environmental goods and services can only ever be partial. And this is so not only because of what we might call strictly 'objective' constraints (i.e., that formally economic production relies on all manner of formally non- or extra-economic production whose complete subordination to the market is simply not possible) but also because of subjective concerns having to do with social struggle over the allocation of biophysical nature (i.e., that quite apart from the physical impossibility of subordinating biophysical processes wholly to the price mechanism, 'society' in the broadest sense will never accept this politically) (O'Connor, 1998). The creation of markets in water, for instance, can give rise to or reinforce the separation of large numbers of people from reliable access to water (Smith, 2004). This in turn can violate commonly held sensibilities concerning rights to water which are perceived to trump commercial, market driven allocation, making the commodification of water a political flashpoint (Page, 2005; Bakker, 2007). This is an example, however, of ways in which 'the economy' is socially embedded via notions of a moral economic order that governs the social allocation of nature as a set of entitlements (Thompson, 1971; Scott, 1976; Booth, 1994). These kinds of arguments present rather fundamental difficulties for the utopian ideal of markets as a sole means of allocating goods and services (Polanyi, 1944).

The lineages of these observations are broadly evident in a recent literature which examines the contradictory and highly specific ways in which non-human nature is made to circulate in the commodity-form. Considerable scholarship has explored the various ways in which highly specific, lively and unruly, material and contested 'natures', including water (Bakker, 2003; Swyngedouw, 2005); fish (McEvoy, 1986; Mansfield, 2003); trees (Prudham, 2003; 2005); wetlands (Robertson, 2006); fossil fuels and minerals (Bridge, 2000; Bridge and Wood, 2005); genes (McAfee, 2003); organic foods (Guthman, 2002; 2004), etc. are extracted, cultivated, refined, processed, represented and made to circulate in the commodity-form, and with all manner of political and ecological implications. A common thread in the literature, echoing Polanyi, is that there is nothing 'natural' about nature's commodification. Rather, considerable work is required on various fronts to circulate nature in the commodity-form.

For instance, one key theme in recent literature concerns the ways in which commodification actually turns on the apparent dissolution of important qualitative differences in the rendering of distinct things equivalent or commensurable. Castree (2003) refers to this as *abstraction*, a process by which systematic representations dissolve the specificity of things (any specific things) in favour of their aggregation into classes of things. A good deal of work along these lines has been inspired by William Cronon's (1991) book *Nature's Metropolis*, and in particular, a chapter on wheat called 'A Sack's Journey'. Cronon traces a series of technological and organisational innovations underpinning the emergence of Chicago as the premier market for wheat in the United States during the 19th century. He examines in particular how the convention of transporting wheat in sacks from individual farms gave way to aggregation, allowing more efficient transport in rail cars, mass storage in grain elevators, and highly fluid forms of exchange including sophisticated futures markets. For Cronon, a key and socially mediated development was the conversion of continuous differentiation in wheat quality into discrete categories or grades of wheat that sold at different prices corresponding to standardised grades. These grades helped dissolve the specificity of wheat and the farms from which it had been shipped in individual, identifiable sacks. Perhaps the chapter's most compelling line of argument is that the expansion of Chicago's wheat market, with all this entailed, could not have occurred had the abstraction of wheat not allowed for it to be aggregated in ways that replaced the sack but still made wheat 'legible' to buyers and traders.

This and work along similar lines suggests that acts of representation and in fact what might be called *social relations of abstraction* are necessary in order for discrete things to be rendered commensurable and exchangeable, particularly where money is involved. A curious feature of abstraction, however, is that difference is both dissolved (as kernels with different characteristics are lumped into the same grade) but also renegotiated and reproduced in legible forms, e.g., as discrete grades of wheat. Without this, the complex circuits of material and symbolic exchange in

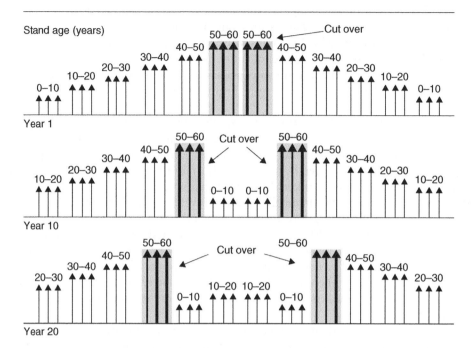

Figure 9.2 The Normalbaum (literally normal tree or forest) idealized as a set of discrete, even-aged forest stands in various stages of regrowth after harvesting. This is the abstracted ideal of 20th century, scientific, sustained yield forestry and has been critical to making forests legible, which in turn enables their rational conversion to wood based commodities. Such scientific representations, whatever else they may accomplish, facilitate the abstraction of timber and indeed whole forests from specific social and ecological contexts, making them commensurate across space and time and thereby enabling exchange and commodification to proceed. Reprinted with permission from Demeritt (2001).

wheat or any other large-scale market could simply not occur. Similar processes and arguments could be inferred from the commodification of many biophysical inputs, from grades of logs and lumber to oil (typically indexed by price and quality to regional variants, e.g., Saudi crude). Moreover, the abstractions that underpin far-flung exchanges tie the commodification of nature to systems of representation more broadly – including weights and measures but also natural science – as regimes of calculation and expertise that more generally make nature and territory 'legible' and governable (Scott, 1998; Mitchell, 2002).

David Demeritt (2001), for instance, examines the development of key techniques for representing forest resources in the context of 20th century American scientific forest management, including via the uptake of the concept of the *Normalbaum* or 'normal forest' from the European tradition of scientific forestry (see figure 9.2). As Demeritt argues (drawing on the conceptual work of Timothy Mitchell and Michel Foucault), these representations allowed the liquidation but also conservation of forest resources in America to become (or at least appear to be) calculable and coherent socio-ecological projects; they thus underpinned the emergence of state-centred forest management as a form of governmentality (literally the conduct of conduct). Equally, however, these representations facilitated the abstraction of timber and indeed whole forests from specific social and ecological context, making them commensurate across space and time and thereby enabling exchange and com- modification to proceed. While these processes of 'statistical picturing' are hardly innocent of power relations (Prudham, 2007b), they also have material effects beyond consolidating managerial expertise and commodification processes. Rather, and as Demeritt also observes, abstraction away from the specificity of forests is also complicit in the production of ecologically simplified forests in the image of the abstraction, while also tending to downplay social contestation of access to and control of forests as social spaces (see Robbins, 2001; Braun, 2002).

Emphasis on the systematic representations that underpin abstraction highlights complex cultural and political processes by which nature as a set of *sign-values* is made to circulate in or attendant with the commodity-form, which in turn is pro- ductive of prevailing conceptions of nature itself on an increasingly widespread if not global scale (Smith, 1984; Braun, 2006). Morgan Robertson (2000) has inter- rogated some of this sort of representational 'work' as it has applied to the circula- tion of wetlands as exchangeable commodities under the US wetland banking system, with a focus on the articulation of environmental science and the commodi- fication of nature. Since the early 1990s, development in wetlands has required a permit from the US Army Corps of Engineers, often granted on the condition of offsets or mitigation. This has propelled the development of systems of commensu- rability in wetland services. Entrepreneurs began building, restoring, or saving wetlands and applying for certification from the Corps in order to then sell the wetland 'credits' to would-be developers. 'Thus was born wetland mitigation banking: the first successful market in ecosystem services defined as such, rather than (as in the case of air- pollution credits) defined in conventional units of weight or volume. Though still a small industry it is experiencing geometrical growth in membership, and has captured the imagination of those who promote market-led environmental policy' (Robertson, 2006, p. 372). Robertson (2006) pays particular attention to the role of scientists in certifying wetlands, and in monitoring the status of wetlands in the programme. Teams of ecologists are enrolled to make scientific judgements about commensurability using what are called Rapid Assessment Meth- odologies (RAMs!). As Robertson writes, 'RAMs function as instruments of *transla- tion between science, policy, and economics. . . .* Early in the development of wetland banking it was recognized that the commodity to be traded must be defined in a way that maintained a consistent identity across space and time. . . . This task must be accomplished before any market can function, not just markets in ecosystem services' (Robertson 2006, p. 373, emphasis added).

All this in mind, it is important to remember that abstraction is not sufficient for commodification to occur, nor is exchange the only nor perhaps even most salient feature of commodification. Consider, for instance, Cronon's narrative about wheat. While he dwells on the construction of new categories of wheat's representation and the concomitant expansion of the Chicago wheat exchange, there is no discussion of processes of farm consolidation, changing agronomic practices, proletarianisa- tion, and rural to urban migration in the context of a rapidly expanding wheat market. Instead, one might well argue that as powerful as Cronon's insights remain, he ends up re-inscribing what Marx called the 'fetish of the commodity' by focusing narrowly on commodities as exchange-values unto themselves (see the next section on fetishism). In a useful review and synthesis, Castree (2003) argues that there are in fact six distinct but inter-related moments in the commodification of nature, including not only abstraction, but also privatisation, alienability, individuation, valuation, and displacement.[7] Picking up on some of these points, Bakker (2005) argues for careful distinctions between privatisation, commercialisation and commodification. These are useful insights provided that they not be seen wholly as separate categories of social action. Privatisation schemes, for instance, are frequently as integral to commodification and the development of far-flung exchange as are processes of representation and abstraction, and these schemes are often sites of contradictory imperatives and intense contestation and social struggle (Mansfield, 2004b; 2007). Moreover, privatisation struggles are pivotal moments tied (directly or indirectly) to processes of accumulation by dispossession (Harvey, 2003; Glassman, 2006) and in this respect are not formally distinct but relational moments in the commodification of nature (Prudham, 2007a).

Commodity Fetishism, Labels and Alternative Commodity Circuits

One of the most commonly noted features of commodities in the contemporary world is that it is by no means obvious even to curious consumers where commodities originate and what kinds of social and environmental inputs went into their production and circulation. From a normative and ethical standpoint, this means that it is not obvious what kinds of activities are being supported and reproduced via the purchase of commodities. As David Harvey (1990, p. 423) put it '[t]he grapes that sit on supermarket shelves are mute; we cannot see the fingerprints of exploitation upon them or tell immediately what part of the world they are from'. Complex relations of transformation, circulation and exchange sever '. . . materially and symbolically the connection between producing exchange and use values . . . masking the qualitative social and environmental relations of production' (Kaika and Swyngedouw, 2000, p. 123). This phenomenon, and specifically, the tendency to reify commodities as things in and of themselves (with a concomitant tendency for commodities to take on values somewhat independent of their production and circulation) was termed the 'fetishism' of the commodity by Marx (1977, p. 165).[8]

This idea of the commodity fetish remains a quite powerful notion for scholars and activists interested in commodification processes. At a basic level, and despite different takes on the idea of fetishism per se, a desire to understand the complex trajectories and valences of commodities has animated a rich literature and social activism concerning the 'lives' of commodities, including their geographies, motivated in part by a sense that the spatio-temporal displacements of commodity provisioning – whether conceptualised in terms of chains, networks, or circuits – are becoming more complex in a globalising world (Winson, 1993; Gereffi and Korzeniewicz, 1994; Hartwick, 1998; Leslie and Reimer, 1999; Robbins, 1999). Much of this work seeks not only to document and understand, but also to transform relations of exploitation in realms of production (e.g., Harvey, 1990; Hartwick, 1998; Hartwick, 2000; Mutersbaugh, 2004). In this sense, commodity chain and commodity circuit analyses offer strong complementarities with life-cycle assessment methodologies developed in the physical and engineering sciences, seeking to document the full range of relations and practices that propel commodities, including ecological inputs and lifetime environmental impacts from production, circulation, and disposal.

All that said, the fetishism idea is not without its critics. One problem is that aggressive invitations to get 'past' or 'behind' the veil of the fetishism of commodities in order to unmask them – as for example explicitly advocated by Hartwick (2000) and Harvey (1990) – run the risk of assuming that the origins of commodities are unambiguous and also that the 'facts' of exploitation and ecological degradation can speak for themselves (Jackson and Holbrook, 1995; Jackson, 1999; 2002; Page 2005). However, highly complex trajectories and displacements of even single commodities in the contemporary international economy (see Dicken, 1998) suggest that 'origins' are multiple and not at all obvious. Indeed, the proliferation of production sites serving mass markets in seemingly generic commodities shows considerable geographical variation, so much so that the geography and politics of production cannot be read backward simply from commodities (Leslie and Reimer, 1999). Moreover, power, agency, and decision-making capabilities are often distributed in complex, dispersed and contradictory ways across networks linking commodity production, distribution and consumption (Marsden et al., 1996; Friedberg, 2004). In some ways, then, commodity chains and circuits do not have clear end points; they merely proliferate, requiring careful analytical and political choices in the conduct of commodity chain analyses and campaigns.

In addition, it is not always apparent what political and ethical commitments, judgements, and actions will or should attend the revealed origins of commodities. Indeed, despite commodity chain analyses that provide a '. . . critique of consumption founded on geographical detective work . . . highlighting the connection between producers and consumers' (Hartwick, 2000, p. 1178), it is not necessarily clear what changes in consumption or production practices ought to follow from this work. Instead, political action requires difficult choices to be made, including between contending forms of social liberation and exploitation among commodity producers, and sometimes between social and ecological dimensions of enhanced sustainability (Mutersbaugh, 2004). Is it socially just, for instance, to choose to reduce food miles by eating locally and truncating food trade if this means depriving distant peasants and farm-workers of their livelihoods in globally integrated food production and distribution circuits (Friedberg, 2004)?

On these and related issues, there is much to draw on from a wide ranging literature that has exploded in the last decade or so concerning the complex geographical and cultural character of commodities and commodity circuits/networks, sometimes referred to generally as the 'commodity cultures' or 'geographies of commodities' literature. This literature is not restricted to questions concerning the commodification of nature, and rather is more broadly concerned with the proliferation of the commodity-form, the complexity of commodity chains/networks, the articulation of culture and economy in and through commodities, and importantly, the complex cultural meanings of commodities and mass consumerism (for useful reviews and commentary, see Jackson, 1999; 2002; Bridge and Smith, 2003; Castree, 2004).

One of the points of contention in this literature is the use (misuse?) of the fetish idea. Some have argued that a focus on fetishism is essentially elitist and pedantic, placing all-knowing scholars (and presumably fair trade activists) above more or less duped consumers (see also Jackson, 2002). Notwithstanding that this is arguably a rather hollow caricature of the fetishism idea as originally formulated by Marx, it at least serves as a useful caution against elitist condemnations of everyday consumption practices. And it leads to the important point that consumers and a politics of (mass) consumption must not be dismissed or disregarded (Miller, 1998; Jackson, 1999). Research on commodity circuits (e.g., Le Heron and Hayward, 2002) and commodity cultures shows instead that consumption is a domain of struggle and contestation, and that forms of cultural learning and of both solidarity and emancipation can also emerge in and through a politics of consumption (Jackson and Holbrook, 1995; Jackson and Taylor, 1996; Jackson, 1999; Johns and Vural, 2000; Sayer, 2003). Through this lens, consumption becomes a site of tremendous political importance, including in forging the very links between otherwise disconnected people (e.g., via the transnationalisation of food and diet) that can easily be overlooked in the rush to get behind commodities and consumption (Cook and Crang, 1995). The commodity cultures literature draws attention to the imagined geographies that can and do circulate with commodities as powerful and productive sources of knowledge about the world (Domosh, 2006). Some of these may well be highly dubious and even manipulative (e.g. think of the utopian Valley of the Jolly Green Giant from whence your vegetables ostensibly emerge, or the smiling campesino Juan Valdez picking *your* perfect coffee bean). And social learning and liberation achieved via the consumption of capitalist commodities will always be fraught (Jackson, 2002).[9] But these imagined geographies are in and of themselves important cultural facets of commodification, and cannot be ignored even if and when they tend to promote homogenous, flatter worlds of 'McDonaldisation'.

All of this only further reinforces that commodification always entails interwoven material and semiotic processes (Robertson, 2000). In fact, debates about the cultures of commodities and the implications of fetishism and commodity displacements highlights an important but sometimes overlooked aspect of commodity fetishism. Increasing displacement from points of social and ecological production together with the sheer proliferation of the commodity-form attendant with commodification implies that the 'meaning' ascribed to commodities becomes potentially more malleable. That is, the very reification of commodities becomes a powerful and productive facet of commodification itself. This is consistent with Marx's provocative description of the proliferation of value in the commodity-form as a process that '. . . transforms every product of labour into a social hieroglyphic (Marx, 1977, p. 167). This almost mystical character of commodity fetishism provides not only an invitation to 'get behind the fetish' as it were, but also to 'get with the fetish' in the sense of coming to terms with the production and reproduction of meaning through commodification. Thus, recognition of the tremendous cultural significance of commodified meanings has led some to talk of fetishism in terms of the dreams, desires, and wish images that come to be attached to and circulate with commodities. As Kaika and Syngedouw put it '[t]he fetish character of commodities often turns them into objects of desire in themselves and for themselves, independent from their use value'. Drawing on the work of Walter Benjamin and Susan Buck-Morss, they continue that it is the '. . . very "estrangement of commodities" that makes them capable of becoming "wish images". Commodities do not only carry their materiality, but also the promise and the dream of a better society and a happier life' (Kaika and Swyngedouw, 2000, p. 123).[10] For them, a specific example is found in the production of coherent notions and wish images of urban modernity which become attached to and signified by highly fetishised technological networks. Somewhat ironically, even though a major facet of these networks is the metabolic transformation of biophysical nature constitutive of the production and reproduction of urban space (e.g., in storm and sanitary sewers, drinking water distribution and storage, energy systems, etc.), the networks themselves come to embody a wish image of the 'urban' defined, in part, as that which is distinct from and an improvement on 'nature'.

The wishes, desires, and dreams of imaginary geographies that circulate with commodities brings us to scholarship and activism seeking explicitly to transform the socio-ecological character of commodity chains and networks in part by taking hold of the sign values of commodities. This is a central facet of ethical, fair and organic production, and trade campaigns seeking more equitable and sustainable material practices in part through the propagation of standards, labels and the like. While these labelling schemes always aim towards some form of greater transparency, as well as a mix of enhanced social justice and ecological sustainability in commodity circuits, they do not eliminate fetishism per se; they rather seek to simultaneously rework both the material and semiotic aspects of commodities (Goodman, 2004).

And in this, consumer education campaigns around better and worse choices of commodity purchases reflect the power that consumers and a politics of consumption can and do have to effect change (see, e.g., Johns and Vural, 2000; Le Heron and Hayward, 2002). Broad-based scholarship and international networks of social activism pursue these goals in part by forging and sustaining connections that span production and consumption, linking disparate human and non-human actors in commodity circuits via mechanisms such as fair, ethical, organic, and sustainable trading regimes, and with wide-ranging implications for the geographies of producing and circulating nature in the commodity-form (McCarthy, 2006). This includes for instance, the development of forest certification schemes which define and seek to support more socially and ecologically sustainable forestry through the certification of wood products, schemes that have had considerable (though contested) impacts in forest commodity networks (Morris and Dunne, 2004; Klooster, 2005; 2006; Stringer, 2006). It also includes a plethora of food labels and certification schemes (e.g., organics) that both reflect and reinforce a widespread cultural and political re-signification of food in recent years, resulting in reworked relations among production and consumption for scholars, activists and 'foodies' alike in conventional and alternative food networks (Watts et al., 2005; Winter, 2003). These dynamics also establish new lines of struggle and contestation as both the form and content of labelling and certification schemes become subject to contending social pressures, on one hand seeking to uphold rigorous standards of social justice and ecological sustainability, and on the other, to hollow these out in favour of light green glosses on conventional, more profit-driven practices (Guthman, 2007).

(De)Commodification Redux

Whatever the outcome of such struggles, it has become clear that the search for alterity in commodity circuits must confront both material and representational practices. Important challenges and dilemmas remain. Can the fetishism of the commodity ever really be enlisted and sustained for the purposes of more socially just and environmentally sound production and consumption relations and practices? Put succinctly, and paraphrasing Guthman (2002), what is the relationship between 'commodified meanings', alternative or otherwise, and 'meaningful commodities' (i.e., more sustainable in a robust sense of the term)? How can resignification schemes overcome the challenge of displacement? Nowhere is the threat of a narrowing of the progressive promise of meaningful commodities more apparent than in organic food commodity circuits, which, absent certain prescribed chemicals and farm practices, look more and more like conventional, industrialised food circuits every day. Transforming the social relations of agricultural production, questioning productivism (including a full range of questionable growing practices), and providing high-quality nutritious and safe foods for *everyone* remain not only in question, but may actually be undermined by increasing market shares for organics (Guthman, 2003; 2004). More and more, the dynamics of organic food markets seem subject to the systemic processes of competitive commodity production under capitalism outlined succinctly by Kloppenburg (2004). As Mutersbaugh (2004) shows, for instance, struggles over the symbolic and material dimensions of certified coffee indicates an ever-present danger that labels will be co-opted, eliminating provisions for genuinely fair trading, including viable economic returns for independent and co-operative peasant producers.

These are not merely ephemeral, contingent and sector specific issues but rather deep, structural challenges to alternative commodity networks. Recognising them need not mean rehearsing tired debates between structure and agency in the evolution of commodity chains, agricultural or otherwise. Alternative commodity circuits have costs associated with them, not least administrative costs associated with certification (including in governance and enforcement). Who will bear the brunt of these costs (Mutersbaugh, 2005)? Is it socially just if only the more affluent consumers of the world can afford alternative commodities? Moreover, it is in the very nature of displacement and commodity fetishism in the context of competitive, capitalist economies that threats are ever present to more just and benign commodity circuits from competitive profit and rent seeking behaviour. Competition between labels and certification standards, for instance, can confuse consumers while placing downward pressure on standards via price-based competition. Even within labels, efforts to sustain and increase profits in commodity production regimes that remain largely capitalist (or are in competition with capitalist commodities in the same sectors) leads to systemic pressures to compromise, presenting a particular challenge to voluntary labelling and certification schemes (see, e.g., Klooster, 2006; Guthman, 2007). These observations are not meant to cast aspersions on efforts to forge alternative, fair, ethical, and more environmentally benign commodity circuits; quite the opposite. They are meant to reflect realistic assessments of the social (not merely technocratic) challenges involved in establishing and sustaining networks of ethical commitment that are frequently transnational in scope (Goodman, 2004). Maintaining these networks requires organising and solidarity, but also new relations of production, representation and governance that allow diverse actors from across commodity circuits – including workers, peasants, environmentalists and consumers – opportunities for meaningful participation in lasting coalitions. These efforts reenforce the need for political relationships in search of alternative commodity circuits to span the same range as those circuits themselves. And this is one more reason for scholars and activists alike to critically engage with the complex dynamics of commodification in a robust and polyvalent sense of the term, from inputs, to production, to distribution and to consumption.

A final word about decommodification. One of the appealing features of the term commodification is its inherently dynamic connotation. This can be interpreted teleologically to imply that everything, eventually, will be commodified, including our own bodies, and the earth, air and water around us. There are depressing trends that indeed point in these directions; yet it is important to recall the observation noted above that commodities, or more accurately, the commodity-form of things is not inherent to them. Commodities are made, not born. The commodity-form, put differently is really just one phase in the complex lives of things and ideas (Appadurai, 1986). Even in the conventional world of commodities produced exclusively for sale by profit seeking capitalist firms, commodification is tenuous, incomplete and ephemeral, not monolithic, complete or necessarily lasting. As Sayer (2003) intriguingly discusses, consumption is a form of de-commodification in so far as it reverses the ontology of things from exchange-value back to use-value. Using the same term, but in a different way, Henderson (2004) has argued that the circuits of value and of commodities in a (more than) capitalist political economy – and thus of commodification – are incomplete and 'leaky'. Even things produced exclusively as exchange-values in order to meet social needs and aspirations via the money economy can have politically charged, unpredictable lives, including mundanely enough in Henderson's discussion, canned food donated as surplus to food banks for relief. One implication then, is that commodified food produced for exchange-value ends up politicising (as opposed to depoliticising) the social allocation of food. A similar line of reasoning might well be applied to myriad environmental concerns linked to the commodification of nature, e.g., the mountains of non-biodegradeable and often toxic waste unevenly distributed across the globe and linked to consumer culture as the detritus of commodification. These represent simultaneously material and semiotic processes of decommodification that draw attention to the limits of commodification as the domination of exchange-value in production, and of some of the limits of displacement in the provision of social needs. Likewise, efforts to achieve fair, ethical, organic or otherwise alternative commodity circuits invoke questions about the limits of commodification, or alternatively, of the degree to which decommodification constrains or bounds the domination of exchange-motivated production. This is not so much about whether or not things are commodities, but the degree to which commodification has taken hold of their social allocation, and what a politics of commodification has to say about that.

ACKNOWLEDGEMENTS

This work, like all writing and scholarship, is the product of collective authorship. I would like to thank a number of specific people who commented on earlier drafts, including the BFP collective, Max Boykoff, Noel Castree, David Demeritt, Emily Eaton, Emma Hemmingsen, Mark Kear, Diana Liverman, Bruce Rhoads, Neil Smith, Tricia Wood, and Tom Young. Any errors, omissions, ambiguities, misrepresentations and the like are my sole responsibility.

NOTES

1. For a discussion of how some of these issues have unfolded in the agrarian and food literatures, for example, see (Goodman, 2002; Guthman, 2002; Whatmore, 2002; Winter, 2003).
2. This critique has been accompanied by a parallel concern with consumption as an end in itself, as opposed to for the provision of need, as for example, with so-called status

or external goods, and more generally, with the emergence of the consumer society and consumption as a measure of social worth – see Sayer (2003).

3. I use this term throughout the essay. At one level, it merely denotes that the provision of discrete objects and ideas has come to occur, at least in significant measure, via commodities produced primarily for sale, and thus that these 'things' are increasingly available in the commodity-form. At another level, it expresses the increasing importance of commodities as vehicles for the circulation and expression of value in a capitalist society, and thus for value itself to take the commodity-form.

4. These terms are productively discussed by Lysandrou (2005) in a paper interrogating globalisation *as* commodification, but he draws on Marx's analysis of the specificity of *capitalist* commodification, particularly in Volume 2 of Capital.

5. I would like to stress here that my point is not to reify the preparation of food as inherently women's work, but rather to simply observe historically that much of this work did indeed fall to women in western households, and that as women have become wage workers in increasing numbers, and as two-wage households have become more common, this has been accompanied by important shifts and evidence of deepening in the commodification of food.

6. One thinks, for instance, of a range of novel synthetic organic and inorganic chemicals produced during the 20th century for a variety of purposes whose toxic legacy, famously chronicled by Rachel Carson (1994), is still unfolding.

7. I do not discuss all of these here, but instead recommend a careful review of Castree's (2003) paper. Briefly, privatisation is the creation of new and exclusive forms of property claims over discrete bits of nature allowing them to be transferred between exclusive owners. Alienability refers to the often taken-for granted physical but also cultural processes whereby it becomes possible to sever bits of nature from sellers. This is related to but not wholly synonymous with ownership. Castree offers the example of internal organs, which may be owned but not easily (or painlessly) sold. Individuation is also closely related, and refers to the physical and cultural process of divorcing discrete things or entities from their social and ecological context. Valuation should also be reasonably familiar but refers to the socially mediated processes whereby value(s) are assigned, including monetisation, as well as (and conversely) how things become vehicles for the circulation of value. Finally, displacement is the most inherently geographical notion at play here, though by no means is it only a geographical process. This refers to the effects of time and space distantiation as commodities undergo complex transformation en route from producers to consumers and in ways that make it difficult for consumers to perceive the social and ecological relations, which underpin commodity production and circulation. There is a close conceptual link with fetishism (see below).

8. Marx writes specifically: 'In order, therefore, to find an analogy, we must take flight into the misty realm of religion. There, the products of the human brain appear as autonomous figures endowed with a life of their own, which enter into relations both with each other and with the human race. So it is in the world of commodities and with the products of men's [sic] hands. I call this the fetishism, which attaches itself to the products of labour as soon as they are produced as commodities, and *is therefore inseparable from the production* of commodities' (emphasis added).

9. Indeed, as Jackson (2002, p.15) notes in a largely sympathetic review, a danger in '. . . literature on commodity cultures has been to become overly fascinated with the spectacle of consumption and its liberating possibilities, to examine discursive and representational aspects of commodities and their meanings without attending to how these are produced, much less to explore in what ways consumption too underpins not just social and cultural difference but culturally inflected social *differentiation*'.

10. On commodity fetishism and desire, see the discussion in Page (2005) concerning water in the commodity-form.

BIBLIOGRAPHY

Appadurai, A. (1986) *The Social Life of Things: Commodities in Cultural Perspective.* Cambridge, UK: Cambridge University Press.

Bakker, K. (2003) *An Uncooperative Commodity: Privatizing Water in England and Wales.* Oxford and New York: Oxford University Press.

Bakker, K. (2005) Neoliberalizing nature? Market environmentalism in water supply in England and Wales. *Annals of the Association of American Geographers*, 95(3), 542–65.

Bakker, K. (2007) The 'commons' versus the 'commodity': alter-globalization, anti-privatization, and the human right to water in the global south. *Antipode*, 39(3), 430–55.

Booth, W. J. (1994) On the idea of the moral economy. *The American Political Science Review*, 88(3), 653–67.

Braun, B. (2002) *The Intemperate Rainforest: Nature, Culture, and Power on Canada's West Coast.* Minneapolis: University of Minnesota Press.

Braun, B. (2006) Environmental issues: global natures in the space of assemblage. *Progress in Human Geography*, 30(5), 644–54.

Bridge, G. (2000) The social regulation of resource access and environmental impact: Nature and contradiction in the US copper industry. *Geoforum*, 31(2), 237–56.

Bridge, G. and Smith, A. (2003) Intimate encounters: Culture – economy – commodity, *Environment and Planning D: Society and Space*, 21, 257–68.

Bridge, G. and Wood, A. (2005) Geographies of knowledge, practices of globalization: Learning from the oil exploration and production industry. *Area*, 37(2), 199–208.

Carson, R. (1994) *Silent Spring.* Boston: Houghton Mifflin.

Castree, N. (2001) Commodity fetishism, geographical imaginations, and imaginative geographies. *Environment and Planning A*, 33, 1519–25.

Castree, N. (2003) Commodifying what nature? *Progress in Human Geography*, 27(3), 273–97.

Castree, N. (2004) The geographical lives of commodities: problems of analysis and critique. *Social and Cultural Geography*, 5(1), 21–35.

Cook, I. and Crang, P. (1995) The world on a plate. *Journal of Material Culture*, 1, 131–53.

Cronon, W. (1991) *Nature's Metropolis: Chicago and the Great West.* New York: W. W. Norton.

Demeritt, D. (2001) Scientific forest conservation and the statistical picturing of nature's limits in the progressive-era United States. *Environment and Planning*, 19, 431–59.

Dicken, P. (1998) *Global Shift: Transforming the World Economy*, 3rd edn. New York: Guilford Press.

Domosh, M. (2006) *American Commodities in an Age of Empire.* New York: Routledge.

Foster, J. B. (2000) *Marx's Ecology: Materialism and Nature.* New York: Monthly Review Press Books.

Friedberg, S. (2004) The ethical complex of corporate food power. *Environment and Planning D: Society and Space*, 22, 513–31.

Gandy, M. (2002) *Concrete and Clay: Reworking Nature in New York City.* Cambridge, MA: MIT Press.

Gandy, M. (2005) Cyborg urbanization: complexity and monstrosity in the contemporary city. *International Journal of Urban and Regional Research*, 29(1), 26–49.

Gereffi, G. and Korzeniewicz, M. (1994) *Commodity Chains and Global Capitalism.* Westport, CT: Praeger.

Glassman, J. (2006) Primitive accumulation, accumulation by dispossession, accumulation by 'extra-economic' means. *Progress in Human Geography*, 30(5), 608–25.

Goodman, D. (2002) Rethinking production-consumption: integrative perspectives. *Sociologia Ruralis*, 42(4), 271–77.

Goodman, M. (2004) Reading fair trade: political ecological imaginary and the moral economy of fair trade foods. *Political Geography*, 23, 891–915.

Guthman, J. (2002) Commodified meanings, meaningful commodities: re-thinking production-consumption links through the organic system of provision. *Sociologia Ruralis*, 42(4), 295–311.

Guthman, J. (2003) Fast food/organic food: reflexive tastes and the making of 'yuppie chow'. *Social and Cultural Geography*, 4(1), 45–58.

Guthman, J. (2004) *Agrarian Dreams: The Paradox of Organic Farming in California*. Berkeley: University of California Press.

Guthman, J. (2007) The Polanyianway? Voluntary food labels as neoliberal governance. *Antipode*, 39(3), 456–78.

Hartwick, E. (1998) Geographies of consumption: a commodity-chain approach. *Environment and Planning D-Society and Space*, 16(4), 423–37.

Hartwick, E. R. (2000) Towards a geographical politics of consumption. *Environment and Planning A*, 32(7), 1177–92.

Harvey, D. (1982) *The Limits to Capital*. Oxford: B. Blackwell.

Harvey, D. (1985) *The Urbanization of Capital: Studies in the History and Theory of Capitalist Urbanization*. Baltimore, MD: Johns Hopkins University Press.

Harvey, D. (1990) Between space and time: reflections on the geographical imagination. *Annals of the Association of American Geographers*, 80(3), 418–34.

Harvey, D. (2003) *The New Imperialism*. Oxford, NY: Oxford University Press.

Henderson, G. (2004) 'Free' food, the local production of worth, and the circuit of decommodification: a value theory of the surplus. *Environment and Planning D: Society and Space*, 22, 485–512.

Heynen, N. C., McCarthy, J., Prudham, S. and Robbins, P. (eds) (2007) *Neoliberal Environments: False Promises and Unnatural Consequences*. London and New York: Routledge.

Jackson, P. (1999). Commodity cultures: the traffic in things. *Transactions of the Institute of British Geographers*, 24(1), 95–108.

Jackson, P. (2002). Commercial cultures: transcending the cultural and the economic. *Progress in Human Geography*, 26(1), 3–18.

Jackson, P. and Holbrook, B. (1995) Multiple meanings: shopping and the cultural politics of identity. *Environment and Planning A*, 27(12), 1913–30.

Jackson, P. and Taylor, J. (1996) Geography and the cultural politics of advertising. *Progress in Human Geography*, 20(3), 356–71.

Johns, R. and Vural, L. (2000) Class, geography, and the consumerist turn: unite and the stop sweatshops campaign. *Environment and Planning A*, 32, 1193–213.

Kaika, M. and Swyngedouw, E. (2000) Fetishizing the modern city: the phantasmagoria of urban technological networks. *International Journal of Urban and Regional Research*, 24(1), 120–38.

Klooster, D. (2005) Environmental certification of forests: the evolution of environmental governance in a commodity network. *Journal of Rural Studies*, 21, 403–17.

Klooster, D. (2006) Environmental certification of forests in Mexico: the political ecology of a nongovernmental market intervention. *Annals of the Association of American Geographers*, 96(3), 541–65.

Kloppenburg, J. R. (2004) *First the Seed: The Political Economy of Plant Biotechnology*, 2nd edn. Madison: University of Wisconsin Press.

Le Heron, K. and Hayward, D. (2002) The moral commodity: production, consumption, and governance in the Australasian breakfast cereal industry. *Environment and Planning A*, 34, 2231–51.

Leslie, D. and Reimer, S. (1999) Spatializing commodity chains. *Progress in Human Geography*, 23(3), 401–20.

Liverman, D. (2004) Who governs, at what scale and at what price? Geography, environmental governance, and the commodification of nature. *Annals of the Association of American Geographers*, 94(4), 734–38.

Lysandrou, P. (2005) Globalisation as commodification. *Cambridge Journal of Economics*, 29(5), 769–97.

MacDonald, K. I. (2005) Global hunting grounds: power, scale, and ecology in the negotiation of conservation. *Cultural Geographies*, 12, 259–91.

Mansfield, B. (2003) From catfish to organic fish: making distinctions about nature as cultural economic practice. *Geoforum*, 34(3), 329–42.

Mansfield, B. (2004a) Neoliberalism in the oceans: 'rationalization', property rights, and the commons question. *Geoforum*, 35, 313–26.

Mansfield, B. (2004b). Rules of privatization: contradictions in neoliberal regulation of North Pacific fisheries. *Annals of the Association of American Geographers*, 94(3), 565–84.

Mansfield, B. (2007). Privatization: property and the remaking of nature-society relations. *Antipode*, 39(3), 393–405.

Marsden, T., Munton, R., Ward, N. and Whatmore, S. (1996) Agricultural geography and the political economy approach: a review. *Economic Geography*, 72(4), 361–75.

Marx, K. (1977) *Capital: A Critique of Political Economy, Volume 1*. New York: Vintage Books.

McAfee, K. (1999). Selling nature to save it? Biodiversity and green developmentalism. *Environment and Planning D-Society and Space*, 17(2), 133–54.

McAfee, K. (2003). Neoliberalism on the molecular scale. Economic and genetic reductionism in biotechnology battles. *Geoforum*, 34(2), 203–19.

McCarthy, J. (2004). Privatizing conditions of production: trade agreements as neoliberal environmental governance. *Geoforum*, 35(3), 327–41.

McCarthy, J. (2006). Rural geography: alternative rural economies – the search for alterity in forests, fisheries, food, and fair trade. *Progress in Human Geography*, 30(6), 803–11.

McCarthy, J. and Prudham, S. (2004) Neoliberal nature and the nature of neoliberalism. *Geoforum*, 35(3), 275–83.

McEvoy, A. F. (1986) *The Fisherman's Problem: Ecology and Law in the California Fisheries, 1850–1980*. Cambridge, CB; New York: Cambridge University Press.

Miller, D. (1998) Coca-cola: a black sweet drink from Trinidad. In D. Miller (ed.), *Material Culture: Why Some Things Matter*. Chicago: University of Chicago Press, pp. 169–88.

Mintz, S. W. (1985) *Sweetness and Power: The Place of Sugar in Modern History*. New York: Penguin.

Mitchell, T. (2002) *Rule of Experts: Egypt, Techno-politics, Modernity*. Berkeley: University of California Press.

Morris, M. and N. Dunne (2004) Driving environmental certification: its impact on the furniture and timber products value chain in South Africa. *Geoforum*, 35, 251–66.

Mutersbaugh, T. (2004) Serve and certify: paradoxes of service work in organic coffee certification. *Environment and Planning D: Society and Space*, 22, 533–52.

Mutersbaugh, T. (2005) Fighting standards with standards: harmonization, rents, and social accountability in certified agro-food networks. *Environment and Planning A*, 37, 2033–51.

O'Connor, J. (1998) *Natural Causes: Essays in Ecological Marxism*. New York: Guilford Press.

Page, B. (2005). Paying for water and the geography of commodities. *Transactions of the Institute of British Geographers*, 30(3), 293–306.

Polanyi, K. (1944) *The Great Transformation: The Political and Economic Origins of Our Time*. Boston: Beacon Press.

Prudham, S. (2003) Taming trees: capital, science, and nature in Pacific slope tree improvement. *Annals of the Association of American Geographers*, 93(3), 636–56.

Prudham, S. (2005) *Knock on Wood: Nature as Commodity in Douglas-Fir Country*. New York: Routledge.

Prudham, S. (2007a) The fictions of autonomous invention: Accumulation by dispossession, commodification, and life patents. *Antipode*, 39(3), 406–29.

Prudham, S. (2007b) Sustaining sustained yield: class, politics, and post-war forest regulation in British Columbia. *Environment and Planning D: Society and Space*, 25(2), 258–83.

Robbins, P. (1999). Meat matters: cultural, politics along the commodity chain in India. *Ecumene*, 6(4), 399–423.

Robbins, P. (2001) Fixed categories in a portable landscape: the causes and consequences of land-cover categorization. *Environment and Planning A*, 33(1), 161–79.

Robertson, M. (2000) No net loss: wetland restoration and the incomplete capitalization of nature. *Antipode*, 32(4), 463–93.

Robertson, M. (2006) The nature that capital can see: science, state, and market in the commodification of ecosystem services. *Environment and Planning D: Society and Space*, 24, 367–87.

Sayer, A. (2003) (De)commodification, consumer culture, and moral economy. *Environment And Planning D: Society and Space*, 21, 341–57.

Sayre, N. (2002) *Ranching, Endangered Species, and Urbanization in the Southwest: Species of Capital*. Tucson: University of Arizona Press.

Scott, J. C. (1976) *The Moral Economy of the Peasant: Subsistence and Rebellion in Southeast Asia*. New Haven: Yale University Press.

Scott, J. C. (1998) *Seeing Like a State: How Certain Schemes to Improve the Human Condition Have Failed*. New Haven: Yale University Press.

Smith, L. (2004) The murky waters of the second wave of neoliberalism: corporatization as a service delivery model in Cape Town. *Geoforum*, 35(3), 375–93.

Smith, N. (1984) *Uneven Development: Nature, Capital, and the Production of Space*. Oxford: Basil Blackwell.

Smith, N. (1996) 'The production of nature. In G. Robertson, M. Mash, L. Tickner, J. Bird, B. Curtis and T. Putnam (eds). *Future/Natural: Nature/Science/Culture*. London: Routledge, pp. 33–54.

Stringer, C. (2006) Forest certification and changing global commodity chains. *Journal of Economic Geography*, 6, 701–22.

Swyngedouw, E. (2005) Dispossessing H2O: the contested terrain of water privatization. *Capitalism, Nature, Socialism*, 16(1), 81–98.

Swyngedouw, E. and Heynen, N. C. (2003) Urban political ecology, justice and the politics of scale. *Antipode*, 35(5), 898–918.

Thompson, E. P. (1971) The moral economy of the English crowd in the 18th century. *Past and Present*, 50, 76–136.

Watts, D. C. H., Ilbery, B. and Maye, D. (2005) Making reconnections in agro-food geography: alternative systems of food provision. *Progress in Human Geography*, 29(1), 22–40.

Whatmore, S. (2002) *Hybrid Geographies: Natures, Cultures, Spaces*. Thousand Oaks, CA: Sage.

Winson, A. (1993) *The Intimate Commodity: Food and the Development of the Agro-Industrial Complex in Canada*. Toronto: Garamond Press.

Winter, M. (2003) Geographies of food: agro-food geographies – making reconnections. *Progress in Human Geography*, 27(4), 505–13.

Wolf, E. R. (1982) *Europe and the People without History*. Berkeley: University of California Press.

Part II   Approaches

Chapter 10

Earth-System Science

John Wainwright

The Origin of Earth-System Science

With over 283 academic references on the ISI database and about 185,000 hits on Google by early 2008, the topic of Earth-System Science has clearly had a major impact. This impact has been significant across a range of disciplines, principally environmental science, ecology, meteorology and atmospheric science and geology (figure 10.1). It has had impacts on disciplines as diverse as psychology, neuroscience and education and a notable feature of these references is the range and integration of different subject areas. Some authors have even used the concept to bridge ideas of science and religion (Primavesi, 2000). By its very (problematic) definition, Earth-System Science brings in a broad range of disciplines and allows them to interact. However, the fact that the term exists outside or across current disciplinary boundaries has often been the source of controversy, uncertainty and suspicion (e.g., Turner, 2002). In this context, is it possible to define how the term came about and to evaluate whether it is – as some have claimed – a new science, or rather the repackaging of some older ideas?

The first specific use of the term 'Earth-System Science' in the literature was by Francis Bretherton (1985) in the *Proceedings of the Institute of Electrical and Electronics Engineers*. At first glance this may appear to be an unusual source for anyone interested in environmental geography, but it must be remembered that the *Proceedings* carries a number of papers on remote sensing – indeed, the title of Bretherton's paper is 'Earth System Science and Remote Sensing'. The underlying rationale for developing Earth-System Science (henceforth ESS) was two-fold. On the one hand, there is an altruistic desire to integrate and mobilise scientific endeavour to tackle pressing problems of anthropic environmental and climate change (note that this paper pre-dates the establishment of the Intergovernmental Panel on Climate Change by three years). While challenging the tendency of scientists to pursue various, typically reductionist, disciplinary research approaches, Bretherton does not make a proscriptive statement of what ESS should be:

Figure 10.1 Citations in the ISI database as of early February 2008 employing the term 'Earth-System Science'. Each of the 283 different references can be cross-referenced into one or more subject areas – the average number of subject areas covered by each paper is 3.4.

The very attempt to articulate an intellectually coherent structure across a field so broad is itself perilous, and the implicit claim to influence over the future advance of knowledge is bold, to say the least. Yet that is the challenge of Earth System Science, and this contribution is intended in that spirit, as a foil to debate and a spur to action. Many of the judgements expressed here may be poorly considered or misleading, and important aspects may have been overlooked. If individuals are stimulated to correct these errors, to fill in pieces of the puzzle, or simply to express opposing views, this survey will have served its purpose (Bretherton, 1985, p. 1119).

Despite that methodological agnosticism, the desire for greater integration does highlight the importance of modelling in order to render the problem of ESS manageable:

> For a situation as complex as the earth, model[l]ing has a critical role. Only by reducing qualitative perceptions to quantitative formulas is it possible to communicate ideas effectively across disciplinary boundaries and to analyze the subtle interactions and feedback loops which control the overall functioning of the system. It may well be that a complete numerical model of the whole system is never constructed and, unless firmly grounded in observation, the results of such a model are assuredly debatable. Nevertheless, experience shows that the attempt to identify its essential features can help focus critical issues, and maintain the perspective and balance which are essential in a program[me] aimed at overall understanding. (Bretherton, 1985, p. 1119)

This model is presented as a pair of systems diagrams (figure 10.2) that aim to represent a way of analysing global change on a decadal to centennial timescale by dividing the Earth system into a physical climate system and a biogeochemical cycling component. Bretherton notes that these two components are actually relatively weakly coupled in the model and goes on to point out four major caveats with respect to the systems diagrams. First, the diagrams – and hence model of ESS – are specific to research objectives, and it is these objectives that control the scale of representation. In the particular case presented, there is a very explicit timescale as well as a global spatial scale. Secondly, the representation is descriptive and not functional and thus does not make claims to completeness. Thirdly, although strongly affected by and having major impacts on the Earth system, humankind is regarded as external to it. Fourthly, there is an assumption that the Earth system can be defined in terms that are deterministic, and thus, predictable, even though parts of the system – e.g., weather and climate – are known to exhibit chaotic behaviour. We will return to the implications of these caveats later in this chapter.

On the other hand, Bretherton's paper provides a methodological statement about the need to employ remote sensing as a way of informing and testing the suggested approach. He highlighted five roles of remote sensing that needed to be developed. First, it provides the necessary global synoptic coverage and shifts emphasis from relatively disconnected point measurements that characterised a number of scientific approaches. Secondly, ESS forces a rethinking of algorithms employed in remote sensing because of the complexity of extracting a signal that can be meaningfully used in the parameterisation and testing of models. Thirdly, the emphasis within ESS is on change and therefore the need for ongoing measurements, with remote sensing being the most cost-effective way of doing so. Fourthly, ESS should promote better practice for integrated data management to understand what is going on in different spectra and thus to characterise different parts of the Earth system simultaneously. Fifthly, there was a need for more training to remove remote sensing from the minority role it had at the time. The extent to which these five roles have been addressed will be considered later.

Of course, such developments could not occur in a vacuum. They would need significant funding initiatives, international cooperation and data exchange, and mechanisms for linking research with governmental and industrial requirements. Bretherton's paper reflected a major US initiative that included efforts from NASA, NOAA and the National Science Foundation, together with inputs from the US

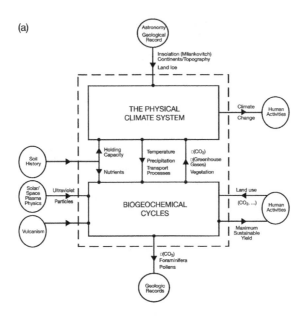

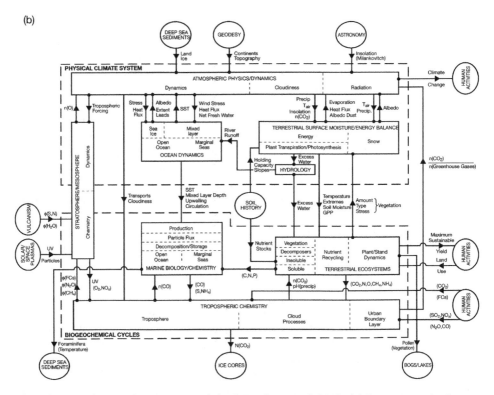

Figure 10.2 The Earth-System Model of Bretherton (1985): (a) is a general schema intended to provide the structure for understanding the drivers of change over decadal to centennial timescales; and (b) is an attempt to provide more detail of the ways the physical climate system and biogeochemical cycles are made up and interact. It was apparently familiarly known as 'the wiring diagram' by the team from the ESS Committee of the NASA Advisory Panel that defined it.

Geological Survey (USGS), Department of Energy, and State Department, as well as data from Defense Department satellites. Notwithstanding the latter, the aim should be 'to build a truly international approach' and in so doing 'must avoid even the appearance of military or economic overtones' (Bretherton, 1985, p. 1126). To evaluate the possibility of these statements, we need first to consider the underlying development of these ideas in institutional, academic and broader contexts.

The Evolution of Earth-System Science

Given this impressive mobilisation of US institutions, it is perhaps surprising that Bretherton is a UK national. A Cambridge-trained applied mathematician, he worked extensively on problems of fluid dynamics, atmosphere and ocean models.[1] He became Director of the US National Center for Atmospheric Research (NCAR) in 1974, taking on an administrative role until 1980, when he became research-active once again in the NCAR Oceanography section. In 1983, he was invited to chair a committee to evaluate how NASA could best restructure to develop an 'Office of Space Science and Applications' that would enable the organisation to carry out Earth Observation activities most effectively. Having tried polite refusal, he was finally convinced to steer what became known as 'The Earth System Sciences Committee'. The committee was composed of 15 other members, from the fields of meteorology, atmospheric chemistry, ionospheric physics, physical oceanographry, marine biology, plant ecology, soils and vegetation interactions, agronomy, cryology, geology, geophysics and space-based instrumentation, together with a representative from NASA. The choice was deliberately restricted to the natural sciences and hydrology was not included as the scientists identified for the committee were unavailable.

The committee first met in early 1984 and agreed on the need to investigate the complex interactions of the Earth system in order to quantify the impacts of human activities in relation to natural variability. Bretherton's mathematical background led to a focus on modelling approaches. The involvement of NASA underpinned the interest in remote sensing, but notwithstanding their input, there were formidable institutional issues inasmuch as NOAA were responsible for weather satellites, and even LANDSAT, which was developed by NASA, had its data distributed by the USGS. It was also important to involve the National Science Foundation, which funds university-based research. Such institutions would need to be convinced their long-term involvement was important despite more rapid changes in staff and in the politicians holding the purse strings. Support would be needed from the general public and professionals in the academic and research institutions to ensure success. At the same time, the international nature of the endeavour would require integration under the umbrella of credible organisations such as the United Nations and World Meteorological Organization. Following an encouraging progress meeting with the NASA Advisory Council, the Committee met in full in June 1984 in Charlottesville, Virginia and quickly agreed on the need to engage other organisations, such as NSF and NOAA, to secure the necessary cooperation.

At the same time, a modelling subgroup met on two occasions to develop what the ESS model should look like. For example, what components it should contain and how they were interconnected? How should processes that operated on very different timescales be interrelated? The outcome of the second meeting in Jackson

Hole, Wyoming, was the system model that was subsequently published in Bretherton (1985) and is reproduced here as figure 10.2. It was subsequently the centrefold of the Overview report of the Committee, published by NASA in May 1986 (Earth System Sciences Committee, 1986). The Overview report was the result of a further full committee meeting at Orcas Island, Washington in June 1985. It presents the overall scientific background to the question of ESS and makes both general and specific recommendations for how progress should be made. A press conference to launch the report was organised with major consequences; the opening statement of the report being 'We, the peoples of the World' as an echo of the opening of the UN Charter. As a result – perhaps combined with a little environmental serendipity – the US Global Change Research Program (USGCRP) was introduced by President Reagan as a Presidential Initiative starting from financial year 1989 (see below). The Earth System Science Committee is now the Earth System Science and Applications Advisory Committee and meets biannually to discuss NASA strategy (ESSAAC, 2008).

The idea of ESS thus had a strong institutional focus within the governmental and non-governmental research organisations in the USA, and a strong disciplinary focus within various approaches to applied mathematics in the environment. To what extent can the science within ESS be considered to be novel? Von Humboldt's 1845 work *Kosmos* states 'the word climate, however, refers to a specific nature of the atmosphere; but this nature depends on the continuous interplay . . . with the heat radiating dry earth which is covered by forest and herbs'[2] (cited in Scheffer et al., 2005). At its simplest level, this statement reflects the representation of climate and biogeochemical cycles inherent in the formation of ESS. Other work in the 1890s by Arrhenius and Hogböm also demonstrated an understanding of similar interactions (Heimann, 1997). Clifford and Richards (2005) suggest that as well as von Humboldt, Huxley's work in physiography also reflects an early forerunner to the holistic approach of ESS. As well as Lovelock's Gaia theory (of which, more will be discussed later), they point to the importance of the development of systems approaches in geography – especially the work of Chorley and Kennedy (1971) and Bennett and Chorley (1980) – grounded in the work of von Bertalanffy and followers in general systems theory, and of parallel (and much earlier) developments in ecology and agronomy and forestry (v. Chorley and Kennedy, 1971, pp. 88–90).

For some, ESS has simply taken (or borrowed or stolen, depending on the perspective) the mantle of systems-based physical geography. Interestingly, this perception may have spread more widely, as noted by the following author, who is based in the Department of Geological and Environmental Sciences, at Stanford University:

> Earth systems science is actually twenty-first century geography: it encompasses the study of the environmental physical and life sciences and engineering, coupled with an analysis of human constructs and political and economic policies. It employs space-age technologies to identify, measure, and manage diverse global databases that serve as a framework and foundation for a coherent discipline. (Ernst, 2000, p. 520)

Another author, this time from the Department of Biological Sciences at Stanford makes similar points:

The subject called 'physical geography' was once taught to general students in many universities. Now it is almost an endangered species; few institutions still have geography departments, and the average contemporary undergraduate – even those studying scientific subjects – would be unable to say what is encompassed by the subject.

For those concerned with the vital matter of environmental quality, physical geography is a core discipline. It attends to the structure and character of the local and the global habitat: landforms, temperature, soils, and climate. It examines the way in which those physical factors determine the pattern of occupancy by living systems – that is, it seeks explanations for the spatial distribution of species of organisms, and of the development, through their interactions, of ecosystems. Finally, it attempts to explain how humans have settled on the land and have used it. Dressed up in a more modern name, physical geography *is* Earth systems science. (Kennedy, 2000, p. 13)

Notwithstanding the explicit borrowing of general systems theory during the quantitative revolution of the 1960s and 1970s, such claims about the intellectual debts of ESS to physical geography do not withstand scrutiny. As an example, Chorley and Kennedy (1971, pp. 82–93) discuss the 'solar energy cascade', which can be seen as a direct parallel to the bicameral atmosphere-biogeochemical cycles approach of ESS. Figure 10.3 illustrates this division very clearly. However, this version of the model is focused on the energy and water components – largely in order to make specific predictions about catchment hydrology. It contains little in the way of biogeochemical linkages with the atmosphere. Such concerns did not feature much in physical geography before the 1990s. One reason, perhaps, was the increasing reductionism of research in physical geography from the 1970s onwards, often with an aim related to environmental management; ESS aimed explicitly to be holistic, to develop understanding, and thus, to guide policy.

The best interpretation is probably one of parallel development from similar backgrounds. Chorley was well aware of the climate-modelling literature (the first edition of *Atmosphere, Weather and Climate* written with Roger Barry was published in 1968), while Bretherton was director of the US National Center for Atmospheric Research (NCAR) from 1974. Climate modelling was part of the NCAR remit from its inception in 1959 (UCAR, 1959) and experience with overseeing the development of the NCAR climate model had a strong influence on Bretherton (*pers. comm.*).

This idea of a holistic approach is part of the underlying rationale for ESS:

Though the specific requirements differ in each case, rational treatment of each such issue [of environmental change] depends on an understanding of many different components of the global environment and the interactions between them, and appreciation of the functioning of the system as a whole. This fundamental knowledge, rather than the isolated issues themselves, is what consititutes Earth System Science (Bretherton, 1985, pp. 1118–9).

Clifford and Richards (2005) criticise this outward holism of ESS. They use complexity theory to suggest that there are many ecosystem features that cannot be accounted for in terms of energetics or biogeochemistry. While undoubtedly true of the original structure of ESS, it is not clear that this critique applies universally. Indeed, many ecological models working within the complexity theory remit *do* deal

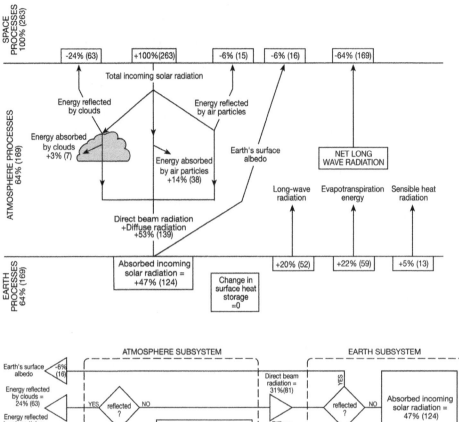

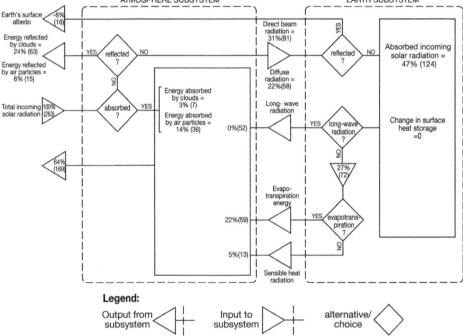

Figure 10.3 The solar energy cascade model of Chorley and Kennedy (1971) given in both diagrammatic and canonical structure form. In the latter, note the division into an atmosphere and an Earth subsystem.

with such ecosystemic characterisics as *emergent properties* of energetics of biogeo-chemical models. Likewise, their complaint about the absence of social interactions within ESS, while perhaps applicable to the work subsequently inspired by ESS, also misrepresents the original aim of ESS, which deliberately excluded the human sphere (see above). The issue here is one of the naïf, initial representation of ESS. If it is to be used in policy formation, the social context cannot be ignored, and the use and abuse of models and their results must be considered (Oreskes et al., 1994; Demeritt, 2001). It is not clear, though, that this use is the same as ESS having 'hegemonising tendencies' (Clifford and Richards, 2005, p. 381). In their argument for a pluralistic approach to social science, Clifford and Richards are remarkably restrictive about what is permissible within social science, and indeed, about the question of whether ESS is one approach or many. The central issue is whether ESS has to be a model of everything, everywhere, all the time. Clifford and Richards argue that ESS must be, and thus argue that it is unscientific on the grounds that it cannot affect closure on any question (in the sense that any research can come to a well-defined result unaffected by the lack of defined boundaries to the research). The initial vision of ESS expressed in Bretherton (1985) is much broader. It notes that different science questions will need to employ different formulations relating to explicit and implicit spatial and temporal scales. No model can live up to the everything, everywhere, all the time goal (see Wainwright and Mulligan, 2003), so this argument is something of a straw man.

A more serious case for the appropriation of physical geography approaches under the banner of ESS comes from the discipline of geology (or Earth Science for locations where that nomenclature has been seen to be more politic). There have been a number of institutions, courses and individuals who have imported the 'systems' into 'Earth science' for a range of motives. For example, the textbook of Merritts et al. (1997, p. 10) called *Environmental Geology: An Earth System Science Approach* suggests that ESS is a natural successor to a sequence of approaches with illustrious protagonists. First came 'The Dawn of Science', with Ptolemy and Aristotle, second 'The Scientific Revolution' of Newton, Steno, Kepler, Galilei and Copernicus, third 'The Age of Earth and Evolution' with Curie, Darwin, Lyell and Hutton, and fourth 'The Plate Tectonics Revolution' with McKenzie, Morgan, Hess and Wegener. The fifth step of 'Earth System Science' is interestingly *not* carried out by scientists with geological track records; in this case, the examples given are Rowland's work on CFC emissions and the ozone hole and Lovelock's work on Gaia theory. Merritts et al. define a single Earth system comprised by various subsystems, or 'spheres' (litho-, pedo-, hydro-, bio- and atmo-), and including humans and their actions. Skinner et al. (1999: first edn 1995) and Ernst (2000, p. 525) also use a very similar terminology, which would not be out of place in any physical geography text since Chorley and Kennedy (1971). The same development is seen in the evolution of the influential text of Press and Siever. The 1982 edition of *Earth* uses the 'Earth Machine' of plate tectonics as a structuring element; systems only explicitly appear on page 40 as 'time-rock units'. Press et al. (2004) use systems terminology throughout. Systems are 'what comes after plate tectonics'. The textbook by Ernst et al. (2000) is related to an elementary course at Stanford called 'Introduction to Earth Systems', developed since 1993 with a philosophy that is 'problem-focused, not discipline-focused' (Ernst et al., 2000, p. vii). The emphasis is on finding 'appropriate ways to *integrate* high-quality disciplinary work from several fields. Although scholars from various disciplines may study the Earth locally – in a tax district, a volcano, a thunderstorm, a patch of forest or a test tube – Earth systems scientists put the accent on "systems", the multiscale interactions of all these small-scale phenomena' (Schneider, 2000, p. 5). The emphasis is clearly on holism via stitching together all the reductionist components.

This appropriation of ESS into geology is intriguing, given Bretherton's (1985, p. 1122) statement that 'though part of the Earth System, basic geology and geophysics are not directly relevant on these timescales [the 10,000 years of heightened anthropic impacts on the environment] except as aids in interpreting drainage and soil patterns'. Arguably, the redefinition of geology as Earth-Systems Science is about repositioning that discipline in a post-oil economy against a background of dwindling and closing departments. It is for this 'brand' of ESS that the arguments of Clifford and Richards (2005) more clearly hold. The same may probably be said with regard to their argument about 'hegemonising tendencies', or at least homogenising tendencies. This point can most clearly be seen by the attempt to standardise undergraduate education in ESS by a series of fixed templates and syllabus suggestions (NASA/ESRA, 2007).

Life on Earth-System Science?

The second major component of Bretherton's (1985) blueprint for ESS is the biosphere. He states that 'global model[l]ing on decades to centuries is dominated by the changes in surface temperature and precipitation and by the sensitivity of photosynthesis and respiration by planets [sic] and phytoplankton to these and to the concentration of CO_2 in the atmosphere' (p. 1124). Dutton (1987, p. 311) further emphasises the role of biological processes, noting the need for 'theoretical and empirical studies necessary to provide a dynamical systems representation of the biological processes and biogeochemical cycles that clearly link the systems together and provide important feedbacks and modifications of the entire planetary environment'. He suggests that it is more likely that local process studies will provide an adequate basis for this work given the lack of theoretical biological work to provide such an underpinning. Theoretical biologists, however, would probably beg to differ.

In parallel to these suggestions, the disciplines of ecology, hydrology and geomorphology at least were already recognising the need for trans- or interdisciplinary work. An early example was Eagleson's seven-paper *magnum opus* on the links between vegetation and hydrology (Eagleson, 1978a–g), which has recently been elaborated in book form (Eagleson, 2002). Ecohydrology has been steadily developing as a research focus (Baird and Wilby, 1999; Newman et al., 2006). Similarly, in the field of geomorphology, Viles (1988) and Thornes (1990) provided collections of papers reflecting the interactions between biological and geomorphic processes. The papers of Viles' biogeomorphology are very much based on specific environments and often limited in terms of large-scale feedbacks. Thornes' work on vegetation and erosion also links back to earlier papers that develop an integrated modelling approach (e.g. Thornes, 1985; 1988) but again with a scale that is essentially that of the hillslope. A third parallel might be seen in the development of landscape ecology. Originally, a term used by the German biogeographer Carl Troll in the 1930s, the idea was developed as a way of investigating the effects of spatial pattern on ecological process among ecologists in the 1980s (Turner et al., 2001). Often this work is focused on the practical needs of environmental management and practitioners sit across the disciplinary boundaries of (bio-)geography and ecology.

Two things characterise these parallels. First, despite the work across disciplines, they still maintain a relatively disconnected approach. In some respects, this disconnectedness relates to the restricted spatial scale of observation (Eagleson's attempts at global ecohydrology notwithstanding), while in others it relates to the lack of development of strong interdisciplinary ties, or ties that are seldom more than bilateral. Secondly, these strands of Earth-surface research were not well connected to work on the atmosphere, which would be central to the ESS project. In part, this disjuncture is due to the decline of exposure to atmospheric sciences. They are decreasingly taught within geography departments – due to their perception as being difficult or too mathematical – and never really had a home within biological science. A notable exception again is the work of Eagleson, coming as it does from a heavily mathematical, engineering perspective in hydrology. However, there have been other key exceptions. Raymo's work on the linkages between plate tectonic activity, weathering and CO_2 release to the atmosphere causing feedbacks that potentially produced the Quaternary glaciations (e.g. Raymo, 1994), shows archetypal ESS interactions, even if it does operate on much longer timescales. Charney (1975), working from a meteorological perspective, pointed out the significant potential feedbacks between vegetation and climate in the Sahel. More recent work has tried to develop this theory, generally from a hydrological or environmental science focus, but including the atmospheric linkages (e.g. Entekhabi et al., 1992). The key outcome seems to be that modelling studies (e.g. Xue and Shukla, 1993; Claussen, 1997; Zeng et al., 1999; Zhou et al., 2007) support the theory, while observations, including remote sensing (e.g. Jackson and Idso, 1975; Wendler and Eaton, 1983) find problems with it.

These examples suggest major weaknesses with the argument that 'ESS *is* physical geography'. Physical geographers have tended to have an overly reductionist focus that has led them to concentrate on very small-scale processes without linking them back to the larger scale. They also tend to lack the appropriate tools for the modelling approach to ESS given the quantitative paradox (that statistics are a requirement while mathematical modelling is considered a minority interest) inherent in the syllabi of many university departments. Church (1998; 2005) has discussed the appropriation of ESS into mainstream geology and its implications for physical geographers, especially geomorphologists, from a similar perspective. That geomorphologists may have missed the boat seems inexcusable to Church, given that the boat was moving at continental drift pace.

To what extent can ecologists be said to have fared any better? Given the bicameral definition of the ESS blueprint, ecological work should inform understanding of the behaviour of the whole Earth system in detail. Nitta (1994) described an early example of how an experimental facility might be used to inform the functioning of the biosphere elements of ESS. Notwithstanding a major conference on using understanding of linkages between plants and the atmosphere over geological timescales in late 2001 (Pataki, 2002), some limited work on forest carbon (White and Nemani, 2003) and the limited ecohydrological and landscape ecological work discussed above, ecology as a discipline seems as unimpressed with ESS as geographers have been. A recent major review of plant response to CO_2 changes (Körner, 2006) totally fails to mention ESS.

As mentioned previously, a number of authors have chosen to highlight Lovelock's Gaia theory as a principle way of linking biosphere and whole-Earth behaviour. Given that it was first defined in 1972, it clearly pre-dates the initiative of the ESS Advisory Panel. Equally, however, it was influenced by the work of NASA – in this case, Lovelock's work to define methods for detecting life on Mars by looking at atmospheric chemistry. In its first formulation, the theory is defined by suggesting:

> that life at an early stage of its evolution acquired the capacity to control the global environment to suit its needs and that this capacity has persisted and is still in active use. In this view the sum total of species is more than just a catalogue, 'The Biosphere', and like other associations in biology is an entity with properties than the simple sum of its parts. Such a large creature, even if only hypothetical, with the powerful capacity to homeostat the planetary environment needs a name; I am endebted to Mr. William Golding [the novelist] for suggesting the use of the Greek personification of mother Earth, 'Gaia'.
>
> As yet there exists no formal physical statement of life from which an exclusive test designed to prove [sic] the presence of 'Gaia' as a living entity. Fortunately such rigour is not usually expected in biology and it may be that the statistical nature of life processes would render such an approach a sterile one. At present most biologists can be convinced that a creature is alive by arguments drawn from phenomenological evidence. The persistent ability to maintain a constant temperature and a compatible chemical composition in an environment which is changing or is perturbed if shown by a biological system would usually be accepted as evidence that it was alive. (Lovelock 1972, p. 579)

The article goes on to produce such statistical evidence, and two further papers (Lovelock and Margulis, 1974; Margulis and Lovelock, 1974) developed the idea in much more detail. A series of books (e.g. Lovelock, 1979; 2000; 2006) did much to popularise the idea.

Kirchner (1991; 2002) has provided some of the strongest critiques of Gaia theory and the slipperiness of its central homeostatic principle. In particular, he differentiates a 'weak Gaia' hypothesis, in which life is said to create a more suitable environment for itself, from a 'strong Gaia' hypothesis in which the entire planet is considered as a single organism. Kirchner (2002) believes that such approaches 'may be useful as metaphors but are unfalsifiable, and therefore misleading, as hypotheses' (p. 393). Lovelock (2000, p. 271) has dismissed the arguments of Kirchner's first paper as 'sophistry, not science' but fails to refute the claims directly. Others have taken on the mantle in trying to test the weak form of Gaia (notably Lenton, 2002; Lenton and Van Oijen, 2002; Lenton and Wilkinson, 2003), although both Volk (2003) and Kirchner (2003) have suggested that they have tended to be selective with the evidence and to focus only on cases where biological activity has tended to stabilise the Earth system.

For present purposes, two important issues arise from this debate. First, proponents of strong Gaia present an argument that is non-scientific – at least in strictly Popperian terms. Given that the ESS blueprint was essentially underpinned by critical rationalist thinking, often with explicit aims of future prediction, the two approaches are incompatible. Weak Gaia, on the other hand, is testable and indeed has rarely been challenged in that it is not too distant from ideas of Humboldt, Huxley and many others since. This form is not incompatible with ESS, but has tended to focus on feedbacks that support the theory. As Kirchner has pointed out, this approach assumes only responses tending towards equilibrium are possible, which is both incorrect and assumes that all systems must reach a stable equilibrium through feedbacks (see Bracken and Wainwright, 2006, for a similar demonstration of this fallacy in geomorphological thinking).

Secondly, there is the extent to which Gaia has a teleological requirement. Such a requirement is clearly evident in the quotation from Lovelock (1972), although his later books tend to present a steady distancing from this more 'New Age' perspective. Teleological Gaia was certainly not a part of the original ESS blueprint, although the environmental problems due to feedbacks from human activity were a central concern. In his latest book, *The Revenge of Gaia*, Lovelock (2006) addresses this same theme directly from a Gaian perspective. He also explicitly equates ESS and Gaia – most clearly in the glossary, where he notes that ESS 'differs from Gaia theory only because it has not had time to digest the mathematical consequences of the union between the Earth and life sciences, the most important of which is that self-regulation requires a goal' (p. 162). It is hard to avoid reading this statement as teleological and thus concluding that Gaia is not the same (nor even a subset of) ESS as commonly perceived. In a parallel argument, Huggett (1999) has also concluded that Gaia is not a good replacement for the concept of biosphere.

Ground Control to Major Tom?

As noted above, one of the key elements of ESS, not least because of its original definition within the NASA Advisory Committee, is that of remote sensing:

> Effective discussion of these [environmental] problems requires an intellectual framework and a long-term program[me] of research and observations which transcend the traditional boundaries of the disciplines in Earth Sciences. The framework must be firmly grounded in the realities of knowledge about the physics, chemistry, and biology of the processes involved, yet must articulate a vision of how this understanding can fit together into a coherent whole. . . . Remote sensing is but one (albeit an expensive one) of several critical tools, and it is vital to keep the vitality of the science and the integrity of the whole endeavo[u]r clearly in view, at the same time as cultivating of the community of interest with more immediate applications of the instrument and data types. (Bretherton, 1985, p. 1119)

A central question then is whether the practitioners of remote sensing have risen to the challenge. Stoms and Estes (1993) provide an early example in the literature of a manifesto for remote sensing to tackle the issue of biodiversity monitoring. This topic has seen a lot of development in the remote sensing literature (e.g. Williams, 1996; Innes and Koch, 1998; Soberon and Peterson, 2004; Duro et al., 2007), although there is little explicit reference to an ESS framework. Bretherton specifically highlighted the need to improve estimates of evaporation from the oceans and evapotranspiration as modulated by vegetation cover over the land surface. An early response was the First International Satellite Land Surface Climatology Project (ISLSCP) Field Experiment (FIFE), which carried out experiments at the Konza Prairie field site in Kansas in 1987 and 1989, with subsequent campaigns up to 1995 (GEWEX, 1995; Hall and Sellars, 1995). There also followed the influential paper of Qi et al. (1994), which attempted to produce an improved empirical method of assessing vegetation cover by taking background soil characteristics into account.

Such approaches were important in planning the new Earth Observing System (EOS) satellite launches, in particular the Moderate Resolution Imaging Spectrora-diometer (MODIS) instruments launched by NASA (Justice and Townsend, 1994; Running et al., 1994). The design of the MODIS (2008) 'products' shows a clear commitment to delivering data that fit into the ESS design criteria, as well as having broader uses. There have been considerable technological inputs into improving the collection and delivery of relevant datasets (e.g. Vetter et al., 1995; Arnavut and Narumalani, 1996; Hyman, 1996; Wanner et al., 1997). NASA has an ongoing commitment to the development of sensors and applications within the scope of ESS (NASA, 2007; ESSAAC, 2008) at least up to 2025 (King and Birk, 2003). Certainly, remote sensing is now routinely used in the parameterisation of general circulation models (e.g. Feingold and Heymsfield, 1992; Webb et al., 2001; Suzuki et al., 2004) that have informed studies of ongoing and potential future anthropic climate change. The ESS linkage of climate and biogeochemistry is explicitly recognised in the Intergovernmental Panel on Climate Change Fourth Assessment (IPCC, 2007). While ESS has certainly played a role in facilitating these developments, it could also be argued that the global scale of climate studies necessitates such a satellite-based approach and so would have occurred (eventually) anyway.

One of the emphases in remote sensing science has been on developing new 'products'. Despite efforts to target a 'market' of ESS and other users, there often tends to be too little integration across disciplinary barriers. The issue is probably still largely a lack of communication from the two sides. This communication aspect is a major issue for the mathematical modelling approach to ESS, which assumes that all aspects of the Earth system are reducible to mathematical description. Many 'products' are still heavily empirical and highly calibrated, and the potential for transferability from one context to another is still poorly understood. The market metaphor is also quite literal in many cases. For example, while many of the MODIS products are freely available, ASTER data produced from the same satellite platform must be bought. The US approach to freedom of information has been useful in the democratisation of data as suggested by the Dublin Agreement on access to environmental information (INFOTERRA, 2000) and indeed the 'we, the people of the World' pronouncement of the ESS Committee (1986). The use of complex models parameterised from these data is not likely to be a broadly democratic process until computer power increases significantly, but at least data can be freely found to assess changing global conditions. The same freedom of information can still not be said of data funded by UK taxes, however.

Paydirt?

One success of the NASA blueprint for ESS was to bring a considerable level of resource into the study of related phenomena. In particular, the US Global Change Research Program (USGCRP) was introduced by President Reagan as a Presidential Initiative starting from financial year 1989 as a direct response to the NASA report, although the introductory document for the programme suggests that the intervention of environmental phenomena ('the discovery of the Antarctic "ozone hole" and the 1988 North American drought') may have helped produce the decision (Committee on Earth Sciences, 1989, p. 1). The USGCRP was set up with three objectives: '1. Establish an Integrated, Comprehensive Monitoring Program[me] for Earth

System Measurements on a Global Scale'; '2. Conduct a Program[me] of Focused Studies to Improve Our Understanding of the Physical, Chemical, and Biological Processes that Influence Earth System Changes and Trends on Global and Regional Scales'; and '3. Develop Integrated Conceptual and Predictive Earth System Models' (pp. 12–13). These objectives follow very directly from the recommendations of Bretherton (1985) and the Earth System Sciences Committee (1986). Funding was initially $133.9 M in the 1989 financial year, increasing to $190.5 M in 1990. By far, the largest proportion of this funding (87 percent in 1989) was intended for basic research. By 1995, the total budget had risen to $1827.7 M, although the actual increase is inflated by inclusion of the NASA space-based observing budget (of $815.5 M), which is not incorporated in the earlier figures. Nevertheless, the value does represent a considerable increase in research funding, which despite some real-term decreases remains very substantial, with $1,505 M in 2006 (all figures from reports on the USGRCP website).

Not wanting to be left out, the UK Natural Environment Research Council (NERC) announced an ESS initiative of its own from 2001, following the appointment of John Lawton as Chief Executive in 1999. Lawton (2001, p. 1965) published a paean to ESS in *Science* noting that:

> ESS as the discipline that deals with our planet as a complex, interacting system. ESS takes the main components of planet Earth – the atmosphere, oceans, freshwater, rocks, soils, and biosphere – and seeks to understand major patterns and processes in their dynamics. To do this, we need to study not only the processes that go on within each component (traditionally the realms of oceanography, atmospheric physics, and ecology, to name but three), but also interactions between these components. It is the need to study and understand these between-component interactions that defines ESS as a discipline in its own right.

It should be noted how exclusionary a vision of ESS is presented here. It certainly does not reflect the role of physical geography in investigating the major components of 'freshwater, rocks, soils'. Replies to *Science* also challenged the failure to mention geology (Carlson, 2001) and the claim that there were no interdisciplinary training programmes in ESS (Ernst, 2000 cited above represents a course in Stanford that has been running since the early 1990s, and there are a number of others: Farmer, 2001).

Nevertheless, this exclusionary approach seems to underlie the implementation of ESS at various levels in NERC. ESS remains a core science theme of NERC in the 2007–2012 strategy document (NERC, 2007). The ESS of NERC is highly reductionist, however: 'Planet Earth is a complex, interconnected system. To build an understanding of the whole system we need to increase our knowledge of its component parts and the ways that these interact. This is called Earth system science' (NERC, 2007, p. 16). It is also strongly focused on ocean and atmosphere processes as well as biogeochemical cycles, and has a relatively restrictive set of 10 major challenges. NERC were unable to specify the extent to which they had directly funded ESS research (*pers. comm.*, July 2007) other than to suggest the QUEST (Quantifying and Understanding the Earth SysTem) thematic programme fell clearly within the topic, and to point to their online database of funded research. The former has a budget of £23 M between 2003 and 2009. It is made up of three themes: 'Contemporary Carbon Cycle' (£3.0 M), 'Natural regulation of atmospheric composition on glacial-interglacial and longer time scales' (£2.6 M), and 'Implications of global environmental changes for the sustainable use of resources' (£1.7 M) with applications for 'Biosphere management for climate mitigation' still outstanding as of March 2008. The latter produced some interesting results (GOTW, 2007). Only a single grant actually mentioned the words 'Earth System Science' and that was awarded £47k for a postdoctoral training network. Widening the search to 'Earth System' did produce another 82 grants funded for a total of £16.9M (excluding those funded through QUEST). Only five of these grants were awarded to geography departments (compared to four of the 29 QUEST-funded projects). Academics from whatever discipline in the UK seem less than keen to promote an ESS narrative, even those who have directly benefited from funding considered to be in the field.

Clearly, there are also political aspects to such decisions and to the relative positionings of different disciplines. As Clifford and Richards (2005, p. 379) point out:

> Positions are now regularly being created with titles like 'Chair in Earth (sometimes Environmental) Systems Science', and in the UK, the Research Assessment Exercise (RAE) now has a sub-panel, which emerged through a rather mysterious lobbying process, confusingly entitled 'Earth Systems and Environmental Sciences'.

For example, the previous professorial appointment to my own in Geography at Sheffield was in ESS. As this paper is being written, the Higher Education Funding Council for England (HEFCE) is carrying out the abovementioned RAE to decide on funding for universities for the period to 2010 at least. The composition of the sub-panel referred to by Clifford and Richards is illuminating: it is chaired by someone from an Earth Sciences department, with a panel made up of four others from Earth Sciences, two from combined Earth Sciences deparments (one Earth and Ocean Science, one Earth, Atmospheric and Environmental Sciences), two from Ocean Science, two from Biology and one each from Environmental Health Sciences, Environmental Sciences and two non-academic members. By comparison, the 'Geography and Environmental Studies' sub-panel (chaired by Richards) has seven panel members from Geography and one each from Geography and Environment; Geography, Earth and Environmental Sciences; Geographical and Earth Sciences; City and Regional Planning; and the Centre for Advanced Spatial Analysis (RAE, 2008). Clearly, environmental science departments needed to decide between the rock of being poorly represented in the 'Earth Systems and Environmental Sciences' sub-panel and the hard place of losing their science in the 'Geography and Environmental Studies' sub-panel. The fallout from such decisions is likely to have long-term impacts on the sorts of research carried out in the UK both within and outwith the ESS umbrella.

At an international scale, the impact of the ESS has been felt by the setting up of a number of initiatives, often within existing organisations. The IPCC has already been mentioned. It was created in 1988 by United Nations resolution as a collaboration between the World Meteorological Organization (WMO) and the United Nations Environment Programme (UNEP) and has strong parallels with the underlying concepts of ESS. The ESSP (Earth System Science Partnership) was set up following the Amsterdam convention of 2001 under the aegis of the International Council for Science (ICSU), to coordinate the efforts of four international research programmes: DIVERSITAS (the international programme of biodiversity science), IGBP (International Geosphere-Biosphere Programme), IHDP (International Human Dimensions Programme on Global Environmental Change), and the WCRP (World

Climate Research Programme) (ESSP, 2001). It has focused, among other things, on supporting research into the impacts of climate change and adaptations to them (Leary et al., 2007). The Amsterdam declaration itself states explicitly that 'The Earth System behaves as a single, self-regulating system comprised of physical, chemical, biological and human components' (Moore et al., 2001), which resonates strongly the wording of the Gaia hypothesis, although Lovelock (2006, p. 25) has argued that this is just paying 'lip service' to Gaia. The Amsterdam wording also runs counter to the *non-necessity* of stable self-regulation as discussed above. Earlier attempts were also made to integrate aspects of ESS and social change by the IGBP (Malone, 1995).

Other changes brought by ESS have been in the field of academic publications. Examples of textbooks have already been given, but there have been a range of new journals. The European Geophysical Society (now Union) established *Hydrology and Earth Systems Sciences* in 1997 and then somewhat schizophrenically *Natural Hazards and Earth System Sciences* in 2001. It is often difficult to see how the contents of either justify the 'and ESS' component of their titles. In 2005, the *Proceedings of the Indian Academy of Sciences (Earth and Planetary Sciences)* was renamed the *Journal of Earth System Science* and there is also an electronic *Journal of Earth System Science Education* (JESSE: http://jesse.usra.edu/) as well as a projected journal *Earth System Science Data and Methods*.

The Ends of Earth-System Science?

When asked to write this chapter, my response to the editors was that I would only do so if I did not have to profess a belief in ESS. Happily, they agreed. While this perspective may apparently make it difficult to write *about* ESS, it reflected a perspective that is still commonly found among colleagues: that most people seem rather unsure about what ESS actually *is*. To some extent, this problem arises because ESS is something of a chimera; every time you ask for a definition, you seem to get a different answer. To some, ESS is (actually, unhelpfully), the study of everything, while to others it is (equally unhelpfully) the study of nothing (scientific, at least). Unfortunately, the success of the core idea in attracting funding has probably compounded this issue by encouraging those seeking a greater share of those funds to redefine ESS in ways more closely aligned to what they are doing. Calls for 'paradigm change' from some quarters of the ESS literature should be seen with a dose of scepticism, even cynicism. There is nothing inherently paradigmatic about ESS in the Kuhnian sense at least. Attempts to standardise undergraduate training in ESS have been criticised as placing constraints on the development of the subject. Such standardisation is not unusual in disciplines such as geology or engineering (or indeed medicine), but is seen as going against the 'free spirit' of geography as a discipline. Church (2005) pointed out that this freewheeling approach has tended to produce geographers who are insufficiently trained to undertake ESS research. However, it is also difficult to see how one could standardise such a (trans-)discipline, given the lack of general agreement on its definition.

One of the central problems in developing such an agreement is that ESS critics often fail to specify which version of ESS they are criticising. There are at least six: what I have termed the ESS blueprint of Bretherton (1985) and the ESS Committee (1986); the post-oil reformulation of the discipline of geology; a general (but implicitly restrictive) interdisciplinary science in the mould of Lawton (2001); a bastardised/plundered physical geography; the implementation of remote sensing in Earth observation; and various Gaian or sub-Gaian homeostatic theories (of whatever strength). Most perceptions probably sit somewhere in a continuum between these versions. In many respects, ESS parallels – and should probably be informed by – developments in complexity science, which is still a highly contested topic (Manson and O'Sullivan, 2006). However, improved understanding can only emerge through debate and dialogue, which have been impeded by the lack of a common scientific language (beyond mathematics, whose use is also contested in some quarters) and the heterogeneous and multidisciplinary nature of ESS enterprise.

As geographers, should we be involved more in this dialogue, or have we already missed the boat on which it is taking place? Certainly, ESS sits in the nexus of the physical and human worlds, which was the traditional definition of the discipline, despite much subsequent drift. In terms of studies of applied environmental change, few geographers would disagree with Bretherton's conclusion that '[d]aunting though these tasks may be, they are matched by the significance of the goal. Humankind is pressing on its environment in unprecendented ways, and we do not understand the implications. We must try, for we may not have a second opportunity' (Bretherton, 1985, p. 1127). The development of an ESS perspective would also speak to an approach using concepts of globalisation as discussed by Davies (2004). If 'one of the most disconcerting aspects of ESS . . . is its apparently homogenizing, normative and nomothetic project, possibly as an unconscious attempt to "make" a more complex world more manageable' (Clifford and Richards, 2005, p. 382), should it not be the role of geographers to enter into a contestation with this normative approach and show the advantages of plurality? To enter the debate, though, we must ensure that geographers are at least ESS-literate (e.g. Church, 2005; Pitman, 2005) and indeed prepared to enter it (Thrift, 2002; Murphy, 2006). Or as Johnston (2006, p. 10) has noted (in response to Pitman, 2005), geographers must 'rid ourselves of the paranoia and inferiority complex' and get on with making a contribution.

If we return to the caveats expressed by Bretherton (1985) about the original blueprint for ESS, then this contribution can address all four. First, geographers are inherently aware of issues of scale and space. Interaction with scientists from other disciplines that also consider these issues (notably ecology) will provide an improved understanding of how Earth systems can be conceptualised in a non-reductionist way and still effect closure to allow scientific investigation. Secondly, geographers are used to looking at the world from a range of different perspectives and spend more time in the field than the average mathematical modeller. They are thus ideally placed to inform 'bottom-up' approaches (i.e., allowing system properties to emerge from its behaviour at a smaller scale) rather than the typical 'top-down' (for some, hegemonising) approaches (i.e., a definition of the system that then structures the resulting behavioural responses). While completeness is not the same as everything, everywhere, all the time, the bottom-up approach is inherently useful (as in complexity theory) for identifying general patterns or at least missing links. Thirdly, geographers have long recognised that humankind *cannot* be regarded as being external to the system. It is part of, interacts with, and strongly affects the Earth system on a range of scales, while also being strongly affected by it. Fourthly, predictability is often no longer an issue in geography, where a concern for rich understanding by diverse means has developed. These points need to be revisited at least if ESS is to mature beyond the vague continuum noted above. For some, this vagueness suggests the end of ESS as a practical (or practicable) endeavour. Certainly, the decline in allocated funding in the United States and the apparent indifference or marginalisation in UK funding may herald such an end. Equally, what we may be observing is simply the immature development of an Earth-System Pre-Science.

ACKNOWLEDGEMENTS

I am extremely grateful that Francis Bretherton took the time to respond in detail to an out-of-the-blue e-mail to answer questions on the origins of Earth-System Science. Other insights have been gratefully received through conversations with a number of colleagues. Any interpretation herein is my sole responsibility.

NOTES

1. This and the following two paragraphs are based on personal communication with Francis Bretherton. He notes that some inaccuracies may have crept in due to the passage of time but that 'it may perhaps provide encouragement or helpful advice to individuals of another generation, who are grappling with the same issues but with even greater urgency'.
2. 'Das Wort Klima bezeichnet allerdings zuerst eine specifische Beschaffenheit des Luftkreises; aber diese Beschaffenheit ist abhängig von dem perpetuirlichen Zusammenwirken . . . mit der wärmestrahlenden trockenen Erde, die . . . mit Wald und Kräutern bedeckt ist.'

BIBLIOGRAPHY

Arnavut, Z. and Narumalani, S. (1996) Application of permutations to lossless compression of multispectral thematic mapper images. *Optical Engineering*, 35, 3442–8.
Baird, A. J. and Wilby, R. L. (eds) (1999) *Eco-hydrology*. Routledge: London.
Bennett, R. J and Chorley, R. J. (1980) *Environmental Systems. Philosophy, Analysis and Control*. London: Methuen.
Bracken, L. J. and Wainwright, J. (2006) Geomorphological equilibrium: myth and metaphor? *Transactions of the Institute of British Geographers NS*, 31, 167–78.
Bretherton, F. P. (1985) Earth System Science and remote sensing. *Proceedings of the IEEE*, 73, 1118–27.
Carlson, M. P. (2001) Earth System Science sentiments. *Science*, 293, 49.
Charney, J. G. (1975) Dynamics of deserts and drought in Sahel, *Quarterly Journal of the Royal Meteorological Society*, 101, 193–202.
Chorley, R. J. and Kennedy, B. A. (1971) *Physical Geography: A Systems Approach*. London: Prentice-Hall.
Church, M. (1998) Think Globally, Learn Locally: Broadening Perspectives of the Earth. Implications for university curricula, and for the requirements for professional registration and accreditation. A report to the Canadian Geoscience Council. http://www.geoscience.ca/papersandreports/church/churchf.html (accessed 20 March 2008).
Church, M. (2005) Continental drift. *Earth Surface Processes and Landforms*, 30, 129–30.
Claussen, M. (1997) Modelling bio-geophysical feedback in the African and Indian monsoon region, *Climate Dynamics*, 13, 247–57.
Clifford, N. and Richards, K. (2005). Earth System Science: an oxymoron? *Earth Surface Processes and Landforms*, 30, 379–83.

Committee on Earth Sciences (1989) *Our Changing Planet: A U.S. Strategy for Global Change Research*. USGCRP, Washington, DC. http://www.usgcrp.gov/usgcrp/Library/OCP-strategy1989.pdf (accessed 8 March 2008).

Davies, W. K. D. (2004) Globalization. A spatial perspective. In J. A. Matthews and D. T. Herbert (eds), *Unifying Geography. Common Heritage, Shared Future*. London: Routledge, pp. 189–214.

Demeritt, D. (2001) The construction of global warming and the politics of science. *Annals of the Association of American Geographers*, 91, 307–37.

Duro, D. C., Coops, N. C., Wulder, M. A. and Han, T. (2007) Development of a large area biodiversity monitoring system driven by remote sensing. *Progress in Physical Geography*, 31, 235–60.

Dutton, J. A. (1987) Predicting the Earth's future. *Acta Astronautica*, 16, 305–12.

Eagleson, P. S. (1978a) Climate, soil, and vegetation. 1. Introduction to water-balance dynamics. *Water Resources Research*, 14, 705–12.

Eagleson, P. S. (1978b) Climate, soil, and vegetation. 2. Distribution of annual precipitation derived from observed storm sequences. *Water Resources Research*, 14, 713–21.

Eagleson, P. S. (1978c) Climate, soil, and vegetation. 3. Simplified model of soil-moisture movement in liquid-phase. *Water Resources Research*, 14, 722–30.

Eagleson, P. S. (1978d) Climate, soil, and vegetation. 4. Expected value of annual evapotranspiration. *Water Resources Research*, 14, 731–9.

Eagleson, P. S. (1978e) Climate, soil, and vegetation. 5. Derived distribution of storm surface runoff. *Water Resources Research*, 14, 741–8.

Eagleson, P. S. (1978f) Climate, soil, and vegetation. 6. Dynamics of annual water-balance. *Water Resources Research*, 14, 749–64.

Eagleson, P. S. (1978g) Climate, soil, and vegetation. 7. Derived distribution of annual water yield. *Water Resources Research*, 14, 765–76.

Eagleson, P. S. (2002) *Ecohydrology. Darwinian Expression of Vegetation Form and Function*. Cambridge: Cambridge University Press.

Earth System Sciences Committee (1986) *Earth System Science. Overview: A Program for Global Change*. Washington, DC: NASA.

Entekhabi, D. Rodriguez-Iturbe, I. and Bras, R. L. (1992) Variability in large-scale water-balance with land surface atmosphere interaction. *Journal of Climate*, 5, 798–813.

Ernst, W. G. (2000) Synthesis of Earth systems and global change. In W. G. Ernst (ed.), *Earth Systems: Processes and Issues*. Cambridge: Cambridge University Press, pp. 519–32.

ESSAAC (2008) Earth System Science & Applications Advisory Committee (ESSAAC). http://science.hq.nasa.gov/strategy/essaac/index.html (accessed 5 March 2008).

ESSP (2001) Earth System Science Partnership Briefing Paper. http://www.essp.org/ (accessed 15 March 2008.

Farmer, C. (2001) Earth System Science sentiments. *Science*, 293, 49.

Feingold, G. and Heymsfield, A. J. (1992) Parameterizations of condensational growth of droplets for use in general-circulation models. *Journal of the Atmospheric Sciences*, 49, 2325–42.

GEWEX (1995) Global Energy and Water Cycle Experiments: International Satellite Land-Surface Climatology Project (ISLSCP). http://www.gewex.org/islscpdata.htm (accessed 20 March 2008).

GOTW (2007) *Grants on the Web*. NERC online database of funded research at http://gotw.nerc.ac.uk (accessed 12 July 2007).

Hall, F. G. and Sellers, P. J. (1995) First International Satellite Land Surface Climatology Project (ISLSCP) Field Experiment (FIFE) in 1995. *Journal of Geophysical Research-Atmospheres*, 100, 25383–95.

Heimann, M. (1997) A review of the contemporary global carbon cycle and as seen a century ago by Arrhenius and Hogbom. *Ambio*, 26, 17–24.

Huggett, R. J. (1999) Ecosphere, Biosphere, or Gaia? What to call the global ecosystem. *Global Ecology and Biogeography*, 8, 425–31.

Hyman A. H. (1996) Information presentation for new sensors: a focus on selected sensors of the Earth Observing System (EOS). *Progress in Physical Geography*, 20, 146–58.

INFOTERRA (2000) Dublin Declaration on Access to Environmental Information. UNEP-Infoterra http://www.unep.org/infoterra/infoterra2000/report1.htm (accessed 15 March 2008).

Innes, J. L. and Koch, B. (1998) Forest biodiversity and its assessment by remote sensing. *Global Ecology and Biogeography*, 7, 397–419.

IPCC (2007) *IPCC Fourth Assessment Report: Working Group I Report 'The Physical Science Basis'* (especially Chapter 7). http://www.ipcc.ch/ipccreports/ar4-wg1.htm (accessed 20 March 2008).

Jackson, R. D. and Idso, S. B. (1975) Surface albedo and desertification, *Science*, 189, 1012–3.

Johnston, R. (2006) Geography (or geographers) and earth system science. *Geoforum*, 37, 7–11.

Justice, C. O. and Townshend, J. R. (1994) Data sets for global remote-sensing – lessons learnt. *International Journal of Remote Sensing*, 15, 3621–39.

Kennedy, D. (2000) Physical geography. In W. G. Ernst (ed.), *Earth Systems: Processes and Issues*. Cambridge: Cambridge University Press, pp. 13–25.

King, R. L. and Birk, R. J. (2003) Science for society: Delivering Earth system science knowledge for decision support in the year 2025. *IGARSS 2003: IEEE International Geoscience and Remote Sensing Symposium, Proceedings – Learning From Earth's Shapes and Sizes*. Toulouse: IEEE, 1035–37, DOI: 10.1109/IGARSS.2003.1294003.

Kirchner, J. W. (1991) The Gaia hypotheses are they testable? Are they useful?. In S. H. Schneider and P. J. Boston (eds), *Scientists on Gaia*. Cambridge, MA: MIT Press, pp. 38–46.

Kirchner, J. W. (2002) The Gaia hypothesis: fact, theory, and wishful thinking. *Climatic Change*, 52, 391–408.

Kirchner, J. W. (2003) The Gaia hypothesis: conjectures and refutations. *Climatic Change*, 58, 21–45.

Körner, C. (2006) Plant CO_2 responses: an issue of definition, time and resource supply. *New Phytologist*. 172, 393–411.

Lawton, J. (2001) Earth System Science. *Science*, 292, 1965.

Leary, N, Kulkarni, J. and Seipt, C. (eds) (2007) *Assessment of Impacts and Adaptation to Climate Change. Final Report of the AIACC Project*. Project No. GFL-2328-2724-4330. The International START Secretariat, Washington, DC. www.aiaccproject.org (accessed 15 March 2008).

Lenton, T. M. (2002) Testing Gaia: the effect of life on Earth's habitability and regulation. *Climatic Change*, 52, 409–22.

Lenton, T. M. and van Oijen, M. (2002) Gaia as a complex adaptive system. *Philosophical Transactions of the Royal Society B*, 357, 683–95.

Lenton, T. M. and Wilkinson, D. M. (2003) Developing the Gaia theory. *Climatic Change*, 58, 1–12.

Lovelock, J. E. (1972) Gaia as seen through the atmosphere. *Atmospheric Environment*, 6, 579–80.

Lovelock, J. E. and Margulis, L. (1974) Atmospheric homeostasis by and for the biosphere – the Gaia hypothesis. *Tellus*, 26, 2–10.

Lovelock, J. E. (1979) *Gaia: A New Look at Life on Earth*. Oxford: Oxford University Press.

Lovelock, J. E. (2000) *Homage to Gaia*. Oxford: Oxford University Press.

Lovelock, J. E. (2006) *The Revenge of Gaia*. London: Allen Lane/Penguin.

Malone, T. F. (1995) Reflections on the human prospect. *Annual Reviews of Energy and Environment*, 20, 1–29.

Manson, S. and O'Sullivan, D. (2006) Complexity theory in the study of space and place. *Environment and Planning A*, 38, 677–92.

Margulis, L. and Lovelock, J. E. (1974) Biological modulation of the Earth's atmosphere. *Icarus* 21, 471–89.

Merritts, D. de Wet, A. and Menking, K. (1997) *Environmental Geology: An Earth System Science Approach*. New York: W. H. Freeman.

MODIS (2008) Moderate Resolution Imaging Spectroradiometer (MODIS). http://modis. gsfc.nasa.gov/ (accessed 20 March 2008).

Moore, B., III, Underdal, A., Lemke, P. and Loreau, M. (2001) The Amsterdam Declaration on Global Change. http://www.sciconf.igbp.kva.se/ (accessed 20 March 2008).

Murphy, A. B. (2006) Enhancing geography's role in public debate. *Annals of the Association of American Geographers*, 96, 1–13.

NASA (2007) *Sensing Our Planet. NASA Earth Science Research Features 2007*. Washington, DC: NASA. http://nasadaacs.eos.nasa.gov/pdf/annual_2007.pdf (accessed 20 March 2008).

NASA/ESRA (2007) *Design Guide for Undergraduate Earth System Science Education. A Product of the NASA/USRA Earth System Science for the 21st Century Program*. http:// essedesignguide.org (accessed 20 March 2008).

NERC (2007) *Next Generation Science for Planet Earth. NERC Strategy 2007–2012. NERC, Swindon*. http://www.nerc.ac.uk/publications/strategicplan/documents/strategy07. pdf (accessed 20 March 2008).

Nitta, K. (1994) Earth environment and closed ecology experiment facilities. *Acta Astronautica*, 33, 155–65.

Oreskes N., Scrader-Frechette, K. and Belitz, K. (1994) Verification, validation, and confirmation of numerical models in the earth sciences. *Science*, 263, 641–6.

Pataki, D. E. (2002) Atmospheric CO_2, climate and evolution – lessons from the past. *New Phytologist*, 154, 10–12.

Pitman, A. J. (2005) On the role of Geography in Earth System Science. *Geoforum*, 36, 137–48.

Press, F. and Siever, R. (1982) *Earth*, 3rd edn. New York: WH Freeman and Company.

Press, F., Siever, R., Grozinger, J. and Jordan, T. H. (2004) *Understanding Earth*, 4th edn. New York: WH Freeman and Company.

Primavesi, A. (2000) *Sacred Gaia: Holistic Theology and Earth System Science*. London: Routledge.

Qi, J., Chehbouni, A., Huete, A. R., Kerr, Y. H. and Sorooshian, S. (1994) A modified soil adjusted vegetation index. *Remote Sensing of Environment*, 48, 119–26.

RAE (2008) Unit of Assessment 17 'Earth Systems and Environmental Sciences' and Unit of Assessment 32 'Geography and Environmental Studies'. http://www.rae.ac.uk/panels/ members/ (accessed 8 March 2008).

Raymo, M. E. (1994) The initiation of Northern Hemisphere glaciation. *Annual Reviews of Earth and Planetary Science*, 22, 353–83.

Running S. W., Justice, C. O., Salomonson, V., Hall, D., Barker, J., Kaufmann, Y. J., Strahler, A. H., Huete, A. R., Muller, J. P., Vanderbilt, V., Wan, Z. M., Teillet, P. and Carneggie, D. (1994) Terrestrial remote-sensing science and algorithms planned for EOS MODIS. *International Journal of Remote Sensing*, 15, 3587–620.

Scheffer, M., Holmgren, M., Brovkin, V. and Claussen, M. (2005) Synergy between small- and large-scale feedbacks of vegetation on the water cycle. *Global Change Biology*, 11, 1003–12.

Schneider, S. (2000) Why study Earth systems science? In W. G. Ernst (ed.), *Earth Systems: Processes and Issues*. Cambridge: Cambridge University Press, pp. 5–12.

Skinner, B. J., Porter, S. C. and Botkin, D. B. (1999) *The Blue Planet. An Introduction to Earth System Science*, 2nd edn. Chichester: John Wiley & Sons, Ltd.

Soberon, J. and Peterson, A. T. (2004) Biodiversity informatics: managing and applying primary biodiversity data. *Philosophical Transactions of the Royal Society of London Series B –Biological Sciences*, 359, 689–98.

Stoms, D. M. and Estes, J. E. (1993) A remote-sensing research agenda for mapping and monitoring biodiversity. *International Journal of Remote Sensing*, 14, 1839–60.

Suzuki, K., Nakajima, T., Numaguti, A., Takemura, T., Kawamoto, K. and Higurashi, A. (2004) A study of the aerosol effect on a cloud field with simultaneous use of GCM modeling and satellite observation. *Journal of the Atmospheric Sciences*, 61, 179–94.

Thornes, J. B. (1985) The ecology of erosion, *Geography*, 70, 222–36.

Thornes, J. B. (1988) Erosional equilibria under grazing, In J. Bintliff, D. Davidson and E. Grant (eds), *Conceptual Issues in Environmental Archaeology*, 193–211, Edinburgh University Press, Edinburgh.

Thornes, J. B. (ed.) (1990) *Vegetation and Erosion*. Chichester: John Wiley & Sons, Ltd.

Thrift, N. (2002) The future of geography. *Geoforum*, 33, 291–98.

Turner, M. G., Gardner, R. H. and O'Neill, R. V. (2001) *Landscape Ecology in Theory and Practice: Pattern and Process*. Springer Verlag: Berlin.

Turner, B. L., II (2002) Contested identities: human-environment geography and disciplinary implications in a restructuring academy. *Annals of the Association of American Geographers*, 92, 52–74.

UCAR (1959) *Preliminary Plans for a National Institute for Atmospheric Research. Prepared for the National Science Foundation.* Second Progress Report of the University Committee on Atmospheric Research. http://www.library.ucar.edu/uhtbin/cgisirsi/20080310155755/SIRSI/0/520/DR000134 (accessed 20th March 2008.).

Vetter, R. V., Ali, M., Daily, D., Gabrynowicz, J., Narumalani, S., Nygard, K., Perrizo, W., Ram, P., Reichenbach, S., Seielstad, G. A. and White, W. (1995) Accessing Earth system science data and applications through high-bandwidth networks. *IEEE Journal on Selected Areas in Communications*, 13, 793–805.

Viles, H. (ed.) (1988) *Biogeomorphology*. Oxford: Blackwell.

Volk, T. (2003) Natural selection, Gaia, and inadvertent by-products: a reply to Lenton and Wilkinson's response. *Climatic Change*, 58, 13–20.

Wainwright, J. and Mulligan, M. (eds) (2003) *Environmental Modelling: Finding Simplicity in Complexity*. John Wiley and Sons: Chichester.

Wanner, W., Strahler, A. H., Hu, B., Lewis, P., Muller, J. P., Li, X., Schaaf, C. L. B. and Barnsley, M. J. (1997) Global retrieval of bidirectional reflectance and albedo over land from EOS MODIS and MISR data: theory and algorithm. *Journal of Geophysical Research – Atmospheres*, 102, 17143–61.

Webb, M., Senior, C., Bony, S. and Morcrette, J. J. (2001) Combining ERBE and ISCCP data to assess clouds in the Hadley Centre, ECMWF and LMD atmospheric climate models. *Climate Dynamics*, 17, 905–22.

Wendler, G. and Eaton, F. (1983) On the desertification of the Sahel zone; 1. ground observations climatic change. *Climatic Change*, 5, 365–80.

White, M. A. and Nemani, R. R. (2003) Canopy duration has little influence on annual carbon storage in the deciduous broad leaf forest. *Global Change Biology*, 9, 967–72.

Williams, B. K. (1996) Assessment of accuracy in the mapping of vertebrate biodiversity. *Journal of Environmental Management*, 47, 269–82.

Xue, Y. K. and Shukla, J. (1993) The influence of land-surface properties on Sahel climate; 1: desertification. *Journal of Climate*, 6, 2232–45.

Zeng, N., Neelin, J. D., Lau, K. M. and Tucker, C. J. (1999) Enhancement of interdecadal climate variability in the Sahel by vegetation interaction. *Science*, 286, 1537–40.

Zhou, L. M., Dickinson, R. E., Tian, Y. H., Vose, R. S. and Dai, Y. J. (2007) Impact of vegetation removal and soil aridation on diurnal temperature range in a semiarid region: Application to the Sahel. *Proceedings of the National Academy of Sciences of the United States of America*, 104, 17937–42.

Chapter 11

Land Change (Systems) Science

B. L. Turner II

Introduction

Land change science (alternatively, land systems science) is a rapidly emerging, interdisciplinary field of study that seeks to understand, explain, and project land-use and land-cover dynamics (Turner, 2002; Gutman et al., 2004; Turner et al., 2007). It has been stimulated by international concern regarding global environmental change, the search for sustainability, and the recognition of the pivotal role of land dynamics in both. Neither the recognition of the human impress on the land (Marsh, 1965[1864]; Thomas, 1956) nor the need for a formal approach to its study, captured in the German geographic concept of *landschaft*, is new. The totality of land changes currently underway and their far-flung consequences (Steffen et al., 2003) are unprecedented, however, spawning a new-found need for integrative studies of land systems dynamics. In this sense, land change or land systems science may be viewed as a reinvention of *landscahft* research with a face decidedly gazing at the environmental sciences at large.

Matching Nature on Land

The human dominion over the terrestrial surface of the earth is well documented. Thirty-to-fifty percent of the land surface has been transformed – radically altered – by human activities (Vitousek et al., 1997), an area roughly the size of South America has been taken to cultivation (Raven, 2002), and virtually no land surface may be considered 'pristine' if co-evolved landscapes, both forest and grasslands, human-induced climate change, and tropospheric pollution are considered (Meyer and Turner, 1994). Land change joins industrial change to elevate human activity more-or-less equivalent to nature in affecting the biogeochemical flows that sustain the biosphere, leading some expert to suggest that humankind has entered the 'Anthropocene' (Crutzen and Stoermer, 2000).

The antiquity of local-to-regional scale land changes of import to society and the environment has long been understood (Thomas, 1956; Redman, 1999). Recent evidence, however, supports interpretations of continental-to-global scale human impacts on land systems dating to prehistory. The extinction of megafauna in Australia and the Western Hemisphere, some 40–30,000 and 10,000 years ago, respectively, apparently involved human-induced landscape changes in concert with climate change (e.g., Martin, 2005). Biota exchanges and concomitant landscape changes amplified during the 'age of exploration' and European colonisation of world also qualifies as global-scale land change (Crosby, 1986). In retrospect, each new technomanagerial phase of humankind has escalated change in the land *structure* of the earth system, with major consequences for ecosystem goods and services and the provisioning of food, fuel, and shelter for humankind (Turner and McCandless, 2004; MEA, 2005). The entire land structure of some regions has long been transformed (e.g., McNeill, 1992; Foster and Aber, 2004; Butzer, 2005; Kirch, 2005), and today the land-cover impacts of deforestation-forestation, cultivation, pasture, arid land degradation and water withdrawal have reached a global dimension in magnitude and spatial reach. Croplands and pasture consume about 40 percent of land surface of the earth (Foley et al., 2005), gained at the expense of forest and arid-land covers (Williams, 2005). An estimated 10–20 percent of arid lands, which cover about 41 percent of the terrestrial surface of the earth (Reynolds and Smith, 2002), is degraded from human activity. Agriculture consumes about 85 percent of annual global water withdrawal – that withdrawal now approaching about 10 percent of renewable resources (Foley et al., 2005) – and rangelands house some 3.3 billion cattle, sheep, and goats (Raven, 2002, p. 954). Including other land uses – for example, settlements, roads, reservoirs, recreation areas – the land covers of the world have been increasingly fragmented, with impacts ranging from access to pollinators to threats of biota extinctions of biota globally (MEA, 2005).

These land changes have reached such a magnitude that they now affect the *function* of the earth system through impacts on albedo (reflectivity) and biogeochemical cycles (but see Ruddiman, 2005). Land-based activities usurp up to 40–60 percent of NPP (Vitousek et al., 1997; Rojstaczer et al., 2001). Synthetic nitrogen production, dominated by fertilizer for agriculture, has superseded nature's flow of nitrogen (Matson et al., 1997) and land uses, largely deforestation and tilling, comprise about 30 percent of the source of anthropogenic carbon in the atmosphere (Watson et al., 2000; Foley et al., 2005). Tropical deforestation, especially in Amazonia, portends to have major consequences on global hydrologic cycle (Zhang et al., 2001).

Institutional Response

These facets of land change were quickly recognised among the international and multidisciplinary sciences addressing climate change (Intergovernmental Panel on Climate Change, IPCC) and earth system science (International Geosphere-Biosphere Programme, IGBP) and led them to call for improved understanding of land dynamics with outputs complementary with their research agendas. The IGBP approached the then budding International Human Dimensions Programme, requesting a joint international project on Land-Use/Cover Change agenda (LUCC). Given that the IGBP already had strong programmes on the biophysical side of the land change, LUCC focused on land-change observation and monitoring (remote sensing), land processes and land-change (spatial) modelling, with the intent that the human subsystem of the land would connect to the environment subsystem through land cover (Gutman et al., 2004; Lambin and Geist, 2005).

The intimate ties between human and environmental subsystems comprising land subsequently led to the merger of the IGBP's ecological research with that of LUCC, as informed by DIVERSITAS (biotic diversity programme), to create the Global Land Project (GLP, 2005). The GLP, in response to sustainability science, as registered in the Earth System Science Partnership (ESSP) and related programmes (Kates et al., 2001), now focuses on land as a coupled human-environment or social-ecological system and expands base research to various synthesis efforts consistent with the needs of decision makers, especially vulnerability-resilience and sustainability activities, moving the GLP closer to such concerns as food security and environmental justice, among others.

Land as a Coupled System

Land systems and their change are product of human-environment interactions and have been at least since humankind mastered the control of fire in the hunt, in the process transforming habitat and fauna (Thomas, 1956; Martin, 2005). These interactions have intensified globally throughout history (Goudie and Vilas, 1997) as society attempts to manipulate the productivity of land for resources, or in earth systems' parlance, ecosystem goods and services (e.g., water, soil nutrients) and to reduce risks to the vagaries of nature (e.g., drought, pest outbreaks, floods). Today, virtually no part of terrestrial surface of the earth remains unclaimed or lacks some form of governance, although governing institutions may be ineffective in their enforcement.

Land systems and their change involve the ambient environmental conditions (e.g., temperate forest biome); the uses of those conditions (e.g., wheat cultivation, suburbia, nature reserves); the consequences of the uses, both human and environmental (e.g., arid land degradation; corporate profits); and the impacts of those consequences on the land systems (e.g., loss of biodiversity, shifts in land uses). Thus land systems are coupled human and environmental subsystems in which both endogenous (e.g., soil conditions and fertilizer applications) and exogenous subsystem dynamics (e.g., global warming or market failures) affect and even change the subsystems. Perhaps owing to this complexity or to the aggregation of phenomena and process required to address land-use/cover change, theory of land system change per se – coupling the two subsystems – has been difficult beyond broad system concepts (Gunderson and Holling, 2001). Despite this lacuna, research on land systems and their change moves forward on all fronts – causes, consequences and system linkages of use and cover change.

Drivers of land use

At the global scale and over the long run, land dynamics appear to track with the PAT variables – population, affluence and technology – in the IPAT identity (Waggoner and Ausubel, 2002; Turner and McCandless, 2004) – as P (population) and A (affluence) serve as surrogates for demand for land and land-resources and T (technology) as the means of fulfilling that demand. At lower spatio-temporal resolutions, however, PAT variables give way, at least quantitatively, to a plethora of political economic and biophysical factors, be they climate change or globalisation, the last of which leads to the loss of spatial congruency between the source of the demand and the land generating product (Lambin et al., 2003). Meta-analyses indicate the power of different combinations of factors to account for land changes, by time and place, including institutions, economy and culture (Agarwal and Yadama, 1997; Barbier and Burgess, 2001; Lambin et al., 2001). This variance has led to such general conclusions as deforestation occurs whenever and wherever the demand for forest use and the power to achieve it exist (paraphrasing Angelsen and Kaimowitz, 1999), especially where 'frontier' forest lands exist (Barbier, 2004). Place-based research points to the role of the specific factors that generate this demand-power function, such as markets (Brown and Pearce, 1994), policy (Binswanger, 1991), transportation-road networks (Cropper et al., 1999) and household lifecycles (Perez and Walker, 2002) impact on deforestation. Yet other research explores older theoretical themes in new ways, such as induced intensification (Laney, 2004), as well as the role of different explanatory approaches for addressing land change (Roy Chowdhury and Turner, 2006).

Land use to land cover and environment

Sustained documentation of the consequences of land uses on land covers continues, although more attention has been given to environmental drawdown than to sustainable activities (e.g., Barrows, 1991; Kasperson et al., 1995; Nepstad et al., 1999; Seto et al., 2000; MEA, 2005; but see Ellis and Wang, 1997; Johnson and Lewis, 2007). Immediate and visible ecological consequences, such as soil erosion, have increasingly shared attention with less visible ecological and earth system ones, including landscape functioning under different levels of habitat loss and fragmentation (Skole and Tucker, 1993; Sala et al., 2000; Higgins et al., 2003; DeFries et al., 2004). Land changes often open the door for invasive species which not only change plant functional relationships but the capacity of the ecosystem to restore itself for some future use, as in swidden or slash-and-burn cultivation, or the economic costs of combating the invasion (Mooney and Hobbs, 2000; Schneider and Geoghegan, 2006). Such 'on-site' consequences are increasingly matched and in regard to the functioning of the earth system, superseded by those cumulative consequences of repeated land uses worldwide (Meyer and Turner, 1994).

Perhaps the best documented global-scale impacts are those of tropical deforestation, largely for cultivation and pasture, on the loss of biodiversity (e.g., Cervigni, 2001; DIVERSITAS, 2002) and on global climate warming, through carbon and radiative dynamics (e.g., Houghton et al., 2000; Zhang et al., 2001; Pielke, 2002; Steffen et al., 2003). Importantly, land-change research has also demonstrated that regional-scale land changes have significant consequences on regional temperature and precipitation regimes (e.g., Pielke et al., 1999), in some cases exceeding the projected changes of global climate. In addition, recent but debated research suggests that urban conglomerations may be affecting warming at regional scales and above (Kalnay and Cai, 2003; Zhou et al., 2004).

Environment to land cover and land use

Research on land-cover change feedbacks on land uses has grown from such base agronomic and climate change issues as, respectively, soil erosion and crop responses (Rosenzweig and Parry, 1994; Lal, 2004) to questions of ecosystem services for land uses (Daily et al. 2000). Land-cover change challenges the capacity of many ecosystems and landscapes to deliver the expected goods and services for specified land management systems, ranging from seed stocks to water filtration (Daily et al., 2000; DeFries et al., 2004; MEA, 2005). Fragmentation of ecosystems or land barriers to the flow biota may disrupt or reduce the movement of biota across landscape and regions, especially along ecoclines or involving keystone species, with implications for the functioning of ecosystems. Increasing research examines this relationship for nature reserves, especially in regard to land changes beyond the reserve (Homewood et al., 2001; Terborgh et al., 2002).

In addition to these immediate feedbacks, land uses are affected by those operating through climate and other atmospheric changes. For example, regional-to-continental scale, ground-level ozone from industrial-urban regions spreads across prime croplands worldwide, interacting with nitrous oxide released from fertilizers to reduce crop yields (Chameides et al., 1994; Matson et al., 1997; Tilman et al., 2002). Climate change in conjunction with land changes also threatens such sensitive land covers as tropical forests (Nobre et al., 1991; Laurance, 1998) as well as the functioning of terrestrial ecosystems and their land covers everywhere (Walther et al., 2002).

Observing-monitoring land change

Perhaps no part of land change science has advanced more than that dealing with observation and monitoring as the use of satellite remote sensing has become increasingly fine-grain in spatial and temporal resolution and employed in novel ways (Walsh and Crews-Meyer, 2002; Fox et al., 2003; Wulder and Franklin, 2006). Seamless global data of different types of land cover can now be produced that address a large number of vegetative attributes (Defries et al., 2000; Loveland, 2000), such as their functional properties (e.g., DeFries et al., 1995), improving datasets for various kinds of global models.

Advances have been made as well in a large array of remote sensing data assessments for specific kinds of land change detection-assessment. Examples include the temporal patterns of landscape burning and their implications for cultivation and burning policies (Laris, 2005); attempts to separate climate from land management impacts on vegetation in order to assess the consequences of stocking strategies (Archer, 2004); detection of 'cryptic' deforestation by way of selective logging (Nepstad et al., 1999; Asner et al., 2005); mapping and monitoring 'hot spots' of biological diversity (Myers et al., 2000), although such efforts perhaps should be directed at populations (Ceballos and Ehrlich, 2006); observing land changes to urbanisation and peri-urban uses (Seto et al., 2002; Seto and Kaufman, 2003); and linking successional states of forest growth to household lifecycles (Moran et al., 1994; McCrackin et al., 1999), participatory mapping (Mapedza et al., 2003), ethnology (Nyerges and Green, 2000) and disasters (Lupo et al., 2001).

While each project tends to design its own land classification suited to the observational instrument and the aims of the project, headway has been made on meta-classification, complete with software, in order to permit individual project products to be compared (Di Gregorio 2005). In addition, land classifications and monitoring are now used as accounting mechanisms for differing governing units (e.g., EEA, 2006).

Modelling land dynamics

The advances in land-change modelling have matched those in remote-sensing observations, driven in large part the demand for spatially (geographically) explicit model outputs (Lambin, 1994; Veldkamp and Lambin, 2001; Agarwal et al., 2002). While little agreement exists regarding model taxonomies, the array of 'integrated' modelling efforts – combining human and environmental variables to address both human and environment outcomes – cross cuts ecology (Liu, 2001), economics (Kaimowitz and Angelsen, 1998; Irwin and Geoghegan, 2001) and the interdisciplinary communities (Liverman et al., 1998; Veldkamp and Lambin, 2001). Both empirical and theoretical models have been directed to projecting land-use/cover changes down to the pixel level (e.g., Veldkamp and Fresco, 1996; Liverman et al. 1998; Bell and Bockstael, 2000; Walker et al., 2004), as have agent-based models (Parker et al., 2003; Manson and Evans, 2007). Modelling efforts address the full array of new GIScience methods (Walsh et al., 1999; Brown et al., 2000; Pijanowski et al., 2002), are applied from frontier to urban settings (Batty et al., 1997; Geoghegan et al., 2005) and explore environmental or land-cover feedbacks on land use (Verburg, 2006). The spatially explicit nature of the land-change modelling has also triggered advances in the measures of the accuracy of the outcomes (e.g., Pontius, 2002).

Coupling and synthesis: future pathways

The research captured in the headings above has not yet reached its maturity, in part owing to the complexity of land system dynamics and the trans-disciplinary nature of integrated assessments, which carry with them an array of analytical problems (Rindfuss et al., 2004). Major advances are expected in each category of research, however, especially regarding observation-monitoring and modelling, if only because of the level of research expenditures devoted to them internationally.

Global environmental change and sustainability science, however, place demands on the land-change community to move rapidly towards synthesis products and assessments (e.g., Turner et al., 2003a); that is, to move from single 'sector' analyses, such as 'hot spots' of xeric land degradation or losses in biotic diversity, to issues of total system resilience-vulnerability (Adger, 2000; Downing et al., 2000; Turner et al., 2003b) and sustainability (Schellnhuber et al., 1997; Berkes et al., 2003; Clark et al., 2004). This orientation, in turn, demands that land be treated as coupled human-environment or social-ecological systems in which the synergy of the sub-systems sets the conditions of the response of both subsystems to external drivers (Cutter et al., 2000; Luers et al., 2003; Turner et al., 2003c), as well as the consequences of the coupled system for the earth system at large (e.g., carbon and nitrogen cycles).

Implications for Geography

From the IPCC to the ESSP, many geographers have been instrumental in the development of global environmental change and sustainability research agenda-setting and research efforts (Kates et al. 2001; Liverman et al., 2004), in part owing to the long-standing, geographic traditions of integrated approaches to and synthesis understanding of earth system processes and human-environment relationships.

Among these activities, land change science, more so than any other endeavour, highlights the full expanse of the geographical sciences (e.g., Lambin and Geist, 2005) because of its spatially explicit treatment of the coupled human-environment system.

The geographical sciences, however, increasingly encapsulate more than the formal discipline of geography. Geographical information science, including remote sensing, now is standard fare in far-flung research communities and the ecological sciences tackle human-environment relationships as they increasingly recognise the intimate role of human activity in environmental processes and outcomes. The coupled system – perhaps the hallmark of the original and contemporary study of, respectively, *landschaft* and human-environment relationships – and its examination in spatially explicit ways are no longer the primary domain of geography, if they ever were. Rather these endeavours are increasingly those of interdisciplinary research institutes (e.g., Postdam Institute for Climate Impact Research and Stockholm Environment Institute), including those directed explicitly to land systems (MacCaulay Institute, UK) and in the United States, newly minted doctoral degree programmes in such elite institutions as Stanford and Columbia, which lack geography programmes.

The immediate future appears to be one in which geographic practitioners of land systems are drawn increasingly into integrative science programmes, while geographic pedagogy, more so than at any other time in the past, opens to practitioners from the beyond the formal discipline. What these developments portend for geography per se are unclear (Turner, 2003). They do point to at least one major conclusion: land systems, however defined, are the topic of engagement by an increasingly large numbers of natural, social and integrated sciences whose shear number of practitioners overwhelms the number of geographers undertaking the topic. The land change/system science of the future will thus decidedly differ from the *landschaft* study of the past.

BIBLIOGRAPHY

Adger, N. (2000) Social and ecological resilience: are they related? *Progress in Human Geography*, 24, 347–64.

Agarwal, C., Green, G. M., Grove, J. M., Evans, T. P. and Schweik, C. M. (2002) *A Review and Assessment of Land-Use Change Models: Dynamics of Space, Time, and Human Choice, Gen. Tech. Rep. NE-297*. Newton Square, PA: U.S. Department of Agriculture, Forest Service, Northeastern Research Station.

Agarwal, A. and Yadama, G. N. (1997) How do local institutions mediate market and population pressures on resources?: forest *Panchayats* in Kumaon, India. *Development and Change*, 28, 435–65.

Angelsen, A. and Kaimowitz, D. (1999) Rethinking the causes of deforestation: lessons from economic models. *The World Bank Research Observer*, 14, 73–98.

Archer, E. R. M. (2004) Beyond the 'climate versus grazing' impasse: using remote sensing to investigate the effects of grazing system choice on vegetation cover in eastern Karoo. *Journal of Arid Environments*, 57, 381–408.

Asner, G. P., Knapp, D. E., Broadbent, E. N., Oliveira, P. J. C., Keller, M. and Silva, J. N. (2005) Selective logging in the Brazilian Amazon. *Science*, 310, 480–81.

Barbier, E. B. (2004) Explaining agricultural land expansion and deforestation in developing countries. *American Journal of Agricultural Economics*, 86, 1347–53.

Barbier, E. B. and Burgess, J. C. (2001) The economics of tropical deforestation. *Journal of Economic Surveys*, 15, 413–33.

Barrows, C. J. (1991) *Land Degradation: Development and Breakdown of Terrestrial Environments*. Cambridge: Cambridge University Press.

Batty, M., Couclelis, H. and Eichen, M. (1997) Urban systems as cellular automata. *Environment and Planning B: Planning and Design*, 24, 159–305.

Bell, K. P. and Bockstael, N. E. (2000) Applying the generalized-moments estimation approach to spatial problems involving microlevel data. *Review of Economics and Statistics*, 82, 72–82.

Berkes, F., Colding, J. and Folke, C. (eds) (2003) *Navigating Social-Ecological Systems: Building Resilience for Complexity and Change*. Cambridge: Cambridge University Press.

Binswanger, H. (1991) Brazilian policies that encourage deforestation in the Amazon. *World Development*, 19, 821–9.

Brown, K. and Pearce, D. (1994) *The Causes of Tropical Deforestation: The Economic and Statistical Analysis of Factors Giving Rise to the Loss of Tropical Forests*. Vancouver: University of British Columbia Press.

Brown, D. G., Pijanowski, B. C. and Duh, J. D. (2000) Modeling the relationships between land use and land cover on private lands in the Upper Midwest USA. *Journal of Environmental Management*, 59, 247–63.

Butzer, K. W. (2005) Environment history in the Mediterranean world: cross-disciplinary investigation of cause-and-effect for degradation and soil erosion. *Journal of Archaeological Science*, 32, 1773–1800.

Ceballos, G. and Ehrlich, P. R. (2006) Global mammal distributions, biodiversity hotspots and conservation. *Proceedings of the National Academy of Sciences U.S.A.*, 103, 19374–79.

Cervigni, R. (2001) *Biodiversity in the Balance: Land Use, National Development and Global Welfare*. Cheltenham, UK: Edward Elgar.

Chameides, W. L., Kasibhatla, P. S., Yienger, J. and Levy, H., II. (1994) Growth of continental-scale metro-agro-plexes, regional ozone pollution, and world food production *Science*, 264, 74–7.

Clark, W. C., Crutzen, P. and Schellnhuber, H.-J. (eds) (2004) *Earth System Analysis for Sustainability*. Dahlem Workshop report no. 91. Cambridge, MA: MIT Press.

Cropper, M., Griffiths, C. and Mani, M. (1999) Roads, population pressures, and deforestation in Thailand, 1976–1989. *Land Economics*, 75, 58–73.

Crosby, A. W. (1986) *Ecological Imperialism: The Biological Expansion of Europe, 900–1900*. Cambridge: Cambridge University Press.

Crutzen, P. J. and Stoermer, E. F. (2000) The 'Anthropocene'. *Global Change Newsletter*, 41, 12–13.

Curran, L. M., Trigg, S. N., McDonald, A. K., Astiani, D., Hardiono, Y. M., Siregar, P., Caniago, I. and Kasischke, E. (2004) Lowland forest loss in protected areas on Indonesian Borneo. *Science*, 303, 1000–3.

Cutter, S., Mitchell, J. T. and Scott, M. S. (2000) Revealing the vulnerability of people and places: a case study of Georgetown County, South Carolina. *Annals of the Association of American Geographers*, 90, 713–37.

Daily, G. C., Söderqvist, T., Aniyar, S., Arrow, K., Dasgupta, P., Ehrlich, P. R., Folke, C., Hansson, A., Jansson, B.-O., Kautsky, N., Levin, S., Lubchenco, J., Mäler, K.-G., Simpson, D., Starrett, D., Tilman, D. and Walker, B. (2000) The value of nature and nature of value. *Science*, 289, 395–6.

DeFries, R., Asner, G. and Houghton, R. (eds) (2004) *Ecosystems and Land Use Change*. Geophysical Mongraph Series Vol. 153. Washington, DC: American Geophysical Union.

DeFries, R. S., Field, C. B., Fung, I., Justice, C. O., Los, S., Matson, P., Matthews, E., Mooney, H., Potter, C., Prenice, K., Sellers, P., Townshend, J. R. G., Tucker, C., Ustin, S. and Vitousek, P. (1995) Mapping the land surface for global atmosphere-biosphere models:

toward continuous distributions of vegetation's functional properties. *Journal of Geophysical Research*, 100, 20867–82.

Defries, R. S., Hansen, M. C., Townshend, J. R. G., Janetos, A. C. and Loveland, T. R. (2000). A Global 1km data set of percent tree cover data derived from remote sensing. *Global Change Biology*, 6, 247–54.

Di Gregorio, A. (revisor) (2005) *Land Cover Classification System: Classification Concepts and User Manuel-Software Version 2.* Rome: Food and Agriculture Organization, U.N.

DIVERSITAS (2002) *Biodiversity, Science, and Sustainable Development.* ICSU Series on Science for Sustainable Development No. 10: International Council of Science.

Downing, T., Olsthoom, A. A. and Tol, R. S. (eds) (2003) *Climate Change and Risk.* New York: Routledge.

EEA (European Environment Agency) (2006) *Land Account for Europe 1990–2000: Towards Integrated Land and Ecosystem Accounting.* EEA Report No. 11/2006. Copenhagen: European Environment Agency.

Ellis, E. C. and Wang, S. M. (1997) Sustainable traditional agriculture in the Tai Lake region of China. *Agriculture, Ecosystems and Environment*, 61, 177–93.

Foley, J. A., DeFries, R., Aner, G. P., Barford, C., Bonan, G., Carpenter, S. R., Chapin, F. S., Coe, M. T., Daily, G. C., Gibbs, H. K., Helkowski, J. H., Holloway, T., Howard, E. A., Kucharik, C. J., Monfreda, C., Patz, J. A., Prentice, I. C., Ramakutty, N. and Snyder, P. K. (2005) Global consequences of land use. *Science*, 309, 570–73.

Foster, D. R. and Aber, J. D. (eds) (2004) *Forests in Time: The Environmental Consequences of 1,000 Years of Change in New England.* New Haven: Yale University Press.

Fox, J., Rindfuss, R. R., Walsh, S. J. and Mishra, V. (eds) (2003) *People and the Environment: Approaches for Linking Household and Community Surveys to Remote Sensing and GIS.* Boston: Kluwer Academic Publishers.

Johnson, D. L. and Lewis, L. (2007) *Land Degradation: Creation and Destruction*, 2nd edn. Oxford: Rowan and Littlefield.

Geoghegan, J., Schneider, L. and Vance, C. (2005) Temporal dynamics and spatial scales: modeling deforestation in the southern Yucatan peninsular region. *GeoJournal*, 61, 353–63.

Gutman, G., Janetos, A., Justice, C., Moran, E., Mustard, J., Rindfuss, R., Skole, D. and Turner, B. L., II (eds) (2004) *Land Change Science: Observing, Monitoring, and Understanding Trajectories of Change on the Earth's Surface.* New York: Kluwer Academic Publishers.

GLP (Global Land Project) (2005) *Science Plan and Implementation Strategy.* IGBP Report No. 53/IHDP Report No. 19. Stockholm: IGBP Secretariat.

Goudie, A. and Vilas, H. A. (1997) *The Earth Transformed.* Boston: Blackwell.

Gunderson, L. H. and Holling, C. (eds) (2001) *Panarchy: Understanding Transformations in Human and Natural Systems.* Washington, DC: Island Press.

Higgins, S. I., Lavorel, S. and Revilla, E. (2003) Estimating plant migration rates under habitat loss and fragmentation. *Oikos*, 101, 345–66.

Homewood, K., Lambin, E. F., Coast, E., Kariuki, A., Kivelia, J., Said, M., Serneels, S. and Thompson, M. (2001) From the cover: long-term changes in Serengeti-Mara wildebeest and land cover: Pastoralism, population, or policies? *Proceedings of the National Academy of Sciences U.S.A.*, 98, 12544–9.

Houghton, R. A., Skole, D., Nobre, C., Hackler, J., Lawrence, K. and Chomentowski, W. (2000) Annual fluxes of carbon from deforestation and regrowth in the Brazilian Amazon. *Nature*, 403, 301–4.

Irwin, E. G. and Geoghegan, J. (2001) Theory, data, methods: developing spatially-explicit economic models of land use change. *Agriculture, Ecosystems, and Environment*, 84, 7–24.

Kaimowitz, D. and Angelsen, A. (1998) *Economic Models of Tropical Deforestation: A Review.* Bogor: Center for International Forestry Research.

Kalnay, E. and Cai, M. (2003) Impact of urbanization and land-use on climate. *Nature*, 423, 528–31.

Kasperson, J. X., Kasperson, R. E. and Turner, B. L., II (eds) (1995) *Regions at Risk: Comparisons of Threatened Environments.* Tokyo: United Nations University Press.

Kates, R. W., Clark, W. C., Corell, R., Hall, J. M., Jaeger, C. C., Lowe, I., McCarthy, J. J., Schellenhuber, H. J., Bolin, B., Dickson, N. M., Faucheaux, S., Gallopin, G. C., Grübler, A., Huntley, B., Jäger, J., Jodha, N. S., Kasperson, R. E., Mabogunje, A., Matson, P., Mooney, H., Moore, B., III, O'Riordan, T. and Svedin, U. (2001) Sustainability science. *Science*, 292, 641–2.

Kirch, P. V. (2005) Archaeology and global change: the Holocene record. *Annual Reviews of Environment and Resources*, 30, 409–40.

Lal, R. (2004) Soil carbon sequestration impacts on global climate change and food security. *Science*, 304, 1623–6.

Lambin, E. F. (1994) Modelling *Deforestation Processes: A Review. Tropical Ecosystem Environment Observations by Satellites.* Trees Series B. Research Report 1. Luxembourg: Office for Official Publications of the European Community.

Lambin, E. and Geist, H. (eds) (2005) *Land Use and Land Cover Change: Local Processes, Global Impacts.* New York: Springer Verlag.

Lambin, E. F., Geist, H. J. and Lepers, E. (2003) Dynamics of land-use and land-cover change in tropical regions. *Annual Review of Environment and Resources*, 28, 205–41.

Lambin, E. F., Turner, B. L., II, Geist, H., Agbola, S., Angelsen, A., Bruce, J. W., Coomes, O., Dirzo, R., Fischer, G., Folke, C., George, P. S., Homewood, K., Imbernon, J., Leemans, R., Li, X., Moran, E. F., Mortimore, M., Ramakrishnan, P. S., Richards, J. F., Skånes, H., Steffen, H., Stone, G. D., Svedin, U., Veldkamp, T., Vogel, C. and Xu, J. (2001) The causes of land-use and land-cover change-moving beyond the myths. *Global Environmental Change: Human and Policy Dimensions*, 11, 2–13.

Laney, R. (2004) A process-led approach to modeling land change in agricultural landscapes: a case study from Madagascar. *Agriculture, Ecosystems and Environment*, 101, 135–53.

Laris, P. (2005) Spatiotemporal problems with detecting seasonal-mosaic fire regimes with coarse satellite data in savannas. *Remote Sensing of the Environment*, 99, 412–24.

Laurance, W. F. (1998) A crisis in the making: response of Amazonian forests to land use and climate change. *Trends in Ecology and Evolution*, 13, 411–15.

Liu, J. (guest ed.) (2001) Special issue: integration of ecology with human demographic, behavior and socioeconomics. *Ecological Modelling*, 140(1–2), 1–192.

Liverman, D., Moran, E. F., Rindfuss, R. R. and Stern, P. C. (eds) (1998) *People and Pixels: Linking Remote Sensing and Social Science.* Washington, DC: National Academy Press.

Liverman, D. M., Yarnal, B., and Turner, B. L., II (2004) The human dimensions of global environmental change. In G. Gaile and C. Wilmott (eds), *Geography in America at the Dawn of the 21st Century.* New York: Oxford University Press, pp. 267–82.

Loveland, T. R. (2000) Development of a global land cover characteristics database and IGBP DISCover from 1 km AVHRR data. *International Journal of Remote Sensing*, 21, 1303–30.

Luers, A. L., Lobell, D. B., Sklar, L. S., Addams, C. L. and Matson, P. A. (2003) A method for quantifying vulnerability, applied to the agricultural system of the Yaqui Valley. *Global Environmental Change Part A*, 13, 255–67.

Lupo, F., Reginster, I. and Lambin, E. F. (2001) Monitoring land-cover changes in West Africa with SPOT vegetation: impact of natural disasters in 1998–1999. *International Journal of Remote Sensing*, 22, 2633–9.

Manson, S. M. and Evans, T. (2007) Agent-based modeling of deforestation in southern Yucata'n, Mexico, and reforestation in the Midwest United States. *Proceedings, National Academy of Sciences (USA)*, 10, 20678–83.

Mapedza, E., Wright, J. and Fawcett, R. (2003) An investigation of land cover change in Mafungausti Forest, Zimbabwe, using GIS and participatory mapping. *Applied Geography*, 23, 1–21.

Marsh, G. P. (1965[1864]). *Man and Nature: or, The Earth as Modified by Human Action.* Cambridge, MA: Belknap Press of Harvard University Press.

Martin, D. L. (2005) *Twilight of the Mammoths: Ice Age Extinctions and the Rewilding of America.* Berkeley: University of California Press.

Matson, P. A., Parton, W. J., Power, A. G. and Swift, M. J. (1997). Agricultural intensification and ecosystem properties. *Science,* 277, 504–9.

McCracken, S., Brondizio, E., Neslon, D., Moran, E., Siqueira, A. and Rodriguez-Pedraza, C. (1999) Remote sensing and GIS at farm property level: demography and deforestation in the Brazilian Amazon. *Photogrammetric Engineering and Remote Sensing,* 65, 1311–20.

McNeil, J. R. (1992) *The Mountain of the Mediterranean World: An Environmental History.* Cambridge: Cambridge University Press.

MEA (Millennium Ecosystem Assessment) (2005) *Ecosystem and Human Well-Being. Vol. 2.* Washington, DC: Island Press.

Meyer, W. B. and Turner, B. L. II (eds) (1994) *Changes in Land Use and Land Cover: A Global Perspective.* New York: Cambridge University Press.

Mooney, H. A. and Hobbs, R. J. (eds) (2000) *Invasive Species in a Changing World.* Washington, DC: Island Press.

Moran, E. F., Brondizio, E., Mausel, P. and Wu, Y. (1994) Integrating Amazonian vegetation, land-use, and satellite data. *Bioscience,* 44, 320–88.

Myers, N., Mittermeier, R. A., Mittermeier, C. G., da Fonseca, G. A. B. and Kent, J. (2000) Biodiversity hotspots for conservation priorities. *Nature,* 403, 853–8.

Nepstad, D. A., Veríssimo, A., Alencar, A., Nobre, C., Lima, E., Lefebvre, P., Schlesinger, P., Potter, C., Mountinho, E. and Cochrane, M. A. (1999) Large-scale impoverishment of Amazonian forests by logging and fire. *Nature,* 398, 505–8.

Nobre, C. A., Sellers, P. J. and Shukla, J. (1991) Amazonian deforestation and regional climate change. *Journal of Climate,* 4, 957–88.

Nyerges, A. E. and Green, G. M. (2000) The ethnology of landscapes: GIS and remote sensing in the study of forest change in West African Guinea savanna. *American Anthropologist,* 102, 271–89.

Parker, D. C., Manson, S. M., Janssen, M., Hoffmann, M. J. and Deadman, P. J. (2003) Multi-agent systems for the simulation of land use and land cover change: a review. *Annals of the Association of American Geographers,* 93, 316–40.

Perez, S. G. and Walker, R. (2002) Household life cycles and secondary forest cover among small farm colonists in the Amazon. *World Development,* 30, 1009–27.

Pielke, R. A., Sr (2002) The influence of land-use change and landscape dynamics on the climate system: relevance to climate-change policy beyond the radiative effect of greenhouse gases. *Philosophical Transactions of the Royal Society A: Mathematical, Physical and Engineering Sciences,* 360, 1705–19.

Pielke, R. A., Sr, Walko, R. L., Steyaert, L. T., Vidlae, P. L., Liston, G. E., Lyons, W. A. and Chase, T. N. (1999) The influence of anthropogenic landscape changes on weather in South Florida. *American Meteorological Society,* 127, 1663–1773.

Pijanowski, B. C., Brown, D. G., Shellito, B. A. and Manik, G. A. (2002) Using neural networks and GIS to forecast land use changes: a Land Transformation Model. *Computers, Environment and Urban Systems,* 26, 553–75.

Pontius, R. G., Jr (2002) Statistical methods to partition effects of quantity and location during comparison of categorical maps at multiples resolutions. *Photogrammetric Engineering and Remote Sensing,* 68, 1041–9.

Raven, P. (2002) Science, sustainability and the human prospect. *Science,* 297, 954–8.

Rindfuss, R. R., Walsh, S. J., Turner, B. L. II, Fox, J. and Mishra, V. (2004) Developing a science of land change: challenges and methodological issues. *Proceedings of the National Academy of Sciences U.S.A.,* 101, 13976–81.

Redman, C. L. (1999) *Human Impact on Ancient Environments.* Tucson: University of Arizona Press.

Reynolds, J. F. and Stafford Smith, D. M. (eds) (2002) *Global Desertification: Do Humans Cause Deserts?* Dahlem Workshop Report 88. Berlin: Dahlem University Press.

Rojstaczer, S., Sterling, S. M. and Moore, N. J. (2001) Human appropriation of photosynthesis products. *Science*, 294, 2549–52.

Rosenzweig, C. and Parry, M. (1994) Potential impact of climate change on world food supply. *Nature*, 367, 133–8.

Roy Chowdhury, R. and Turner, B. L. II (2006) Reconciling agency and structure in empirical analysis: smallholder land use in the southern Yucatán, Mexico. *Annals of the Association of American Geographers*, 96, 303–22.

Ruddiman, W. (2005) *Plows, Plagues, and Petroleum: How People Took Control of Climate.* Princeton: University of Princeton Press.

Sala, O. E., Chapin, F. S., Armesto, J. J., Berlow, E., Bloomfield, J., Dirzo, R., Huber-Sanwald, E., Huenneke, L. F., Jackson, R. B., Kinzig, A., Leemans, R., Lodge, D. M., Mooney, H. A., Oesterheld, M., Poff, N. L., Sykes, M. T., Wlaker, B. H., Walker, M. and Wall, D. H. (2000) Biodviersity: global biodiversity scenarios for the year 2100. *Science*, 287, 1770–4.

Schellnhuber, H. J., Black, A., Cassel-Gintz, M., Kropp, J., Lammel, G., Lass, W., Lienenkamp, R., Loose, C., Lüdeke, M. K. B., Moldenhauer, O., Petschel-Held, G., Plöchl, M. and Reusswig, F. (1997) Syndromes of global change. *GAIA*, 6, 19–34.

Schneider, L. C. and Geoghegan, J. (2006) Land abandonment in an agricultural frontier after bracken fern invasion: linking satellite, ecological and household survey data. *Agricultural Resources Economic Review*, 11, 1–11.

Seto, K. C. and Kaufmann, R. K. (2003) Modeling the drivers of urban land use change in the Peral River Delta, China: integrating remote sensing with socioeconomic data. *Land Economics*, 79, 106–21.

Seto, K. C., Kaufmann, R. K. and Woodcock, C. E. (2000) Landsat reveals China's farmland reserves, but they're vanishing fast. *Nature*, 406, 121.

Seto, K. C., Woodcock, C. E., Song, C., Huang, X., Lu, J. and Kaufamnn, R. K. (2002) Monitoring land-use change in the Pearl River Delta using Landsat Tm. *International Journal of Remote Sensing*, 23, 1985–2004.

Skole, D. and Tucker, C. (1993) Tropical deforestation and habitat fragmentation in the Amazon: satellite data from 1978–1988. *Science*, 260, 1905–10.

Steffen, W., Sanderson, A., Tyson, P., Jäger, J., Matson, P., Moore, B., III, Oldfield, F., Richardson, K., Schellnhuber, H.-J., Turner, B. L., II and Wasson, R. (2003) *Global Change and the Earth System: A Planet under Pressure, IGBP Global Change Series.* Berlin GR: Springer-Verlag.

Terborgh, J., Van_Shaik, C. P., Davenport, L. and Rao, M. (eds) (2002) *Making Parks Work: Strategies for Preserving Tropical Nature.* Washington, DC: Island Press.

Thomas, W. M., Jr (1956) *Man's Role in Changing the Face of the Earth.* Chicago: University of Chicago Press.

Tilman, D., Cassman, K. G., Matson, P. A., Naylor, R. and Polasky, S. (2002) Agricultural sustainability and intensive production practices. *Nature*, 418, 671–7.

Turner, B. L., II (2002) Toward integrated land-change science: advances in 1.5 decades of sustained international research on land-use and land-cover change. In W. Steffan, J. Jäger, D. Carson and C. Bradshaw (eds), *Challenges of a Changing Earth: Proceedings of the Global Change Open Science Conference, Amsterdam, NL, 10–13 July 200.* Heidelberg, GR: Springer-Verlag, pp. 21–6.

Turner, B. L., II (2003) Contested identities: human-environment geography and disciplinary implications in a restructuring academy. *Annals of the Association of American Geographers*, 92, 52–74.

Turner, B. L., II, Geoghegan, J. and Foster, D. R. (eds) (2003a) *Integrated Land-Change Science and Tropical Deforestation in the Southern Yucatán: Final Frontiers.* Oxford: Clarendon Press.

Turner, B. L., II, Kasperson, R. E., Matson, P. A., McCarthy, J. J., Corell, R. W., Christensen, L., Eckley, N., Kasperson, J. X., Luers, L., Martello, M. L., Polsky, C., Pulsipher, A. and Schiller, A. (2003b) A framework for vulnerability analysis in sustainability science. *Proceedings of the National Academy of Sciences, U.S.A.*, 100, 8074–79.

Turner, B. L., II, Matson, P. A., McCarthy, J. J., Corell, R. W., Christensen, L., Eckley, N., Hovelsrud, G., Kasperson, J. X., Kasperson, R. E., Luers, L., Martello, M. L., Mathiesen, S., Naylor, R., Polsky, C., Pulsipher, A., Schiller, A., Selin, H. and Tyler, N. (2003c) Illustrating the coupled human-environment system for vulnerability analysis: three case studies. *Proceedings of the National Academy of Sciences, U.S.A.*, 100, 8080–5.

Turner, B. L., II and McCandless, S. (2004) How humankind came to rival nature: a brief history of the human-environment condition and the lessons learned. In W. C. Clark, P. Crutzen and H.-J. Schellnhuber (eds), *Earth System Analysis for Sustainability*, Dahlem Workshop report no. 91. Cambridge, MA: MIT Press, pp. 227–43.

Turner, B. L., II, Lambin, E. F. and Reenberg, A. (2007) The emergence of land change science for global environmental change and sustainability. *Proceedings, National Academy of Sciences, U.S.A.*, 104, 20666–71.

Veldkamp, T. and Lambin, E. (guest eds) (2001) Special issue: predicting land-use change. *Agriculture, Ecosystems and Environment*, 85(1–3): 1–292.

Veldkamp, A. and Fresco, L. O. (1996) CLUE: a conceptual model to study the conversion of land use and its effects. *Ecological Modelling*, 85, 253–70.

Verburg, P. H. (2006) Simulating feedbacks in land use and land cover change models. *Landscape Ecology*, 21, 1171–83.

Vitousek, P. M., Mooney, H. A., Lubchenco, J. and Melillo, J. M. (1997) Human domination of earth's ecosystems. *Science*, 277, 494–500.

Waggoner, P. E. and Ausubel, J. H. (2002) A framework for sustainability science: A renovated IPAT Identity. *Proceeding of the National Academy of Sciences U.S.A.*, 99, 7860–65.

Walker, R., Drzyga, S., Li, Y., Qi, J., Arima, E. and Vergara, D. (2004) A behavioral model of landscape change in the Amazon Basin: the colonist case. *Ecological Applications*, 14(Supplement), S299–312.

Walsh, S. J. and Crews-Meyer, K. A. (eds) (2002) *Linking People, Place, and Policy: A GIScience Approach*. Boston: Kluwer Academic Publishers.

Walsh, S. J., Evans, T. P., Welsh, W. F., Entwisle, B. and Rindfuss, R. R. (1999) Scale-dependent relationships between population and environment in Northeastern Thailand. *Photogrammetric Engineering and Remote Sensing*, 65, 97–105.

Walther, G.-R., Post, E., Convey, P., Menzel, A., Parmesan, C., Beebee, T. J. C., Fromentin, J.-M., Hoegh-Guldberg, O. and Bairlein, F. (2002) Ecological responses to recent climate change. *Nature*, 416, 389–95.

Watson, R. T., Noble, I. R., Bolin, B., Ravindranath, N. H., Verardo, D. J. and Dokken, D. J. (2000) *Land Use, Land-Use Change, and Forestry*. A Special Report of the IPCC. Cambridge: Cambridge University Press.

Williams, M. (2005) *Deforesting the Earth: From Prehistory to Global Crisis*. Chicago: University of Chicago Press.

Wulder, M. and Franklin, S. (eds) (2006) *Understanding Forest Disturbance and Spatial Pattern: Remote Sensing and GIS Approaches*. Boca Raton: CRC Press.

Zhang, H., Henderson-Sellers, A. and McGuffie, K. (2001) The compounding effects of tropical deforestation and greenhouse warming on climate. *Climatic Change*, 49, 309–38.

Zhou, L., Dickinson, R. E., Tian, Y., Fang, J., Li, Q. and Kaufmann. R. K. (2004) Evidence for a significant urbanization effect on climate in China. *Proceedings of the National Academy of Sciences U.S.A.*, 101, 9540–4.

Chapter 12

Ecology: Natural and Political

Matthew D. Turner

Introduction

The term 'ecology' has many meanings. It is often used loosely as a synonym for 'environment' or 'nature', the latter arguably one of the most complex, difficult, and meaning-rich terms in the English language (Williams, 1980). As such, the juxtaposition of 'natural' and 'political' ecology raises issues of the division (Latour, 1993; Castree, 2005), mutual construction (Ellen and Fukui, 1996; Demeritt, 1998; Castree and Braun, 2001; Bakker and Bridge, 2006), and hybridity (Swyngedouw, 1999; Whatmore, 2002) of 'nature' and 'society'. Recent biophysical and social research has questioned whether what we think of as 'natural' or 'wild' can be rightly seen as such (Cronon, 1996; Neumann, 1998; Braun, 2002), while others have argued that the gradations of naturalness ignored by some social commentators are important (e.g., Vale, 1998).

More specifically, ecology is defined as the scientific study of the relationships among biological organisms and with their physical environment. Often the term is used as well to refer not only to a field of study but to the actual interrelationships being studied. This distinction is important in this chapter and to avoid confusion, 'ecology' will refer to the science and 'ecological relations' will refer to ecologists' foci of study. This chapter is concerned less with advances in ecology but more with the implications of ecology and ecological relations for the study of society-environment relations. 'Ecology' as a term connotes complex interrelatedness and as such, has proven a popular label for a suite of scholarly approaches, primarily in anthropology and geography, for analysing society-environment relations (cultural ecology, human ecology, political ecology). I do not intend to survey these approaches here – this has already been done admirably by others (Ellen, 1982; Zimmerer, 1996; Turner, 1997; Robbins, 2004). In this chapter, I will focus on the relationships between ecology and ecological relations (science and subject of study) with the politics of the environment and particularly how they have been addressed by the diverse interdisciplinary field of political ecology. More specifically, I explore two questions:

1. Under what conditions are the goals of political ecology scholarship advanced
    by a serious engagement with ecological relations?
2. How may ecology as a science be implicated in understandings of environmental
    politics by political ecologists?

Political Ecology's Engagement with Ecological Relations

Political ecology is a maturing, rapidly expanding field in geography and to a lesser
extent, allied disciplines such as anthropology, rural sociology and development
studies. Its major lineage in geography developed out of the uneasy marriage
between cultural ecology and agrarian political economy – a marriage that emerged
from an interest in the political and economic roots of land degradation in rural areas
of the developing world (Watts, 1983; Blaikie, 1985; Blaikie and Brookfield, 1987).
Geography's political ecology, as originally framed by Piers Blaikie and Harold
Brookfield, engaged critically with dominant environmental analyses (those strongly
shaped by neo-Malthusian and 'Tragedy of the Commons' concepts) and sought to
combine a detailed understanding of rural producers' use of the natural resources as
conditioned by their 'access to resources' which is, in turn, shaped by changes in the
biophysical environment and the broader political economy. The explanatory focus
was the dialectical relationship between social and environmental change with a
particular emphasis on the connection between poverty and environmental mismanagement. Since the early 1990s, there has been a movement away from its structural
roots. In addition there has been a diversification of political ecological scholarship,
reflecting the rapid growth of this field, which has attracted scholars to its promise
for drawing connections between: social and ecological change, the environment and
social justice, global and local change, and political interests and the construction of
dominant views of environment. Outside of the land-use and land cover change field,
a large fraction of people-environment geography now self-identifies as 'political
ecology.' As a result, it has become difficult to specify a common theoretical or methodological framework for political ecology in geography (but see Forsyth, 2003;
Zimmerer and Bassett, 2003; Robbins, 2004; Neumann, 2005).

One could describe much of political ecology work as analyses of society-
environment relations, contextualised by history and place, with a particular emphasis on environmental and social justice implications of broader political economic
change. Such a broad description hides many differences within the field. A minority
of contemporary political ecology work remains focused on understanding the
relationship between social and environmental change (Walker, 2005). A much
larger fraction of contemporary political ecology is not concerned with environmental change *per se* but with the politics surrounding the use of, and struggles for,
access to natural resources. In the sections below, I will discuss the experience and
future potential of engagements with ecological relations for these two areas of
inquiry within political ecology.

Different encounters with ecological relations

The history of approaches to the study of people-environment relations is one that
has been plagued, from the start, with analytical conundrums associated with the
drawing of boundaries between 'human society' and 'nature' as well as the identification of mediators between environmental and social change, once boundaries are

drawn (Ellen, 1982; Latour, 1993; Castree 2005). Environmental determinism, historical possibilism; cultural ecology, human systems ecology, political economy of natural resources, environmental history, resource economics, environmental security, and various versions of political ecology all draw boundaries between human society and everything else in different ways with significant implications for what researchers see. Mediating concepts, currencies, or mechanisms such as Julian Steward's cultural core; energy and nutrient flows in human systems ecology; evolutionary concepts of 'adaptation' in cultural ecology; and notions like capital (de)appreciation in resource economics – are all different ways in which the mediation between social and environmental change have been conceptualised.

When first introduced, Blaikie and Brookfield's 'access to resources' concept provided a novel mediating link (Ribot and Peluso, 2003) between social change (processes, powers, and institutions affecting how people can make effective use to resources) and ecological change (changes in the physical availability of resources). In this way, it provided a means through which to think about the relationship between environmental and social change without reducing the social to the ecological (Watts, 1983) or ignoring the ecological relations influencing resource properties. While ecological relations (as defined above) were not fully incorporated in the broad schema of political ecology (beyond the recognition that resources are distributed unevenly across rural landscapes), chapters in Blaikie and Brookfield (1987) do seriously engage with the complexities of measuring soil erosion and understanding land degradation processes.

In reflecting on political ecology's engagement with ecological relations, I do not seek to replay earlier debates about where the 'ecology' is in political ecology (well summarised by Walker, 2005) or the mirrored set of arguments of where the 'politics' are in political ecology (see Watts, 1990). It can be said that such debates are necessary for political ecology to establish its identity as a field of study and stake its claim on an interdisciplinary intellectual terrain. Still, I adopt here a much more modest but potentially more illuminating position here – the appropriate level of engagement should be determined by the questions being asked about particular places at particular historical moments.[1] By engagement, I refer to the level of understanding (and incorporation into the study) of the ecological relations of relevance to the people-environment relation being studied.[2] By adopting this position, I leave open the possibility of research approaches that mix methods and epistemologies. Such mixed approaches, while less pure, have attracted significant interest and debate within people-and-environment studies and political ecology in particular over the past decade.

Increasing analytical engagement with ecological relations involves the successive recognition and understanding of: ecological heterogeneity, ecological dynamics, responsiveness to human resource use, longer temporal scales of response, and the embeddedness of the ecological parameter within a wider set of ecological relations (figure 12.1). All analysts are aware of the complexity of ecological relations. Where they differ is the degree to which they engage with these complexities in their analyses of people-environment relations. I locate some examples of different levels of engagement in figure 12.1 with the general level of engagement increasing as the number of the level increases (1–7).

By presenting this simple diagram (figure 12.1), I hope to stimulate greater reflection by political ecologists of what is lost and gained by adopting different levels of engagement with ecological relations for understanding people-environment

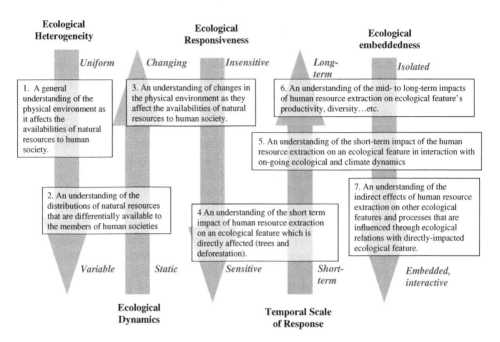

Figure 12.1 Levels of engagement with the complexity of ecological relations (1–7) across five dimensions of conceptual difference: ecological heterogeneity, dynamics, responsiveness (to human actions), temporal scale and the degree to which the ecological feature is seen as embedded within a broader set of relations.

questions. In so doing, the discussion below cautions against a naïve embrace of the complex web of social and ecological relations by the 'big-picture' analyst – such experiments will generally lead to ideographic descriptions or violent reductions of complexity through various forms of systems analysis. At the same time, the discussion cautions against knee-jerk invocations of the incommensurate epistemologies of social and ecological analysis – thus sparing the political ecologist from moving into uncomfortable ontological/epistemological spaces that run across and through people-environment relations. This position does not derive from monistic vision but an embrace of the analyst's agency. While difficult, we, as analysts, can place different logics and epistemologies in parallel looking at congruencies and divergences without being captured by any one. It is such integrative work where arguably many advances in people-environment study will come – including those from within political ecology.

It is important to recognise at the outset that engagement with ecological relations is not without costs. Political ecologists may not have the necessary training or time to perform ecological research themselves or the contacts or inclination to collaborate with biophysical scientists. These costs are important and have worked to shape the questions posed by political ecologists performing social-ecological change research. An important but unexamined cost of greater engagements is how they complicate the exposition of research results. While political ecologists seek to provide richer and more complex narratives than simple declensionist or cornucopian story lines, adding both ecological and social complexity may place too many demands on a tractable story line. If we treat ecological and social systems as open, interactive and not determined by the other, the embracement of complexity will necessarily be associated with the uncertainty. Recognition of this uncertainty works against constructing a compelling narrative. The mix of different types of information of contrasting spatial and temporal resolutions that are created through engagement with both social and ecological relations make it difficult to maintain a narrative thread. Simply having the interest, training, and resources to engage seriously with both social and ecological complexity does not produce compelling research results. Political ecologists who have attempted to engage seriously with the fuller complexity of social and environmental change may find themselves publishing the ecological and social portions of their analysis as separate products, directed at different audiences. Therefore, political ecologists should think seriously about these limits of full engagement with ecological relations in designing their research.

Current political ecological research spans the range of engagements presented in figure 12.1. Research in the two foci areas of contemporary political ecology – environmental politics and social-ecological change – tend to be found somewhere on either end of this range. The majority of political ecology research treats ecological relations as a backdrop to environmental politics – a backdrop that either: (i) broadly defines what is and what is not possible in terms of human activity (#1 of figure 12.1) or (ii) produces the spatiotemporal variation in resource availability that helps shape the nature of resource-related conflict (#2 and 3 of figure 12.1). The ecological response to human land-use has generally been less of a concern (engagements greater than #3). The political ecological research on social-ecological interactions has generally engaged more with ecological relations, ranging from the documenting the direct, short-term effects of resource extraction (#4 of figure 12.1) to mid- to long-term impacts of human extraction on ecological variables (#6 of figure 12.1). These are general observations of prior work in political ecology. More detailed explorations of promise and pitfalls of deeper engagements with ecological relations in these two focal areas are presented below.

Environmental politics: conflicts over natural resources

A major focus of political-ecological research is the environmental politics surrounding the claiming, using and managing of the natural resources. As described above, this body of work has engaged much less with ecological relations than the much smaller body of work focused on understanding social-environmental change. The costs of engagement are similar to those described above while the benefits are less clear. Environmental politics is historically embedded and shaped by ideology and social relations. Any material influence is necessarily given political meaning by these social features.

Many efforts to tie the material world to conflicts over natural resources, ranging from the works of Semple (1915), Rappaport (1984), Homer-Dixon (1994) and Diamond (2006), have ignored or downplayed how material constraints or resource scarcity is strongly mediated if not produced by social relations. Political ecology as an approach developed (Watts, 1983; Blaikie, 1985; Neumann, 1998), and has recurrently reinforced (Peluso and Watts, 2001), its identity through its critical engagement with simple but highly influential treatments of environmental governance and the etiology of resource conflict. In many ways, political ecology has developed as a subfield around the important idea that resource-related conflict is inherently social and that changing material (ecological) conditions influence environmental politics only through the divergent meanings attached to that change by individuals and groups with divergent powers. How might a greater engagement with ecological relations contribute to a political ecologist's understanding of environmental politics? The fact that a tree has microrhizal associations allowing it to better capture nutrients released from vegetative burning or decay would seem to have little role to play in the conflicts that surround one group's interest in cutting the tree down and another group's interest in letting it stand for production or preservation purposes. Under what circumstances would greater engagement with ecological relations contribute to an understanding of the unfolding of politics surrounding natural resources? I will explore these questions by focusing on two areas of particular interest to political ecologists: differentiation of wealth and power, and environmental governance.

Differentiation of wealth and power

Only those ecological relations that surround resources valued by human society are likely to attract attention from those studying environmental politics. A classic political ecology perspective conceptualises environmental politics as being constituted by struggles over 'access to resources.' Framed in this way, 'conflicts over resources' are not simply here-and-now struggles over the resources made available by productive ecologies but they are, in fact, socially mediated. Politics are between people not between people and trees. Resource-related conflicts among people, while having a material basis, are often expressed through the invocation of principles governing social conduct: fairness, justice, past agreements, and historical precedent. Therefore, it is difficult to disentangle rigorously the material, ideological, and political roots of any conflict.

This said, we can recognise that environmental change and ecological dynamics do affect the nature of environmental politics. On a broad level, areas attracting the interests of international capital (extractable resources) or conservation (biodiversity) have quite different political ecologies than other areas. Areas of greater resources elicit greater investments into governance structures at the level of communities, districts, and national governments. As a result, governance structures, the potential for competing interests, and power differentials will be affected by the human valuation of nature's objects (nature as 'resource') and the magnitude of resources available for extraction.

At the level of the human community, early political ecology work has shown that the temporal dynamics (Watts, 1983) and spatial heterogeneity (Blaikie, 1985) of biological productivity, as mediated through social relations of production, play important roles in the process of social differentiation of wealth and of ecological vulnerability. High spatiotemporal variability of resource availability, by limiting capital investment and primitive accumulation, could be seen to work against differentiation. However, more predictable cycles of temporal variability allow the rich to speculate across these cycles to their advantage and to the disadvantage of the poor. While most of this work has focused on climatic variability, one could imagine environmental changes leading to resource production dynamics that differ in their predictability and spatiotemporal variability. By changing temporal resonances with markets and spatial resonances with capital investment, such environmental changes could, through changes in the distributions of wealth and power, affect environmental politics.

The spatial distribution of biological productivity also may play an important role in differentiation. Resource enclosures of all sorts have contributed to changes in social relations of production and more uneven distributions of wealth and power. In this way, the degree to which resources are sufficiently aggregated to elicit enclosure moves by the more powerful will affect the nature of resource control within communities, districts and countries. Environmental change can lead to an increase or decrease in the spatial aggregation of resource availability with important social implications. Shifts in the relative importance of rainfed, floodplain, or irrigated land as cropland or pastures have an important effect on the distribution of power within agropastoral communities. Shifts in the spatial aggregation of the extracted resource by those who hunt, gather, or log forests will influence the distribution of wealth. One can even think of transitions towards cash-cropping in this way. Access to the resource is now determined not only by resource tenure but also effective local access to chains supplying regional and international markets, which are often monopsonistic and socially-embedded. Under such circumstances, the ability to accumulate wealth from a widespread resource is limited to a few.

Environmental governance

Political ecologists see the distribution of power and wealth as a major factor shaping natural resource politics. Another factor is the exercise of this power through formal and informal governance structures. The effectiveness of different institutional forms to monitor and regulate resource use will vary depending on the characteristics of the resource. In short, different institutional forms, by creating different spaces for conflict and negotiation, will change the nature of environmental politics. The properties of a resource and of the biophysical processes that affect the resource, may require deviations from the requirements or assumptions embedded within institutional forms. Such deviations can take the form of a social group adopting a different institutional form *or* the less-than-effective functioning of the ill-fitted form. In both cases the nature of environmental politics will differ.

Examples come from two popular institutional forms: common property resource management (CPRM) and market-based resource management. Many natural resource management programmes in the developing world are applications of highly influential work on common property institutions (e.g., Ostrom, 1990). This work sees the management of resources held in common by a social group as a collective action problem. Common property institutions should clarify the boundaries of the resource managed by the social group, which should in turn be clearly circumscribed. Once a closed socio-ecological system is created in this way, the goal is to establish a set of rules that provide the proper mix of incentives to individuals so that they utilise the resource for the common good (along with the design of monitoring and enforcement capabilities). Applications of this approach have often sought to improve management of the CPR by creating territorial boundaries around the resources in question – in many cases, representing the first step in the process of privatisation (Mansfield, 2004). In situations where resources are highly mobile or ephemeral, such boundaries are unworkable – resource location is best seen as a constantly shifting mosaic – laying claim to only a portion of this mosaic would increase vulnerability of the social group.[3] Under these circumstances, more socially-porous boundaries around resources may be preferable – replacing a politics framed by rigid rules and boundaries with one associated with negotiation, favours, and mutual obligations. The highly dynamic ecology demands such porosity – if territorial boundaries are drawn, they will be circumvented, with such circumventions referred to as corruption. In this way, environmental politics is strongly shaped by high spatiotemporal variability of resource availability.

A major trend in environmental governance has been the increased reliance on market mechanisms (Daily and Ellison, 2002). Environmental services are priced and increasingly traded within government created and managed markets. To be bought and sold, the features of a natural process, ecological community, or a wild plant or animal population need to be necessarily abstracted. Moreover, the matrix from or into which the service is bought or sold is ignored – one wetland, forested patch, or biodiverse grassland is treated as equal to another in its category. The characteristics of ecological processes may resist such categorisations and in so doing change the appropriate scales at which these markets are established and the different interests and debates implicated in environmental politics. These politics may increasingly rely on ecological science to stabilise facts and create tradeable indicators of ecosystem functioning (Robertson, 2006). The actors, discursive strategies and institutions implicated in environmental politics are thus affected by the ecological relations.

Land management as political instrument

To understand environmental politics, the political ecologist, even in the cases described above, would need to understand only up to the third level of engagement with ecological relations (figure 12.1). It could be argued that there is no need to delineate the reasons for increased aggregation, spatiotemporal variability, or the strong embeddedness in place of ecological features or processes for understanding how these changes have influenced environmental politics. Certainly a more complete story could be told with information about the anthropogenic and non-anthropogenic processes leading to such resource characteristics but is it necessary to strive for such completeness to understand environmental politics? Can we think of environmental politics as simply reactive to the environmental change or is the intentional reworking of ecological relations to some end part of this politics? The simple answer to this question is that yes, politics are mediated through ecological relations to serve particular interests. Political actors' (mis)understandings of ecological response and their reactions to ecological change as it unfolds may very much be tightly intertwined in the politics surrounding resource use and control. In such situations, greater engagement with ecological relations beyond level 3 may be necessary to understanding the unfolding of environmental politics.

Relationship between ecological and social change

Arguably, the early explanatory focus of geographical political ecology was to investigate the relationship between social and ecological change. This is an ambitious intellectual project especially when one considers the checkered history of environmental determinism, cultural ecology, and human ecology in people-and-environment research. In different ways, these prior approaches sought integration of social and environmental change by favouring ecological or social logics in their choice of common currencies (cash, labour, energy, nutrients, etc.), strong materialist treatments of society, or strong socialisations of 'nature'. Moreover, there was a tendency for cultural adaptation and its various functionalist variants to result in very static, teleological treatments of 'environment' and 'society' interaction. The question of integration remains a major conundrum within people-and-environment study – how can we simultaneously treat 'ecological relations' and 'social relations' as following multiple, divergent logics while embedded within interacting, open systems? Political ecological research has contributed more sophisticated treatments of the multiple-scaled social relations that surround natural resource use to the society-environment tradition in geography. However, while improving on previous approaches, it has generally failed to produce full, balanced depictions of dynamic ecology-society interaction. As described above, a major weakness has been a relatively shallow engagement with the complexity of ecological relations.

In this section, I avoid replaying arguments about whether a 'political ecology' approach needs to engage more deeply with ecological relations. Prescriptive declarations either way do little to clarify the costs and benefits of greater engagement with ecological relations. The position taken here is that the appropriate level of engagement (figure 12.1) depends on the research question(s) and the specifics of land-use ecology (sensitivity of the ecological parameters to human extraction pressures). Each of these issues will be explored in the following two sections.

Research questions and political ecological engagements with ecological relations

Just as it is not feasible or desirable for political ecologists to study all aspects of society, it is not feasible to study all aspects of the biophysical environment when performing social-ecological change research. How political ecologists frame their studies has a significant effect on the appropriate level of engagement with ecological relations. No matter how one divides 'human society' from the 'biophysical environment', all human activities influence the environment. One's study is concerned with the interaction of what human activity(ies) with what ecological parameters? How is the 'environment' categorised and 'change' viewed in terms of spatial and temporal scales? These questions relate directly to how society-environment interaction and environmental change is framed and conceptualised. This has a strong effect on the research questions posed in political ecology research which in turn influence the appropriate level of engagement with ecological relations (figure 12.1).

For example, deforestation has been a major topic of political ecological studies of social-ecological change. Not only is this a major environmental problem in many parts of the world, it is most commonly framed by political ecologists and others in ways that require much less engagement with ecological relations. Deforestation, as typically treated, is short-term change in land cover resulting from the removal of trees by humans. Compared to other environmental changes (e.g., soil fertility decline, species composition shifts in vegetation, or wildlife population declines), it is much more tractable without significant fieldwork or engagement with ecological relations. The documentation of change is straightforward and the evidence pointing to an anthropogenic cause is clear. It is analogous to describing vegetation removal associated with anthropogenic prairie fires – the short-term losses are clear. What is complicated is how these removals of vegetation affect soils, seed stocks and microclimates, which in turn, will influence the vegetation that will replace what has been removed. While in the prairie case, research focused on ecological impact would most likely engage with the ecological relations implicated in post-fire succession, the environmental change narrative in deforestation research tends to stop at initial vegetative loss.

These disjunctures are surprising given the emphasis within policy and scientific circles, illustrated by terms such as sustainability and degradation, on understanding the longer-term persistence of an ecosystem's productive potential. They are not surprising however when we consider temporal framing issues. Changes that are short with respect to human lifetimes and planning horizons are less likely to be ignored by researchers. Changes that are seen as long are more easily ignored. This is due to conceptual, policy and methodological issues. It is more difficult to study processes that transcend human lifetimes. These changes are less tractable and actionable by policymakers. Returning back to the deforestation example, the time needed for forest structures to be reestablished is longer than those for the reestablishment of herbaceous cover on a burned prairie. But is this the appropriate criteria? Are we not interested in post-deforestation ecological changes for their own sake not simply with an eye towards the reforested landscape? Adopting the deforested landscape as the endpoint implicitly treats subsequent ecological change as long-term and therefore emphasises structural recovery or return to 'climax' as the post-disturbance change of concern.

Endpoints of change are necessarily established in all research. The choice of endpoints reflects not only how social-environment interaction is conceptualised and framed both spatially and temporally (as described above) but also the environmental change's policy relevance, economic importance, and feasibility of study. Political ecologists are more likely to incorporate the effects of resource extraction on ecological relations that have a direct influence on economy and politics. This results from their interest in the recursive relations between political economy and ecology over time. Therefore, 'ecological relations' captured within political ecology research are more likely to be those within which human-defined resources are implicated. Second-order effects of resource extraction that have little potential for strong feedback to human economic activity (effects of resource extraction on ecological populations of little economic importance) are less likely to be studied (e.g., at level 7 of figure 12.1). In this way, political ecology research of social/ecological change has an inherent anthropocentric emphasis.

It is important to acknowledge that the framing of research is not solely governed by conceptual issues – more practical concerns come into play. Despite the importance of soil degradative processes to the original framers of political ecology (Blaikie, 1985; Hecht, 1985; Blaikie and Brookfield, 1987), anthropogenic changes in the physical and chemical properties of soils have not continued to attract attention from political ecological analysts of social-ecological change (with exception of Gray, 1999). In these ways, the preponderance of social-environmental change research performed by political ecologists and other social scientists is focused on environmental changes, such as deforestation, that are more tractable by outside observers. Therefore, the level of engagement with ecological relations may remain limited due to the difficulties, especially in light of the background and training of political ecologists, of performing certain types of environmental change research.

Land-use ecology and political ecological engagements with
ecological relations

Despite the early calls for greater engagement with the biophysical world, the vast majority of political ecological research shares with other social scientific counter-

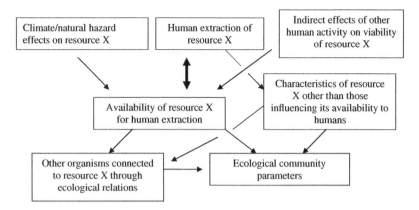

Figure 12.2 Land-use ecology: resource extraction in isolation or in relation.

parts the use of very simple conceptual models of land-use ecology ranging from 'stock depletion' to 'tipping bucket' models. The 'stock depletion' model treats ecological response as linearly-related to human resource use. Ecological relations are a depletable stock – an expansion of population, cash cropping, timber extraction by multinationals . . . etc. will necessarily lead to commensurate levels of environmental transformation. Dependence on this model is often implicit – political ecological analyses whose 'chains of explanation' trace changes only to the point of documenting 'greater pressures on the environment' rely on this model. Without clearly stated caveats, references to 'greater pressures on the environment' are viewed as instances of 'environmental change' by most audiences. Somewhat more sophisticated treatments incorporate a threshold concept, above which increases in human pressures will have a disproportionate impact on the environment (the bucket tips).

Political ecology, while making quite different claims about the social roots of environmental change, shares with some of the social scientific approaches that it critiques (neo-Malthusianism, hazards theory, IPAT model, environmental security . . . etc.), conceptual models of ecological response to human activities. In this way much political ecology work has reinforced the dominant views that: (i) the resource is isolated from the broader web of ecological relations; and (ii) ecological response is proportional to the magnitude of resource extraction. As shown in figure 12.2, a singular focus on how resource extraction influences the availability of the resource may ignore other factors affecting the resource's availability; features of the resource which do not affect its availability to human society; and broader effects of resource extraction on other organisms and ecological community parameters. Cases when the resource is strongly tied to other organisms and processes, more fuller engagements with ecological relations may be necessary to truly understand the environmental effects of resource extraction. Even if the extracted resource is the sole concern of the researcher, it's availability may be determined just as if not more by climate or indirect effects of human activities (via the dynamics of other organisms and physical processes) than by the primary extraction activity itself. Moreover, the direct effects on resource X may be small compared to the indirect effects of resource X extraction on other populations, processes or community parameters. For example, the removal of a few predators will have a disproportionate effect on the dynamics of prey populations. In short, assumptions of the ecological isolation of resource extraction need to be interrogated.

In terms of impact on ecological relations, the ways in which resources are extracted are often seen as relatively unimportant compared to the aggregate magnitude of resource extraction. Under what situations (research question, land-use ecology, and variation of extraction practices) is the manner in which resources are extracted likely to be important in terms of impact on ecological productivity? By 'manner of resource extraction', I refer to the following: the spatial pattern of extraction; the seasonality of extraction; the portion of ecological population that is extracted (e.g., demographic cohort or development stage of ecological population); or the morphological parts of the organism extracted (e.g., roots, leaves, stem, fruits, seeds of plants; horn, hair, wool, body . . . etc of animals).

Over a wide range of extraction pressures, *how* a resource is extracted may have more important effects on ecological relations than the aggregate level of extraction. This may be true not only due to variable sensitivities to the manner of extraction of the biological populations from which resources are extracted but also due to the indirect effects of resource extraction on broader biophysical systems and biological populations not directly the focus of extractive pressure. An important example of the former is that of the ecological effects of livestock grazing. On the annual grasslands of Sahelian West Africa, where livestock husbandry retains certain levels of mobility, how animals are grazed is more important than the aggregate stocking rate at spatial scales ranging from a village territory to the whole region. This results from: (i) grasslands are more sensitive (in terms of productivity and species composition) to the timing (duration and seasonality) of grazing than to the aggregate level of animals stocked over a yearly cycle; and (ii) grazing management results in a wide variation in the magnitude and timing of grazing pressures experienced from one grazing site to another. Grasslands are primarily sensitive to grazing during the rainy season. Given the nature of soils and the many 'natural' factors leading to declines in grass cover during the dry season, grazing by domestic livestock during the dry season has limited effect. In such a system, one needs to not only understand the factors that contribute to grazing management variability but also the details of ecological response to grazing during the rainy season. For example, defoliation experiments have found that short-term response to grazing varies in magnitude and direction during the period of active vegetative growth. In these experiments, simulated grazing bouts 2–3 weeks apart produce the most positive and negative responses to grazing on same-year productivity.

In sum, greater engagements with ecological relations are required to understand social/ecological change when ecological features of interest display non-linear, heterogeneous ecological responses to variation in how resources are extracted. Heterogeneous response may work at the individual to population levels. At the level of the individual organism, heterogeneous response is most common when extracting resources from living organisms (lopping of trees, grazing of vegetation, harvesting of fruits . . . etc.). At the level of the population, more complex response patterns are most common where strong dependencies exist between the extracted organism and others (e.g., figure 12.2). For those studying social-environmental change, serious consideration of land-use ecology is important in evaluating the appropriate level of engagement with ecological relations.

Political Ecology's Engagement with the Scientific Practice of Ecology

As already seen in the discussion above, environmental change and politics are very much shaped by our ideas about ecological relations. Ecology as a broad field, including not only academic ecology but also the applied ecologies of natural resource extraction and management (forestry, grazing management, agroecology, wildlife ecology, conservation biology, fisheries science . . . etc.), plays an important role in how we understand ecological relations. In many ways, the ecological sciences mediate our understandings of ecological relations and therefore are implicated in environmental politics and our understanding of anthropogenic environmental change. Increasingly, political ecologists are critically engaging with scientific practices and the depictions of ecological relations they produce. In this section, I do not attempt to review this rapidly expanding literature (see Forsyth, 2003; Taylor, 2005), but instead outline how greater engagement in the practices of ecological scientists may enrich the two core themes of political ecology.

Ecology and the relationship between environmental and social change

Those scholars concerned with the relationship between social and environmental change have, through their own engagements with ecological relations, critically engaged with the assumptions, methods, and theories of ecological scientists. Such critical engagements have raised questions about the use of inappropriate models of ecological dynamics in under-studied regions of the world by resource managers, scientists and policymakers (Fairhead and Leach, 1996; Bassett and Zuéli, 2000; Forsyth, 2003). Moreover, this work has raised questions about the social content in environmental diagnoses and prescriptions of environment (Fairhead and Leach, 1996; Neumann, 1998; Bassett and Zuéli, 2000; Taylor, 2005). By 'social content' I refer to the observations that: (i) most applied ecologies have developed in close relation to a resource extraction activity bundled with a particular social organisation and set of technologies;[4] (ii) due the limited availability of data, many diagnoses of environmental problems are based less on biophysical fact and more on cursory observations of: a. local social conditions that are seen as leading to environmental mismanagement (population density, property regime, resource-related conflict) and; b. poor environmental condition (defined by visual appearance); (iii) local understandings of the geographical and historical contexts of changing resource availability have most often been ignored as unscientific; and (iv) knowledge claims about the ecological relations are far from politically neutral in effect or intent – resource management and policy institutions may make scientific claims that reinforce their control over resources and people within the target area.

Due to the social content in environmental scientific work, political ecologists' place-based orientation for understanding the relationship between social and environmental change will often lead to a critical engagement with scientific claims that implicate local resources and people. The degree to which that engagement delves into the underlying methods, assumptions, and practices of scientist and resource managers depends not only on the political ecologist's adherence to critical realist or social constructivist epistemologies but also on her intended audience. A critical realist stance would favour delving into method and assumptions and would also be more useful in constructing a critique that resonates within policy and scientific circles (audience). In this way, the degree of engagement with scientific practice by those studying social and environmental change depends in part on how they position themselves with respect to resource management science.

Ecology and environmental politics

Those studying environmental politics have increasingly treated these as a 'politics of knowledge' with a major arena of conflict and competition over the question of what counts as valid ecology. In one way, this emphasis is linked to the movement away from political ecology's structural roots and the post-structural embrace of discourse, difference, and identity politics. Early political ecology may have tended to conceptualise environmental politics as straightforward exercises of power in response to material interests. Today, these interests are not seen to be less important but mediated discursively and materially through the practices of knowledge production and circulation. Since such mediation is incomplete, contested, and indeterminate, it is important to understand the knowledge politics of ecology.

A major theme in political ecology's treatment of environmental politics has been the social justice implications of the exercise of power by governments, multinational corporations, and powerful individuals to claim/manage natural resources from less powerful social groups. This politics is infused with a knowledge politics as the knowledge claims of the less powerful, schooled, and wealthy are discounted by those with more power. Scholars that share political ecology's critical engagement with ecological science have, by questioning the neutrality and generalisability of ecological science, contributed to a revisionist 'indigenous knowledge' literature to that which had been dominated by treatments that sought to translate lay people's knowledge into western scientific frameworks. Instead, these scholars have argued that it is actually highly problematic to reify 'western scientific knowledge' from all other knowledge forms. All knowledge bears the imprint of the context in which it is produced. Different knowledge systems have strengths and weaknesses and one should not rush to incorporate one into another but to treat each with respect looking for areas of overlap and divergence (Nadasdy, 1999; Goldman, 2007). Such treatments resonate strongly with efforts to democratise science and development practice.

A number of trends have changed how power is exercised within conservation and development contexts and in so doing have exposed the importance of how ecological knowledge is produced, invoked, and circulated. As part of the neoliberal turn in natural resource governance away from government towards civil society and the market, a whole set of code words describing conservation and development programmes have been invoked such as participatory, community-based, devolution, decentralisation ... etc. These programmes have arguably increased the reliance on scientific management as a means by which control by the powerful is maintained. The most common check on devolved authority to communities and citizen groups is the scientific oversight provided by international NGOs and government (Nadasdy, 1999). In this way, the politics of ecological knowledge has become an important locus through which conflict over access to resources occurs.

Conclusions

Ecology as both science and subject of scientific inquiry are of interest to political ecologists concerned with understanding environmental politics or social-environmental change. Political ecology is a highly diverse field of inquiry. Political ecological research is focused on different subjects of study which influence its varying levels of 'engagement' with the heterogeneity, dynamics, responsiveness, and embeddedness of ecology as subject of study (ecological relations). In general, political ecological research has shown limited engagement in ecological relations – reflecting the real costs of such engagements. Rather than promote a particular recipe for political ecological research, this chapter has explored how the questions we ask influence the appropriate level of the engagement with ecological relations. In so doing, the hope has been to avoid a relativist trap (e.g., all levels of engagement are fine) and delineate where further engagement promises to reveal either a hidden politics or more complex social-environmental change.

Political ecologists have increasingly contributed critical assessments of the social and environmental work done by ecological science. The silences, emphases, and framings embedded within the assessments, monitoring, and solutions of environmental scientists, international conservation organisations, and state officials etc. have been shown to have significant social and environmental consequences. In analysing these politics of knowledge, political ecologists have emphasised how scientific arguments have been used to silence those (often the disempowered) who have developed their own understandings of environmental change in ways that deviate from the western scientific programme.

While each has been treated separately in this chapter, one could argue that in fact to engage critically with the truth claims of ecology, political ecologists themselves need to engage more with ecological relations. Otherwise, their criticisms of ecology as practiced within their study areas may enjoy a following among social scientists but have little effect on ecological scientific practice. Through greater engagement with ecological relations, political ecologists could reveal the assumptions, methodological lapses and social content of ecological assessments and in so doing, gain a greater appreciation of the difficulties of ecological inquiry. Such engagements are not without cost (as outlined above). It is important for researchers to be deliberate about these choices, reflecting on their audiences and research subjects and how these in turn influence the benefits of greater engagements across the social and biophysical scientific divide.

NOTES

1. I adopt this stance, not out of ignorance of the need for political ecology, as a relatively young field of inquiry, to define itself, but to explore the benefits (and costs) of further engagement with ecological relations as affected by the research questions being asked by political ecology researchers. Such an exploration is not possible if one adopts a prescriptive view of what political ecological analysis should or should not contain.
2. 'Engagement' does not necessarily require ecological fieldwork. Still, it should be noted that the relative importance of context increases with greater engagement, which in turn increases the chance that fieldwork is necessary. In this way, the need for ecological fieldwork (performed by political ecologist or by collaborators) tends to increase as the level of engagement increases.

3. Interestingly, this is an argument used by CPR theorists against full privatisation as the solution to the 'tragedy of the commons'. Applications of CPR theory often work to erect boundaries using existing jurisdictions which often are smaller than the effective production spaces of rural producers. In so doing, such programs make resource access boundaries between social groups less porous. A major new emphasis is matching the scale of environmental governance (through systems of multi-scaled management) to the spatial scales of processes influencing the availability of the resource.
4. Forestry is obvious example of this. It developed within a particular institutional context and with particular production imperatives. Much of scientific forestry, based on models of whole tree removal, has little relevance to management concerns in many forested areas of the world where the major extractive flows from the forest are through lopping of branches and collection of non-timber forest products.

BIBLIOGRAPHY

Bakker, K. and Bridge, G. (2006) Material worlds? Resource geographies and the 'matter of nature'. *Progress in Human Geography*, 30(1), 5–27.

Bassett, T. J. and Zuéli, K. B. (2000) Environmental discourses and the Ivorian Savanna. *Annals of Association of American Geographers*, 90(1), 67–95.

Blaikie, P. (1985) *The Political Economy of Soil Erosion in Developing Countries*. London: Longman.

Blaikie, P. and Brookfield, H. (eds) (1987) *Land Degradation and Society*. London and New York: Methuen.

Braun, B. (2002) *The Intemperate Rainforest : Nature, Culture, and Power on Canada's West Coast*. Minneapolis: University of Minnesota Press.

Castree, N. (2005) *Nature*. New York: Routledge.

Castree, N. and Braun, B. (eds) (2001) *Social Nature: Theory, Practice, and Politics*. Malden, MA: Blackwell.

Cronon, W. (ed.) (1996) *Uncommon Ground: Rethinking the Human Place in Nature*. New York: WW Norton and Company.

Daily, G. C. and Ellison, K. (2002) *The New Economy of Nature : The Quest to Make Conservation Profitable*. Washington, DC: Island Press/Shearwater Books.

Demeritt, D. (1998) Science, social constructivism and nature. In B. Braun and N. Castree (eds), *Remaking Reality*. New York: Routledge.

Diamond, J. M. (2006) *Collapse : How Societies Choose to Fail or Succeed*. New York: Penguin.

Ellen, R. (1982) *Environment, Subsistence and System*. Cambridge: Cambridge University Press.

Ellen, R., and Fukui, K. (eds) (1996) *Redefining Nature: Ecology, Culture and Domestication*. Oxford: Berg.

Fairhead, J. and Leach, M. (1996) *Misreading the African Landscape*. Cambridge: Cambridge University Press.

Forsyth, T. (2003) *Critical Political Ecology: The Politics of Environmental Science*. London: Routledge.

Goldman, M. (2007) Tracking wildebeest, locating knowledge: Maasai and conservation biology understandings of wildebeest behavior in northern Tanzania. *Environment and Planning D: Society and Space*, 25(2), 307–31.

Gray, L. C. (1999) Is land being degraded? A multi-scale investigation of landscape change in southwestern Burkina Faso. *Land Degradation and Development*, 10, 329–43.

Hecht, S. B. (1985) Environment, development, and politics: Capital accumulation and the livestock sector in Eastern Amazonia. *World Development*, 13, 663–84.

Homer-Dixon, T. F. (1994) Environmental scarcities and violent conflict. *International Security*, 19(1), 5–40.

Latour, B. (1993) *We Have Never Been Modern*. Cambridge: Harvard University Press.

Mansfield, B. (2004) Neoliberalism in the oceans: 'rationalization', property rights and the commons question. *Geoforum*, 35, 313–26.

Nadasdy, P. (1999) The politics of TEK: power and the 'integration' of knowledge. *Artic Anthropology*, 36(1–2), 1–18.

Neumann, R. P. (1998) *Imposing Wilderness: Struggles over Livelihood and Nature Preservation in Africa*. Berkeley: University of California Press.

Neumann, R. P. (2005) *Making Political Ecology, Human Geography in the Making*. London: Hodder Arnold.

Ostrom, E. (1990) *Governing the Commons: The Evolution of Institutions for Collective Action*. Cambridge: Cambridge University Press.

Peluso, N. L. and Watts, M. (eds) (2001) *Violent Environments*. Ithaca, NY: Cornell University Press.

Rappaport, R. (1984) *Pigs for the Ancestors: Ritual in the Ecology of a New Guinea People*, 2nd edn. New Haven: Yale University Press.

Ribot, J. C. and Peluso, N. L. (2003) A theory of access. *Rural Sociology*, 68(2), 153–81.

Robbins, P. (2004) *Political Ecology*. London: Blackwell.

Robertson, M. M. (2006) The nature that capital can see: science, state, and market in the commodification of ecosystem services. *Environment and Planning D: Society and Space*, 24, 367–87.

Semple, E. C. (1915) *Influences of Geographic Environment*. New York: Henry Holt.

Swyngedouw, E. (1999) Modernity and hybridity: Nature, regeneracionismo, and the production of the Spanish waterscape, 1890–1930. *Annals of the Association of American Geographers*, 89, 443–65.

Taylor, P. J. (2005) *Unruly Complexity: Ecology, Interpretation, Engagement*. Chicago: University of Chicago Press.

Turner, B. L. (1997) Spirals, bridges and tunnels: engaging human-environment perspectives in geography. *Ecumene*, 4(2), 196–217.

Vale, T. R. (1998) The myth of the humanized landscape: an example from Yosemite National Park. *Natural Areas Journal*, 18, 231–36.

Walker, P. A. (2005) Political ecology: where is the ecology? *Progress in Human Geography*, 29(1), 72–83.

Watts, M. J. (1983) On the poverty of theory: natural hazards research in context. In K. Hewitt.(ed.), *Interpretations of Calamity*. London: Allen and Unwin, pp. 231–62.

Watts, M. J. (1990) Is there politics in regional political ecology. *Capitalism, Nature, Socialism*, 4, 123–31.

Whatmore, S. (2002) *Hybrid Geographies: Natures, Cultures, Spaces*. London and Thousand Oaks, CA: SAGE.

Williams, R. (1980) Ideas of nature. In *Problems in Materialism and Culture*. London: Verso, pp. 67–85.

Zimmerer, K. (1994) Human geography and the 'New Ecology': the prospect and promise of integration. *Annals of the Association of American Geographers*, 84(1), 108–25.

Zimmerer, K. (1996) Ecology as cornerstone and chimera in human ecology. In C. Earle, K. Mathewson and M. S. Kenzer (eds), *Concepts in Human Geography*. Lanham, MD: Rowman & Littlefield, pp. 161–88.

Zimmerer, K. S. and Bassett, T. J. (eds) (2003) *Political Ecology: An Integrative Approach to Geography and Environment-Development Studies*. New York: Guilford Press.

Chapter 13

Quaternary Geography and the Human Past

Jamie Woodward

A New Quaternary Geography

Research published in the 1990s reshaped fundamental ideas about Quaternary geography and the tempo of global environmental change. Data from the Greenland ice cores and from North Atlantic marine sediments show that the last cold stage was punctuated by a remarkable series of abrupt and high-amplitude changes in climate and oceanographic conditions (Bond et al., 1993; Dansgaard et al., 1993). These discoveries are driving a new research agenda focused on the causal mechanisms and how ecosystems and geomorphological processes on the surrounding continents reacted to such rapid and repeated oscillations in Quaternary climate (Fuller et al., 1998; Allen et al., 1999; Walker and Lowe, 2007). Some researchers have returned to previously well-studied sites to scrutinise records at much higher resolution in order to examine the sensitivity and response of terrestrial environments during this period (e.g., Tzedakis et al., 2002). It is now clear that these findings have major ramifications for all components of Physical Geography – for geomorphology, biogeography and climatology and how we conceptualise the interactions between them. These findings have also led to new research questions and new approaches in the study of human-environment interactions during the Quaternary Period; especially the Palaeolithic archaeology of the last cold stage (Woodward and Goldberg, 2001; Gamble et al., 2004; Mellars, 2006; Tzedakis et al., 2007).

Much earlier in the 20th century, in the absence of reliable stratigraphic frameworks for both the Quaternary and archaeological records, it proved difficult to tackle even very basic questions about the nature of the relationship between Quaternary environmental change and the human past. Peake (1922, p. 6), for example, outlined some of the central issues at this time:

> The problem before us is twofold. Firstly we have to consider whether there was one ice age or several, and in the latter case how many we must account for. The other question, upon which opinion is rapidly hardening, is the relation of the different Palaeolithic periods to the glacial phases.

Quaternary research has always been a meeting place for geographers, geoscientists and archaeologists. These disciplines have long been fascinated by the pace, amplitude and drivers of environmental change, by the nature of ice age ecosystems and by the challenges presented to early human societies by a changing Quaternary geography. It is the emergence of new and more finely resolved windows into the past – as well as improvements in dating methods for both geological and archaeological records – that have, in recent decades, radically transformed the way we think about Quaternary environments, landscape change and past human activity. There is now abundant evidence for rapid and repeated reorganisations of Quaternary ecosystems over centennial to millennial timescales and these would have impacted significantly upon resource availability and human subsistence strategies (see Mithen, 1999).

It is against this background of a newly energised Quaternary geography that this chapter aims to explore some of the practical and theoretical issues associated with locating the Palaeolithic archaeology of the last glacial stage within a precise environmental framework. This is a key research goal despite the traditional tensions within archaeology between 'environmentalist' and 'internalist' theories. The former emphasise ecological relationships and the determining or limiting effect of basic biological and environmental factors, while the latter derive their main inspiration from the social sciences and emphasise what they call the inherent dynamic of social relations (Bailey, 1983). In this context, Mithen (1999, p. 478) has set out a robust defence of the study of long-term human-environment interactions:

> People are not detached from natural environments but are part of them; the natural environment provides opportunities and constraints on human behaviour, and it is in turn changed by that behaviour. When human-environment interactions can be studied over the longer term, it is possible to explore how people adapted to environmental change and this provides us with basic information about the nature of the human condition.

Bailey (1983) and Mithen (1999) argue that these approaches need not be in conflict and, in any case, it can be argued that we need to establish the environmental context of the archaeological period under discussion before that debate can begin. This chapter is not advocating a purely deterministic approach to the investigation of human activity in Late Pleistocene environments. Rather, by recognising the new and exciting opportunities presented by recent advances in Quaternary science, it seeks to explore how we may define new questions and test existing ideas about human-environment interactions in light of the reality of a highly dynamic Quaternary geography and the developing potential for improved dating frameworks. Dating control is a key theme throughout this chapter. Without reliable dating frameworks, it is not possible to establish the pace of environmental change or to compare records – both environmental and cultural – from different contexts.

A good deal of the material covered here builds directly upon the pioneering contributions of Nick Shackleton and Willard Libby (figure 13.1) who revolutionised approaches to the Quaternary record through, respectively, the study of the oxygen isotope record in deep sea sediments (Shackleton, 1967; Shackleton and Opdyke, 1973) and the development of radiocarbon dating (Libby, 1955).

Figure 13.1 Nick Shackleton (1937 to 2006) and Willard Libby (1908 to 1980) are arguably the two most influential figures in Quaternary science in the second half of the 20th century. Nick Shackleton is shown picking forams for oxygen isotope analysis in the Godwin Laboratory in Cambridge (University of Cambridge Newsletter, December 2004). Willard Libby is shown in his Laboratory at UCLA in 1968 (University of California history digital archives).

The Upper Palaeolithic Revolution

Advances in radiocarbon dating are an important part of this story because, as far as the archaeological record of the Upper Palaeolithic is concerned, this is the main method for dating this period and for building correlations between archaeological sites over the past 50,000 years or so (Mellars, 2006). This is a small portion of Quaternary time, but it incorporates dramatic changes in both the environmental and archaeological records. It is a key period in human history that saw the demise of the Neanderthals (the end of the Middle Palaeolithic) and the establishment of anatomically modern humans (Cro-Magnons) as the sole human species. In Europe this is known as the Upper Palaeolithic revolution (Mellars, 1994).

The Upper Palaeolithic is marked by a set of fundamental cultural shifts that set it apart from the Middle Palaeolithic. Stone tool production shows greater innovation and a much wider range of tool types. New raw materials enter the human tool kit – with elaborate use of bone and antler – and this period saw the development of more effective social networks (Gamble 1986; 1999). Perhaps the most striking aspect of the Upper Palaeolithic record in Europe are the remarkable cave paintings of France and Cantabrian Spain from sites such as Lascaux, Chauvet and Altamira (Bahn and Vertut, 1997) (figure 13.2). Questions surrounding the causes (whether environmental or cultural, or a combination of both) and precise timing of the Neanderthal extinction, and the pace of modern human (Cro-Magnon)

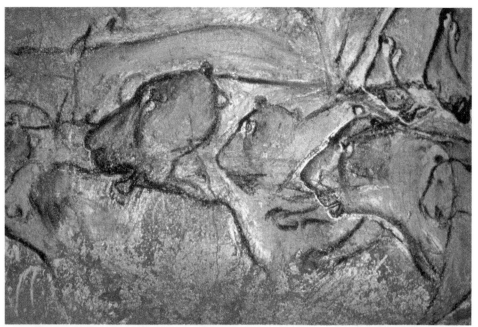

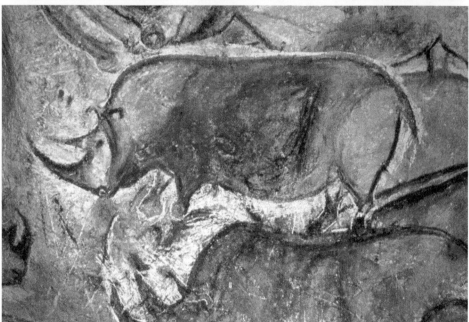

Figure 13.2 Upper Palaeolithic cave paintings from Chauvet Cave in southern France. The upper image shows a pride of lions and the lower image shows a rhino from a group of seventeen in Chauvet Cave. Both photographs by Jean Clottes and reproduced with permission from the French Ministry of Culture and Communication (Direction régionale des affaires culturelles Rhône-Alpes). AMS radiocarbon dating has allowed the pigments and charcoal from such images to be dated directly and this can provide valuable insights into the nature of ice age ecosystems. Upper Palaeolithic art in Europe has a very distinctive geography – it is mainly concentrated in the Dordogne, Cantabrian Spain and the Rhone Valley.

dispersal across Europe, have generated much debate (Stringer and Gamble, 1993). A key area of controversy is the interpretation of radiocarbon dates because many of these cultural changes took place towards the practical upper limit of this dating method (Mellars, 2006).

To explore some of the problems involved in charting the interactions between environmental change and human activity over the course of the last cold stage, this chapter will focus upon examples from Western Europe and the Mediterranean within the period between c. 50,000 and 10,000 years ago. We now know that this period includes Heinrich Events 1 to 5 when massive discharges of icebergs from the Laurentide Ice Sheet chilled the surface of the North Atlantic and created a bitterly cold and dry climate across the surrounding land masses. This period also includes the global Last Glacial Maximum (c. 20–22 ka) when the major continental ice sheets in North America and Eurasia reached their maximum extent. To understand the full significance of the data obtained from the North Atlantic marine sediment record and the Greenland ice cores, it is instructive to consider some early ideas about the glacial record and the first major paradigm shift in Quaternary science that took place in the 1970s.

The Alpine Model of Quaternary Glaciation

In the late 19th and early 20th centuries, much effort was centred on establishing the number of Quaternary glaciations and the antiquity of humans (Peake, 1922 and see Grayson, 1990; Gamble, 1994). This period saw some of the earliest interaction between geologists and archaeologists (Goudie, 1976). For much of the 20th century, the glacial record of the Quaternary was synonymous with the framework put forward by Albrecht Penck and Eduard Brückner published in 1909 and this model gained widespread support after the First World War (see Peake, 1922; Bowen, 1978). This scheme was based on geomorphological fieldwork in the northern forelands of the Alps where they recognised a series of glacial and fluvial landforms (primarily moraines and river terraces) and associated sediments representing four main periods of Quaternary glaciation (table 13.1). These glacial periods were named Gunz, Mindel, Riss and Würm after the river valleys that contained these deposits. The Alpine scheme is based on a discontinuous terrestrial record that

Table 13.1   Penck and Brückner's (1909) model of four major Quaternary glaciations and interglacials (based on table 2.1 in Bowen, 1978). The values on the right are estimates of the length of the interglacials relative to the post-glacial (Holocene) period. The Mindel-Riss interglacial became known as the Great Interglacial

Stage	*Landform or process*	
Post-Würm (Holocene) interglacial	*incision*	1
Würm Glaciation	Niedterrassen (Low Terrace)	
Riss-Würm interglacial	*incision*	3
Riss Glaciation	Hochterrassen (High Terrace)	
Mindel-Riss interglacial	*incision*	12
Mindel Glaciation	Younger Deckenschotter	
Günz-Mindel interglacial	*incision*	3
Günz Glaciation	Older Deckenschotter	

contains large gaps. The interglacial periods, for example, are represented, not by sediments, but by long phases of incision in the river valleys (table 13.1). A key weakness of this model was the lack of a reliable time frame for the events it represented (Bowen, 1978) although Penck and Brückner (1909) did provide estimates for the length of the interglacial periods relative to the Holocene or post-glacial period (table 13.1). This framework became a cornerstone of Quaternary research and records from around the world were correlated with the Alpine model (see Bowen, 1978 for a detailed discussion).

The Marine Realm: Oxygen Isotopes and the Glacial Record

The Penck and Brückner model persisted for so long partly because there were no convincing alternatives and partly because it could not be challenged effectively in the absence of reliable dating frameworks for long records of change. This situation changed in the 1960s and 1970s as attention shifted to the study of the continuous Quaternary records in the deep ocean basins and the use of oxygen isotope analysis.

Oxygen has three stable isotopes (^{16}O, ^{17}O and ^{18}O) and because the lighter isotope evaporates more easily, atmospheric water vapour contains more ^{16}O and less ^{18}O than the parent sea water. This process is called fractionation and it means that continental ice sheets and glaciers are enriched in ^{16}O and, as ice sheets grow, the oceans become relatively enriched in the heavier isotope ^{18}O. The oxygen isotope ratio of ocean water is recorded in the calcium carbonate shells of tiny organisms called forams. When they die they form part of the marine sediment record. These simple creatures and these physical principles were the key to unlocking the glacial record of the Quaternary.

Some species of forams produce shells with a composition that is in isotopic equilibrium with the water that they inhabit and the oxygen isotopes can be measured using a mass spectrometer. This means that a long core of foram-rich marine sediment can provide a record of long-term shifts in the isotopic composition of the oceans. A cold stage or glacial ocean is enriched in the heavier isotope (^{18}O) because huge amounts of ^{16}O are locked within the continental ice sheets. Conversely, a warm stage or interglacial ocean contains more of the lighter isotope (^{16}O) because ice sheet melting returns ^{16}O to the oceans. Shackleton (1967) had already shown that the oxygen isotope record from Quaternary marine sediments was primarily a record of changes in global ice volume and not a record of changes in ocean temperature as had been argued previously (Emiliani, 1955). Thus, oxygen isotope measurements can provide valuable insights into long-term changes in the global hydrological cycle.

Shackleton and Opdyke (1973) worked on a 16-m sediment core (V28–238) recovered from the Solomon Plateau on the floor of the equatorial Pacific and measured the oxygen isotope ratio of foram samples for the entire length of the core. Their oxygen isotope curve is shown in figure 13.3. The troughs in this curve mark those periods when global ice volume reached its maximum extent during glacial stages and these are marked with even numbers. The odd numbers represent interglacial periods when global ice volume was much reduced and eustatic sea level was high. These are commonly referred to as marine isotope stages (MIS) and MIS 5, for example, is the last interglacial.

This record revolutionised the study of the Quaternary because Shackleton and Opdyke (1973) were the first to set an oxygen isotope curve (and the long-

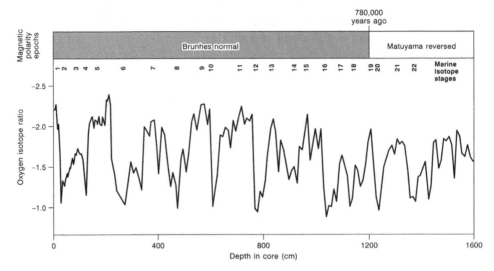

Figure 13.3 The oxygen isotope record from ocean core V28–238 modified from Shackleton and Opdyke (1973). The Brunhes-Matuyama magnetic reversal coincides with Marine Isotope Stage 19 and this event provided the basis for estimating the age of the glacial and interglacial stages shown in this core. Shackleton and Opdyke used a value of 730 ka for this event, but this was later refined to 780 ka in the light of more recent data. Note the rapid transition from cold stages to warm stages – these are known as Terminations as shown on figure 13.5. The global Last Glacial Maximum corresponds to MIS 2 and the post-glacial or Holocene period corresponds to MIS 1.

term record of continental glaciation it represented) within a robust dating framework. To achieve this they utilised long-term changes in the Earth's magnetic field.

Marine isotope stage 19 in core V28–238 coincides with the Bruhnes-Matuyama magnetic reversal, the last time that the Earth's magnetic field flipped. All of the sediments above MIS 19 have normal polarity. The age of this reversal event was known and it allowed a timescale to be developed for the entire record based on the reasonable assumption of a constant rate of sedimentation for deep ocean sediments. The ages of all sections of the core were interpolated from this datum. This showed that the Bruhnes Epoch alone contained eight full glacial-interglacial cycles (figure 13.3) and this accounted for much less than half of Quaternary time. In contrast, the Penck and Brückner model had just four for the entire two million years or so of the Quaternary Period (table 13.1).

The continuous record of glacial and interglacial cycles from core V28–238 had a profound impact on Quaternary science and it provided a yardstick against which all other records could be evaluated. At the end of the decade, Bowen published a paper in *Progress in Physical Geography* entitled 'Geographical Perspective on the Quaternary' that drew much of its inspiration from the V28–238 record. It began with the following statement (Bowen, 1979, p. 167):

A revolution has taken place in Quaternary research that is in effect comparable to that of plate tectonic theory in the geological sciences as a whole. Its implications extend deep into the fundamental methodological foundations of all geographical sciences concerned with matters related to the past. It relates predominantly to a change in scale and complexity which, although hinted at earlier this century, could hardly have been imagined at the opening of the post-World War II period.

As Bowen argued so forcefully, the oxygen isotope record from the marine archive had major implications for all components of Quaternary geography – it made geomorphologists, glaciologists and ecologists rethink the tempo of Quaternary landscape and vegetation dynamics – it also led to new ideas in human evolution and adaptation (Gamble, 1986). In short, this work forced a radical rethink about the complexity and dynamics of Earth system change during the course of the Quaternary Period and it signalled the end of the Alpine framework. The record was soon replicated in all the major marine basins and a series of marine oxygen isotope records were later used by Shackleton and co-workers to show that the rhythms of the ice ages were controlled by astronomical parameters as predicted by Milankovitch much earlier in the century (Hays et al., 1976; Imbric and Imbric, 1979).

Dating the Terrestrial Records: The Radiocarbon Method

Radiocarbon dating was developed by Willard Libby and his team at the University of Chicago in the years immediately after the Second World War. Libby was awarded The Nobel Prize for Chemistry in 1960 'for his method to use carbon-14 for age determination in archaeology, geology, geophysics, and other branches of science'. Radiocarbon (^{14}C) is continually produced in the upper atmosphere and it enters all living organisms via the carbon dioxide cycle. On the death of a plant or animal, the uptake of radiocarbon ceases and the radiocarbon store in the organism continues to decay, but without replenishment. So death sets the radiocarbon dating clock ticking so that with a few assumptions, it is possible to establish the amount of residual radioactivity per gram of carbon in a fossil sample and, using modern standards and the measured half-life of radiocarbon (5,570 ± 30 years), it becomes possible to calculate a date for the death of the sample (Libby, 1955; Bowman, 1990).

The measurement of radiocarbon requires sensitive and specialist laboratory equipment because for every one million million atoms of stable carbon (^{12}C) in a living organism, there is just a single atom of ^{14}C (Lowe and Walker, 1997). The sensitivity of the method has been significantly enhanced through the use of Accelerator Mass Spectrometry (AMS) as this allows ^{14}C atoms to be detected and counted directly in contrast to conventional dating which only detects those atoms that decay during the time interval allotted for an analysis. AMS offers several advantages because the measurement time is much quicker and only very small samples of carbon (1 mg or less compared to 5 to 10 g for conventional dating) are needed for dating (Gowlett et al., 1997; Bell and Walker, 2005). AMS represented a key breakthrough for studies of the Middle and Upper Palaeolithic because it allowed small samples of charcoal to be dated instead of bone samples – the latter are susceptible to contamination by more recent carbon from percolating groundwater. This process can top up the amount of residual radiocarbon in a bone sample to give a spuriously young age.

Another recent breakthrough has seen the application of the AMS approach to obtain radiocarbon determinations directly from cave paintings by dating small samples of the pigments and fragments of charcoal that form the images on the cave walls (e.g., Valladas, 2003) (figure 13.2). Previously, the chronology of the cave paintings was loosely based on the style of the images. This work offers the potential to provide a much more rigorous basis for the development of a detailed chronology of Upper Palaeolithic art across Europe, but some of the AMS results and their interpretation have been contested (Pettitt and Bahn, 2003).

A key assumption of the method is that the ratio of radiocarbon (^{14}C) to stable carbon (^{12}C) has remained constant in the Earth's atmosphere so that the measurement of residual radiocarbon in a given sample provides a reliable indication of its true age. However, it is now well established that radiocarbon production in the upper atmosphere has fluctuated markedly during the Quaternary (see Bard et al., 1990; Mellars, 2006) and radiocarbon dates therefore have to be calibrated because radiocarbon years are not directly equivalent to calendar years. In theory, radiocarbon dating can be used to date organic materials up to 50,000 years old, but in practice many researchers do not place much faith in dates older than about 40,000 years because the ages can be distorted by sample contamination. Furthermore, the development of calibration curves and algorithms for such old samples is still in its infancy.

Since the radiocarbon method was pioneered by Libby in the 1940s, it has seen a series of fundamental changes in the measurement and interpretation of results. The key changes are largely responses to the problems associated with age calibration and sample contamination, and these factors are especially acute for radiocarbon determinations beyond six half-lives. However, in a stimulating and sanguine review, Mellars (2006) argues that recent methodological advances have dramatically reduced both of these sources of error. First, new pretreatments for the purification of bone collagen have effectively removed the problem of contamination by more recent carbon. Second, a new calibration model based on data from various sites around the world now provides the best available means of calibrating radiocarbon dates over the last 50,000 years (e.g., Hughen et al., 2004). This calibration shows that a radiocarbon date of 35,000 years BP is equivalent to a calendar age of approximately 40,500 years BP. It is therefore of crucial importance when reporting dating results to make a clear distinction between radiocarbon years and calendar years. The systematic displacement of radiocarbon ages from true calendrical ages has very clear implications for any comparison between radiocarbon-based chronologies from archaeological sites and proxy climate records such as the Greenland ice cores or other geological archives that have been dated by other methods (Woodward, 2003; Mellars, 2006). If the purification of bone samples and calibration back to 50,000 BP become routine over the next few years, this will present exciting opportunities to test ideas about the nature of the Middle to Upper Palaeolithic transition. Mellars (2006) has already begun to put forward a case for a much more rapid transition and a more rapid dispersal of modern humans across Europe.

Quaternary Geography: Transects Across Europe

Figure 13.4 shows transects across Europe from the Mediterranean Sea to the Arctic Ocean under interglacial and glacial conditions. Each transect shows, in broad terms, the major ecosystems present across the continent at the extremes of warm and cold stages – the odd and even numbered stages, respectively, on figure 13.3. The cold stage geography of Europe shows a large mid-latitude ice sheet fringed by a belt of polar desert and steppe tundra. It shows a few trees on the southern slopes

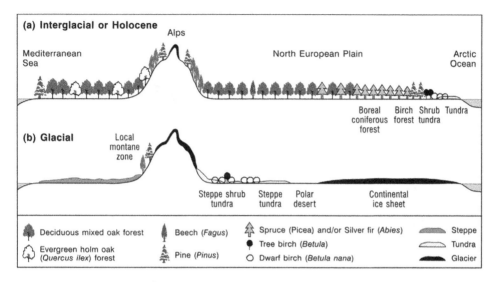

Figure 13.4 Transects across Europe from south to north showing schematic representations of the vegetation belts associated with glacial and interglacial stages (modified from van der Hammen et al., 1971). We know from the long pollen records in the Mediterranean that trees survived in the south throughout the Pleistocene. Without the presence of long-term refugia in the south, northern tree populations could not be re-established during interglacials.

of the Alps and the Mediterranean region is shown as a steppe environment. Each ecosystem would have supported a range of fauna including large herbivores such as mammoths, reindeer and horses, and each presented a range of opportunities and challenges for human societies (Gamble, 1986). These and other fauna have been recorded from the bone assemblages recovered from the excavation of Middle and Upper Palaeolithic sites across the region. These animals are also represented in the cave paintings of France and northern Spain where the galleries provide wonderful insights into the geography of ice age Europe and showcase the creativity and skill of Upper Palaeolithic human groups (Figure 13.2). The cold stage geography of Europe will be discussed in more detail below.

The Quiet Revolution: Rapid Climate Change

Shackleton continued to work on much longer cores throughout the 1970s and 1980s and he extended the oxygen isotope record back for the entire Quaternary and deep into pre-Quaternary time. This work provided the first indications of a Quaternary geography that was far more dynamic than anyone had previously contemplated. Figure 13.5 shows four long proxy climate records. Figure 13.5a is an oxygen isotope curve for the entire Quaternary from ODP Site 677. This is a much longer record than the one from core V28–238 shown in figure 13.3 as it shows changes in global ice volume over the last 2.5 million years. This remarkable pattern of environmental change was compiled by Shackleton and Crowhurst (1996) and it includes several magnetic reversals back to MIS 104. There is an important step change around 900,000 years BP (MIS 22) known as the mid-Pleistocene Revo-

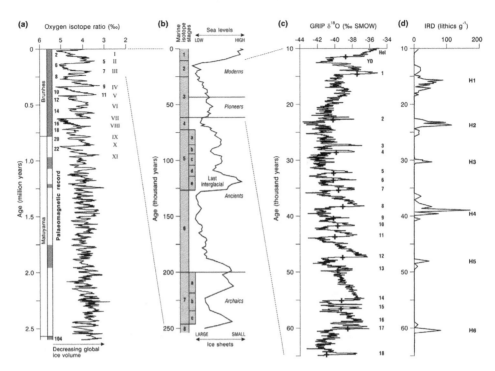

Figure 13.5 (a) The long oxygen isotope record from ODP Site 677 in the tropical Pacific Ocean showing changes in global ice volume and shifts from glacial to inter-glacial conditions for the entire Quaternary Period. Note that this record extends back beyond 2.5 million years and the record contains several magnetic reversals. Compare to figure 13.3. Roman numerals mark Terminations I to XI (modified from Shackleton and Crowhurst 1996). (b) The marine oxygen isotope record for the last 250,000 years in more detail showing the subdivisions within the interglacials of MIS 5 and MIS 7. The terms used by Gamble (1994) for the Palaeolithic occupation of Europe are also shown. Towards the end of MIS 3 the Neanderthals (Ancients) were replaced by Cro-Magnons (Moderns). (c) Abrupt changes in temperature from 60 to 10 ka shown in a high-resolution ice core record from Greenland (after Dansgaard et al., 1993). (d) The record of ice-rafted debris (IRD) from marine sediments in the North Atlantic between 60 and 10 ka. These discrete layers of coarse material mark the Heinrich Events of the last cold stage (modified from Roucoux et al., 2005 and sources therein).

lution when the glacial cycles change to a dominant 100,000-year cycle and there is a much larger contrast in ice volume between cold and warm stages.

The next curve (figure 13.5b) shows the last 250,000 years in more detail along with the terms from Gamble (1994) for the human occupation of Europe in the Palaeolithic. The Ancients are the Neanderthals and the Pioneers are the final Nean-derthal groups who disappeared from the archaeological record in Europe in MIS 3. Figure 13.4 showing the transects across Europe discussed above was published in 1971 and it can be argued that it reflects a more general tendency for some Quaternary researchers to focus on the palaeogeography at the *extremes* of glacial and interglacial conditions. To some extent the oxygen isotope records reinforced this view as Porter (1989, p. 245) has argued:

In Quaternary research, it is all too easy to view the glacial ages simplistically as a succession of glacial and interglacial culminations during which the extent and volume of glacier ice were at a maximum or minimum.

In reality, and this is shown very clearly by the marine oxygen isotope records (figures 13.3 and 13.5a,b), conditions during the Quaternary Period were, for much of the time, *intermediate* between these extremes and, after all, the peaks of glacial and interglacial periods were relatively short-lived. Porter (1989) argued that this was an important consideration when examining geomorphological and ecological processes during the Quaternary Period. It is clear, however, that for much of the Quaternary, global ice volume was much greater than present-day values. Figure 13.5b shows a relatively slow build-up of continental ice during the course of marine isotope stages 2 and 6 with extended periods of very harsh conditions ending in a brief period of rapid ice sheet melting known as a Termination. However, the high-resolution data from the North Atlantic and from the Greenland ice cores have dispelled any notions of long-term ice sheet stability and glacial monotony as is shown in figure 13.5c,d. One of the most remarkable discoveries of the 1990s was that the last cold stage was punctuated by centennial- to millennial-scale variations in climate and these are clearly recorded in the ice core records (figure 13.5c) with significant fluctuations in air temperatures in Greenland throughout this period. These are known as Dansgaard-Oeschger cycles and they represent air temperature shifts of the order of 15°C (Dansgaard et al., 1993).

Another key discovery of the last two decades was the presence of ice-rafted debris in the marine sediment record across the North Atlantic (figure 13.5d). These sediments show that the North American (Laurentide) ice sheet was highly dynamic throughout the last cold stage as large discharges of icebergs periodically flowed out across the North Atlantic and cooled the ocean surface. These are known as Heinrich Events and their impact on the climate system has been recorded in a variety of proxy records across the European continent (Bell and Walker, 2005; Anderson et al., 2007). As the drifting ice melted, it lowered the salinity of surface waters and this is clearly recorded in the oxygen isotope signal from foram species that lived in the upper part of the water column (Bond et al., 1993). The recognition of Heinrich Events showed the potential scale and rapidity of cryosphere-ocean-atmosphere interactions during the last cold stage.

For the second time within two decades, revelations from the marine record have forced Quaternary scientists to revise their ideas about long-term ice sheet dynamics and the drivers of environmental change, and to ask new questions of the terrestrial records. Indeed, the extract from Bowen (1979) cited above is just as relevant almost 30 years on as the combined impact of these findings alongside the Greenland ice core records has been profound across both the Quaternary science and archaeological communities. A direct result of these revelations is that most research is now done at much higher resolution than before, with more finely resolved sampling and better dating control. The impact of these changes for the study of long-term human-environment interactions will be discussed below.

A Mediterranean Perspective: High-Resolution Records

Another important development in European Quaternary research in the last two decades has seen a significant increase in the volume of work conducted south of the Alps in the Mediterranean region. There are good reasons for this. It can be argued that the Mediterranean has the best set of long-term terrestrial Quaternary records in Europe if not the rest of the world (Woodward, 2009). The region contains distinctive tectonic settings with long-term sediment sinks spanning multiple glacial-interglacial cycles and, in some cases, all of the Quaternary. The long lake sediment records, for example, can be compared directly with the marine archive (Tzedakis et al., 1997) and because the region lay south of the major European ice sheets, many of these sedimentary records are continuous and well preserved. An added advantage is the fact that the geology of the region offers many opportunities for dating and often at better resolution than in other parts of Europe. The widespread occurrence of limestone, for example, has produced karstic features and secondary carbonates that can be dated using uranium-series methods and this has produced new insights into the glacial records for example (Woodward et al., 2004; Hughes et al., 2006). The presence of explosive volcanic centres has spread volcanic ash (tephra) over wide areas (Narcisi and Vezzoli, 1999; Wulf et al., 2004) and this material can be dated directly. Tephras can be used to correlate between records that are many hundreds of kilometres apart and they have even been found in Upper Palaeolithic rockshelter sediment records in Greece (Farrand, 2000) and Montenegro (Brunnacker, 1966).

Parts of the Mediterranean region formed important refugia for tree species during cold stages of the Pleistocene. When climate ameliorated and trees were able to expand their ranges from refugial centres, the long pollen records show that they were able to do this very rapidly (figure 13.6). In contrast to areas much further to the north, this created a much more dynamic Pleistocene geography. Allen et al. (1999) have examined the long lake sediment record from Lago Grande di Monticchio in southern Italy. This record covers the last 102,000 years and it shows a series of rapid environmental changes during the last cold stage that correlate well with the Greenland ice core records. This is a sensitive, high-resolution record that allows centennial to millennial scale climate variability to be examined. Rapid vegetation changes took place in this region during the last cold stage over timescales of less than 200 years. This a key terrestrial archive of environmental change in southern Europe for the last glacial cycle. Allen et al. (1999) show very clearly that the terrestrial biosphere was a full participant in these rapid fluctuations and they conclude that:

> the closely coupled ocean-atmosphere system of the Northern Hemisphere during the last glacial extended its influence at least as far as the central Mediterranean region

The marine sedimentary record in the Mediterranean is also a very distinctive archive of environmental change that is linked directly to the North Atlantic via water exchange at the Straits of Gibraltar. The Mediterranean Sea is a relatively small body of water in the global ocean system, but it is very sensitive to environmental change and the high sedimentation rates in the basin form excellent records of change (Cacho et al., 1999). It is now well established that the impact of Heinrich Events is clearly recorded in the western Mediterranean basin because cold North Atlantic waters entered the basin via the Straits of Gibraltar and the regional climate became cooler and drier during these periods. Figure 13.6a shows two parameters from the marine archive in the western Mediterranean that record the impact of Heinrich Events very clearly. These cooling and drying episodes placed great stresses on terrestrial ecosystems and the pollen records from several sites in the Mediterranean show that tree cover contracted rapidly during these periods. Three long pollen records from basins in contrasting settings in Greece are shown in figure

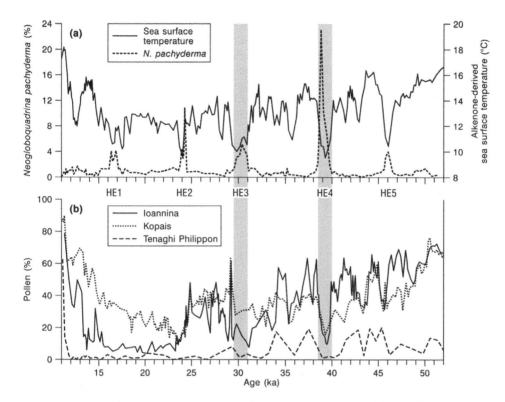

Figure 13.6 High-resolution marine and terrestrial records from the Mediterranean basin between c. 50 and 10 ka (see figure 13.5). The upper figure (a) shows the marine record from core MD95-2043 in the Alboran Sea showing discrete cooling events in the western Mediterranean Sea in response to Heinrich Events in the North Atlantic (modified after Cacho et al., 1999). The lower figure (b) shows the response of the temperate tree populations at three sites in Greece (table 13.2). See text for discussion (modified from Tzedakis et al., 2004).

13.6b and the impact of Heinrich Event 4 is especially clear at Ioannina and Kopais (table 13.2). A key challenge is to establish the impact of Heinrich Events on resource availability and human survival strategies at these times.

In the same kinds of deep limestone caves that contain the ice age art mentioned earlier, important high-resolution records of climate change have been recovered from speleothems in the Mediterranean region (figure 13.7). Speleothems are the product of calcium carbonate precipitation from groundwater. This process takes place very slowly over long periods of time and they record the changing oxygen isotopic composition of the groundwater. Speleothems can be dated using the uranium-series method and this provides a robust chronological framework. These caverns and their hydrology are also sensitive environmental systems and they have recorded the impact of Heinrich Events, for example, in the most easterly parts of the Mediterranean region over 4,000 km from the North Atlantic Ocean. An example from Soreq Cave in Israel is shown in figure 13.7. It shows marine isotope stages 1 to 6 with evidence of rapid environmental change within MIS 5 as well as rapid and high-amplitude change between 50,000 and 10,000 years BP. This record is important because it shows that the Heinrich Events in the North Atlantic

Table 13.2 Geographical attributes for three long pollen records in Greece (modified after Tzedakis et al., 2004). Ioannina is west of the Pindus Mountains and is the highest site with much higher rainfall than both Kopais and Tenaghi Philippon. Kopais is the driest site with higher summer and winter temperatures and greater losses of moisture to evaporation. Tenaghi Philippon is the most northerly location and is prone to incursions of cold continental air masses from the north and east. These topographic and meteorological factors combine to create limiting factors for tree growth in the drier parts of Greece to the east of the Pindus Mountains divide

Site	Latitude and Longitude	Elevation	MAP	T_{jan}
Tenaghi Philippon	41°10′N and 24°20′E	40 m	600 mm	3.4°C
Ioannina	39°45′N and 20°51′E	470 m	1200 mm	4.6°C
Kopais	38°26′N and 23°03′E	95 m	470 mm	9.0°C

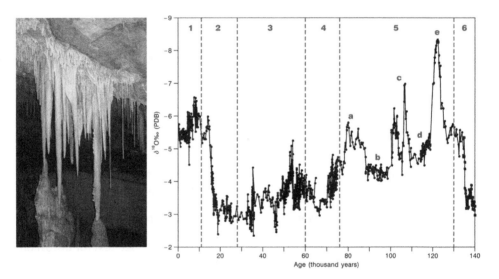

Figure 13.7 Speleothems and palaeoclimate data from speleothem records in the Mediterranean. The photograph shows speleothems from Campanet Cave in Mallorca. The diagram shows an oxygen isotope curve from speleothems in Soreq Cave in Israel based on the work of Bar-Matthews et al. (2000). The record shown here goes back to MIS 6 and evidence for rapid climate change is especially clear in marine isotope stages 5, 3 and 2 (modified after Bar-Matthews et al., 2000).

impacted on climatic conditions across the entire Mediterranean region (Bar Matthews et al., 1999) and it also shows that some were felt more strongly than others. Such proxy climate records from the Mediterranean are important because the region contains many Middle and Upper Palaeolithic sites and the region formed a refuge for humans during the last cold stage.

Environmental Archives: Resolution and Sensitivity

All Quaternary archives of change that provide us with insights into past environments and processes can usefully be assessed and compared in terms of their temporal resolution and their sensitivity to environmental change (Lewin, 1980; Allen et al., 1999; Woodward and Goldberg, 2001). One approach is illustrated in figure 13.8. Temporal resolution is a measure of the completeness and precision of the stratigraphic record at a given site – or within a particular sequence – and the dating control available for that record. A sequence with many erosional gaps and few dates would constitute a low resolution record and this would provide only a very limited window into the past. In contrast, however, some depositional environments involve more or less continuous sedimentation and this provides a sound basis for the development of a reliable and consistent record of environmental change, especially where sedimentation rates are high and the preservation of pollen and other proxies is good. These tend to be low energy settings such as lake and marine environments where sub-aerial erosion is absent and sediments can accumulate, undisturbed, for an extended period of time.

Environmental sensitivity is less easy to quantify, but it is a useful concept and a key characteristic of any environmental system (such as a lake or marine basin, a cave or river catchment system) that produces a long-term record of environmental change (see Wright, 1984). This property relates to the archive's ability to respond to and record an environmental change. Its sensitivity may determine whether it records local, regional or global signals in a consistent and predictable way.

Figure 13.8 shows that lake sediments and speleothems can provide well-dated, high-resolution records of change and these are typically associated with systems that are sensitive to change – they are commonly responsive to external climate fluctuations and they record them in a reliable and consistent manner. This sensitivity can be tested by contemporary process studies (Bar Matthews et al., 1999). Some lake systems accumulate sediments with annual laminations that can be counted. These contexts provide a basis for the development of extremely robust and detailed chronologies and they can be used to test the integrity of other dating methods (e.g. Allen et al., 1999). In contrast, coarse-grained clastic cave sediments (such as the ones shown in figure 13.9) plot at the opposite end of this continuum. These are angular scree sediments that can be produced by a range of mechanisms including frost action or even seismic shaking (Bailey and Woodward, 1997).

A key point to make here is that much of the Middle and Upper Palaeolithic record in Europe has been recovered from coarse-grained cave and rockshelter sediments, and from coarse-grained river sediments (Gamble, 1986; Woodward and Goldberg, 2001). Fluvial sediments are the product of flood events and the flood regime can respond in a sensitive way to environmental change – but coarse-grained river sediments are often deposited very quickly and their temporal resolution is limited in comparison to other records. The archaeological records, therefore, are typically limited in resolution and it has become increasingly difficult to make effective comparisons between the cultural and environmental records even for the most recent cold stage. The resolution and quality of many proxy climate records has become far superior to the existing archaeological datasets. Mithen (1999, p. 480) has made this point within a discussion of Mesolithic archaeology and changing Late glacial environments in Britain:

> there is in fact an increasing dislocation regarding the fine chronological resolution with which palaeoecologists can reconstruct local environmental history, and the much cruder chronological resolution with which archaeologists have to work.

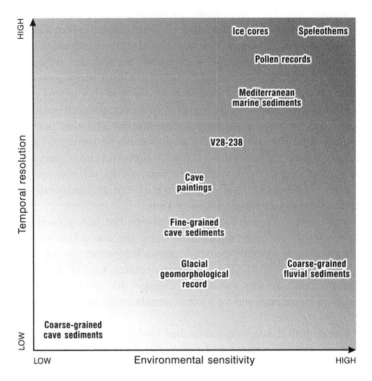

Figure 13.8 The various environmental archives discussed in this chapter character-ised on the basis of their temporal resolution and environmental sensitivity. See the text for further explanation. Plotting the precise location of each archive is somewhat subjective and one could discuss whether cave paintings should be plotted at all, but the purpose of this exercise is to highlight the variability in the nature and quality of data retrieved from different Quaternary archives. Plotting a range of archives using these axes and then justifying their locations can be a very useful exercise for students.

The problem is compounded because rockshelter sedimentary records lack preci-sion and are commonly very 'noisy' with complex stratigraphies (e.g., Bailey and Woodward, 1997) and a typical example with a very wide range of sediment particle sizes is shown in figure 13.9. Furthermore, human occupation can disturb the sedi-ment record and alter the fine sediment matrix through physical and chemical processes. This means it can be very difficult to decouple the cultural and environ-mental signals in rockshelter and cave sediment records (Woodward and Goldberg, 2001). On a positive note, some clastic rockshelter and cave sediment records in the Mediterranean with Middle and Upper Palaeolithic cultural assemblages have been shown to record the influence of some rapid climate change events (e.g., Courty and Vallverdu, 2001; Karkanas, 2001). However, such contexts are unusual and it is difficult to make secure correlations if dating control is limited.

Mithen (1999, p. 481) has argued that archaeologists can only feel frustrated at the relatively poor degree of chronological resolution that appears possible from their data. This mismatch in resolution and dating control means that establishing

Figure 13.9 A section in a rockshelter sediment record showing the wide range of particle sizes (from large boulders to fine clays) and complex stratigraphies that can be encountered. This photograph shows the deep trench at Crvena Stijena in western Montenegro. This is a large limestone rockshelter that contains over 20 m of Quaternary sediments and includes rich Middle Palaeolithic deposits. The photograph contrasts very coarse-grained rockshelter sediments in the central and upper part of the photograph with the lower right section where well-bedded fine-grained alternations of light and dark sediments associated with hearth features are present.

the environmental context of a culture or occupation phase can face many problems, given that climate during the last cold stage is now known to have fluctuated abruptly over timescales of centuries and even decades. Tzedakis et al. (2007) have proposed a novel method for circumventing this problem. They have directly mapped the radiocarbon dates of interest from archaeological contexts onto the high-resolution palaeoclimatic record from the marine sediments of the Cariaco Basin off Venezuela as the latter has been used to develop a radiocarbon calibration curve for the last 50,000 years (Hughen et al., 2004). This approach has provided new perspectives on the environmental context of 'late' surviving Neanderthal groups from Gorham's Cave in Gibraltar.

Quaternary Geography: Sensitivity and Thresholds

Figure 13.10 shows how the tree populations in the three areas of Greece shown in figure 13.6b might have responded to environmental stresses such as the drying and cooling associated with Heinrich Events. The response to such a stress is very different between the three regions and this is a function of local environmental conditions. A key point here is that we should not expect the same response to rapid climate change events in all parts of the landscape as some populations already lie close to their tolerance limit. The schematic representation of temperate tree abundance shown in the lower part of figure 13.10 shows the variable response between each region to climatically induced stress. Tzedakis et al. (2004) use the example of temperate tree abundance, but the variable response could equally be glacier mass balance, karst spring discharge, river sediment yield, or the availability of a key plant or animal resource for a group of Middle Palaeolithic foragers. Some systems may have switched on and off while others showed fluctuations in some measure of abundance or yield. This figure could even represent the population shifts in Neanderthal groups in their refuges in southern Europe before their final demise (see Gamble et al., 2004).

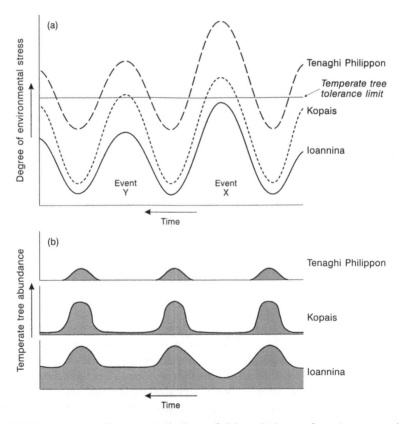

Figure 13.10 A schematic representation of (a) variations of environmental stress encountered by tree populations and (b) the response of the local temperate tree population at the three sites under discussion (modified after Tzedakis et al., 2004). Additional data on each of the three sites are given in table 13.2.

A key point here is that figure 13.10 implies a dynamic and spatially variable Quaternary geography associated with rapid climate change events during the last cold stage and this is in marked contrast to the rather static geography associated with the last cold stage as portrayed in figure 13.4. The response of ecosystems and landscape processes to rapid climate change will be modulated by local and regional environmental conditions and an appreciation of these environmental factors is clearly very important. The ability of human groups to cope with these changes will determine their success in the long term.

More generally, Bowen (1979) has proposed a basic working philosophy for the study of the Quaternary that represents a combination of *geological appraisal* for sequence and *geographical evaluation* for spatial reconstruction, coupled with the particular problems and techniques serving it – be they palaeobotanical, palaeoclimatological or geomorphological (or archaeological in this case). He goes on to argue that the time-space 'event sequence' forms the vehicle for ordering the view of the world on this basis. If we consider this approach in relation to figure 13.8, any assessment of the temporal resolution of an environmental archive is essentially a geological appraisal and any attempt to assess the sensitivity of a system will require a geographical evaluation of the lake basin, rockshelter or marine environment in question. At the same time a key aspect of any geographical evaluation must try to factor in the role of environmental thresholds and the potential for a spatially variable response of natural systems to rapid climate change as illustrated in figure 13.10. This problem is analogous to the complex response model put forward by Stan Schumm in the 1970s. He argued that different parts of river basins may respond in radically different ways to an environmental change by either aggrading or incising channel beds for example (Schumm, 1977)

Quaternary Geography and the Human Past

An important challenge is the development of new interdisciplinary approaches that will allow the cultural data from key Middle and Upper Palaeolithic sites to be examined in relation to the high-resolution proxy climate records for the last cold stage. One way of getting around the deficiencies inherent in the records from individual rockshelter and cave sites is to integrate the data from many sites over much larger spatial and temporal scales. Gamble et al. (2004) have compiled a database of over 2,000 radiocarbon dates from across Western Europe and this has allowed them to explore, in very broad terms, population dynamics across Europe from Britain to the Mediterranean between 30,000 and 6,000 BP. In this example, the radiocarbon dates come from archaeological sites across Western Europe and all of them have been calibrated to facilitate comparison with the GRIP ice core record (figure 13.11). This has allowed, for the first time, a regional scale analysis of human response to changing ecological conditions.

The radiocarbon dates have been used as a proxy for Upper Palaeolithic population history. Figure 13.11 shows the importance of southern Europe as a refuge for humans during the last cold stage but it also points to extreme cold tolerance by human populations. The analysis by Gamble et al. (2004) suggests that climate affects population contraction rather than expansion and they also argue that the dispersal of modern humans across Europe took place within wide climatic tolerances. These people had strategies to cope with extreme conditions so that explaining such events by general trends of warming or cooling is not possible.

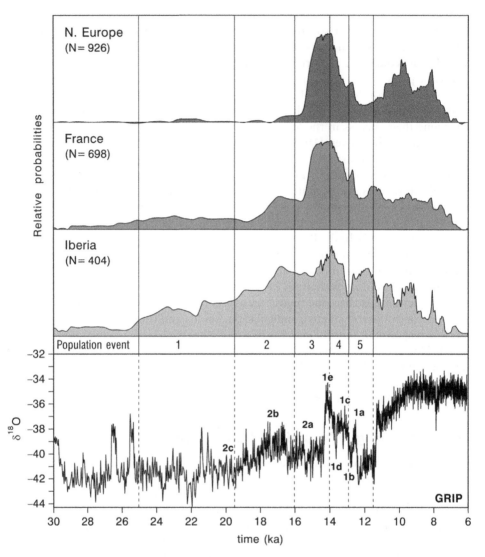

Figure 13.11 Palaeolithic human geography: using radiocarbon dates as a proxy for population expansion and contraction in three regions of Europe. This figure shows the radiocarbon database of Gamble et al. (2004) plotted by region and shown in relation to the GRIP ice core record for 30 to 6 ka. The population events (1 to 5) discussed by Gamble et al. (2004) are also shown. See text for further explanation. GRIP = Greenland Ice Core Project (modified from Gamble et al., 2004).

Conclusions

With the demonstration that much of the last cold stage was punctuated by a remarkable series of rapid and high-amplitude environmental changes, Quaternary geoscience has entered an exciting and challenging new era. The information on rapid change comes primarily from archives such as the Greenland ice cores, marine sediments in the North Atlantic, and, in Europe, from long pollen and speleothem records in the Mediterranean basin. A key challenge is to explore the relationship between the Middle and Upper Palaeolithic records in Europe and this new palaeoclimatic framework. However, much of what we know about the human past comes from material preserved in rockshelter and cave sediment records. These records are discontinuous and dating control is often inadequate (Woodward and Goldberg, 2001). New approaches are therefore needed to establish the environmental context of the archaeological record of the last cold stage. Important progress has already been achieved and the next decade will see further advances. If calibration of the radiocarbon timescale back to 50,000 years BP becomes routine practice, it may soon be possible to explore population dynamics across the Middle and Upper Palaeolithic transition in the same way that Gamble et al. (2004) have done for the period between 30,000 and 6,000 years BP. Also, as a more robust dating framework emerges for Upper Palaeolithic cave paintings (using direct AMS dating of the materials used to produce the images), it would be fascinating to explore the relationship between the faunal elements they depict and the records of rapid climate and ecosystem change. The geography of the last cold stage was highly dynamic – both temporally and spatially – with evidence for rapid change in geomorphological processes and ecosystems. The reality of the new records of rapid and high-amplitude climate change means that a geographical perspective on the Quaternary is now more relevant than ever.

ACKNOWLEDGEMENTS

The author would like to thank Noel Castree for the invitation to write this chapter, John Lewin for helpful comments on the manuscript and Nick Scarle for drawing the figures.

REFERENCES

Allen, J. R. M., Brandt, U., Brauer, A., Hubberten, H.-W., Huntley, B., Keller, J., Kraml, M., Mackensen, A., Mingram, J., Negendank, J. F. W., Nowaczyk, N. R., Oberhänsli, H., Watts, W. A., Wulf, S. and Zolitschka, B. (1999) Rapid environmental changes in southern Europe during the last glacial period. *Nature*, 400, 740–43.

Anderson, D. E., Goudie, A. S. and Parker, A. G. (2007) *Global Environments through the Quaternary*. Oxford: Oxford University Press.

Bahn, P. G. and Vertut, J. (1997) *Journey through the Ice Age*. London: Weidenfeld and Nicholson.

Bailey, G. N. (1983) Concepts of time in Quaternary prehistory. *Annual Review of Anthropology*, 12, 165–92.

Bailey, G. N. and Woodward, J. C. (1997) The Klithi deposits: sedimentology, stratigraphy and chronology. In: G. N. Bailey (eds) *Klithi: Palaeolithic Settlement and Quaternary Landscapes in Northwest Greece. Volume 1: Excavation and Intra-site Analysis at Klithi*, Cambridge: McDonald Institute for Archaeological Research, pp. 61–94.

Bar-Matthews, M., Ayalon, A., Kaufman, A. and Wasserburg, G.J. (1999) The Eastern Mediterranean palaeclimate as a reflection of regional events:Soreq cave, Israel. *Earth and Planetary Science Letters*, 166, 85–95.

Bar-Matthews, M., Ayalon, A., and Kaufman, A. (2000) Timing and hydrological conditions of sapropel events in the Eastern Mediterranean, as evident from +speleothems, Soreq cave, Israel. *Chemical Geology*, 169, 145–56.

Bard, E., Hamelin, B., Fairbanks, R. G. and Zindler, A. (1990) Calibration of the ^{14}C timescale over the past 30,000 years using mass spectrometric U–Th ages from Barbados corals. *Nature*, 345, 405–10.

Bell, M. and Walker, M. J. C. (2005) *Late Quaternary Environmental Change: Physical and Human Perspectives*, 2nd edn. Harlow, Essex: Pearson Education.

Bond, G., Broecker, W., Johnsen, S., McManus, J., Labeyrie, L., Jouzel, J. and Bonani, G. (1993) Correlations between climate records from North Atlantic sediments and Greenland ice. *Nature*, 365, 143–47.

Bowen, D. Q. (1978) *Quaternary Geology: A Stratigraphic Framework for Multidisciplinary Work*. Oxford: Pergamon Press.

Bowen, D. Q. (1979) Geographical perspective on the Quaternary. *Progress in Physical Geography*, 3, 167–86.

Bowman, S. (1990) *Radiocarbon Dating*. Berkeley: University of California Press.

Brunnacker, K. (1966) Die sedimente de Crvena Stijena, *Galsnik-Sarajevo*, 21, 31–65.

Cacho, I., Grimalt, J. O., Pelejero, C., Canals, M., Sierro, F. J., Flores, J. A., and Shackleton, N. (1999) Dansgaard-Oeschger and Heinrich event imprints in Alboran Sea paleotemperatures. *Paleoceanography*, 14, 698–705.

Courty, M.-A. and Vallverdu, J. (2001) The microstratigraphic record of abrupt climate change in cave sediments of the Western Mediterranean. *Geoarchaeology: An International Journal*, 16, 467–99.

Dansgaard, W., Johnsen, S. J. Clausen, H. B., Dahl-Jensen, D., Gundestrup, N. S., Hammer, C. U., Hvidberg, C. S. Steffensen, J. P., Sveinbjörnsdottir, A. E., Jouzel, J. and Bond, G. (1993) Evidence for general instability of past climate from a 250-kyr ice-core record. *Nature*, 364, 218–20.

Emiliani, C. (1955) Pleistocene temperatures. *Journal of Geology*, 63, 538–75.

Farrand, W. R. (2000) *Depositional History of Franchthi Cave: Sediments, Stratigraphy and Chronology*: Fascicle 12. Bloomington and Indianapolis: Indiana University Press.

Fuller, I. C., Macklin, M. G., Passmore, D. G., Brewer, P. A., Lewin, J. and Wintle, A. G. (1998) River response to high-frequency climate oscillations in southern Europe over the past 200 k.y. *Geology*, 26, 275–78.

Gamble, C. S. (1986) *The Palaeolithic Settlement of Europe*. Cambridge: Cambridge University Press.

Gamble, C. S. (1994) The peopling of Europe 700,000–40,000 years before the present. In B. Cunliffe (ed.), *Prehistoric Europe: An Illustrated History*. Oxford: Oxford University Press, pp. 5–41.

Gamble, C. S. (1999) *The Palaeolithic Societies of Europe*. Cambridge: Cambridge University Press.

Gamble, C. S., Davies, W., Pettitt, P. B. and Richards, M. B. (2004) Climate change and evolving human diversity in Europe during the last glacial. *Philosophical Transactions of the Royal Society of London, Series B*, 359, 243–54.

Goudie, A. S., (1976) Geography and prehistory. *Journal of Historical Geography*, 2 197–205

Gowlett, J. A. J., Hedges, R. E. M. and Housley, R. A. (1997) Klithi: the AMS radiocarbon dating programme for the site and its environs. In G. N. Bailey (ed.), *Klithi: Palaeolithic Settlement and Quaternary Landscapes in Northwest Greece. Volume 1: Excavation and Intra-site Analysis at Klithi*. Cambridge: McDonald Institute Monographs, pp. 27–40.

Grayson, D. K. (1990) The provision of time depth for paleoanthropology. In L. F. Laporte (ed.), *Establishment of a Geologic Framework for Paleoanthropology*. GSA Special Paper, 242, 1–13.

Hays, J. D., Imbrie, J. and Shackleton, N. J. (1976) Variations in the Earth's orbit: pacemaker of the ice ages. *Science*, 194, 1121–31.

Hughen, K., Lehman, S., Southon, J., Overpeck, J., Marchal, O., Herring, C. and Turnbull, J. (2004) ^{14}C activity and global carbon cycle changes over the past 50,000 years. *Science*, 303, 202–7.

Hughes, P. D., Woodward, J. C., Gibbard, P. L., Macklin, M. G., Gilmour, M. A. and Smith, G. R. (2006) The glacial history of the Pindus Mountains, Greece. *Journal of Geology*, 114, 413–34.

Imbrie, J. and Imbrie, K. P. (1979) *Ice Ages: Solving the Mystery*. Cambridge, MA: Harvard University Press.

Karkanas. P. (2001) Site formation processes in Theopetra cave: a record of climatic change during the Late Pleistocene and Early Holocene. *Geoarchaeology: An International Journal*, 16, 373–99.

Lewin, J. (1980) Available and appropriate timescales in geomorphology. In R. A. Cullingford, D. A. Davidson and J. Lewin (eds), *Timescales in Geomorphology*. John Wiley and Sons: Chichester, pp. 1–8.

Libby, W. F. (1955) *Radiocarbon Dating*, 2nd edn. Chicago: University of Chicago Press.

Lowe, J. J. and Walker, M. J. C. (1997) *Reconstructing Quaternary Environments*, 2nd edn. London: Addison-Wesley-Longman.

Mellars, P. A. (1994) The Upper Palaeolithic Revolution. In B. Cunliffe (ed.), *Prehistoric Europe: An Illustrated History*. Oxford: Oxford University Press, pp. 42–78.

Mellars, P. A. (2006) A new radiocarbon revolution and the dispersal of modern humans in Eurasia. *Nature*, 439, 931–35.

Mithen, S. J. (1999) Mesolithic archaeology, environmental archaeology and human palaeo-ecology. *Quaternary Proceedings No. 7*, 477–83.

Narcisi, B. and Vezzoli, L. (1999) Quaternary stratigraphy of distal tephra layers in the Mediterranean – an overview. *Global and Planetary Change*, 21, 31–50.

Peake, H. J. E. (1922) The ice age and man. *Man*, 22, 6–11.

Penck, A. and Bruckner, E. (1909) *Die Alpen im Eiszeitalter*. Tauchnitz, Leipzig.

Pettitt, P. B. and Bahn, P. G. (2003) Current problems in dating Palaeolithic cave art: Candamo and Chauvet. *Antiquity*, 77, 134–41.

Porter, S. C. (1989) Some geological implications of average Quaternary glacial conditions. *Quaternary Research*, 32, 245–61.

Roucoux, K. H., de Abreu, L., Shackleton, N. J. and Tzedakis, P. C., (2005) The response of NW Iberian vegetation to North Atlantic climate oscillations during the last 65 kyr. *Quaternary Science Reviews*, 25, 1637–53

Shackleton, N. J. (1967) Oxygen isotope analyses and Pleistocene temperatures re-assessed. *Nature*, 215, 15–17.

Shackleton, N. J. and Opdyke, N. D. (1973) Oxygen isotope and palaeomagnetic stratigraphy of equatorial Pacific core V28–238: oxygen isotope temperatures and ice volume on a 10^5 and 10^6 year scale. *Quaternary Research*, 3, 39–55.

Shackleton, N. J. and Crowhurst, S. (1996) Timescale calibration – ODP 677. IGBP PAGES/ World Data Center–A for Paleoclimatology Data Contribution Series # 96-018. NOAA/ NGDC Paleoclimatology Program, Boulder, CO, USA.

Schumm, S. A. (1977) *The Fluvial System*. New York: John Wiley.

Stringer, C. and Gamble, C. S. (1993) *In Search of the Neanderthals*. New York: Thames and Hudson.

Tzedakis, P. C., Andrieu, V., de Beaulieu, J.-L., Crowhurst, S., Follieri, M., Hooghiemstra, H., Magri, D., Reille, M., Sadori, L., Shackleton, N. J. and Wijmstra, T. A. (1997) Comparison of terrestrial and marine records of changing climate of the last 500,000 years. *Earth and Planetary Science Letters*, 150, 171–76.

Tzedakis, P. C., Lawson, I. T., Frogley, M. R., Hewitt, G. M. and Preece, R. C. (2002) Buffered tree population changes in a Quaternary refugium: evolutionary implications. *Science*, 297, 2044–47.

Tzedakis, P. C., Frogley, M. R., Lawson, I. T., Preece, R. C., Cacho, I. and de Abreu, L. (2004) Ecological thresholds and patterns of millennial-scale climate variability: the response of vegetation in Greece during the last glacial period. *Geology*, 32, 109–112.

Tzedakis, P. C., Hughen, K. A., Cacho, I. and Harvati, K. (2007) Placing late Neanderthals in a climatic context. *Nature*, 449, 206–208.

Valladas, H. (2003) Direct radiocarbon dating of prehistoric cave paintings by accelerator mass spectrometry. *Measurement Science and Technology*, 14, 1487–92.

Van der Hammen, T., Wijmstra, T. A. and Zagwyn, W. H. (1971) The floral record of the late cenozoic of Europe, In K. K. Turekian (ed.), *The Late Cenozoic Glacial Ages*. New Haven, Yale University Press, pp. 391–424.

Walker, M. J. and Lowe, J. J. (2007) Quaternary Science 2007: a fifty-year retrospective. *Journal of the Geological Society of London*, 164, 1073–92.

Woodward, J. C. (2003) Geoarchaeology. In H. Viles and A. Rogers (eds), *The Student's Companion to Geography*, 2nd edn). Oxford: Blackwell, pp. 34–42.

Woodward, J. C. (2009) *The Physical Geography of the Mediterranean*. Oxford: Oxford University Press (in press).

Woodward, J. C. and Goldberg, P. (2001) The sedimentary records in Mediterranean rock-shelters and caves: archives of environmental change. *Geoarchaeology: An International Journal*, 16 (4), 327–354.

Woodward, J. C., Macklin, M. G. and Smith, G. R. (2004) Pleistocene glaciation in the mountains of Greece. In J. Ehlers and P. L. Gibbard (eds), *Quaternary Glaciations – Extent and Chronology. Volume 1: Europe*. Amsterdam: Elsevier, pp. 155–173.

Wright, H. E. (1984) Sensitivity and response time of natural systems to climatic change in the Late Quaternary. *Quaternary Science Reviews*, 3, 91–131.

Wulf, S., Kraml, M., Brauer, A., Keller, J. and Negendank, J. F. W. (2004) Tephrochronology of the 100 ka lacustrine sediment record of Lago Grande di Monticchio (southern Italy). *Quaternary International*, 122, 7–30.

Chapter 14

Environmental History

Georgina H. Endfield

The Meaning Of Environmental History

Environmental history tells the story of society's interaction with the physical environment. It represents an increasingly important approach with which to explore the many ways in which humans and the environment affect each other at a range of temporal and spatial scales, and is concerned with a number of important themes. The first is the way in which nature is organised, functions and operates, the identification of the physical attributes of past environments, and how these have changed over time. The second is the interaction between the socio-economic realm and the environment. Finally, the third theme is concerned with the our intellectual encounters with nature, that is to say the different ways in which humans perceive, value and record nature though myths, law, ethics, custom and perception and other symbolic mechanisms (Worster, 1988).

It is only over the last three decades or so, however, that environmental history, at least as a discrete discipline, has been formally recognised. Moreover, its emergence and development has been somewhat chequered: its remit, scope, goals and purpose have been dissected and disputed, its intellectual, disciplinary, regional and conceptual origins and evolution contested, and its definition debated. To some extent this degree of scrutiny is a function of the subject's diverse and often contested ancestry and its multiple traditions. Indeed, as a subject area, environmental history is thought to have developed along a variety of distinctive evolutionary intellectual and practical trajectories (Williams, 1994; Baker, 2003).

Environmental history is often regarded as a predominantly American enterprise (Williams, 1994). As Crumley (1994, p. 21) highlighted, in the United States, 'environmental historians initially focused on leaders of the conservationist and preservationist movements and on the relationship of the frontier and wilderness to American culture and politics'. Environmental history as an institutional form and intellectual project, however, was only consolidated in the United States in the 1960s and 1970s, as something of a radical branch of history and as a function of growing environmental consciousness around this time (McNeill, 2003).

It has been suggested, however, that the North American historian's apparently 'new' and arguably radical take on this subset of their academic discipline after the 1960s and 1970s actually served to reopen existing debates on what, for many, were fairly old issues (Williams, 1994, p. 3). Geographers have long dealt with the environmental problems of past and present (Baker, 2003, p. 75), and the degree of overlap has encouraged strong links to be drawn between historical geography and environmental history. As McNeill (2003, p. 9) suggests 'the subject matter' of both (sub) disciplines 'is essentially the same, and the differences are mainly matters of style, nuance and technique'.

The roots of British environmental history, for example, are in fact thought to have stemmed directly from geography-based studies of natural and landscape history, drawing on archaeological, palaeecological and historical evidence. Much environmental history research is and has been conducted under the banners of historical geography, which has remained a vibrant field in Britain (MacKenzie, 2004, p. 376). Practitioners in the distinctive subfield of cultural geography, moreover, have made particularly significant contributions to the interpretation of landscape as texts.

The emergence of environmental history in other parts of the world may have had similar origins. In Europe generally, but perhaps especially in Scandinavia, landscape studies have long focused on integrating information similarly derived from both scientific and cultural evidence. Likewise, although environmental history in Australia was argued to be a burgeoning 'new' field in the mid-1990s, Joe Powell (1996, p. 257) was among the first to suggest that Australia's historical geographers had in fact long been covering much of the ground 'pegged out' for environmental history at that stage.

It is clear that environmental history is 'many things to many people' (McNeill, 2003, p. 6). As a result, there have evolved many different 'species' of the subject (Stewart, 1998), each tending to be relatively narrowly defined, legitimising one or other disciplinary perspective on the topic (Sorlin and Warde, 2005). Given these competing traditions, it is not surprising that there is a lack of genuine coherence, numerous definitions and an unwieldy breadth in environmental history (Weiner, 2005) as well as persistent institutional and disciplinary cleavages.

The different ancestral routes of environmental history, the disputes and debates over its origins and its competing definitions, however, have resulted in an immensely diverse – and, thanks to a rather challenging degree of intellectual scrutiny in recent years – a dynamic field of enquiry. But precisely what kinds of issues and concerns have been studied under the remit of environmental history and how have these changed over time? How is environmental history practised? What are the current and new directions in the field? Moreover, what contributions can and have geographers made to the study of environmental history?

Agency, Method and Multidisciplinarity in Environmental History

The role of human and non-human agency in environmental change has traditionally been, and continues to be, a fundamental concern within the study of environmental history. Geographers have long investigated the way in which humans have shaped and transformed the landscape and how they situate themselves in nature. Anthropogenic modification of the environment has featured conspicuously in geographical work that now could be classed as environmental history. By raising 'environmental consciousness' and highlighting the degree to which human forces were capable of shaping the natural world, for example, George Perkins Marsh's (1864) *Man and Nature* represents one fundamental text to which both geographers and historians lay claim as an important influence in their understanding of environmental history. Such themes were developed in the 1956 multidisciplinary volume *Man's role in changing the face of the earth*, edited by W. L. Thomas.

Perhaps the most important individual in this vein, however, has been that of Carl Sauer, founder of the Berkeley School of Geography and who contributed a decisive chapter on 'The agency of man on the earth' to Thomas's edited volume. Sauer's particular interests lay in the fashioning and transformation of landscapes by human culture, or the production of 'cultured landscapes', ideas which were to be and are still being developed by later geographers. The edited volume entitled *The Earth Transformed* (Turner et al., 1990), for example, is regarded as the direct successor to Thomas's 1956 publication (Williams, 1994), while various authors have explored the way in which societies have impacted upon their landscapes at different points in time (see for example, Matthewson, 1993; Nicholson and O'Connor, 2000).

But the environment is of course itself an active player in human affairs. In recent decades, a number of American environmental historians have highlighted the 'earth as an agent' (Worster, 1988, p. 289) and 'nature as a historical actor' (Merchant, 1989, p. 7). Such notions, however, are far from new (McNeill, 2003, p. 13), and this is again a subject to which geographers have long devoted their attention. Climate in particular has featured prominently as a molding force throughout history. During the period of colonial expansion, for instance, issues of acclimatisation or the ability with which societies could adapt to different environments became a key political, scientific and economic dilemma. Geographical discussions of climate then also focused on centring patterns of variation in levels of human civilisation within a regional climatic framework (Livingstone, 1991, p. 2002). When combined, these themes provided an enduring moral and ethno-climatological frame of reference that would permeate geographical debates and indeed colonial policy into the 1900s, and would help fuel the development of damaging racial ideologies in the first half of the century.

After the Second World War, in a context of anti-colonialism, a drive towards quantification in geography and as a backlash against the naïve precepts of such climatic determinism, there was a shift from 'reductionist' thinking towards complex socio-economic explanations of cultural variation, development and change. More recent decades have witnessed the emergence of approaches to environmental history, which posit ecological or environmental change as the consequence of cultural choice and human action rather than environmental circumstance. Moreover, much of the environmental literature of the 1990s began to situate human values as central to environmental transformation (Rothman, 2002, p. 491).

Other agencies have also featured as central characters in environmental history texts. In Alfred Crosby's now classic global scale environmental history, *Ecological imperialism*, biological vectors, disease pathogens, weeds and domestic animals, or the 'portmanteau biota', are charged with facilitating the process of conquest and change in the new world. Crosby argues that the 'unconscious teamwork' of these biota facilitated European imperial success, though his thesis has since been criticised for displacing the role of human agency in the crises and hence any direct blame for the tragedies that befell the subjects of the new colonies. (McNeill, 2003, p. 33). Moreover, more recent multidisciplinary work discussed in the next section, is serving to challenge other aspects of the theory more fundamentally.

Arguably what's now being termed a 'new' environmental history adopts a nuanced take on agency, highlighting nature as a 'co-creator' of the past, and seeing nature and society as interdependent, interactive and 'causally bound up in a narrative of decline and potential renewal' (Walton, 2001, p. 903). Acknowledging these interdependencies is of course important, but the real difficulty for the environmental historian lies in disentangling these different agents of change from the signals that each leaves in the historical record and identifying the respective role that each may have played individually or in combination in modifying the environment. Problems arise because past human influences are normally cumulative, with different stages of activity being superimposed upon each other, and upon any impacts and changes associated with climate (Russell, 1997), adding a further degree of complexity. An additional obstacle is the ubiquity of humanised landscapes, thus rendering it very difficult to distinguish or decipher the independent influence of non-human agents of change.

A wide range of methods to explore issues of agency do exist. Written records, traces left on the landscape and analysis of sedimentary records, or so called geo-archives, however, can provide some insight into the environments of the past and how and why they may have been changed and might also shed light on the nature of the relationship that humans might have had with their environment at different points in the past. Moreover, by combining the latest scientific and archaeological techniques with information gathered from documents, other areas of archaeology, art, and ethnography, it is possible to identify the nature, type and specific impacts of different diseases on society in the past. Detailed consideration of the potential advantages, resolution and limitations of a variety of different approaches to such 'reconstructions' are outlined in detail elsewhere (see, for example, Roberts and Manchester, 2007) but have traditionally been tools employed by physical and historical geographers, historical ecologists, palaeobotanists and archaeologists (McNeill, 2003).

To some extent geographers might be well placed, perhaps uniquely so, to combine both social and physical scientific methodologies *within* the boundaries of their discipline to similarly disentangle the relative role of humans and climate in environmental change in a variety of geographical contexts. Davies et al., (2004), for example, have demonstrated the immense potential of integrating both human geography and physical geography approaches and data (palaeolimnological and archival material) to disentangle complex drought-society relationships in west central Mexico over the last thousand years.

The combination of different methodological approaches, however, drawn from multiple disciplines can perhaps shed the most light on the intractable problem of agency, as the two following examples drawn from studies of Scottish environmental history reveal. Geographers and archaeologists, for example, have effectively combined their expertise to explore issues of agency in environmental modification. Edwards et al., (2007), for example, employed a variety of biotic and sedimentological evidence to explore the respective impacts of climate and anthropogenic influence on woodland cover in Scotland in the Late Holocene, while recent work by Dodgshon and Olsson (2006), combining historical geographical and biological (palaeobotanical) perspectives, has focused on the degree to which humans have been responsible for the creation of the distinctive heather moorland landscapes of the Scottish Highlands. Such investigations highlight how interdisciplinary approaches might well serve to advance our knowledge of the environmental history of hitherto little understood yet culturally important places and the role of different agents of change in their creation.

Changing Directions in Environmental History

The plurality of intellectual routes through which environmental history is thought to have evolved has resulted in some interesting trends in the geographical and spatial focus of research within the discipline. American environmental history, for example, was for a long time focused primarily on the history of the American West and the idea of the frontier. Preoccupation with these themes is thought to have stemmed originally from a number of benchmark and often cited publications, Turner's (1893) *The Significance of the Frontier in American History*, Walter Prescott Webb's (1931) *The Great Plains* and Malin's (1967) *The Grassland of North America*, among them. Interpretations of the history of the Great Plains and the American West generally, the idea of the frontier and the concept of wilderness have remained pervasive themes in American environmental history in the last few decades (Nash, 1967; Cronon, 1992a). There was, however, something of a shift from a focus on 'wilderness' to the idea of 'ordinary nature' in the late 1980s and 1990s. These ideas were coupled with the much more of a 'constructivist' spirit in the study of human-environmental relations. Cronon (1991, p. 69) noted, for example, that wilderness was in fact a 'product of civilisation', a social construction in itself.

Cronon's work was also instrumental in shifting the geographical focus of environmental history research more fundamentally. Research on landscape and the urban environment is long established in urban history, historical geography, and archaeology, but environmental history per se has traditionally been very rural in orientation (Grove, 2001, p. 264). Cronon's (1992b) *Nature's Metropolis: Chicago and the Great West*, however, explored the environmental implications of the urban centre (Baker, 2003, p. 81), and heralded a new wave of urban environmental histories, typified by Mike Davis's (1998) *The Ecology of Fear*, in which he tackles the geographical specificities of vulnerability to extreme weather events and natural hazards in the city and explores a variety of different narratives of the way Los Angeles has been understood and conceptualised. A number of specifically urban environmental histories have followed, addressing critical problems such as public water supply, waste disposal and links and interactions between social and physical ecologies in variety of urban contexts (see, e.g., Tarr, 1996; Melosi, 2000; Rome, 2001; Schott et al., 2005).

There has also been an increasing internationalisation in environmental history. There are now thriving Chinese, Australian, New Zealand, Latin American and Indian schools of Environmental History. It is worth highlighting the coincident emergence of a very strong 'southern' research agenda incorporating African, Latin American, Asian and Australasian environmental histories (Baker, 2003, p. 80). There are now many examples of place or country specific environmental histories of the 'global south', including Elvin (2004) for China, Arnold and Guha (1995) or Grove et al. (1998) for Southeast Asia and McCann (1999) and Dovers et al., (2003) for Africa. As Grove (2001) anticipated, and as work by Moon (2005) has demonstrated, the centre of gravity of environmental history research is also shifting eastwards towards Russia and the Middle East.

According to the 'Introduction' of the recently published *Encyclopaedia* on the subject, however, environmental history should be, by definition, a global endeavour. After all, environmental issues, be they climate change, land degradation, deforestation or pollution, are trans-national and unfold without respect to political or administrative borders. Most research, however, remains wed to the idea of the nation state or has been conducted at the case study level, often with a political boundary, nation or region as a geographical delineation. Although there is obvious scope for more regionally focused environmental histories, there have been calls for more cross-border or cross-national regional, comparative and global studies of environmental history (Steinberg, 2004, p. 266; Lekan, 2005). The last few years has seen an increase in the number of such syntheses (e.g., Beinart and Coates, 1995; McNeill, 2000; Hughes, 2001; Richards, 2003).

A focus on environmental features, including plants, animals, mountain ranges, rivers or forests and climate, and of processes such as climate change, deforestation, soil erosion or pollution, might help broaden the fields of vision beyond the spatial confines of the political or administrative boundary. Rising concerns over the implications of global warming and fears of escalating human vulnerability to natural calamities and extreme events together form one such highly topical issue which 'cuts to the heart of the debate about the relationship between nature and culture' (Steinberg, 2004, p. 275). The study of climate history and historical climatology,[1] long-term climate reconstruction, and explorations of the impact of past climate change and particularly extreme weather events on communities thus represent important growth areas in environmental history.

Advances in dendroclimatology and the analysis and dating of materials held in ice, sea and lake cores, have all been used to glean invaluable insight into longer-term climate trends in different parts of the world. Building on earlier climate history work of the French *Annales* School of History, and associated most notably with the work of Braudel and Le Roy Ladurie (1972), interests in climate-society interactions over the historical time period have also grown significantly over the last few decades (see, for example, Lamb, 1982; Grove, 1988; Barriendos, 1997; Pfister and Brazdil, 1999; Jones et al., 2001; Brazdil et al., 2005). This pioneering work is not only providing detailed regional climate histories but is also affording important insights into how societies have been affected by, coped with and have responded to climatic variability and anomalous weather events and weather-related events in the past. Investigations of past climate change and extreme weather and weather-related events are in turn encouraging a growing interest in the timing, frequency and implications of historical natural disasters, from weather related events to historical earthquakes, volcanic eruptions and storms (Kempe and Rohr, 2003).

Environmental processes and products have also formed the focus of interesting transboundary environmental histories. Work on histories of deforestation (Williams, 2004), of particular species (Griffiths, 2001) and of global commodities (e.g., *Salt* [1998] or *Cod* [2002], both by Kurlansky) serve as examples of themed studies through which more global or synthetic environmental histories might be explored. Recent 'eco-biographical' work targeting attention on specific geographical and topographic features, landmarks and water courses are also beginning to make a significant impact in cross-boundary environmental histories. There have been a number of environmental histories of mountain regions (e.g., McNeill, 2004) and water courses. Evenden's (2004) exploration of the contestation between dam development and salmon conservation on the Fraser River, British Columbia, represents a case in point, as does Mark Cioc's (2006) investigation of the historical political, economic and ecological dimensions of the Rhine from the early nineteenth century.

Some of the avenues which have been heralded as 'new' for environmental historians have already been subject to the geographers' gaze. Recent reviews of work on smell and aural histories have highlighted a niche for sensory environmental histories (Coates, 2005), though there is also a body of geographical work that addresses the themes of work and music (Jones, 2005), art and sound (Butler, 2006) and smell and taste (Law, 2001). Environmental historians have also begun to explore the association of environment and international trade, and the complexities of the relationship between production and consumption (Klingle, 2003, p. 94). Again, their arguments might be usefully informed by the considerable geographical literature on these themes (e.g., Hughes, 2006; Klooster, 2006).

Frontiers in Environmental History

There are very many other new possible avenues of environmental history research. The following subsections focus on a number of vibrant themes, selected for three key reasons. First, they reflect- but might also benefit from recent multidisciplinary developments; second, they have the potential to take the study of environmental history beyond the spatial confines of the political boundary or nation state; and third, they impinge upon and incorporate some of the other relatively new departures in environmental history to which geographers are making important contributions.

The environmental history of the oceans and seas

Beyond a focus on pollution, environmental historians have, until recently, tended to neglect water as a medium for investigation (McNeill, 2003). The oceans in particular have remained 'outside of history' (Bolster, 2006). Contemporary concerns over depleted fish stocks, the destruction of marine habitats, especially coral reefs, and the threat of ship-borne biological invasions have served to highlight the vulnerability of the oceans. It is now being recognised that these problems are part of a much longer history of human impact (Igler, 2005; Van Sittert, 2005). In this context, it is perhaps not surprising that the sea is now being recognised as one of the new frontiers for environmental historians (Bolster, 2006).

Geographers have again already made some valuable contributions to this theme. Philip Steinberg, for example, has adopted a political geography perspective to examine how nations and peoples have viewed and used the oceans in his 2001 'Social construction of the oceans'. A recent special issue of the *Journal of Historical Geography*, moreover, has highlighted the potential for more work on the imaginative geographies, conceptualisations and representations of the sea, on the oceans as sites of biological, economic and cultural exchange and scientific investigation, on the geography of seafaring and maritime disaster and the sea as a militarised space (Lambert et al., 2006).

Some of the most exciting developments in the field of maritime history have come as a result of multidisciplinary efforts. The History of Marine Animal Populations research programme (HMAP), for example, a joint initiative between the Universities of Southern Denmark, New Hampshire (USA) and Hull (UK) has brought together marine ecologists and maritime historians to investigate how, why and through what human or natural mechanisms there have been changes to the biodiversity of the world's oceans over the last 2000 years (Holm et al., 2001; Holm, 2003). Historical perspectives are revealing long-term ecological declines in species richness and abundance in some regions as a result of the combined impacts of past fishing extractions, habitat destruction and pollution (Holm, 2005). Such information will prove pivotal in the formulation of appropriate policy and regulation of fisheries and marine habitats.

In a similar vein, a recently completed EU funded project has brought together scholars from Spain, the UK, the Netherlands and South America and from a wide variety of disciplinary backgrounds to construct a Climatological Database of the World's Oceans, 1750–1850 (CLIWOC), from English, Dutch and Spanish maritime and mariners' records, including ships logbooks and trading documents (García-Herrera et al., 2005). Analysis of the precise time and place specific information on wind direction, wind force, weather conditions, sea state and sea ice reports held in the log books is providing useful insight into climate variability associated with the North Atlantic Oscillation and the Southern Oscillation. The project is also illuminating a number of additional historical themes of interest to maritime environmental history. Analysis of the muster books of the East India Company reveals that the most frequent entries by naval officers relate to the health of the crew and mortality. These data can be used to trace developments in ventilation, hygiene and diet on board ship and to measure the effects of environmental conditions on the health of the crews. It may also be possible to employ these sources to explore the relationship between infectious disease and climate variability in the past (Wilkinson, 2005).

Health and disease in environmental history

Issues of land, health and sickness were, to some extent, central to the early conservation thought and practice of pioneering environmental historians like Aldo Leopold (Mitman, 2005). Yet histories of disease, health and public health response have not really featured as central tenets in recent environmental history discourse (Luckin, 2004). Historical geographers and medical historians have examined changing conceptualisations of the healthiness of places over time according to advances in medical knowledge and of acclimatisation of plants, animals and people (Livingstone, 1991; 2002; Anderson, 1996; Harrison, 1996). Geographers have also been at the forefront of monitoring spatial diffusion of disease and particularly epidemic disease at the local, regional and global levels (Smallman-Raynor et al., 2003; Cliff et al., 2004). The study of epidemic disease, however, is beginning to creep into the new suite of urban environmental histories. Craddock's exploration of epidemic disease in nineteenth and twentieth century San Francisco provides insight into the way in which scourges of smallpox and tuberculosis were understood, conceptualised and addressed, but also demonstrates developments in the field of epidemiological thinking over time (Craddock, 2000). Attention is also now shifting to the way in which health can be and has been 'influential historically to the development, image and identify of everyday places' (Kearns and Andrews, 2005, p. 2697).

Environmental histories are also contributing to our understanding of disease diffusion with rising concerns over the threat of new and deadly infections, which are being transferred from continent to continent with relative ease in an increasingly globalised world. As John McNeill (1999, p. 175) suggests, however, such threats are not necessarily new and it is possible to find corollaries in past centuries. The importing of sugar to Atlantic America in the seventeenth century, for example, heralded a new chapter in ecological transformation and created a set of environmental conditions that were conducive to the propagation of yellow fever.

Other long-standing debates over the relationship between disease and empire, however, have also recently been reopened with an ecological twist. It is generally accepted that 'virgin soil epidemics' – the introduction of Old World diseases to a people and land with little or no resistance (Crosby, 1986) – triggered massive depopulation across the New World throughout the 16th and early 17th centuries. To take Mexico as an example, wave after wave of epidemic disease swept across the country causing unprecedented life loss among indigenous populations. It has recently been suggested, however, that the most serious epidemics to strike central Mexico in the sixteenth century may have been caused by the same haemorrhagic fever (Acuña-Soto et al., 2000). Perhaps most controversially, it has been posited that this fever may have had a New World etiological agent, which was stimulated by a combination of extreme drought, post-conquest changes in agricultural practices and modifications to local settlement and infrastructure (Acuña-Soto et al., 2002). The expanding network of tree-ring studies is also playing a pivotal role in exploring these new disease histories. Dendroclimatological investigations have resulted in the identification of a period of 'megadrought' in the 1550s, one of the most severe droughts in North American history, which may have also interacted with prevalent ecological and sociological conditions, magnifying the human impact of infectious disease in central Mexico (Cleaveland et al., 2003; Dias et al., 2002; Acuña-Soto et al., 2002).

War, environmental and militarised landscapes

Opportunities for exploring the environmental impacts and consequences of conflict, nature's effects on war and the landscapes of battle have recently been highlighted (Tucker and Russell, 2004). The nature of the relationship between warfare and environment are often quite complex and mediated by social, economic and political structures and intervention. Bennett's work on the environment of post conquest Fiji, for example, illustrates how a state of emergency might have proffered opportunities for dramatic changes to land legislation, specifically the appropriation of land by the Crown which in turn had significant environmental consequences for the land and forest reserves and associated social implications (Bennett, 2001) Historical geographers have also been among those responsible for opening up warscapes as spaces of investigation. Clout (2006) has recently drawn attention to the considerable research completed on the landscapes of the Second World War in Western Europe. Geographers have also spearheaded analysis of the relationship between military conflict, civil strife and the (re)emergence of disease in epidemic proportions (Cliff and Smallman-Raynor, 2004).

The Future of Environmental History: Prospects for Progress?

Much of the work conducted within environmental history has tended to focus on environmental problems that pose a challenge to humanity, or which require some kind of action to ensure social well-being, be it management, recovery or restoration. Having purchase on real and often pressing environmental problems, concerns and threats has undoubtedly served the development of the discipline well and particularly in recent years, when the sensitivity of the relationship between humans and their environment has become increasingly and often tragically clear. Certainly, the new raft of studies exploring the environmental histories of city, water and waste, disease, hygiene and climate change, some of the themes which have been referred to in this chapter, may have been stimulated in part by the contemporary environmental zeitgeist. Perhaps for this reason, however, most environmental history research to date has been very much declensionist (McNeill, 2003), at once grounded in and justified by fears of environmental crisis and the need for intervention.

But in as much as environmental historians can and indeed should be focusing on research that is both topical and policy relevant, and though some aspects of environmental history may also have been born of, and have helped to promote environmental consciousness, a case might be made for studying more optimistic (or in McNeill's (2003) terms, 'ascensionist') environmental histories. Work that focuses on technological adaptation through time, or the dynamics of human resilience to and ability to cope with environmental transformations are arguably just as important as developments in environmental awareness and indeed protection. Pessimistic climate prediction, to take one example has tended to obscure the history of human adaptation and resilience and the exploration of the institutions and cultural coping strategies that help people adapt to climate changes in the past (Fraser et al., 2003).

Moreover, as Cronon (1993) suggested, to have real relevance for the future, environmental histories must reach beyond rhetoric and affect the views of policymakers in real and tangible ways. This is perhaps where the greatest challenge for practitioners lies. Deriving policy relevant insights about the contemporary and future world from past interpretations of human-environmental interaction is problematic (McNeill, 2005, p. 178). The fact that past societies differ markedly from those in the modern world makes simple analogies or parallels with the historical past unrealistic (Meyer et al., 1998). Knowledge of successes and failures in adaptation to past environmental challenges, however, can increase the ability to respond to the threats of long-term future changes.

Obtaining this knowledge requires us to undertake empirical reconstruction of environmental change over more recent time-scales, that is to say, material environmental histories, and a willingness to relate these changes to the cultural record. Moreover, it is essential to try to obtain a better understanding of how societies have conceptualised these changes and endeavoured to make themselves effectively more resilient to them. As Butzer (2005) has recently illustrated, human perception and, in particular, ecological behaviour are absolutely pivotal to understanding cause and effect. Such 'pluri-disciplinary' or at least multi-dimensional investigations should be seen and indeed are regularly discussed as constituting the very essence of environmental history, but clearly there is still scope for much more cross-disciplinary collaboration (Butzer, 2005). Thus as a self-conscious area of enquiry, it might well be the case that 'environmental history has pushed and shoved its way' into centre stage (Steinberg, 2004, p. 265) and is arguably now more important than ever before. It is perhaps only through integrated, holistic and, above all, truly multidisciplinary investigations, however, that it can make a real scholarly and practical contribution.

NOTE

1. I am using the term here to refer to the broad range of studies of climate's past, involving the reconstruction of climate variability at a range of timescales as well as the exploration of the impact of climate change on societies. See Brazdil et al. (2005) for a good outline of the development of Historical Climatology.

BIBLIOGRAPHY

Acuña-Soto, R. Stahle, D. W, Cleaveland, M. K. and Therrell, M. D. (2002) Megadrought and megadeath in 16th century Mexico. *Emerging Infectious Diseases*, 8(4), 360–62.

Anderson, W. (1996) Disease, race and empire. *Bulletin of the History of Medicine*, 70(1), 62–67.

Arnold, D. and Guha, R. (1995) *Nature, Culture, Imperialism. Essays on the Environmental History of South Asia.* Oxford: Oxford University Press.

Baker, A. (2003) Environmental geographies and histories. In A. Baker (ed.) *Geography and History. Bridging the Divide.* Cambridge: Cambridge University Press, pp. 72–108.

Barriendos, M. (1997) Climatic variations in the Iberian peninsula during the Late Maunder Minimum (AD 1675–1715). An Analysis of data from rogation ceremonies. *Holocene*, 7, 105–11.

Beinart, W. and Coates, P. A. (1995) *Environment and History. The Taming of Nature in the USA and South Africa.* London: Routledge.

Bolster, J. W. (2006) Opportunities for marine environmental history. *Environmental History*, 11(3), 567–97.

Brazdil, R., Pfister, C., Wanner, H., von Storch, H and Luterbacher, J. (2005) Historical climatology in Europe – the state of the art. *Climatic Change*, 70(3), 363–430.

Butler, T. (2006) A walk of art: the potential of the sound walk as practice in cultural geography. *Social and Cultural Geography*, 7(6), 889–908.

Butzer, K. W. (2005) Environmental history in the Mediterranean world: cross-disciplinary investigation of cause-and-effect for degradation and soil erosion. *Journal of Archaeological Science*, 32, 1773–1800.

Cioc, M. (2006) *The Rhine. An Eco-Biography, 1815–2000.* Washington, DC: University of Washington Press.

Cleaveland, M. K., Stahle, D. W., Therrell, M. D. Villanueva-Diaz, J. and Burns, B. T (2003) Tree-ring reconstructed winter precipitation and tropical teleconnections in Durango, Mexico. *Climatic Change*, 59(3), 369–88.

Cliff, A. D. and Smallman-Raynor, M. R. (2004) *War Epidemics: A Geography of Infectious Diseases in Military Conflict and Civil Strife, 1850–2000.* Oxford: Oxford University Press.

Cliff, A. D., Haggett, P. and Smallman-Raynor, M. R (2004) *World Atlas of Epidemic Diseases.* London: Arnold.

Clout, H. (2006) Beyond the landings: the reconstruction of Lower Normandy after June 1944. *Journal of Historical Geography*, 32(1), 127–48.

Coates, P. A. (2005) The strange stillness of the past: towards and environmental history of sound and noise. *Environmental History*, 10(4), 636–65.

Craddock, S. (2000) *City of Plagues: Disease, Poverty and Deviance in San Francisco*. Minneapolis: University of Minnesota Press.

Cronon, W. (1991) The trouble with wilderness: or getting back to the wrong nature. In Cronon, W. (ed.), *Uncommon Ground*. London: W.W Norton and Company, pp. 69–90.

Cronon, W. (1992a) A place for stories: nature, history and narrative. *The Journal of American History*, 78, 1347–76.

Cronon, W. (1992b) *Nature's Metropolis: Chicago ad the Great West*. New York: Norton.

Cronon, W. (1993) The uses of environmental history. *Environmental History Review*, 0(Fall), 17(3), 1–22.

Crosby, A. W. (1986) *Ecological Imperialism: The Biological Expansion of Europe, 900–1900*. Cambridge: Cambridge University Press.

Crumley, C. (1994) Historical ecology: a multidimensional ecological orientation. In C. Crumley (ed.), *Historical Ecology: Cultural Knowledge and Changing Landscapes*. Santa Fe, NM: School of American Research Press, pp. 1–16.

Davies, S. J, Metcalfe, S. E, MacKenzie, A. B, Newton, A. J, Endfield, G. H and Farmer, J. G. (2004) Environmental changes in the Zirahuén Basin, Michoacán, Mexico, during the last 1000 years. *Journal of Paleolimnology*, 31, 77–98.

Davis, M. (1998) *The Ecology of Fear. Los Angeles and the Imagination of Disaster*. New York: Metropolitan Books.

Dias, S. C., Therrell, M. D., Stahle, D. W and Cleaveland, M. K (2007) Chihuahua (Mexico) winter-spring precipitation reconstructed from tree–rings, 1647–1992. *Climate Research*, 22, 237–44.

Dodgshon, R. A and Olsson, G. A. (2006) Heather moorland in the Scottish Highlands: the history of a cultural landscape, 1600–1880. *Journal of Historical Geography*, 32(1), 21–37.

Dovers, S., Edgecomb, R. and Guest, B. (2003) *South Africa's Environmental History: Cases and Comparisons*. Athens: Ohio University Press.

Edwards K. J., Langdon P. G. and Slugden, H. (2007) Separating climatic and possible human impacts in the early Holocene: biotic response around the time of the 8200 cal. yr BP event. *Journal of Quaternary Science*, 22(1), 77–84.

Elvin, M. (2004) *The Retreat of the Elephants: An Environmental History of China*. New Haven: Yale University Press.

Evenden, M. (2004) *Fish versus Power. An Environmental History of the Fraser River*. Cambridge: Cambridge University Press.

Fraser, E. D. G., Mabee, W. and Slaymaker, O. (2003) Mutual vulnerability, mutual dependence. The reflexive relation between human society and the environment. *Global Environmental Change*, 13, 137–44.

García-Herrera, R., Wilkinson, C., Koek, F. B., Prieto, M. R, Calvo, N. and Hernández, E. (2005) Descriptions and general background to ships' logbooks as a source of climatic data. *Climatic Change*, 73(1–2), 13–36.

Griffiths, T. (2001) *Forests of Ash: An Environmental History*. Cambridge: Cambridge University Press.

Grove, J. M. (1988) *The Little Ice Age*. London: Routledge.

Grove, R. H. (2001) Environmental history. In P. Burke (ed.), *New Perspectives on Historical Writing*. Cambridge: Polity Press, pp. 261–82.

Grove, R, Damodaran, V. and Sangwan, S. (eds) (1998) *Nature and the Orient: the Environmental History of South and Southeast Asia*. Oxford: Oxford University Press.

Harrison, M. (1996) The tender frame of man: disease, climate and racial difference in India and the West Indies. *Bulletin of the History of Medicine*, 70(1), 68–93.

Holm P., Smith, T. D. and Starkey, D. J. (2001) *The Exploited Seas: New Directions for Marine Environmental History*. St John's, NL: History Association / Census of Marine Life.

Holm, P. (2003) History of marine animal populations: a global research program of the Census of Marine Life. *Oceanologica Acta*, 25, 207–11.

Holm, P. (2005) The human impacts on fisheries resources and abundance in the Danish Wadden Sea, c.1520 to the present. *Helgoland Marine Research*, 59(1), 39–44.

Hughes, A. (2006) Learning to trade ethically: knowledgeable capitalism, retailers and contested commodity chains. *Geoforum*, 37(6), 1008–20.

Hughes, J. D. (2001) *An Environmental History of the World. Humankind's Changing Role in the Community of Life.* London: Routledge.

Igler, D. (2005) Longitudes and latitudes. *Environmental History*, 10(1, Special Anniversary Forum), 30–109.

Jones, K. (2004) Music in factories: a twentieth-century technique for control of the productive self. *Social and Cultural Geography*, 6(5), 723–44.

Jones, P. D., Ogilvie, A. E. J., Davies, T. D and Briffa, K. R. (eds.) (2001) *History and Climate: Memories of the Future?* London: Kluwer Academic.

Kempe, M. and Rohr, C. (2003) Coping with the unexpected- natural disasters and their perception. Special issue of *Environment and History* 9 (2), Editorial.

Kearns, R. A. and Andrews, G. J. (2005) Everyday health histories and the making of place: the case of an English coastal town. *Social Science and Medicine*, 60(12), 2697–2713.

Klingle, M.W. (2003) Spaces of consumption in environmental history. *History and Theory*, 42, 94–110.

Klooster, D. (2006) Environmental certification of forests in Mexico: the political ecology of a nongovernmental market intervention. *Annals of the Association of American Geographers*, 96(3), 541–65.

Kurlansky, M. (1998) *Cod: A Biography of the Fish That Changed the World.* New York: Penguin.

Kurlansky, M. (2002) *Salt: A World History.* New York: Penguin.

Lamb, H. H. (1982) *Climate, History and the Modern World.* London: Routledge.

Lambert, D. Martins, L. and Ogborn, M. (2006) Currents, visions and voyages: historical geographies of the sea Journal of Historical Geography, 32, 479–93.

Law, L. (2001) Home cooking: Filipino women and geographies of the senses in Hong Kong. *Ecumene*, 8(3), 264–83.

Le Roy Ladurie, E. (1972) *Times of Feast, Times of Famine: A History of Climate Since the Year 1000.* London: George Allen and Unwin.

Lekan, T. (2005) Globalising American environmental history. *Environmental History*, 10(1, Special Anniversary Forum), 30–109.

Livingstone, D. N. (1991) The moral discourse of climate: historical considerations of race, place and virtue. *Journal of Historical Geography*, 17(4), 413–34.

Livingstone, D. N. (2002) Race, space and moral climatology: notes toward a genealogy. *Journal of Historical Geography*, 28(2), 159–80.

Luckin, B. (2004) At the margin: continuing crisis in British environmental history. *Endeavour*, 28(3), 97–100.

MacKenzie, J. M. (2004) Introduction to Special 10th Anniversary Issue. *Environment and History*, 10(4), 371–78.

Malin, J. C. (1967) *The Grassland of North America: Prolegomena to Its History.* Kansas: Lawrence.

Marsh, G. P (1869) *Man and Nature, or Physical Geography as Modified by Human Action.* New York: Scribner Press.

Matthewson, K. (ed.) (1993) *Culture, Form and Place, Essays in Cultural and Historical Geography.* Geosciences and Man 32 (Louisiana State University).

McCann, J. (1999) *Green Land, Brown Land, Black Land: An Environmental History of Africa, 1800–1990.* Portsmouth: Heinemann.

McNeill, J. R (1999) Ecology, epidemics and empires: environmental change and the geopolitics of Tropical America, 1600–1825. *Environment and History*, 5(2), 175–84.

McNeill, J. R. (2000) *Something New Under the Sun: An Environmental History of the Twentieth Century.* New York: Penguin.

McNeill, J. R. (2003) Observations on the nature and culture of environmental history. *History and Theory*, 42, 5–43.

McNeill, J. R. (2004) *Mountains of the Mediterranean World. Studies in Environment and History.* Cambridge: Cambridge University Press.

McNeill, J. R. (2005) Diamond in the rough: is there a genuine environmental threat to security: a review essay. *International Security*, 30(1), 178–95.

Melosi, M. V. (2000) *The Sanitary City: Urban Infrastructure in America from Colonial Times to the Present.* Baltimore: Johns Hopkins University Press.

Merchant, C. (1989) What is environmental history? In C. Merchant (ed.), *Major Problems in American Environmental History.* Lexington: DC Heath and Company, pp 1–31.

Meyer, W.B., Butzer, K. W., Downing, T. E., Turner, B. L., II, Wenzel, G. W. and Westcoat, J. L. (1998) Reasoning by analogy. In S. Raynor and E. L. Malone (eds), *Human Choice and Climate Change, No. 3, Tools for Policy Analysis.* Columbus, OH: Batelle Press, pp. 218–89.

Mitman, G. (2005) In search of health: landscape and disease in American environmental history. *Environmental History*, 10(2), 184–210.

Moon, D. (2005) The environmental history of the Russian steppes: Vasilii Dokuchaev and the harvest failure of 1891. *Transactions of the Royal Historical Society*, 6th series 15, 149–17.

Nash, R. (1967) *Wilderness and the American Mind.* New Haven: Yale University Press.

Nicholson, R. A. and O'Connor, T. P. (eds) (2000) *People as an Agent of Environmental Change.* Symposia of the Association of Environmental Archaeology. Oxford: Oxbow Books.

Pfister, C. and Brazdil, R. (1999) Climatic variability in sixteenth century Europe and its social dimension: a synthesis. *Climatic Change*, 43(1), 5–53.

Powell, J. M. (1996) Historical geography and environmental history: an Australian interface. *Journal of Historical Geography*, 22(3), 253–73.

Roberts, C. and Manchester, K. (2007) *The Archaeology of Disease*, 3rd edn. New York: Cornell University Press.

Schott, D., Luckin, B. and Massard-Guilbaud, G (eds) (2005) Resources of the City: Contributions to an Environmental History of Modern Europe. Historical Urban Studies Series. Aldershot: Ashgate Publishing.

Smallman-Raynor, M., Nettleton, C. and Cliff, A. D. (2003) Wartime evacuation and the spread of infectious diseases: epidemiological consequences of the dispersal of children from London during WWII. *Journal of Historical Geography*, 29(3), 396–421.

Richards, J. F. (2003) *The Unending Frontier: Environmental History of the Early Modern World.* Berkeley: University of California Press.

Rome, A. (2001) *The Bulldozer in the Countryside Suburban Sprawl and the Rise of American Environmentalism.* Cambridge: Cambridge University Press.

Rothman, H. (2002) Conceptualising the real: environmental history and American studies. *American Quarterly*, 54(3), 485–97.

Russell, E. W. B. (1997) *People and the Land Through Time. Linking Ecology and History.* London: Yale University Press.

Sauer, C. O. (1956) The agency of man on the earth. In W. L. Thomas, Jr (ed.), *Man's Role in Changing the Face of the Earth.* Chicago: University of Chicago Press, pp. 45–69. (Also Summary remarks: retrospect, pp. 1131–35.)

Sorlin, S and Warde, P. (2005) The problem of the problem of environmental history – a re-reading of the field and its purpose. *Environmental History*, 12(1), 9–34.

Steinberg, T. (2004) Fertilising the tree of knowledge: environmental history comes of age. *Journal of Interdisciplinary History*, 35(2), 265–77.

Stewart, M. A. (1998) Environmental history: profile of a developing field. *The History Teacher*, 31(3), 351–68.

Tarr, J. (1996) *The Search for the Ultimate Sink: Urban Pollution in Historical Perspective*. Ohio: Akron Press.

Thomas, W. L., Jr. (ed.) (1956) *Man's Role in Changing the Face of the Earth*. Chicago: University of Chicago Press.

Tucker, R. P and Russell, E. P. (2004) *Natural Enemy, Natural Ally: Toward an Environmental History of War*. Corvallis: Oregon State University Press.

Turner, B. L., II, Clarke, W. C, Kates, R. W., Richards, J. F., Mathews, J. T. and Meyer, W. B (eds) (1990) *The Earth as Transformed by Human Action: Global and Regional Changes in the Biosphere over the Last 300 Years*. Cambridge: Cambridge University Press.

Turner, F. J. (1893) The significance of the frontier in American history. *The Annual Report of the American Historical Association for the year 1893*, 199–227.

Van Sittert, L. (2005) The other seven tenths. *Environmental History*, 10(1, Special Anniversary Forum), 30–109.

Walton, J. (2001) City natures. *International Journal of Urban and Regional Research*, 25(4), 903–6.

Webb, W. P. (1931) *The Great Plains*. New York: Blaisdell Publishing Company.

Weiner, D. R. (2005) A death defying attempt to articulate a coherent definition of environmental history. *Environmental History*, 10(3), 404–20.

Wilkinson, C. (2005) The non-climatic research potential of ships logbooks and journals. *Climatic Change*, 73(1–2), 155–67.

Williams, M. (1994) The relations of environmental history and historical geography *Journal of Historical Geography*, 20, 3–21.

Williams, M. (2004) *Deforesting the Earth: from Prehistory to Global Crisis*. Chicago: University of Chicago Press.

Worster, D. (1988) Appendix: doing environmental history. In D. Worster (ed.), *The Ends of the Earth: Perspectives on Modern Environmental History*. Cambridge: Cambridge University Press, 289–307.

Chapter 15

Landscape, Culture and Regional Studies: Connecting the Dots

Kenneth R. Olwig

Introduction

The nexus of *landscape, culture and regional studies* is critical to contemporary attempts to rethink the nature/culture dichotomy and the resultant divide between physical and cultural geography. The power of this nexus lies in its formation of a totality, or *gestalt*, that is greater than the sum of its parts. It is, however, largely left to the imagination to connect the dots of •landscape•culture•region•nature• environment, and form a mental picture of what this connectivity means in terms of social practice and power relations. The way we connect dots of this kind depend on our world picture, or to use a fancier terms, our *cosmographic* picture of the world as it is tied to our *cosmological* understanding of the world.

Concepts such as nature and culture are related to a given society's worldview, and Geography – as both a school and university subject – has long played a central role in educating people to *see* the world in terms of given world pictures. I have italicised *see* because this is not just a question of the way people think, but of the way people and disciplines perceive and know the world, which is to say their *epistemology*. It may seem paradoxical that the most exciting state-of-the-art writing about this nexus is often *historical, conceptual* and *interdisciplinary* in approach. The reason for this is that the verities of modernism's world-picture, which have long been taken for granted, are now being questioned by scholars whose focus is not just upon the 'modern' present, but upon the past, when the idea of the modern took shape. This work is being undertaken in important measure by non-geographers who are, in effect, questioning fundamental assumptions of Geography as a modern science by showing how its spatialised discourse has its origin in older discourses of landscape and nature, which modernist geography has long since relegated to the scrap heap of history (W. J. T Mitchell, 1994; Olwig 1996a). This naturally goes against the grain of geographers who define themselves as modernists, and this may explain why many of the key figures in this critique are not geographers, even if their concerns lie at the heart of geography.

According to the French anthropologist and student of science Bruno Latour (1993), modernism, as it emerged from the Renaissance and Enlightenment, bore the promise that science, by isolating nature as an object of study, would be able to transform and control that nature to the benefit of society. The contemporary environmental crisis, however, suggests to Latour that this separation and objectification of nature in relation to culture may be a source of environmental problems, rather than their solution. A solution for scholars like Latour is thus to question the philosophical foundations of modernity itself, and to do this many scholars have gone back into history to re-examine the way that modernism's world picture, which we now take for granted, was originally constructed. Thus, if we want to know why it is that geographers often claim that their discipline forms a bridge between the natural sciences and social sciences, and why it is that geographers nevertheless often do not cross that bridge, one needs to go back to the origins of the modernist tendency to split nature from culture. An important clue to how this dichotomisation occurred can be found in the work of students of landscape, who have shown how the construction of a pictorial concept of landscape in the Renaissance, which was based on foundational work by geographers of both the Renaissance and ancient Greece, played a central role in the development of modernism's world picture. This world picture, in turn, helped shape the modern concept of region in relation to culture and environment.

Landscape is often loosely thought of as meaning more or less the same as environment or nature. It would be more correct to say, however, that the way one defines landscape has a great deal to say about the way one connects the dots between culture and region and between environment/nature. Discussions of the concept of landscape are complicated by the fact that it embodies two somewhat contradictory meanings, each of which connects the concepts of landscape, culture and region in different ways. The two meanings are well expressed in the definition of *landscape* in Dr. Samuel Johnson's classic 1755 dictionary (note that neither definition specifically mentions either nature or environment) (Johnson, 1755[1968]):

1. 'A region; the prospect of a country'
2. 'A picture, representing an extent of space, with the various objects in it'

In Johnson's dictionary definition (1) is the oldest, whereas definition (2) is the most modern. Definition (2) is arguably a quintessential expression of the modern worldpicture. Since I am concerned here with the contemporary critique of modernism, I will begin by examining definition (2) before returning to definition (1) as the possible source of a 'non-modern' alternative to understanding landscape, which does not dichotomise culture and nature. This reversal of the temporal course of history contradicts, of course, the mentality of modernity, which casts history as a progressive linear movement that continuously relegates the past to obsolescence, and privileges the present (Olwig, 2002b).

Landscape Two: Modern, Scenic, Pictorial Space

The tendency to conceptualise the world in terms of an extent of space, with the various objects in it, is a characteristic of the rise of a movement of thought in the Renaissance and Enlightenment that can be said to mark a beginning of the ideology modernism. As one student of the Renaissance writes:

> The systematic origins of Renaissance art and of the Copernican astronomy can be found in a movement of thought which may be properly called a 'Ptolemaic renaissance'. . . . When scientific 'pictures' of the world came to be constructed according to these same principles, modern astronomy and geography began their rise. (Gadol, 1969, p. 157)

The re-discovery of the cartographic techniques of Ptolemy (1991), an ancient Greek astronomer and geographer (or 'cosmographer,' as he was then known), facilitated the creation of the modern map, in which locations are plotted upon the grid of an absolute space. What was less well known, until the topic began to be explored by scholars with an understanding of art history, including the geographer Denis Cosgrove (1984; 1988; 1993), was that the rise of modern surveying, cartography and geography, also made possible the development of the techniques of central point perspective which were used to create landscape pictures that were expressive of what became the modern world picture. These pictures represented the world in terms of an extent of space, with the various objects in it. What we then see is that, at the same time surveying and cartography were being developed by the 'modern' geographers of the Renaissance for the practical purpose of both understanding and transforming the material world, these same techniques were also being used (often by the same individuals) to create an equally modern world picture by which to comprehend and form that world. It was in the Renaissance and Enlightenment that modern *space* was discovered and deified as the 'sensorium' of God, as Isaac Newton put it (Newton, 1717, p. 380), meaning the seat of the mind of God. What made this space modern was that *Nature* was seen to operate according to eternal laws instituted by *God*. In this way natural science, with its knowledge of *Nature's* laws, was able to lay claim to powers once reserved for religion.

'Sensorium',' at Newton's time, meant 'the brain or a part of the brain regarded as the seat of the mind,' but in literal terms it refers specifically to 'the parts of the brain that are concerned with the reception and interpretation of sensory stimuli' (Merriam-Webster Dictionary, 2000). Perspective functioned as a kind of spatial sensorium through which a particular central and focused world picture could be formed. This spatial framing of phenomena facilitates their treatment in terms of measurable spatial relationships governed by natural laws, such as those of physics. This framing, furthermore, facilitates a shift in meaning of 'land' in landscape from a cultural phenomenon, a land in the sense of country or *pays*, to a natural object, land understood as a physical phenomenon like soil. It thus makes landscape an object of scientific interest, as in landscape ecology. This approach also implies that human cultural phenomena can be treated in the same way, thus reducing cultural phenomena to a subset of the natural.

To understand the way the conceptualisation of landscape as pictorial, scenic space helped create a larger world picture one must take a closer look at the way these pictures functioned during the Renaissance and the Enlightenment. Pictorial landscape images played an important role in the Renaissance development of the modern theater with its perspective scenery, framed on a stage. 'Theater' thereby came to provide a metaphor by which everything from war ('theater of war'), medicine ('operating theater') and even the globe (atlases were called theaters and theaters could be called 'The Globe', which was the name of Shakespeare's London playhouse). The stage, with its landscape scenery, created a stratified space with nature as its foundation. Its structure consisted of superimposed layers of earth, flora and fauna, and, finally a layer of culture. The continuing power of the metaphor of the theatre can be seen in the work of Bernard le Bovier de Fontenelle (1657–1757), who is regarded as a central figure in promoting the modern conception of science. Fontenelle writes in *Conversations on the Plurality of Worlds* from 1686:

> I have always thought that nature is very much like an opera house. From where you are at the opera you don't see the stages exactly as they are; they're arranged to give the most pleasing effect from a distance, and the wheels and counter-weights that make everything move are hidden out of sight. You don't worry, either, about how they work. Only some engineer in the pit, perhaps, may be struck by some extraordinary effect and be determined to figure out for himself how it was done. That engineer is like the philosophers. But what makes it harder for the philosophers is that, in the machinery that Nature shows us, the wires are better hidden – so well, in fact, that they've been guessing for a long time at what causes the movements of the universe. . . . Whoever sees nature as it truly is simply sees the backstage area of the theater (Fontenelle 1990[1686], p. 12).

For Fontelle the space of the theater functioned as a kind of 'sensorium' of nature, providing a metaphorical framework within which its workings could be understood.

Landscape, in this pictorial, scenic and theatrical sense described by Fontenelle, provided a world-picture functioning at a metaphoric meta-level that facilitated the ability of scientists to reflect upon physical phenomena from their position as distanced observers, thus objectifying nature and separating it from the human spectator. This notion of landscape, thus, helped shape the taken for granted world picture that lay behind much scientific research from the Renaissance through the Enlightenment, though the science of geography had yet to be established as a formal university discipline (Glacken, 1967). Landscape emerges as a focus of scientific geographical interest with the development of modern geographical science as a university discipline in the early 19th century, particularly through the work of Alexander von Humboldt (Minca, 2007), and in the course of the 20th century landscape became a central concept in the geographical discipline which was becoming firmly entrenched at all levels of education and research (Sauer, 1925). The fact, however, that geography was both a form of science and a key subject in schools, inculcating world-pictures into the heads of children, means that the two dimensions of the discipline cannot be explored in isolation from each other. It is thus necessary to consider the ideological and educational role of landscape in geography before considering its situation today.

The Landscape of the Theatre of State

One of the subjects of particular interest in the ongoing re-examination of the assumptions of modernism is the role of the state. The rise of modernism went hand in hand with the rise of the centralised state, which has been seen to be a central agent of modern development. The issue of the state is important when examining the ideas of geographers because they were so thoroughly implicated in the construction and functioning of the modern state. They not only mapped the regional organisation used to make the state and its territory governable, they also produced the textbooks used to educate the populace to see themselves as natural-born citizens of the state, as the geographers Anssi Paasi and Jouni Häkli have amply shown

(Häkli, 1998; Paasi, 2008). Today, however, the state has been weakened by the rise of modern globalist ideology, and with it supra-national organisations such as the European Union and multi-national corporations. As the power of the modern state weakens, we see the re-emergence of historical regions, such as Catalonia or Wales, that pre-existed the rise of the modern state (Kaplan and Häkli, 2002). Such historical regions, which are often roughly equivalent to the Dutch *landschap*, are usually not defined by the sharp map-drawn boundaries of a state bureaucracy, but by fuzzier cultural practices such as the speaking of a language or dialect, as the geographer Tomas Germundsson (2006) has argued.

During the Renaissance many regents sought to consolidate their power in the form of centralised states under their absolute control. The metaphor of the 'theater of state' provided an official cultural means of envisioning the organisation of the state, its spatial enframement and regionalisation, and the material and social progress that was promised by the centrally organised state. The boundaries of the state, as in the case of Britain, or Sweden, were often naturalised and legitimised through the argument that they naturally should follow a mapable environmental barrier, such as a coastline. In England the argument was even made, using the vehicle of the theatre and its landscape scenery, that such bounded territories formed natural human entities, linking hitherto separate nations (e.g. the English, Welsh and Scotts), into a single 'British' natural region, where people shared a temperament and race determined by their physical environment (Olwig, 2002a, pp. 148–75).

In the course of the Enlightenment the ability to perceive the landscape as spatial scenery came to be seen as a mark distinguishing a person belonging to the educated elite, and hence a person capable of running an estate within a modern state (Barrell, 1987). The emergence of popular democracy and the nation state in the late 18th and early 19th centuries also saw the emergence of popular education, and by the end of the 19th century children from all classes were being educated to see the world from the perspective of landscape scenery and cartographic space. The problem with this form of education, however, was that the landscape model has the tendency to almost subliminally give the impression that the natural foundation of the landscape determines its subsequent cultural overlay. This characteristic helped further a research and educational agenda known as 'environmental determinism' in which differing landscape environments were seen to foster different types of societies – e.g., freedom-loving people in cold mountains, slavish people on warm irrigated planes, etc. etc. A problem with this approach is that, while it creates a framework within which to examine the relationship between culture and nature, it also polarises society and nature, creating a tendency to see the one (normally nature) as the determinant of the other (normally culture) (Olwig, 1996a).

Environmental determinism was not only questionable as science, it also facilitated nationalistic indoctrination, so that children would be taught that their national character and culture was an outgrowth of the landscapes demarcated by their national territory – an ideology known as 'blood and soil.' Social groups (such as Jews or the Romany) who were not seen to be rooted in the native soil of the nation, could then be made the object of nationalistic discrimination and scapegoating (Livingstone, 1994). The ideological excesses of W.W. II gave, in retrospect, this form of geography a bad name, and landscape was largely abandoned, in name, by mainstream Anglo-American human geography, though it maintained scenic landscape's focal spatial 'sensorium'. Postwar mainstream human geography thereby redefined itself as a modern science of space in which regions were defined in terms of the relations between phenomena within an encompassing space, and the unmodern study of the historical interaction between society and nature, as expressed in landscape was largely abandoned (Smith, 1989; Olwig, 1996a). Landscape remained a viable concept within physical geography, but the transformation of human geography into the modern science of space created a situation where human and physical geography became divorced from each other.

The scenic landscape continued to be a focus of study in continental Europe, particularly in eastern Germany, and this helped give impetus to the emergence of the discipline of *Landscape Ecology* in the late 20th century (Brandt and Vejre, 2004). *Landscape ecology*, which has attracted the interest of some physical and cultural geographers, has perpetuated the landscape paradigm, as it involved from the Renaissance, and maintained its modernism and ties to state planning (Groening, 2007). Landscape ecologists thus tend favor the use of the latest GIS computer technology to build planning models. The scenic approach to landscape is thus alive and well at the turn of the 21st century, and this helps explain why it remains an object of interest to critics of the modernist landscape world picture.

Post-modern versus Circulating Landscapes

It was, as noted, largely through the early work of Denis Cosgrove (1984; 1988; 1993) that geographers were made aware of the relationship between Renaissance scientific developments in the area of surveying, cartography and perspective in the creation of a modern world picture that, in turn, influenced the shaping of the land. Cosgrove's work was, in this sense, both modernist and materialist with a structural Marxist inspired dialectic, between scientific development, ideology and the progressive material transformation of nature. Cosgrove did not, however, follow up in-depth on the rich implications of his work for the understanding of the relationship between human and physical geography, or for geography's approach to environmental issues. Instead, he turned toward a more 'post-modern' position on landscape that was in keeping with the growing contemporary critique of modernism (Daniels and Cosgrove, 1988) – though he later modified stance concerning the substantivness of landscape (Cosgrove, 2004). From this perspective landscape is a form of visual representation, not the things represented. Landscape was thus conceived to be a 'simulacrum', to use the term of the French philosopher Jean Baudrillard (1988), which is to say a form of representation that takes on a life on its own. In his later work Cosgrove also was associated with James Duncan, who regarded landscape as a form of text to be 'read', and who was largely concerned with ideological issues. While these issues are important, for example with regard to the role of the scenic spatial approach to landscape in fostering imperialism (W.J.T. Mitchell 1994), they do not speak particularly to geographers concerned with society-environment issues, or with the relations between physical and human geography (Demeritt, 1994).

A process described by Latour as 'circulating reference' (Latour, 1999; Olwig, 2004) provides a way of comprehending the relationship between the landscape as spatial representation and the material world that recognises the tendency of representations to become simulacra while, at the same time, granting them a role in the creation of scientific knowledge. From this perspective the application of a graphic representational system, such as the map, or visual perspective, to material phenomena may well produce a useful perspective on such phenomena.

The problem, however, is that a continuing form of circulation develops so that one continually moves between the form of representation, the material world to which it refers, and back to the form of representation, in a process in which each modifies the perception of the other so that the they eventually blur. This leads to a situation in which scientists loose sight of the way that their knowledge may be shaped and constrained by taken-for-granted representations. On the basis of field-work following a team of soil scientists in Brazil, Latour concluded that between their graphic representations and the material world to which they refer 'there is neither correspondence, nor gaps, nor even two distinct ontological domains, but an entirely different phenomenon: circulating reference' (Latour, 1999, p. 24). Latour's studies of the soil scientists' research process led him to observe the following: 'Remove both maps, confuse cartographic conventions, erase the tens of thousands of hours invested in . . . [their] atlas, interfere with the radar of planes, and our . . . scientists would be lost in the landscape and obliged once more to begin all the work of exploration, reference marking, triangulation, and squaring performed by their hundreds of predecessors' (p. 29). The consequence of this disorientation is that 'Lost in the forest, the researchers rely on one of the oldest and most primitive techniques for organising space, claiming a place with stakes driven into the ground to delineate geometric shapes against the background noise, or at least to permit the possibility of their recognition. Submerged in the forest again, they are forced to count on the oldest of the sciences, the measure of angles, a geometry whose mythical origin has been recounted by Michel Serres' (pp. 41–42). 'Yes', he finally exclaims, 'scientists master the world, but only if the world comes to them in the form of two-dimensional, superimposable, combinable inscriptions' (p. 29).

Today, with the development of mapping and modeling tools such as GIS and GPS, the forms of representation noted by Latour are more powerful than ever, and this has fostered a resurgence of scientific interest in landscape in fields such as landscape ecology.[1] In this situation, when an anthropologist like Latour points out that what these scientists are actually seeing is vitally shaped by circulating reference, the reaction of natural scientists can be quite negative, even vituperative, and this has helped give rise to what has become known as the 'science wars' (Latour, 1999). Such wars, of course, do little to further cooperation between physical and cultural geographers. A possible way around the problem of circulating reference with regard to landscape and its relationship to culture, region and nature, lies in a re-examination of the foundations of the modern notion of landscape as scenic, pictorial, space. This brings us back to Dr Johnson's definition (1): 'A region; the prospect of a country.'

Landscape One: Historical Region, the Place of Community

In his first definition of landscape Dr Johnson equated landscape with the prospect of a country, understood as region and place. He defines both region and country as 'the place which any man inhabits' and as the inhabitants themselves, as when Shakespeare wrote: 'All the *country*, in a general voice, Cry'd hate upon him' (S. Johnson, 1755[1968]: region). Country/region, in this sense, means the same as 'land,' as in Scot*land*, the country, region or place which is the habitation of the Scots. *Land*, like *country*, can also refer to a human community, as when we refer to the 'land' rising up against an oppressor.

If landscape is understood to be an historically evolved human region, the place of a polity and society, then the role of representation in the study of this region changes. Landscape then ceases to be a form of space, represented on a map or diagram, and rather is understood to be a non-representational phenomenon, which is to say a phenomenon that exists through practice rather than on the basis of a representation. Such a phenomenon, however, can also be represented in various ways, and those representations can influence its perception and understanding. Such a representation can be a visual prospect, but it can also be, for example, a representative political body. This understanding of landscape thus eliminates the privileged perspective of space as the sensorium of God, or Nature, or Science, or the State, and interest turns toward the properties of perception itself. How people perceive landscape, and how this affects the way we behave toward the perceived landscape, become central questions when landscape is understood this way.

Though it is difficult for any geographer living in the modern era to avoid being influenced by the conception of landscape as scenic space, a number of geographers have focused their work upon landscape as the historically constituted place of a polity and society. In Europe the focus upon landscape as place, and its perception, is particularly associated with the influential school of the French geographer Vidal de la Blache which took its point of departure as the historical region of the *pays*, the root of the French word for landscape *pay*sage, and focused upon the way these places were stamped with the culture of a particular regional polity (Buttimer, 1971). In America it was the Berkley school of historical and cultural geography, founded by Carl Sauer, which took this approach (Sauer, 1969). For this school it is vital to understand how differing perceptions have effected the shaping of, particular regional landscapes, and with it their nature and environment (Lowenthal, 1961; Glacken, 1967; Tuan, 1974). An important inspiration for this school was the work of the 19th American geographer George Perkins Marsh whose 1864 book *Man and Nature: Or, Physical Geography as Modified by Human Action* (1965) sought to change the perception of society's relationship to its environment and thus contributed to the development of the subsequent conservation and environmental movements (Lowenthal, 2000). My own approach to landscape, and to society-nature issues, has grown out of this school, and this is why I have focused in my work upon the historical development of actual regional landscape polities in relationship to differing perceptions of those regions, and the consequences of these perceptions for both society and its material surroundings (Olwig, 1984; 2002a).

Pre-Modern and Non-Modern Contemporary Landscape Polities

If one conceptualises landscape in terms of region and culture, it makes sense to focus on the Dutch landscape paintings of the *landschap* regions that originally lent their name to the genre (Merriam-Webster, 2000, landscape), rather than on the central point perspective paintings of Renaissance Italy, which were usually representations, in both form and content, of an idealised *Nature*. The Dutch paintings were prospects of landscape regions, called *landschap*, and the English therefore called them by the equivalent English word, 'landscape'.[2] For the Dutch painters the important thing was the character of the regional landscapes as they had been formed by local custom and history. The low-lying, often federated, lands of this part of Europe were engaged at this time in a struggle against an emperor based in Spain who wished to subsume these low lands under the imperial power of a centralising state. For this reason the maintenance of a common identity was a life and death matter for these societies, and *landschap*, because it united community and place, was, according to the geographer Tom Mels, a vital expression of that identity (Mels, 2006).

Polities of the kind depicted in the Dutch paintings still exist. Examples of contemporary regional landscape polities can be found along the coast of the shallow waters of the Wadden Sea that links modern Denmark, Germany and the Netherlands. Farms here are family-based, and many of the dikes are still maintained through collective local effort. Living with this environment has historically been a question of 'going with the flow.' One does not build sharp barriers against the water. Rather one allows the water to follow around, through, and even over, the land according to weather conditions. Modern society tends to define environment in terms of physical objects external to a society, but this idea is difficult for the people of such a polity to understand because their community has created, through a gradual process of diking and drainage, the very soil upon which they live, and which they share with various forms of wildlife, including great quantities of migratory birds. This soil is not seen as being external to their landscape polity, but rather it is seen to be a 'substantive' part and parcel of their social and legal practices, and hence their habitus.[3] The word *substantive*, as used here, should not be confused with *substantial*, because it is meant here in the legal sense of 'creating and defining rights and duties', as in the branch of law that deals with the prescription of 'the rights, duties, and obligations of persons to one another as to their conduct or property' (Olwig, 1996b; Merriam-Webster, 2000). Their landscape polity is thus built upon the culture of a 'moral economy', to borrow a term from the English historian E.P. Thompson, which involves substantive cooperative community effort, particularly with regard to the maintenance of its dikes and drainage systems, which places a legal and moral burden upon members of the community (Knottnerus, 1992; Thompson, 1993). Landscape is thus a practice, something that you do, as an individual, and as part of a community, and as part of a material habitus. Landscape, therefore, is not primarily a scene of an objectified nature that you observe from a distance, or perform upon as a stage, but a part of the nature of your community and the customs and values that are deemed normal, and hence natural. Another way of describing this kind of community's relationship to its material habitus could be to borrow the concept of 'actant' from Latour, who has argued that the natural environment is an active participant in the social relations that determine our environmental agenda, though not a determinant of those relations (Latour, 1993).

There is an ongoing pressure in the Wadden Sea to 'modernise' by constructing ever-higher centrally planned dikes that will separate the water from dry land rigidly. The problem with this, as residents of areas with similar environmental issues (such as Venice, New Orleans or even the Thames basin) well know, is that there does not seem to be a technological fix that can maintain a sharp separation between water and land. When unusual flooding conditions occur, the disaster for the residents in such 'modernised' areas will become correspondingly greater. The diked landscapes of the Wadden Sea area, with their moist coastal grasslands, are a paradise for migratory birds, and this attracts bird-watchers, who would like to limit human use of the area. The dikes, however, are often maintained through human use as part of the moral economy that holds together the fabric of their community, but this is little appreciated by environmentalists. The result is a some-times nasty clash between environmentalists, who seek to maintain a sharp divide between society and nature, and the farmers, who think of themselves as part and parcel of the land they have made, as shown by the anthropologist Werner Krauss (Krauss, 2006).

The 'non-modern' sense of landscape as used to apply to places or regions that express the habitus of a polity/community may be subaltern to the modern scenic sense, but it is by no means dead, neither in the Wadden Sea nor, for example, in New Orleans. In an article in the *New York Times* we thus find issues of law, custom, culture, region and environment linked in a reference to the 'social landscape' of New Orleans where we are told that: 'The 'second line' clubs of New Orleans, which lead the distinctive black tradition of Sunday parading, say they feel threatened by new police fees, the latest sign of conflict between old customs here and the altered social landscape left by Hurricane Katrina' (Nossiter, 2006). The landscape described in the *Times* is not an extent of scenic pictorial space, with the various objects in it, but a complex of social, political and natural practices and processes, that the police wish to control. But if the term landscape were to be interpreted in scenic terms, with a foundational natural landscape that *determines*, the social landscape upon it, then the *Times*' statement could provide a problematic legitimisation for police suppression. It is possible that much of the antipathy often expressed towards American landscape geography by British scholars (Jackson, 1989; Duncan, 1990) may owe to the fact that British scholars have tended to focus upon landscape as scenic space, and thus may not appreciate the broad and complex meaning of land-scape as used, for example, by the Berkeley school. The term 'Anglo-American' is often applied to both British and American geography, as if a common lan-guage created a scholarly unity, but the fact is that American geography, like the American people, derives from many different national sources, and it is clear, when reading the work of Carl Sauer, that his understanding of landscape has been informed by a broad personal and intellectual continental European background which contrasts markedly with the narrow British understanding of landscape taken by his critics.

The Present-Day Situation

Today, there would seem to be something of a revival of the idea of landscape as region and place – a revival not in the simple sense of an unreconstructed resur-rection of the old but in the sense of a new way of examining older concerns. This revival is far from uniform, in terms of its intellectual details or its relative strength, but it is real nonetheless. As Lesley Head (2007, p. 837) puts it, many geographers interested in region and place have 'entered a post-humanist moment and want to talk about the agency of [things like] wolves and trees'. This, she continues, is generative of 'the idea of landscape as a bioculturally collaborative product . . .' (p. 840). To give a flavor of this new work, I want to list three devel-opments within British and North American geography that are proving espe-cially influential at the present time. First, as Matthew Turner's chapter shows, there is a rediscovery of the agency of nature within 'political ecology', an enormously important branch of geography in North America. Though some have suggested that political ecology has lost sight of the material world, the so-called 'new' or 'non-equilibrium ecology' has been one impetus for some political ecologists to rediscover how local ecologies are integral to human practices at a range of scales (see Walker, 2005; Head, 2007). Second, in the USA there is post-Sauerian, post-Cosgrovian tradition of landscape research associated with Marxist Don Mitchell and students at Syracuse University. Mitchell sees landscape very much as a physical and material thing – not only a product the social processes of whose construction must be deciphered, but also something that can reinforce or alter human practices. So, for Mitchell, the materiality of landscapes is not immaterial to the practices giving rise to their construction, maintenance and decomposition over time (see Mitchell, 1996; 2007; Kirsch and Mitchell, 2004). Finally, where neither of these two kinds of work especially accents the human body, a third new area of research does: so-called 'non-representational geography', which is very much a British phenomenon. This corpus of work, inspired greatly by Nigel Thrift's writings, aims to 'reembody' human actors. For too long, so non-representational geographers argue, we have analysed people as seeing and speaking beings – that, as re-presenters of worlds – rather than as fleshy, multi-sensual, practical ones. Their aim has been to disclose the range of not-only-visual, not-only-oral encounters we have with landscape with a view to depicting a more fully 'human' human geography, as well as a more environmentally-embedded one (see, for instance: Lorimer, 2005; Wylie, 2007).

There are important intellectual differences between these new 'joined-up' approaches to landscape. For instance, the muscular conception of power and social structure one finds in much of Mitchell's writing is apparently absent from the 'radical empiricism' so characteristic of studies by non-representational geographers. Even so, there is a family resemblance in all three cases, based on a commitment to see landscape as very much about process, practice and action.

Conclusion

It is impossible for one person, with a particular disciplinary approach, to explicate the many different ways landscape, region, culture and environment are linked in fields as far ranging as cultural ecology, bioregional studies, physical geography, landscape ecology etc., especially in an essay of limited length. The approach presented here sheds light on a modern world-picture that continues to frame, and bifurcate, differing approaches to landscape, culture, region and the environment. Readers coming from differing backgrounds should be able to apply this approach to their own discipline, and discover for themselves how it may shape their disciplinary world-picture. Perhaps this can help them to think outside the box of a modernism that has divided nature from culture and, thereby, physical from human geography. Geography – insofar as landscape, region, and culture are concerned – has entered a new intellectual phase wherein environment/nature are no longer seen as things existing 'out there' waiting to be represented or else worked upon by humans like a tabula rasa.

ACKNOWLEDGEMENT

I would like to thank Noel Castree for making some useful additions to this chapter.

NOTES

1. Within physical geography, that way of seeing space as container for objects whose relations with each other are governed by universal laws of nature has become the dominant one, though of late there has also been some interest among physical geographers in uniqueness of place and the merits of a more ideographic understanding of the earth sciences as historical (even hermeneutic) sciences concerned with the unique and historically contingent evolution of particular landforms that perhaps harks back to your first landscape tradition (e.g., Baker and Twidale, 1991; Frodeman, 1995; Beven, 2000).
2. Merriam-Webster (2000) gives the following etymology for landscape: 'Dutch *landschap*, from Middle Dutch *landscap* region, tract of land (akin to Old English *landscipe* region, Old High German *lantscaf*, Old Norse *landskapr*), from land + -scap -ship; akin to Old High German *lant* land and to Old High German -*scap* -ship – more at LAND, -SHIP.'
3. The word *habitus* derives from a Latin word meaning 'condition, appearance, attire, character, disposition, habit,' (Merriam-Webster, habit), and it thus belongs to a constellation of words like character that are related to custom (habit) as well to morality (moral habitus). The concept of 'habitus' as an expression of social practice is particularly identified with the French anthropologist/sociologist Pierre Bourdieu (1977).

BIBLIOGRAPHY

Baker, V. and Twidale, C. (1991) The reenchantment of geomorphology. *Geomorphology*, 4, 73–100.

Baudrillard, J. (1988) Simulacra and Simulations. *Selected Writings*, Stanford, Stanford University Press: 166–184.

Bourdieu, P. (1977) *Outline of a Theory of Practice*. Cambridge: Cambridge University Press.

Barrell, J. (1987) The public prospect and the private view: the politics of taste in eighteenth-century Britain. In J. C. Eade, (ed.), *Projecting Landscape*. Australia: Humanities Research Centre Australian National University, pp. 15–35.

Beven, K. J. (2000) Uniqueness of place and process representations in hydrological modelling. *Hydrology and Earth System Sciences*, 4, 203–213.

Brandt, J. and Vejre, H. (2004) Multifunctional landscapes – motives, concepts and perspectives. In J. Brandt and H. Vejre (eds), *Multifunctional Landscapes: Theory, Values and History*, Vol. 1. Southhampton: WIT Press, pp. 3–31.

Buttimer, A. (1971) *Society and Milieu in the French Geographic Tradition*. Chicago: Rand McNally.

Cosgrove, D. (1984) *Social Formation and Symbolic Landscape*. London: Croom Helm.

Cosgrove, D. (1988). The geometry of landscape: practical and speculative arts in sixteenth-century Venetian land territories. In D. Cosgrove and S. Daniels (eds), *The Iconography of Landscape*. Cambridge: Cambridge University Press, pp. 254–76.

Cosgrove, D. (1993) *The Palladian Landscape: Geographical Change and Its Cultural Representations in Sixteenth-Century Italy*. University Park: The Pennsylvania State University Press.

Cosgrove, D. (2004) Landscape and landschaft. Lecture delivered at the 'Spatial Turn in History' Symposium German Historical Institute, 19 February 2004. *Ghi Bulletin*, 35, 57–71.

Duncan, J. S. (1990) *The City as Text: the Politics of Landscape Interpretation in the Kandyan Kingdom*. Cambridge: Cambridge University Press.

Daniels, S. and Cosgrove, D. (1988) Introduction: iconography and landscape. In D. Cosgrove and S. Daniels (eds), *The Iconography of Landscape*. Cambridge: Cambridge University Press, pp. 1–10.

Demeritt, D. (1994) The nature of metaphors in cultural geography and environmental history. *Progress in Human Geography*, 18(3), 163–85.

Fontenelle, Bernard le Bovier de (1990[1686]) *Conversations on the Plurality of Worlds*. Berkeley: University of California Press.

Frodeman, R. (1995) Geological reasoning: geology as an interpretive and historical science. *Bulletin of the Geological Society of America*, 107: 960–68.

Foucault, M. (1979) [1975] *Discipline and Punish: The Birth of the Prison [Surveiller et punir]*. Harmondsworth: Penguin.

Gadol, J. (1969) *Leon Battista Alberti: Universal Man of the Early Renaissance*. Chicago: University of Chicago Press.

Germundsson, T. (2006) Regional cultural heritage versus national heritage in Scanias disputed national landscape. In K. R. Olwig and D. Lowenthal (eds), *The Nature of Cultural Heritage and the Culture of Natural Heritage. Northern Perspectives on a Contested Patrimony*. New York: Routledge, pp. 19–35.

Glacken, C. J. (1967) *Traces on the Rhodian Shore: Nature and Culture in Western Thought from Ancient Times to the End of the Eighteenth Century*. Berkeley: University of California Press.

Groening, G. (2007) The landscape must become the law – or should it? *Landscape Research*, 32(5), 595–612.

Häkli, J. (1998) Discourse in the production of political space: decolonizing the symbolism of provinces in Finland. *Political Geography*, 17(3), 331–63.

Head L. (2007) Cultural ecology: the problematic human and the terms of engagement, *Progress in Human Geography*, 31: 837–46.

Jackson, P. (1989) *Maps of Meaning: An Introduction to Cultural Geography*. London: Unwin Hyman.

Johnson, S. (1755[1968]). *A Dictionary of the English Language*. London: W. Strahan.

Kaplan, D. and Häkli, J. (eds) (2002) *Boundaries and Place: European Borderlands in Geographical Context*. Lanham, MD: Rowman & Littlefield.

Kirsch, S. and Mitchell, D. (2004) In the nature of things. *Antipode*, 36(5), 687–705.

Knottnerus, O. S. (1992) Moral economy behind the Dikes: class relations along the Frisian and German North Sea Coast during the Early Modern Age. *Tijdschrift voor sociale geschiedenis*, 18(2/3), 333–52.

Krauss, W. (2006) The natural and cultural landscape heritage of Northern Friesland. In D. Lowenthal and K. R. Olwig (eds), *The Nature of Cultural Heritage and the Cultural of Natural Heritage: Northern Perspectives on a Contested Patrimony*. London: Routledge, 37–50.

Latour, B. (1993) *We Have Never been Modern*. Cambridge, MA: Harvard University Press.

Latour, B. (1999) *Pandora' s Hope Essays on the Reality of Science Studies*. Cambridge, MA: Harvard University Press.

Livingstone, D. N. (1994) Climate's moral economy: science, race and place in post-Darwinian British and American geography. In N. Smith and A. Godlewska (eds), *Geography and Empire*. Oxford: Blackwell, pp. 132–54.

Lorimer, H. (2005) 'Cultural geography: the busyness of being 'more-than-representational', some recent work. *Progress in Human Geography*, 29(1), 1–12.

Lowenthal, D. (1961) Geography, experience, and imagination: towards a geographical epistemology. *Annals of the Assnociation of American Geographers*, 51, 241–60.

Lowenthal, D. (2000) *George Perkins Marsh: Prophet of Conservation*. Seattle: University of Seattle Press.

Marsh, G. P. (1965) [1864] *Man and Nature: Or, Physical Geography as Modified by Human Action*. Cambridge, MA: Belknap Press.

Mels, T. (2006). The Low Countries' connection: landscape and the struggle over representation around 1600. *Journal of Historical Geography*, 32, 712–30.

Merriam-Webster (2000) *Webster's Third New International Dictionary of the English Language, Unabridged*. Springfield, MA: Merriam-Webster.

Minca, C. (2007) Humboldt's compromise, or the forgotten geographies of landscape. *Progress in Human Geography*, 31(2), 179–93.

Mitchell, D. (1996) The lie of the land: migrant workers in the Californian landscape. Minneapolis: University of Minnesota Press.

Mitchell, D. (2007) Work, struggle, death, and geographies of justice: the transformation of landscape in and beyond California's Imperial Valley. *Landscape Research*, 32(5), 559–577.

Mitchell, W. J. T. (1994) Imperial landscape. In W. J. T. Mitchell (ed.), *Landscape and Power*. Chicago: The University of Chicago Press, pp. 5–34.

Newton, I. (1717) *Opticks: Or, a Treatise of the Reflections, Refractions, Inflexions and Colours of Light*. London: Royal Society.

Nossiter, A. (2006) Another conflict confronts altered social landscape of New Orleans. Times Digest: New York Times. New York: 3.

Olwig, K. R. (1984) *Nature's Ideological Landscape: A Literary and Geographic Perspective on its Development and Preservation on Denmark's Jutland Heath*. London: George Allen & Unwin.

Olwig, K. R. (1996a) Nature – mapping the 'ghostly' traces of a concept. In C. Earl, K. Mathewson and M. S. Kenzer, (eds), *Concepts in Human Geography*. Savage, MD: Rowman and Littlefield, pp. 63–96.

Olwig, K. R. (1996b) Recovering the substantive nature of landscape. *Annals of the Association of American Geographers*, 86(4), 630–53.

Olwig, K. R. (2002a) *Landscape, Nature and the Body Politic: From Britain's Renaissance to America's New World*. Madison: University of Wisconsin Press.

Olwig, K. R. (2002b) Landscape, place and the state of progress. In R. D. Sack (ed.), *Progress: Geographical Essays*. Baltimore: Johns Hopkins University Press, pp. 22–60.

Olwig, K. R. (2004) This is not a landscape: circulating reference and land shaping. In H. Sooväli, H. Palang, M. Antrop and G. Setten (eds), *European Rural Landscapes: Persistence and Change in a Globalising Environment*. Dordrecht: Kluwer, pp. 41–66.

Paasi, A. (2008) Finnish landscape as social practice: mapping identity and scale. In K. Olwig and M. Jones (eds), *Nordic Landscapes: Region and Belonging on the Northern Edge of Europe*. Minneapolis: University of Minnesota Press.

Ptolemy, C. (1991) *The Geography*. New York: Dover.

Sauer, C. O. (1925) The morphology of landscape. *University of California Publications in Geography*, 2(2), 19–53.

Sauer, C. O. (1969) *Land and Life: A Selection from the Writings of Carl Ortwin Sauer*. Edited by John Leighly. Berkeley: University of California Press.

Smith, N. (1989) Geography as museum: private history and conservative idealism in *The Nature of Geography*. In S. D. Brunn and J. N. Entrikin (eds), *Reflections on Richard Hartshorne's The Nature of Geography*. Washington, DC: AAG, pp. 89–120.

Thompson, E. P. (1993) *Customs in Common*. London: Penguin.

Tuan, Y.-F. (1974) *Topophilia: A Study of Environmental Perception, Attitudes, and Values.* Englewood Cliffs, NJ: Prentice-Hall.

Walker, P. A. (2005) Political ecology: where is the ecology? *Progress in Human Geography* 29: 73–82.

Wylie, J. (2007) *Landscape.* London: Routledge.

Chapter 16

Ecological Modernisation and Industrial Transformation

Arthur P. J. Mol and Gert Spaargaren

Introduction

The emergence and maturation of the idea of ecological modernisation in the 1980s should be understood in reaction to the demodernisation ideas of the environmental movement in the 1970s on the one hand, and the not very successful curative approaches of environmental state authorities in Europe on the other. In the early eighties, the environmental movement in northwestern Europe was facing an internal debate on the effectiveness of its strategy and the adequacy of its ideology. In Germany the political party The Greens was the platform of a debate between 'realos' (realists) and 'fundis' (fundamentalists), while in The Netherlands the debate centred on the radicalisation of the anti-nuclear and squatter movement in the early 1980s. At the same time, environmental state authorities in Europe were facing failures in coping with the environmental crisis, basically because their end-of-pipe, curative and command-and-control strategies were widely perceived as unsuccessful in making serious advances in combating the environmental crisis, while at the same time the political climate turned increasingly towards deregulation and liberalisation, rather than stronger state involvement. In this context, and with ideas of sustainability rising on the political agendas after the 1987 Brundtland report, ecological modernisation became increasingly used as the academic equivalent of the more popular notion of sustainable development (cf. Mol and Spaargaren, 1992).

Historical Development of Ecological Modernisation Ideas

The notion of ecological modernisation (hereafter EM) was first proposed in Germany by the political scientist and politician Martin Jänicke at the end of the 1970s in the Berlin city parliament. As a social theory, EM developed from the mid-1980s onwards primarily in a small group of Western European countries, most notably Germany, the Netherlands, the United Kingdom and somewhat later the Scandinavian countries. It developed especially within the disciplines of sociology, political sciences and geography, but later found application in other social and interdisciplinary sciences. The social scientists Joseph Huber, Martin Jänicke, Volker von Prittwitz, Udo Simonis and Klaus Zimmermann (Germany), Gert Spaargaren, Maarten Hajer and Arthur P.J. Mol (the Netherlands), Albert Weale, Maurie Cohen, Joseph Murphy, Andrew Gouldson and David Gibbs (United Kingdom) and Michael Skou Anderson, Pekka Jokinen, Lennart Lundqvist, (Denmark, Finland and Sweden, respectively), among others, have made substantial European contributions to the early or later stages of the development of the EM theory. In addition, various empirical studies using this theoretical framework have been carried out in various other countries and regions.

Throughout the relatively short time of its existence, there has been considerable diversity and internal debate among the various contributors to the EM theory. These differences spring from national backgrounds (with authors referring to various empirical references and interpretations, as I will illustrate below) and theoretical roots,[1] but also chronology. Though an extensive analysis and overview of ecological modernisation literature up to now is outside the scope of the present contribution, I believe it makes sense to distinguish at least three stages in the development and maturation of the EM theory. The first contributions, for instance those by Joseph Huber (1982; 1985; 1991), were characterised by: a heavy emphasis on the role of technological innovations in bringing about environmental reforms, especially in the sphere of industrial production; a rather critical attitude towards the (bureaucratic and inefficient) state, as found in the early writings of Martin Jänicke (1986); a very optimistic, perhaps naïve, attitude towards market actors and market dynamics in environmental reforms (later on glorified by neo-liberal scholars); a system-theoretical perspective with a relatively underdeveloped concept of human agency and social struggle; and a concentration on national or sub-national studies. Some of the more critical remarks on the EM theory still refer to these initial contributions.

Building upon several of these limitations, ecological modernisation studies in the second period, from the late 1980s onward, showed less emphasis on and a less deterministic view of technological innovations as the motor behind ecological modernisation. These contributions gave evidence of a more balanced view of state and market dynamics in ecological transformation processes, as illustrated by the work of Albert Weale (1992) and the later Martin Jänicke (1991; 1993). During this phase, the institutional and cultural dynamics of ecological modernisation were given more weight, as well as the role of human agency in environment-induced social transformations. The emphasis remained on national or comparative studies of industrial production in OECD countries. Critical remarks on the concept of ecological modernisation in this period – articulated by scholars both inside and outside the ecological modernisation tradition (cf. Mol, 1995; Blowers, 1997; Blühdorn, 2000) – focused on its Eurocentrism, since the EM theory had been developed primarily in the context of a small group of Western European countries. In addition, comments pointed out its limited definition of the environment (Spaargaren and Mol, 1992), its overly optimistic expectations of environmental reforms in social practices, institutional developments and environmental debates, and its disregard for lifestyles and consumption practices.

The third period, from the mid-1990s onwards, encompasses innovations in three fields. First, studies on industrial production were increasingly complemented by work done on ecological transformations related to consumption processes (cf. Spaargaren and van Vliet, 2000; Spaargaren, 2003). This of course resembles the wider attention in the social sciences to consumption, often referred to as the 'consumerist turn'. Second, the Eurocentrism criticism of the second period resulted in various national studies on environmental reforms in non-EU countries (newly industrialising countries in especially Southeast and East Asia and the transitional economies in Central and Eastern Europe, but also, for instance, the USA, Canada, Australia and Japan), leading to mixed conclusions on the relevance of this theoretical framework for understanding the processes of environmental reform. Finally, on the wave of international relations and globalisation studies, growing attention was paid to the global dynamics of ecological modernisation (e.g. some contributions in Mol 2001; Sonnenfeld and Mol, 2002; Spaargaren et al., 2006). In the last section we will turn back to the periodisation of EM studies, by asking whether a new fourth phase has started following some of the recent debates around globalisation, materiality and informationalisation.

In spite of national, temporal and theoretical differences, all these contributions can still be gathered together under the umbrella of the EM theory, not only because they identify themselves as such, but also because they have in common: (i) that environmental deterioration is conceived of as a challenge for sociotechnical and economic reform, rather than the inevitable consequence of the current institutional structure; (ii) the emphasis on the actuality and necessity of transformation of modern institutions in the fields of science and technology, the nation-state and global politics, and the (global) market, to achieve environmental reform, and (iii) a position in the academic field that is distinct from the more or less strict neo-Marxists, as well as from counter-productivity and post-modernist analyses. Against this shared background, I will outline the central theoretical notion that lies behind the variety of contributions to EM theory, as well as its core features.[2]

Fundamentals of Ecological Modernisation

The basic idea of EM is that at the end of the second millennium modern societies witness a centripetal movement of ecological interests, ideas and considerations in their institutional design. This development crystallises in a constant ecological restructuring of modernity. Ecological restructuring refers to the ecology-inspired and environment-induced processes of transformation and reform in the central institutions and social practices of modern society.

Within the so-called 'EM theory' this ecological restructuring is conceptualised at an analytical level as the growing autonomy, independence or differentiation of an ecological rationality vis-à-vis other rationalities (cf. Mol, 1995; Spaargaren, 2003).[3] In the domain of states, policies and politics the emergence of an ecological rationality emerged already in the seventies and early eighties, and 'materialised' or 'institutionalised' in different forms. The construction of governmental organisations and departments dealing with environmental issues dates from that era. Equally, environmental (framework) laws, environmental impact assessment systems and green political parties date back to that period. The same is true in the domain of ideology and the life world. A distinct 'green' ideology – as manifested by, for instance, environmental NGOs, environmental value systems and environmental periodicals – started to emerge in the 1970s. Only in the 1980s, however, this 'green' ideology assumed an independent status and could no longer be interpreted in terms of the old political ideologies of socialism, liberalism and conservatism, as argued by among others Paehlke (1989) and Giddens (1994).

However, the crucial transformation that makes the notion of the growing auton-omy of an ecological rationality especially relevant, is of more recent origin. After an ecological rationality has become relatively independent from the political and socio-ideological rationalities (in the 1970s and 1980s), this process of growing independence began to extend to the economic domain in the 1990s. And since, according to most scholars, this growing independence of the ecological rationality from its economic counterpart is crucial to 'the ecological question', this last step is the decisive one. It means that economic processes of production and consumption are increasingly analysed and judged, as well as designed and organised from both an economic *and* an ecological point of view. Some profound institutional changes in the economic domain of production and consumption have become discernable in the 1990s. Among these changes are the widespread emergence of environmental management systems, environmental accountancy and environmental reporting in companies; the introduction of economic valuation of environmental goods via the introduction of eco-taxes, among other things; the emergence of environment-inspired liability and insurance arrangements; the increasing importance attached to environmental goals such as natural resource saving and recycling among public and private utility enterprises; and the articulation of environmental considerations in economic supply and demand, for instance by eco-labels and environmental cer-tification schemes. Within ecological modernisation ideas these transformations are analysed as *institutional* changes, indicating their semi-permanent character. Although the process of ecology-induced transformation should not be interpreted as linear, evolutionary and irreversible, as was common in the modernisation theo-ries in the 1950s and 1960s, these changes have some permanency and would be difficult to reverse.

Some environmental sociologists and commentators in the environmental reform tradition go even one step further. They suggest that environmental considerations and interests not only activate institutional transformations in contemporary indus-trial societies, but even evolve into a new Grand Narrative. The traditional Grand Emancipatory Narratives of modernity (e.g., the emancipation of labour, the dissolu-tion of poverty) place us in history as human beings who have a definite past and a more or less predictable future. Now these traditional narratives have ceased to perform as overarching 'storylines', some believe the ecology (or, alternatively, sus-tainability) will emerge as the new sensitising concept through which modern society orientates itself in its future development. Environment/sustainability – or rather environmental/sustainability considerations and interests – is then the leading notion, the structuring principle, the *leitmotiv* for a new round of institutional transformations in what can be labelled (in a variation on Hobsbawm) the 'Age of Environment'. That still needs to be proven.

Ecological Modernisation as Environmental Reform

Most EM studies focus on actual environmental reform dynamics in specific social practices and institutions. One of the constantly returning debates with respect to EM relates to the empirical evidence for environmental reform. It has been well established by political scientists, geographers and others that public and private institutions around the world have become more openly and positively oriented towards the natural environment, at least as expressed in treaties, policies, plans, organisations and financial resources. At the same time, environmental economists and others have argued and illustrated convincingly that economic institutions have partly included environmental considerations, for instance in markets, pricing, investments, and research and development. Sociologists have illustrated how civil society transformed through environmental challenges, as new social movements emerged, households reorganised, processes of distinction changed, and norms, values and discourses have been reframed. But what have been the enduring consequences of such socio-institutional accomplishments for the physical environment (less pollution, more nature reserves, less energy use)? To what extent and how has the adoption of environmental perspectives and rationalities in economic, political and socio-cultural institutions resulted in material improvements? And how do we measure these physical improvements: per capita, per GDP, per country/region, per time unit?

In the majority of the EM studies, ecological modernisation is interpreted as a social theory, providing us with ideas and heuristics on understanding, interpreting and advancing environment-induced social change. This means that social transformations that are induced, motivated or triggered by environmental considerations are the object of ecological modernisation reflections, and not so much actual physical improvements. But one can of course not completely delink the former from the latter: eco-modernised transformations in economy, politics and society must have some positive environmental consequences. An EM perspective on environmental reform can be categorised in five themes. Empirical studies in the ecological modernisation tradition often address more than one theme, also because these five themes logical hang together in a common perspective. Related to these themes are numerous other studies and approaches on processes and dynamics of environmental reform, which do not coin themselves as ecological modernisation, but develop perspectives and empirical evidence very much in line with it (see next section).

First there are studies on the new role of science and technology in environmental reform. First, science and technology are no longer only analysed and judged for their role in causing environmental problems (so dominant in the 1970s and early 1980s), but also valued for their actual and potential role in bringing about environmental reforms and preventing environmental crises. Second, environmental reforms via traditional curative and repair technologies are replaced by more preventive socio-technological approaches that incorporate environmental considerations from the design stage of technological and organisational innovations. Finally, the growing uncertainties with regard to scientific and expert knowledge and complex technological systems in bringing about environmental reforms do not lead to a denigration of science and technology in environmental reform, but rather in new environmental and institutional arrangements. Exemplary EM studies in this theme are those of Joseph Huber, both his earlier works (1985; 1991) as well as his more recent studies (2000; 2004). Related studies are to be found in the tradition of cleaner production, of industrial ecology, studies on the environmental reform of large socio-technical or network-bound systems, and to some extent so-called transition studies. In these fields a primarily technological emphasis dominates in environmental reform, although these perspectives are widening towards economic and political dimensions of socio-environmental reform, as can be witnessed from the more recent contributions to among others the *Journal of Cleaner Production* and the *Journal of Industrial Ecology*.

A second theme are studies focused on the increasing importance and involvement of economic and market dynamics, institutions and agents in environmental reform. Producers, customers, consumers, credit institutions, insurance companies, utility sectors, and business associations, to name but a few, increasingly turn into social carriers of ecological restructuring, innovation and reform (in addition to, and not so much instead of, state agencies and new social movements; cf. Mol, 1995; Mol and Spaargaren, 2000). This goes together with a focus on changing state–market relations in environmental reform, and on a growing involvement of economic and market institutions in articulating environmental considerations via monetary values and prices, demand, products and services, and the like. Exemplary empirical studies are those of Michael Skou Andersen (1994) on green taxes in Europe, those on specific categories of industries (e.g., Revell, 2007, on SMEs) and those by environmental or ecological economists, for instance in the journal *Ecological Economics*. Studies in the tradition of the 'environmental Kuznets curve', and the fierce debates around that concept, can be interpreted as empirical cases on this theme. But also various (non-economic) empirical studies on the role of shareholders in green investments, the emergence of green accounting and company environmental reporting, and the role of insurance companies during the Kyoto protocol negotiations are typical studies that resonate ecological modernisation ideas on the growing importance of economics and markets in environmental reform.

A third theme in ecological modernisation relates to the changing role, position and performance of the 'environmental' state (often referred to as political modernisation in Europe [cf. Jänicke, 1993; Tatenhove et al., 2000], or regulatory reinvention in the US [cf. Eisner, 2004]). The traditional central role of the nation-state in environmental reform is shifting, leading to new governance arrangements and new political spaces. First, there is a trend towards more decentralised, flexible and consensual styles of national governance, at the expense of top-down hierarchical command-and-control regulation. Second, there is a larger involvement of non-state actors and non-state arrangements in environmental governance, taking over conventional tasks of the nation-state and conventional politics (e.g., privatisation, public–private partnerships, conflict resolution by business-environmental NGO coalitions without state interference, and the emergence of subpolitics). Finally, supra-national and global environmental institutions and governance arrangements to some extent undermine the conventional role of the sovereign nation-state or national arrangements in environmental policy and politics. Under this theme one finds especially contributions by political scientists and geographers, among others in the journals *Environmental Politics*, *Geoforum*, *Environment and Planning C* and *Journal of Environmental Policy and Planning*. Typical studies are, for instance, Hills (2005) on Hong Kong, Jänicke and Weidner (1997) with a comparative study on national environmental policies, and Jokinen (2000) on the EU. While ecological and political modernisation perspectives have arguably been very timely in investigating shifts in environmental governance, by now the same themes can be found is a very diverse and voluminous body of literature on environmental governance and environmental partnerships.

Fourth, the modification of the position, role and ideology of social movements (vis-à-vis the 1970s and 1980s) in the process of ecological transformation emerges as a theme in ecological modernisation. Instead of positioning themselves on the periphery or even outside the central decision-making institutions on the basis of de-modernisation ideologies and limited economic and political power, environmental movements seem increasingly involved in decision-making processes within the political and, to a lesser extent, economic arenas. Legitimacy, accountability, transparency and participation are the new principles and values that provide social movements and civil society the resources to gain a more powerful position in environmental reform processes. Within the environmental movement this transformation goes together with a bipolar or dualistic strategies of cooperation and conflict, and internal debates on the tensions that are a by-product of this duality. Typical studies in the ecological modernisation traditions are David Sonnenfeld's (1996) study on the environmental movement in South-East Asia and Leonardus Rinkevicius (2000) on the environmental movement in Lithuania, while increasingly various strands of social movement studies resembles ideas of ecological modernisation in discussing changing tactics, strategies, alliances and discourses of the environmental movement (e.g., in the works of John Dryzek et al. 1997; 2003).

And finally, EM studies concentrate on changing discursive practices and frames, and the emergence of new ideologies in political and societal arenas. Neither the fundamental counter-positioning of economic and environmental interests nor a total disregard for the importance of environmental considerations are accepted any longer as legitimate positions. Intergenerational solidarity in the interest of preserving the sustenance base seems to have emerged as the undisputed core and widely shared principle, although differences remain on interpretations and translations into practices and strategies. The classical ecological modernisation study in this theme is the dissertation of Maarten Hajer (1995), while many have followed with case studies and further explorations of the ecological modernisation discourse, such as recently Davidson and Mackendrick (2004) and Keil and Desfor (2003). Numerous studies focusing on the sustainable development discourse resemble ecological modernisation ideas and frames.

Industrial Transformation and Beyond

EM perspectives have been applied in studying environmental reforms in a variety of sectors and geographies, focusing of different indicators and processes of environment-induced social change. Usually, in studying empirical socio-ecological transformations various themes (as mentioned above) come together. Arguably the three most studied transformations within EM are industrial reforms (of individual companies, sectors, industrial networks, industrial parks and regions, industrial products or chains, self regulations, certifications, technological change, R&D, reporting and auditing), environmental policies (policy integration, new instruments, policy styles, prevention, partnerships and alliances, vertical relations between local-national-international policy systems) and utility provisioning (greening of network-bound system such as those related to water, energy, waste and transport; consumer and citizen involvement in these systems, socio-technological change, new management and organisational styles, differentiation and monitoring, pricing, demand side management).

A significant number of concepts have emerged that are used to study empirical processes that are not too far beyond the research agenda of EM. Industrial transformation is arguably the most overarching concept that brings together empirical studies in these three – and various other – fields of environmental reform. The International Human Dimensions Program on Industrial Transformation started originally with a more narrow focus on industrial processes and products, but has considerably widened its scope in the new millennium. To a significant extent, industrial transformation builds upon and incorporates older ideas and research agendas of cleaner production, but includes higher levels of aggregation and a wider variety of non-industrial sectors and activities. In that sense industrial ecology seems to be close to industrial transformation, be it that it still has – similar to cleaner production – a primarily technical/natural science outlook. Innovations studies and transition models are less dominated by the natural and technical sciences, and share with ecological modernisation a primarily social science approach to understanding environmental reforms.

But the boundaries between these research traditions are increasingly blurred. Cleaner production and industrial ecology studies now incorporate social science perspectives and analyses in their domain (see for instance the contents of the journals that carry these names), and all traditions range from small-scale or micro-level analyses of environmental reform to large-scale macro level analyses of socio-technical systems, even beyond the nation-state container. Still, most studies in these traditions resonate starting points and characteristics of ecological modernisation: a focus on environmental reform through (socio, technical, economic, cultural) innovations, an inclusion of science and technology in a social science analysis, a balanced perspective of the prevailing economic order.

Ecological Modernisation and Its Critics

From various (theoretical) perspectives and from the first publications onwards, the growing popularity of ecological modernisation studies and ideas has met debate and criticism.[4] Coming from subdisciplines that had been preoccupied with explaining the continuity of environmental crises and deterioration, such a move to environmental reform perspectives cannot but meet (fierce) debate. The debates and criticism on EM have been summarised and reviewed in a number of publications.[5] Here I want to categorise these various critiques and debates in three baskets.

First, several objections have been raised during the short history of EM, which have been incorporated in more recent versions of the theory. While these objections made sense in referring to the first or second period of EM studies (cf. Sonnenfeld and Mol, 2002), within the third generation approaches they are no longer adequate. This is valid, for instance, regarding criticism on technological determinism in EM , on the productivist orientation and the neglect of consumption and the consumer, on the lack of power and inequality in ecological modernisation studies and on its Eurocentricity. Not withstanding what I would call the 'outdated' character of these critiques at the turn of the millennium, several scholars continue repeating them up until recently (e.g., Carolan [2004] on the productivist orientation; Murphy and Bendell [1997] on technological determinism; Gibbs [2006] on missing power relations).

Second, there are a number of critiques on EM perspectives that find their origin in radically different paradigms and approaches. Neo-Marxist criticism by Schnaiberg et al. (2002), Pepper (1999) and Blowers (1997) emphasise consistently the fundamental continuity of a capitalist order that does not allow any environmental reform beyond window dressing. Deep ecology inspired scholars argue against the reformist agenda of EM, as it opts for a light green reform agenda, in stead of a deep green fundamental and radical change of the modern order, sometimes even towards post-modernity (e.g., Toke, 2001). Human-ecologists, inspired by neo-Malthusianism and sometimes in remarkable alliances with neo-Marxists (cf. York and Rosa, 2003), blame an EM perspective for the neglect of quantities, not in the last place population growth and ever growing consumption quantities. And post-modernists, such as Ingolfur Blühdorn (2000), argue that ecology is just the last modernist storyline in a post-modern world that does no longer allow such frames in making sense of contemporary developments. Consequently, EM perspectives are blamed to be overly optimistic/naïve, not showing environmental reforms and/or ill-equipped for 'real' radical and structural changes of the modern order towards sustainability. It is not so much that these objections are completely incorrect. From their starting points and the basic premises of these schools of thought, the points raised against ecological modernisation are internally logic, consistent and coherent. But their claims are too narrow, limited and one-sided, when they claim that no environmental reform can be witnessed and refuse to interpret anything new under the sun as long as we continue to have capitalism, population growth or modernity. While EM scholars would not deny that in various locations, practices and institutions environmental deterioration is still continuing and even prevailing, they object to the conclusion of these critics that no environment-induced transformation can be identified in contemporary modern societies.

Third and finally, there is a category of comments and debates, which are less easy either incorporated or put aside if we want to analyse and understand environmental reform in late modern society. These issues have to do with the nation-state or national society centredness of EM, the strong separation between the natural/physical and the social in ecological modernisation, and the continuing conceptual differentiation in state, market and civil society actors and institutions. Here it is especially the changing character of modern society – especially through processes of globalisation – that makes that new, early twenty-first century environmental reform dynamics are not always easily fitting ecological modernisation conceptualisations of the 1990s. This is not too dissimilar to the fact that the environmental reform dynamics of the 1990s did not fully fit the conceptualisations of the 1970s environmental reform studies.

New horizons of Ecological Modernisation Debates

It is particular the latter category of debates and criticism that is challenging current EM perspectives. In this last section we want to explore four innovations and challenges EM studies are facing at the moment, following among others these debates, to turn finally to the question whether we are entering a fourth phase of ecological modernisation, or alternative we are in need of a new, fundamentally different perspective on environmental reform.

With the wider attention in the social sciences to the role of citizen-consumers in social development and a stronger focus on consumption in sustainability studies, a growing interest among EM scholars can be witnessed in how citizen-consumers (can) contribute to environmental reforms. While there is wide consensus that the conventional attitude-behaviour models are no longer adequate, various innovative citizen-consumer oriented approaches and conceptualisations are at the moment under construction: Micheletti's political consumerism, Spaargaren's social practices model, Anheier's global civil society studies, etc. With increasing attention to civil society and citizen-consumers in environmental reform, the assessment of their role is also severely under debate, ranging between captive consumers, responsible citizens, and powerful change agents. The main debate seems to be on how structures and systems relate to capable actors in environmental reforms, to what extent are knowledgeable and capable citizen-consumers essential in ecological modernisation (can reform not take place behind the back of citizen-consumers?), and is a citizen-consumer orientation not blaming the wrong polluter?

The second challenge for EM perspective relates to globalisation. It has become common knowledge that globalisation is not just an additional layer that needs to be included in our analyses to understand environmental deterioration or reform. In contrast, globalisation processes are fundamentally challenging our notions, concepts and models of social change, and that is not different for ecological modernisation than for any other field. While EM scholars have worked with and upon globalisation, the question whether the fundamental changes brought about by globalisation can be incorporated into ecological modernisation ideas, and if so how, remains yet unanswered. Of course, geography as a discipline is especially equipped to contribute to this relation between EM and globalisation, among others, through its central focus on space. Beck's work on second modernity, world risk society and cosmopolitanism; Urry's sociology of network and flows; Held's elaborations on cosmopolitan democracy; and Anheier's studies of global civil society are framings that I see as useful building stones for assess how environmental reform models should be interpreted in the global age and to what extent we could still label them EM. Our recent volume on *Governing Environmental Flows* (Spaargaren 2006) is no more than a first attempt to redefine EM in a era of globalisation.

Related to globalisation, and following the work of Latour, Callon, the sociology of science and technology, and debates on the materiality of the social, new debates on the relation between the physical objects and social relations can be expected. EM studies have traditionally taken a rather realist perspective on materiality, and have entered into critical debate with strong social constructivists and developed a sceptical attitude towards lifting the borders between the social and the material. With a general turn in the social sciences to complexity theory, hybridity, and materiality one can expect new debates emerging or the relation between the social and the material, and new concepts being developed that cope with that, such as actants. It is also here that geography, with its social and physical sub-disciplines, can be expected to make a contribution to these new horizons of ecological modernisation.

Finally, the information revolution and the emergence of the knowledge society has been poorly understood and studied in the environmental social sciences, including the EM literature. With the growing importance of information flows, the Internet, transparency and accountability, and monitoring capacity the horizons of environmental reform are changing. New lines of inquiry in the informationalisation of environmental reform – whether that is in civil society activism, multinational corporations or the transnational state system – will be on the future agenda, together with critical inquiries into the digital divide, the validation of information flows, and the new power relations of media conglomerates.

One of the often-asked questions is, whether these new debates and trajectories can and should be considered still as part of the EM project, or rather as a divergence from it, towards new conceptualisations. The answer depends very much on where one wants to put the core of EM ideas. Is it within the fundamental ideas of an ecological rationality being developed, introduced and institutionalised in social practices and institutional developments, as the Wageningen School of EM has always emphasised? Is it within more substantial notions and models of environmental change in politics, technology or economics, as emphasised by Martin Jänicke and Joseph Huber, among others? Or is it with the physical changes in terms of environmental indicators, industrial ecology, factor 4 or factor 10, as emphasised by some of the EM critics? The fact that this debate is still vivid is perhaps the best proof that EM remains a relevant category in the various social science disciplines studying the environment.

NOTES

1. These theoretical traditions range from system theoretical analyses by, for instance, Joseph Huber (1985, 1991), more institutional analysis by, for instance, Mikael Skou Andersen (1994) and Arthur Mol (1995), up to discourse analyses by, for instance, Maarten Hajer (1995) and, to a lesser extent, Albert Weale (1992).
2. A full historical analysis and overview of developments in EM literature up till now is beyond the scope of this chapter. See for such overviews the volumes edited by Spaargaren et al. (2000) and Mol and Sonnenfeld (2000), and special issues of the Journals *Environmental Politics* (2000, no.4), *Geoforum* (2000, no. 31), *Journal of Environmental Policy and Planning* (2000, no. 4), and *American Behavioural Scientist* (2002, no.9). The more American-oriented journals *Society and Natural Resources* and *Organization & Environment* contain regularly ecological modernisation studies and debates.
3. See Dryzek (1987) for an early development of the idea of an emerging ecological rationality, although by then not (yet) in a framework of ecological modernisation.
4. Cf. Mol and Spaargaren (2000). This growing importance of EM perspectives is even acknowledged by its critics, who often do not challenge the analytical and descriptive qualities of this theory for West European societies but rather its normative undertones. While contemporary environmental policies and reforms may indeed be 'based' on or reflect ideas of EM, they should be criticized for that, as such attempts to solve the environmental crisis suffer from various problems, according to these critics.
5. For evaluations and critiques on the idea of EM as the common denominator of environmental reform processes starting to emerge in the 1990s, see for instance: Hannigan (2006), Christoff (1996), Blowers (1997), Gouldson and Murphy (1997), Tatenhove et al. (2000), Blühdorn (2000), Buttel (2000), Mol and Spaargaren (2000), Pepper (1999), Schnaiberg et al. (2002), Gibbs (2006).

BIBLIOGRAPHY

Andersen, M. S. (1994) *Governance by Green Taxes. Making Pollution Prevention Pay.* Manchester: Manchester University Press.

Blowers, A. (1997) Environmental policy: ecological modernization and the risk society? *Urban Studies*, 34(5–6), 845–71.

Blühdorn, I. (2000) Ecological modernisation and post-ecologist politics. In G. Spaargaren, A. P. J. Mol and F. Buttel (eds), *Environment and Global Modernity*, London: Sage.

Buttel, F. H. (2000) Ecological modernization as social theory. *Geoforum*, 31(1), 57–65.

Carolan, M. (2004) Ecological modernization: what about consumption? *Society and Natural Resources*, 17(3), 247–60.

Christoff, P. (1996) Ecological modernisation, ecological modernities. *Environmental Politics*, 5(3), 476–500.

Davidson, D. J. and Mackendrick, N. A. (2004) All dressed up with nowhere to go: the discourse of ecological modernization in Alberta, Canada. *Canadian Review of Sociology and Anthropology*, 41(1), 47–65.

Dryzek, J. S. (1987). *Rational Ecology: Environment and Political Economy*. Oxford and New York: Blackwell.

Dryzek, J. S. (1997) *The Politics of the Earth: Environmental Discourses*, Oxford: Oxford University Press.

Dryzek, J. S., Downes, D., Hunold, Ch., Schlosberg, D. and Hernes, H.-K. (2003) *Green States and Social Movements. Environmentalism in the United States, United Kingdom, Germany, and Norway*. Oxford: OUP.

Eisner, M. A. (2004) Corporate environmentalism, regulatory reform, and industry self-regulation; toward genuine regulatory reinvention in the United States. *Governance: An International Journal of Policy, Administration, and Institutions*, 17(2), 145–67.

Gibbs, D. (2006) Prospects for an environmental economic geography: linking ecological modernization and regulationist approaches. *Economic Geography*, 82(2), 193–216.

Giddens, A. (1994) *Beyond Left and Right: The Future of Radical Politics*. Cambridge: Polity Press.

Gouldson, A. and Murphy, J. (1997) Ecological modernization: economic restructuring and the environment. *The Political Quarterly*, 68(5), 74–86.

Hajer, M. A. (1995) *The Politics of Environmental Discourse: Ecological Modernisation and the Policy Process*. Oxford: Clarendon Press.

Hannigan, J. (2006) *Environmental Sociology*, 2nd edn. London and New York: Routledge.

Hills, P. (2005) Environmental reform, ecological modernization and the policy process in hong kong: an exploratory study of stakeholder perspectives. *Journal of Environmental Planning and Management*, 48(2), 209–40.

Huber, J. (1982) *Die verlorene Unschuld der Ökologie. Neue Technologien und superindustrielle Entwicklung*. Frankfurt: Fisher.

Huber, J. (1985) *Die Regenbogengesellschaft. Ökologie und Sozialpolitik*. Frankfurt am Main: Fisher Verlag.

Huber, J. (1991) *Unternehmen Umwelt. Weichenstellungen für eine ökologische Marktwirtschaft*. Frankfurt am Main: Fisher.

Huber, J. (2000) Towards industrial ecology: sustainable development as a concept of ecological modernization. *Journal of Environmental Policy and Planning*, 2, 269–85.

Huber, J. (2004) *New Technologies and Environmental Innovation*. Cheltenham: Edward Elgar.

Jänicke, M. (1986) *Staatsversagen. Die Ohnmacht der Politik in die Industriegesellschaft*. München: Piper.

Jänicke, M. (1993) Über ökologische und politieke Modernisierungen. *Zeitschrift für Umweltpolitik und Umweltrecht*, 2, 159–75.

Jänicke, M. and Weidner, H. (eds) (1997) *National Environmental Policies: A Comparative Study of Capacity Building*. Berlin and New York: Springer.

Jokinen, P. (2000) Europeanisation and ecological modernisation: agri-environmental policy and practices in Finland. *Environmental Politics*, 9(1), 138–70.

Keil, R. and Desfor, G. (2003) Ecological modernisation in Los Angeles and Toronto. *Local Environment*, 8(1), 27–44.

Mol, A. P. J. (1995) *The Refinement of Production: Ecological Modernisation Theory and the Chemical Industry*. Utrecht: Jan van Arkel/International Books.

Mol, A. P. J. (2001) *Globalization and Environmental Reform*. Cambridge, MA: MIT.

Mol, A. P. J. and Sonnenfeld, D. S. (eds) (2000) *Ecological Modernisation Around the World: Perspectives and Critical Debates*. London: Frank Cass.

Mol, A. P. J. and Spaargaren, G. (2000) Ecological modernization theory in debate: a review. *Environmental Politics*, 9(1), 17–49.

Murphy, D. F. and Bendell, J. (1997) *In the Company of Partners: Business, Environmental Groups and Sustainable Development Post-Rio*. Bristol: The Policy Press.

Paehlke, R. C. (1989) *Environmentalism and the Future of Progressive Politics*. New Haven and London: Yale University Press.

Pepper, D. (1999) Ecological modernisation or the 'ideal model' of sustainable development? Questions prompted at Europe's periphery. *Environmental Politics*, 8(4), 1–34.

Revell, A. (2006) The ecological modernisation of SMEs in the UK's construction industry. *Geoforum* (forthcoming).

Revell, A. (2007) The ecological modernization of SMEs in the UK's construction industry, *Geoforum*, 38, 1, pp. 114–126.

Rinkevicius, L. (2000) Ecological modernisation as cultural politics: transformations of civic environmental activism in Lithuania. *Environmental Politics*, 9(1), 171–202.

Schnaiberg, A., Weinberg, A. S. and Pellow, D. N. (2002) The treadmill of production and the environmental state. In A. P. J. Mol and F. H. Buttel (eds), *The Environmental State Under Pressure*. London: JAI/Elsevier, pp. 15–32.

Sonnenfeld, D. (1996), *Greening the Tiger? Social Movements' Influence on the Adoption of Environmental Technologies in the Pulp and Paper Industries of Australia, Indonesia and Thailand*. PhD Thesis, Santa Cruz: University of California (Sociology).

Sonnenfeld, D. A. and Mol, A. P. J. (eds) (2002). Symposium on 'Globalization, Governance, and the Environment'. *American Behavioral Scientist*, 45(9), 1311–1461.

Spaargaren, G. (2003). Sustainable consumption: a theoretical and environmental policy perspective. *Society and Natural Resources*, 16(8), 687–702.

Spaargaren, G. and Mol, A. P. J. (1992) Sociology, Environment and Modernity. Ecological modernization as a theory of social change, *Society and Natural Resources*, 5, 5, pp. 323–345.

Spaargaren, G. and van Vliet, B. (2000) Lifestyles, consumption and the environment: the ecological modernisation of domestic consumption. *Environmental Politics*, 9(1), 50–76.

Spaargaren, G., Mol, A. P. J. and Buttel, F. H. (eds) (2006) *Governing Environmental Flows. Global Challenges for Social Theory*. Cambridge, MA: MIT.

Tatenhove, J. van Arts, B. and Leroy, P. (eds) (2000) *Political Modernisation and the Environment: The Renewal of Policy Arrangements*. Dordrecht: Kluwer.

Toke, D. (2001) Ecological modernisation: a reformist review. *New Political Economy*, 6(2), 279–91.

Weale, A. (1992). *The New Politics of Pollution*. Manchester: Manchester University Press.

York, R. and Rosa, E. A. (2003) Key challenges to ecological modernization theory. *Organization and Environment*, 16(3), 273–87.

Chapter 17

Marxist Political Economy and the Environment

George Henderson

Introduction

It is both easy and difficult to place Marxist thought (and politics) in relation to the environment. Easy because from a Marxist perspective, we are never removed from the environment; difficult because Marxism is not an environmentalism in any traditional sense of the word. The ambiguity is doubled because environmentally oriented social movements on the Left have succeeded in politicising environmental problems to an unprecedented degree, and even though an active engagement with Marxist thought is maintained by many participants, especially outside the United States, it is not always clear how much their environmentalism is able to draw from Marxism (e.g., Benton, 1996a; Panitch and Leys, 2006; see Goldman [2005] for a study of how environmental politics are by no means reducible to the Left). It is these ambiguities that ground the present chapter. My purpose will be to convey an appreciation of how environmental problems have been approached by Marxists – many of which are to be found in geography – and, conversely, how explicit attention to these problems presents an opportunity to reconstruct Marxism. I will begin by introducing Marx's thoughts on what he called our species being, an important concept strongly linked to his notion of human labour and its mixing with non-human matter. Then follows a discussion of how species being is joined to capitalism and why capitalism gives rise to environmental questions and issues. The latter half of the chapter offers an in-depth look at three reconstructions of Marxist theory that have gained particular currency in the field of geography: the notions of a second contradiction of capitalism, the production of nature, and the plane of immanence.

Species Being

Let us begin with a provocation planted by Marxist geographer David Harvey, in one of his classic works, *Justice, Nature, and the Geography of Difference*: 'There's nothing unnatural about New York City'. Now that jars. Its common sense seems blinded to the common sense that worries about the impact of human presence on the earth, the violation of 'natural laws,' and the multiplication of environmental risks and hazards.[1] In order to make sense of Harvey's remark, what sort of environmentalism it outfits, and why Harvey as a Marxist would utter it, we need to look closely at how Marx comes to his concept of the environment. He comes to it with a question: How do we interact with the environment and why do we interact with it the way we do? I will begin with two observations.

First, Marx was a thinker of the actual material situations that frame people's lives and that, wittingly or not, they are in the process of changing. He was interested in what people do and what they think, given the materials and social settings they have to work with. Through the ages people have devised different truths about the world, because they have lived in the world in different ways, and lived together in that world in different ways too. People's consciousness of themselves and the surroundings with which they interact is inextricable from how they practically experience each other and the world and from the intentions they carry forward to the world beyond themselves. It hardly needs saying that Marx focused most intently on the *social* qualities of those practical experiences and intentions. And this for the simple reason that individuals are not the sole authors of their own experiences, intentions, and thoughts. Some readers will understand this last statement in terms of the influence of family and friends, of ethnic or race relations, of religious belief, and other shaping forces. Marx was not inured to these but had a special interest in the difference made by our (in)access to the resources, tools, and technology used to sustain life in capitalist societies, versus other kinds of societies (e.g., slave, feudal or mixed, but also societies to come). That is, it is crucial to consider how life is framed by questions of who controls the fruits of labour, who controls the tools necessary to perform labour, and who controls the concept of labour itself? The 'controls' are socially not individually authored. They are also malleable. Yet they are *not* obvious, precisely because 'common sense' ossifies the answers.

Second, at the same time it appears Marx was a thinker of universals. He was interested in what generalisations might be drawn from the diversity of actual concrete situations. For example, in so far as life among other human beings, in conjunction with a non-human environment, is a common feature across time and space, should not we have an interest in what compels us in a general sort of way to interact with others and with an external environment? That is, we can not simply, naively, step into the heart of a given social/historical situation, into the specifics of a given mode of interaction, without also wondering about what conspires to bring about an interaction and compose it in the first place. That human beings are *in and of* nature was a first rule for Marx that held true everywhere.

These two observations intersect in a revealing and instructive way in a chapter Marx wrote on the labour process for the first volume of *Capital*. A previous chapter of this book has already established that in capitalism working classes do not own means of production; they are owners of the capacity to work (labour power) which, in order to sustain their lives, they sell for a wage to capitalists: 'Hence, what the capitalist sets the labourer to produce, is a particular use-value, a specified article'. It is these use values, in the form of commodities, that workers buy back in order to meet their needs. But 'the fact that the production of use-values, or goods, is carried on under the control of a capitalist and on his behalf, does not alter the general character of that production' (Marx, 1967, p. 177). For this reason the labour process has to be considered apart from any particular instance of it.

'Labor is, in the first place, a process in which both man and Nature participate, and in which man (sic.) of his own accord starts, regulates, and controls the material re-actions between himself and Nature' (Marx 1967, p. 177; cf. Marx, 1967, pp. 42–43) Though there is a kind of anthropocentrism here, it is not a simple one: man 'opposes himself to Nature as one of her own forces, setting in motion arms and legs, head and hands, the natural forces of his body, in order to appropriate Nature's productions in a form adapted to his own wants'. When Marx writes of people initiating their participation with nature of their own accord, this does not presuppose a binary between people and nature. On the contrary, it speaks to the idea of a single Nature that includes people, a diversity within a unity. At the same time, this nature is a not a static, pre-given whole. 'By thus acting on the external world and changing it, he at the same time changes his own nature. He develops his slumbering powers and compels them to act in obedience to his sway' (Marx, 1967, p. 177). There are two ideas here. One is that human beings have no unchanging essence – we are not creatures who are doomed to repeat and live out precisely the same relations and practices from time immemorial. Nature has an *emergent* quality. The other concerns the intentionality and creativity of the human being. Marx formulates his idea, that man is that part of nature who compels his part toward an enlarged, creative capacity, because he wants to draw a distinction between the reactive and the active. Unlike one notion of animal being which emphasises animals as simply responding to stimuli or acting on unreflective instinct, Marx's notion of the human animal is something else altogether. 'A spider conducts operations that resemble those of a weaver, and a bee puts to shame many an architect in the construction of her cells. But what distinguishes the worst architect from the best of bees is this, that the architect raises his structure in imagination before he erects it in reality. At the end of every labour process we get a result that already existed in the imagination of the labourer at its commencement' (Marx, 1967, p. 178). The bee and spider cannot help what they do; human bodies in their interactions with their environments realise and give form to an idea.

But it is just at this point where the argument takes an interesting turn, with enormous political implications:

> He (sic.) not only effects a change of form in the material on which he works, but he also realizes a purpose of his own that gives the law to his modus operandi, and to which he must subordinate his will. And this subordination is no mere momentary act. Besides the exertion of the bodily organs, the process demands that, during the whole operation, the workman's will be steadily in consonance with his purpose. This means close attention. The less he is attracted by the nature of the work, and the mode in which it is carried on, and the less, therefore, he enjoys it as something which gives play to his bodily and mental powers, the more close his attention is forced to be (Marx, 1967, p. 178).

Here it happens that the realisation of a purpose, while capable of bringing joy, is not a realm of pure freedom and autonomous activity – as if the idea of what one wants to produce and how to do it could come to fruition of its own. Instead, human beings are never outside the relational world. Throughout a labour process, through which one aspect of nature transforms another, and which means a good deal more than 'work,' something is demanded of us: vigilance, attention, acquisition and practice of competencies. That is, our freedom, our capacity to enhance ourselves, comes about interactively and can be no other way. To be in nature at all, to be an aspect of nature, is to be in a world of relations among multiple phenomena. And this, Marx suggests, is the source of our joy as human beings, our particular *species being*. Elsewhere Marx writes: 'The labor-process . . . is the necessary condition for effecting exchange of matter between man and Nature; it is the everlasting Nature-imposed condition of human existence, and therefore is independent of every social phase of that existence, or rather, is common to every such phase' (Marx, 1967, p. 184). One has to read only a little more of Marx to pick up on the innuendo that the labour process within capitalism has indeed become a demeaned and diminished thing: 'the less he is attracted by the nature of the work . . . the more close his attention is forced to be'.

But how does this take a political turn? There is a qualitative distinction between what nature forces people to do, in a generic sense, and what workers are compelled to do under capitalism. 'Nature does not invent capitalists on the one side and workers on the other . . .' Marx moves back and forth, as thinker of specific historical situations, on the one hand, and of universals, on the other hand, to make a point. Workers, even under capitalism, are doing a *kind* of thing (nature imposed) they would be doing anyway. Yet under capitalism, this compulsion is mediated by capitalists, who compel workers to sell their labour power *as the primary vehicle through which to engage with nature*. (Marx would not consider the consumption of nature through visits to parks, scenic overlooks, or drives through the countryside as engagements with nature, in his fully blown, relational sense of the word.) One might say that Marx means to point to an inversion whereby the capitalist assumes the role of nature. Where nature would impose a real necessity; capital interlopes, imposing a false one – nature does not create capitalists on one side and workers on another. Under capitalist social relations, workers experience what nature would impose as something capitalists impose instead, in a bastardised, constricted way.

But what is so wrong with the mediation of social relations? Indeed mediation is not really the problem. (The problem is capitalist social relations.) Here we have to go back to the idea that in transforming nature we transform ourselves, and back to the potentially positive valence this has for Marx. 'In the labor-process . . . man's (sic.) activity, with the help of the instruments of labor, effects an alteration, designed from the commencement, in the material worked upon. The process disappears in the product; the latter is use-value, Nature's materials adapted by a change of form to the wants of man. Labor has incorporated itself with its subject: *the former is materialised, the latter transformed*' (Marx, 1967, p. 180, emphasis added). The materialisation of labor is none other than the manner in which we make ourselves in new ways, whilst at the same time remaking portions of the world in new ways. And this is not somehow a process outside nature: it is one aspect of nature acting upon (incorporating itself into) another, producing in the end a quite different nature than was there before: there is nothing unnatural about New York City.

Does Harvey mean there is nothing problematic about New York City? Not at all. Under capitalism the labour process is the mechanism whereby capitalists control what workers do (in nature) and assume the products of labour (transformed nature) as their own. At the same time the waste products are often placed outside the economy as so-called externalities for society and the state to deal with (Katz, 2001). And all this, for Harvey, is what New York City expresses. There is nothing 'unnatural' about this, it is simply (!) what nature becomes in its capitalist guise. The vision is a broad one and allows for all manner of political struggles, inclusive of labour struggles, to be viewed as environmental issues. Grasping these points goes some way towards explaining Marxism's historical ambivalence about fixating on environmental concerns (see Benton, 1996b). They seem a step removed from where the real action takes place, seem more the provenance of a politically fickle middle class, and are easily co-opted by bourgeois forces wanting to deflect attention from themselves. In particular, environmental problems are framed ideologically as something 'we' are all in together (Waterstone, 1993; Enzenberger, 1996). The sections below will have more to say about this.

The Strange Case of Value versus Wealth: $M \dots C \dots M + \Delta$ and 'delta' Ecologies

The foregoing might give the impression that capitalists primarily aim to orchestrate production and command nature (inclusive of labour power) to do their bidding. It might be thought that Marx excoriated capital based on its objective to master the environment of humans and non-humans, all to accumulate wealth. This would not be quite right. Within capitalism, Marx argued, the objective is not to accumulate material wealth but to continually generate so-called surplus value. This distinction is a crucial one. It is because of the need for surplus value that there can never be enough material wealth. It is why arguments that capitalist societies can be taught to contain their material acquisitiveness, can learn that enough is enough, fail (Postone, 1999). (They must, according to James O'Connor, be badgered into containment by social movements. See below.) Under capitalism, Marx argued, the point is not to accumulate goods, and even less to fulfill all needs. The point is to expand value. But what *is* this 'value'? This is not an easy question to answer, but we can get at least some insight by approaching the issue simply, which is what Marx does at the very beginning of *Capital*.

Unlike the first section of this chapter in which I began with the idea of a labour process that brings us, our intentions, and our ideas into an active relation with a more-than-human world, the analysis of capitalism as such begins with the commodity. The commodity, for Marx, stands at the center of the more-than-human world that capitalism has wrought. Why? Partly, because it is commodities with which we most directly provision ourselves. We obtain clothing, food, and shelter after these have been placed on the market as commodities. And partly because it is primarily commodities that we are labouring to produce when we work at our jobs. A little less clear perhaps, but also quite obvious, is that commodities are not only produced, and not only consumed; they are exchanged for other commodities, with money as the means of exchange. We need to follow this thread. The site of exchange is what we call the market, and all-important there is the matter of who is bringing what to market and for what purpose. Workers bring their labour power (a commodity) and in exchange for their exercising it they typically are paid a money wage, which they then use to purchase needful things, other commodities. Note that this is an exchange of qualitatively different things: labour power for money, money for needful things. This is a useful exchange of differences precisely because it cumulates in life support. The notation Marx uses to describe the exchange of qualitative differences is $C \dots M \dots C$ (Commodities, Money, [other] Commodities). Now, supposing there are just two social classes (a reduction that Marx makes to illustrate his case), this leaves capitalists, the ones who pay the workers, to consider. What capitalists bring to market is money. With this money they purchase commodities (labour power and means of production) by which other commodities will be made (suffused with the value that labour imparts to them), and they will sell these finished commodities for money (preserving that value). $M \ldots C \ldots M$ denotes this series of exchanges. But note that the exchange does not end with something qualitatively different from its beginning. It begins and ends in money, a perfectly useless exchange: it is not clear yet why anyone would do it. So, Marx surmises, the only reason to engage in the exchange of same for same is if more of the same was actually the goal: more money, profit, a surplus value. Profit is built into the very conception of the capitalist (and of capital, money invested to make profit). The notation Marx uses to describe this series of exchanges, inclusive of the extra increment, the 'delta', the surplus value, is $M \ldots C \ldots M + D$. The series ends with a quantitative difference rather than a qualitative one. Or rather the point is that the series does not end. A portion of surplus value is reinvested when money capital is thrown back into production on a continuing basis: $M \ldots C \ldots M + D \ldots M \ldots C \ldots M + D \ldots$. What accounts for the 'D' in a well-developed capitalism? In short it is that the mass of labour power expended by workers is capable of making more goods than it actually needs for its own replenishment.

Important questions follow. I have selected from the classic problems that Marxism presents to capitalism: the relative position of labour vís a vís capital, the problem of surpluses in capitalism, and the possibility of limits to growth. We can frame them as follows, showing how each bridges to environmental issues:

Don't workers buy back what they make? Aren't profits and wages perfectly compatible? No, answers Marx. If this were the case there would be no profits for capitalists. Workers are compensated for their *efforts* (their labour power) not for the actual value of the goods they produce: the value of labour power is distinct from the (greater) value of what labour produces (the much vaunted fruits of labour). This virtually ensures, as labour history verifies, an ongoing struggle between the two classes over the fate of that surplus value. It also ensures struggles over what is considered adequate compensation for supplying labour power on a continuing basis. That is to say wages must ensure 'social reproduction', access to the bundle of needs and wants that are central to so much political struggle (Katz, 2001; Mitchell et al., 2004). Clearly, capitalists will not pay more for minimal efforts expended over the maximum time: they will not pay for workers' time regardless of how long a worker takes to get a job done. And they will consider carefully whether a wage needs to cover the cost of dining at expensive restaurants or health care. Marx argued that, all things being equal, wages are paid out on the basis of the average amount of time it takes to make a certain amount of goods under average conditions of production and social reproduction, *in so far as these are the outcome of specific struggles.*[2] Environmental questions matter here: what wage is adequate compensation for living next to toxic wastes? For working with hazardous chemicals? For taking a job in a place designated specifically as environmentally unregulated (see Pulido, 1994; Katz, 2001)? For living amidst blighted and abandoned urban landscapes (Bunge, 1971)? Whatever the wage, in its rising/falling state, it cannot be allowed to threaten the conditions for the expansion of value (Harvey, 1982).

If wages are less than profits but profits are based on sales, doesn't this imply more goods on the market than workers' wages to purchase them? What keeps capitalist economies going? Yes, basically, to the first, and two short answers to the second: devaluation and expansion. Losses (devaluation) are endemic to capitalism, but so is expansion: expansion of markets into new geographical territories so that more goods may be off-loaded, expansion of the working class, so that more people will become consumer-producers sopping up the ready-made, expansion of the kinds of things that become commodities, so that the source of profits will be augmented. The last of these, what Marx called primitive accumulation and David Harvey has recently termed accumulation by dispossession is of renewed importance as an environmental issue – witness the commodification and patenting of genetic materials and privatisation of common water resources, not to mention fossil fuels (Castree, 2003; Harvey, 2003; Swyngedouw, 2006, for a wider, very useful discussion of the commodification of nature). There is no sense, as yet, that devaluation and expansion harmlessly balance out. Quite the contrary, the history of capitalism in the 20th and 21st century is fraught with crisis. If anything, the surprise is the degree to which capitalism accommodates itself to crisis, perhaps beyond what Marx imagined. Looming especially large are financial and credit instruments by which to extend production-consumption circuits well beyond current earnings. (See Box 17.4 below for an appreciation of the important role finance capital plays in 'nature-based' economies such as agriculture). That these sometimes come crashing down (e.g. the spectacular failure of the subprime mortgage market in the United States in 2007–2008), and that they are currently a central part of the economic imaginary regarding solutions to the climate crisis (e.g., the brisk market in so-called carbon credits) alerts us to the heightened role financial instruments play.

If the expansion of surplus value is manifested as the proliferation of goods (and the transformation/exploitation of nature that is involved), will we see a point where capitalism will have produced 'enough' goods and consumed enough resources in the process? Individual workers may well decide to scale back their consumption (though scaling back is problematic for reasons already noted). Capital, capitalists, as such, cannot, at least not in the aggregate. Capital is forced to expand: capital is that which expands. Money is not traded for money, it is traded for more money. It is not interested in things as they are useful for provisioning life, nor in things by which the measure of a sufficient life might be made. It is interested in more-of-the-same money. And it must embark on this venture spatially and temporally. This is the systemic nature of capitalism, beyond what individual, well-meaning, philanthropic capitalists may intend. Capital simply *is* the expansion of economic value throughout the field of those interacting aspects of nature (people, raw materials, etc.) that comprise the matter of capitalism. Capitalists as much as workers are caught up in these interactions. But these interactions are in process, so to speak. In Marx's view capitalist social relations set in motion the possibility for their own erasure. How this might happen is the subject of huge debate, well beyond the scope of this chapter (though see the discussion on O'Connor below).[3]

Can the self-expansion of value move forward indefinitely; does it encounter any obstacles or limits set by nature – doesn't nature matter in a more than trivial way? Although we can deny (and should deny) that there is an unvarnished anthropocentrism at work in Marx's ideas, different aspects of nature (the human) seem to be given more agency than others (see Benton, 1996b). The effect is to make it seem that people are indeed the active player on a field of more pliant non-humans, though we cannot for a moment forget that the human being's becoming was only through relations with other entities. Elmar Altvater (2006; cf. Benton, 1996b; Coronil, 2000) offers an interesting effort to insert this idea more directly into the concept of value. He writes about a 'fossil capitalism'. Capitalism, he advises, has to be understood not only through the value chain in general $(M \ldots C \ldots M + D)$, but through specific flows of matter and energy characteristic of capitalist development and change at any time (cf. Benton, 1996b). Fossil fuel, he argues, has been the perfect energy source for industrial capitalism because of the physical properties of fossil fuels themselves. (See Box 17.1).

Box 17.1   The congruence between fossil fuels and industrial capitalism

Petroleum in particular has a high Energy Return on Energy Input (EROEI). That is, 'only a small amount of energy needs to be invested in order to harvest much greater amounts of energy, because the entropy of petroleum is very low and its energy concentration is very high, yielding a high energy surplus' (Altvater 2006: 39). Moreover, there is a specific congruence between the physical properties of fossil fuel and the 'requirements of the capitalist process of accumulation' (Altvater, 2006, p. 41). First, fossil energy eases the transformation to capitalism in places where it did not exist before. This energy resource can be transported almost anywhere, which means capitalist production is free to move around globally, other things being equal. Local resource limitations are no longer a fetter. Second, unlike solar radiation, a major source of energy in the past, fossil energy is available non-stop, year round, and at 'constant intensity' irrespective of seasons and biological rhythms. Production (and productivity) can be accelerated and the friction of distance reduced by temporal speed-up: the times and spaces of nature can be altered by the times and spaces of capital (see David Harvey [1989] on time-space compression). Third, fossil energy can be used and diverted flexibly, so that its consequences are multiplied. It allows for the expansion of electricity-based motors and illumination, as well as the internal combustion engine and its diffusion. These developments have magnified the effect that capital is seemingly freed from natural constraints. This freedom is more relative than absolute, Altvater cautions. But still it is impressive. 'Although something like capitalist social forms occasionally could be found in ancient societies (in Latin American and Asia as well as in Europe), they could not grow and flourish without fossil energy. The entropy of the available energy sources was too high, and the EROEI too low, to allow significant surplus production. Therefore, growth was limited, and in fact the annual growth rate was close to zero before the industrial revolution of the late 18th century. But in the course of the industrial revolution economic growth rates jumped from 0.2 percent to more than 2 percent a year until the end of the 20th century; world population increased faster than ever before' (Altvater, 2006, p. 42). While Altvater is careful to note that fossil energy is not the sole or even primary cause of growth (he points additionally to capitalist social relations as such, European rationality and colonialism, and the disembedding of the market from society), he is emphatic on the point that fossil fuel-led growth is not possible forever. In that sense too the advantages of fossil capitalism are purely relative. 'The limits of growth are among the conditions of life and the laws of evolution on planet Earth, and are a direct consequence of the limits of the resources – and especially fossil resources – which fuel growth' (p. 43). To return then to the notion of EROEI, it decreases 'in step with the exhaustion of global oil reserves' (p. 39). And yet oil extraction may still be profitable.

The discussion in Box 17.1 indicates there is a difference between what is rational economically and what is rational in energy terms (cf. Dryzek, 1987). This difference points to the very calculus Marx argued was distinctive about capitalism in the first place: it aims to increase value, even when there are diminishing returns in other arenas. Capitalism will notice a decreasing reserve in its resource base only if value decreases. If not, these two spheres are free (for a time) to move in opposite directions. It seems then that while it is important simultaneously to theorise value and the matter through which it materialises, they are not reducible to each other as the same thing. As to the prospects of other energy sources, none compare to the (dis) advantages of fossil energy, Altvater argues. For example, 'today, and possibly forever, it is impossible to power the machine of capitalist accumulation and growth with 'thin' solar radiation-energy. It simply lacks the advantages mentioned above, i.e., the potential of time and space compression, which 'thick' fossil energy offers' (Altvater, 2006, p. 45). Fossil energy as useful as it has been is an obstacle to future development, as capitalism understands that term 'development'. Fossil energy's supply is limited; its harmful ecological effects too severe. Political machination, oil wars, and new oil discoveries only forestall the day when fossil capitalism will come to a close. How close we are to that day is subject to intense debate. For example, Marxists such as Michael Watts disagree that the recent Iraq Wars are essentially 'oil wars' announcing the approach of peak oil production. Nor are periods of supposed oil shortage in the 20th century premonitory: 20th-century capitalism's problem was rather one of oil glut and declining prices (Retort, 2006). Marxists such as Daniel Buck (2006), following Storper and Walker (1989) warn against making oil and capitalism synonymous with each other: oil is just one kind of energy framework for capitalism; it will be followed by others.

Where does Marx stand on all this? On the one hand, it seems he was less concerned with environmental problems as we now understand them (e.g., ecological destruction, pollution, and so on); his concern was reserved more for what happens in the human world under capitalism than the non-human world, although this cannot be known for certain given the little he said and the openness with which he said it (Soper, 1996). On the other hand, as we go about construing the congruent relations between capital and the physical properties of the substances it articulates with, the translation of purposes and ideas into material forms through the labour process should *not* be made to seem too smooth. This is a theme oddly underdeveloped by Altvater. 'Energy invested' does not automatically find its mark, even under high EROEI. Altvater skips over a long history of technological blunder and serendipity at play in the fields of oil, blunder and serendipity that made available the very properties of oil that are so 'congruent' with capital (Bowker, 1994; cf. Bakker and Bridge, 2006, p. 9; cf. Latour, 1993; cf. Braun, 2000). The smooth translation of value through the circulatory, reproductive process of capital is much belied by volume 2 of *Capital,* an indication of Marx's dialectical method: begin with inadequate concepts and serially refine them by charting what other concepts (e.g., mismatches of supply and demand, overinvestments of capital, etc.) they bump into (see Harvey, 1982; Castree, 1996; Arthur, 2002; Henderson, 2004).

Some telling questions are now out of the bag: if buildings exist in the heads of architects (or oil derricks and drills in the heads of engineers), does this mean that non-human actors nowhere intervene, interrupt, cause adaptation and revision, to say nothing of plain old failed attempts to control nature? And if the non-human has agency how shall that agency be thought of? Geographers interested in these issues have cautioned against lapsing towards a deterministic view of nature. As put in Bakker and Bridge's (2006) account, 'If one adopts the ontological position that non-human entities have active capacities . . . then one needs a way to express the physicality and causality of the non-human without straying into object fetishism, or without attributing intrinsic qualities to entities/categories whose boundaries are 'extrinsic' (i.e., that are defined, at least in part, socioculturally). To the skeptic, then, the resurgence of the material after a decade of social constructionism should be reason for pause, since it raises specters of worn-out dualisms, resurgent physicalism, object fetishism and environmental determinism' (p. 8). What many geographers seek to explain, then, is how non-human entities gain agency not through their innate properties, but 'through the way they are embedded in a wider set of socio-technical relations' (Castree, 1995, p. 13).[4] Having introduced certain classic Marxist problems and bridging these to environmental matters, I want next to examine how the question of environmental agency has been taken on in Marxist inspired research on the environment. I have chosen three influential takes on these issues: nature as both a sustaining and limiting 'outside' to capitalism; nature as 'produced' by capitalism; nature and capitalism as in utter need of conceptual repair and rethinking. These are to some degree overlapping themes but I will present them sequentially in the next three sections.

Ecological Marxism: From Environmental Destruction to the 'Second Contradiction' of Capitalism

The above heading alludes to a 'second contradiction' – but what about the 'first'? Traditionally Marxism identifies the contradiction between forces and relations of production as what generates the crises and conflicts, the historical and geographical 'oomph', that provide potential openings for a socialist or communist transition, or at a minimum, grist for 'anti-capitalist', 'alternative economy' movements. By relations of production is meant the large-scale question of who owns, controls or manages means of production, and who does not own, control, or manage those means. In most of *Capital*, Marx assumed, for purposes of argument and political strategy, a two-class system: capitalists who own means of production and workers who do not. Relations of production also refers to intra-class relations: the cooperative and competitive tendencies among capitalists and the same contradictory tendency among workers. (The term is also widened to refer to access and control over the preponderance of finance capital, service industries, and commodity distribution. Furthermore, the term has been transported into production at the household scale, where it notices the gender division of labour.) Forces of production is a more amorphous, flexible term that refers to the assemblage of given technologies, scientific knowledge, labour skills, productive capacity of machinery, and even natural processes (e.g., physical properties and potentials of water, metals, wood, heat, and so on), that enter into commodity production. Marx argued that forces and relations of production were themselves related to each other in contradictory enabling and constraining ways, as when he proposed that a given technological improvement in agriculture threw agricultural labourers out of work, while some other technological improvement allowed for workers to mass together under one roof in an urban factory, while further labour-saving improvements would lead to a 'surplus army of the unemployed'. (The idea of a growing surplus of workers, induced specifically by capitalist forces, was Marx's famous answer to Thomas Malthus's equally famous theory of poverty as an expression of poor families having increased their number beyond the capacity of the natural resource base that sustains them. Malthusianism remains a popular idea, especially in the 'developed' world; it is one most Marxists reject out of hand [see the important works of Harvey, 1974; Watts, 1983].) Forces of production are therefore instrumental in the emergence of an industrial working class. But the idea is that this is an inherently contradictory situation. Not only is there a struggle over who gets to keep surplus value. Marx held out the possibility that working people and forces of production could become a joint force capable of rendering capitalists, or more properly the specific social role a capitalist plays, obsolete.

In contrast to the 'first contradiction' between forces and relations of production, ecological Marxism (see footnote 11 in O'Connor [1997] for background on this term) focuses on a 'second contradiction', between capitalist production relations (including productive forces) and conditions of capitalist production. The proliferation of terms here is a bit unfortunate but not impossible to sort out. Conditions of capitalist production refers to the ability of capitalist social relations and forces of production to be sustained and reproduced; it is the question of what must be accomplished so that capital can be reproduced.

Marx himself, according to O'Connor, defined three types of production conditions. First are natural factors external to capital appropriated for use within capital. Marx termed these 'external physical conditions'. 'Personal conditions of production,' second, is a term reserved to describe workers' labour power, and the wherewithal to reproduce it. The third condition is 'the communal, general condition of social production' (O'Connor, 1997, p. 160). This consists of such things as the means and capacity to communicate and enter into association with other human beings by virtue of the wider human made landscape. In other words, what must be in place for capital to be produced at all are viable, functioning natural/ecological systems, human mental and bodily being, and sociality itself. The concern for the environment by ecological Marxism is part of a broad remit; it makes no easy distinction between natural and social environment, wild and urban environment and so on. O'Connor emphasises that 'neither human labour power nor external nature nor infrastructures, including their space/time dimensions, are produced capitalistically, although capital treats these conditions of production as if they are commodities or commodity capital' (p. 164). In so doing, production conditions regularly become stressed or despoiled, generating counter-movements. These movements form the real substance of the second contradiction. Most especially, the state steps in to regulate capital's treatment of production conditions or social movements push back against capital, or both. 'This means that whether or not raw materials and needed labour skills and useful spatial and infrastructural configurations are available to capital in requisite quantities and qualities and at the right times and places depends on the political power of capital, the power of social movements that challenge particular capitalist forms of production condition, [and] state structures that mediate or screen struggles over the definition and use of production conditions' (p. 165).

Unlike the direct confrontation between capital and labour in the production process, which is viewed as an act of private exchange outside politics, the second contradiction of capitalism is politicised by definition, precisely because capital encounters conditions that it cannot make for itself, but that matter a great deal to ordinary people – a domain of intensely held values that capitalist value does not make or properly represent via money. The second contradiction illuminates a universe of use values, in other words, that when recruited into capitalist exchange relations incites a good deal of push back. The second contradiction, in the form of counter-movements, exposes capitalism as a system that does not provide all good things, and ruins a good many. 'Listen to yourselves! Look at what your own movement show you! Take yourselves seriously!' would perhaps be O'Connor's injunction.

The problem of course is that these common use values (healthy ecosystems, functioning human bodies, a functioning society) are treated as if they come rolling right off the conveyor belt: they are treated like commodities for sale or as if they can only be sustained if commodity production is sustained, and are thereby laid waste. Examples are many and varied: the warming of the atmosphere, acid rain, salinisation of water tables, toxic wastes, soil erosion, urban congestion and soaring rents, hazardous and dehumanising workplaces, decrepit infrastructure (O'Connor, 1997, p. 166). Where certain strands of environmentalism might construe these as scarcities induced by population growth – the Malthusian Myth – O'Connor argues that capitalism in effect produces its own scarcity and its own environmental problems, in so far as it degrades its own conditions.[5] See Box 17.2 on how this myth relates to water resources.

The degradation caused by capital produces an underproduction of capital, O'Connnor argues. 'We can . . . introduce the possibility of capital underproduction once we add up the rising costs of reproducing the conditions of production' (O'Connor, 1997, p. 166). Indeed vast sums are expended on health care, environmental remediation, policing of the social environment, research and development monies to develop substitutes for degraded or depleted natural substances, oil and other resource wars.

> No one has estimated the total revenues required to compensate for impaired or lost production conditions and/or to restore these conditions and develop substitutes (much less how much of these 'costs' actually fall on capital. It is conceivable that total revenues allocated to protecting or restoring production conditions may amount to one-half or more of the total social product. . . . Is it possible to link these unproductive expenditures . . . to the vast credit and debt system in the world today? To the growth of fictitious capital? To the fiscal crisis of the state? To the internationalization of production? The traditional Marxist theory of crisis interprets credit/debt structures as the result of capital overproduction. An ecological Marxist approach might interpret the same phenomena also as the result of capital underproduction and unproductive use of capital produced (O'Connor, 1997, p. 166).

These are interesting issues to be sure. But why do they matter? It is not so that traditional and ecological Marxists can each crunch the numbers to see who is more correct. Nor is it to suggest that these two independent lines of inquiry should forever run in parallel. As a Marxist – traditional *or* ecological – O'Connor's point is that the crises which capitalism generates are also forces that goad it to restructure and rebuild. (Crisis is itself a use value, one might cynically venture, to be folded back into the circulatory apparatus of capital). At the same time, he is arguing that the sources of crisis are more plentiful than Marxists have realised and the crises generated through the second contradiction have enhanced potential for generating and nurturing the possibility for a socialist transition, or at the very least a height-ened awareness of capitalism as exorbitantly problematic. Because the second con-tradiction already calls upon the state to mediate capital's domination of production

---

Box 17.2   Water and the Malthusian Myth

In a classic essay of 1974 on natural resources and the ideology of science, David Harvey argued that the environmental sciences of the postwar period were fixated erroneously on 'population pressures'. Drawing upon Marx, he argued that so-called surplus populations could not be understood apart from global capitalist processes through which lands and waters were enclosed and through which labour was induced to urbanise. Barring catastrophe, he argued, there is no essential relationship between population growth and resource scarcity, as Malthusians argue. Consider the case of water. Eric Swyngedouw (2006) argues that universally available potable water would be possible were it not for the social, political, and economic relationships that structure (in)access to it. That is, access to potable water is structured by access to capital. 'In Mexico City, for example, 3 percent of households have 60 percent of all urban potable water, while 50 percent make do with 5 percent. In Guayaquil, Ecuador, 65 percent of urban dwellers receive 3 percent of the produced potable water at a price that is at least two hundred times higher (20,000 percent) than that paid by a low-volume consumer connected to the piped urban water network' (pp. 199–200). Meanwhile the capital used for private development of water (its development is increasingly privatised) has been restructured, coming more from international sources, a trend promulgated by the World Bank. And yet what these private companies have begun to discover is that they cannot conduct business profitably without public assistance and financing. The road to more water runs through capital, but capital has to run through the state. Swyngedouw concludes that 'transforming H20 into a useful "thing" requires remodelling and reorganising the socio-hydrological cycle so that it serves particular socio-physical ends (irrigation, recreation, sanitation, etc.) The resulting hydro-social cycle is embedded in and organised through the commodification of water' (p. 206). It is not an independent resource base to be drawn down by 'overpopulation'. For other scholars working on natural resource issues, the above is not to say that capitalism cannot have a 'Malthusian effect'. See for example Bernstein and Woodhouse (2006).

---

conditions – and therefore exposes capitalism as a predatory system, whose natural necessity is radically unnecessary – periods of restructuring intensify that very exposure. It becomes apparent that capitalism requires more planning and regulation, not less, and cannot abide by the neoliberal reforms sweeping the globe. Social movements geared to politicising the conditions of production are thereby handed an opportunity to constantly call capitalism and desires for capitalism into question. Thus, it is the very forms of socialisation already widely embraced to deal with the second contradiction (e.g., 'political bipartisanship in relation to urban redevelopment, educational reform, environmental planning' (O'Connor, 1997, p. 168)) that point a way forward out of capitalism. Through the crises generated by the second contradiction – O'Connor hypothesises a possible 'second path to socialism'. Such a second path would in fact pose a challenge to socialism as such, necessitating that it be as equally green as it is red (a major theme of the prominent annual socialist review, *Socialist Register* 2007).

The Production of Nature: Nature as an Accumulation Strategy

But what if we turn the tables and instead of focusing on how capitalism erodes its conditions, (re)visit how it alters them. Here is a statement by the geographer Neil Smith: 'The social provision of sustenance has always involved a certain "production of nature". In capitalist societies, however, the production of nature mutates from an incidental and fragmented reality to a systematic condition of social existence, from a local oddity to a global ambition' (Smith, 2006, p. 21). What does this mean? Not what one might initially think. It will take some space to say why, including what makes his argument different from O'Connor's.

The notion of production of nature is an idea different from traditional views of nature generated in capitalism, which Smith refers to as *external* and *universal* nature. On the one hand, it is convenient to consider nature as external to human beings – to propose that nature is precisely what *is* external. This view reaffirms the drive to dominate, control, and commodify nature; to take charge of it by developing technological and scientific means to bring it to heel for human use. Nature is approached as if it were a 'repository of biological, chemical, physical and other processes that are outside the realm of human causation or creation, and the repository too of identifiable objects – subatomic and molecular, specific organisms and species, terrestrial "bodies", and so forth' (Smith, 2006, p. 22). On the other hand is a notion of universal nature that also finds a place in capitalism. The idea that nature does not lie outside the human being but rather includes the human being: 'the entire world – human and non-human – is subject to natural events and processes' (p. 22). This belief generates strong Romantic countercurrents to industrial capitalism and the ideology of domination. The desire to get back to nature, to seek solace and personal repair away from cities, or, just as common to redesign and plan cities and landscapes in better harmony with nature is part of this countercurrent. What many Marxists insist on is that capital seizes the countercurrents and commodifies them. It is possible then to see a role for the ideas of external and universal nature in capitalism.

The idea of the production of nature is quite apart from this. It falls outside this binary of external versus universal nature in two ways. First, it denies that human beings dominate and control nature in any sort of secured way. To say nature is produced is not to speak of domination and control. Second, while it shares with the idea of universal nature an insistence that humans are not outside nature, it suggests that it is not enough to speak of natural processes that human beings are simply a part of. Smith offers a telling passage from Marx and Engel's *The German Ideology* which begins to get at the notion of 'produced nature':

> So much is this activity, this unceasing sensuous labor and creation, this production, the basis of the whole sensuous world as it now exists, that, were it interrupted only for a year, Feuerbach would not only find an enormous change in the natural world, but would very soon find that the whole world of man and his own perceptive faculty were missing. . . . The nature that preceded human history is not by any means the nature in which Feuerbach lives, it is nature which today no longer exists anywhere (quoted in Smith, 2006, p. 23).

Smith argues that one finds here a validation of neither the external nor the universal notions of nature. One finds instead the partial truth of each. 'The externality and universality of nature are real enough, but these are not to be taken as ontological givens,' Smith writes (p. 23). The growing presence of the human being and the expansion of the capitalist world creates a world in which a putative non-human nature simply has no place apart from humans and their traces, while a universal nature that would seamlessly blend human with non-human does not account well for the manner of relationship among nature's parts (humans included). What do these ideas mean? On the one hand, the so-called natural world and its processes and laws now occur within a field that has been altered by human activity, from traces of industrial chemicals in biota globally to anthropogenic climate change. Such examples, in addition to more mundane ones like flying planes or driving cars, indicate that a human presence is afoot producing new admixtures (socionatures, as some call them) of so-called nature: producing entities never produced before with the aegis of natural 'laws' and with material substances that are of but not reducible to 'nature'. It is not possible, when we explore such examples, to pry apart the operations of 'natural' laws from the logics of instrumental, scientific reason. Put another way, Marx renders humans a constitutive part of nature rather than a passive part subject to transcendental laws (Braun, 2006, pp. 195–96).

On the other hand, the idea of an external nature is supplanted by the notion of *externalising* processes. Recall that for Marx humans beings, in order to be human, must engage with the material world around them, a world they are ontologically part of but that will not support them without intervention and reconstruction. In practical terms human beings make objects that reflect their needs and desires, objects that reflect who these human subjects are, and again allow them to become who they are. We become human *through* our objects. (Every object is a subject for another object, wrote Marx.) The fruit of our labour embodies potentially the best of who we are and can be, while not exhausting our potential for further growth and transformation. So, thought Marx, in the best of all worlds this is a joyful, creative, collective endeavor; in the worst, shear drudgery, repetition, and alienation. Under capitalism the process of externalisation is indeed problematic, as those who produce value (commodities) are enjoined to produce (using means of production they do not own) a surplus, which must be forked over to a class other than themselves. Direct producers in other words produce an 'external' nature – they externalise themselves, as one part of nature producing another – which then becomes the property of another, thus rendering the nature they have produced a very poor reflection indeed of what they might become.

And here Smith reminds us of an important consequence of the notion that human beings are part of nature: the changing nature of human beings themselves. Over time, labour within capitalism involves what Marx called formal and real subsumption. In the first case, and speaking historically, labour is not too much changed by capitalist work. The way to get producers to produce more is to work them longer and harder, while not changing the work itself too much. This simple fact of working for capital is what Marx called formal subsumption. But production soon reaches its limit by way of this strategy. Once workers worked hard and longer, more products can only come by finding more workers. The only option left is to change the work itself: develop new technologies, new scientific know-how, new materials, new machinery, through which workers – now with new 'skills' – might produce more in the same amount of time as before and with the same amount of effort. Thus is born the real subsumption of labour. In the first volume of *Capital* Marx devotes several long and rich chapters to these processes. The result is that labourers become instruments of machinery rather than the reverse. Their very bodies become attuned to and shaped by their labours; they make parts of things according to a detailed division of labour, rather than whole things; they find that machines meant to save labour for capitalists do not save labour for them. Defying, at least in part, O'Connor's notion that conditions of production preexist capitalist production, the body's 'natures', ever pliable and moldable, change. Not, of course, without incident and resistance.[6]

The point of the production of nature idea is that it is not sufficient to think of nature as that which is non-human, nor sufficient to say that nature includes the human. Nature takes on particular forms over time and space; therefore, the spread of capitalism over time and space involves a particular production(s) of nature(s) under capitalism. But by production of nature does Smith just mean that nature is controlled and dominated? Is the production of nature another way of saying capital is masterful, knowing, controlling and determinant? I have already hinted at the answer. It is that capitalistically produced nature shares important features of capitalism: where capitalism seeks to dominate and control the conditions under which value and surplus value are to be produced and appropriated, it finds its domination and control incomplete and fissured, with a host of unintended consequences and contradictions in train. 'Just as capitalists never entirely control the production process, its results, or the global capitalism it generates, so capitalist society does not entirely control nature' (Smith, 2006, p. 25). There are strong parallels here with O'Connor's thesis of a second contradiction of capitalism; and strong parallels too with the ideas of Ted Benton, another prominent 'green' thinker concerned to reconstruct the Marxist tradition. Benton (1996b) has long argued that Marx's idea of the labour process was overly general: some labour processes utterly transform the non-human natures they encounter, but others merely regulate or channel those natures.

What Smith emphasises though is the further dialectical quality of the second contradiction. Even though vast quantities of resources are vacuumed up by capital, only for conditions of production to be threatened, nature becomes increasingly capitalised – there is a large and growing trade in pollution credits, for example – and *more elements* of nature are found for capital's circuits, from plant and animal DNA to human body organs to seeds and water. Capitalism, or *particular sectors of capital,* finds ways to make money on that which seems to pose an obstacle to it.[7] As Benton (1996b) argues there is in fact no necessary 'second' contradiction at all, or at least this may not be the best way to capture the capital-nature relation.

As David Harvey (1974) has noted there is nothing essential about any element of non-human nature that makes it a resource. Something becomes a resource in capitalist society as a factor of social, technological, and scientific change. What we have learned in the meantime is that how to think of the physical properties, the materiality as such, of these natural resources remains something of a mystery. They are not determined by social/technological/scientific change and they are not unaffected by it either. There is, as Noel Castree (2005) writes, a 'both/and' case to be explored, a case that will simply look different as different natures come to matter differently in different capitalist formations. Consider his work on the 'war against the seals', which treats seals as agents within a tense network of actors – see Box 17.3.

Unlike Altvater who is primarily attuned to the advantages fossil energy gave industrial capitalism, and unlike Smith who is simply more interested in the natures that become *internal* to capital, Castree's seals, though nearly depleted, neither

> **Box 17.3   Nature as actor: the north Pacific fur seal**
>
> During a 40-year period, beginning in the 1870s, north Pacific fur seals, valued
> for their pelts, were hunted by competing Americans, Canadians, Russians, and
> Japanese sealers to near extinction. How could seals have possibly been agents
> under these conditions? Castree insists we notice both the *possibilities* for value
> expansion as well as the *obstacles* posed by the north Pacific fur seal. The pos-
> sibilities posed included the sheer density of the pelts that 'created a market for
> garments made from those pelts' (p. 160). It included the very large number (more
> than 3 million) of seals to be exploited, a measure of the seals' ecological success.
> That the seals migratory lives stretched across the Pacific Ocean and included
> both land and sea habitats, invited the success of sealers from the countries named
> above who then utilised a variety of hunting strategies (land- and sea-based). And
> yet the nature of seals also posed obstacles. The difficulty of maintaining a count
> of seals while they were in their ocean-going migratory phase made it impossible
> to know the proportion of living to killed seals. It was also not easy to tell male
> from female seals, resulting in a large number of pregnant seals hunted down.
> Seal nature then could be seen as having had a hand in developing the interna-
> tional regulatory apparatus that followed the over hunting. As Castree notes
> though, just why these material/natural features of seals mattered (among their
> many other features) is 'relative to the demands made on them by the mode of
> production' (Castree, 2005, p. 160).

disappear into capital, nor are autonomous from it. They are both/and, in a context
where different natures are going to matter differently from situation to situation
(cf. Benton, 1996b).

There is some precedent for theoretically inserting the specific properties of
nature into theorisations of capital. Marxist writers have for a long time noted that
certain natural processes pose obstacles to capital and induce it to innovate. (This
is not to suggest that such obstacles are the only goad to change. Inter-capitalist
competition as such plays a strong role in fostering the search for new resource
frameworks through which to accumulate capital. See Storper and Walker [1989];
Buck [2006].) I have indicated one such natural obstacle already, the body of the
worker, the obstacles to which have been reckoned with by the passage from formal
to real subsumption. Another 'classic' case is that of agricultural production. See
Box 17.4.

The inventiveness of the commodification process, with respect to even a force-
full nature, seems to know no end, as Neil Smith suggests (see also the prescient
essay by Enzenberger [1996], originally published in 1974, especially his discussion
of the eco-industrial complex). To the above examples we can add carbon credits,
wetlands credits, even woodpecker credits – in sum, the buying and selling of the
right to pollute or degrade nature. All are startling innovations. These credits are
sophisticatedly packaged to take their place alongside any other security and allow
polluters to play the part of environmentalist. Like the environment itself, it seems
environmental problems have no essence. An environmental problem is not a
problem if it can be traded for a benefit elsewhere. Such is the imposition of
value flows globally when 'a $40 ton of unproduced Costa Rican carbon is entirely

---

Box 17.4   Nature as actor? Food and forest production

In much agricultural production, the natural environment and natural forces loom large. As a generalisation, agriculture is space extensive, it relies on soils requiring a high degree of maintenance, its suffers inclement weather, it involves waiting out the growing season until crops can be harvested and profits made, and so on. How does capital take on such risky business? Indeed, why would it? In an influential book, Goodman et al. (1987) explored these sorts of questions. Building on the work of Karl Kautsky, Susan Mann, James Dickinson and other Marxist theorists of the 'agrarian question', they argued that capital has developed two basic strategies to deal with agriculture. Over time, productive activities that once were the remit of the farm household or village economy were *appropriated* and redeveloped on an industrial basis, e.g., farm implements and machinery or food processing and preserving. A strategy of *substitution* was also devised, whereby 'natural' material inputs were replaced by industrially produced inputs: fossil fuels replaced draught animals, and to a degree people, as a power source; manufactured, chemical fertilizers replaced manures and mulches; factory produced food additives replaced whole foods. Appropriation and substitution speak to the power of industrial capital to get around certain nature-imposed obstacles (cf. McMichael 2006 on 'metabolic rift'). Finance capital has played an important role too. Again, as a generalisation, the history of capitalist agriculture is of increasing capital intensiveness: farmers have had to purchase more and more inputs, and frequently more land, in order to remain successful. They have not always had the money to do so, and still often do not. This has enlarged the role of finance capital in capitalist agriculture. Over the course of the 20th century, banks, insurance companies, and the securities 'industry' have all developed financial instruments (e.g., crop loans, farm mortgages, bonds) as a way to capitalise on the obstacles nature poses (see Henderson, 1999). This does not mean that environmental constraints or the matter of nature are left behind. It does mean that these can worked-up to the same old ends: making money. Thus, in the timber industry of the Pacific Northwest of the United States, that Scott Prudham (2005) has written about in *Knock on Wood*, the long growing season of Douglas Fir, and no less the specific woody traits of that species matter to the regional geography of production, to labour relations in the forest, and to the scale of timber firms. But precisely through these natural dynamics, and the environmental effects that the industry produces (e.g. reduction of spotted owl habitat) new inducements to capital also result: commercial forest tree breeding, business-friendly eco-regulation. None of this is free from contradiction; capital does not get the playground to itself.

---

equitable – the commodity equivalent – with a $40 ton of produced carbon from the Houston oil industry. . . . And whether carbon is or is not released into the atmosphere becomes, literally, a matter of capitalist equivocation' (Smith, 2006, p. 29). To these developments Smith provocatively asks, what will happen when the credit system collapses and cannot impose itself as that which it is not? (p. 34 – see also p. 25).

The Limits to Marxism's Nature?

We have in contention two urges. One of these sees the need for a natural outside to capital, a set of conditions of capitalist possibility that capitalists cannot make for themselves, but which they are free to spoil (until state and social movements intervene): the earth from which are drawn use values, the bodies of workers from which is drawn labour power, and sociality itself from which is drawn the capacity to be cooperative at all (cf. Benton, 1989; Altvater, 1991; Swyngedouw, 1999; Bridge, 2000). The other urge sees a fundamentally altered nature within which no natural pre-conditions are left untouched. There are 'natural processes', sure enough but nowhere on the planet do these exist without an admixture from human species being (cf. Harvey, 1996). The ontological positions taken by these urges are not easily reconciled. At risk of splitting the difference it is worth noting the limited truth told by each. Thus, to put a fine point on the matter, it is important to say that when labourers are crushed and beaten, sometimes shot in the street or poisoned by coal dust or effectively locked inside the factory gates, it matters to notice this in and of itself. And surely it is important that ecosystems are destroyed and communities laid waste? We can read O'Connor as saying there is a history of capitalism which involves forms of, well, death; deaths that are not going to be brought back to life by a nature-producing capitalism for which crisis is a use value during spasms of economic restructuring. (That is to say, perhaps the political significance of O'Connor's thesis is revealed more strongly through a change of perspective: before the significance of ruined production conditions becomes evident to capital, might they not be made evident to us?) However, if these deaths are to be struggled against, and if these deaths were once forms of life that capital alone did not make, must we invoke a natural order to be preserved and conserved as the alternative? I will get to that question in a moment. What of the truth told by the 'production of nature' thesis? Smith poses the issue himself: there is no getting around the fact that we are always standing in the middle of a stream that we have had a hand in making. There is no way to live in a wholly external nature. The reality we are faced with is the one we have a hand in making. Yet, as he asks, what would a better politics of produced nature be like if we are not to leave it up to the astounding acts of recovery and sleight of hand capitalism seems so far to be capable of? If produced nature is indeed the project, who participates and how, whom does it serve and whom not?

Marxism is materialism. Everything worth noting; everything we possibly could note is traceable to being the productive parts of nature that we human beings are; to what we must do to play those parts; and to the specific kinds of collectives (and thus politics) of humans and non-humans that emerge from the ways the parts are played out. What more can be said? How can this conversation be moved along? Recent encounters between geographical Marxism and what some call 'immanentist' philosophy are instructive. I will draw upon some recent work by geographer Bruce Braun (2006) to make the case. See also Chapter Two of this book, written by Braun.

The idea of immanence will have some immediate appeal. It posits that 'things' and 'entities' are, in part, the effects of the processes that constitute them, rather than the other way around. These processes and the particular ways they come together congeal in ways that are of highly variable temporal duration and spatial extent – sometimes very durable, sometimes of seemingly well-bounded extent. But whether the appearance is that of 'things' like trees or glaciers, or mosquitoes or bees, it would be a mistake to find nature in such 'natural kinds'. By the same token, as Braun notes, it is a mistake to under-appreciate that these relative permanences are in certain respects constitutive of how nature is organised. (A view of nature as pure flow, pure process, in which anything is possible at any time, is not sustainable. Rather, there are limits 'to the open-ended actualisation of being' [Braun, 2006, p. 210].)

The immediate appeal of an immanentist outlook may well stop here, however. For as Braun insists the philosophers who propound this outlook (e.g., Michel Serres, Bruno Latour, Gilles Deleuze and Felix Guattari), view nature as organised in only a very particular sort of way. To wit, the immanentist view of nature is that its various parts/processes are not founded upon something more real than the parts/processes themselves. Nothing underlies the so-called 'plane of immanence': all things that are real are equally real. Evolution is not more real than any given organism; plate tectonics is not more real than any given mountain range. And there is nothing in nature that is determined by something outside it, supplemental to it, whether God, Spirit or Law. Reality is not 'stratified'. Moreover, the world is forever becoming 'otherwise': we are not witness to a replay of the same processes entering into the same interactions that they always have entered into. That matters are not settled beforehand imparts a political quality to the notion of immanentism: 'Whatever organisation exists at any given moment must be understood as an effect of the forces and practices that cause things to hold together in a particular way, even if to us the 'things' of the world appear stable and unchanging. . . . [And yet] at any given moment the rules of combination are precise: they have to do with *this* practice, *this* connection, *this* bifurcation – which is why becoming is the domain of ethics and politics, not blind chance, and why politics must always begin 'in the middle of things' (Braun, 2006, pp. 204, 211, emphasis in original). Why must politics now be a constituent feature of the 'natural'? It must be so in a world with humans. That is, a politics of nature, of humans in nature, of humans as more-than-human, cannot be about compliance to putative natural laws. These smack too much of the eternal, of forces outside time and history. As Braun (following another philosopher, Manuel DeLanda) stipulates, the immanentist conception of nature takes time seriously. Time is not about a replay of the eternal same; it is directional, irreversible. 'The earth is *in the making* . . . in ways that cannot simply be undone' (p. 206). It is open, not closed, and thus we are positioned ethically rather than with respect to discoverable, timeless essences and rules for living. (Note though how Altvater draws a somewhat contrasting conclusion: that nature is irreversible – fossil energy capitalism will come to a close – should inform our 'rules' for living! [Altvater, 2006].)

Braun argues that immanentism disturbs the ecological Marxism of O'Connor because of the latter's emphasis on a pre-given nature, *as well as* the Marxism of geographers such as Neil Smith and David Harvey (whose work Braun's essay addresses), whose production of nature thesis actually shares with immanentism a notion of nature as an open-ended becoming. For Braun the problem has to do with centering the production of nature too much on capital. In the production of nature thesis, recall, the social-nature produced under capitalism takes on the qualities of capital itself. (Although Smith may be more flexible than Braun gives him credit for, in so far as Smith argues that 'second nature' has become a *context* for the operations of 'first nature'). Social-nature comes to serve the dictates of commodity production and exchange. Its geographical particularities reflect the uneven development that is one of capitalism's hallmarks. Even its dystopic qualities (e.g., anthropogenic climate change) are recuperated into the circulation of finance capital (e.g., carbon credits). As Braun reads Harvey, 'the future of socio-nature is very quickly reduced to a movement *internal* to the temporal rhythms and spatial orderings of capitalism' (Braun, 2006, p. 213), and this despite periodic attempts by Harvey to offer more complex accounts of produced nature. More compelling for Braun, because they too cry out for investigation and interpretation, are the myriad social-natures that are not accountable to capitalism as such, but have differing and highly variable sources, be these the techno-sciences behind genetic engineering, the different nature ideologies of, say, North America and Europe that inform differential creations of landscape, or the geopolitical concerns over sovereignty and security that inform new notions of biological life (see Braun, 2007; cf. Braun, 2000; Whatmore, 2002; Latour, 2004).

One could respond to this critique of a too capital-centric produced-nature that Harvey and Smith, like Marx, are at the end of the day only concerned with those sites and situations where the capitalist mode of production prevails. But Braun cautions that a 'value-theoretical understanding is valid in the general or universal sense that Harvey presumes, *only so far as such conditions that make capitalist calculation possible are extended and "hold together" over space and time'* (Braun, 2006, p. 214, emphasis in the original). These are very carefully chosen words. The issue at stake for Braun has indeed everything to do with the conditions that make capitalist calculation possible; it is not about replacing capitalist calculation as such. For him, these conditions are not determined by capital alone but are instead irreducibly heterogeneous. The influential work of Timothy Mitchell (2002) provides him with a case in point. In Mitchell's richly detailed case study of the emergence of 'the economy' in Egypt – the emergence, that is, of the belief in a distinct, autonomous sphere endowed with causative powers – a wide range of actors were actually at work: capitalists, yes, but also (as it happens) the anopheles mosquito, the malaria causing faciparum parasite, the chemistry of ammonium nitrate, the hydraulic force of the river, family networks, imperial connections, and more. None of these reduce to capital, or the economy. None were wholly determined nor incorporated within it. Their interactions do, however, 'make possible a world that somehow seems the outcome of human rationality and programming' (Mitchell, 2002, p. 30).

The appeal of Mitchell's account, Braun argues, is its resistance to abstraction. Explanatory power and, no less, anti-capitalist politics, are reduced once a certain level of abstraction is reached in the attempt to explain events. Otherwise the temptation may be to attribute causal powers to the abstraction itself. To be avoided is any trace of the idea that there are two ontologies at work: one ontology for 'practices and things', as Braun puts it (Braun 2006: 215), and another for underlying, causative logics. One might respond that, just the same, we need accounts that spell out why interactions come to be ordered the way they are. For Harvey, interactions within the capitalist world are ordered by the exigencies of value, since it is value that those who rule capitalism must tend to. It is value that rules the rulers, even. But now we can hear the response of immanentist philosophy: value must itself be assembled and provided a territory, and once assembled it does not fully internalise those elements which enter into it (they remain potentials, as it were). Value, therefore, cannot account for itself; it must be accounted for:

> If, at the dawn of the 21st century, capital has become an axiomatic it is so not because it has magical powers, but because of the fine weave of practices – from the congresses

of the WTO to the boardrooms of the World Bank, from intellectual property laws to labor legislation in developing countries, from the laboratories of US universities to the field mapping of genetic materials in Mexico and Costa Rica – that produce a territory for the 'law' of value to operate on, and where profit-oriented economic rationalities can both occur and moreover, contribute to the 'decoding' of the social-ecological assemblage defined by these networks. To the extent that networks are constituted in such a manner, then an analysis of capitalism, its institutions and its imperatives is clearly on the table. And to the extent that these 'hold together', then Harvey is entirely correct to call attention to how the drive for profit can draw new places and new ecologies in relation, and how capitalism can 'reterritorialize' the earth (Braun, 2006, p. 217).

Braun has, then, treated two problems, that of Marxism's capital-centred production of nature and that of a pre-given capital logic capable of making the world a reflection of itself. He proposes, via Deleuze, Guattari, and others in the immanentist tradition of thought that the 'plane of immanence' is the way to handle these two problems. The plane of immanence requires we pose once again the question of an outside to capital. (Although we do this differently than O'Connor: Braun would seem to disagree with O'Connor that conditions of capitalist production are not produced by capital, as if the dividing line could be drawn easily.) Why? Because we refuse to see capital as a reality 'underlying' other entities or as self-organising according to a predetermined law. The plane of immanence also places 'man' in the same domain of action as non-human entities. That which appears to be a human accomplishment is merely an effect, a joint accomplishment of humans and non-humans alike: an accomplishment that occurs without sublation, even while the opposite seems to be the case.[8] (See Box 17.5.)

The advantage of the plane of immanence as deployed by Braun – see Box 17.5 for comparisons with the 'Second contradiction of capitalism' and 'production of nature' approaches – is its refusal to accept natural orders that not only constrain unnecessarily our notions of who and how humans and non-humans might 'become', but that have been all too useful in the exploitation of humans and non-humans alike. At the same time the plane of immanence does not suggest that all things are possible at all times – it understands the, let us say, stickiness of assemblages. Some Marxists worry that these post-structural approaches lose the specificity of capitalism by dissolving it into its context or argue that Marx already went some way in accounting for them (e.g., Kirsch and Mitchell, 2004). But even these scholars, many of them at least, appreciate that Marx left themes undeveloped and that it is ultimately to the good to keep to a range of sources and ideas. Whether Marxism is amenable to Braun's read of capitalism also depends on how we read Marx. Doubtless, by the time he wrote *Capital* Marx was primarily interested in theorising capitalism as such, whilst also holding to the notion that given capitalist formations were rarely *only* capitalistic. This put him in a jam. It is by no means easy to capture the relationality he wished to convey, a capitalism that was expanding, yet mutually defined in specific circumstances by the forces with which it contended. So, that capitalism *here* differs from *there;* that different factions of capital (money capital versus productive capital versus commodity capital) have variable spatial and temporal mobility; that these mobilities have to be locally established, and heterogeneously so, to use Braun's term – these are realities that geographers are well positioned to appreciate. Braun's critique taps into another enduring theme of Marxist geography. Geographical Marxists have, as a group, always been leery of

Box 17.5   The 2nd Contradiction of Capitalism, Production of Nature, and
Plane of Immanence: Propositions and Politics

The '2nd contradiction of capitalism'

- Capital subsists on conditions (e.g. ecosystems, labor power) outside itself, which are treated like commodities. The emphasis on making surplus value, forces capital to degrade these very conditions: capitalism produces scarcity.
- Social movements and/or the state are forced to step in as counter-movements to regulate capital's excesses, on a more rather than less frequent basis. Capitalism is shown to be unable to subsist on its own.
- Politics: If capitalism is destructive of entities that society acknowledges are valuable and useful, can there emerge a social movement or political collective powerful enough to dislodge capitalism and sustain social reproduction and environmental conditions?

Production of nature:

- Capitalism intervenes in the nature-based conditions upon which it depends; little pure nature is left. Nature is thus said to be produced. But production of nature is incomplete. Capitalism and nature are locked up in each other, refusing each other a pure state of being.
- Production of nature nonetheless emphasizes that *more elements* of nature are found for capital's circuits, from plant and animal DNA to human body organs to seeds and water. Even degraded nature is grist for capital's mills (e.g., pollution credits).
- Politics: The clock cannot be turned back. We will continue to live in a world where nature is produced. Who participates in that process and how, whom does that process serve and whom not? Why should these questions be settled on capitalism's terms?

The plane of immanence

- Entities (things), whether 'natural' or 'social' (the distinction may be artificial) are not the building blocks of processes, but the effect of processes. Entities and relations among them have emergent qualities: they are in a continual state of becoming, though they are not free to form any which way.
- Capital (the 'law of value'), while real enough, is irreducibly heterogeneous: it does not fully internalize the elements that enter into it. To the extent it has a territory within which to operate, capital seems, but only seems, the central organizing force of the heterogeneous processes that comprise it.
- Politics: Radical efforts to address environmental problems must not reduce them to 'capitalism', nor can they address the injustices of 'capitalism' by invoking a pre-given state of nature.

positing nature in deterministic terms. Talk of 'limits' is always approached carefully, the preference being for how present accomplishments might be leveraged towards a radical future.

Coda

In Marxist scholarship today, there is indeed renewed interest in reading the diversity of Marx's writings. This is a theme worth ending on. A major schism between some Marxists and some 'greens' is over the latter's accusation that actually existing socialisms, at least prior to 1989, have been overly productivist, i.e., allowing industrial development to run roughshod over the earth, placing 'man' at the center of the economy. Was this Marx's view? Again, complexity abounds. Ample evidence indicates that Marx thought industrial capitalism, with its galloping productive capacity and its centralising and socialising functions, were creating the conditions of possibility for an entirely new kind of society, even while leaving a wake of destruction. But if the question is whether Marx posited people as sole creators of wealth, and the absolute center of 'value', the answer must be a firm 'No'. His 'Critique of the Gotha Programme' (Marx, 1978, pp. 525–26) is definitive on this score. Addressing a group of socialists' plans for a new economy, he heaps uncompromising scorn on the notion that labour alone creates wealth. Wealth is a joint accomplishment of humans and non-humans he reminded them. That wealth is the creation of labour alone he considered bourgeois tyranny. And we can or should see why. It is capitalism, with its history of stripping away means of production only to leave workers their labour power that posits 'man' as the creator of wealth. It is capitalism, with its Spartan view of work and its injunction to work harder and longer, that places 'man' at the center, even as it struggles to undo the accomplishments of 'man' as soon as threatening gains have been made. But it was precisely this conceit that capitalism cannot make a reality (cf. Negri, 1992; Althusser, 1997; Dussel, 2001). The capitalist abstraction of 'man' fails; always has, always will.

NOTES

1. It must be noted that environmental themes in Marxist geography, whilst introduced early in the Marxist turn (e.g., Harvey 1974; cf. Harvey, 1996), remained for some time a latent interest, more implied than rigorously explored, and playing second fiddle to research on urbanisation and industrial location (FitzSimmons, 1989). Strongly Marxist-influenced political ecology in geography would be an important exception (e.g., Watts 1983; Peet and Watts, 1996).
2. This idea of an average, together with the distinction between labour power and labour as such, is what sets his ideas apart from other 'labour theories' of value.
3. One basic idea of Marx's is that the contradictions generated within capital (e.g., between capital and labour) would be resolved by social powers that capital itself has a hand in creating. For example, in so far as labour power is developed mutually along with other means of production, labour develops the capacity to take over production as such, recognising the superfluity of 'capital'. See Marx's chapter on 'Machinery and modern industry' in the first volume of *Capital*. Marx was clearly fascinated by and worked to foment the new kinds of collective identities that capitalism fostered and depended upon. His hope was that these collectivities arising as part and parcel of capital would emerge as social forces potentially transformative of capital. For an excellent discussion of the

nature of these collectivities, and especially of the argument that the famed 'proletariat' are not 'workers', but in fact exceed the limits of 'identity politics' in any form see Thoburn (2003).

4. Just so. But still up for grabs in a research area that Marxists by no means own is to what sort of relations these very sociotechnical relations are themselves related. There are, for example, impulses to *collapse* capitalism into broader socio-technical relations (e.g., Whatmore, 2002), to argue that capitalism *lends a distinctive taint* to dominant sociotechnical relations (see Castree, 2002; Kirsch and Mitchell, 2004) and to *rethink* what capitalism and sociotechnical relations are *cum* nature (e.g., Braun, 2006).

5. This is a point made widely by 'green' thinking Marxists both before and after O'Connor – see Benton, 1996b; Enzenberger, 1996; Leff, 1996; Perelman, 1996; Skirbekk, 1996; Altvater, 2006; Bernstein and Woodhouse, 2006; Löwy, 2006; McMichael, 2006; Swyngedouw, 2006.

6. I am stressing the difference that capitalist work makes to the body of labourers. One finds other themes of difference in Marx, though not satisfyingly developed. Particularly important are the differences produced before or along with capitalist work – differences of race, gender, sexuality, that are forced into articulation with capitalist work (and capitalist consumption). See Spivak's (1985) 'Scattered speculations on the question of value'. Smith frames such differences in terms of the concept of external nature: 'By corollary, of course, this external conception of nature becomes a powerful ideological tool for justifying . . . forms of social difference and inequality as 'natural' rather than social in their social genesis' (Smith, 2006, p. 23). That is to say, the idea of nature as outside the human can be used by some human beings to cast marginalised other human beings as *belonging* to the realm of controllable (or uncontrollable, excessive, overabundant) nature, or as being closer to nature and further from culture (as enslaved to the body and its emotions, for example).

7. Readers should note Smith is cannier than some of the critics of the production of nature thesis allow. Read Bakker and Bridge (2006) and compare with the Smith (2006) essay summarised here.

8. Why though does it appear that human actions, human projects, human objects, seem only human? How is the ruse perpetuated? Why is it not patently obvious that this is wrong? There can be no easy answer, although it is easy to see how such an effect could be generated, first by the qualities of species being: the requirement to 'always-already' be in contact with nature with hand and brains. To forfeit this is to forfeit life: As Marx wrote, we are always objects for another subject. Second, class processes perhaps serve to reinforce anthropocentric ways of thinking. Here I mean class in the broadest sense of the word: the elite cloistering of technologies (economic, scientific, political) that perpetuate the notion of nature as external. One could add that, after all, a great deal depends on the illusion that things are under control. Daily life would seem to depend upon a high degree of faith in our ability to use certain 'natural' forces in order to keep other natural forces at bay.

BIBLIOGRAPHY

Albo, G. (2006) The limits of eco-localism and scale, strategy, socialism. In L. Panitch and C. Leys (eds), *Socialist Register 2007: Coming to Terms with Nature*. New York: Monthly Review Press, 337–63.

Althusser, L. (1997) The object of capital. In L. Althusser and E. Balibar (eds), *Reading Capital*. London: Verso, pp. 73–198.

Altvater, E. (1991) *The Future of the Market: An Essay on the Regulation of Money and Nature After the Collapse of 'Actually Existing Socialism'*, trans. P. Camiller. London: Verso.

Altvater, E. (2006) The social and natural environment of fossil capitalism. In L. Panitch and C. Leys (eds), *Socialist Register 2007: Coming to Terms with Nature*. New York: Monthly Review Press, pp. 37–59.

Arthur, C. J. (2002) *The New Dialectic and Marx's Capital*. Boston: Brill.

Bakker, K. and Bridge, G. (2006) Material worlds? Resource geographies and the 'matter of nature'. *Progress in Human Geography*, 30(1), 5–27.

Benton, T. (1989) Marxism and natural limits. *New Left Review*, 178, 51–86.

Benton, T. (ed.) (1996a) *The Greening of Marxism*. New York: Guilford.

Benton, T. (1996b) Marxism and natural limits: an ecological critique ad reconstruction. In T. Benton (ed.), *The Greening of Marxism*. New York: Guilford, pp. 157–83.

Bernstein, H. and Woodhouse, P. (2006) Africa: eco-populist utopias and (micro) capitalist realities. In L. Panitch and C. Leys (eds), *Socialist Register 2007: Coming to Terms with Nature*. New York: Monthly Review Press, pp. 147–69.

Bowker, G. (1994) *Science on the Run: Information Management and Industrial Geophysics at Schlumberger, 1920–1940*. Cambridge, MA: MIT Press.

Braun, B. (2000) Producing vertical territory: geology and governmentality in late-Victorian Canada. *Ecumene*, 7(1), 7–46.

Braun, B. (2006) Towards a new earth and a new humanity. In N. Castree and D. Gregory (eds), *David Harvey: A Critical Reader*. Malden, MA: Blackwell, pp. 191–222.

Braun, B. (2007) Biopolitics and the molecularization of life. *Cultural Geographies*, 14(1), 6–28.

Bridge, G. (2000) The social regulation of resource access and environmental impact: production, nature, and contradiction in the US copper industry. *Geoforum*, 31, 237–56.

Brunnengräber, A. (2006) The political economy of the Kyoto protocol. L. Panitch and C. Leys (eds), *Socialist Register 2007: Coming to Terms with Nature*. New York: Monthly Review Press, pp. 213–30.

Buck, D. (2006) The ecological question: can capitalism prevail? In L. Panitch and C. Leys (eds), *Socialist Register 2007: Coming to Terms with Nature*. New York: Monthly Review Press, pp. 60–71.

Bunge, W. (1971) *Fitzgerald: Geography of a Revolution*. Cambridge, MA: Schenkman Publishing Company.

Castree, N. (1995) The nature of produced nature: materiality and knowledge construction in Marxism. *Antipode*, 27(1), 12–48.

Castree, N. (1996) Birds, mice and geography. *Transactions of the Institute of British Geography*, 21(3), 342–62.

Castree, N. (2002) False antithesis? Marxism, nature and actor-networks. *Antipode*, 34(4), 111–46.

Castree, N. (2003) Commodifying what nature? *Progress in Human Geography*, 27(3), 273–97.

Castree, N. (2005) *Nature*. New York: Routledge.

Coronil, F. (2000) Towards a critique of globalcentrism: speculations on capitalism's nature. *Public Culture*, 12(2), 351–74.

Dussel, E. (2001) *Towards an Unknown Marx: A Commentary on the Manuscripts of 1861–63*. London: Routledge.

Enzenberger, H. M. (1996) A critique of political ecology. In T. Benton (ed.), *The Greening of Marxism*. New York: Guilford, pp. 17–49.

FitzSimmons, M. (1989) The matter of nature. *Antipode*, 21(2), 106–20.

Gare, A. (1996) Soviet environmentalism: the path not taken. In T. Benton (ed.), *The Greening of Marxism*. New York: Guilford, pp. 111–28.

Goldman, M. (2005) *Imperial Nature: The World Bank and Struggles for Social Justice in the Age of Globalization*. New Haven: Yale University Press.

Goodman, D., Sorj, B. and Wilkinson, J. (1987) *From Farming to Biotechnology: A Theory of Agro-Industrial Development*. New York: Basil Blackwell.

Harriss-White, B. and Harriss, E. (2006) Unsustainable capitalism: can capitalism prevail? In L. Panitch and C. Leys (eds), *Socialist Register 2007: Coming to Terms with Nature*. New York: Monthly Review Press, pp. 72–101.

Harvey, D. (1974) Population, resources, and the ideology of science. *Economic Geography*, 50(2), 256–77.

Harvey, D. (1982) *The Limits to Capital*. Oxford: Blackwell.

Harvey, D. (1989) *The Condition of Postmodernity*. Cambridge: Blackwell.

Harvey, D. (1996) *Justice, Nature, and the Geography of Difference*. Oxford: Blackwell.

Harvey, D. (2003) *The New Imperialism*. New York: Oxford University Press.

Henderson, G. (1999) *California and the Fictions of Capital*. New York: Oxford University Press.

Henderson, G. (2004) Value: the many-headed Hydra. *Antipode*, 36(3), 445–60.

Katz, C. (2001) Vagabond capitalism and the necessity of social reproduction. *Antipode*, 33(4), 709–28.

Kirsch, S. and Mitchell, D. (2004) The nature of things: dead labor, nonhuman actors, and the persistence of Marxism. *Antipode*, 36(4), 687–705.

Latour, B. (1993) *We Have Never Been Modern*. Cambridge, MA: Harvard University Press.

Latour, B. (2004) Politics of nature: how to bring the sciences into democracy. Cambridge, MA: Harvard University Press.

Leff, E. (1996) Marxism and the environmental question: from the critical theory of production to an environmental rationality for sustainable development. In T. Benton (ed.), *The Greening of Marxism*. New York: Guilford, pp. 137–56.

Löwy, M. (2006) Eco-socialism and democratic planning. In L. Panitch and C. Leys (eds), *Socialist Register 2007: Coming to Terms with Nature*. New York: Monthly Review Press, pp. 294–309.

Martinez-Alier, J. (2006) Social metabolism and environmental conflicts. In L. Panitch and C. Leys (eds), *Socialist Register 2007: Coming to Terms with Nature*. New York: Monthly Review Press, pp. 273–93.

Marx, K. (1967) *Capital*, volume 1. New York: International Publishers.

Marx, K. (1973) *Grundrisse: Foundations of the Critique of Political Economy*. New York: Penguin.

Marx, K. (1978) Critique of the Gotha Programme. In R. Tucker (ed.), *The Marx-Engels Reader*, 2nd edn. New York: Norton, pp. 525–41.

McMichael, P. (2006) Feeding the world: agriculture, development and ecology. In L. Panitch and C. Leys (eds), *Socialist Register 2007: Coming to Terms with Nature*. New York: Monthly Review Press, pp. 170–94.

Mitchell, K., Marston, S. and Katz, C. (eds) (2004) *Life's Work: Geographies of Social Reproduction*. Malden, MA: Blackwell.

Mitchell, T. (2002) *Rule of Experts: Egypt, Techno-Politics, Modernity*. Berkeley: University of California Press.

Negri, A. (1992) *Marx beyond Marx: Lessons on the Grundrisse*. London: Pluto Press.

O'Connor, J. (1997) The second contradiction of capitalism. In J. O'Connor, *Natural Causes: Essays Ecological Marxism*. New York: Guilford, pp. 158–77.

Panitch, L. and Leys, C. (eds) (2006) *Socialist Register 2007: Coming to Terms with Nature*. New York: Monthly Review Press.

Peet, R. and Watts, M. (eds) (1996) *Liberation Ecologies: Environment, Development, Social Movements*. New York: Routledge.

Perelman, M. (1996) Marx and resource scarcity. In T. Benton (ed.), *The Greening of Marxism*. New York: Guilford, pp. 64–80.

Postone, M. (1999) Contemporary historical transformations: beyond postindustrial and neo-Marxist theories. *Current Perspectives in Social Theory*, 19, 3–53.

Prudham, S. (2005) *Knock on Wood: Nature as Commodity in Douglas-Fir Country*. New York: Routledge.

Pulido, L. (1994) Restructuring and the contraction and expansion of environmental rights in the United States. *Environment and Planning A*, 26(6), 915–36.

Retort (2006) *Afflicted Powers: Capital and Spectacle in a New Age of War*, 2nd edn. New York: Verso.

Skirbekk, G. (1996) Marxism and ecology. In T. Benton (ed.), *The Greening of Marxism*. New York: Guilford, pp. 129–36.

Smith, N. (1991) *Uneven Development: Nature, Capital, and the Production of Space*. Cambridge, MA: Blackwell.

Smith, N. (2006) Nature as accumulation strategy. In L. Panitch and C. Leys (eds), *Socialist Register 2007: Coming to Terms with Nature*. New York: Monthly Review Press, pp. 16–36.

Soper, K. (1996) Greening prometheus: Marxism and ecology. In T. Benton (ed.), *The Greening of Marxism*. New York: Guilford, pp. 81–99.

Spivak, G. (1985) Scattered speculations on the question of value. *Diacritics*, 15(4), 73–93.

Storper, M. and Walker, R. (1989) *The Capitalist Imperative: Territory, Technology, and Industrial Growth*. New York: Basil Blackwell.

Swyngedouw, E. (1999) Modernity and hybridity: nature, *Regeneracionismo*, and production of the Spanish waterscape, 1890–1930. *Annals of the Association of American Geographers*, 89(3), 443–65.

Swyngedouw, E. (2006) Water, money, and power. In L. Panitch and C. Leys (eds), *Socialist Register 2007: Coming to Terms with Nature*. New York: Monthly Review Press, pp. 195–212.

Thoburn, N. (2003) *Deleuze, Marx, and Politics*. New York: Routledge.

Waterstone, M. (1993) Adrift on a sea of platitudes: why we will not resolve the greenhouse issue. *Environmental Management*, 17(2), 141–52.

Watts, M. (1983) *Silent violence: food, famine, and peasantry in northern Nigeria*. Berkeley: University of California Press.

Wen, D. and Li, M. (2006) China: hyper-development and environmental crisis. In L. Panitch and C. Leys (eds), *Socialist Register 2007: Coming to Terms with Nature*. New York: Monthly Review Press, pp. 130–46.

Whatmore, S. (2002) *Hybrid Geographies: Natures, Cultures, Spaces*. Thousand Oaks: Sage.

Chapter 18

After Nature: Entangled Worlds

Owain Jones

No-one yet has made the crossing from nature to society, or vice versa, and no-one ever will. There is no such boundary to be crossed. (Ingold, 2005, p. 508)

Introduction

Although there is but one world in common, somehow it has long been common to suppose that the world is in fact divided in two: into a world of nature and another, one of culture. For more than four centuries this nature/culture dualism has shaped knowledge, politics, and ethics in the West – with often debilitating consequences.

From this long-established perspective, the title of my chapter, 'After Nature', might be understood as referring to the pursuit of nature, as if nature were something elusive or endangered that I am seeking or lamenting. This is a very common rhetoric at a time when human impact on the environment all around us seems greater than ever. Bill McKibben (1990) has written movingly about the 'end of nature' now that global climate change means there is no place left on earth free from the mark of human influence.

In contrast to that vision of what comes after nature, this chapter understands what is entailed by 'after' rather differently. My focus here is on the end of that binary understanding of a world divided cleanly in two. In its place, this chapter explores a number of new analytical approaches in geography, and elsewhere, that seek to abolish this binary division. Despite important differences among them, these approaches are all 'after nature' in the sense that they reject the idea of nature as an ontologically pure realm that exists outside, and apart from, a separate one of human knowledge, culture, and society. Instead they address life in ways that recognise it as an ongoing outcome of complex interplays, or entanglements, between all manner of processes and elements – bio-physical, economic, cultural, technological, human and non-human. It should also be clear that not only is the 'nature' side of the nature/culture dualism being called into question, but so too is the 'cultural' or 'social' as well. In dissolving the divide between them, the way we see both is transformed.

It might seem perverse, even dangerous, to abandon the idea of nature as a set of pristine spaces and autonomous processes and structures at this point in history. After all, as McKibben and other environmentalists continually remind us, we face a deepening global environmental crisis. Even if human-generated climate change is perhaps still open to a degree of scientific uncertainty, it is not the only point of concern about human impacts on the environment around us. Deforestation, the over fishing of the oceans, and the degrading of soil and water resources are other very pressing matters of concern. Nature seems to be in trouble, and many environmentalists insist that it is precisely our failure to respect its autonomy and material limits that have gotten us into this mess. At the same time, scientists continue to reveal the amazing depths of 'nature' in ever more detail. Wildlife films show astonishing pictures of 'life on earth'. The Hubble telescope reveals wondrous formations in deep space. A new generation of particle accelerators are hunting for the primary sub-atomic particles of existence itself. In such circumstances is it responsible, let alone reasonable, to be calling into question the very idea of nature itself?

'Yes' is the response contained in a growing body of research in Geography and cognate disciplines that rejects the nature/culture divide as an obstacle to forging a world in common. The 'modern constitution', as Latour (1993) terms it, has made us blind to the everyday realities of life on earth, and thus unable to grasp how it comes about, how it goes on, and how it might be shaded one way or the other. Systems such as those taking food from field to market involve elements of 'nature' and 'culture' so closely woven together that separating them out into 'natural' dimensions to be studied by physical geographers and social ones of concern to human geographers is neither practical nor analytically possible. And yet, under the modern constitution, that is exactly what we have sought to do. Distractions about policing the arbitrary and inevitably leaky borders between 'natural' foods and artificial ones, for example, have hobbled more constructive interventions in the food system about what makes for better and more ethically acceptable foods. Discussions over such things as GM foods have been constrained by the nature/culture dualism that breaks down discussion into issues of scientific fact and risk, on the one hand, and political desirability or consumer choice on the other.

Abandoning these longstanding habits of thought opens up an exciting conceptual landscape in which the world is no longer fixed by some timeless and essential nature, but instead is understood as the ongoing outcome of myriad entanglements of elements and processes spanning both sides of the supposed divide of old between nature and culture.

Being 'after nature' thus involves relational ways for thinking about the environment, nature, and society. Emphases on relationality, flows, networks, and ecologies have come into Geography from a variety of sources, which I review in the first part of this chapter. There are important conceptual and political differences, for example, between the approaches to relationality offered by actor-network theorists and Marxist dialecticians. These ethical and political implications are what I turn to in the second part of the chapter. They require new approaches to disciplinary divisions and methodologies, and to our ways of thinking about politics and ethics. The project of moving beyond, or after nature is thus an ambitious one, and the stakes are high. Some suggest that abandoning the idea of nature and the nature culture/

dualism brings risks and that we 'should not be too quick, in our renewed under-standing of its contingency, to now jettison any concept of nature at all' (Castree and Head, 2007, p. 2). But surely it has to be admitted that the modern constitution has failed us in certain key ways, not least in terms of environmental security and how we live with what has been termed nature and how we understand ourselves. Constitutional reform is needed now!

Understanding the World 'After' Nature

Pressure on the modernist nature/culture dualism has come from various sources. Somewhat ironically perhaps, some of the most important influences have been developments in the natural sciences, which have exposed the implausibility of any sharp break between the social and the natural. Donna Haraway (1992a, p. 193) suggests that 'biology and evolutionary theory . . . [have] reduced the line between humans and animals to a faint trace'. Increasing understanding of complex processes of interchange between systems (e.g., ecologies and biospheric exchange systems) shows that no clear divide exists between the living community of the world and its material environment. The fundamental condition of all life is relational rather than autonomous and independent.

Putting the increasingly detailed evidence being gathered by environmental archaeologists alongside histories of societies shows that what were once hard-to-explain rises and falls of social systems can often be attributed to long-term varia-tions and cycles in natural systems, such as shifting climate zones or the fallout from extreme events like super volcano eruptions. The interpenetration of natural and social systems has always been far greater than we have allowed. History has been a narrow, partial story of the strictly social. But if we look closely enough

> What we will find there . . . is not so much an interplay between two kinds of history, human and non-human, as a history comprised by the interplay of diverse human and non-human agents in their mutual relationships (Ingold, 2005, p. 506).

In addition to these emerging knowledges of relational life processes spun around the world are knowledges of nature-culture entanglements on a different scale – that of the body. Scientific advances in genetics and neurology challenge ideas dating back to Rene Descartes and the origins of the Enlightenment, of a conscious, ratio-nal, language-using mind as something set apart from, and in command of, its material body. Instead contemporary biological science emphasises the complex relationships between mind and body, while also holding out the prospect of human enhancement through pharmaceutical and other interventions at the molecular level that call into question any simple idea of a fixed and essential human nature (Rose, 2001). As Thrift (2008, p. 252) points out in relation to non-representational theory, 'affect is a challenge to what we regard as the social because it involves thinking about waves of influence which depend on biology to an extent that is rarely rec-ognised or theorised in the social sciences'. In turn the body is not set apart from, or at odds with space/environment, but a dwelt inter-emergent production.

Challenges to the intellectual divide between society and nature are also emerging thick and fast from a range of other disciplines across the social sciences and humanities, including environmental ethics, environmental history, sociology, poli-tics and geography and also in art and literature. Within these are a series of theoretical approaches to the 'after natural' that for all their differences also share some basic traits. For example, they are generally processual views of the world: they see reality as unfolding or becoming, rather than being composed of fixed and stable structures to be revealed. They often draw upon a minority philosophical legacy, which has been at odds with Cartesian foundations from the start (see Watts, 2005). Marxism, poststructuralism, phenomenology, expanded notions of ecology ('new ecology'), are just some of the approaches drawn upon.

All bring valuable insights to the after nature project, yet also differ in their angle of their 'attack' on the binary world view. I sketch out each one in turn. As I do so I consider some areas of similarity and difference between them, and some of their comparative strengths and weaknesses. But the point is not to see these positions as simply exclusive or contradictory. These approaches can (to varying degrees) be drawn upon pragmatically in pluralist research frameworks. Theories are tools, possibly useful as we approach the *complexity* of the after nature world. We might need many to work effectively. Just as we see the world as an increasingly entangled, complex process of becoming, so our forms of knowledge (need to) become entangled and creative in response. Inter-disciplinary and inter-theoretical approaches are becoming increasingly common within, and between the natural and social sciences, and arts and humanities.

New ecologies

Since it was developed by the 19th-century German biologist Ernst Haeckel, ecology has been regarded as one of the natural sciences and a branch of biology. It studies interactions between organisms and their environment including interactions within and between species, groups, and individuals. It has multiple forms centred on population, community, ecosystem, behavioural, and spatial dynamics. This diversity is instructive because once the principle of establishing connections and relationships within the natural world was established many differing paths of inquiry presented themselves as vital.

Mostly, but not exclusively, ecology has focused on the nature side of the nature/culture dualism. However, given its focus on relationships, it has increasingly moved towards a thoroughly relational view of life. In so doing it has drawn in other disciplines such as geography, genetics, chemistry, and physics in order to effectively map webs of life. In effect ecology, in its tracing out of those vital connections, has bumped up against the limits of the nature/culture world view.

Ideas of 'new ecology' seek to trace life in its various forms and connections across the nature/culture divide *and* to question long established ideas of nature as separate and a stable world of equilibrium, harmony, and balance. In their place new ecology stresses 'disequilibria, instability, and even chaotic fluctuations in biophysical environments, both natural and human impacted' (Zimmerer, 1994, p. 108). This raises fundamental questions about the limits to ecosystem predictability, management and control. More emphasis is placed upon the importance of spatial, scalar, and temporal variation, and on complexity and uncertainty within ecosystems. This involves a rejection of the view of nature as a separate realm into which human life simply intrudes, and inevitably corrupts, distorts, and lessens, along with corresponding ideas of 'first nature' and 'second nature'.

Along with a greater emphasis on disequilibria in new ecology there is a (related) assumption that systems are more open than closed, with leakage and flux between systems occurring constantly. There is concern for spatial and temporal dynamics developed in detailed and situated studies of humans and non-humans in places, using, in particular, historical analysis as a way of explaining environmental change across time and space. Of particular interest are mechanisms though which the states and 'directions' of systems might be transformed. Within this the challenge is 'to regard human actors as always already *part of* complex and changeable biophysical systems' (Castree, 2005, p. 235, original emphasis).

This recognition opens up new geographies for ecologies, habitats, and biodiversity to be found (and created) in the world's proliferating urban spaces. With their webs of non-human life, in mosaics of abandoned and planned spaces such as gardens, parks, allotments, derelict land, transport network verges, car parks, roof tops and underground systems, cities can offer much richer habitats than intensively farmed, but apparently green, rural landscapes. Maxeiner and Miersch (2006) point out that Berlin is in fact the biodiversity 'hotspot' of Germany, being home to 141 species of birds and more types of wild flowers per square kilometre than just about anywhere else in the country. This comparative abundance may in part be the result of drastic declines in biodiversity outside cities, but it shows the extent to which 'nature' and 'society' are intimately tied together in ways that often escape our notice. Norman (2006, p. 16) goes so far as to state that 'the city and not the countryside is the true home of nature' and that 'the bigger the city, the more ecological niches it offers to nature'.

Recognising that the quintessentially social spaces of the city are in fact ecologically rich, or potentially so, opens up a new kind cosmopolitics. No longer are we concerned solely with multiculturalism but with a wider politics of conviviality encompassing humans and non-humans alike. Insofar as the century is going to be an urban one, in which the majority of the world's human population lives in cities, the quality of such urban hybridity is a vital matter. Such entanglements, where both humans and non-humans might collectively flourish, have often been neglected in scientific and political agenda focused on either one or other side of the nature/culture divide.

Social constructionism

The social constructionist approach mounts a concerted attack on the nature/culture dualism and raises key questions about knowledges and practices of nature. It sprang up as a response to realist approaches, which took at face value the vision of 'external nature' as presented by the modern constitution.

Social construction calls into question the very existence of a stable, objective, knowable realm of nature in the first place – 'nature has come to be seen as never simply, or not even, natural' (Castree, 2001, p. 16). This is not to deny the existence of material reality, but rather that it is always known through socially contingent and geographical variant languages and discourses which arise from and shape specific cultures, societies, and economies. Even spaces of 'wilderness', where pure nature (one might think) still abounds, are creations of powerful discourses of Western modernity (Cronon, 1996).

The social construction of nature works both discursively and materially (Castree, 2001). Discourse is about a world beyond discourse which is shaped by and shaped those discourses. 'Artefactual natures' are the myriad things that are 'purposefully engineered' by socio-economic systems. What types of artefactual natures are being developed, by whom, and with what consequences are vital questions to be asked in this era of globalised capitalism (Demeritt, 2001). We can never talk simply of 'nature' or, the particular things and spaces to which that term refers, such as forests, oceans, and climate systems, without recognising how culture and economy are always at work shaping these objects.

Social construction dissolves the nature/culture divide by swamping the apparently natural with flows of constructive power and practice from the other side of the divide. This one-way traffic has attracted critical comment. Whatmore (2003, p. 165) complains that 'the world is rendered as an exclusively human achievement in which "nature" is swallowed up in the hubris of social construction'. This is not an adequate position because, as already declared, if the 'nature' of the nature/culture dualism is unsustainable, then so too is the 'culture'. If these two realms are always entangled together, we need new ways of thinking about the world that do not rest on *either* wing of the old foundation.

New dialectics

Marxism had a central interest in nature and nature-culture relationships from its inception, not least in how capitalism alienates society from nature and renders it as property. Latterly, various attempts have been made to project Marxism into ecologism (or visa versa) so as to make 'green Marxism'. At the heart of the Marxist approach is dialectical materialism in which nature is seen as embedded in social processes as both a cause and effect.

Particularly as developed by David Harvey, new dialectics attempts to develop these key trajectories of Marxism and embrace a more fully relational and fluid view of the world (see Braun, 2006 for detailed review). As such, it makes a sustained attack on the nature/society dualism *and* the dualisation of space and time. Here there is a relational view of things very similar to that propounded by Actor Network Theory (ANT) (see below). Things are not understood as separate, or in possession of innate, stable identities; rather their natures are relationally inscribed through the networks that constitute them. Harvey, from his Marxist base, suggests that the imperatives of capitalism are now central to organising and transforming the networks of the modern world, and that this fundamental and problematic process needs to be confronted. Humans and non-humans in 'socio-ecological formations', as Harvey puts it, 'become the "arteries" through which an invisible process of ceaseless [economic] value expansion operates' (Castree, 2005, p. 233). His materialist theory sees (capitalist) society making nature in its image and actively transforming elements and organisms, for example highly bred, even genetically modified, poultry raised to grow so rapidly that they cannot express any meaningful kind of (natural) life/behaviour. Nature, in turn, dialectically reworks society (think of crises in capitalist industrial food chains). There is often a resistance from natural elements as they are forcefully enrolled into capital accumulation networks, which sets-up the dialectical dynamic (see George Henderson's chapter, this volume).

Castree (2005) feels that the impact of new dialectics has been limited by the more general turn away from Marxism in geography and the social sciences. Also new dialectics is thought by some to slip inadvertently back towards a dualised view of nature/culture embedded in dialectical reasoning. As Braun (2006, p. 202) summarises, 'a number of scholars – like Sarah Whatmore, Donna Haraway and Bruno Latour – suggest that dialectics is too crude a method for understanding the heterogeneous processes that constitute the environment'. Another criticism is that socio-ecological relations are ultimately defined as conflictual, whereas other approaches are keen to seek out potentially positive sum relationships between economy and ecology (see 'Entangling Politics' below).

Felix Guattari's (2000) take on nature-society relations (which Braun [2006] sees as bearing much affinity with the work of Harvey) makes it plain that it is globalised capitalism which is denuding cultural, psychological, and ecological diversity to the extent that we are witnessing 'ecocide' on a global scale. New dialectics does an important job in bringing to our attention the force, and ubiquity, with which capitalist networks (re)shape the world. However the new dialectical analysis of how to act (critically) in these circumstances seems limited by its stance of fundamental opposition. The debate over the nature of critical Left geography between Amin and Thrift (2005; 2007), on the one side, and Harvey (2006) and Smith (2005) on the other, shows that for some the central focus on Marxism is a limiting rather than empowering for critical knowledge, and this may be the case for its knowledges of after-natural worlds. ANT shares much with new dialectics but seeks more multi-faceted, multi-scaled and flexible responses which might work within what is now considered to be a more heterogeneous and open capitalism than heretofore, and which offer spaces of possibility for creative and life-enhancing after natural entanglements.

Actor network theory

ANT is perhaps the most prominent 'after nature' approach considered in this chapter. It mounts a root-and-branch attack on the two world model, insisting on a 'symmetrical' view across the previously inscribed divides of nature/society, object/subject, structure/agency, fact/value, and more besides. Indeed, attacking these dualisms, which underpin the modern constitution, is the very *raison d'être* of ANT. Whereas other approaches (e.g., new ecology) have come to question the nature/culture divide in the course of their developing inquires, ANT *begins* with a philosophical agenda set against the modern constitution. One of its chief proponents repeatedly insists that we must now completely by-pass the distinctions of old. Instead of treating them as starting points or foundations for inquiry (as so many do), 'we need to rethink the whole assemblage from top to bottom and from beginning to end' (Latour, 2004a, p. 227).

To that end ANT emphasises the networks of heterogeneous connections that can be seen, if only we care to look for them. The work of the late Jonathan Murdoch (1995; 1997; 1998; 2001; 2003) has been particularly significant in interpreting ANT within geography and sociology. He shows just how challenging ANT is for a range of disciplines in terms of understanding space and what and who counts as actors and agency. For example, power generation and consumption could be considered as a vast network of technological devices, information, politics, biogeochemical processes, people and so on. A power station itself could be seen in exactly the same way. (At what scale we read networks and where they meaningfully stop and start are some of the challenges posed by ANT). It is the *network* which has power, agency and affect – not the individual elements in it. Such networks shape space itself; think of the advent of rail networks and the Internet. Fixed notions of space (e.g., local and global) are problematised as *topological* mappings of space. Relations and manifold, multiscalar networks are instead understood as making up the fabric of unfolding life.

In effect all elements of any network are potential actors with agency, or rather 'actants' with 'actency' – terms devised to de-centre the human subject. For Whatmore (1999, p. 26) agency is 'a relational achievement, involving the creative presence of organic beings, technological devices and discursive codes'. This redefinition of agency redistributes it on both sides of the nature/society divide rather than as an exclusive property of humans alone. This recognition that, as Haraway (1992b, p. 297) puts it, 'the actors are not all "us"', is central to ANT. Its force is not to deny the uniquely distinctive capacities of humans, but rather to greatly expand the notion of agency. Latour (2004a) has pointed out that the term agency needs to be redistributed in order to account for the differing ways the global population of things can act creatively. He stresses that:

> No science of the social can even begin if the question of who and what participates in action is not first opened-up, even though it might mean letting elements enter, that, for lack of a better term, we call nonhumans (2004a, p. 226).

ANT is particularly interested in devices which connect and can effectively transmit agency/power from one part of a network to another. How are actants enrolled and held in place? What manner of translations and translating devices are needed to allow differing types of actants to pass power up-and-down the line?

The weaving together of diverse elements into networks has been termed 'heterogeneous engineering'. Thus, hybridity is another central motif of ANT and has been developed alongside it as an approach itself (Whatmore, 2002). From this perspective hybridity is about more than just the conjoining of differing elements to make new compounds; hybridity is a characteristic of the elements themselves. The elements of any network are themselves composed of other hybrid elements that can be decomposed and recombined without end. This fluid, anti-foundational vision of ontology is common to other, related poststructuralist approaches. Bodies, of various kinds, humans, organic/non-organic non-humans, all merge into each other, or, more precisely perhaps, were never separated out. Pure, discrete corporealities are rare. The human gut contains a complex bacterial flora without which we could not digest our food. Likewise, machines are packed with 'natural' elements and minerals that enable them to function. Haraway's (1992a) cyborg (organic, informational, technological, human/nonhuman) is one such figure. The capacities of hybrid bodies or assemblages are greater than the sum of their parts. Deleuze and Guattari (1987) discuss how rider, horse, and riding technology combine to form a new entity with new spatial attributes and life making potentials. But as Latour (1993) insists our divided disciplines and methods have made us blind to the traffic that routinely criss-crosses the nature/society border. Old dualistic habits of mind allow monstrous formations to proliferate unnoticed and un-policed.

Murdoch (2003) has made the case for deploying ANT with its sensitivity to the non-human and hybridity within rural studies:

> The idea that the countryside is simply a social construction, one that reflects dominant patterns of social relations, cannot adequately account for the 'natural' entities found within its boundaries. There is something beyond the 'social' at work as the countryside *displays a material complexity that is not easily reducible to even the most nuanced social categories* [ ] to paraphrase Sarah Whatmore (1999) the countryside is 'more than human' (p. 264, emphasis added).

On these grounds Murdoch argues that the concept of hybridity, which is as yet 'not in common usage' in rural studies, has 'the potential to capture the socio-natural complexity of the countryside more easily than traditional modes of representation' (p. 264). Significantly, the suggestion is that taking non-humans seriously should not be a specialised and discrete form of study, but rather that non-humans are likely to be actively present, and thus, deserving of theoretical and empirical attention, *in just about any consideration of any rural phenomena*. Therefore a general sensitivity to hybridity is needed.

ANT is a persuasive, even arresting view of the world. It seems undeniable that most everyday processes of life involve a whole host of actants from right across the spectrum of existence (ideas, texts, chemicals, machines, organisms, processes, finances and so on). However, ANT has been questioned in a number of ways. Its view of dispersed power leaves it open to the accusation that it cannot account for uneven power relations and the victimisation or injustice that stems from them. Thrift (1999) suggests that it fails to deal very well with ideas and realities of place, and Bingham and Thrift (2002: 299) argue that it misses 'the sizzle of the event' – in other words, the affective dimensions of interaction between actants. It certainly has a strongly technological inflection, which seems to under-represent living things and their characteristics (Whatmore and Thorne, 2000). Its treatment of thing-identity as purely relational has also been questioned (Philo and Wilbert, 2000).

So ANT can be seen as an effective way of diagramming the world 'after nature', but other approaches are needed to flesh-out how these mappings develop. ANT establishes the first principles of interconnectedness and distributed agency, but then how life feels and functions for bodies in networks, and the kinds of spaces, places and experiences they find themselves in need to be dealt with. The next two approaches to be discussed might help in that respect.

Dwelling

Dwelling is 'the thesis that the production of life involves the unfolding of a field of relations that *crosscuts* the boundary between human and non-human.' (Ingold, 2005, p. 504). Springing from the later work of Heidegger and phenomenology of Merleau-Ponty, it offers a ground from which life (human and non-human) can be rethought away from Cartesian-derived dualisms. Dwelling takes

> The immersion of the organism-person in an environment or life world as an inescapable condition of existence. [ ] The world continually comes into being around the inhabitant, and its manifold constituents take on a significance through their incorporation into a regular pattern of life activity (Ingold, 2000, p. 153).

It is thus focused upon the ever-active and becoming push of life and centred on the body-in-environment which is always sensing, engaging, doing, and remembering. No great distinction is made between human and (organic) non-human life. Ingold (2000), Thrift (2005), and Harrison et al. (2004) use dwelling, and Von Uexküll's (1957) closely linked notion of 'life-world', to stress the 'intelligences' and embodied practices of everyday human and non-human life. As Whatmore and Hinchcliffe (2003, p. 5) put it dwelling is about 'the ways in which *humans and other animals* make themselves at home in the world through a bodily register of ecological conduct' (emphasis added).

Unlike social constructionism, dwelling stresses the physical, relational, sensual, performative orchestration of body and space/environment. It differs from ANT in that it retains a topographical interest in place, and bodies in place, whereas ANT takes a more markedly topological approach that focuses on newly measured spatially extended networks. Also the biology of the body is interpreted differently. Dwelling is more focused on living organisms as a grouping than ANT (but is interested in technologies as means of dwelt life). Given the remarkable vitality and complexity of living things, set against human-generated technologies (still pretty crude albeit with rudimentary vitality and complexity), this seems important.

Recent sociologies of nature suggest that dwelling overcomes conflicts between (dualised) 'realist' and 'idealist' approaches to nature and environment (Macnaghten and Urry, 1998). Franklin (2002) places it at the heart of a new anthropology of nature focused on 'practice, practical knowledge, things, technologies and embodied sensual experience' (p. 80). He pushes dwelling towards a more animated, turbulent vision of the world unfolding in a burgeoning, far-reaching (in time and space) interfolding of processes:

> The basic building blocks of this anthropology are unmediated perceptual knowledge, practical experience and knowledge of the world, the technologies that link humans and non-humans, the aesthetic and sensual composition of experience and the cultural choices that are made in reference to these (pp. 71–72).

Recent uses of dwelling in geography problematise the dwelt/authentic – undwelt/inauthentic life dualism present in Heidegger. Also problematised is the dualism between bounded space and network, showing how places can and need to be understood as continuously articulated in both senses (Cloke and Jones, 2001). Wylie (2003, p. 146) suggests that Ingold's approach 'offers a potentially fruitful means of reconfiguring cultural geographies of landscape within the ambit of em-bodied practice and performativity', which scramble a range of settled dualisms.

Understandings of dwelling have undoubtedly been bound-up with the notions of home, local and (rural) rootedness and accordingly can be accused of endorsing a naïve, pre-modern and quasi-romantic politics, as Ingold (2005) himself admits. It apparently sits uneasily with the mobile, speeded-up, stretched-out nature of much contemporary (urban) life. But in such life, and in all life, specific time/space deepened experience, articulated through bodies and biophysical processes, remains inevitable. Dwelling offers a kind of mirror image of the strengths and weaknesses of ANT. It focuses on body-in-(immediate)-environment and thus perhaps struggles to deal with the topological consecutiveness of networks in which bodies find themselves. It is for precisely this reason that Thrift (1999) uses ANT and dwelling in combination (thus illustrating the multi-theoretical approach flagged-up earlier). Thrift does so in his development of 'ecologies of place', and it is to such new ideas of place that we turn.

Places (and landscapes) as entanglements

After nature entanglements can be usefully grounded in developing notions of places (and landscapes). Places and landscapes have been variously seen as the outcome of combinations of elements of culture and nature in *local* relations – as in US geographer Carl Sauer's famous idea of the 'cultural landscape'. In more humanistic approaches developed in Geography during the 1970s, subjective, imaginative notions of life in local place were emphasised. These configurations have been dogged by somewhat one-dimensional notions of place as fixed, bounded space, and problematic ideas of who and what belongs to the 'authentic' chemistry of any given place. With topographic sensibilities very much to the fore, nature and culture were seen as co-present yet separate categories, mixed in dense, kaleidoscopic local arrangements.

Such topographical notions of place and landscape are being challenged and reworked by more topologically infused ideas that figure them as temporary outcomes of processes, networks and hybridity. Thrift (1999), Harrison et al. (2004) and Massey (2005) jettison any notion of place as bounded, static, or exclusively social spaces. They instead understand places as temporal processes where all manners of trajectories – people, non-humans, economies, technologies, ideas and more – come together to assemble enduring (but also changing/open to change) distinctive patterns which are still fully networked into the wider world. As Amin and Thrift (2002, p. 30) summarise,

> Places . . . are best thought of not so much as enduring sites but as moments of encounter, not so much as 'presents', fixed in space and time, but as variable events; twists and fluxes of interrelation.

These comings together operate at differing velocities, rhythms, and trajectories, 'where spatial narratives meet up or form configurations, conjunctions of trajectories which have their own temporalities' (Massey, 2005, p. 139). It is not just a case of social flows whirling through, and tangling with, more fixed grounds of nature. Both are on the move. If a long enough view is taken, even land itself can be seen swirling across the surface of the globe through the movements of plate tectonics. Other processes operate in flows and rhythms more amenable to immediate human apprehension, such as weather patterns or planetary, tidal, and seasonal cycles and the corresponding rhythms of animal and plant life.

All manner of entities thus bring their agency to the formation of place, which is in turn rendered 'local' by the dwelt processes of living bodies. Cities, oceans, fields and parks, even seemingly inanimate objects, like a desk, can all be seen in this way. Along with ideas of dwelling and ANT that Thrift (1999) incorporates into his 'ecologies of place' are the further entanglements of human memories, longing and affect, and even hauntings, thus widening again the repertoire of life to be considered in its own unfolding. Seen in this way place can be a ground on which to hold all these rich entanglements together. In turn this more dynamic notion of place might offer a way to work through the politics and ethics an acknowledgement of relational demands. Massey (2005) advocates an open, non-foundational, non-purifying/ed notion of place, which is sensitive to 'other' notions of 'other' places.

Methodologies, Ethics and Politics Suitable for an 'After Natural World'

The approaches set out above attempt in differing ways to draft new constitutions for knowledges of life on earth. They are very challenging because they question the very foundations of Western knowledge upon which much of our academic *and* everyday assumptions are based. Settled ideas and practices of disciplines and methodologies, ethics and politics are all are thrown into question. The implications of these unsettlements are now briefly sketched out.

Entangling disciplines and methodologies

An after-natural worldview challenges settled disciplinary divisions and methodologies in Geography. If particular formations in the world, say networks of food production and consumption, involve a whole set of interacting processes and elements spanning conventional divides between the natural and social, then the forensic skills of both the natural and social sciences are needed. But they need to be employed, first in the acknowledgement that they are studying a symptom, an organ in a larger body (rather than a entire body), and second, in ways which can communicate with other investigations to build more holistic accounts ('epistemic fungibility'). Thus, the after nature approach seeks to dissolve rigid disciplinary boundaries and to promote new multidisciplinary, interdisciplinary or even transdisciplinary ways of working (Nowotny et al., 2001). There are at least some signs that this key insight is getting through to those who drive research agendas. For example the Research Councils UK 'Rural Economy and Land Use' (RELU) programme insisted that applying teams had to demonstrate a multidisciplinary approach to various rural environmental processes ranging from food chains, pollution and water catchment area management.

Explorations of various forms of interdisciplinarity are emerging in Geography (Evans and Randalls, 2008; Harrison et al., 2008; Lau and Pasquini, 2008), Anthropology (Strathern, 2004) and beyond. But it is early days in terms of what these disciplinary reconfigurations will look like and how (indeed, whether) they will work to generate new after-natural knowledges. This is about more than just placing disciplines side-by-side (Whatmore, 1999). It is also about changing the ontological and political basis for scientific research, admitting and then embracing the idea of knowledges as interventions in the 'creative becoming of the earth' (Braun, 2003, p. 175). But it is likely that differing entanglements will

> require a [new] procedural methodology, taking seriously the particularities of the sites, the unpredictability of circumstances, the uneven patterns of landscapes and the hazardous nature of becoming. (Henaff, 1997, p. 72)

With regard to methodologies employed by disciplines, after-natural approaches do not entail a complete abandonment of the many forms of expertise that have developed to describe aspects of the world. Rather, they redefine how, where, and in what ways expertise is deployed:

> There are many ways to know the world and many forms of expertise have developed to describe aspects of the world. To backtrack [ ] is not to abandon the expertise that has developed in investigating and exploring it. (Harrison et al., 2004, p. 7).

The approaches set out above, particularly new ecology, ANT, place and dwelling (if taken in connection with affective performativity), promote various new kinds of interdisciplinary engagement. In turn those frameworks of after-natural knowledge include – rather than sit outside – questions of ethics and politics.

Entangling ethics

Environmental ethicists have long contended that our narrowly anthropocentric frameworks of human rights and ethical values are both unethical in themselves *and* a key driver of unsustainable practices. This is one of the starkest examples of the divide at work. Human life is (supposedly at least) sacred; all other life is anything but: a resource (at best) with no rights and no intrinsic value. Recent disputes around animal rights form a small but highly significant skirmish on this ethical borderline. But beyond such minor shifts, the prevailing Cartesian dualism 'has stalled the process of getting on with a viable environmental philosophy' (Grange, 1997, p. 11). Reconfiguring ethics to somehow apply to the whole-again after nature world, is now a primary task. As Buell (1995, p. 2) puts it:

> If, as environmental philosophers contend, Western metaphysics and ethics need revision before we can address today's environmental problems, then the environmental crisis involves a crisis of the imagination the amelioration of which depends on finding better ways of imagining nature and humanity's relation to it.

This is, however, a challenge which requires 'an act of considerable moral imagination for those raised in the heart of the monster, the Western dualism of moral insiders and outsiders' (Cheney, 1999, p. 144). The stakes are indeed high:

> Through exclusively social contracts, we have abandoned the bond that connected us to the world. . . . What language do the things of the world speak that we might come to an understanding of them contractually?. . . . In fact, the Earth speaks to us in terms of forces, bonds and interactions . . . each of the partners in symbiosis thus owes . . . life to the other, on pain of death. (Serres, 1995, cited by Whatmore, 1999, p. 26)

Whatmore (2002), Thrift (2005) and others are keen to develop a new ethico-political practice based on affective inter-corporeal ethics of care. ANT and new ecology offer useful mappings of relationality, which can be pored over in the processes of ethical readjustment. Dwelling points to moments of embodied becoming in which ethical resonances might be embedded. While Marxism takes up the principles of relational ontology and unfixity, it seems to remain ethically and politically wedded to an anthropocentrism that other relational approaches are ready to give up, or at least call into question.

Entangling politics

Many political implications and questions stem from the newly broken grounds of after nature. How do we value and engage with each other, with other environments and non-humans? Are we freed from nature as an essential ground for being as hoped/argued for by feminist, anti-racist and queer theory?

We have already seen how ANT and related approaches have been critiqued for providing an uncertain ground on which to conduct politics. The first response to this charge is the claim that revealing the relational, hybrid nature of everyday formations is a political act *in itself* insofar as it unsettles conventional modernist dualisms. Just as Latour (2004b) suggests that political ecology and social science cannot even begin without first demolishing those metaphysical divisions, neither can any kind of meaningful politics. In *We Have Never Been Modern* Latour (1993) first set out his notion of 'the parliament of things' as the central motif for a new kind of 'cosmopolitics' in which things and hybridity are forgrounded and political settlement no longer depends on appeal to any transcendent realm of Nature (Latour, 2004b; 2007; Latour and Weibel, 2005; Stott, 2008). Science has to change from a hegemonic voice of modernity (representing nature in modern politics as an external force which limits human choice) to a more polyphonic voice in chorus with other discourses, which presents the panoply of non-human life in a reconfigured politics. Latour continues to pursue the idea of the parliament of things, but it might now be multiple, mobile parliaments – communities incorporating humans and nonhumans and building on the experiences of the sciences as they are actually practiced.

Latour develops the idea of 'multinaturalism', to sit along side that of multiculturalism. Membership of these complex collectives is not determined by outside experts claiming absolute (scientific) authority, but by 'diplomats' or spokespersons, who are treated with caution but who speak for otherwise mute things in order to ensure that the collective equitably involves both humans and non-humans. The right to represent nature is expanded to include not just scientists but a whole range of spokespeople for example ethicists, poets, farmers and architects. Nature becomes internalised to 'social' systems.

Latour (2007) has termed these ideas 'Dingpolitik'. There is, as he puts it, (p. 1) 'a pixilation of politics'. Just as we have said that methodologies need to be site, or entanglement, specific, then so do politics. New political practices form through and around networks, collectives, and 'matters of concern', where the more-than-human is recognised and takes voice, where political assemblies are therefore multiple, and where there are not grand narratives of succession, but many streams of politics flowing at once. Collectives are flexible are open to pragmatic and incremental adjustments towards better futures of human and non-human flourishing.

Related forms of politics and governance are being proposed through which the active agencies of things/nature can represent themselves or at least be better represented. Murdoch (2005) sets out some principles of 'ecological planning' whereby the processes and forces of nature are built into the very fabric of planning. Whatmore and Hinchcliffe (2003) – in the context of nature in the city – call for a cosmopolitan politics of conviviality. They draw, as does Latour, upon the cosmopolitics of Stengers which is

> a politics of knowledge in which the admission of non-humans into the company of what counts invites new alignments of scientific and political practices and more democratic distributions of expertise (Whatmore, 2005, p. 93).

This kind of cosmopolitics

> learn[s] how new types of encounter (and conviviality) with nonhumans, which emerge in the practice of the sciences over the course of their history, can give rise to new modes of relation with humans, i.e., to new political practices (Paulson, 2001, p. 112).

If Latour's politics is very much about tracing technical networks and the things that comprise them, other related approaches to politics 'after nature' focus on affect, emotion, and the body within networks. Thrift (2005; 2008), for instance, discusses 'various kinds of practical affective politics' designed to stimulate kindness, compassion, cooperation, play as new forms of political (and ethical) action. The aim is to engineer increased capacitates for affirmative life drawing upon 'an ethics of intelligence' that includes wonder, modesty, trial, and error (pragmatism), proliferation, and embracing of plurality, uncertainty and difference. Together these make up a 'non-representational geographic ethics of knowing'. Thrift (2005, p. 472).

The challenge for after nature politics and ethics is to take these micro-dynamics of the body, and bodies in action in space, and see how they connect to the larger entanglements of relational life, spun out around the world in ecologies, and networks that are at once cultural, economic, technological, organic, human, and non-human. The aim is not simply to represent these networks, but to creatively intervene in them as they unfold in always novel arrangements, seeking to build capacity of human and non-human flourishing.

Final Thoughts: Destructive or Creative Entanglements?

If we understand the world as a huge entanglement of entanglements in ceaseless motion, within which bodies, places, landscapes, ecologies and cultures form for a while, before dissolving or moving on, then we must consider, as best we can, those various entanglements in terms of their desirability. Depleted resources, degraded land, fished out oceans, and environmental injustice, these are undesirable entanglements in action. Cities with rich and wild ecologies, sustainable agricultural systems that produce food and ecologies simultaneously; these are desirable entanglements in action.

We need to learn how our maps of those entanglements might also become positive interventions for good. This vision of theory and research as creative interventions *in* the world as much as maps of it lies at the heart of emerging non-representational geographies of nature. For example Hinchcliffe (2007) pleads in the conclusion of *Geographies of Nature*:

> rather than offering interpretations of nature, or analytical concepts, the injunction must be to join the doings, to experiment, to engage in the doings of environments, to environ them in better ways (p. 191).

Such a perspective is difficult to enact if we retain the modernist framework of seeing nature as separate and given, and cultural contact with it as inevitably corrosive. Just as Latour suggests that we have been blind to the proliferation of monstrous formations spanning the nature/culture dualism, we have also been blind to the many often commonplace burgeonings of relational life in the same zone.

Enriching entanglements within which life (human and non-human) can and does flourish are all around us, if we only know how to look. Some of the richest habitats in the UK for biodiversity have been adaptive landscapes, such as chalk downland grazed by domestic animals and coppiced woodlands (Adams, 1996). Here economic production has gone *hand in hand* with the production of rich ecologies. Other happy nature-culture mixings are evident in urban settings. There are persistent claims in the UK that urban bees are fairing better than their rural counterparts and that they make better honey (Norman, 2006)! This is because, somewhat counter intuitively, there is more 'nature' – a wider range of plant biodiversity – for them to exploit in urban areas than in the modernist countryside. Wildlife can be found in the heart of our biggest cities, not only in the mosaic of urban green spaces, but also in apparently unlikely habitats such as building facades and rooftops.

These ideas of rich, productive co-habitation can be expanded to a more global scale and to ideas of biodiversity. Whatmore (2002, pp. 90–91) recounts how The United Nations Food and Agriculture organisation dedicated the 1993 World Food Day to the business of linking the primary ambition of 'food for all' to the equally important aim of sustaining 'the biological diversity of our planet' (FAO, 1993). Their report stated:

> Humanity's place in nature is still not widely understood. Human influences on the environment are all pervasive; even those ecosystems that appear most 'natural' have been altered directly or indirectly during the course of time. Starting some 12,000 years ago, our forebear, as farmers, fisherman, hunters and foresters, have created a rich diversity of productive ecosystems. (FAO, 1993, p. 1, in Whatmore, 2002, pp. 90–91).

Whatmore contrasts this relational view with the definitions and ontologies of the United Nation's Convention on Biodiversity in which nature is still seen as a pure, separate realm once intact, but increasingly under threat from human activity. Whatmore concludes that there is no 'state of nature' only richly inhabited ecologies in which the 'precious metal of bio-diversity' is intimately bound up with the diversities of cultural practices (Whatmore, 2002, pp. 115–16).

If we try to hold nature and culture apart and regard the relationship between them as an inevitably zero sum game, then we are in big trouble. As the first views of earth from space showed, we live on one world, in a gossamer biosphere gathered to the planet's flank by gravity, and shielded from harmful solar emissions (as we now know) by its magnetic field which is generated by the molten iron core.

Some people worry about where we stand – about the foundations of our knowledges. If nature as a separate, fixed ground is lost, what do we stand upon when deciding how to know, judge, and act? The answer is that we are never standing: we are always on the move along with the unfolding world around us. It is constantly being made, unmade, and remade (even if it is at speeds as various as the drifting of continental plates and the pulse of fibre optic communication). This is a slippery context in which to practice knowledge, politics, ethics and sustainable societies. But it is a hopeful slipperiness. In flux is possibility. Process ontologies such as those discussed above espouse knowledges and politics of experimentation through which new entanglements might flourish. In a world that is after nature and after culture, where entanglement is all, we must, as the song goes, 'accentuate the positive, eliminate the negative, and don't mess with Mister In-between'.

ACKNOWLEDGEMENTS

Thanks to David Demeritt and Noel Castree for their patience, support and guidance.

BIBLIOGRAPHY

Adams, W. M. (1996) *Future Nature. A Vision for Conservation*. London: Earthscan Publications.

Amin, A. and Thrift, N. (2002) *Cities: Reimagining the Urban*. Cambridge: Polity.

Amin, A. and Thrift, N. (2005) What's Left? Just the Future. *Antipode*, 37, 220–38.

Amin, A. and Thrift, N. (2007) Being Political, *Trans Inst Br Geogr*, NS 32 112–15.

Bingham, N. and Thrift, N. (2000) Some new instructions for travellers: the geography of Bruno Latour and Michel Serres. In M. Crang and N. Thrift (eds), *Thinking Space*, London: Routledge, pp. 281–301.

Buell, L. (1995) *The Environmental Imagination, Thoreau, Nature Writing and the Formation of American Culture*. Cambridge, MA: Harvard University Press.

Braun, B. (2003) Nature and culture: on the career of a false problem. In J. S. Duncan, N. C. Johnson and R. H. Schein (eds), *A Companion to Cultural Geography*. Oxford: Blackwell, pp. 149–79.

Braun, B. (2006) Towards a new earth and a new humanity: nature, ontology, politics. In N. Castree and D. Gregory (eds), *David Harvey. A Critical Reader*. Oxford: Blackwell Publishing, pp. 191–222.

Castree, N. (2001) Socializing nature: theory, practice and politics. In N. Castree and B. Braun (eds), *Social Nature: Theory, Practice and Politics*. Oxford: Blackwell, pp. 1–21.

Castree, N. (2005) *Nature*. London: Routledge.

Castree, N. and Braun, B. (eds) (2001) *Social Nature: Theory, Practice and Politics*. Oxford: Blackwell.

Castree, N. and Head, L. (2007) Culture, nature and landscape in the Australian region. *Geoforum*, 39(13), 1255–7.

Cheney, J. (1999) The journey home. In A. Weston (ed.), *An Invitation to Environmental Philosophy*. Oxford: Oxford University Press.

Cloke, P. and Jones, O. (2001). Dwelling, place, and landscape: an orchard in Somerset. *Environment and Planning A*, 33, 649–66.

Cronon, W. (ed) (1996) *Uncommon Ground: Rethinking the Human Place in Nature*. New York: W. W. Norton.

Demeritt, D. (2001) Being constructive about nature. In N. Castree and B. Braun (eds), *Social Nature: Theory, Practice and Politics*. Oxford: Blackwell, pp. 22–40.

Deleuze, G. and Guattari, F. (1987) *A Thousand Plateaus. Capitalism and Schizophrenia*. London: The Athlone Press.

Evans, J. and Randalls, S. (2008) Interdisciplinarity and geography: views from the ESRC/ NERC studentship programme. *Geoforum*, 39, 581–92.

Franklin, A. (2002) *Social Nature*. London: Sage.

Grange, J. (1997) *Nature: An Environmental Cosmology*. Albany: State University of New York Press.

Guattari, F. (2000) *The Three Ecologies*. London: The Athlone Press.

Haraway, D. J. (1992a) A manifesto for cyborgs: science. technology, and social feminism in the 1980s. In L. J. Nicholson (ed.), *Feminism/Postmodernism*. London: Routledge.

Haraway, D. J. (1992b) The promises of monsters: a regenerative politics for inappropriate/d others. In L. Grossberg, C. Nelson and P. A. Treichler (eds), *Cultural Studies*. New York: Routledge, pp. 295–337.

Harrison, S., Massey, D. and Richards, K. (2008) Conversations across the divide. *Geoforum*, 39, 549–51.

Harrison, S., Pile, S. and Thrift, N. (2004) *Patterned Ground: Entanglements of Nature and Culture*. London: Reaktion Books.

Harvey, D. (1996) *Justice, Nature, and the Geography of Difference*. Oxford: Blackwell.

Harvey, D. (2006) The geographies of critical geography. *Transactions. Institute of British Geographers*, NS 31, 409–12.

Henaff, M. (1997) Of stones, angels and humans: Michel Serres and the Global City. *SubStance*, 83, 59–80.

Hetherington, K. (1998) *Expressions of Identity*. London: Sage

Hinchcliffe. S. (2007) *Geographies of Nature: Societies, Environments and Ecologies*. London: Sage.

Ingold, T. (2000). *The Perception of the Environment. Essays in Livelihood, Dwelling and Skill*. London: Routledge.

Ingold, T. (2005) Epilogue: towards a politics of dwelling. *Conservation and Society*, 3(2), 501–8.

Latour, B. (1993) *We Have Never Been Modern*. Hemel Hempstead: Harvester/Wheatsheaf.

Latour, B. (2004a) Non-Humans. In S. Harrison, S. Pile and N. Thrift (eds), *Patterned Ground: Entanglements of Nature and Culture*. London: Reaktion Books.

Latour, B. (2004b) *Politics of Nature: How to Bring the Sciences into Democracy*. Cambridge, MA: Harvard University Press.

Latour, B. (2007) From Realpolitik to Dingpolitik – or How to Make Things Public. http://www.bruno-latour.fr/articles/article/96-DINGPOLITIK2.html (accessed 03 07 2007).

Latour, B. and Weibel, P. (2005) *Making Things Public. Atmospheres of Democracy*. Karlsruhe: SKM.

Lau, L. and Pasquini, M. (2008) 'Jack of all trades'? The negotiation of Interdisciplinarity within geography. *Geoforum*, 39, 552–60.

Macnaghten, P. and Urry, J. (1998) *Contested Natures*. London: Sage Publications.

Massey, D. (2005) *For Space*. London: Sage.

Maxeiner, D. and Miersch, M. (2006) The urban jungle. In J. Norman (ed.), *Living for the City*. London: Policy Exchange, pp. 52–67.

McKibben, B. (1990) *The End of Nature*. London: Penguin Books.

Murdoch, J. (1995) Actor-networks and the evolution of economic forms: combining description and explanation in theories of regulation, flexible specialisation, and networks. *Environment and Planning A*, 27, 731–58.

Murdoch, J. (1997) Inhuman/nonhuman/human: actor-network theory and the potential for a non-dualistic and symmetrical perspective on nature and society. *Environment and Planning D, Society and Space*, 15, 731–56.

Murdoch, J. (1998) The spaces of actor-network theory. *Geoforum*, 29(4), 357–74.

Murdoch, J. (2001) Ecologising sociology: actor-network theory, co-construction and the problem of human exemptionalism. *Sociology*, 35, 111–33.

Murdoch, J. (2003) Co-constructing the countryside: hybrid networks and the extensive self. In P. Cloke (ed.), *Country Visions*. Harlow: Pearson, pp. 263–82.

Murdoch, J. (2005) *Post-Structuralist Geography: A Guide to Relational Space*. London: Sage.

Norman, J. (2006) Introduction living for the city. In J. Norman (ed.), *Living for the City*. London: Policy Exchange, pp. 13–20.

Nowotny, H., Scott, P. and Gibbons, M. (2001) *Re-Thinking Science: Knowledge and the Public in an Age of Uncertainty*. Cambridge: Polity Press.

Paulson, W. (2001) For a cosmopolitical philology: lessons from science studies. *SubStance*, 96(30.3), 101–19.

Philo, C. and Wilbert, C. (2000) Animal spaces, beastly places: an introduction. In C. Philo and C. Wilbert (eds), *Animal Spaces, Beastly Spaces*: New Geographies of Human-Animal Relations. London: Routledge, pp. 1–34.

Rose, N. (2001) The politics of life itself. *Theory, Culture, and Society*, 18, 1–30.

Serres, M. (1995) *The Natural Contract*. Ann Arbor: Michigan University Press.

Smith, N. (2005) What's left? Neo-critical geography, or the flat pluralist world of business class. *Antipode*, 37, 887–99.

Strathern, M. (2004) *Commons and Borderlands: Working Papers on Interdisciplinarity, Accountability and the Flow of Knowledge.* Wantage: Sean Kingston Publishing.

Stott, P. (2008) *A Parliament of Things.* http://parliamentofthings.info/index.html (accessed 22 March 2008).

Thrift, N. (1999) Steps to an ecology of place. In D. Massey, P. Sarre and J. Allen (eds), *Human Geography Today.* Oxford: Polity, pp. 295–352.

Thrift, N. (2005) From born to made: technology, biology and space. *Transactions. Institute of British Geographers,* NS 30, 463–76.

Thrift, N. (2008) *Non-Representational Theory: Space, Politics Affect.* London: Sage.

Watts, M. (2005) Nature: culture. In P. Cloke and R. Johnston (eds), *Spaces of Geographical Thought.* London: Sage.

Whatmore, S. (1999) Rethinking the 'human' in human geography. In D. Massey, P. Sarre and J. Allen (eds.), *Human Geography Today.* Oxford: Polity, pp. 22–41.

Whatmore, S. (2002) *Hybrid Geographies: Natures, Cultures, Spaces.* London: Sage.

Whatmore, S. (2003) Culturenatures: introduction: more than human geographies. In K. Anderson, M. Domash, S. Pile and N. Thrift (eds), *Handbook of Cultural Geography.* London: Sage, pp. 165–67.

Whatmore, S. (2005) Generating materials. In M. Pryke, G. Rose and S. Whatmore (eds), *Using Social Theory.* London: Thousand Oaks, CA and New Delhi: Sage Publications with the Open University, pp. 89–104.

Whatmore, S. and Hinchcliffe, S. (2003) Living cities: making space for urban nature. *Soundings. Journal of Politics and Culture,* Jan. This version downloaded as PDF; www.open.ac.uk/socialsciences/habitable_cities/habitable_citiessubset/.

Whatmore, S. and Thorne, L. (2000) Elephants on the move: spatial formations of wildlife exchange. *Environment and Planning D: Society and Space,* 18, 185–203.

Wylie, J. (2003) Landscape, performance and dwelling: a Glastonbury case study. In P. Cloke (ed.), *Country Visions.* Harlow: Pearson, pp. 136–57.

Zimmerer, K. (1994) Human geography and the 'new ecology'. *Annals of the Association of American Geographers,* 1, 108–25.

Part III   Practices

Chapter 19

Remote Sensing and Earth Observation

Heiko Balzter

Earth Observation has become more and more prominent in many aspects of contemporary environmental research. Earth Observation describes the process of remotely measuring certain properties of the Earth system, like atmospheric gas concentrations, aerosol optical depth, land cover, forest fires or glacier velocity. Observations from aircraft and satellites provide spatially explicit and often multi-temporal geographical datasets that can be used for detecting, monitoring and mapping environmental changes. The increasing accessibility of Geographical Information Systems (GIS) have made it possible to base decision-making processes increasingly on remotely sensed observations since imagery can be used to provide data layers in GIS databases.

International organisations and national departments increasingly take up remotely sensed data in their everyday procedures. An example is the United Nations Food and Agriculture Organisation (FAO), which deploys remote sensing techniques for early warning systems, environmental monitoring and rapid assessment of crop growth conditions, as well as for natural resource inventory, monitoring and management at local, national, regional and global scales. One such application is the Global Forest Resources Assessment from 1990, 2000 and 2005. For the 2010 assessment a comprehensive global remote sensing survey is planned (Ridder, 2007). Since 1988, the FAO has also operated the Africa Real Time Environmental Monitoring Information System (ARTEMIS). This system acquires and processes hourly estimates of rainfall and vegetation index (NDVI) images in (near) real time using Meteosat and NOAA data. ARTEMIS covers the whole African continent. The 10-day and monthly data products are used to identify areas in which food security might be at risk or where desert locust control measures may be necessary.

Data from Earth Observation is now central to environmental research and to a number of international science programmes. The International Geosphere-Biosphere Programme (IGBP) views Earth Observation as vital. Dr Will Steffen, Executive Director of IGBP, told the Earth Observation Summit in Washington in 2003 that 'the Earth has a number of "Achilles' heels" that are particularly sensitive to human activities. Examples include the Gulf Stream in the North Atlantic, the ozone hole over Antarctica and the Amazon basin in Brazil. [...] IGBP sees the Earth Observation System as a very positive step towards producing the globally consistent and reliable data needed for all nations to deal with global change' (IGBP, 2003). In its capacity to deliver essential monitoring information, Earth Observation supports a wide range of international conventions, such as the UN Framework Convention on Climate Change and its implementation in the Kyoto protocol, the RAMSAR convention, the UN Convention to Combat Desertification and the UN Convention on Biodiversity. This chapter aims to describe selected principles and applications of remote sensing with a focus on land applications.

Physical Principles of Remote Sensing

Remote sensing methods are based on instruments measuring electromagnetic energy received from a remote target (e.g., the land surface). The electromagnetic spectrum encompasses a range of wavelengths. Here, we will describe methods utilising visible light (wavelengths between 0.4 and 0.75 μm), infrared radiation (0.75 to 10^3 μm) and microwaves (10^3 to 10^5 μm). We distinguish active from passive remote sensing systems. Active systems, such as Synthetic Aperture Radar (SAR), illuminate the target by transmitting electromagnetic energy pulses and recording the return to the sensor. Passive systems, like imaging spectrometers and multispectral scanners, record the intensity of electromagnetic radiation originating from an independent source, usually the sun. Dependence on solar illumination consumes less power than active sensors but limits the imaging opportunities in the high-latitude winter season (polar night) and at night-time.

Optical/near-infrared

When observations of the land surface are required for environmental applications, the imaging process is complicated by the atmospheric pathway that the electromagnetic radiation (the sunlight) has to travel before being reflected from the target and again on its way back. Absorption of radiation in certain parts of the electromagnetic spectrum through atmospheric constituents (water vapour, etc.) can change the spectral signal. Scattering processes, such as those caused by aerosols, can alter the direction of the electromagnetic waves, and thus, influence the recorded signal. Such distortion needs to be atmospherically corrected to get the characteristic reflectance of the target (Leroy and Roujean, 1994; Los et al., 2005).

In the blue, green, red and near-infrared spectral bands, we can distinguish different land surface types based on their characteristic spectral reflectance properties, also called spectral signatures. Green plants absorb photosynthetically active radiation primarily in the red spectrum, while reflectance in the near-infrared spectrum is almost stable, which leads to a characteristic green reflectance from vegetation canopies. These differences have been used to derive vegetation indices, the most common of which is the Normalised Difference Vegetation Index (NDVI). It is based on the reflectance in the near-infrared (NIR) and red (RED) spectrum:

$$NDVI = \frac{NIR - RED}{NIR + RED}$$

The NDVI has been found to increase with increasing green biomass.

Based on vegetation indices, biophysical parameters of the vegetation canopy can be estimated from optical/near-infrared remote sensing. The leaf-area index (LAI) of a vegetation canopy is defined as the ratio of the area of all leaves to that of the ground surface. LAI can be estimated from remote sensing by applying radiative transfer modelling techniques. Another biophysical parameter that can be derived from Earth Observation is the fraction of absorbed photosynthetically active radiation (fAPAR), which describes the amount of radiation absorbed by plants during photosynthesis.

Synthetic Aperture Radar (SAR)

SAR is an active remote sensing technique in the microwave domain. Electromagnetic energy pulses are transmitted from the satellite to the surface. The pulse is then scattered by the target and a certain fraction of it is scattered back in the direction of the satellite and received by the sensor. The backscatter intensity depends both on the viewing geometry and target characteristics, particularly the water content and the structure (dielectric constant, orientation distribution and number density of the scattering elements, e.g., branches and leaves). SAR signals can be transmitted and received at defined polarisations, either horizontal (H) or vertical (V). In the case of a fully polarimetric SAR, the instrument can thus record horizontally transmitted and horizontally received (HH) backscatter as well as VV, HV and VH backscatter. Since the backscatter intensity at these different polarisations varies as a function of target structure and viewing geometry a fully polarimetric system can be used to estimate fractions of backscatter caused by fundamental scattering mechanisms: rough surface scattering from the ground, double-bounce scattering (e.g., trunk/ground interactions) and volume scattering (e.g., from multiple scattering in a tree crown). The contributions of these three basic scattering mechanisms provide information on the properties of the imaged target, e.g., how rough the ground surface is, whether the microwaves penetrate the canopy and are scattered from the stems or whether most of the radiation is scattered from a dense canopy. The wavelength of the SAR plays an important role in determining which scattering elements contribute most to the signal. At longer wavelengths like L- or P-band branches and tree trunks contribute more to the backscatter, while at shorter wavelengths like X- or C-band, the leaves and needles are important scattering elements (figure 19.1). This means that multi-wavelength observations can give important structural information about the target.

If two SAR acquisitions were taken at a suitable spatial baseline, the complex correlation coefficient between the two SAR acquisitions can be calculated pixel by pixel. This technique is called SAR interferometry because the resulting interferogram image shows characteristic fringe patterns caused by the interference of the electromagnetic waves from the two acquisitions. The interferometric coherence is the magnitude of the complex correlation coefficient between the two SAR images and indicates how similar the SAR signals are. The interferometric phase is an angular measurement and is related to the vertical height of the scattering phase centre, the location from which most of the signal originates. The scattering phase centre is the integral of the returns from a large ensemble of scatterers, which include ground, stems, branches, and leaves or needles. Which types of scatterers interact most strongly with the radar wave depends on the wavelength, polarization, incidence angle and vegetation density. Interferometric techniques can be used for

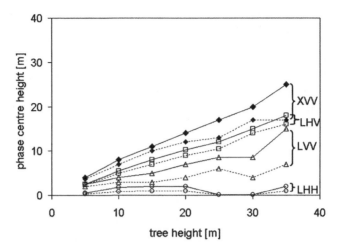

Figure 19.1 Scattering phase centre height at X-band VV polarization and L-band HH, HVand VV polarization (H = horizontal, V = vertical) from the coherent microwave model CASM for tree densities of 1000 (dotted lines) and 2000 ha^{-1} (solid lines) at 45° incidence angle. The maximum phase centre separation is between the phase centres of X-VV and L-HH. Reprinted from Balzter, H., Rowland, C.S., and Saich, P (2007b): 'Forest canopy height and carbon estimation at Monks Wood National Nature Reserve, UK, using dual-wavelength SAR interferometry', *Remote Sensing of Environment*, 108, 224-239, copyright (2007), with permission from Elsevier.

estimating forest canopy height because from the phase at different wavelengths and polarizations the underlying terrain height and vegetation canopy height can be derived (Balzter et al., 2007a; Balzter et al., 2007b). Because SAR is an active remote sensing technique, it is able to acquire imagery at night-time, including at high latitudes during the polar winter.

Light detection and ranging (LiDAR)

Light detection and ranging is an active technique based on emitting infrared radiation from the sensor to the surface and then measuring the time it takes until a signal is received. If the target is a forest site then the signal will stem from a range of vegetation components at different heights in the canopy, and a full pulse of returns can be recorded by more advanced sensors. Of particular interest is often the first and last return since they are correlated to the highest and lowest point within the LiDAR footprint. Airborne laser scanning (or Imaging LiDAR) is a very accurate method for mapping topographic and vegetation height. It has been extensively applied by the Environment Agency in the UK for flood defence monitoring along the coastline and major rivers, for mapping vegetation canopies (Hill et al., 2002; Hinsley et al., 2002; Gaveau and Hill, 2003; Patenaude et al., 2004; Hill and Thomson, 2005; Hinsley et al., 2006) as well as building structures in urban areas (Zhou et al., 2004; Sohn et al., 2005; Luo and Gavrilova, 2006). Ground-based LiDAR systems can be deployed closer to the target and have mm to cm range resolutions (Henning and Radtke, 2006; Van der Zande et al., 2006; Danson et al., 2007).

Profiling LiDAR instruments like SLICER (Ni-Meister et al., 2001; Lefsky et al., 2002; Kotchenova et al., 2004) or ICESAT-GLAS (Zwally et al., 2002; Harding and Carabajal, 2005; Lefsky et al., 2005; Lefsky et al., 2006) do not produce images but have the advantage of recording full waveforms of signal returns, which can be used to estimate forest canopy height and more generally the 3D target structure.

Environmental Applications

These basic techniques of Earth Observation are now being applied to address a vast array of practical concerns ranging from environmental monitoring to mapping the three-dimensional pattern of major cities for flight simulator software, to inter-planetary observation and monitoring compliance with international conventions on nuclear non-proliferation. However, most of the research done by geographers and environmental scientists has focused on applying Earth Observation methods and data to understand various processes of change in land cover and the earth surface.

Fires

Biomass burning is a major contributor to global carbon dioxide fluxes from the land to the atmosphere (Patra et al., 2005). Wildfires also dramatically decrease the surface albedo (Eugster et al., 2000), with potential feedbacks to the climate system. With climate change, the frequency and intensity of fire regimes in some parts of the world are increasing. Because of their economic impacts on the forestry sector, as well as on air quality and ecosystems, wildfires are often classified as disasters.

Remote sensing is used to routinely monitor the occurrence of wildfires. Three basic approaches can be distinguished: (i) the detection of 'hot spots' (active fire detections) using thermal infrared data (Kant et al., 2000; Stroppiana et al., 2000; Barducci et al., 2002; Schultz, 2002; Lasaponara et al., 2003; Li et al., 2003; Soja et al., 2004; San-Miguel-Ayanz et al., 2005; Csiszar et al., 2006); (ii) 'burned area' mapping, typically using near-infrared or middle infrared wavelengths (BourgeauChavez et al., 1997; Fraser et al., 2000; Diaz-Delgado and Pons, 2001; Brivio et al., 2003; Sukhinin et al., 2004; Balzter et al., 2005; George et al., 2006; Gerard et al., 2003); and (iii) the measurement of fire radiative power, to infer fire radiative energy which can be related to carbon release (Wooster et al., 2003).

The method of active fire detection can serve as a near-real time early warning system, but it has the drawback that it can fail to detect fires under dense clouds or between satellite overpasses. Depending on the sensor characteristics, false detections caused by hot surfaces or sun glint can also be a problem. Active fire data are generally not an accurate estimate of burned area. Examples of global active fire datasets are the World Fire Atlas at the European Space Agency (http://dup.esrin.esa.int/ionia/wfa/), the Fire Information for Resource Management System (FIRMS, http://maps.geog.umd.edu/firms/) or the World Fire Web at the European Commission's Joint Research Centre.

Burned area mapping approaches are based on various types of change detection methods. Some authors have used the changes in Normalised Difference Vegetation Index (NDVI) before and after a fire to map the burned area (Fraser et al., 2000). Others have used the Short-wave Infrared Index (NDSWIR) to map burned areas.

This index has been shown to detect historic fire scars as old as 10–15 years (Gerard et al., 2003; Balzter et al., 2005; George et al., 2006). An example of a burned area map of Central Siberia is shown in figure 19.2. This map was used in an integrated full greenhouse gas accounting system in the European project SIBERIA-2.

The method for estimating fire radiative power and integrating it to fire radiative energy is based on the 4 μm and 11 μm mid-infrared channels. The basic concept is that an active fire radiates heat, which is being measured by the 4 μm channel and contrasted to the background radiances that would have been expected without a fire. The choice of the 4 μm wavelength is based on the fact that the 4 μm spectral radiance is proportional to the integrated total radiance at all wavelengths for fires burning at different temperatures. Thus, based on measuring a single channel, it is possible to estimate the total heat emitted from the fire. The radiative power estimates have to be integrated over time and over space within the fire scar to give fire radiative energy estimates. Since the combustion of a quantified amount of carbon generates a known amount of heat release, fire radiative energy can be used to estimate the amount of burned biomass (Wooster et al., 2005). This approach

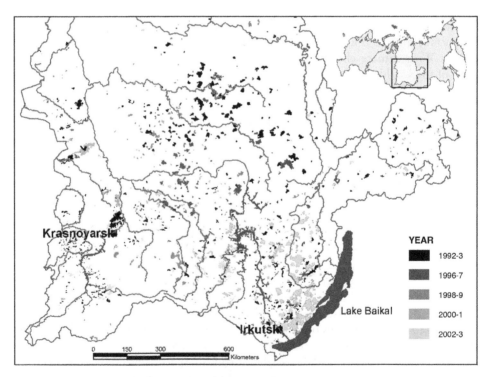

Figure 19.2 Forest disturbance map of Central Siberia. The shading indicates the years in which the largest proportion of the polygon burnt. Redrawn from George, C., Rowland, C., Gerard, F. and Balzter, H. (2006) Retrospective mapping of burnt areas in Central Siberia using a modification of the normalised difference water index. *Remote Sensing of Environment*, 104, 346–59, copyright (2006), with permission from Elsevier.

avoids the use of empirical burning efficiency and vegetation density factors used in traditional approaches.

Land cover and land-use change

Remote sensing can provide operational information on land-use and land-cover state and change. Land use describes the human land management practices, while land cover means the biophysical properties of the land surface. The mean surface reflectance at each wavelength recorded by the sensor is called spectral signature. Many land cover types have characteristic spectral signatures that can be used to discriminate between them. The more similar two spectral signatures are, the higher the risk of confusion between land cover classes. There is a general trade-off between the number of land cover classes (thematic detail) and classification accuracy. The desire for more classes generally leads to a lower spectral separability between some of the classes. One way to tackle this problem is to define a hierarchy of land cover classes, in which the first level of the classification discriminates a low number of very accurately classified land cover types, while the next level subdivides the classes further with less accuracy. We can distinguish supervised and unsupervised classification algorithms. On the one hand, in a supervised approach, the spectral signatures of each land cover type are defined *a priori* either from a digital library of spectral signatures or by visually identifying areas of known land cover and estimating the signatures from the image data. In an unsupervised approach such as *k*-means or Isodata, no prior knowledge of the land cover types in the image is assumed, and an algorithm automatically cagetorises the image data into a given number of classes. Unsupervised approaches tend to find better class separability because they merge and split classes to maximise the difference between their spectral signatures. On the other hand, an unsupervised classification needs careful interpretation. Since the classes are purely based on spectral characteristics, they do not necessarily map onto the thematic requirements of a land cover map.

In the UK, land cover mapping from satellite is used in routine environmental monitoring. Fuller et al. (1994) mapped the entire land area of Great Britain using Landsat imagery (30-m resolution) as part of the Countryside Survey 1990. The survey was later repeated in 2000, leading to a new updated map that used an advanced algorithm (Fuller et al., 2002). Whereas the 1990 map is a pixel-based map, the UK Land Cover Map 2000 is a polygon database, in which each polygon or land parcel is linked to a comprehensive attribute data table, containing information on the most likely, second most likely land cover class and so on. The maps have been used by over 500 registered data users and are a prime example of the operational use of satellite remote sensing. At European level, a standardised land cover map is also available. The CORINE 1990 and CORINE 2000 land cover maps with their 44 classes were generated based on amalgamations of nationally produced land cover maps (e.g., the UK Land Cover Map), but the minimum mapping units were increased to 25 ha, which limits their use for site-specific habitat management.

It can be hard to reconcile land maps using different remote sensing classification schemes, since map producers often have their own locally specific understandings what each land cover class actually means in the real world. This subjectivity in remote sensing image classification and interpretation can lead to substantial semantic confusion (Comber et al., 2005). Whenever land cover information of different dates derived from different methodologies and interpretation systems is used together, the inconsistent class definitions can cause problems in understanding the resulting land cover change maps. However, formal methods exist for dealing with inconsistent classification systems (Comber et al., 2004).

Remotely sensed data have also been applied in studies modelling the processes of land-use and land-cover change (De Almeida et al., 2005). Since they provide a spatial and temporal data source, they enable new insights into space-time dynamics of the land surface and its driving factors.

Forest biomass

Forest biomass is a parameter of great economic importance to forest enterprises, natural resource managers and climate scientists. It can be indirectly estimated from stand parameters and canopy height using sets of published species-specific allometric equations. Forest canopy height can be estimated from stereophotogrammetry, LiDAR, and Synthetic Aperture Radar (SAR). On a landscape-scale forest parameters can be mapped at high accuracy and a spatial resolution of 0.5–2 m using Imaging LiDAR. The basic principle is that a model of the underlying unvegetated terrain is subtracted from the signals received from the canopy top to derive a vegetation canopy height model. The terrain model can be generated from the last return data of the LiDAR, but it generally needs to undergo several filtering and interpolation steps.

Detailed maps from airborne remote sensing have a high spatial resolution and are used in land and habitat management, forestry and agriculture. In this way, remote sensing can help update forest inventory geographic information systems like those used by the British or the Russian Forestry Commissions. Forest biomass is also directly correlated to carbon content of the vegetation. This information is useful for informing government agencies about aboveground carbon stocks, which are needed for national reports to the Secretariat of the United Nations Framework Convention on Climate Change under the Kyoto Protocol to combat climate change.

Researchers interested in global change often require large-scale coverage that can only be obtained from a satellite-borne instrument. The first spaceborne profiling LiDAR instrument, the Geoscience Laser Altimeter System (GLAS), was launched on-board ICESAT on 12 January 2003. While the mission focus is to measure ice sheets, clouds and aerosols, the instrument can also be exploited for land and vegetation applications. GLAS fires 40 pulses per second and records the intensity of the reflected radiation as a function of travelling time, resulting in approximately 70-m footprints, which are spaced about 170 m apart along the track. Between orbital paths, footprints are between 2.5 km (near the poles) and 15 km (near the equator) apart (University of Texas, 2003). Each pulse results in a full waveform measurement, which provides a profile of the illuminated footprint in the third dimension (figure 19.3). Biomass indicators that can be derived from the full-intensity waveform are terrain elevation, canopy height and crown depth (Harding and Carabajal, 2005).

Whereas LiDAR technology essentially measures the time that the radiation takes to reach the target and back at a specified point or footprint location, Synthetic Aperture Radar (SAR) is based on active microwave radiation. Since microwaves

ID 258798423 shot 6

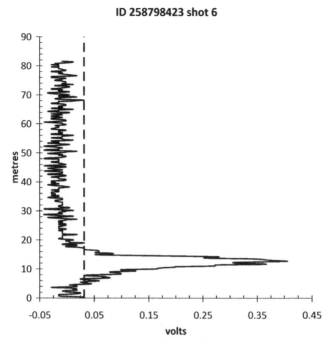

Figure 19.3 Full waveform of an ICESAT-GLAS footprint (cloud-free acquisition) in Siberia. The full waveform can be used to derive forest canopy structure and terrain height. The intensity (volts) indicates from which height in the footprint most of the reflected radiation originated. The relative travelling time of the pulse is a measure of height. The thickness of the vegetation canopy can be determined by finding the maximum and minimum height values that exceeded the noise level (vertical dashed line), in this example 17 m–5 m = 12 m, which was verified from forest inventory data. Acknowledgement: Claire Burwell (University of Leicester) GLAS data from Zwally, H. J., Schutz, R., Bentley, C., Bufton, J., Herring, T., Minster, J., Spinhirne, J., and Thomas, R. (2006) *GLAS/ICESAT L1A Global Altimetry Data V028*. Boulder, CO: National Snow and Ice Data Center. Digital media

partially penetrate through vegetation tissue, SAR can measure vegetation volume and structure in three dimensions. Three basic approaches of biomass mapping from SAR can be distinguished: backscatter, coherence and phase-based approaches. *Radar backscatter* is the energy that is received by the SAR sensor after being scattered by the imaged target area. As described above, radar backscatter from forests originates from three basic physical scattering mechanisms: volume scattering within the canopy, rough surface scattering from the ground and double-bounce scattering from trunk-ground interactions. Backscatter intensity has been used to estimate woody biomass of forest (Kasischke et al., 1997), forest biomass (Le Toan et al., 1992), forest aboveground dry biomass (Rignot et al., 1994) and timber volume (Balzter et al., 2002a). Forests with higher biomass generally show higher backscatter, but the biomass–backscatter relationship saturates at a wavelength-dependent biomass level. This saturation problem restricts SAR applications for biomass

Figure 19.4 X-band VV-polarised SAR backscatter image acquired by the airborne E-SAR system over Monks Wood National Nature Reserve, East Anglia, UK in 2000. Spatial resolution = 3 m. Data were provided by the German Aerospace Centre (DLR).

mapping to some extent (Imhoff, 1995). Additional errors may arise from effects of vegetation structure, which can influence radar backscatter to a similar magnitude as biomass dependence.

While radar backscatter images like the one shown in figure 19.4 can be acquired from a single data acquisition, *coherence-based approaches* using SAR interferometry (InSAR) require at least two SAR acquisitions of the same area at a suitable spatial baseline (the distance separating the viewing positions of the sensors at the time of the acquisitions). These techniques exploit the fact that the coherence between the two SAR images tends to be lower for areas with high forest biomass due to temporal and volume decorrelation processes. In boreal forest, this method works best in winter, when the entire biome is frozen and the scattering elements (branches, needles, stems and ground) do not change in water content or geometric properties between the two image acquisitions. Because SAR is an active sensor, it can acquire data during the boreal winter night-time when the use of passive sensors is limited. Superior performance of wintertime coherence for boreal forest stem volume retrieval was reported by Pulliainen et al. (2003). Forest stem volume was also retrieved from ERS-1 and ERS-2 C-band tandem coherence (one day between acquisitions) under winter conditions in Finland and Sweden by Askne and Santoro (2005), and from L-band JERS-1 wintertime coherence (44 days between acquisitions) in Siberian taiga forest (Eriksson et al., 2003).

Phase-based approaches measure the location of the scattering phase centre within the canopy layer, from where most of the backscatter is originating. The scattering phase centre in a forest is the 'integral of the returns from a large collection of scatterers, which include ground, stems, branches and leaves or needles'

(Balzter et al., 2007b) and describes the vertical location from which most of the back scattered radiation originates. Canopy gaps smaller than the radar resolution cell will influence the interferometric height of the scattering phase centre, which will be a measurement of the area-weighted average of the canopy top and the proportion of open ground seen by the radar (Hagberg et al., 1995). Some studies have exploited the wavelength-dependent penetration depth of microwaves to estimate canopy top and underlying terrain height, e.g., with the airborne dual-wavelength SAR system TOPOSAR containing a single-pass X-band and a repeat-pass P-band sensor system (Andersen et al., 2004), or from the single-pass X-band and repeat-pass L-band E-SAR system (Balzter et al., 2007b). The method of polarimetric interferometry (Cloude et al., 2001) makes use of different polarisations to estimate contributions of different scattering mechanisms and of interferometry to locate the scattering phase centre heights of these mechanisms. A limitation of this method is that it relies on sufficient scattering phase centre separation to derive canopy height estimates.

An example of a large-scale mapping project using two wavelengths (C-band interferometry from ERS-1 and 2, and L-band backscatter from JERS-1) is the Siberian forest cover map (Balzer et al., 2002b), which classified open water, smooth surfaces (agriculture) and four forest growing stock volume classes (<20, 20–50, 50–80 and >80 m³/ha) over an area of about a million km² at 50-m spatial resolution. The map was used in operational forest management by over 40 Russian forest enterprises.

Vegetation phenology

The seasonal cycle of vegetation greening during the photosynthetically active growing seasons followed by dormancy during winter or the dry season is commonly described as vegetation phenology. Obtaining spatially explicit quantitative information on vegetation phenology is important because the year-to-year differences in the carbon flux from terrestrial metabolism have almost been as large as variations in the growth rate of atmospheric CO_2 concentrations (Houghton, 2000). A statistical framework for the analysis of AVHRR time-series data was presented by de Beurs and Henebry (2005). The same authors relate changes in NDVI to annual growing degree days (GDD), a statistic which is defined as the average of the daily maximum and minimum temperatures compared to a base temperature. GDD are linked to the metabolic activity of plants and can indicate the onset of flowering and other phenological events. A number of studies have found that phenological indicators from remote sensing are correlated with climatic indicators. For the Amazon basin, the seasonal amplitude (amount of temporal change within one season) of NDVI was observed to increase during El Niño periods with concurrent low rainfall anomalies, and to decrease during wet La Niña episodes (Asner et al., 2000). For Europe, spring phenology correlates with winter temperature anomalies and the winter North Atlantic Oscillation (NAO) index (Stockli and Vidale, 2004). For Siberia, the timing of leaf appearance in spring shows a strong correlation with sea surface temperatures over the equatorial Pacific of the previous summer, which are related to El Niño–Southern Oscillation patterns (Vicente-Serrano et al., 2006).

Delbart et al. (2005) studied the timing of the onset of greening-up and leaf senescence over Central Siberia and compared three spectral indices from the SPOT-VEGETATION sensor. They conclude that in the boreal biome, NDVI-based greening-up dates are affected strongly by snow effects influencing the spectral signatures in the red and near-infrared bands. This means that NDVI time-series based methods cannot be used to determine the onset of the greening-up of the vegetation alone, but instead indicates the timing of the snow melt, when the NDVI signature increases in spring. In contrast to NDVI, the normalised difference water index (NDWI) only showed a small snow effect. The estimated dates of leaf colouring in autumn show a lesser accuracy than the greening-up dates (Delbart et al., 2005). Delbart et al. (2006) report an average advance of spring greening-up over Siberia of 3.5 days between 1982 and 2004. This trend varies spatially across regions and temporally: From 1982 to 1991, the start of the greening-up advanced by 7.8 days, but between 1991 and 1999, only random variation is observed while from 2000 to 2004, there was a trend towards a later greening-up by as much as 7 days (Delbart et al., 2006). Similar observations were made from an analysis of fAPAR time-series data (fraction of absorbed photosynthetically active radiation) over Siberia by Balzter et al. (2007c) illustrated in figures 19.5 and 19.6.

Analysis of NDVI-derived biophysical parameters from 1981 to 1991 indicated a greening trend and an increase in growing season length in the Northern Hemisphere (Myneni et al., 1997) that was attributed to a biosphere response to climate change. After the Mount Pinatubo volcanic eruption in June 1991, a drop in NDVI was observed that was probably caused by reduced vegetation photosynthetic activ-

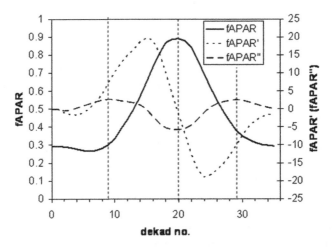

Figure 19.5 Illustration of the phenological signal in the second order slope of the FASIR fAPAR time series data for one pixel location (pixel 74, line 253; 57.03°N 95.01°E) and one year (1995). Left y axis: fAPAR; right y axis: first and second local derivatives of fAPAR. The typical 'camel back' appearance of peak, trough, and second peak in the second order derivative (dashed line) corresponds to the start, peak, and end of growing season indicators shown as vertical broken lines, which are determined based on the local maxima and minimum of the local second derivative. Reprinted from Balzter, H., Gerard, F., Weedon, G., Grey, W., Combal, B., Bartholomé, E., Bartalev, S. and Los, S. (2007c) Coupling of vegetation growing season anomalies with hemispheric and regional scale climate patterns in Central and East Siberia. *Journal of Climate*, 20(15), 3713–29, doi: 10.1175/JCLI4226. (C) Copyright 2008 American Meteorological Society (AMS).

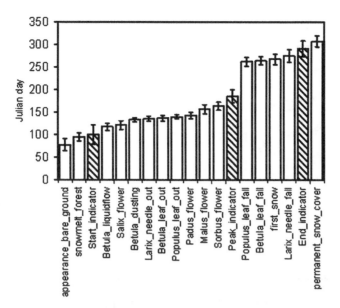

Figure 19.6 Comparison of phenological indicators from field observations in Siberia at 3 sites over 5 years with the remotely sensed indicators 'Start', 'Peak' and 'End' of growing season. Remotely sensed indicators are highlighted with a diagonal fill pattern. Shown is the mean Julian day of the events averaged over the years and sites. Bars show plus/minus one standard deviation. Reprinted from Balzter, H., Gerard, F., Weedon, G., Grey, W., Combal, B., Bartholomé, E., Bartalev, S. and Los, S. (2007c) Coupling of vegetation growing season anomalies with hemispheric and regional scale climate patterns in Central and East Siberia. *Journal of Climate*, 20(15), 3713–29, doi: 10.1175/JCLI4226. (C) Copyright 2008 American Meteorological Society (AMS).

ity because of a dimming and cooling effect from volcanic aerosols (Slayback et al., 2003). This drop was greater at high latitudes and particularly pronounced in the boreal biome, where Slayback et al. (2003) found NDVI trends between 1982 and 1999 to be almost zero, despite the evidence for a Northern Hemisphere greening trend observed by Bogaert et al. (2002) using averaged NDVI values over the growing season. Differences in remote sensing data processing and statistical methods make direct comparisons between studies very difficult. Nevertheless, the increasing length of available time series of remote sensing data offers tremendous new opportunities to detect changes in the biosphere.

Atmospheric gases

So far, this chapter has covered various remote sensing methods to characterise terrestrial carbon pools and fluxes, including forest biomass, land cover and fire. However, ultimately atmospheric greenhouse gas concentrations contribute to the observed increase in global average temperatures (known as global warming). Recently, a range of novel approaches has been developed to estimate atmospheric greenhouse gas concentrations in the air column between the satellite sensor and the ground. In 2002, the European Space Agency launched the ENVISAT satellite,

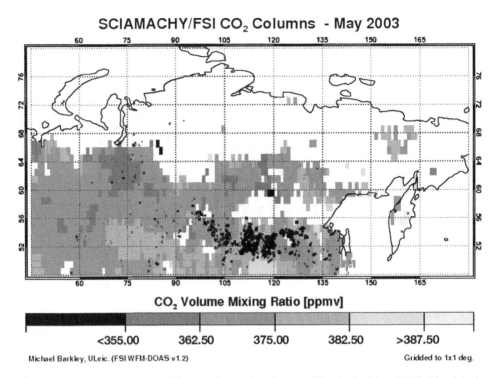

Figure 19.7 CO$_2$ volume mixing ratio retrievals over Siberia in May 2003. The black dots represent fire counts from MODIS thermal anomaly data, which are clearly concentrated to the north of the elevated CO$_2$ region. Figure provided by M. Barkley and P. Monks, University of Leicester.

carrying a whole range of instruments, including SCIAMACHY (SCanning Imaging Absorption spectroMeter for Atmospheric CartograpHY). SCIAMACHY is a passive ultraviolet, visible/near-infrared (240–2380 nm wavelengths) hyper-spectral spectrometer. To retrieve concentrations of the most important greenhouse gas – CO$_2$ – from space observations, Barkley et al. (2006a) developed the Full Spectral Initiation (FSI) WFM-DOAS retrieval algorithm. It extends previous work on the WFM-DOAS algorithm by Buchwitz et al. (2006), and is aimed at SCIAMACHY measurements in the near-infrared (NIR) domain. Different spectral wavelength intervals can be used to retrieve concentrations of a range of atmospheric gases, including carbon dioxide, methane and carbon monoxide from the NIR (Buchwitz et al., 2004; Buchwitz et al., 2005), water vapour from the near-visible (Noel et al., 2004), and ozone from the UV (Coldewey-Egbers et al., 2005). These gases show specific absorption features in the electromagnetic spectrum as the radiation from the sun passes through the atmosphere before reaching the sensor. The algorithm by Barkley et al. (2006b) generates absorption spectra using the radiative transfer model SCIATRAN (Rozanov et al., 2002). It accounts for the expected climatological conditions at the time of the satellite observation, the latitude of the ground pixel, the solar zenith angle and aerosols (Barkley et al., 2006b). A monthly composite of the atmospheric CO$_2$ volume mixing ratio for May 2003 over Russia retrieved by this method is shown in figure 19.7. Also shown in figure 19.7 are the

Figure 19.8 LiDAR image of Leicester city centre derived from first return height data. Individual tall buildings, housing areas, a train line and lamp posts are clearly visible. The brightest colours represent the tallest height. LiDAR Data Copyright Environment Agency Geomatics.

detected heat emissions from the MODIS instrument on the Terra satellite for the same month (black dots). In the region 48°–52°N | 110°–120°E an anomalously high-atmospheric CO_2 concentration was observed, which is located just south of a region of major catastrophic forest fire outbreaks that occurred in Siberia in May 2003 (indicated by the black dots). While the causality of the two observations has not yet been proven, the example illustrates the potential for joint exploitation of terrestrial and atmospheric satellite observations.

Urban mapping

In addition to these natural resource monitoring and management applications, remote sensing in general and imaging LiDAR in particular has a range of applications in human geography and the social sciences. From the detailed high-resolution LiDAR height maps, three-dimensional urban models can be generated (Batty and Hudson-Smith, 2005). These models provide an increasing level of detail, and when combined with optical photographs of building textures can be used to create virtual cities. This technology has applications in urban planning, the games industry, architecture and visual building impact assessment through line-of-sight simulations, as well as building material and surface structure assessment. Figure 19.8 shows an airborne LiDAR image of the recorded first return heights of Leicester city centre.

Concluding Remarks

The selected applications of remote sensing described above can serve to illustrate the broad range of Earth-System Science that has only become possible over the last years because a very complex system of satellite and airborne sensors are available. Some environmental changes have only been identified because of satellite data. Other processes have been better quantified using satellite-derived information, and the uncertainties of these processes have been reduced as a direct result of the sensing capability. For geographers, access to comprehensive spatial and temporal information on the state of the physical, chemical and biological environment provides exciting new opportunities to better understand the many rapid changes that are currently reshaping our planet. Earth Observation also opens up previously unimaginable opportunities to visualise and analyse three-dimensional space-time processes of urban change of relevance for spatial planning. Finally, remote sensing is not just a tool for monitoring. Its use for environmental monitoring can also act as a force in shaping the environment it observes by driving land management and policy decisions (Robbins, 2001). This reciprocal relationship makes remote sensing a fascinating subject to study for Human Geographers, too.

The international community is coordinating Earth Observation needs and the required technological infrastructure through the Global Earth Observation System of Systems (GEOSS). This idea was conceived following the first Earth Observation Summit in Washington, DC in July 2003, which established the intergovernmental Group on Earth Observations (GEO). At the second Earth Observation Summit in Tokyo in April 2004, a Framework Document for GEOSS was adopted. At the third Summit in Brussels in February 2005, the GEOSS 10-Year Implementation Plan was endorsed. The societal needs for a global observing system were identified as follows:

- 'Reducing loss of life and property from natural and human-induced disasters.
- Understanding environmental factors affecting human health and well-being.
- Improving management of energy resources.
- Understanding, assessing, predicting, mitigating and adapting to climate variability and change.
- Improving water resource management through better understanding of the water cycle.
- Improving weather information, forecasting and warning.
- Improving the management and protection of terrestrial, coastal and marine ecosystems.
- Supporting sustainable agriculture and combating desertification.
- Understanding, monitoring and conserving biodiversity'. (Group on Earth Observations, 2006).

In conclusion, the current Earth Observation capabilities will further develop, and we will see new information services emerge.

BIBLIOGRAPHY

Andersen, H.-E., McGaughey, R., Reutebuch, S., Schreuder, G., Agee, J. and Mercer, B. (2004) Estimating canopy fuel parameters in a Pacific Northwest conifer forest using multifrequency polarimetric IFSAR. Proceedings of the ISPRS conference, Istanbul.

Andreae, M. O. (1997) Emissions of trace gases and aerosols from southern African savanna fires. In B. W. van Wilgen et al. (eds), *Fire in Southern African Savannas: Ecological and Atmospheric Perspectives*. Johannesburg: Witwatersrand Univiversity Press, pp. 161–84.

Askne, J. and Santoro, M. (2005) Multitemporal repeat pass SAR interferometry of boreal forests. *IEEE Transactions on Geoscience and Remote Sensing*, 43, 1219–28.

Asner, G. P., Townsend, A. R. and Braswell, B. H. (2000) Satellite observation of El Nino effects on Amazon forest phenology and productivity. *Geophysical Research Letters*, 27, 981–4.

Balzter, H., Baker, J. R., Hallikainen, M. and Tomppo, E. (2002a) Retrieval of timber volume and snow water equivalent over a Finnish boreal forest from airborne polarimetric Synthetic Aperture Radar. *International Journal of Remote Sensing*, 23, 3185–208.

Balzter, H., Talmon, E., Wagner, W., Gaveau, D., Plummer, S., Yu, J. J., Quegan, S., Davidson, M., Le Toan, T., Gluck, M., Shvidenko, A., Nilsson, S., Tansey, K., Luckman, A. and Schmullius, C. (2002b) Accuracy assessment of a large-scale forest cover map of central Siberia from synthetic aperture radar. *Canadian Journal of Remote Sensing*, 28, 719–37.

Balzter, H., Gerard, F. F., George, C. T., Rowland, C. S., Jupp, T. E., McCallum, I., Shvidenko, A., Nilsson, S., Sukhinin, A., Onuchin, A. and Schmullius, C. (2005) Impact of the Arctic Oscillation pattern on interannual forest fire variability in Central Siberia. *Geophysical Research Letters*, 32, L14709.1–L14709.4.

Balzter, H., Luckman, A., Skinner, L., Rowland, C. and Dawson, T. (2007a) Observations of forest stand top height and mean height from interferometric SAR and LiDAR over a conifer plantation at Thetford Forest, UK. *International Journal of Remote Sensing*, 28, 1173–97.

Balzter, H., Rowland, C. S. and Saich, P. (2007b) Forest canopy height and carbon estimation at Monks Wood National Nature Reserve, UK, using dual-wavelength SAR interferometry. *Remote Sensing of Environment*, 108, 224–39

Balzter, H., Gerard, F., Weedon, G., Grey, W., Combal, B., Bartholome, E., Bartalev, S. and Los, S. (2007c) Coupling of vegetation growing season anomalies with hemispheric and regional scale climate patterns in Central and East Siberia. *Journal of Climate*, 20(15), 3713–29; doi: 10.1175/JCLI4226.

Barducci, A., Guzzi, D., Marcoionni, P. and Pippi I. (2002) Infrared detection of active fires and burnt areas: theory and observations. *Infrared Physics and Technology*, 43, 119–25.

Barkley, M. P., Friess, U. and Monks, S. (2006a) Measuring atmospheric CO_2 from space using full spectral initiation (FSI) WFM-DOAS. *Atmospheric Chemistry and Physics*, 6, 3517–534.

Barkley, M. P., Monks, P. S., Friess, U., Mittermeier, R. L., Fast, H., Korner, S. and Heimann, M. (2006b) Comparisons between SCIAMACHY atmospheric CO_2 retrieved using (FSI) WFM-DOAS to ground based FTIR data and the TM3 chemistry transport model. *Atmospheric Chemistry and Physics*, 6, 4483–98.

Batty, M. and Hudson-Smith, A. (2005) Urban simulacra: London. *Architectural Design*, 178, 42–47

Bogaert, J., Zhou, L., Tucker, C. J., Myneni, R. B. and Ceulemans, R. (2002) Evidence for a persistent and extensive greening trend in Eurasia inferred from satellite vegetation index data. *Journal of Geophysical Research-Atmospheres*, 107, ACL 4-1–ACL 4-11.

BourgeauChavez, L. L., Harrell, P. A., Kasischke, E. S. and French, N. H. F. (1997) The detection and mapping of Alaskan wildfires using a spaceborne imaging radar system. *International Journal of Remote Sensing*, 18, 355–73.

Brivio, P. A., Maggi, M., Binaghi, E. and Gallo, I. (2003) Mapping burned surfaces in sub-Saharan Africa based on multi-temporal neural classification. *International Journal of Remote Sensing*, 24, 4003-18.

Buchwitz, M., de Beek, R., Bramstedt, K., Noel, S., Bovensmann, H. and Burrows, J. P. (2004) Global carbon monoxide as retrieved from SCIAMACHY by WFM-DOAS. *Atmospheric Chemistry and Physics*, 4, 1945–60.

Buchwitz, M., de Beek, R., Noel, S., Burrows, J. P., Bovensmann, H., Bremer, H., Bergamaschi, P., Korner, S. and Heimann, M. (2005) Carbon monoxide, methane and carbon dioxide columns retrieved from SCIAMACHY by WFM-DOAS: year 2003 initial data set. *Atmospheric Chemistry and Physics*, 5, 3313–3329.

Buchwitz, M., de Beek, R., Noel, S., Burrows, J. P., Bovensmann, H., Schneising, O., Khlystova, I., Bruns, M., Bremer, H., Bergamaschi, P., Korner, S. and Heimann, M. (2006) Atmospheric carbon gases retrieved from SCIAMACHY by WFM-DOAS: version 0.5 CO and CH4 and impact of calibration improvements on CO2 retrieval. *Atmospheric Chemistry and Physics*, 6, 2727–51.

Cloude, S., Papathanassiou, K. P. and Pottier, E. (2001) Radar polarimetry and polarimetric interferometry. *Ieice Transactions on Electronics*, E84C, 1814–22.

Coldewey-Egbers, M., Weber, M., Lamsal, L. N., de Beck, R., Buchwitz, M. and Burrows, J. P. (2005) Total ozone retrieval from GOME UV spectral data using the weighting function DOAS approach. *Atmospheric Chemistry and Physics*, 5, 1015–25.

Comber, A., Fisher, P. and Wadsworth, R. (2004) Integrating land cover data with different ontologies: identifying change from inconsistency. *International Journal of Geographical Information Science*, 18, 691–708.

Comber, A. J., Fisher, P. F. and Wadsworth, R. A. (2005) You know what land cover is but does anyone else? An investigation into semantic and ontological confusion. *International Journal of Remote Sensing*, 26, 223–28.

Csiszar, I. A., Morisette, J. T. and Giglio, L. (2006) Validation of active fire detection from moderate-resolution satellite sensors: the MODIS example in northern Eurasia. *IEEE Transactions on Geoscience and Remote Sensing*, 44, 1757–64.

Danson, F. M., Hetherington, D., Morsdorf, F., Koetz, B. and Allgower, B. (2007) Forest canopy gap fraction from terrestrial laser scanning. *IEEE Geoscience and Remote Sensing Letters*, 4, 157–60.

De Almeida, C. M., Vieira Monteiro, A. M., Camara, G., Soares-Filho, B. S., Coutinho Cerqueira, G., Lopes Pennachin, C. and Batty, M. (2005) GIS and remote sensing as tools for the simulation of urban land-use change. *International Journal of Remote Sensing*, 26, 759–74

De Beurs, K. M. and Henebry, G. M. (2005) A statistical framework for the analysis of long image time series. *International Journal of Remote Sensing*, 26, 1551–73.

Delbart, N., Kergoat, L., Le Toan, T., L'Hermitte, J. and Picard, G. (2005) Determination of phenological dates in boreal regions using normalized difference water index. *Remote Sensing of Environment*, 97, 26–38.

Delbart, N., Le Toan, T., Kergoat, L. and Fedotova, V. (2006) Remote sensing of spring phenology in boreal regions: a free of snow-effect method using NOAA-AVHRR and SPOT-VGT data (1982–2004). *Remote Sensing of Environment*, 101, 52–62.

Diaz-Delgado, R. and Pons, X. (2001) Spatial patterns of forest fires in Catalonia (NE of Spain) along the period 1975–1995 – analysis of vegetation recovery after fire. *Forest Ecology and Management*, 147, 67–74.

Eriksson, L. E. B., Santoro, M., Wiesmann, A. and Schmullius, C. C. (2003) Multitemporal JERS repeat-pass coherence for growing-stock volume estimation of Siberian forest. *IEEE Transactions on Geoscience and Remote Sensing*, 41, 1561–70.

Eugster, W., Rouse, W. R., Pielke, R. A., McFadden, J. P., Baldocchi, D. D., Kittel, T. G. F., Chapin, F. S., Liston, G. E., Vidale, P. L., Vaganov, E. and Chambers, S. (2000) Land-atmosphere energy exchange in Arctic tundra and boreal forest: available data and feedbacks to climate. *Global Change Biology*, 6, 84–115.

Fraser, R. H., Li, Z. and Cihlar, J. (2000) Hotspot and NDVI differencing synergy (HANDS): a new technique for burned area mapping over boreal forest. *Remote Sensing of Environment*, 74, 362–76.

Fuller, R. M., Groom, G. B. and Jones, A. R. (1994) The land-cover map of Great-Britain – an automated classification of Landsat thematic mapper data. *Photogrammetric Engineering and Remote Sensing*, 60, 553–62.

Fuller, R. M., Smith, G. M., Sanderson, J. M., Hill, R. A. and Thomson, A. G. (2002) The UK land cover map 2000: construction of a parcel-based vector map from satellite images. *Cartographic Journal*, 39, 15–25.

Gaveau, D. L. A. and Hill, R. A. (2003) Quantifying canopy height underestimation by laser pulse penetration in small-footprint airborne laser scanning data. *Canadian Journal of Remote Sensing*, 29, 650–57.

George, C., Rowland, C., Gerard, F. and Balzter, H. (2006) Retrospective mapping of burnt areas in Central Siberia using a modification of the normalised difference water index. *Remote Sensing of Environment*, 104, 346–59.

Gerard, F., Plummer, S., Wadsworth, R., Sanfeliu, A. F., Iliffe, L., Balzter, H. and Wyatt, B. (2003) Forest fire scar detection in the boreal forest with multitemporal SPOT-VEGETATION data. *IEEE Transactions on Geoscience and Remote Sensing*, 41, 2575–85.

Group on Earth Observations (2006) About GEO. http://www.earthobservations.org/about/about_GEO.html (accessed 14 June 2007).

Hagberg, J. O., Ulander, L. M. H. and Askne, J. (1995) Repeat-pass SAR interferometry over forested terrain. *IEEE Transactions on Geoscience and Remote Sensing*, 33, 331–40.

Harding, D. J. and Carabajal, C. C. (2005) ICESat waveform measurements of within-footprint topographic relief and vegetation vertical structure. *Geophysical Research Letters*, 32.

Henning, J. G. and Radtke, P. J. (2006) Ground-based laser imaging for assessing three-dimensional forest canopy structure. *Photogrammetric Engineering and Remote Sensing*, 72, 1349–58.

Hill, R. A. and Thomson, A. G. (2005) Mapping woodland species composition and structure using airborne spectral and LiDAR data. *International Journal of Remote Sensing*, 26, 3763–79.

Hinsley, S. A., Hill, R. A., Gaveau, D. L. A. and Bellamy, P. E. (2002) Quantifying woodland structure and habitat quality for birds using airborne laser scanning. *Functional Ecology*, 16, 851–57.

Hinsley, S. A., Hill, R. A., Bellamy, P. E. and Balzter, H. (2006) The application of LiDAR in woodland bird ecology: climate, canopy structure, and habitat quality. *Photogrammetric Engineering and Remote Sensing*, 72, 1399–1406.

Houghton, R. A. (2000) Interannual variability in the global carbon cycle. *Journal of Geophysical Research-Atmospheres*, 105, 20121–30.

IGBP (2003) IGBP calls for better observation of Earth's 'Achilles' heels. http://www.igbp.kva.se/page.php?pid=282, 5/8/2003 (accessed 18 March 2008).

Imhoff, M. L. (1995) Radar backscatter and biomass saturation – ramifications for global biomass inventory. *IEEE Transactions on Geoscience and Remote Sensing*, 33, 511–18.

Kant, Y., Prasad, V. K. and Badarinath, K. V. S. (2000) Algorithm for detection of active fire zones using NOAA AVHRR data. *Infrared Physics and Technology*, 41, 29–34.

Kasischke, E. S., Melack, J. M. and Dobson, M. C. (1997) The use of imaging radars for ecological applications – a review. *Remote Sensing of Environment*, 59, 141–56.

Kotchenova, S. Y., Song, X. D. Shabanova, N. V. Potter, C. S., Knyazikhin, Y. and Myneni, R. B. (2004) Lidar remote sensing for modeling gross primary production of deciduous forests. *Remote Sensing of Environment*, 92, 158–72.

Lasaponara, R., Cuomo, V., Macchiato, M. F. and Simoniello, T. (2003) A self-adaptive algorithm based on AVHRR multitemporal data analysis for small active fire detection. *International Journal of Remote Sensing*, 24, 1723–49.

Le Toan, T., Beaudoin, A., Riom, J. and Guyon, D. (1992) Relating forest biomass to SAR data. *IEEE Transactions on Geoscience and Remote Sensing*, 30, 403–11.

Lefsky, M. A., Harding, D. J., Keller, M., Cohen, W. B., Carabajal, C. C., Espirito-Santo, F. D., Hunter, M. O. and de Oliveira, R. (2005) Estimates of forest canopy height and aboveground biomass using ICESat. *Geophysical Research Letters*, 32.

Lefsky, M. A., Harding, D. J., Keller, M., Cohen, W. B., Carabajal, C. C., Espirito-Santo, F. D., Hunter, M. O., de Oliveira, R. and de Camargo, P. B. (2006) Estimates of forest canopy height and aboveground biomass using ICESat. *Geophysical Research Letters*, 33.

Lefsky, M. A., Cohen, W. B., Harding, D. J., Parker, G. G., Acker, S. A. and Gower, S. T. (2002) Lidar remote sensing of above-ground biomass in three biomes. *Global Ecology and Biogeography*, 11, 393–99.

Leroy, M. and Roujean, J. L. (1994) Sun and view angle corrections on reflectances derived from NOAA AVHRR data. *IEEE Transactions on Geoscience and Remote Sensing*, 32, 684–97.

Li, Z. Q., Fraser, R., Jin, J., Abuelgasim, A. A., Csiszar, I., Gong, P., Pu, R. and Hao, W. (2003) Evaluation of algorithms for fire detection and mapping across North America from satellite. *Journal of Geophysical Research-Atmospheres*, 108, ACL 20-1–ACL 20-14.

Los, S. O., North, P. R. J., Grey, W. M. F. and Barnsley, M. J. (2005) A method to convert AVHRR Normalized Difference Vegetation Index time series to a standard viewing and illumination geometry. *Remote Sensing of Environment*, 99, 400–11.

Luo, Y. and Gavrilova, M. L. (2006) 3D building reconstruction from LiDAR data, p. 431-439 Computational Science and Its Applications – ICCSA 2006, Pt 1, Vol. 3980.

Moulin, S., Kergoat, L., Viovy, N. and Dedieu, G. (1997) Global-scale assessment of vegetation phenology using NOAA/AVHRR satellite measurements. *Journal of Climate*, 10, 1154–70.

Myneni, R. B., Nemani, R. R. and Running, S. W. (1997) Estimation of global leaf area index and absorbed par using radiative transfer models. *IEEE Transactions on Geoscience and Remote Sensing*, 35, 1380–93.

Ni-Meister, W., Jupp, D. L. B. and Dubayah, R. (2001) Modeling LiDAR waveforms in heterogeneous and discrete canopies. *IEEE Transactions on Geoscience and Remote Sensing*, 39, 1943–58.

Noel, S., Buchwitz, M. and Burrows, J. P. (2004) First retrieval of global water vapour column amounts from SCIAMACHY measurements. *Atmospheric Chemistry and Physics*, 4, 111–25.

Patenaude, G., Hill, R. A., Milne, R., Gaveau, D. L. A., Briggs, B. B. J. and Dawson, T. P. (2004) Quantifying forest above ground carbon content using LiDAR remote sensing. *Remote Sensing of Environment*, 93, 368–80.

Patra, P. K., Ishizawa, M., Maksyutov, S., Nakazawa, T. and Inoue, G. (2005) Role of biomass burning and climate anomalies for land-atmosphere carbon fluxes based on inverse modeling of atmospheric CO2. *Global Biogeochemical Cycles*, 19, GB3005, 1–10.

Phillips, M., Page, S., Saratsi, E., Tansey, K. and Moore, K (2007) Diversity, scale and green landscapes in the gentrification process: traversing ecological and social science perspectives. *Journal of Applied Geography* (doi: 10.1016/j.apgeog.2007.07.003).

Pulliainen, J., Engdahl, M. and Hallikainen, M. (2003) Feasibility of multi-temporal interferometric SAR data for stand-level estimation of boreal forest stem volume. *Remote Sensing of Environment*, 85, 397–409.

Ridder, R. M. (2007) *Global Forest Resources Assessment 2010: Options and Recommendations for a Global Remote Sensing Survey of Forests*. FAO Forestry Department, Forest Resources Assessment Programme Working Pages 141, Rome: FAO.

Rignot, E., Way, J. B., Williams, C. and Viereck, L. (1994) Radar estimates of aboveground biomass in boreal forests of interior Alaska. *IEEE Transactions on Geoscience and Remote Sensing*, 32, 1117–24.

Robbins P. (2001) Fixed categories in a portable landscape: the causes and consequences of land-cover categorization. *Environment and Planning A*, 33, 161–79.

Rozanov, V. V., Buchwitz, M., Eichmann, K. U., de Beek, R. and Burrows, J. P. (2002) SCIATRAN – a new radiative transfer model for geophysical applications in the 240–2400 nm spectral region: the pseudo-spherical version. *Remote Sensing of Trace Constituents in the Lower Stratosphere, Troposphere and the Earth's Surface: Global Observations, Air Pollution and the Atmospheric Correction*, 29, 1831–35.

San-Miguel-Ayanz, J., Ravail, N., Kelha, V. and Ollero, A. (2005) Active fire detection for fire emergency management: potential and limitations for the operational use of remote sensing. *Natural Hazards*, 35, 361–76.

Schultz, M. G. (2002) On the use of ATSR fire count data to estimate the seasonal and interannual variability of vegetation fire emissions. *Atmospheric Chemistry and Physics*, 2, 387–95.

Slayback, D. A., Pinzon, J. E., Los, S. O. and Tucker, C. J. (2003) Northern hemisphere photosynthetic trends 1982–99. *Global Change Biology*, 9, 1–15.

Sohn, H. G., Yun, K. H., Kim, G. H. and Park, H. S. (2005) Correction of building height effect using LiDAR and GPS. High Performance Computing and Communications, Proceedings, Vol. 3726, 1087–95.

Soja, A. J., Sukhinin, A. I., Cahoon, D. R., Shugart, H. H. and Stackhouse, P. W. (2004) AVHRR-derived fire frequency, distribution and area burned in Siberia. *International Journal of Remote Sensing*, 25, 1939–60.

Stockli, R. and Vidale, P. L. (2004) European plant phenology and climate as seen in a 20-year AVHRR land-surface parameter dataset. *International Journal of Remote Sensing*, 25, 3303–30.

Stroppiana, D., Pinnock, S. and Gregoire, J. M. (2000) The Global Fire Product: daily fire occurrence from April 1992 to December 1993 derived from NOAA AVHRR data. *International Journal of Remote Sensing*, 21, 1279–88.

Sukhinin, A.I., French, N. H. F., Kasischke, E. S., Hewson, J. H., Soja, A. J., Csiszar, I. A., Hyer, E. J., Loboda, T., Conrad, S. G., Romasko, V. I., Pavlichenko, E. A., Miskiv, S. I. and Slinkina, O. A. (2004) AVHRR-based mapping of fires in Russia: new products for fire management and carbon cycle studies. *Remote Sensing of Environment*, 93, 546–64.

Tansey, K., Gregoire, J. M., Stroppiana, D., Sousa, A., Silva, J., Pereira, J. M. C., Boschetti, L., Maggi, M., Brivio, P. A., Fraser, R., Flasse, S. Ershov, D., Binaghi, E., Graetz, D. and Peduzzi, P. (2004) Vegetation burning in the year 2000: global burned area estimates from SPOT VEGETATION data. *Journal of Geophysical Research-Atmospheres*, 109, D14, CiteID D14S03.

University of Texas (2003) GLAS Instrument Description. http://www.csr.utexas.edu/glas/Instrument_Description (accessed 14 June 2007).

Van der Zande, D., Hoet, W., Jonckheere, L., van Aardt, J. and Coppin, P. (2006) Influence of measurement set-up of ground-based LiDAR for derivation of tree structure. *Agricultural and Forest Meteorology*, 141, 147–60.

Vicente-Serrano, S. M., Delbart, N., Le Toan, T. and Grippa, M. (2006) El Nino-Southern Oscillation influences on the interannual variability of leaf appearance dates in central Siberia. *Geophysical Research Letters*, 33, L03707.

Wooster, M. J., Zhukov, B. and Oertel, D. (2003) Fire radiative energy for quantitative study of biomass burning: derivation from the BIRD experimental satellite and comparison to MODIS fire products. *Remote Sensing of Environment*, 86, 83–107.

Wooster, M. J., Roberts, G., Perry, G. L. W. and Kaufman, Y. J. (2005) Retrieval of biomass combustion rates and totals from fire radiative power observations: FRP derivation and calibration relationships between biomass consumption and fire radiative energy release. *Journal of Geophysical Research-Atmospheres*, 110, D24 311, 1–24.

Zhou, G. Q., Song, C., Simmers, J. and Cheng, P. (2004) Urban 3D GIS from LiDAR and digital aerial images. *Computers and Geosciences*, 30, 345–53.

Zwally, H. J., Schutz, B., Abdalati, W., Abshire, J., Bentley, C., Brenner, A., Bufton, J., Dezio, J., Hancock, D., Harding, D., Herring, T., Minster, B., Quinn, K., Palm, S., Spinhirne, J. and Thomas, R. (2002) ICESat's laser measurements of polar ice, atmosphere, ocean, and land. *Journal of Geodynamics*, 34, 405–45.

Chapter 20

Modelling and Simulation

George L.W. Perry

Introduction

Models and the practice of modelling have been the subject of ongoing debate in geography. Modelling 'has arguably become the most widespread and influential research practice in the discipline of geography, as indeed within the sciences more generally' (Demeritt and Wainwright, 2005, p. 206). The geographical literature is replete with reviews of various approaches to modelling and debates as to the merits, or otherwise, of modelling itself (e.g., see Macmillan, 1989b; Canham et al., 2003; Wainwright and Mulligan, 2004). In this chapter, I aim to provide a picture of the 'state-of-the-art' in the modelling of human-environment interactions, with a focus on simulation models and their evaluation. The focus is on the place of models and the nature of modelling as intellectual activities, rather than on the mechanics of model-building. The chapter is divided into two broad sections; the first focuses on current perspectives on modelling in geography and the second uses a series of case studies to illustrate how modelling is being practised.

Fundamental concerns for effective model-building and analysis are: (i) ensuring that the entity under investigation is appropriately represented and (ii) obtaining the data required to parameterise the models. These two issues relate to some of the crucial decisions of model-making: how detailed should a model be? How much causal (process) representation does it need to incorporate? At what scales in time and space should it operate? The problem of determining optimal model complexity, in terms of representational and empirical adequacy, is a recurrent theme of this chapter. A second underlying theme is that of complexity and complexity science (Medd, 2001; O'Sullivan, 2004). Over the last decade interest in and insights from 'complexity' and 'complexity science' have led to significant shifts in modelling socio-environmental systems. It is important to distinguish between complicated and complex systems. In complicated systems many components interact in a linear, or somehow predictable, manner (e.g., a multi-component, yet inherently predictable, system such as an aeroplane), whereas complex systems may comprise but few components, but (indirect) interactions between those components result in unexpected behaviours at the system-level (so-called 'emergence'). The central lesson of complexity science is that the dynamics of seemingly complex/complicated entities can be reproduced by simple models. In other words, complex problems do not necessarily require complicated answers[1]. That computer-based simulations are the main tool of complexity science has led to an examination of the place of simulation in science more generally and in particular to the question of 'where do simulation models lie in relation to theories and to experiments?' (see Humphreys, 1995/96; Dowling, 1999).

Approaches and Issues

What, why and how?

Models are idealised simplifications of some phenomenon or system. If modelling is nothing more than a process of simplified representation then nearly all conceptual activities might be described as modelling and verbal descriptions and cartographic maps could be called models. In this context, simplification entails paring back the representation of an entity until it contains only what is relevant to a given problem (a process often termed 'abstraction') as well as deliberate distortion to aid understanding (e.g., economists may assume perfectly rational decision making). As a result, models are *inherently* false and are known to be so. Thus, as the basic empiricist argument against scientific realism emphasises, considering a model to be 'true' is perilous (Morton, 1993; Oreskes et al., 1994; Beven, 2002). Nevertheless, and crucially, as Beven (2002) and Frigg and Hartmann (2006) point out, models may still be *approximately* true.

Given that all models are simplifications, one of their key traits is their level or degree of detail. This is usually thought of in terms of the number of parameters a model contains or processes it represents. As Batty and Torrens (2001) and Mulligan and Wainwright (2004) emphasise, parsimony is central to good modelling: we seek the simplest model that serves our purpose adequately. This does not mean that the absolutely simplest model is always the best solution; rather, we seek the simplest model that also serves the purpose we require of it. It is worth noting, however, that although modellers have tended to adopt this 'parsimony principle', observations of the 'real' world suggest that it is not simple, nor do simple answers seem consistently more useful than complicated ones. In practice, most (environmental) modellers tend to adopt what Beven (2002) terms 'pragmatic realism', that is they: (i) attempt to make their models as realistic as possible, and (ii) consider that even if current models are limited they will, over time, become ever more faithful mimics of the entity being represented.

Types of models

Methodologically, models are often classified as being conceptual, analytical (mathematical), empirical-statistical or simulation in form (table 20.1). Conceptual models are simply verbal, narrative or graphical descriptions of the system of interest, and the interactions and interdependencies between its components, while analytical (mathematical) models are distillations of conceptual models into the formalisms of mathematics. Empirical-statistical and simulation models are often distinguished by how they treat causality. Empirical-statistical models (e.g., regression approaches)

Table 20.1 A typology of modelling approaches

Model type	Description
Conceptual	Description of some system or process using narrative or graphical tools.
Analytical (mathematical)	Formal description of some system or process using the language of mathematics; can take many different forms including both deterministic and stochastic approaches. Note, however, that the term 'mathematical' is somewhat misleading as models almost invariably contain, to a greater or lesser degree, mathematical elements in some guise (Guisan and Zimmermann, 2000).
Empirical (statistical)	Models based on observed data (usually, but not necessarily, quantitative); includes statistical models.
Simulation	In a loose sense simulation simply involves 'building a likeness' (Kleindorfer et al., 1998). In general, however, it is usually taken to mean computer-based or *in silico* (see page 341) activity. Simulation modelling encompasses a multitude of activities ranging from the numerical solution of analytically intractable systems of equations to attempts to produce faithful *in silico* mimics or surrogates of specific 'real' world systems and processes (Winsberg, 2003; Küppers and Lenhard, 2005).

Note: Falling outside this typology are 'hardware' models, that is, scaled physical reconstructions such as flumes and wind tunnels.

are based on observations and focus on prediction of a system's dynamics; they do *not* consider why a change will occur, only what the nature of the change will be. Conversely, simulation models tend to consider the dynamics of the system and the processes that explain those dynamics; they consider what the response of the system might be to change and what processes explain that response. Thus, simulation models are often also referred to as 'mechanistic' or 'process-based' (Guisan and Zimmermann, 2000). In many cases the boundaries between the methodologies are blurred; for example, nearly all simulation models contain mathematical elements and some empirical component.

Another view is to consider models as being either 'top-down' or 'bottom-up' (Grimm, 1999). Bottom-up modelling is an atomistic approach, motivated by the belief that the dynamics and organisation of complex systems arise from, and can be explained by, interactions between the units that comprise that system. In environmental geography, agent-based models (ABMs) epitomise bottom-up modelling (Parker et al., 2003; Brown et al., 2004; Brown, 2006). In ABMs, the agents are autonomous, goal-seeking entities. Although agents often represent individuals, they may also represent aggregate structures such as family units, tribes, settlements or business organisations. Schelling's (1978) segregation model provides a famous example of a bottom-up, agent-based approach. In Schelling's model, householders are divided into two groups and have preferences regarding how many of each type of neighbour they prefer to live next to. 'Unhappy' households move to new sites in an effort to improve their situation. Over time the model produces broad-scale patterns of segregation, arising purely from decisions made by individual house-holders; macro-level patterns (segregation) 'emerge' from micro-level (individual) decisions.

By contrast, top-down modelling focuses on aggregate entities (e.g., entire populations) and on representing system-level relationships between aggregate variables with the goal of finding relationships between those variables. As such, it involves the application of general frameworks to particular problems (Grimm, 1999). The classical models of population dynamics, such as the exponential ($dN/dt = rN$) and logistic ($dN/dt = rN[1 - N/K]$) models, represent top-down approaches. These models assume that while all populations behave in the same general way, as is encoded in the functional form of the equation, the specific nature of their behaviour will vary from case to case, and this is specified by the exact parameter values used.

A final way to classify models is according to their use. Models serve three broad purposes in environmental geography: (i) predicting the future state of some system or phenomenon, (ii) making inferences about how a system or phenomenon is structured and changes, and (iii) integrating and synthesising knowledge and data from disparate sources. Bankes (1993) identifies two basic purposes of modelling:

1. *consolidation*: modelling based on compiling all available information about a system with the goal of creating a realistic and faithful surrogate of it. In this context prediction will be important, whether to test the realism of the model or to inform management and policy decisions about the actual system being modelled; and
2. *exploration*: modelling in the face of epistemic uncertainty, where the model is used experimentally to reduce this uncertainty by investigating the consequences of various assumptions about the modelled object. The goal of such modelling is heuristic.

This classification does *not* represent a rigid either-or division. Exploration and consolidation are synergistic. Improving our *understanding* of a process or system should enable us to predict its behaviour better (or determine whether it has the quality of predictability). Likewise, *reliable prediction* may lead to better under-standing (Brown et al., 2006).

Consolidation: models for prediction

The desire to predict a system's or phenomenon's behaviour is a common motivation for modelling. Making predictions and testing them is central to the 'conventional' deductive-nomological model of scientific inquiry. Predictive models take many forms, from simple deterministic analytical models to complicated stochastic simulation models. In geography, predictive modelling is often equated with empirical-statistical models (e.g., regression models); indeed statistical modelling is probably the most commonly applied and most criticised form of modelling used by geographers (Macmillan, 1989a). As outlined above, empirical-statistical models are for-malised descriptions based on observed characteristics of the entity of concern. While they may describe the links between components in a system, they do not consider the underlying mechanisms. This approach has often been denounced for yielding acausal and astructural 'black-boxes' that provide little heuristic insight (e.g., see Sayer, 1992).

Although empirical models are usually seen as focused strongly, if not solely, on prediction, they can also be used in an explanatory sense. The general intent of most empirical modelling is establishing a relationship between some variable x and a suite of predictor variables; establishing this relationship allows *indirect* causal relationships to be established (Mac Nally, 2000). Furthermore, there is increasing interest in applying statistical frameworks and tools, such as information-theoretic model selection and Bayesian statistics, to bridge the gap between exploration and prediction (Hobbs and Hilborn, 2006). In any case, the users of a prediction may be concerned solely with the reliability of the prediction. In such cases, a black-box approach may even be more appropriate than a complicated process-based model that explains the underlying processes responsible for driving the system being predicted (Demeritt and Wainwright, 2005). Furthermore, such models may also be suggestive of mechanism and help to generate new hypotheses.

Irrespective of how predictive modelling is best conducted there is, undoubtedly, a pressing need for reliable prediction to inform (environmental) public policy and decision making (Sarewitz et al., 1999; Clark et al., 2001; Pielke, Jr., 2003). Nevertheless, the goal of accurate prediction has, itself, been questioned. Clark et al. (2001, p. 657) take the pragmatic stance that '"Forecastable" ecosystem attributes are ones for which uncertainty can be reduced to the point where a forecast reports a useful amount of information'. However, Oreskes (2003) comments that the very factors that often lead us to modelling (limited understanding of/empirical information about a complex and/or complicated system) restrict the use of models for quantitative prediction. She argues that successful prediction in science has been limited to short duration, repetitive systems of low dimensionality, and that, even in such cases, successful prediction has often been reliant on trial and error. Conversely, socio-ecological systems may play themselves out over long durations, be non-repetitive, exhibit emergent or path-dependent behaviours, and be of high dimensionality – all traits that seem to preclude prediction (Batty and Torrens, 2001).

Unpredictability is also the key lesson of chaos theory. In chaotic (non-linearly deterministic) systems infinitesimally small differences in initial conditions will, in the long-term, result in completely different dynamics and system-states. These differences in initial conditions are much smaller than could ever be measured, and so, in a practical sense, chaotic systems do not even possess the quality of predictability (Gleick, 1987). Concerns over the ability to make reliable or meaningful predictions have, for example, been at the centre of the debate over the siting of the US high-level nuclear waste repository at Yucca Mountain, Nevada. 'Science', including, but not limited to, modelling, has played a central role in attempting to assess the performance of Yucca Mountain as a waste disposal site and billions of dollars (US) have been spent on this process (Ewing and Macfarlane, 2002). With a regulatory framework demanding safety assessments spanning tens of thousands of years (!), 'geoscientists in this project are challenged to make unprecedented predictions . . . ' in a context where epistemic uncertainty is high and the policy implications of those predictions even higher (Long and Ewing, 2004, p. 364). In such situations, where science and politics are intertwined and interdependent, there are important issues at stake about how the predictions scientists make are best interpreted and used (Macfarlane, 2003).

Exploration: models for learning

Besides prediction, models are vehicles for learning about the 'real' world. This is particularly true of simulation models. Recently, simulation models have become seen as systems that are open to examination in similar ways to other 'traditional' experimental systems (e.g., see Humphreys, 1995/96; Dowling, 1999; Winsberg, 2001; 2003; Peck, 2004). Certainly, the application of simulation modelling in some disciplines falls between traditional theorising and experimentation (Humphreys, 1995/96; Dowling, 1999). This approach opens up the possibility that following Dowling (1999), simulation models provide a means of 'experimenting on theories'. 'Experimental' simulation modelling seeks to mimic systems *in silico*[2]. The *in silico* form has the advantage that it can be manipulated in ways the 'real' world cannot; global climate change models are obvious examples of this (Frigg and Hartmann, 2006). Using models in this manner is a two-step process: we learn about the model and then transfer knowledge about the model to the target system. In practice, however, analysis often concentrates predominantly on the model. Nevertheless, it must be remembered that the model is a tool designed to help understand the real world; the (often understated) difficulty with detailed models is maintaining that connection (O'Sullivan, 2004; Frigg and Hartmann, 2006).

Models for integration: adaptive and participatory approaches

Models have become important tools for aiding in the decision-making process (e.g., forecasts of air quality are used to inform decisions about public health). Such modelling has often been viewed as the domain of the 'expert' and has been isolated from the rest of the decision-making process. Recently, this has begun to change as models are seen as integrative tools. Adaptive environmental management and assessment (AEMA) is an iterative process of structured learning through modelling, field experimentation and system monitoring (Walters, 1986). AEMA uses models to aid in the synthesis and integration of data and understanding, and to identify and reduce uncertainty. For example, Walters et al. (2000) used a series of conceptual and simulation models to filter various alternatives for restoring the flow regime affected by the Glen Canyon Dam in the Grand Canyon. Their models considered multiple spatio-temporal scales from localised algal responses to long-term patterns of sedimentation. They were used to: (i) highlight key areas of uncertainty in the system, and (ii) identify components of the system potentially amenable to controlled field experimentation. Model outcomes demonstrated the potential inability of the current monitoring framework to detect ecosystem responses to either experiment or management. Thus, models form(ed) part of an iterative and adaptive process, in which knowledge and understanding are constantly refined and management practices adapted to reflect this.

Models are also used to facilitate communication both between researchers in different disciplines and between the various stakeholders involved in environmental decision making. Castella et al. (2005) provide an interesting example of this approach. Castella et al. used a range of tools including a narrative model, an ABM, a role-playing game (derived from the ABM) and a GIS in an attempt to understand human-environment interactions and LUCC following Vietnam's *doi mois* economic reforms of the 1980s. The ABM explicitly considered: (i) farmers' decision-making strategies, (ii) the institutions that control resource use and access, and (iii) the dynamics of the biophysical and socio-economic components of the system. LUCC scenarios were developed with local land users using the role-playing game and the model, and were refined by repeated interactions between the researchers and the land users. The role-playing game helped the researchers to improve their understanding of farmers' decision making and how the actors deal with the risks engendered by uncertainty; it built trust and facilitated communication, and hence, model development.

Evaluating models: confrontation and experimentation

Verification and validation of models are much contested issues. Verification focuses on assessment of a model's structure (i.e., is the model free of logical, mathematical or coding errors?), whereas validation addresses on how exactly a model reproduces observed system dynamics (i.e., a model's predictions are confronted with observational data to assess its empirical adequacy). While some researchers believe that validation is central to modelling, others have argued that it is a logical impossibility (see Rykiel, 1996). Both verification and validation are, in essence, concerned with evaluating a model's adequacy against some criteria; what is 'adequate' will vary with a model's purpose. I will use the term 'evaluation' to encompass this broad(er) range of processes.

Models and their outcomes can be evaluated in many ways (table 20.2; Gardner and Urban (2003)). Kleindorfer et al. (1998) distinguishes objectivist, or founda-

Table 20.2 Some common methods of model evaluation and analysis, and their purpose

Method	Description and purpose
Structural	– Error propagation: Analysis of error in model output(s) as a function of the uncertainty associated with each parameter input to the model. – Sensitivity analysis: Identification of components of a model most sensitive to uncertainty and error in parameterisation.
Confrontational	– Visual 'diagnostics': Visual comparison of empirical observations and model predictions (i.e. by graphs). – Visual inspection for systematic bias, etc. – Statistical methods: Summary of differences between observations and predictions. – Quantitative comparison of predictions and observations (via correlation, regression and residual analysis, t-tests, difference measures, etc.). – Assessment of spatio-temporal trends in model performance and error.
Experimental	– Pattern-oriented modelling: Use of multiple observed patterns to evaluate and refine models and select between alternate representations (this will include structural and confirmatory evaluation). – 'Social' validation: Accepting a model as legitimate on the basis of consensus that it is valid by its users (this may or may not include structural and confirmatory evaluation).

Note: These methods are not mutually exclusive and most models are evaluated using a combination of the three.

tionalist, approaches to model evaluation from relativist and anti-foundationalist approaches. 'Classical' objectivist approaches to model analysis hinge on the 'confrontation' of a model with data, with the aim of establishing resemblance between the model's predictions and observations of the 'real' world; they emphasise the empirical verification of models and their outcomes. The tools used for establishing resemblance include graphical and visual diagnostics (e.g., time-series, residuals plots) and statistical (e.g., correlation and regression analyses, t-tests, summary difference measures) analyses (Mayer and Butler, 1993). Confrontational evaluation tends to emphasise an 'either-or' perspective: either the model and the predictions it generates are unambiguously valid, or they are rejected as unambiguously indefensible, with little in-between (Oreskes et al., 1994; Kleindorfer et al., 1998).

Contemporary philosophy of science emphasises several problems with the objectivist view that there is any unambiguous and impartial foundation for evaluating models and theories through some kind of self-evident and unproblematic confrontation with empirical data (Kleindorfer et al., 1998). First, recent discussions of model evaluation focus on the problems in seeing a model as 'true' (Rykiel, 1996; Oreskes, 1998; Brown et al., 2006). But second, even those embracing the idea of falsification as an alternative to the idea of validation must confront the problem of underdetermination. Observational data, it is argued, do not provide unambiguous grounds for evaluating theories as infinitely many hypotheses *might* explain a given dataset, even if only a small subset of these are actually plausible. This means that just because a model's predictions match empirical observations to some acceptable level, a model *cannot* be deemed either 'true' or 'correct'. A subset of the underdetermination problem is equifinality where there may be 'multiple model representations that provide acceptable simulations for any environmental system' (Beven, 2002, p. 2417). Finally, even the observed data used in the validation process carry assumptions, and so their place as a unique or truthful description of a system or phenomenon is itself questionable (Oreskes et al., 1994; Kleindorfer et al., 1998).

Even if their truth cannot be demonstrated incontrovertibly, models do have utility for elucidating how a system 'works' and for isolating where epistemic uncertainty is highest. Thus, and in keeping with a more exploratory approach to modelling, alternative modes of model evaluation have been developed, which tend to focus on what has been learned rather than on assessing the degree to which observations match model predictions. The adoption of more experimental approaches towards simulation modelling is premised on the belief that if models are experiments they should be evaluated as such (Dowling, 1999; Peck, 2004). One such approach is pattern-oriented modelling (POM – Wiegand et al., 2003; Grimm et al., 2005). POM involves the use of multiple observed spatio-temporal patterns with the aim of optimising model structure (by identifying components of the model central to aspects of observed behaviour), reducing parameter uncertainty, and testing and exploring alternate model representations (Grimm et al., 2005). Another more experimental approach is what Castella et al. (2005) call 'social validation' in which a model's users collectively agree that a model is a legitimate representation of the system (cf. Küppers and Lenhard, 2005); again, this is very different from the traditional emphasis on resemblance between observations and predictions. Castella et al. argue that social validation is crucial in participatory modelling, stating (p. 27) 'a model can only be used as a mediating tool for concerted action once it has been perfectly understood and is considered by decision makers to be legitimate'; this echoes Kleindorfer et al. (1998) who argue that model validation should be an open process involving model builder(s) and other stakeholders. Evaluating models in this way represents a significant departure from the objectivist methods typically used in the natural sciences. Development of alternative ways to evaluate models of all types remains fertile, if contested, ground.

Case Studies: Land-Use and Cover Change (LUCC)

Modelling LUCC is of active interest across geography and many other disciplines[3]. To illustrate the points raised in previous sections, I will consider some of the approaches taken to modelling LUCC. I do not intended to provide an exhaustive overview of activity in the field, but rather to provide an overview of the types of approaches that have been adopted. I will consider models in terms of the typology introduced in table 20.1, with the *caveat* that models typically span multiple of these categories; for example, simulation models usually contain analytical and empirical-statistical components. Finally, although LUCC is an obvious example of socio-ecological modelling, there are many other areas of environmental geography where models are routinely applied, including urban planning, climate change and its implications, resource models of water use and agricultural production, transport planning, reconstruction of palæo-environments, and prediction of the distribution of species and ecological communities (past, present and future), among other applications. The chapters in Wainwright and Mulligan (2004) provide a number of examples of specific modelling applications across the broad field of environmental geography.

Analytical models

Analytical models of LUCC focus on changes in the abundance of different land uses or conditions (e.g., economic values). These 'distributional models' (*sensu* Baker, 1989) are non-spatial and focus on how much change is taking place rather than where change is occurring. Transition (Markov) matrices are a commonly used type of distributional model. In Markov models, locations in the landscape are classified as being in one of n discrete categories. Repeatedly multiplying a $n \times n$ matrix, which describes the probability of transitions between each category, by a vector, which contains the abundance of each category in the landscape, results in a projection of change in the abundance of the various categories present in the landscape into the future under various restrictive assumptions. This approach has often been used in modelling LUCC (e.g., Turner, 1987; Hall et al., 1991; Romero-Calcerrada and Perry, 2004) because it is intuitive, conceptually simple and relatively easily parameterised (e.g., via time-series of remotely sensed imagery). However, in their simplest form, Markov models assume stationarity (constant rates of change in space and time) and ignore spatial neighbourhood effects.

A discipline where analytical modelling of land-use change has been much applied is economics. I will consider this economic framework here as much contemporary simulation modelling of LUCC (especially the agent-based approach) has been developed as a *reaction* to the microeconomic approach and its assumptions. The standard economic approach to land-use change is the 'bid-rent model' in which parcels of land (characterised by their location and other attributes) are allocated to the use earning the highest rent. This framework, based on rational utility theory, was originally developed by von Thünen for urban areas where property owners seek to optimise their location by trading off access to the urban centre with land rents. The model is equilibrial and spatially homogeneous, and (perhaps unsurprisingly) fails to reproduce observed patterns of city growth adequately; instead it produces concentric rings reflecting the balance between land value and transportation costs (Bockstael, 1996; Irwin and Geoghegan, 2001; Brown, 2006).

A few recent microeconomic models have departed from some of these restrictive assumptions and have adopted a spatially explicit perspective. For example, Bockstael (1996) and Irwin and Geoghegan (2001) describe a spatially explicit model of the economics of land-use conversion in the Patuxent watershed in northeast Maryland, USA. This region is a heterogeneous mix of rural and urban land uses and is undergoing rapid urbanisation, precisely the type of situation that confounds non-spatial bid-rent models. In Bockstael's model, land owners make decisions about whether or not to change land use at a given site on the basis of the future stream of returns to the parcel given how it is currently used (taking into account conversion costs). Because knowledge surrounding these decisions is imperfect, this decision-making process is framed as discrete probability choices. If there are n categories of land use, then there are n^2 decisions that land owners could potentially make. Bockstael (1996) reduces this to just one choice: whether or not to convert a land parcel from being undeveloped to developed. Thus, the model requires two pieces of empirical information: (i) the value of each parcel of land under any possible uses and (ii) the probability of conversion given those land values and associated conversion costs. To estimate these, Bockstael used an empirical model of land values (what economists term a 'hedonic pricing model') in which spatial factors such as neighbourhood conditions were included as drivers of land value, alongside more usual economic determinants of land value such as parcel size and access to transport infrastructure. The outcome of this model is a *static* map of probabilities of change. Using this framework, the implications of different public policy scenarios can be explored, as they influence the hedonic model, and the resultant probability maps compared. Subsequent extensions to the model (see Irwin and Geoghegan, 2001) made it temporally dynamic by incorporating a term that describes the optimal *timing* of the decision to convert land.

Although analytical approaches grounded in microeconomic theory have proven useful, they represent a different direction to that taken by geography and other disciplines (Drechsler et al., 2007). One of the key criticisms of such microeconomic models is the assumption that those involved in represent *Homo economicus* – the perfectly rational and informed decision maker. Furthermore, the emphasis in econometrics has largely been on temporal change and on equilibrial conditions (although spatial econometric tools are being developed – Irwin and Geoghegan, 2001). Again, these research directions are somewhat different to those taken in geography where the emphasis on space and disequilibrial conditions makes the use of analytical models problematic.

Empirical-statistical models

Empirical-statistical models, and in particular, a multitude of regression-derived approaches, have been widely applied for modelling LUCC. These regression approaches have been criticised on heuristic and methodological grounds; Brown et al. (2004, p. 401) identify some general problems with empirical-statistical models. First, statistical models of LUCC often assume that rates of change are stationary either in space or time or in both. Second, there are scale-related issues arising from the ecological fallacy and the modifiable area unit problem. Finally, the way in which change is represented is restricted by the limited way that relationships between predictor and dependent variables can be represented mathematically. In essence, the question to ask is 'how much can an empirical-statistical model illuminate process and causality?'

Millington et al. (2007) used empirical-statistical models in an effort to both understand and predict LUCC in the SPA 56 (central Spain). The SPA 56 is a heterogeneous and dynamic landscape comprising a range of land uses including agriculture, urban, peri-urban, recreation and forestry; it is designated a special protection area under the EU's 'Bird Directive' (Natura-2000 scheme). As in much of Mediterranean Europe this area has seen considerable land abandonment since the 1960s, largely driven by the decline of the traditional rural economy and rural-to-urban migration. Using satellite imagery, categorical maps and census information, Millington et al. (2007) derived statistical models of LUCC in the SPA 56. They employed multinomial logistic regression models, whose predictions were evaluated on the basis of pixel-by-pixel comparisons and by comparing the accuracy of the statistical models with a null model of zero change in the landscape (figure 20.1). The multinomial models suggested that the transformation of agricultural land to scrubland will continue into the future. Millington et al.'s predictive models

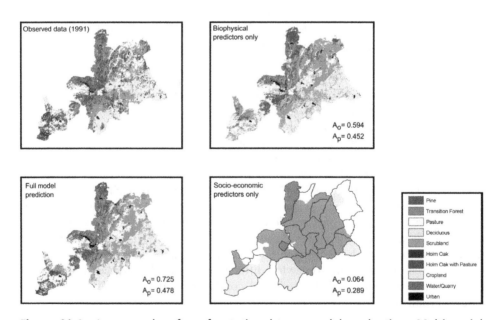

Figure 20.1 An example of confrontational-type model evaluation. Multinomial regression models containing different predictor sets (a full 'saturated' model, a model using only biophysical predictors and a model using only socio-economic predictors) were used to predict landscapes in the SPA-56, Central Spain. The predictions (for 1991) are compared with observed data (from 1991) on the basis of overall composition (A_o) and pixel-by-pixel (A_p) accuracy (proportional); analyses conducted by James Millington.

only perform better than the null model of no change over longer time periods, where they predicted approximately 70 percent of the landscape correctly on a pixel-by-pixel basis. Although they suggest how the landscape might change in the future should the *status quo* be maintained, these models epitomise the predictive 'black-box' approach frequently critiqued by geographers and others (e.g., Sayer, 1992; Mac Nally, 2000). Nevertheless, statistical tools are being developed that help explore relationships between a suite of predictor variables and the observed data. For example, hierarchical partitioning (used by Millington et al., 2007) estimates the contribution of each predictor to the total variance both in isolation and in conjunction with all other variables. Using such methods shifts the emphasis from producing the 'best' predictive model to isolating the variance explained by each predictor (Mac Nally, 2000). Such approaches are far better suited to hypothesis formulation than is the (often blind) search for the single 'best' predictive model.

Simulation models

Simulation models are used for prediction (e.g., forecasting of response to change using 'what if . . . ?' scenarios), synthesis and integration of data, and heuristic insight. They range in representational detail from very simple cellular-automata models to detailed agent-based representations of the decision-making process in spatio-temporally dynamic landscapes. There is a tension in simulation modelling of LUCC between models emphasising the ecological heterogeneity of the landscape at the expense of representing the actors engaged in decision making and *vice-versa*. This divide between landscape and actor has perhaps arisen due to the different foci of the various disciplines modelling LUCC (Veldkamp et al., 2001). In the social sciences the emphasis is on understanding the micro-level motivations of decision makers, whereas in ecology it is more on aggregate macro-level patterns of land use and habitat, with the hope that the socio-economic drivers of change are subsumed within the transition rules or probabilities. However, as Bockstael (1996) points out this means that the nature of these drivers is not transparent; public versus private and exogenous versus endogenous effects, for example, cannot easily be disentangled.

The two most widely adopted types of simulation model are grid-based and agent-based. Grid-based models (sometimes also called cellular or raster models) have been much used for spatial modelling of LUCC, especially, but not exclusively, by ecologists. In such models, the landscape is typically conceived of as a 2D $m \times n$ lattice, whose cells are internally homogeneous, with their state described by either a categorical (e.g., habitat type) or continuous (e.g., land value) variable. The cell size used will vary depending on the problem being addressed and may range from sub-meter (e.g., individual plants) to km+ (e.g., broadscale landscape pattern). Representations of change in grid-based models take a variety of forms including transition matrix approaches, simple quantitative neighbourhood rules, or more complicated hybrid semi-qualitative approaches (Perry and Enright, 2006).

Jenerette and Wu (2001) used a grid-based model to explore patterns of urban LUCC near Phoenix, Arizona. They employed a spatially explicit Markov approach in which transitions were a function of neighbourhood conditions. They developed models at two spatial grains: 250 × 250 m (coarse) and 75 × 75 m (fine). Jenerette and Wu used a parameterisation based on observed transitions and another one selected to optimise the models' fit to the observed data using genetic algorithms. Thus, Jenerette and Wu's predictions of urban change in the region combine statistical extrapolation with simulation modelling. Jenerette and Wu (2001) deemed the performance of the coarse-scale model over the period 1975–95 to be satisfactory. However, the fine-scale model did not perform as well, which Jenerette and Wu attributed to a mismatch in the scales at play in the system and in the model. Jenerette and Wu (2001) also experienced problems with the models' temporal (re) scaling. Estimation of the transitions between different land uses was based on observed data separated by a 20-year interval. These data had to be downscaled to annual transitions, but this downscaling failed when urbanisation was 'non-accretive' (i.e., occurred in entirely new parts of the landscape).

Agent-based models (ABMs) explicitly simulate interactions between autonomous goal-seeking entities, especially, in the case of LUCC, in some sort of dynamic landscape. Over the last decade ABMs have received increasing attention as tools for exploring human-environmental interactions and change (e.g., see Parker et al., 2003). One reason that they have been so eagerly adopted is dissatisfaction with the analytical rational-choice models traditionally used by economists. It has even been argued that bottom-up modelling (of which ABMs are a conspicuous component) represents a new 'generative' approach to social (Epstein, 1999) and landscape sciences (Brown et al., 2006).

An interesting use of ABMs of LUCC, in its broadest sense, is the reconstruction of human-environment interactions. One of the best known of such applications is the 'Artifical Anasazi' model. The Anasazi were a Puebloan (meso-American) group who occupied parts of the south-west of the USA. The Anasazi developed a rich culture in and around Long House Valley (NE Arizona) from about 1800 BC. before a rapid collapse triggered abandonment of these sites c.1300 AD. Detailed reconstructions of palæoecological and palæoclimatic conditions, based on dendrochronology and analysis of Packrat middens, have enabled estimates of annual maize production and hydrological dynamics, which have been used to parameterise the model. ABMs of this social system have been developed covering the period 300–1300 AD; in these models, the individual households are the agents (Dean et al., 2000; Axtell et al., 2002; Gumerman et al., 2003). The 'Artifical Anasazi' ABM follows the fate of individual families in the valley with households fissioning (as female agents age and marry) and moving in the landscape in response to water availability and food production. Early versions of the 'Artificial Anasazi' model (Dean et al., 2000) included few differences between individual actors and limited heterogeneity in the physical environment. Although this version of the model showed qualitative similarities to reconstructed population and settlement dynamics, quantitatively it was very different in that it predicted much larger populations and individual settlements than seems likely from the archæological record. More recent versions of the model (Axtell et al., 2002; Gumerman et al., 2003) incorporating more spatial heterogeneity in the landscape and variation in individual agent's characteristics provide a closer fit to the available data. In a spatial sense, the model now mirrors the known (from the archæological record) location of settlements, and it also mirrors, with one crucial exception, the expansion and rapid collapse of the population, in the face of deteriorating environmental conditions, in particular drought and changes in the water table (figure 20.2). The crucial exception is that the 'Artificial Anasazi' model predicts continued occupancy of the valley after it is believed that Long House Valley was completely abandoned. Thus, the modelling

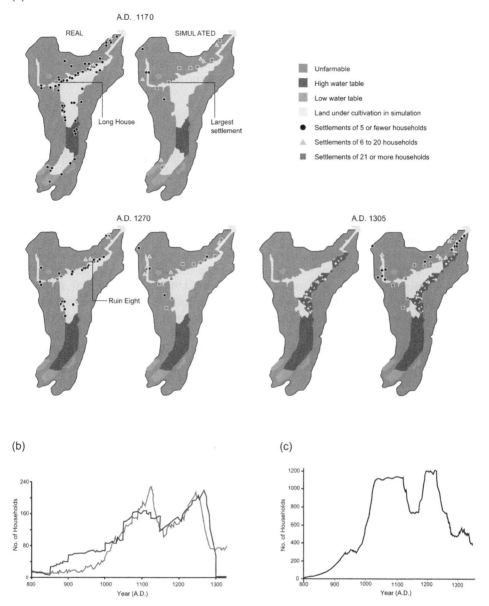

Figure 20.2 Evaluation of the Artificial Anasazi model: (a) comparison of landscape occupancy in the Artificial Anasazi model and as reconstructed from the archæological record; (b) time-series comparisons of number of households as observed (blue) and predicted (red) by the models of Axtell et al. (2002) and Gumerman et al. (2003); and (c) predictions of number of households in Long Valley in an earlier form of the model Dean et al. (2000) with limited spatial and inter-agent heterogeneity (note different y-axis scaling); figure drafted by Nicky Perry, (after Kohler, 2005). Original artwork by Lucy Reading-Ikkanda for Scientific American Magazine and reproduced with permission.

exercise suggests that although the environment may well have been the key control on the socio-environmental dynamics of this system, outside factors, such as longer-distance familial or other social ties, also influenced the Anasazi's behaviour, and may explain the total abandonment of the landscape that the model fails to predict.

Putting it all together

The examples discussed above lead to a series of questions about model representation and evaluation. All of the case studies are concerned with the broad question of what drives LUCC in some landscape, but the various models vary markedly in how they conceptualise and represent the landscape and the processes driving change in it. This variety suggests that there is not a single 'best' modelling approach. Rather some approaches will be more or less useful than others depending on the task at hand. Whatever their purpose, all models must wrestle with the challenges of balancing detail with parsimoniousness and determining the appropriate spatio-temporal scales to consider. In a now famous paper, Levins (1966) suggested that all model builders are forced to trade-off generality, precision, and realism. He believed that, at best, a single model could only achieve two of those three criteria. Arguably, the different approaches to modelling listed in table 20.1 and described above each focus on a different one of Levin's objectives (figure 20.3). While recognising that the boundaries are blurred, it might be said that analytical models focus on generality and precision, empirical models on precision and reality, and (mechanistic) simulation models on generality and realism (Guisan and Zimmermann, 2000).

The case studies also highlight the different approaches taken to model analysis and evaluation. The analytical models described by Bockstael (1996) and Irwin and

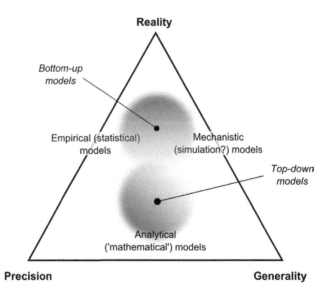

Figure 20.3 A classification of methodological approaches to modelling in relation to the 'goal' of the modelling activity (after Guisan and Zimmerman, 2000); reproduced with kind permission of Elsevier Press.

Geoghegan (2001) and the empirical-statistical models of Millington et al. (2007) rely, largely, on confrontational approaches in that they compare 'real' world observations with model predictions. The tools used in this confrontation vary but include visual comparison of predicted and observed spatial patterns, comparative statistical measures (r^2 and likelihood methods) and pixel-by-pixel comparisons (e.g., the kappa statistic, κ). Jenerette and Wu's 2001 model of urbanisation in Phoenix was evaluated by comparison of the model's predictions with various measures of spatial pattern in the landscape. As their model was stochastic they used Monte Carlo methods (i.e., where did 'real' world observations fall in relation to model estimates?) and avoided pixel-by-pixel confrontation. The agent-based 'Artifical Anasazi' models are evaluated through both confrontation and experiment. The population dynamics produced by the models are visually compared to population changes inferred from archæological reconstructions, and are experimentally evaluated by the researchers 'tinkering' (*sensu* Dowling, 1999) with the model until some adequate resemblance is reached (similar to pattern-oriented modelling).

The case studies also highlight the difficulties in establishing an adequate typology of models and modelling, whether based on methodology or purpose. Methodologically, *all* of the models considered above blur the boundaries between analytical, empirical-statistical and simulation modelling. For example, based on the outcomes of the (empirical-statistical) models developed by Bockstael (1996) and Millington et al. (2007), maps of possible future change may be produced using stochastic simulation. A typology based on purpose is no clearer: all of the examples presented above contain elements of consolidative, integrative and exploratory modelling, and all in some way attempt to improve understanding and to make predictions.

Evaluating the Role of Models in Environmental Geography

A discussion of models and modelling in geography would be incomplete without some mention of the debates about their place in the discipline[4]. During geography's (so-called) 'quantitative revolution', quantitative modelling was embraced as a methodology, peaking in the aftermath of Chorley and Haggett's seminal *Models in Geography* (1967). While models and modelling remain key components in much geographic research (especially in physical geography), geographers continue to debate the appropriate place and use of modelling. Critics of modelling range in position from those who view it as being a worthwhile, but typically poorly done, enterprise, through to those who see it as having little or no place in geography (Flowerdew, 1989). In the following discussion, I will focus on the criticisms put forward by human geographers. This is not because physical geographers all agree about the use and role of modelling, but rather because their debate(s) tend to be rather narrower and methodological (e.g., concerning the appropriateness, or otherwise, of specific techniques and representational assumptions).

In essence, the debate over modelling in geography is an extension of the long-running debate over the usefulness or otherwise of positivism and the scientific method in the discipline (Rhoads, 1999; Demeritt and Wainwright, 2005). Haines-Young (1989) identifies three common critiques of science and positivism in, but not limited to, geography. First, some human geographers complain that modelling, based on abstract quantitative theorising, cannot address the fundamental questions of human geography relating to uniqueness of place, individuality, imagination, morals and æsthetics. Second, there is the realist perspective that truly understanding some entity requires deeper understanding of its structure and the properties that change it and enable it to change. Realists such as Sayer (1992) have argued that the language of mathematics is unable to do this (recall the discussion above of empirical-statistical models as acausal and astructural). Finally, there is the postmodern 'attack' that science and modelling do not hold privileged positions as guarantors of objectivity or truth compared with other approaches, and quantitative modelling is just one of many means of geographic description (Cosgrove, 1989). Science, and indeed knowledge, it is argued, are socially constructed and, as such, are products of the social milieu in which they are created and embedded.

Another, and related, criticism levelled at geographic modelling is that it fails to address the important questions of geography. For example, Harvey (1989) argues that geography is a historical discipline and that the language of mathematics and the positivist approach are ill-suited to the development of theory in this domain (see the realist perspective above). He argues that modelling is limited to repetitive events (cf. Oreskes, 2003, view that prediction is only possible for repetitive systems). Harvey questions what modelling can teach and has taught us about the important historical-geographical shifts that he believes should be the focus of human geography; he states (p. 212) 'those who have stuck with modelling . . . have largely been able to do, I suspect, by restricting the nature of the questions they ask' and bemoans (p. 213) the 'sad degeneration and routinisation of modelling into mere data crunching, numerical analysis and statistical inference *instead of careful theory building*' (my italics). Here lies the crux of the debate: to what extent can models and modelling contribute to effective theory building in geography?

Conclusions

Modelling occupies a central place in geography and related disciplines, and it continues to receive considerable attention in the geographic literature. Although important questions remain about the ontology and epistemology of models and modelling, models are increasingly used in environmental geography to make predictions, to improve understanding, to synthesise and integrate data and to aid in communication. Recent developments in modelling are inextricably intertwined with developments in technology. As new analytical approaches have been developed, new sources of data become available, and computer power has increased and become more readily available, it has become possible to implement ever more detailed ('realistic'?) models. However, detailed and more realistic 'mimics' are not a panacea for the long-standing challenges of identifying appropriate representation and scale. Detailed representation is beguiling, but 'models of this sort may provide an unjustified sense of verisimilitude' (Levin et al., 1997, p. 335). While the pragmatic realist might see ever more detailed models as ever-truer representations, the fact remains that the 'truer' a model, the harder it is to establish its 'truth' (Oreskes, 2003). Likewise, while detailed models may be more empirically adequate, they may be premature and mask a lack of understanding of the entity being modelled (Frigg and Hartmann, 2006). Alongside the development of effective tools for model evaluation, finding the appropriate level of representational detail remains the key challenge for modellers and modelling.

ACKNOWLEDGEMENTS

Thanks to Peter Perry, David O'Sullivan and the editors for commenting on earlier drafts of this chapter. James Millington performed the statistical analyses depicted in figure 20.1. Nicky Perry drafted the figures.

NOTES

1. Of course, 'complex' also has an everyday meaning, which implies that an entity is not simple and comprises many parts; in essence it is 'complicated'. This everyday use of complexity is commonly used in the modelling literature. For example, detailed models are often described as being 'complex'.
2. *In silico* refers to entities or analyses that solely exist or are performed entirely within a computer.
3. By 'land-cover' I mean the nature of the land surface (e.g. forest, urban, etc.); this does not necessarily imply how the land is 'used', which is encompassed by the more anthropocentric term, 'land use'.
4. In this section, I will draw on contributions to Macmillan's *Rebuilding Geography* as they provide a relatively accessible introduction to what is, at times, a somewhat dense and daunting body of literature.

BIBLIOGRAPHY

Axtell, R. L., Epstein, J. M., Dean, J. S., Gumerman, G. J. and Swedlung, A. C. (2002) Population growth and collapse in a multiagent model of the Kayenta Anasazi in long house valley. *Proceedings of the National Academy of Sciences (USA)*, 99, 7275–79.

Baker, W. L. (1989) A review of models of landscape change. *Landscape Ecology*, 2, 111–34.

Bankes, S. (1993) Exploratory modeling for policy analysis. *Operations Research*, 41, 435–49.

Batty, M. and Torrens, P. M. (2001) Modeling complexity: the limits to prediction, *Cybergeo*, 12th European Colloquium on Quantitative and Theoretical Geography.St-Valery-en-Caux, France, 7–11 September 2001. St-Valery-en-Caux, France, 7–11 September 2001, Article 201. http://www.cybergeo.eu/index1035.html (accessed: 17 March 2008).

Beven, K. (2002) Towards a coherent philosophy for modelling the environment. *Proceedings of the Royal Society of London (Series A)*, 458, 2465–84.

Bockstael, N. E. (1996) Modeling economics and ecology: the importance of a spatial perspective. *American Journal of Agricultural Economics*, 78, 1168–80.

Brown, D. G. (2006) Agent-based models. In H. Geist (ed.), *The Earth's Changing Land: An Encyclopaedia of Land-Use and Land-Cover Change*. Westport, CT: Greenwood Publishing Group, pp. 7–13.

Brown, D. G., Aspinall, R. and Benett, D. A. (2006) Landscape models and explanation in landscape ecology: a space for generative landscape science? *Professional Geographer*, 58, 369–82.

Brown, D. G., Walker, R., Manson, S. and Seto, K. (2004) Modeling land use and land cover change. In G. Gutman, A. C. Janetos, C. O. Justice, E. F. Moran, J. F. Mustard, R. R. Rindfuss, D. L. Skole, B. L. Turner, M. A. Cochrane (eds), *Land Change Science: Observing, Monitoring and Understanding Trajectories of Change on the Earth's Surface*. New York: Springer, pp. 395–409.

Canham, C. D., Cole, J. J. and Lauenroth, W. K. (eds) (2003) *Models in Ecosystem Science*. Princeton, NJ: Princeton University Press.

Castella, J. C., Tran Ngoc Trung, T. N. and Boissau, S. (2005) Participatory simulation of land-use changes in the northern mountains of Vietnam: the combined use of an agent-based model, a role-playing game, and a geographic information system. *Ecology and Society*, 10, 27. http://www.ecologyandsociety.org/vol10/iss1/art27/ (accessed 18 March 2008).

Chorley, R. J. and Haggett, P. (eds) (1967) *Models in Geography*. London: Methuen.

Clark, J. S., Carpenter, S. R., Barber, M., Collins, S., Dobson, A., Foley, J. A., Lodge, D. M., Pascual, M., Pielke, R. J., Pizer, W., Pringle, C., Reid, W. V., Rose, K. A., Sala, O., Schlesinger, W. H., Wall, H. and Wear, D., 2001. Ecological forecasts: an emerging imperative. *Science*, 293, 657–60.

Cosgrove, D. (1989) Models, description and imagination in geography. In B. Macmillan (ed.), *Remodelling Geography*. Oxford: Basil Blackwell, Ltd, pp. 230–44.

Dean, J. S., Gumerman, G. J., Epstein, J. M., Axtell, R. L., Swedlund, A. C., Parker, M. T. and McCarroll, S. (2000) Understanding Anazasi culture change through agent-based modeling. In T. A. Kohler and G. J. Gumerman (eds), *Dynamics in Human and Primate Societies: Agent-Based Modeling of Social and Spatial Processes*. New York: Oxford University Press, pp. 179–205.

Demeritt, D. and Wainwright, J. (2005) Models, modelling, and geography. In Castree, N., Rogers, A. and Sherman, D. (eds), *Questioning Geography*. Oxford: Blackwell, pp. 206–25.

Dowling, D. (1999) Experimenting on theories. *Science in Context*, 12, 261–73.

Drechsler, M., Grimm, V., Mysiak, J. and Watzold, F. (2007) Differences and similarities between ecological and economic models for biodiversity conservation. *Ecological Economics*, 62, 232–41.

Epstein, J. M. (1999) Agent-based computational models and generative social science. *Complexity* 4, 41–60.

Ewing, R. C. and Macfarlane, A. (2002) Yucca mountain. *Science*, 296, 659–60.

Flowerdew, R. (1989) Some critical views of modelling in geography. In Macmillan, B. (ed.), *Remodelling Geography*. Oxford: Basil Blackwell Ltd., pp. 245–52.

Frigg, R. and Hartmann, S. (2006) Models in science. In Zalta, E. N. (ed.), *The Stanford Encyclopaedia of Philosophy*. Stanford University, USA: Center for the Study of Language and Information). http://plato.stanford.edu/archives/spr2006/entries/models-science/ (accessed 18 March 2008).

Gardner, R. H. and Urban, D. L. (2003) Model validation and testing: past lessons, present concerns, future prospects. In Canham, C. D., Cole, J. J. and Lauenroth, W. K. (eds), *Models in Ecosystem Science*. Princeton, NJ: Princeton University Press), pp. 184–204.

Gleick, J. (1987) *Chaos: Making a New Science*. (London: William Hienmann Ltd).

Grimm, V. (1999) Ten years of individual-based modelling in ecology: what have we learned and what could we learn in the future? *Ecological Modelling*, 115, 129–48.

Grimm, V., Revilla, E., Berger, U., Jeltsch, F., Mooij, W. M., Railsback, S. F., Thulke, H.-H., Weiner, J., Weigand, T. and DeAngelis, D. L. (2005) Pattern-oriented modeling of agent-based complex systems: lessons from ecology. *Science*, 310, 987–91.

Guisan, A. and Zimmermann, N. E. (2000) Predictive habitat distribution models in ecology. *Ecological Modelling*, 135, 147–86.

Gumerman, G. J., Dean, J. S. and Epstein, J. M. (2003) The evolution of social behavior in the prehistoric American southwest. *Artificial Life*, 9, 435–44.

Haines-Young, R. (1989) Modelling geographic knowledge. In Macmillan, B. (ed.), *Remodelling Geography*. Oxford: Basil Blackwell Ltd, pp. 22–39.

Hall, F. G., Botkin, D. B., Strebel, D. E., Woods, K. D. and Goetz, S. J. (1991) Large-scale patterns of forest succession as determined by remote sensing. *Ecology*, 72, 628–40.

Harvey, D. (1989) From models to Marx: note on the project to 're-model' contemporary geography. In B. Macmillan (ed.), *Remodelling Geography*. Oxford: Basil Blackwell Ltd) pp. 211–16.

Hobbs, N. T. and Hilborn, R. (2006) Alternatives to statistical hypothesis testing in ecology: a guide to self teaching. *Ecological Applications*, 16, 5–19.

Humphreys, P. (1995/96) Computational empiricism. *Foundations of Science*, 1, 119–30.

Irwin, E. G. and Geoghegan, J. (2001) Theory, data, methods: developing spatially explicit economic models of land use change. *Agriculture, Ecosystems and Environment*, 85, 7–23.

Jenerette, G. D. and Wu, J. (2001) Analysis and simulation of land-use change in the central Arizona-Phoenix region. *Landscape Ecology*, 16, 611–26.

Kleindorfer, G. B., O'Neill, L. and Ganeshan, R. (1998) Validation in simulation: various positions in the philosophy of science. *Management Science*, 44, 1087–99.

Küppers, G. and Lenhard, J. (2005) Validation of simulation: patterns in the natural and social sciences. *Journal of Artificial Societies and Social Simulation*, 8(4). http://jasss.soc.surrey.ac.uk/8/4/3.html (accessed 18 March 2008).

Levin, S. A., Grenfell, B., Hastings, A. and Perelson, A. S. (1997) Mathematical and computational challenges in population biology and ecosystems science. *Science*, 275, 334–43.

Levins, R. (1966) The strategy of model building in population biology. *American Scientist*, 54, 421–31.

Long, J. C. S. and Ewing, R. C. (2004) Yucca Mountain: earth-science issues at a geologic repository for high-level nuclear waste. *Annual Review of Earth and Planetary Sciences*, 32, 363–401.

Mac Nally, R. (2000) Regression and model-building in conservation biology, biogeography and ecology: the distinction between – and reconciliation of – 'predictive' and 'explanatory' models. *Biodiversity and Conservation*, 9, 655–71.

Macfarlane, A. (2003) Underlying Yucca Mountain: the interplay of geology and policy in nuclear waste disposal. *Social Studies of Science*, 33, 783–807.

Macmillan, B. (1989a) Modelling through: an afterword to Remodelling Geography. In B. Macmillan (ed.), *Remodelling Geography*. Oxford: Basil Blackwell Ltd, pp. 291–313.

Macmillan, B. (ed.) (1989b) *Remodelling Geography*. Oxford: Basil Blackwell Ltd.

Mayer, D. G. and Butler, D. G. (1993) Statistical validation. *Ecological Modelling*, 68, 21–31.

Medd, W. (2001) What is complexity science? Towards an ecology of 'ignorance'. *Emergence*, 3, 43–60.

Millington, J. D. A., Perry, G. L. W. and Romero-Calcerrada, R. (2007) Regression techniques for examining land use/cover change: a case study of a Mediterranean landscape. *Ecosystems*, 10, 562–78.

Morton, A. (1993) Mathematical models: questions of trustworthiness. *The British Journal for the Philosophy of Science*, 44, 659–74.

Mulligan, M. and Wainwright, J. (2004) Modelling and model building. In J. Wainwright and M. Mulligan (eds), *Environmental Modelling: Finding Simplicity in Complexity*. Chichester: John Wiley & Sons, Ltd, pp. 7–68.

Oreskes, N. (1998) Evaluation (not validation) of quantitative models. *Environmental Health Perspectives*, 106(Suppl. 6), 1453–60.

Oreskes, N. (2003) The role of quantitative models in science. In C. Canham, J. Cole, W. Lauenroth (eds), *Models in Ecosystem Science*. Princeton, NJ: Princeton University Press, pp. 13–32.

Oreskes, N. and Shrader-Frechette, K. and Belitz, K. (1994) Verification, validation, and confirmation of numerical models in the earth sciences. *Science*, 263, 641–46.

O'Sullivan, D. (2004) Complexity science and human geography. *Transactions of the Institute of British Geographers*, 29, 282–95.

Parker, D. C., Manson, S. M., Janssen, M. A., Hoffmann, M. J. and Deadman, P. (2003) Multi-agent systems for the simulation of land-use and land-cover change: a review. *Annals of the Association of American Geographers*, 93, 314–37.

Peck, S. L. (2004) Simulation as experiment: a philosophical reassessment for biological modeling. *Trends in Ecology and Evolution*, 19, 530–34.

Perry, G. L. W. and Enright, N. J. (2006) Spatial modelling of vegetation change in dynamic landscapes: a review of methods and applications. *Progress in Physical Geography*, 30, 47–72.

Pielke, R. A., Jr (2003) The role of models in prediction for decision making. In C. Canham, J. Cole and W. Lauenroth (eds.), *Models in Ecosystem Science*. Princeton, NJ: Princeton University Press, pp. 111–39.

Rhoads, B. L. (1999) Beyond pragmatism: the value of philosophical discourse for physical geography. *Annals of the Association of American Geographers*, 89, 760–71.

Romero-Calcerrada, R. and Perry, G. L. W. (2004) The role of land abandonment in landscape dynamics in the SPA 'Encinares del río Alberche y Cofio', Central Spain, 1984–1999. *Landscape and Urban Planning*, 66, 217–32.

Rykiel, E. J. (1996) Testing ecological models: the meaning of validation. *Ecological Modelling*, 90, 229–44.

Sarewitz, D., Pilke, R. and Byerly, R. (1999) Prediction: a process, not a product. *Geotimes*, 44(4), 29–31.

Sayer, A., (1992) *Method in Social Science*, 2nd edn. London, UK: Routledge.

Schelling, T. (1978) *Micromotives and Macrobehavior*. New York: Norton.

Turner, M. G. (1987) Spatial simulation of landscape changes in Georgia: a comparison of three transition models. *Landscape Ecology*, 1, 29–36.

Veldkamp, A., Verburg, P. H., Kok, K., de Koning, G. H. J., Priess, J. and Bergsma, A. R. (2001) The need for scale sensitive approaches in spatially explicit land use change modeling. *Environmental Modeling and Assessment*, 6, 111–21.

Wainwright, J. and Mulligan, M. (eds) (2004) *Environmental Modelling: Finding Simplicity in Complexity*. Chichester: John Wiley & Sons, Ltd.

Walters, C. J. (1986) *Adaptive Management of Renewable Resources*. New York: McMillan Publishing Co.

Walters, C. J., Korman, J., Stevens, L. E. and Gold, B. (2000) Ecosystem modelling for evaluation of adaptive management policies in the Grand Canyon. *Conservation Ecology*, 4(2). http://www.ecologyandsociety.org/vol4/iss2/art1/ (accessed 18 March 2008).

Wiegand, T., Jeltsch, F., Hanski, I. and Grimm, V. (2003) Using pattern-oriented modeling for revealing hidden information: a key for reconciling ecological theory and application. *Oikos*, 100, 209–22.

Winsberg, E. (2001) Simulations, models, and theories: complex physical systems and their representations. *Philosophy of Science*, 68, S442–S454.

Winsberg, E. (2003) Simulated experiments: methodology for a virtual world. *Philosophy of Science*, 70, 105–25.

Chapter 21

Integrated Assessment

James Tansey

Introduction

It has become commonplace to argue that interdisciplinarity within and between the social and natural sciences is important for advancing human understanding of social and natural systems. This version of interdisciplinarity looks a bit like cross cultural dialogue between disciplinary tribes who have developed their own specialised lexicon and are seeking to innovate at the margins with other disciplines. In contrast, Integrated Assessment (IA) adopted interdisciplinarity as a central organising principle. The impetus for this embrace of interdisciplinarity came from the strong problem orientation of this cluster of methods. The early history of Integrated Assessment focused on addressing the complexities of representing and modelling global climate change but the methods developed by practitioners were soon extended to include a range of problem domains including regional air pollution, land use planning, and urban development.

The identification of global climate change over the last two decades laid bare the fundamental limitations of narrow disciplinary approaches to understanding the complex interactions within and between biogeophysical systems operating on a planetary scale. There are broader efforts to understand coupled systems on a global scale that do not label themselves 'Integrated Assessment' but they often lack policy orientation or a strong problem focus. The central problematic that led to the birth of the term 'Integrated Assessment' was the task of connecting atmospheric models, to models of ocean circulation and terrestrial biotic and abiotic systems in a manner that made them policy relevant. Over time it became increasingly important also to represent social systems as well. Policymakers wanted answers to questions about the likely costs of climate change and, more recently, researchers have sought to understand how adaptation, innovation, and technological diffusion inter-relate to create the possibility of distinct development pathways over the coming century. As will become clearer below, the participatory turn in IA in the late nineties, which parallels in many ways the participatory turn in risk assessment and environmental assessment, spawned a number of highly innovative approaches to modelling and consultation. While much of this work has been undertaken by researchers who might classify themselves as human or physical geographers, there has been surprisingly little overlap with mainstream geography literature. In practice, IA has made very significant progress in connecting social and natural sciences through regional or even global modelling efforts, without intersecting with practitioners interested in participatory GIS or participatory planning systems.

A range of definition of IA can be found in the literature (see, for example, van der Sluis, 1997) of which the most inclusive is:

> Integrated assessment is an interdisciplinary process of combining, interpreting, and communicating knowledge from diverse scientific disciplines in such a way that the whole set of cause-effect interactions of a problem can be evaluated from a synoptic perspective'. (Rotmans and Dowlatabadi, 1998)

Thus, there are two defining features of IA. First, an IA must reach beyond the bounds of a single discipline and consider more than one sector or one aspect of the problem under consideration. Second, it must have as a central purpose 'to inform policy and decision-making, rather than to advance knowledge for its intrinsic value' (Weyant et al., 1996). In contrast to the majority of research, in which problems are identified endogenously through debates among scholars, IA practitioners aspire to be more pragmatic and problem oriented. This pragmatism may count as making a virtue out of necessity, since most of the problems they address involve open systems, characterised by uncertainty and indeterminancy: for instance one of the central challenges for climate modelling has been to account for fluxes of CO_2 that represent as much as 25 percent of the total carbon budget.

Until recently, IA was primarily characterised by the construction of large-scale integrated models – Integrated Assessment Models or IAM's. As a result of methodological innovation models are only one of a number of tools now available to IA practitioners (Rotmans, 1998). Broadly speaking, four methods are currently used in IA research, either alone or in combination (Rotmans and Dowlatabadi, 1998):

1. Computer-aided integrated assessment models, which are used to analyse complex systems;
2. Simulation gaming, which involves the 'representation of a complex system by a simpler one with relevant behavioural similarity' (Rotmans and Dowlatabadi: 294);
3. Scenarios, which are used as tools for identifying and exploring a range of possible futures, in order to assess their desirability or feasibility, or, particularly when applied in the business context to identify possible adaptive strategies (for instance, Kasemir et al., 2000);
4. Qualitative IA, which eschews the use of formal models and strongly resembles expert systems, applied to the task on future oriented assessment.

Despite these innovations, computer simulation modelling still remains at the heart of most IA, so this chapter will begin by reviewing the practices and problems of IA modelling. Though often treated simply as technical challenges, the uncertainty of IA model predictions and the difficulties of validating them are also bear centrally on how best to use IA in policy. After reviewing those debates about the role of IA in policy, the chapter addresses the participatory turn that is emerging in response before closing with some concluding thoughts about the future of IA.

Integrated Assessment Modelling (IAM)

IAM is often mischaracterised as seeking to create tightly specified predictive models. Most practitioners are keenly aware of the limitations of models and the fundamental uncertainty involved in forecasting the future. They insist, however, that not every future is equally possible in biogeophysical terms, and thus, that the goal of IA ought to be to identify and assess 'not implausible' futures (Yohe et al., 1999). While prediction may still represent a distant but ideal goal, there are many intermediate goals that are also extremely important in seeking to inform policy responses to the challenge of climate change.

Rotmans and Dowlatabadi (1998) identify four central benefits of IAM. First, IA can illuminate the feedbacks and relationships between linked systems all too often studied in isolation. For instance, IAM can represent interactions between climate change and other global issues such as ozone depletion (which affects CO2 uptake in the Antarctic Ocean), desertification (which can transfer more dust into the atmosphere), and acid rain. IAMs have also been used to represent the impacts of a range of non-climate driving forces that exist alongside increasing greenhouse gas emissions; for instance, land use change contributes directly to climate change and can also exacerbate the vulnerability of human populations and ecosystems. Second, IA can represent the contingency of global environmental systems; human choices about development pathways have fundamental impacts on the viability of adaptation and mitigation strategies and on the vulnerability of populations. The entire IPCC process is driven by a scenario based approach where distinct development pathways over a century generate large differences in emissions, land use, energy use, population etc. IAMs can be used to represent the high level interactions between development pathway choices, their impacts on the global climate, and then in turn the feedbacks of those climate changes on development pathways.

Third, IA can be used for exploratory purposes, or as a kind of hypothesis machine. The development of plausible integrated models allows researchers to develop hypotheses based on their observations of the order of magnitudes of fluxes and responses and the sensitivity of the model to variations in input choices. Often these models simplify more complex disciplinary models to reduced form models that allow for more successful integration. Integrated models allow researchers to explore the dependencies between natural systems and to identify critical uncertainties about relationships between them. For instance, Sigman and Boyle's (2000) classic study of the role of the southern oceans in drawing down carbon dioxide levels during interglacial periods basically synthesises a large number of disciplinary studies, connecting up plausible fluxes and stocks into a simple mass balance framework. As a result, they were able to suggest that the traditional explanations for the decline in CO2 during the interglacial periods (temperature changes, changes in salinity, changes in photosynthesis) are not of a sufficient magnitude to explain a drop in atmospheric CO2 from roughly 280ppm to 80–100 ppm. Instead they discuss changes to the acidity of oceans, complex geochemical reaction in the deep waters of the southern oceans and physical changes resulting from changes in the distribution of sea ice. Drawing together evidence about the order of magnitude of effects necessary to change carbon dioxide concentrations, this integrated modelling exercise generated a number of hypotheses that can be tested through other research.

Fourth, IA can play an important role in helping to translate and communicate uncertainty by showing how differences in worldviews influence the framing of scientific problems and choices about how social systems respond to climate change. In this way, and linked to the third point above, IA can help identify and prioritise critical gaps in scientific knowledge.

A very large number of Integrated Assessments have been undertaken and a large number of models have been created and improved over many years of work. Over time, key submodels are added, refined and their links to other models are improved. Some of the classic examples include:

- Mackenzie Basin Impact Study, which examined the interactions between climate change, regional development and management responses in Canada's Mackenzie Basin.
- POLESTAR was developed in the US, specifically to explore the interactions between natural and socio-political systems. Driving forces that underpinned the various scenarios included population growth, development strategies and societal responses.
- GCAM, an IAM which represented the interactions between atmospheric composition, human active, climate and sea level and ecosystems.
- IMAGE 2.0 focuses on the interactions between socio-economic processes, land cover, atmospheric and climate process, ecological and economic impacts and interventions in a series of broader 'Pressure-state-impact-response-loops'.

Looking internationally, Rotmans and Dowlatabadi identified a major divide in the underlying approaches between European approaches, more focused biosphere-climate oriented models, and North American ones more focused on macro-economic models. Of course, there are exceptions, but the broad difference does reflect interesting national differences in the framing of the climate problem as a whole.

In summarising some of the limitations and drawbacks of IAM, Rotmans and Dowlatabadi point to a number of problems. Overly complex model structures result in researchers learning more about the model than reality. Many IAMs have unacceptably high levels of aggregation and do not include random behaviour. In addition, there is often inadequate treatment of uncertainty within climate models, limited verification and validation and limitations on the methods used within the model. All these factors reduce the credibility of models within the wider scientific community. Moreover, despite all the caveats attached to models, they are still often treated as predictive truth machines, rather than heuristic tools or abstractions from much more complex systems.

Many of these limitations and drawbacks reflect the underlying challenge of climate change: it demands that researchers seek to represent systems where they know that they do not understand many of the dynamics of the system. Judged by the standard canons of science, this is the wild west of research: frontier work with crude tools cobbled together into an odd looking toolkit. The standard response from IA modellers is that by the time we are able to fully specify the problem, it will be too late to take any meaningful action.

Nonetheless IA practitioners have given serious consideration to issues of calibration, validation, and uncertainty. The challenge in all three domains is that the systems are open and that there is no baseline or benchmark against which to compare results. Calibration seeks to reproduce recent historical conditions to test whether models pass the test of plausibility. The fundamental challenge is that models that seek to examine impacts fifty or one hundred years into the future do not have a long enough historical record against which to generate a comparison.

Practical and conceptual validation of models introduces further challenges. Practical validation tests the validity of model outcomes, whereas conceptual validation tests whether the model structure is consistent with current knowledge. Validation and calibration become more difficult the more integrated the model becomes. There is good evidence (Rotmans and Dowlatabadi, 1998) that simple models often perform as well as more complex models and much of the work of validation occurs at the level of submodels. The task of validating an integrated model involves testing for perverse or implausible outcomes either by direct comparison to the systems being studied or through comparison with other models. Comparisons across models have produced highly divergent results, and even if outputs are comparable, modellers must show that they get similar answers via the same causal mechanisms.

Finally, uncertainty analysis may rely on forms of sensitivity testing. One modelling team developed an innovative system to utilise networked personal computers using a small programme downloaded by the user. The programme uses the processor when the computer is not being used to undertake model runs where the model parameters are altered minutely. This distributed computing approach engaged hundreds of thousands of willing participants and allowed the team to test the sensitivity of their model under a scenario where carbon dioxide doubles (Stainforth et al., 2005). The results of this study have been used to suggest that temperature changes of up to 11 degrees centigrade are possible with a doubling of carbon dioxide, but this is a misreading of the results. The exercise simply shows that under some combinations of input variables, the *model* is capable of generating changes of that magnitude. The extreme finding would need to be validated on a practical and conceptual level to show that these parameterisation changes were in fact plausible in the real world.

Over the last decade we have seen very rapid advances in IA models, driven by improvements in the quality of the underlying science, improvements in the integration of submodels and, inevitably, increases in computing power. The distributed solution described above generated the processing power of 1–2 supercomputers.

Prediction, Policy and IA

Ultimately, however, all complex systems and IA models face a shared problem: they are expected to predict, but are only able to suggest. This dilemma is shared by all assessment methods, including environmental, risk and cumulative assessment. This is a deeper and more intractable challenge that results from the way that science is used to underpin a range of assessment methods. There are many examples where conventional scientific research has been responsible for identifying the presence of biogeophysical 'limits to growth'. Iconic examples include research on the bioaccumulation of DDT through the food web, dramatised by Rachel Carson's *Silent Spring* (1962), or the dramatic discovery of a 'hole' in the ozone layer over the poles by scientists from the British Antarctic Survey. Even so, these dramatic interventions by scientists are relatively rare, compared to the more mundane everyday use of science in society. Assessment regimes depend on the application of scientific research at the project and strategic levels to the prediction of the potential impacts of human activities on natural systems over extended periods. The policy process and legal system demand a great deal from science and in most instances, these demands violate the principles of credible scientific prediction.[3]

The most common mistake is to confuse scientific explanation and scientific prediction. While scientists are often in a position to explain what is occurring at a particular point in time in, for instance, an aquatic ecosystem, they cannot scientifically predict what will occur if the system conditions are changed. Such prediction is only possible in a closed system where external perturbations can be controlled. In most instances, the science used in environmental assessment is not the 'normal science' of the traditional physics laboratory or ecological experiment, but a form of professional consultancy. In these cases of professional consultancy or 'post-normal science', values and framing assumptions, exert a significant influence over the results.[4] Andrew Stirling (1999) warns that while there is a stunning array of hybrid techniques including risk assessment, cost-benefit analysis and multi-criteria analysis, 'a proliferation of candidate understandings is not necessarily a sign of imminent enlightenment ... many of the analytic approaches aspire to develop a nice, clean 'analytical fix' for the messy (and intrinsically political) business of decision-making on sustainability'. These techniques are not scientific in the conventional sense of the word; rather they are hybrid approaches that represent an 'uneasy marriage' between science and policy.[5] To put it bluntly, the mere fact that a scientist is speaking or contributing to an environmental assessment does not make the findings scientific.

In broad terms, the sources of uncertainty regarding environmental decision making are the result of[6]:

1. the resilience of ecosystems and the fact that it is rarely possible to identify where their thresholds lie;
2. the uncertainty regarding the value of changes to ecosystems, both at the intrinsic level and with regards the functions and service they supply to human systems; and
3. the uncertainty regarding the future supply of ecosystem functions.

It is misleading to suggest that the problems related to the role of science in the policy process simply represent a crisis of overconfidence (Jasanoff and Wynne, 1998). In many instances, the political struggle for power and influence is inseparable from the science.[7] Science becomes an instrument in the pursuit of power, and decisions informed by interests and values are given a polished veneer of objectivity through the selective framing of studies and through the partial revelation of findings. The myth that is perpetuated in this process is that science involves the production of certain truth, whereas in practice, scientific research often generates greater complexity by highlighting more precisely the limits of our knowledge.

Scientific practice is a systematic and rigorous analysis of the natural world, but the legitimacy of a given policy intervention or environmental assessment requires science to be embedded in a much broader process. Ultimately the challenge is that the legal and political system in developed countries requires greater clarity than normal or post-normal science can deliver.

One solution to the inevitably value laden character of IA has been to open up technical models to greater public involvement and stakeholder scrutiny. The justification is that this allows policymakers and stakeholders to be explicit about the impacts of value judgements embedded in models. More recently, demands for stakeholder and public input to model development have emerged, in parallel to similar calls in environmental assessment and risk assessment. This participatory turn in IA is discussed below.

The Participatory Turn in IA

The late 20th century witnessed a decline in the public's perception of the credibility and authority of science in society. A range of explanations have emerged, including science's culpability in the creation of many of the problems (made explicitly in Beck's *Risk Society*), a general decline in deference for authority and rising educational levels and access to information. In response, the environmental and risk assessments underpinning public policy often now require significant public input for substantive as well as procedural reasons. Substantive justifications recognise that stakeholders may be able to provide useful input to scientific assessments. Procedural justifications recognise that participation may improve the legitimacy of decision-making processes. Participatory IA has been shaped by both demands. Substantive justifications go beyond the usual claims that stakeholders have useful information to bring to bear on modelling exercises. Once we admit that IAMs are often driven by socio-economic scenarios, then human choice has a direct influence over the future. To take a trivial example, public acceptance of alternative energy systems such as wind power will have an impact on development pathways. Broadly speaking, the argument runs that non-experts should be engaged in making choices about the range of social futures represented in the models. In the strongest cases, described below, the *desirability* of modelled futures is one of the key performance criteria.

As IA has evolved in this direction it has come to share more in common with scenario based approaches made famous by Shell over the last three decades (Van Heijden, 2005). Recognising that their business operates under conditions of great uncertainty, Shell began to challenge their own assumptions about the business environment and then used these variations in input assumptions to create different scenarios, informed in many cases by quantitative data. Shell utilises narrative-based approaches that develop clear storylines for their scenarios and these make the outputs accessible to a wider audience. This tradition informed the backcasting approach described below. Narrative driven approaches have also emerged under the banner of qualitative IA, which undertakes only limited modelling and focuses effort instead on creating compelling visions of the future. Narrative based approaches have the advantage of telling the story of the future in an attractive and accessible manner; they create a coherent storyline that can hold influence over policymakers who may struggle with numerical outputs.

Some of the most important contributions to this domain have sought to combine qualitative and quantitative methods, using focus group discussions and other qualitative social scientific methods to characterise the input variables for models that generate outputs and provide feedback to participants. These approaches seek to combine the quantitative sophistication of numerical models with the accessibility of qualitative visioning and scenario based approaches. These qualitative processes range from single-day events involving members of the public, to six-month processes involving civil servants in the Dutch government. A number of examples are summarised in table 21.1.

One example of how qualitative and participatory models can be combined with quantitative modeling tools is the recent Canadian effort to use a scenario tool – known as Quest – to develop a regional strategy for sustainable development. It was based on an explicitly normative approach approach to forecasting and future scenario development called 'backcasting'. Rather than beginning with the present state of affairs and projecting forwards, either through a formal predictive approach or through a less formal effort to identify alternative pathways, backcasting defines one or more normative endpoints first and then works backwards to identify the steps that would necessary to achieve the endpoint. It is similar to normative visioning exercises, although most applications have populated backcasting scenarios with quantitative data and have often used formal models to support the exercise. One of the ambitions behind this initiative was to not only sketch out a normatively desirable future and represent it using integrated modeling frameworks, but also to provide an opportunity for participants in the exercise to modify their positions based on the interaction between discrete choices.

The term 'backcasting' describes an approach used in the soft path energy studies, which emerged following the oil shocks of the 1970s. Soft path energy studies differed from traditional approaches to energy planning by focusing on demand-side management, energy efficiency, alternative non-fossil fuelled and decentralised supply technologies, behavioural change, rather than focusing on conventional centralised supply-side options. These studies took experts' articulation of a desirable future and analysed how feasible such goals were. The purpose of the analyses was to shed light on the policy and resource implications of different sectoral end-points by describing the trajectories required to connect the current state-of-play with the desired future.

The conceptual basis of backcasting lies in the recognition that the distant future is inherently unknowable, particularly in problem contexts like sustainability. Human choice and behavioural change can shape a desirable future, which is not necessarily the most likely based on past and present conditions (Robinson, 2003). Policy choices in such contexts are oriented by goals that require substantive change from current trends. These discontinuities are not typically resolved by forecasting approaches that are concerned with extrapolating what is most likely (Morgan et al., 1999). Rather than focusing on the likelihood of various version of the future, backcasting explores the feasibility of desirable futures. Backcasting forces a plurality of futures by asking 'what would it take to make this happen?' (Robinson, 1988; Dreborg, 1996).

According to Quist (2003), backcasting is composed of four principal steps: strategic problem orientation; articulation of values and generation of desirable future scenario(s); backcasting of trajectories; and identification of interventions to implement or initiate backcast trajectories. Backcast trajectories are typically described in terms of first-order economic, social, technological and institutional milestones and changes. These in turn inform the types of policy measures and behavioural shifts upon which the trajectories would be founded. Backcasting provides a framework for identifying the interventions or actions required to imple-

Table 21.1 A methodological comparison of participatory IA studies

Steps	COOL (van de Kerkhof et al., 2002)	Sustainable technology development (Weaver et al., 2000)	SusHouse (Young et al., 2001; Green and Vergragt, 2002; Quist, 2003)	GBFP-Strategies (Tansey et al., 2002)
1. Strategic problem context	– climate change – 80 percent reduction of GHGs by 2050 – national scale (Holland)	– sustainable technology adoption by 2030–50 – national scale (Holland)	– sustainable households by 2050 – household scale (EU)	– sustainability in 2040 – regional scale (British Columbian watershed)
2. Value articulation and futures definition	– two different future images (market dynamics and social adaptability)	– future visions based on solutions to the 'factor 20' challenge	– 'design-orienting scenarios' – products, services and their impacts and benefits	– value-based choices – future visions created by integrated assessment model – iteration for learning about trade-offs
3. Trajectory backcasting	– technology adoption pathways – major obstacles / opportunities and ways of overcoming / exploiting them	– technology adoption pathways with co-evolutional cultural and socio-economic conditions – descriptions at time intervals	– social innovation pathways and their acceptability to consumers – economic and environmental impacts	– projected by integrated assessment model – described by a range of environmental, economic, and social indicators
4. Identification of Interventions	– addressing single most significant challenge – key actors	– research programmes, innovation networks and social alliances, business opportunities	– design options for products, services, systems and social arrangements	– policy formulation

Other aspects

Participants in backcasting	Government, business, civil society, ENGOs	Government, business, research bodies, public interest groups	Non-governmental stakeholders	Government
Research duration	2 years	8 years	2.5 years	6 months

ment, or more modestly, to initiate the trajectories which would lead to the desired future (step 4). Interventions are often discussed as part of the trajectory backcasting step, for example, by identifying targets or obstacles to be overcome (Quist, 2003).

The Quest integrated assessment tool, developed to support the backcasting process, is composed of a series of linked sectoral submodels. While the submodels are generally relatively simple, the tool as a whole is much more sophisticated as it is horizontally integrated across submodels and is driven by a user-friendly interface. It is described in detail in two recent papers (Tansey et al., 2002, Carmichael et al., 2005).

This approach is described in detail because it is one of the few attempts to fully operationalise a participatory integrated assessment model combined with a robust participatory process. The geographic focus is on a region facing significant population growth and development. Forcing these two domains into close proximity revealed a number of lessons that speak to the wider challenges of making integrated assessment policy relevant.

First, participatory processes seek to be open with regards to the shape and form that the future may take. At the extreme, visioning exercises and even backcasting approaches typically take public values and opinions as sovereign. In a pure participatory application, any future, considered desirable, should be accepted on face value. Once one introduces a model into the process, the dynamic changes since, by definition, the model identifies critical thresholds and boundary conditions that introduce a measure of constraint. It is not technically feasible to create a model where every combination of every variable is possible. Even if the model is presented as a heuristic device that generates highly contingent outputs, it can change the character of the deliberations among participants. An integrated model is a form of disembodied expertise and it is not possible to negotiate with the model, to check its assumptions or to learn more about the dynamics it represents. Moreover, even if the assumptions are correct, the participatory process may generate a version of the future that is inconsistent with what are understood to be real biogeophysical and economic constraints. For instance, the group may decide that over two years, all houses should be converted to high-density, climate-neutral units. While technically feasible at the outer limits of possibility, this decision would have vast implications for the rest of the economy. The problem is that group dynamics may create a very strong commitment to a set of objectives that a model suggests has wider and negative implications.

Second, with respect to the political process, it is widely understood that decision-making is not driven by a comprehensive rational analysis. To think of IAMs simply as more refined tools for creating input for reasoned political choices is naïve. Decisions are made through the alignment of complex networks of political actors. Scientific research is often carefully framed using terms of reference and selective funding mechanisms. Even the IPCC is subject to close scrutiny, as research on the political process involved in the preparation of the summary for policymakers (Shaw and Robinson, 2006) shows. Scientific research and models in particular may be recruited or used for rhetorical purposes. In the translation of knowledge from the realm of the laboratory into the realm of the political system, contingency and uncertainty are squeezed out. While more structured processes such as citizen's juries hold the promise of subjecting expertise to close scrutiny such processes are still typically treated as inputs to the institutionalised decision-making mechanisms of representative democracies. Participatory IA suffers from the same problem as other participatory approaches: it seeks to solve problems that lie outside its sphere of influence. In other words, in many cases the problem is not so much about process or model design, rather it is about the structure of decision making in modern democracies.

Conclusion

IA model development has been enabled by rapid advances in computing power over the last two decades, which has allowed a new generation of problems to become technically tractable. IAMs generate provocative hypotheses across disciplinary domains and start from the assumption that the climate system is open and poorly understood. Current limits on modelling efforts are probably more a function of limits in our understanding of the underlying systems as they are a function of the limits of computing horsepower. IA will always remain highly dependent on focused disciplinary efforts to create and refine our knowledge of specific systems, while holding other interconnected system constant. The kernel of truth at the heart of IA is that nothing can remain constant in the climate system.

The introduction of participatory methods into IA has followed the same logic and grammar as was followed in a range of other areas. The benefits and limitations are understood and on some level have simply been repeated in the context of IA. The frontiers lie in at least two directions. First, if participatory processes are to be authentic and useful, they must be allowed to have some influence over actual policy formulation. We trust juries to make judgements about the guilt or innocence of criminals and we underwrite this effort with a significant commitment of resources. Participatory processes in IA or indeed in environmental assessment are rarely given such gravity.

Second, our notion of participation might also benefit from becoming less technocratic, methodical and more playful and accessible. A relatively simple model of the global climate, developed as a role playing computer game for the BBC was downloaded 500,000 times by individuals around the world. Open source systems, including open source journal systems make knowledge more accessible and subject to scrutiny and refinement. Perhaps the ultimate transition will be towards an integrated Wiki-Earth, collating the expertise of social and natural scientists from around the world and utilising the computing power of a vast distributed network to generate a vast array of plausible futures.

NOTES

1. I am grateful to one reviewer for pointing out that this framing is consistent with what has been labelled 'mode 2' science, as defined by Gibbons, Michael; Camille Limoges, Helga Nowotny, Simon Schwartzman, Peter Scott and Martin Trow (1994). *The New Production of Knowledge: The Dynamics of Science and Research in Contemporary Societies*. London: Sage.
2. The issue of IA vs. IAMs is discussed in more detail in an earlier paper see Rothman, D. and Robinson, J. (1996), Sustainable Development Research Institute Discussion Paper Series.. For examples of IA projects that have *not* been centred around the development of a single large-scale model (see Cohen, 1997).

3. A strong technical argument is made in Oreskes, N. K., Shrader-Frechette, K. and Belitz, K. (1994) Verification, validation and confirmation of numerical models in the earth sciences. *Science.*, 263, 641–46.
4. For a review of the different forms of science, see Funtowicz, S. O. and Ravetz, J. R. (1985) Three types of risk assessment: a methodological analysis. In C. Whipple and V. Covello (eds), *Risk Analysis in the Private Sector.* New York: Plenum.
5. The implications for Environmental Assessment are explored in Farrell, A., VanDeever, S. D. and Jaeger, J. (2001) Environmental assessment: four under-appreciated elements of design. *Global Environmental Change*, 11, 311–33.
6. Young, R. A. (2001) *Uncertainty and the Environment.* Northhampton, MA: Edward Elgar.
7. There is an extensive literature on this topic including Schrecker, T. (1984) *Law Reform Commission of Canada.* Ottawa, pp. 112 and Jasanoff, S. and Wynne., B. (1998.) Science and decisionmaking. In S. Rayner and E. Malone (eds), *Human Choice and Climate Change, Vol. 1.* Columbus: Battelle Press, pp. 1–87.

BIBLIOGRAPHY

Beck, U. (1992) *Risk Society: Towards a New Modernity.* London: Sage.

Carmichael, J., Tansey, J. and Robinson, J. (2004) Georgia basin quest: an integrated assessment modelling tool. *Global Environmental Change*, 14(2), 171–83.

Dreborg, K. (1996) Essence of backcasting. *Futures*, 28(9), 813–28.

Green, K. and Vergragt, P. (2002) Towards sustainable households: a methodology for developing sustainable technological and social innovations. *Futures*, 34, 381–400.

Jasanoff, S. and Wynne, B. (1998) Science and decisionmaking. In S. Rayner and E. Malone (eds), *Human Choice and Climate Change*, Vol. 1. Columbus: Battelle Press, 1–87.

Kasemir, B., Dahinden, U., Swartling, A., Schüle, R., Tabara, D. and Jaeger, C. (2000) Citizens' perspectives on climate change and energy use. *Global Environmental Change – Human and Policy Dimensions*, 10(3), 169–84.

Morgan, M. G., Kandlikar, M., Risbey, J. and Dowlatabadi, H. (1999) Why conventional tools for policy analysis are often inadequate for problems of global change. *Climatic Change*, 41, 271–81.

Oreskes, N. K., Shrader-Frechette, K. and Belitz, K. (1994) Verification validation, and confirmation of numerical models in the Earth sciences. *Science*, 263, 641–46.

Parson, E. A. and Fisher-Vanden, K. (1997) integrated assessment models of global climate change. Edited by R. H. Socolow, D. Anderson and J. Harte. *Annual Review of Energy and the Environment*, 22, 589–628.

Quist, J. N. (2003) *Greening Foresighting Through Backcasting: More Than Looking Back from the Future.* Sheffield: Greener Management International.

Robinson, J. (2003) Future subjunctive: backcasting as social learning. *Futures*, 35, 839–56.

Rotmans, J. (1998) Methods for IA: the challenges and opportunities ahead. *Environmental Modelling and Assessment*, 3(3), 155–79.

Rotmans, J. and Dowlatabadi, H. (1998) Integrated assessment modelling. In S. Rayner and E. L. Malone (eds), *Human Choice and Climate Change, Vol. 3: The Tools for Policy Analysis.* Washington, DC: Battelle Press, pp. 292–377.

Shaw, A. and Robinson, J. (2004) Relevant but not prescriptive: science policy models within the IPCC. *Philosophy Today*, 48(5/5; Supplement: Toward a Philosophy of Science Policy: Approaches and Issues), 84–95.

Sigman, D. M. and Boyle, E. A. (2000) Glacial/interglacial variations in atmospheric carbon dioxide. *Nature*, 407, 859–69.

Stirling, A. (1999) On science and precaution in the management of technological risks. Final report of a project for the EC Forward Studies Unit under the auspices of the ESTO Network, Sussex: SPRU.

Tansey, J., Carmichael, J., VanWynsberghe, R. and Robinson, J. (2002) The future is not what it used to be: participatory integrated assessment in the Georgia Basin. *Global Environmental Change*, 12(2), 97–104.

Van Der Heijden, K. (2005) *Scenarios: The Art of Strategic Conversation*, Chichester: Wiley and Sons.

van de Kerkhof, M., Hisschemoller, M. and Spanjersberg, M. (2002) Shaping diversity in participatory foresight studies: experience with interactive backcasting in a stakeholder assessment on long-term climate policy in The Netherlands. *Greener Management International*, 37(Spring), 85–99.

van der Sluis, J. (1997) Anchoring Amid Uncertainty. PhD thesis, Universiteit Utrecht, Utrecht, The Netherlands.

Weaver, P., Jansen, L., van Grootveld, G., van Spiegel, E. and Vergragt, Ph. (2000) *Sustainable Technology Development*. Sheffield: Greenleaf Publisher.

Weyant, J., Davidson, O., Dowlatabadi, H., Edmonds, J., Gruff, M., Richels, R., Rotmans, J., Shukla, P., Cline, W., Fankhauser, S., Tol, R. and Parson, E. A. (1996) *Climate Change 1995: Economic and Social Dimensions of Climate Change*. Edited by J. Bruce, H. Lee and E. Haites. Cambridge: Cambridge University Press, pp. 367–96.

Yohe, G., Jacobsen, M. and Gapotchenko, T. (1999) Spanning 'not-implausible' futures to assess relative vulnerability to climate change and climate variability. *Global Environmental Change*, 9, 233–249.

Young, C. W., Quist, J., Toth, K., Anderson, K. and Green, K. (2001) Exploring sustainable futures through 'design orienting scenarios': the case of shopping, cooking and eating. *Journal of Sustainable Product Design*, 1(2), 117–29.

Chapter 22

Ethnography

Kevin St. Martin and Marianna Pavlovskaya

Introduction

Ethnography has recently emerged as a powerful descriptive and explanatory approach within critical human geography. It has also become important to environmental geographers, particularly political ecologists, who increasingly employ it in fieldwork projects in both the global North and South. Traditionally, ethnography was closely associated with anthropology but it has also long intersected with geography, especially its cultural ecology tradition (Livingstone, 1992, chap. 8). In addition, since the expansion of humanistic approaches in human geography in the 1970s, ethnography emerged as central to cultural geography and its critical response to positivist and structural forms of explanation. The current *tour de force* of ethnography in critical human, and increasingly environmental, geography is, however, most clearly a product of the turn in social science towards critical social and cultural theory, especially feminism, post-structuralism and post-colonialism (e.g., Aunger, 2004; Noblit et al., 2004; Madison, 2005).

Ethnography is the direct observation and documentation of some group or community, their practices and habits, and, primarily, aspects of their culture. Generally, participant observation, or living among other people for a prolonged period, provides the foundation for writing a detailed anthropological monograph about the culture or community studied (e.g., the great ethnographies of Malinowski, Boas or Mead). It seeks to explain social and cultural phenomena via a holistic understanding that comes from the researcher's immersion and time spent in the field. While this sort of ethnography is certainly still practiced, 'ethnography' has come to mean considerably more. Indeed, the ethnography being adopted by geographers today cannot reduce to a single method or a single form of writing around which a research project is organised.

While geographers often label their work 'ethnographic' as a way to characterise the extended and immersed nature of one's research in place, their research is increasingly likely to include a plurality of qualitative methods beyond participant observation (e.g., in-depth interviewing, focus groups, oral history, archival research or map biographies) and to break from traditional correspondence theories of knowledge that privilege direct and allegedly objective observation (for an overview see Crang, 2002; 2003; Cloke et al., 2004). While geographers rarely engage with traditional forms of ethnographic research, it is their continued (and renewed) presence in 'the field' – interacting with research subjects and places, searching for multiple ways in which realities are constituted by both the researcher and the subjects of research – that aligns them with contemporary ethnography (cf. Madison, 2006). In this sense, the emergent interest in ethnography signals a shift in method-ological possibility across a variety of subfields in geography. This is particularly true for those subfields that were once distant from ethnography (e.g., economic, urban, political or environmental geography) but are now open to ethnographic approaches as a way to operationalise epistemological innovations such as feminism or post-structuralism.

Environmental geography is focused on understanding the interactions between environmental and human processes rather than other societies or cultures *per se* and, as a result, has rarely relied upon traditional ethnographic methods or modes of explanation that privilege observation of and interaction with subjects. Recent changes in ethnographic practice have, however, made it more amenable to the varied objectives of environmental geography. Yet, the adoption of ethnographic methods by environmental geographers also implies a change in environmental geography. While the broad interests of environmental geographers remain the same (i.e., understanding human/environment interactions), the mode of understanding has changed given, among other things, critical social theory approaches that stress local knowledges, micropolitical processes, identity politics, the positionality of various actors and agents and the social construction of nature generally (Castree, 2003). While the traditional interests of environmental geography and those of ethnography have served to distance them from each other, the gap is now closing as both are transformed in new directions that make their combination in critical, interdisciplinary and multi-method approaches both possible and useful.

In this chapter, our primary goal is to consider the nature of and the potential for ethnography in environmental geography. While we are interested to see envi-ronmental geographers adopt ethnographic methods, we will not discuss individual methods themselves because there is already a large geographical literature on the subject (Crang, 2003; Cloke et al., 2004). We begin by briefly looking into ethnog-raphy's origins in anthropology, its relationship to geography, and its recent trans-formation via critical social theory. We then discuss the potentials of ethnography and what ethnographic methods can offer to environmental geographers: we do so by identifying some of the important theoretical and empirical questions that these methods promise to illuminate. Finally, we examine issues related to the practice of ethnography as it relates to environmental geography; in particular, we address the question of politics and the rise of mixed methods in geographic research.

Ethnography's Transformations

Like geography, anthropology has a complicated heritage in which the heroics of 'discovery' and travel are mixed with the colonial practices of gathering information about peoples and territories that were to become subject to imperial power (Blunt and Rose, 1994). As such, much early anthropological and geographical work, using ethnographic methods, generated representations of 'primitive' societies in distant (global South) locations in need of European 'civilisation'. In the global North, early ethnographies were also about 'others', albeit in select urban areas (e.g., poor neighborhoods) or rural locations (e.g., Appalachia).

While ethnographies of peripheral peoples and places are still common (and, indeed, vitally important when re-cast as distinctly post-colonial projects), ethnographic research now encompasses studies of governing elites, environmental NGOs, transnational development agencies, and complexly positioned subjects across scales and sites in both the global North and South.

> Instead of a royal road to holistic knowledge of 'another society,' ethnography is beginning to become recognizable as a flexible and opportunistic strategy for diversifying and making more complex our understanding of various places, people, and predicaments through an attentiveness to the different forms of knowledge available from different social and political locations. (Gupta and Ferguson, 1997, p. 37)

Ethnography has been transformed such that 'the field' for ethnographic research has been not only expanded but 'decentred'; ethnography has been broadened and blurred beyond participant observation in a single community to a suite of methods applicable across a variety of sites and open to a variety of social science disciplines (Gupta and Ferguson, 1997).

Ethnography's transformation is largely due to the influence of contemporary critical theory that works to deconstruct assumed subject positions, blur the boundaries between centres and peripheries, disrupt the distancing of 'others' and reveal the intermingling of local and global processes. As such, critical theory initially provided strong theoretical tools to critique ethnography's role in the construction of colonial and neocolonial subjects as well as the European appropriation of resources (including human labour and knowledge). More recently, however, it has served to recast ethnography itself as a key method for producing intersubjective and situated understandings of other people and their environments and has even repositioned ethnographic research as a tool for local interventions that counter global hegemonic power.

While ethnography has had a long presence in geography, it is fair to say that at each point of intersection with the discipline it was understood differently and offered different potentials. Initially, ethnography was very popular with geographers, particularly those aligned with the cultural ecology tradition, in the first half of the 20th century. Its popularity waned, however, due to its association with an overly ideographic regional geography (Livingstone, 1992, chap. 9; Cloke et al., 2004). The post-WWII rise of spatial scientific methods relegated ethnography, along with the description of specific places and peoples, to the margins of human geography. Ethnography was seen as primarily a descriptive approach that was unable to explain geographic (particularly spatial) phenomena. It was not until the advent of new humanistic approaches in the 1970s that ethnographic methods became a mainstay of cultural geography and its critical response to the excesses of both positivist and structural approaches (e.g., Ley and Samuels 1978). Ethnography would be recast as able to provide unique insight into human 'lifeworlds', how people actually experienced and related to places and environments.

The humanistic approaches of the 1970s and their interest in ethnography were influential but more as critique than as a new model for human geography research (Cloke et al., 1991; Livingstone, 1991). The specificity of what ethnography revealed worked well to contradict structural assumptions and to reinsert people's lives into geography but it was not clearly linked to either a social theoretic or policy agenda in geography. Ethnographic methods remained marginal within human geography until at least a decade later when, aligned with feminist and post-structural critical social theory, they re-emerged as not only explanatory but, potentially, as ways to create knowledge that would inform change. As in anthropology, ethnography became relevant to critical explorations of urban, economic, political, and environmental processes across a variety of sites.

The adoption of a transformed (and transformative) ethnography is clearly seen within feminist geographic scholarship generally (e.g., Katz 1992) and feminist environmental geography in particular (e.g., Rocheleau 1995). As opposed to humanists who argued for the inclusion and better representation of people using ethnographic methods, feminists question the very possibility of any unbiased scientific representation within a discipline founded upon masculinist practices and ways of knowing. Traditional geographic fieldwork, after all, presumes a heroic (male) geographer traveling to observe other natural and social worlds while maintaining an objective distance from the subjects observed (Rose 1993). To transform the masculinist character of geographic fieldwork (cf. Sundberg 2003), feminist scholars argue for ethnographic methods as a way of co-producing knowledge with subjects and enacting progressive change relative to both the researcher and the researched.

Following Donna Haraway's (1991) concept of situated knowledge, feminist scholars reconceptualise practices of knowledge production as neither objective nor neutral but, using Cindi Katz's (1992) precise term, as 'oozing with power' (p. 496). Feminist scholars call for research that is self-reflective and conscious of its effects on the people it studies and represents. In addition, they insist upon a research practice aligned with a politics of emancipation and social change. A transformed and broadly defined ethnography, often in combination with other methods, facilitates both as is clear in the case of much environmental research that not only explicates environmental injustices but works to facilitate social change (e.g., Routledge, 2002; Sundberg, 2004; Wolford, 2006).

Much contemporary work in human geography is inspired by critical social theories such as feminism and is, increasingly, informed by ethnographic methods. Such research continues to engage a range of important and challenging issues that are likely to be of interest to environmental geographers. These include, for example, the meaning of 'the field' and its masculinist character (Rose, 1993; Hyndman, 2001; Sundberg, 2003); the gendered politics of fieldwork (Katz, 1994; Staeheli and Lawson, 1994), the ethical concerns and unequal modalities of power between academics and research subjects (England, 1994); the politics of team research (Hanson and Pratt, 1995); and the relationship of ethnographic methods to traditional (quantitative) research methods (Lawson, 1995) or to new techniques such as Geographic Information Systems (GIS) (Pavlovskaya and St. Martin, 2007).

That many of the above examples are from environmental geography indicates that they address issues increasingly important to the latter as it too begins to rely more on feminist and post-structural understandings amenable to and informed by ethnographic research methods (e.g., Schroeder, 1999; St. Martin, 2001; Robbins, 2003; Wolford, 2006). Environmental geography, with its pragmatic focus on environmental policy and its reliance upon positivist methods, was well insulated from both 1970s humanism and the broader 'cultural turn' in human geography that would soon follow and to which ethnography was so important. Today, however, ethnography's influence and use spans the breadth of human geography's subfields (e.g., urban, economic, political, environmental, as well as cultural).

Ethnography's Potentials

What makes ethnography, along with other qualitative methods, a powerful research method that is increasingly central to much human geography? How is it able to provide insights that are largely hidden from secondary data or statistical analysis? What issues does ethnography address that other methods cannot address? In this section we will discuss the potentials of ethnography that, we believe, have profoundly transformed the production of scholarship in environmental geography: its emphasis on explanation, its engagement with discursive practices and everyday life, its ability to understand the production of environmental subjectivities and governmentality, its attention to issues of power and its insight into the constitution of geographic scale.

Explanation instead of generalisation

A particular strength of ethnography is that it seeks to explain the phenomenon observed. This is very different from quantitative research, which seeks to either detect patterns and regularities or test a hypothesis that makes a generalised statement about a relationship between variables. Note that interpreting regularities or developing hypotheses also requires qualitative work, which is rarely acknowledged (cf. Pavlovskaya 2006). While qualitative research, including ethnography, is often limited to single sites or a limited number of cases, its power emerges from its ability to construct an explanation based on an intimate and profound understanding of the phenomena, social group or place in question. Information gathered via different ethnographic methods (e.g., archives, interviews, participant observation) is triangulated or checked for consistency allowing the connections between processes, events and phenomena to emerge (Nightingale, 2003). That depth of understanding allows for showing and explaining complexity, tracing connections between people and environments and working across scales.

One example from geography is Cindi Katz' fieldwork in Sudan. Information gathered via interviews and participant observation over two decades allowed her to write rich ethnographies focused on the environmental knowledge of village youth and their families. She calls these accounts 'topographies' (Katz, 2001) and sees them as a means to embed local environmental knowledge and practices within local and global political economies, politics, warfare, and gender, ethnic, and race relations. As a result, a story about a particular place in Sudan becomes a way to understand, 'on the ground', the transformations of society and the environment that are produced by globalisation. While not representative statistically, such an account is representative theoretically (Pavlovskaya, 2006). To allow for this theoretical rigor, ethnographic explanation does not separate nature from economy or from culture. Instead, its thick 'topographies' assemble together the relevant driving forces, including discursive productions, and show their interplay in the constitution of a process, a group of people or an environment.

Discourse as a maker of the world

Understandings of environmental processes and practices are, today, incomplete without a consideration of not only natural but economic, social, cultural, political and other events. This interdisciplinary approach, now clearly the trajectory of environmental geography, is aligned with the now commonplace understanding of 'nature' as socially produced (Braun and Castree, 1998; Demeritt, 2002; Castree, 2003). Understandings of nature as a social construction and as, in part, an outcome of environmental discourse have yielded compelling research on how scientific knowledge, environmental policy, colonial representations and theories of sustainability and/or economic development have come to produce particular environments and landscapes. In addition, this work has linked those productions of, for example, forests (Braun, 2002; Robbins, 2003; Agrawal, 2005), climate (Demeritt, 2001), fisheries (St. Martin, 2001), soils (Engle-Di Mauro, 2006) or bedrock (Braun, 2000) to particular manifestations of power in economic, social or cultural realms.

The production of nature – via the practices that environmental discourses engender and the positionalities and subjectivities it creates – is, however, enacted and performed by people in particular places. How nature is 'made' is, then, accessible not only through an analysis of discourse but through, and perhaps necessarily so, field-based research. Ethnography, in this case, provides a means by which to understand how discourse is effectively performed and it, unlike analyses of discourse in print, opens the door to the micropolitics of environmental knowledge production, management and resource use (e.g., Sletto, 2005). Ethnography is central to a movement beyond the analyses of environmental discourse *per se* to an understanding of environmental governmentality, an understanding of the people, mechanisms, dynamics and power relations produced through and within particular environmental regimes.

The power of ethnographic research combined with analyses of environmental knowledge/discourse is nowhere more powerfully demonstrated than in the work of Agrawal (2005) on 'environmentality'. Through a mixture of archival, interview and participant observation techniques, Agrawal demonstrates how knowledge of forests and forest practices in the Kumaon region of India (from colonial times to the present) produces both environments and environmental subjects. The use of a broadly defined ethnographic approach provides a rich understanding of the production of Indian forests via discourse (e.g., via colonial accounting methods), both historical and contemporary struggles over the forest resources, and how local people come to see themselves in relation to the environment.

Subjectivities and actors

Ethnography gives meaning to those positions afforded by particular discourses (e.g., citizen, worker, capitalist, patriarch, housewife, fisherman, farmer, rancher, etc.) and helps to answer how they are experienced relative to economic, political, gender and environmental systems of power and oppression. In this way, ethnography is widely used by Marxist and feminist researchers to examine how people not only experience but resist neoliberal capitalism, globalisation, gender inequality and environmental injustice (e.g., Rocheleau, 1995; Little, 1999; St. Martin, 2007). Resistance is possible, particularly from a post-structural perspective, because an individual's subjectivity is neither fixed nor without agency. Subjectivity is also constitutive of reality and, insofar as it is always also 'becoming', offers a potential site not only for observation but for intervention and change via, among other things, participatory research methodologies (Cameron and Gibson, 2005; Gibson-Graham, 2006; Kindon et al., 2007).

Understanding the contradictory production of environmental subjects is particularly important as emerging neoliberal regimes worldwide create conditions for new types of environmental governmentality (McCarthy and Prudham, 2004; Heynan et al., 2007). The enclosure of remaining common resources, withdrawal of state maintenance of resources, reliance on market solutions to environmental crises and enrollment of local communities in conservation and resource management are all opportunities for not only increased capital accumulation but for the simultaneous disciplining of resource users into neoliberal subject positions.

But these neoliberal pressures are never complete and subjects act upon new environmental developments, policies and practices in different and always contradictory ways, which, in turn, create openings for resistance. As Agrawal suggests,

> The relationships of subjects to the environment [. . .] need to be examined in their emergence [. . . and t]o pursue such a making of environmental subjects, it would be necessary to give up the concept of subjects and interests that are always already given by their social-structural locations and instead examine how they are made. (Agrawal, 2005, p. 211)

To discover and understand how environmental subjectivities are emerging in particular social and geographic locations is possible with ethnographic methods and environmental geographers increasingly use them to address just how complex socio-economic processes related to environmental regulation are actually enacted (and sometimes subverted) by environmental subjects in the context of, for example, the privatisation of common resources (St. Martin, 2007); expansion of cash-crop agriculture (Katz, 2001; Wolford, 2006); access to indigenous environmental knowledge (Nightingale, 2003); and establishment of protected areas and access to them (West et al., 2003).

Juanita Sundberg, for example, has studied the 'identities-in-making' of women involved in the gathering of medicinal herbs in the Maya Biosphere Reserve in Guatemala by being 'attentive to how disciplining discourses and practices are invoked, enacted, (re)configured, subverted and transformed by individuals who chose to be '*for* some worlds and not others" (Sundberg, 2004 p. 47, quote from Haraway, 1997, p. 37, emphasis in original). Clearly, as Agrawal (2005) has suggested, researchers need to examine knowledge, politics, institutions *and* subjectivities, the latter being accessible, understood and, potentially, alterable via ethnographic methods.

Understanding power

In the past, environmental research was seen as primarily empirical and distanced from power struggles as it was mainly concerned with informing pragmatic policy decisions. The exception was research by political ecologists who, since the 1970s, brought to environmental geography the concerns of political economy such as the enclosure of common property, uneven distributions of access to resources and class exploitation as it relates to land and environment (Peet and Watts, 1996; Robbins,

2004). Clearly focused on questions of power and politics since its inception, political ecology has also often relied upon ethnographic methods to access the lived experiences of peasants and/or 'land managers' subject to political-economic transformations and global structural forces.

Many of political ecology's major concerns are finding their way into more mainstream environmental research and policy development. The new focus on community impacts of regulations, community participation in environmental issues and attempts to mitigate environmental injustices are all recent developments that clearly intersect with political ecology research. In the case of fisheries in the United States, for example, new federal regulations make clear that the impacts of fisheries regulations must be assessed relative to both 'fishing communities' and to questions of environmental justice (Olson, 2005; St. Martin, 2006). While often contradictory, these mainstream adoptions of political ecology concerns nevertheless suggest a broadening of the environmental field such that questions of uneven and unjust impacts, if not power, might be acknowledged and addressed. With this broadening there is also an expansion of ethnographic methods as a way to address these issues.

Where power is understood as the result of political economic structures, ethnography has played an important role as the method by which political ecologists can closely examine the effects of power on local people, their livelihoods and their environments. The question of power and its relationship to the environment is, however, not just a question of forces from above and their impacts locally but one of struggles across scales involving a host of individual moments, actors and enactments (e.g., Sletto, 2005). Since Foucault, understanding power requires not just an analysis from the top down but

> [. . .] an ascending analysis of power, starting, that is, from its infinitesimal mechanisms, which each have their own history, their own trajectory, their own techniques and tactics, and then see how these mechanisms of power have been – and continue to be – invested, colonised, utilised, involuted, transformed, displaced, extended etc., by ever more general mechanisms and by forms of global domination. (Foucault, 1980)

In this conception of power, ethnography's role is again vital. Here it not only gives us access to impacts but works to explain power itself, how it emerges through and within daily interactions and how it is aligned with power mechanisms, practices and dominations. Furthermore, similar to the above concerning subjectivity, ethnography's explication of power can also be recast as an intervention into power, its maintenance, its disruption or its redirection.

Scale, global/local

Understanding scale and the relationships between processes operating at different scales remains a major research task of geography. This is also true of environmental geography. Much of this work has focused on the downward effects of 'macro' scale processes associated with power – global, national and regional – where places and communities were most often seen as 'recipients' of those global processes be they economic, environmental or cultural (Hart, 2004). This is true for both traditional environmental geography that relies upon quantitative impact analyses and for political ecology that reveals impacts through a variety of methods including those that are ethnographic. As new understandings of scale as socially constructed rather than fixed have emerged (Smith, 1993), however, the local scale has become more than a recipient of impacts, it has also become the 'ground of globalisation' (Katz, 2001) and the site where processes operating at a variety of scales are manifest 'in location' (Gupta and Ferguson, 1997). As a result of this recasting of scale, it becomes clear that ethnography can offer insight into and give meaning to a range of processes once thought inaccessible via methods limited to assessments of only 'the local'.

Using ethnographic methods, environmental geographers are breaking new ground and producing nuanced and grounded understandings of global (or local/global) processes. For example, environmental geographers have addressed the varying impacts of global climate change on places and people possessing different economic power and differing in terms of gender and race (Leichenko and O'Brian, 2006), explored the differential impacts of hazards, both natural and technological (Steinberg and Shields, 2007), documented the varying environmental outcomes of economic globalisation as well as resistance to it (Wolford, 2006), and, building upon political ecology traditions, compellingly revealed the diverse consequences of the global move towards neoliberal forms of natural resource management (Robertson, 2004).

Ethnography's Practices

Much of the above assumes a 'new' ethnography. How that new ethnography might be actually practiced relative to environmental geography will be discussed below. We focus on an ethnography transformed by critical social theory. We discuss the implications of such an approach to the politics of research and as a dynamic method that mixes and merges with other geographic methodologies (e.g., GIS) and approaches to knowledge production.

Politics and participation

In environmental geography, being effective demanded a rigorous collection and rational analysis of environmental and social data as a way to achieve better resource management practices. Today, environmental geographers are more acutely aware of the non-instrumental and, perhaps, unplanned or unseen effects of their research and knowledge production (Castree, 2003). This is especially true as they strive to integrate more local social and cultural processes into their research and are encountering the well-known problematics of representation. For example, the very identities of research subjects and of researchers, the motivations and expected behaviours of resource dependent communities, etc. are constituted '*in* the action of knowledge production, not *before* the action starts' (Haraway, 1997, p. 29, emphasis in original, quoted in Sundberg, 2004, p. 46). How identities are constituted is complicated and made political by the unequal power relations between researchers and their subjects.

The positionality of researchers relative to the researched has been widely debated and addressed in the literature (see Crang, 2003 for a recent overview) and ethnographic research, perhaps, most clearly illustrates the power dynamics of research due to its history of representing others as well as its overt embodied nature. Geographers have also attempted to address the problematic of unequal power between researcher and subject. They have, for example, volunteered skills to assist in community struggles as a form of compensation and a means to building reciprocal relations (see Cloke et al., 2004). During her ethnographic work in a Mayan community in Guatemala, Sundberg (2004) had to promise to help with labeling medicinal plants in English and 'volunteered to assist in every way possible' (p. 50). In his research on environmental activism in India, Paul Routledge (2002), similarly, was only allowed to 'observe' the activities of an NGO that was protesting new large-scale hotel developments after promising to participate in their action. Routledge, clearly breaking from the ethical canon of ethnographic research, spoke with hotel developers and managers disguised as a Western businessman interested in the booming tourist industry and its (often illegal) hotel construction.

These examples demonstrate that the traditional model of a 'detached observer' is increasingly irrelevant and that the dynamics of research need to be clearly exposed (rather than submerged). Such research suggests that the subjectivities of both academics and research participants are mutually affected, transformed and (re)constituted during fieldwork as well as the analytical and writing stages, follow-up visits and so on (Routledge, 2002; Sundberg, 2004; see Hyndman, 2001 on fieldwork as unbounded).

While impossible to avoid, the politics of ethnographic representation can be directly addressed where participation is explicitly incorporated into the ethnographic method. That is, researchers can acknowledge the co-production of identities and environmental knowledge by both subjects and researchers and, using participatory forms of research, see that the desires and needs of both are addressed through the research process.

> Rather than viewing ethnographic intervention as a disinterested search for truth in the service of universal humanistic knowledge, we see it as a way of pursuing specific political aims while simultaneously seeking lines of common political purpose with allies who stand elsewhere [. . .] (Gupta and Ferguson 1997, p. 37)

Participatory action research (PAR) is an approach where researchers not only recognise the effects of their research but they design projects around the possible transformations (e.g., of identity, politics, environments) they would like to enact. PAR relies upon a host of qualitative methods such as workshops, personal interviews, participant observation, team research, etc. that are clearly aligned with an explicitly political and participatory ethnography (Kindon et al., 2007).

Our research on the economic and environmental transformations of fisheries in New England uses a PAR approach to facilitate a 'community becoming' and a potential for community-based management of fisheries resources (St. Martin and Hall-Arber, 2007). Using ethnographic methods we engaged members of fishing communities into a cooperative investigation of fisher's local environmental knowledge, territoriality and sense of community. Fishers were recruited to conduct in-depth interviews with other fishers in their communities. In addition to eliciting rich narratives of community and environmental histories (the forte of ethnographic methods), the interviews also worked to generate a new subjectivity among the participants, one that emphasised their positionality vis-à-vis community, shared environmental knowledge and common territories of resource utilisation. The ethnographic approach in a PAR context proved vital as a means to foster a potential for community-based management practices and to counter the individualist neo-liberal subject given by dominant forms of resource management.

Mixing methods

> Any serious decentering of 'the field' has the effect, of course, of further softening the division between ethnographic knowledge and other forms of representation flowing out of archival research, the analysis of public discourse, interviewing, journalism, fiction, or statistical representations of collectivities. (Gupta and Ferguson, 1997, p. 38)

The emergence of quantitative methods in the 1960s undermined ethnographic research and related qualitative methods and associated them with purely descriptive and, hence, unscientific work. Today, the scientific authority of qualitative methods has been re-established and they are being widely adopted. They are, however, increasingly seen as not single methods (e.g., participant observation) but as part of a suite of methods (qualitative and quantitative), any of which might be used on a given project. The combination of methods is made possible by not only the broadening of ethnographic and other qualitative approaches across disciplines but by the re-thinking of quantitative methods and, even, GIS as tools for postpositivist research (Lawson, 1995; Sheppard, 2005; Pavlovskaya and St. Martin, 2007).

Methods once seen as epistemologically incompatible are being successfully combined within 'mixed method' research paradigms that often include ethnography (Creswell, 2003). The success of such approaches is due to their ability to produce knowledge that otherwise would not be possible to create. This is particularly important in light of the simultaneous expansion of secondary data, mainly in digital form (e.g., remotely sensed data, census information, consumer databases, etc.), and its growing prominence in various types of analysis, including environmental policy.

Environmental geography is well suited for mixed-methods approaches. This is clearly demonstrated by political ecologists who are combining, for example, geomatics techniques with ethnographic methods (cf. Turner and Taylor, 2003). Hong Jiang (2003) argues for the integration of satellite imagery analysis with ethnographic accounts of landscape change. Combining these methods produced insights into environmental and cultural change in Inner Mongolia that would not have been revealed by either method alone. Paul Robbins' research (2003; see also Robbins and Maddock, 2000) interrogates professional foresters' and villagers' concepts of 'forest' in India. Using remotely sensed images as well as in-depth interviews, his research not only reveals but explains the dissonance between both groups' categorisation of forests.

Conclusion

The current popularity of ethnography in human geography is a result of the renewed attention to human subjectivity characteristic of many realms of human geography including, recently, environmental geography. Where in the past, the power of ethnography existed in its ability to comprehensively describe and thereby appropriate other peoples and resources, its strengths today suggest a number of ways that it can inform an environmental geography that is itself moving beyond the instrumental analysis of environmental impacts. These include ethnography's abilities to theoretically (rather than statistically) explain social and environmental phenomena, to explicate 'on the ground' just how environments and environmental subjects are constituted (and constitute each other) via practices of discourse and knowledge production, to document the dynamics and impacts of power as it is experienced and performed by people within particular environments, and to effectively examine a host of processes (e.g., social, economic, environmental, as well as cultural) as they are manifest 'in location' rather than relegated to scales other than the local.

To access the potentials of ethnography, environmental geographers are fundamentally rethinking the objectives of research and 'fieldwork'. They are acknowledging the ways in which academic research constitutes environments and are beginning to use their research, via participatory ethnographic methods, as vehicles for social/environmental change. They are also pragmatically mixing methods to better complement the mixed and interdisciplinary strengths of environmental geography itself. To address questions of social practices and meanings relative to the environment, ethnographic approaches are merging with statistical, GIS, survey and other methods long familiar and effectively used by environmental geographers. Environmental geography, as it hones its unique interdisciplinary contribution to understandings of nature/society relations, will increasingly rely upon the power of ethnography to explain those relations and, indeed, to transform them.

BIBLIOGRAPHY

Agrawal, A. (2005) *Environmentality: Technologies of Government and the Making of Subjects*. Durham, NC: Duke University Press.

Aunger, R. (2005) *Reflexive Ethnographic Science*. New York: Altamira Press.

Blunt, A. and Rose, G. (eds) (1994) *Writing Women and Space: Colonial and Postcolonial Geographies*. New York and London: Guilford Press.

Braun, B. (2000) Producing vertical territory: geology and governmentality in late Victorian Canada. *Ecumene*, 7(1), 7–46.

Braun, B. (2002)*The Intemperate Rainforest: Nature, Culture, and Power on Canada's West Coast*. Minneapolis: University of Minnesota Press.

Braun, B. and Castree, N. (eds) (1998) *Remaking Reality: Nature at the Millennium*. London: Routledge.

Cameron, J. and Gibson, K. (2005) Participatory action research in a poststructuralist vein. *Geoforum*, 36(3), 315–31.

Castree, N. (2003) Commodifying what nature? *Progress in Human Geography*, 27(3), 273–97.

Cloke, P., Philo, C. and Sadler, D. (1991) *Approaching Human Geography an Introduction to Contemporary Theoretical Debates*. New York and London: The Guilford Press.

Cloke, P., Cook, I., Crang, P., Goodwin, M., Painter, J. and Philo, C. (2004) *Practising Human Geography*. London: Sage publications.

Crang, M. (2002) Qualitative methods: the new orthodoxy? *Progress in Human Geography*, 26(5), 647–55.

Crang, M. (2003) Qualitative methods: touchy, feely, look-see? *Progress in Human Geography*, 27(4), 494–504.

Creswell, J. W. (2003) *Research Design: Qualitative, Quantitative, and Mixed Methods Approaches*. London: Sage Publications.

Demeritt, D. (2001) The construction of global warming and the politics of science. *Annals of the Association of American Geographers*, 91(2), 307–37.

Demeritt, D. (2002) What is the 'social construction of nature'? A typology and sympathetic critique. *Progress in Human Geography*, 26(6), 766–89.

England, K. (1994) Getting personal: reflexivity, postionality and feminist research. *The Professional Geographer*, 46, 80–89.

Engle-Di Mauro, S. (2006) From organism to commodity: gender, class, and the development of soil science in Hungary, 1900–1989. *Environment and Planning D: Society and Space*, 24, 215–29.

Foucault, M. (1980) *Power/Knowledge: Selected Interviews and other Writings, 1972–1977*. New York: Pantheon Books.

Gibson-Graham, J. K. (2006) *A Postcapitalist Politics*. Minneapolis: University of Minnesota Press.

Gupta, A. and Ferguson, J. (1997) Discipline and practice: 'the field' as site, method, and location in anthropology. In A. Gupta and J. Ferguson (eds), *Anthropological Locations: Boundaries and Grounds of a Field Science*. Berkeley: University of California Press, pp. 1–46.

Hanson, S. and Pratt, G. (1995) *Gender, Work, and Space*. New York: Routledge.

Haraway, D. (1991) *Simians, Cyborgs and Women: The Reinvention of Nature*. London: Free Association Books.

Haraway, D. (1997) *Modest_Witness@Second_Millenium.FemaleMan©_Meets_OncoMouse™. Feminism and Technoscience*. New York: Routledge.

Hart, G. (2004) Geography and development: critical ethnographies. *Progress in Human Geography*, 28(1), 91–100.

Herbert, S. (2000) For ethnography. *Progress in Human Geography*, 24(4), 550–68.

Heynan, N., McCarthy, J., Prudham, S. and Robbins, P. (eds) (2007) *Neoliberal Environments: False Promises and Unnatural Consequences*. New York: Routledge.

Hyndman, J. (2001) The field as here and now, not there and then. *Geographical Review*, 91(1/2), 262–73.

Jackson, P. (2000) Ethnography. In R. J. Johnston, D. Gregory, G. Pratt and M. Watts (eds), *The Dictionary of Human Geography*. Oxford: Blackwell, pp. 238–39.

Jiang, H. (2003) Stories remote sensing images can tell: integrating remote sensing analysis with ethnographic research in the study of cultural landscapes. *Human Ecology: An Interdisciplinary Journal*, 31(2), 215–32.

Katz, C. (1992) All the world is staged: intellectuals and the projects of ethnography. *Environment and Planning D: Society and Space*, 10, 495–510.

Katz, C. (1994) Playing the field: questions of fieldwork in geography. *The Professional Geographer*, 46, 67–72.

Katz, C. (2001) On the grounds of globalization: a topography for feminist political engagement. *Signs: Journal of Women in Culture and Society*, 26(4), 1213–35.

Kindon, S, R. Pain and M. Kesby (eds) (2007) *Connecting People, Participation and Place: Participatory Action Research Approaches and Methods*. London: Routledge.

Lawson, V. (1995) The politics of difference: examining the quantitative/qualitative dualism in post-structuralist feminist research. *The Professional Geographer*, 47(4), 449–57.

Leichenko, R. and O'Brien, K. (2006) *Double Exposure: Global Environmental Change in an Era of Globalization*. New York: Oxford University Press.

Ley, D. and Samuels, M. S. (1978) Introduction: contexts of modern humanism in geography. In D. Ley and M. S. Samuels (eds), *Humanistic Geography: Prospects and Problems*. Chicago: Maaroufa Press, Inc.

Little, P. E. (1999) Environments and environmentalisms in anthropological research: facing a new millennium. *Annual Review of Anthropology*, 28, 253–84.

Livingstone, D. N. (1992) *The Geographical Tradition: Episodes in the History of a Contested Enterprise*. Oxford: Blackwell Publishers.

Madison, S. (2005) *Critical Ethnography: Methods, Ethics, and Performance*. Thousand Oaks, CA: Sage Publications.

McCarthy, J. and Prudham, S. (2004) Neoliberal nature and the nature of neoliberalism. *Geoforum*, 35, 275–83.

Nightingale, A. (2003) A feminist in the forest: situated knowledges and mixing methods in natural resource management. *ACME: An International E-Journal for Critical Geographies*, 2(1), 77–90.

Noblit, G. W., Flores, S. Y. and Murillo, E. G. (eds) (2004) *Postcritical Ethnography: Reinscribing Critique*. Cresskill, NJ: Hampton Press.

Olson, J. (2005) Re-placing the space of community: A story of cultural politics, policies, and fisheries management. *Anthropological Quarterly*, 78(1), 247–68.

Pavlovskaya, M. (2006) Theorizing with GIS: a tool for critical geographies? *Environment and Planning A*, 38(11), 2003–20.

Pavlovskaya, M. and St. Martin, K. (2007) Feminism and GIS: from a missing object to a mapping subject. *Geography Compass*, 1(3), 583–606.

Peet, R. and Watts, M. (eds) (1996) *Liberation Ecologies, Environment, Development, Social Movements*. [ew York: Routledge.

Robbins, P. (2003) Beyond ground truth: GIS and the environmental knowledge of herders, professional foresters, and other traditional communities. *Human Ecology*, 31(2), 233–53.

Robbins, P. (2004) *Political Ecology: A Critical Introduction*. Malden, MA: Blackwell Publishing.

Robbins, P. and Maddock, T. (2000) Interrogating land cover categories: metaphor and method in remote sensing. *Cartography and Geographic Information Science*, 27(4), 295–309.

Robertson, M. (2004) The neoliberalization of ecosystem services: wetland mitigation banking and problems in environmental governance. *Geoforum*, 35, 361–73.

Rocheleau, D. (1995) Maps, numbers, text, and context: mixing methods in feminist political ecology. *The Professional Geographer*, 47(4), 458–66.

Rose, G. (1993) Feminism and geography: the limits of geographical knowledge. Minneapolis: University of Minnesota Press.

Routledge, P. (2002) Travelling east as Walter Kurtz: identity, performance, and collaboration in Goa, India. *Environment and Planning D: Society and Space*, 20, 477–98.

Schroeder, R. (1999) *Shady Practices: Agroforestry and Gender Politics in The Gambia*. Berkeley: University of California Press.

Sletto, B. I. (2005) A swamp and its subjects: conservation politics, surveillance and resistance in Trinidad, the West Indies. *Geoforum*, 36, 77–93.

Sheppard, E. S. (2005) Knowledge production through critical GIS: geneology and prospects. *Cartographica*, 40(4), 5–21.

Smith, N. (1993) Homeless/global: scaling places. In J. Bird, B. Curtis, T. Putnam, G. Robertson and L. Tucker (eds). *Mapping the Future, Local Culture, Global Change*. New York: Routledge, pp. 87–119.

Staeheli, L. A. and Lawson, V. (1994) A discussion of 'Women in the field': the politics of feminist fieldwork. *The Professional Geographer*, 46(1), 96–102.

Steinberg, P. and Shields, R. (eds) (2007) *The Urban After Katrina: Place, Community, Connections and Memory*. Athens: University of Georgia Press.

St. Martin, K. (2001) Making space for community resource management in fisheries. *Annals of the Association of American Geographers*, 91(1), 122–42.

St. Martin, K. (2007) The difference that class makes: neoliberalization and non-capitalism in the fishing industry of New England. *Antipode*, 39(3), 527–49.

St. Martin, K. and Hall-Arber, M. (2007) Environment and development: (re)connecting community and commons in New England fisheries. In S. Kindon, R. Pain and M. Kesby. (eds), *Connecting People, Participation and Place: Participatory Action Research Approaches and Methods*. Lonson: Routledge.

Sundberg, J. (2003) Masculinist epistemologies and the politics of fieldwork in Latin Americanist Geography. *The Professional Geographer*, 55(2), 180–90.

Sundberg, J. (2004) Identities in the making: conservation, gender, and race in the Maya Biosphere Reserve, Guatemala. *Gender, Place and Culture: A Journal of Feminist Geography*, 11(1), 43–66.

Turner, M. and Taylor, P. (2003) Critical reflections on the use of remote sensing and GIS technologies in human ecology research. *Human Ecology*, 31(2), 177–82.

West, P., Igoe, J. and Brockington, D. (2006) Parks and peoples: the social impact of protected areas. *Annual Review of Anthropology*, 35, 251–77.

Wolford, W. (2006) The difference ethnography can make: understanding social mobilization and development in the Brazilian Northeast. *Qualitative Sociology*, 29, 335–52.

Chapter 23

Analysing Environmental Discourses and Representations

Tom Mels

Reservations about the Natural Environment

One of the most striking features of modern environmental experience is that it takes place in a world suffused with discursive forms. Images and texts attract tourists to a natural park, a map leads them from the parking lot to a walk on a wilderness path, commentaries are provided at strategic locations to guide the experience of nature, and an exhibition room at the entrance provides visitors with brochures, plans, stories, and films. What I find most interesting about this is not so much the pervasiveness of discursive forms, nor their technological sophistication. Instead, it is that they tend to leave many people with a somewhat amorphous sense of discomfort.

The very awareness of discursive forms awakens a feeling that the environment presented 'as-it-really-is' may not be all that natural, but the exact expression of an abstract system of manipulable, authoritative discourses. With every more cognizant look at imagery, maps, and texts, intriguing questions of the social production of knowledge and reality and the disciplining of experience come to one's mind. With every visit, national parks, exhibition rooms and wilderness trails stand out as spatialities at which environmental knowledges are produced rather than merely found. I say spatialities, because they are more than the physical sites at which knowledge is presented and encountered. They are places in a network of sites (universities, bureaucracies, studios and desks) from which the natural parks are conjured up through interpretive practices by particular people in particular social and occupational positions. Rather than submitting to an exhaustive and consistent story of 'the environment', these practices tend to unleash a stream of discourses and counter-discourses. It seems to me, then, that the power-laden tension between the reifying tendencies of discourses (their fixity and claim to meet reality-as-it-is), the elusiveness of the environment (its instability and shifting guises in different discourses), and the spatiality of those discourses (their spatio-historical emergence within a networked hierarchy of social sites) may help explain feelings of discomfort (see figure 23.1).

Figure 23.1   Welcome to Stenshuvud! This information board provides a vivid example of the spatiality of environmental discourse. With the help of texts and carefully selected imagery (scenic paintings, flora and fauna, a green map, the national park symbol), the authorities communicate a particular discourse about one of Sweden's national parks. This discourse is situated within a local, national and international context of environmental history readings and political discourse. Site-specific conservation practices specify some of the ways in which the official discourse is not only about valuing or discarding earlier environmental practice (e.g., fruit growing, coppicing, fishery), but also involves prioritizing certain future material relations and processes in the field (e.g., zoning, grazing, footpaths, conservation measures). In a sense the information board and the nearby information centre, offers a discursive spatial fix of what remains a landscape of contested social meaning. (Source: Author)

What I suggest is that the myriad things, processes, and relations we call environment, how they work, and how we should act towards them, are inherently discursive problems. They refer to various ways in which the reality of the biogeophysical world is at all times mediated before we speak or think about and act upon it. The stones, trees, marches, mountains, sounds, currents and waves are media in which cultural values and meaning are always already invested when encountered by humans. The mounting supply of journals, books, and conferences devoted to environmental discourses may be seen as a measure of the degree to which geographers as much as anthropologists, historians, philosophers, political scientists, sociologists, and others recognise this theme. It also shows that the discomfort to which I referred is not easily taken away, but rather something many academics have embraced or learned to live with.

And it is here that an additional discomfort arises. For in the same breath as I mention this broad scholarly acceptance of the importance of discursive ordering, I do not want to ignore difficulties arising from its attendant tendencies towards discursive dematerialisation and political relativism. Yet before I rush to hasty conclusions, it remains essential to bear in mind that there is a plethora of approaches to discourse analysis. For that reason my chapter reflects on how different approximations of discourse and its materiality feed into a variety of methodological and ecopolitical implications.

The chapter opens with a rather brief discussion of what is meant by 'environmental discourse' and 'representation' and how these are hinged together, with an emphasis on theoretical debate within geography. These are complex issues, which I can only discuss parsimoniously here, since I also want to spend some time on the question of how these theoretical insights are mobilised in actual research projects. The second part of the chapter is devoted to giving a range of illustrative examples of how geographers in their research go about unpacking environmental discourses. For reasons of consistency, I will focus on Marxist, post-structuralist, and political ecology work on sustainability and conservation. Their differences and commonalities illuminate the complex formation of environmental discourse as a geography of matter, meaning and power.

Environmental Discourses and the Spatiality of Power Systems

When geographers refer to discourses they tend to have more expansive things in mind than the colloquial reference to speech or language generally. If there is anything special about geographer's contribution to the understanding of environmental discourse it must be their attention to the *spatialities* of discourse. Since its theoretical breakthrough in the discipline during the 1980s, the term discourse has frequently been associated with a broad range of more or less strategic forms of representation (maps, imagery, narratives) mobilised within the ongoing struggle over spaces and places. Environmental discourses draw attention to how the production, circulation and justification of meaning within particular constellations of power permeate all social practices and thereby always enter into the constitution of the biogeophysical environment.

A useful starting point for thinking about the relations involved in environmental discourse is offered by the concept of 'regional discursive formations', first introduced by Richard Peet and Michael Watts. This describes 'certain modes of thought, logics, themes, styles of expression, and typical metaphors' that tend to 'run through the discursive history of a region, appearing in a variety of forms, disappearing occasionally, only to reappear with even greater intensity in new guises. A regional discursive formation also disallows certain themes, is marked by absences, repressions, marginalised statements, allowing some things to be mentioned only in highly prescribed, 'discrete', and disguised ways'. For Peet and Watts, these regional discursive formations 'originate in, and display the effects of, certain physical, political-economic, and institutional settings' (Peet and Watts, 1996b, p. 16). Regional discursive formations are also part of an extensive relational geography of scale, because they articulate and develop a society's wider 'environmental imaginary' in which discourses of nature are a principal element. Such an awareness of the politics and changing spatial situatedness involved in the production of knowledge and the shaping of practices is typical for discourse analysis within geography. And, *pace*

Peet and Watts, I would argue that this gestures beyond the regional level to a discursive spatiality, which refuses to privilege or essentialise any particular scale.

Peet and Watts' portrayal of discourses as part of an environmental imaginary is illuminating for at least three basic reasons, which resonate with more extensive claims about discourse and representation within geography. While discussing this below, I will argue that discourse bears important similarities and differences with its conceptual cousins ideology and hegemony, some of the erstwhile preferred notions among critical minds.

In the first place, Peet and Watts explicitly hook up discourse to shifting relations of social power. Power may here be understood in terms of situated, relational practices of dominance and resistance around both meaning and matter. Rather than seeing power as strictly centred ('power is possessed by a particular social class') or universal ('power is everywhere'), this emphasises the particularities of who exercises it in conjunction with why and how it operates in specific biogeophysical environments. As David Harvey explains, there are good reasons to couple discourse with power, most basically 'because words like 'nature' and 'environment' convey a commonality and universality of concern that can all too easily be captured by particularistic politics. 'Environment' is, after all, whatever surrounds or, to be more precise, whatever exists in the surroundings of some being that is *relevant to* the state of that being at a particular place and time (Harvey, 1996, p. 118). The social situatedness from which such relevance is defined varies considerably, and this will affect the shape of discourses as modalities of power.

Discourses often come as specific packages, as 'formations' of representations, narratives, storylines, concepts, metaphors, and conventions – constituting, if you like, a multimedia dialectic, in which a more or less coherent worldview is communicated in mutually confirming (or contradictory) guises of maps, images, and texts (Mels, 2002). Assemblages or chains of references of this kind tend to circumscribe particular interests and organisations of, for instance, bureaucratic, military, legal or corporate control. Discourses and their constitutive representations in that sense codify and substantiate particular social power relations and intervene in the material reconstitution of the environment, actively producing the 'very reality they appear to describe' (Said, 1978, p. 94).

Yet, like ideology, the power of discourse need not lie in deliberate maneuvers, but can also operate in a more subterranean fashion as 'broad taken-for granted frames of reference, including practical knowledge that results in embodied material practices of engaging with the world. Discourses contain common sense ways of knowing, valuing, and doing – for example, knowing what one likes without knowing how to explain why, or seeing any reason to do so' (Duncan and Duncan, 2004, p. 38). Importantly, the power of discourse is relative and relational. Like hegemony, discourses tend to be contested and struggled-over in ways that mediate geographically specific interests of class, gender, and ethnicity. Shifts in discursive power relations can appear when, for instance, local activists appropriate the discursive techniques of elites, present counter-discourses which map out 'lost' social relationships, or contest the homogenisation or naturalisation of space, property relations, plans and policies.

In the second place, and by extension, Peet and Watts' formulation leaves an opening to deeper philosophical issues about plural knowledge-claims and worlds, epistemologies and ontologies. It has often been argued that this recognition of plurality explains why many academics nowadays prefer to speak of discourse rather than its neighboring concepts ideology and hegemony. That these are neigh-boring concepts is easily comprehended with a closer look at definitions. To put it 'very schematically . . . an ideology is a system (with its own logic and rigour) of representations (images, myths, ideas or concepts, depending on the case) endowed with a historical existence and role within a given society' (Althusser, 1996, p. 231). More specifically, it circumscribes the elaboration of representation 'into a system-atic idealizing of existing conditions, those conditions that make possible the eco-nomic, social, and political primacy of a given group or class' (Lefebvre, 1968, p. 68). Ideologies 'refract (rather than reflect) reality via preexisting representations, selected by the dominant groups and acceptable to them' (p. 69). Ideologies can be envisaged as one of a dynamic range of cultural practices, immersed in the material and mental reality of human subjects, by which powerful classes preserve consent to its primacy within an existing social order or 'hegemony'.

It is not very important whether one agrees with these definitions or not, but what matters is that they do allow me to draw out two points. First, conventional notions of hegemony and ideology tend to allow for a residual believe in unmedi-ated access to the material world: a pure point of overview replacing misleading refractions of reality (ideology) by demystified reflection of the authentic reality of social and natural processes (scientific knowledge). In one of its key conventional uses within Marxism, ideology is thought of pejoratively as a distorted set of ideas, as a 'false' consciousness, which fails to recognise the real circumstances of social life (Williams, 1977, p. 103). Second, hegemony and ideology have traditionally been identified with a more centred notion of power, shaped by the interests of the bourgeoisie or other elite groups.

Discourse analysis won terrain in academic writing in the latest round of debate around what many saw as positivistic inclinations buried in traditional notions of ideology critique. Many scholars who have adopted the notion of discourse analysis insist that there is no extra-discursive, immediate access to reality. They often invoke the French thinker Michel Foucault whose employment of the term discourse (and its intimate relationships to power) entailed a profound disagreement with the epis-temological and ontological status of ideology within Marxism (Foucault, 1980, p. 118). While discourses have truth-effects, Foucault denied that they could be assessed as ideologies because that would suggest some veridical reference to a pre-discursive reality. From this reading, discourse analysis signals a rejection of what is seen as the epistemological realism lurking behind the ideology/science distinction. Scien-tists, business, green movements, the media and others produce environmental discourses which become received 'truths' because of social processes and position-ings, never because they are 'asocial' reflections of the biogeophysical things, spaces or mechanisms they describe. By extension, some claim that discourses structure society at large and that there is no easily identifiable social interest or class with full control over their shape, contents and functioning. This also conveys the idea that discourse (and hence power) tends to be a situation-specific, struggled-over, dispersed, relational and often concealed effect rather than a universal, stable, centred and always overt resource. By such account, attempts like Peet and Watts' to wed Foucaultian notions of discourse and power with historical materialist notions of hegemony and ideology seem contentious.

Perhaps somewhat ironically, the move towards discourse and representation has reactivated ontological and epistemological quarrels that were also central to earlier theoretical disputes about ideology critique. A key objection to a focus on discourse has been the ostensibly increasing remove from the materiality of the biogeophysical environment. Some might argue that relativistic distrust to any mode of representation is inappropriate in a time of increasingly sharpened political stakes of environmental issues. For how can we say anything substantive and meaningful about pending or existing environmental catastrophes when retreating into a detached world of endless signification and interwoven discursive reflection? Yet again, is any such comparison between our stories and knowledges about the material environment based on an erroneous belief that we can break through discourse to reclaim some unmediated reality?

These recurrent questions concerning the dichotomisation of 'cultural' discourse and the 'natural' realm of environment/nature are important enough and widely reviewed (Castree and Braun, 2001). From these debates one can identify various degrees to which discourse analysis within environmental geography is prepared to go beyond the domain of discourse to study biogeophysical environments. Discourse analysis is certainly not limited to the kind of dematerialised constructionism, which shelves any reference to an extra-discursive world. In Marxian and some post-structuralist quarters, it remains common to emphasise that discourses do not just relate to other discourses nor to a universal play of power. Instead, they relate to the range of material processes through which people shape the environment and to specific expressions of power within particular social formations.

Congruent with such an attempt at transcending ingrained oppositions between materialism and idealism (and this is exactly the dualism which Marx described as false consciousness!), 'discursive relations and representational practices are constitutive of the very ways that nature is made available to forms of economic and political calculation and the ways in which our interventions in nature are socially organized' (Castree and Braun, 1998, p. 16). What matters most to geographers engaged in this kind of discourse analysis, is not necessarily the degree of correspondence to reality (always a mediation) but by whom and how discourse is produced, how it works, and what is does. Mapping out the ascendance through which some environmental discourses have come to posses their present power in society may help to challenge taken-for-granted truths and reflexively shape alternative and emancipatory ways forward.

In the third place, and by extension, although Peet and Watts envision discourses as constituent parts of a society's environmental imagination, the academic reception of discourse analysis remains selective. While human geographers have been at pains to map discourses of various kinds in both theory and practice, physical geographers have as yet spent far less thought on this issue (cf. Castree, 2005, chap. 4). To some degree this may be unsurprising since discourse analysis takes human meaning as its prime objective, rather than inanimate objects studied by physical geographers. Occasionally, of course, physical geographers have utilised discursive material, such as qualitative data from interviews, for estimating quantitative environmental changes. Most physical geographers nevertheless hold on to a kind of correspondence theory of truth, in which science provides access to what they regard as an ontologically independent world and thus produces increasingly accurate referential knowledge. On a principal level, however, physical geographers too are discursively situated and they play an active role in shaping environmental discourse. I will revisit these issues in the section that follows, where I will try to address some of the myriad ways in which discourse analysis has been mobilised in research practice.

Geographies of Environmental Discourse

My aim here is to tease out a variety of ways in which geographers have appre-
hended their ambitions of unpacking environmental discourses. For reasons of
conciseness, I centre my discussion on research related to conservation and sustainability from within three broad varieties of discursive strategies. Looking at Marxist,
post-structural, and political ecology approaches to environmental discourse, I will
show how their theoretical positioning of discourse reverberates in methodology
and ecopolitics. Let me say from the outset that I am aware that terms, such as
Marxism and post-structuralism, signal metaphilosophical perspectives, while political ecology refers to a disciplinary field, and that there are arguably as many commonalities between the three approaches as there are differences within them. Yet,
rather than teasing out niceties of taxonomies, subdisciplines and philosophical
angles, my intention is primarily to discuss a range of geography's engagements with
environmental discourse.

Marxism: regulating corporate discourse

While there is a range of Marxist approaches to discourses, a frequently recurring
thread is that they are seen as devises of abstraction vital to capitalism's production
of nature. If environments are produced as commodities by labor power applied
under specific conditions, they are also liable to be represented in ways that efface
and reify the struggles, processes and relationships that go into their making
(Henderson, 1999; Walker, 2001). In that sense environments can be theorised in
politico-economic terms as 'dead labor': material and conceptual reifications of what
are really social relationships and struggles (Mitchell, 2003).

Some of the key characteristics of Marxist engagement with environmental discourse can be extracted from a study by Gavin Bridge and Phil McManus. Their
approach owes much to regulation theory, which tries to comprehend the societal
framework of capitalism as a system full of contradiction and conflict that nevertheless manages to attain periodic stability. Rather than resorting to transhistorical
imperatives of social reproduction, regulationists analyze capitalism in more
contingent terms of geographically and historically embedded, institutionally sanctioned modes of socio-spatial control and organisation. Adopting and adapting
components of this line of thought, Bridge and McManus (2000) argue 'that
regulation of the forestry and mineral sectors in contemporary market economies
is increasingly achieved through the deployment and co-optation of narratives of
sustainability' (p. 11). According to their reading, environmental discourses are
moments in the mode of social regulation: they are simultaneously a guiding framework for and outcome of the institutional structures and material practices that
make possible the reproduction of the conditions for capital accumulation. Sustainability narratives are of particular importance to industries with an unsavory environmental reputation, because they can negotiate and deflect accumulation crises
by disenfranchising opposition, co-opting green language, creating coalitions of
support, smoothing over contradictions and facilitating access to new deposits.

This is not to say that discourses and their regulatory mechanisms stand in any
seamless, functional relationship with accumulation systems. Rather, the authors
accentuate that these relationships tend to be contextual, contingent, politicised,
contradictory, and highly negotiated. Simultaneously, their concern lies with the

shifting ways in which the institutions of capital accumulation disseminate and nor-
malise discourses that codify and legitimise prevailing social relationships of environmental practices within capitalist societies (Bridge and McManus, 2000, p. 20).

Exemplifying their approach with a case study of the forest industry in Canada's
British Columbia, Bridge and McManus (2000, p. 27) lay bare how the 'discursive
framework of forestry . . . increasingly focuses on manipulating considerations of
time and space to ensure the perpetuation of the industry'. In the latest decades of
crisis in the province's industry, this is accomplished by, e.g., rhetorically rescripting
and resituating the forest in a space and time of long-term sustained yield, by making
the industry seem compatible with the international discourse of sustainability, by
appeals to public and national interests, and by sowing doubt about more radical
notions of sustainability. While the rhetorical greening of industry signals a shift in
the mode of social regulation (i.e., institutions and discursive practices), it does
so without any fundamental adaptations of the regime of accumulation (i.e., technologies and the organisation of production), or the accumulation system (i.e.,
production-consumption connections). Notwithstanding important contextual differences within and between sectors, the US gold mining industry offers a similar
example of how corporate discourses effectively regulate environmental practice and
sidetrack opposition.

What I find noteworthy here is that Bridge and McManus seek to understand
environmental transformations in terms of contested representations and discourses,
but emphasise how those discourses play a vital *ideological* role in capital's search
for regime stability. I say ideological because they prefer a notion of discourse in
which power is largely (though not exclusively) situated in corporate hands, to
discourse in a more outspread Foucaultian sense. The authors argue that material
and discursive appropriation of the environment tends to serve the interests of
powerful economic classes. Importantly, this remains a *contradictory* process
whereby ongoing environmental degradation and commoditisation stand in sharp
contrast with corporate espousal of sustainable development jargon.

Bridge and McManus claim that critical analysis of this contradiction needs to
bring out the couplings and synergetic relationships between the regime of accumulation and the mode of social regulation (including its changing discursive moments).
Such analysis can only be successful if we refrain from collapsing these conceptual
components of capitalist economies together. Thus, their discourse analysis covers a
vital but *limited* space in their critique of corporate capital. Discourse becomes an
important yet restricted ideological 'moment' that does its work within the mode of
social regulation but is almost absent in the analysis of the organisational and technological qualities of the regime of accumulation. In the mind of post-structuralists
this would arguably be a far too 'clean' separation, as I will show next.

Post-structuralism: a forest genealogy

According to most post-structuralists, discourses and established categories of
knowledge do not in the first place bear testimony to some ultimate factual reality,
but are rather associated with a solidification of meaning and reality serving interests
of social control. Discourses percolate through the social power struggles of disciplinary institutions; they work as modes of socialisation, and tend to facilitate self-
disciplinary practices. This raises questions about how, by whom, and with what
consequences discourse and categories are made solid and taken-for-granted.

Recent work by the Canadian geographer Bruce Braun may serve to illustrate the critical purchase of post-structuralist strategies to interrogate discursive practices. Central to his inquiry on the Clayoquot Sound, a heavily forested area in British Columbia, is the method of *genealogy*, inspired by Foucault's Nietzschean approach to history. This seeks to detonate the ostensibly obvious nature of things, the search for origins and timeless essences. It is to splinter notions of unity, to expose the heterogeneity and discontinuity of what seems to be consistent and continuous, in order to grapple with 'the historical, cultural, and political conditions through which objects attain legibility' (Braun, 2002, p. 3). Braun's genealogy pays careful attention to specific configurations of power/knowledge, bringing out the capacity of institutionally sanctioned epistemologies to present certain categories and narratives as trustworthy and real. Behind the preservation of such discursive coherence – which is instrumental to the ability to maintain social power – lies a hidden social history of exclusion, forgetting and silencing.

And so Braun turns to the language of industrial foresters, scientists, environmental groups, experts, and various forms of scientific categorisation, nature writing and photography with the intention of tracing the (subjugated) histories, (buried) epistemologies and morphologies of different environmental discourses. In Braun's treatment, each discourse not only adds layers of partial meaning to the environment, but these meanings are in their turn subjected to further deconstruction.

From a critical analysis of official documents, Braun argues that environmentalists and the logging lobby have at first sight constructed radically different discourses about the same old-growth forest. Where the forest company advances a scientistic account of the forest as a set of manageable resources, the environmentalists view the area through a more romantic veil as a pristine, sparsely peopled wilderness. However, for all their further differences, both of these environmental discourses make strong claims to transcend their discursive domain and capture reality as-it-really-is. Both trade on a widely reproduced nature/culture dualism. Their shared view of the Sound as pure nature entails a near denial of the historical presence of indigenous peoples, thereby (perhaps unintentionally) harking back on colonial discourses and practices of displacement. In these discourses, the Nuu-chah-nulth are doubly excluded from the environment, being either seen as a cultural aberration within nature or as a traditional anomaly within Canadian modernity.

Braun's approach to environmental discourse owes much of its depth to his systematic attention to the ways in which past environmental discourses and epistemologies reappear historically. As it turns out (and this echoes a general argument within post-colonial theory) much spadework for the currently dominant expert discourses on Canadian wilderness was done in the 19th century. These past discourses are kept alive not in the first place by immediate reference, but by their reproduction in social memory through imagery, storylines, and habits of thought, which in their turn are inscribed in the landscape itself. On a more theoretical level, Braun's work is attuned to a post-structuralism in which we cannot in any meaningful sense break through discourse to describe what a particular environment is like. Instead of allowing some discourse to speak authoritatively in the name of the environment, this calls for attention to closures in *all* discourses, be they hegemonic or disruptive, scientific or lay (Braun, 2002, p. 262).

I think it is important at this point to measure the distance between the post-structuralist method of critique presented here (revealing and challenge binaries, buried within discourses) and a more Marxian critique, which refuses to separate discourses from ideology and the material social conditions they speak of. Instead of reaching back to the emancipatory idiom of traditional ideology critique, which characterises Bridge and McManus' efforts, Braun follows a more Foucaultian road to criticise discursive power. His discussion also tunes in to the theoretical language of Deleuze and Guattari and envisages the environment metaphorically in term of heterogeneous assemblages, as fluxes of material de- and re-territorialisations. About those assemblages, he argues, we need to enquire the processes of their becoming, simultaneously 'opening space for thinking, doing and being otherwise. It is a politics with a purpose, but without any certain or final outcome' (Braun, 2002, p. 267).

Leaving aside further questions about its practicality for progressive ecopolitical change, this politics non-authoritatively returns questions about the materiality of what our environmental discourses are about to the materiality of discourse itself. In other words, it confirms that environmental politics demarcates a material geography of socially situated knowledges. Still, this does little to alter the clearly privileged attention to various modes and languages of representation that pass through Braun's and other post-structuralists' research on human-environment relations. Critics may say that this has little of substance to offer when questions about the biogeophysical aspects of environmental change appear (cf. Gandy, 1996). One may indeed ask if this does not ultimately reduce and relativise the environment and ecopolitics to habits of epistemology. The question arises how the insight that colonial discursive privileges serve systems of social domination and rationalise unjust material appropriations of land can be coupled with claims that meaning remains ultimately undecidable. After all, the insight itself remains an expression of meaning.

Political ecology: multimethod triangulation

Asking questions about the privileging of representation in research leads to scholars who proffer a political ecology approach to environmental discourse and its materiality. While the term 'political ecology' circumscribes a heterogeneous and interdisciplinary field of research rather than a metaphilosophical vantage point, it has been important for thinking about discourse and environment within geography. It is no random decision to spend some time on political ecology after discussing a Marxist analysis, which emphasises how discourses play a vital *ideological* role in capital's search for regime stability (Bridge and McManus), and a *genealogical* non-identity thinking which seeks to denaturalise all claims to environmental truth (Braun). Simplified, with Marxism, political ecology shares an interest in environmental practice and justice, but also tends to probe further beyond the epistemology offered by a critique of capitalism. With post-structuralism it shares an interest in discourse, but in many cases sees them as materially constrained, experientially based, and 'grounded in the social relations of production and their attendant struggles' (Peet and Watts, 1996a, p. 263). The work discussed here proposes a realist (not genealogical) *denaturalizing* confrontation of (post)colonial geography with discourses and a multimethod debunking of misconceived discourses.

Roderick Neumann's recent work, based on periods of fieldwork in Tanzania and a triangulation of methods (archival research, observations, household surveys and interviews), follows what I read as a 'realist' line of thought concerning environmental discourse. His intention is to explore the ways in which a European and increasingly commoditised aesthetics of unpeopled wilderness came to reinvent African environments and was mobilised to remove and displace indigenous people.

The political ecologist submits that these discourses not only fit comfortably with the authoritarianism of (post)colonial wildlife conservation, but are problematic for at least two additional reasons. First, a tragic irony was that the biogeophysical complexity of the region depended on the very traditional human land uses which were now terminated with reference to a 'purified' nature discourse. The result was that real natural processes sometimes contradicted lofty preservation efforts (Neumann, 1998, p. 28). Second, indigenous peoples, such as the Meru peasant society, did not share this dualistic environmental discourse. For them, Mount Meru was both a vital material resource in everyday life and a physical manifestation of their history and identity, not some aesthetic capital (p. 178). These discourses underlie Meru interpretations of justice and morality and, by extension, rationalise acts of peasant resistance against conservation laws.

It is evident that Neumann's approach to discourse analysis shares important traits with, for instance, Braun's post-structuralism. One of the more obvious congruities is that both view conservation as more than a question of control over material resources. It is also a matter of politics in which privileged (post)colonial ideals and naturalised discourses of nature are socially enforced and imposed upon the material world. Indeed, Neumann's conclusion from an analysis of popular texts and photographs, that 'discursive constructions have important material consequences' in biodiversity conservation would readily be accepted by post-structuralists or Marxists (Neumann, 2004, p. 833). Even so, Neumann presents an approach to representation and discourse (epistemology) as selectively connected to the material history of the environment – a history which Neumann's book lays bare and denaturalises in an ontological realist manner that defies any radical undecidability of meaning. Quite unlike Braun's more undecided stance, Neumann emphasises that he is 'not arguing that global biodiversity conservation constitutes a discourse (although it may) or that the threat of biodiversity loss is not "real" but some sort of linguistic fabrication', and he asserts in a footnote 'that biodiversity, in all its forms, has been historically diminished by human activities, is presently increasingly threatened, and that this is economically, culturally, and ecologically a negative outcome' (ibid.: 823).

Another example from political ecology research on environmental discourses in West Africa contrasts more sharply with Braun's approach, not only philosophically but also methodologically. Instead of 'purer' forms of discourse analysis, Thomas Bassett and Zuéli Koli Bi (2000) place it in a whole constellation of complementary methods (cf. Batterbury et al., 1997). Field research and analytical techniques were mobilised to collect information on land use and vegetation, while environmental perceptions were elicited from farmers and pastoralists through focus group discussions, interviews, and survey-research in a savanna landscape in northern Côte d'Ivoire. Out of this impressive collection of data, Bassett and Koli Bi tease out the disjunctions between two sets of environmental discourses.

The first discourse comprised the global and national desertification narratives underlying, for instance, the Ivorian government's National Environmental Action Plan (NEAP) and mandated by the World Bank as a condition for further loans. It presents an alarming process of environmental degradation as the result of overgrazing, bush fires, and mismanagement by peasants and pastoralists. One of the preconditions for sustainable environmental management, so the plan says, is that land rights give way to a 'modern' freehold tenure system.

The second discourse runs counter to this 'official' discursive formation of West Africa by way of place-specific perceptions of land users. In contrast to the hegemonic desertification story, this discourse describes how the growing number of livestock led to a decline in grass cover followed by an extension of trees and shrubs. Bush fires were less aggressive due to a changing fire regime, combining early dry-season fires with stronger grazing pressure and an expansion of cropland.

The purpose of mapping these discourses was comparative, not in the conventional deconstructive sense of bringing out silences and gaps, but as a stage in the process of making accurate scientific judgements. Hence, the next step in the research project: 'To assess whether local perceptions of environmental change were congruent with scientific findings, we reviewed the specialist literature on human-induced modifications of savanna vegetation' (Bassett and Koli Bi, 2000, p. 71). This was then further mapped with an examination of aerial photographs, quantifications with the help of Geographic information systems, and on the ground species inventories. Somewhat simplified, the findings of virtually all of these analyses supported the farmer-herder discourse and ran counter to the dominant desertification narrative guiding current environmental policymaking. Although there was no clear sign of desertification, 'heavy grazing and early fires have significantly reduced the quality of the savanna for livestock raising' (p. 90). Given the government's prioritisation of livestock development, this would advise policymakers to encourage rangeland rehabilitation rather than the currently prevailing concern with reforestation.

On the basis of my capsule summary, I think it is interesting to point at ways in which this project differs methodologically, philosophically and ecopolitically from Marxist and post-structural approaches. In the hands of Bassett and Koli Bi, discourse analysis is located in a wider array of multi-scale research methods which complement each other in order to distinguish actual from imagined environmental problems. And so, the authors' own research suggests 'that the dominant environmental narrative for the Côte d'Ivoire is misconceived' and that 'environmental analysts and planners are occupied with an imaginary environmental problem' (Bassett and Koli Bi, 2000, pp. 69, 90). In line with this, farmers' and herders' discourses are marginalised in planning, while their 'understanding of environmental change is more nuanced and sophisticated than the dominant narrative' (p. 91). Post-structuralists would probably be unwilling to arbitrate between discourses in this way. They would also resolutely reject a grounded, multimethod, materialist approach, asserting that 'those who embrace constructivist approaches to "nature" but stop short of accepting the radical undecidability of meaning often end up making arguments that are too rigorous, or too "clean," in their separation of ontology and epistemology' (Braun and Wainwright, 2001, p. 61). From a Marxist point of view, the turn to natural science and grounded methods will be misguided as long as researchers fail to unravel the ideological role of discourse in, e.g., the mode of regulation and capital's search for regime stability in African societies.

I am not merely inclined to agree that these issues are underplayed by Bassett and Koli Bi's treatment of discourse, but also think that it can be explained with reference to the academic framework from which it emanates. Their research agenda seeks to contribute to an increasingly common goal in political ecology: traversing the sociocultural and biogeophysical processes within human geography by way of a multimethod triangulation technique (cf. Zimmerer, 1996; Forsyth, 2003). The argument is that accentuating 'local knowledge, environmental history, multi-scale politics, and socially differentiated resource-management practices, requires inten- sive field study and multiple research methodologies' (Bassett and Koli Bi, 2000, p. 68). This illustrates a strong (critical) realist turn in political ecology in which evaluations of the biogeophysical processes shaping human-environmental dynam- ics depend on an understanding of both human discourses and physical geography (Zimmerer and Bassett, 2003, p. 3).

Importantly, this approach goes to some length beyond the deep-seated anthro- pocentrism of many geographical investigations of environmental discourse. On the level of ecopolitics, it is critical of the lack of attention to geographical contexts typical of mainstream sustainability discourses – not least those materialised in the guise of the World Bank's embracement of technocratic, neoliberal ideology and its way of 'assisting dozens of African governments to develop NEAPs which, in assembly-line fashion, are being produced according to a blueprint' (Bassett and Koli Bi, 2000, p. 68). In concord with the philosophical realism of their research design, concrete suggestions for policy reforms could be extracted from the results, which is an additional difference with what tends to be the case with post-structural and Marxist approaches.

'The' Environment Is No More

In my view, one of the presently most imperative challenges for environmental geographers is to decipher the work and logic of discourse by keying it to the destructive logic of capitalist nature but without resorting to some crude, unreflexive realism. Environmental discourses are power systems, which seek to systemise, capture and fix what is constantly mediated, in process, and getting away. As soon as it looks as if all the shapes are in place and audiences convinced, the environment has somehow always already made its escape, only to return in different guises. This is one of the reasons why the struggle over environmental discourse has become a profoundly political matter. Although the work discussed in this chapter offers illustrative rather than exhaustive insights into geographical approaches to dis- course, I would suggest that it does motivate some tentative general conclusions.

First, geographers tend to link the history of discursive ordering and representa- tional practices to the material appropriation of the world. Struggles over discourse and representation are crucial in the geographically uneven struggle over the envi- ronment and what counts as environmental issues in science and society. Bringing out a variety of power struggles and taken-for-granted assumptions and reifications is thus not necessarily a hyper-hermeneutic diversion from ostensibly more impor- tant material practices. On the contrary, this is just as important, precisely because discourses and representations help arrange, codify and challenge the practices that make up environmental politics.

Second, differences in approach to environmental discourse tend to emanate from philosophically distinct ways of imagining knowledge to be related to the biogeo- physical world. Different research strategies of denaturalisation (genealogy or more realist) demand different ways of working with scientific data and other knowledges. The broad compasses of '-isms' and 'posts' translate into a variety of methodological maneuvers – ranging from deep-seated deconstruction in which discourse leaves no room for 'facts' or 'science', to multimethod triangulations, which can corroborate, verify or falsify discourses.

Third, this shallow or deep 'space' of discourse within the research process also reflects ecopolitical differences, including where the political is located in research and beyond, and how political struggles over the environment are or ought to be structured and contested. The way discourses are located within power systems, ranging from universally sprawled to specifically centred bends analysis in different directions, e.g., deconstructing colonial environmental discourses by way of discourse; or denaturalizing them with a multimethod realist strategy; falsifying the scientific base of policymaking by way of science; and bringing out the contradictions of discourse as corporate ideology. Different approaches nevertheless raise some shared concerns: which ensembles of ideas are regarded as legitimate?; whose knowledge becomes widely accepted?; which discourses serve to sustain particular power systems?; and how are these knowledges reproduced and transformed into sets of practices? Questions such as these belong to the current standard arsenal, which many human geographers haul to the forefront of environmental discourse analysis.

Most generally, the examples illuminate the complex formation of environmental discourse as a geography of physicality, meaning and power, with imbrications in colonialism, capitalism and various social struggles. To grasp the real power of 'the' environment we cannot ignore the ways in which competing environmental discourses are constituted and reproduced within a set of material relationships, activities and socio-spatial power systems. In other words, the value of discourse analysis is seriously limited if it does not provide ways of explaining the physical and social power relations that determine the privileged or subordinated position of particular discourses.

ACKNOWLEDGEMENTS

I am grateful for several rounds of discourse at Joensuu University, Finland, in particular with Bruce Braun and Ari Lehtinen. Many thanks to Noel Castree and David Demeritt for edifying comments. The usual caveats apply.

BIBLIOGRAPHY

Althusser, L. (1996) *For Marx*. London: Verso.
Bassett, T. J. and Koli Bi, Z. (2000) Environmental discourses and the Ivorian savanna. *Annals of the Association of American Geographers*, 90(1), 67–95.
Batterbury, S., Forsyth, T. and Thomson, K. (1997) Environmental transformation in developing countries: hybrid research and democratic policy. *The Geographical Journal*, 163, 126–32.
Braun, B. (2002) *The Intemperate Rainforest: Nature, Culture and Power on Canada's West Coast*. Minneapolis: University of Minnesota Press.
Braun, B. and Wainwright, J. (2001) Nature, poststructuralism and politics. InN. Castree and B. Braun (eds), *Social Nature: Theory, Practice, Politics*. Oxford: Blackwell, pp. 41–63.
Bridge, G. and McManus, P. (2000) Sticks and stones: environmental narratives and discursive regulation in the forestry and mining sectors. *Antipode*, 32(1), 10–47.
Castree, N. (2005) *Nature*. London: Routledge.

Castree, N. and Braun, B. (1998) The construction of nature and the nature of construction. In B. Braun and N. Castree (eds), *Remaking Reality: Nature at the Millenium*. London: Routledge.

Castree, N. and Braun, B. (eds) (2001) *Social Nature: Theory, Practice and Politics*. Oxford: Blackwell.

Duncan, J. and Duncan, N. (2004) *Landscapes of Privilege*. London: Routledge.

Forsyth, T. (2003) *Critical Political Ecology: The Politics of Environmental Science*. London: Routledge.

Foucault, M. (1980) *Power/Knowledge: Selected Interviews and Other Writings 1972–1977*. New York: Pantheon Books.

Gandy, M. (1996) Crumbling land: the postmodernity debate and the analysis of environmental problems. *Progress in Human Geography*, 20, 23–40.

Harvey, D. (1996) *Nature, Justice and the Geography of Difference*. Oxford: Blackwell.

Henderson, G. (1999) *California and the Fiction of Capital*. New York: Oxford University Press.

Lefebvre, H. (1968) *The Sociology of Marx*. London: Penguin.

Mels, T. (2002) Nature, home and scenery: the official spatialities of Swedish national parks. *Environment and Planning D: Society and Space*, 20, 35–54.

Mitchell, D. (2003) Dead labor and the political economy of landscape – California living, California dying. In K. Anderson, M. Domosh, S. Pile and N. Thrift (eds), *Handbook of Cultural Geography*. London: Sage, pp. 233–48.

Neumann, R. P. (1998) *Imposing Wilderness: Struggles for Livelihood and Nature Preservation in Africa*. Berkeley: University of California Press.

Neumann, R. P. (2004) Moral and discursive geographies in the war for biodiversity in Africa. *Political Geography*, 23, 813–37.

Peet, R. and Watts, M. (1996a) Conclusion: towards a theory of liberation ecology. In R. Peet and M. Watts (eds), *Liberation Ecologies: Environment, Development, Social Movements*. London: Routledge, pp. 260–69.

Peet, R. and Watts, M. (1996b) Liberation ecology: development, sustainability, and environment in an age of market triumphalism. In R. Peet and M. Watts (eds), *Liberation Ecologies: Environment, Development, Social Movements*. London: Routledge, pp. 1–45.

Said, E. (1978) *Orientalism*. New York: Pantheon Books.

Walker, R. (2001) California's golden road to riches. *Annals of the Association of American Geographers*, 91, 167–99.

Willems-Braun, B. (1997) Buried epistemologies: the politics of nature in (post)colonial British Columbia *Annals of the Association of American Geographers*, 87, 3–31.

Williams, R. (1977) *Marxism and Literature*. Oxford: Oxford University Press.

Zimmerer, K. S. (1996) *Changing Fortunes: Biodiversity and Peasant Livelihood in the Peruvian Andes*. London and Berkeley: University of California Press.

Zimmerer, K. S. and Bassett, T. J. (2003) Approaching political ecology. In K. S. Zimmerer and T. J. Bassett (eds), *Political Ecology: An Integrative Approach to Geography and Environment-Development Studies*. New York and London: The Guilford Press, pp. 1–25.

Chapter 24

Deliberative and Participatory Approaches in Environmental Geography

Jason Chilvers

Participatory Environmental Geographies

In the environmental sphere it is now very fashionable to be doing, or at least talking about, 'participation'. This never used to be the case of course. Environmental policy has traditionally been determined by scientists and other policy elites in exclusive, technocratic processes. Indeed, it is unlikely that this chapter would have featured in this *Companion* had it been written even a decade ago. Emerging participatory practices seek to empower voices often marginalised in science-policy processes. Their development in environmental geography forms part of a wider 'participatory turn' across the discipline, which has been particularly visible in development geography (e.g., Binns, 1997), social geography (e.g., Pain, 2004) and GIS (e.g., Craig et al., 2002).

Ideas of citizen participation are central to green political ideology and have gained greater prominence with the rise of the sustainability agenda and global attempts to implement Local Agenda 21 post-Rio (Macnaghten and Jacobs, 1997). This has coincided with increasing realisation of the 'post-normal' nature of environmental risk issues where 'facts are uncertain, values in dispute, stakes high and decision urgent' (Funtowicz and Ravetz, 1993, p. 740). Uncertainty, indeterminacy (see Chapter 6, this volume) and intense epistemic and ethical differences demand that an extended range of actors, knowledges and values are incorporated into environmental policymaking (Eden, 1996), as well as core expert domains of environmental science, appraisal, and management (Irwin, 1995). Such engagements overlap with partnership approaches linked to new forms of environmental governance beyond the state (see Chapter 32, this volume). These global trends overlay participatory developments in environmental planning in the global 'north' (Healey, 1997), and international development in the global 'south' (Chambers, 1997).

Across these domains the search for deliberative and participatory alternatives has been driven by a common realisation that 'top-down' technocratic approaches are deficient, often due to their apparent irrelevance and insensitivity to local contexts. Yet as the idea of participation has swept through successive policy arenas it has prompted something of a critical backlash. An increasing number of authors are arguing that participation can also disempower, exclude, oppress, be manipulated by powerful interests, and act as a smokescreen behind which decision-making institutions conduct business as usual (e.g., Cooke and Kothari 2001). This decidedly 'janus-faced' nature of new participatory governance (Swyngedouw, 2005) fuels polarised disputes between proponents and critics, which remain unresolved for the time being.

The contested and wide-ranging terrain of participatory environmental geography presents obvious challenges for mapping out the current and future state of the field. Any such review would be incomplete if it ignored related work across the social sciences by sociologists, anthropologists, psychologists, planners, political scientists and others. Indeed, a key characteristic of most geographers working in these very public and participatory worlds is the solidarities they have forged within a wider 'epistemic community' of researchers, practitioners, policymakers and activists (see Chilvers, 2007). And while social scientists have undoubtedly led the way, this wider community increasingly includes physical geographers and other natural scientists.

In surveying this multidisciplinary field the chapter argues that empirical research within it has developed in three main streams, which precede each other to some extent, sometimes running in parallel and sometimes converging. The first stream is committed to developing participatory practices and innovative deliberative methodologies in research and policy contexts. The second stream, which centres on evaluating the quality of participatory processes and outcomes, has shadowed evolving practices but received relatively less attention. The third stream encompasses emerging critical studies of participation, which offer more wide-ranging and reflexive accounts of its construction, performance, and discourse. Before charting these streams in turn, it is important first to map out the theoretical and conceptual landscape within which they are situated, which in itself represents a contested and continually evolving body of work.

Deliberating Environmental Participation: Concepts, Rationales and Critiques

Participation is a highly contested term that means different things to different people. To add to the confusion, participation is often called (or equated to) many different things, such as 'engagement', 'empowerment', 'involvement', 'consultation', 'deliberation', 'dialogue', 'partnership', 'outreach', 'mediation', 'consensus building' and 'civic science'. The list goes on. The most popular means of clarifying this situation remains Shelly Arnstein's (1969) ladder of participation, which defines eight steps of increasing devolution of power *from* decision makers *to* citizens, moving through 'non-participation' (including information provision), 'degrees of tokenism' (including consultation), to 'degrees of citizen control' in influencing proposals and decisions. It is the latter category that most commentators take to mean participation proper, where participants often engage in deliberation over extended periods of interaction, discussion and debate.

The conceptual origins of deliberation lie in theories of deliberative democracy developed by critical and political theorists tapping roots that stretch back to the *polis* of ancient Greece (Dryzek, 2002). Deliberative theorists view democracy as an inclusive forum where reasoned debate transforms judgements in the face of publicly convincing arguments that appeal to the 'public good' rather than individual self-interest (Bohman and Rehg, 1997; Mason, 1999; Dryzek, 2002). Jürgen Habermas (1984; 1987) remains the most authoritative deliberative theorist, most notably through his works on communicative rationality and the 'ideal speech situation', a discursive ethics that emphasises the central role of language and undistorted communication as the basis for reaching non-coerced mutual understanding and agreement through the force of the best argument alone. Habermas's communicative rationality seeks to counter the dominance of instrumental rationality and formal expertise over lifeworld (cultural) rationality grounded in understandings of everyday life, an asymmetry that is particularly ingrained in the environmental sphere.

Over the years a range of arguments have been advanced promoting the virtues of deliberation and participation. Fiorino (1990) differentiates between normative, instrumental, and substantive rationales. From a normative (or ethical) viewpoint participation is simply the right thing to do. Deliberative democratic theories hold that citizens have a right to influence decisions that affect their own lives, based on principles of citizen empowerment, equity, and social justice (Renn et al., 1995; Bohman, 1996). Instrumental (or practical) rationales contend that participation is a better way to achieve particular ends, such as, increasing public trust and the legitimacy of governing institutions (see Irwin, 2006), enhancing the acceptance and implementation of environmental policy (Eden, 1996) or reducing conflict surrounding decisions (Renn et al., 1995). Whether such 'benefits' are actually realised in practice is a different matter.

Substantive rationales claim that participation leads to better ends, in both the quality of environmental science (Funtowicz and Ravetz, 1993) and environmental decisions (Coenen et al., 1998). Work in geography, science and technology studies (STS), and risk research has argued that inclusion of wider non-scientific knowledges, values and meanings can lead to environmental science and policy that is of greater analytical rigour (Stirling, 1998) and social intelligence or robustness (Leach et al., 2005). Substantive rationales encompass both epistemological arguments for incorporating other 'ways of knowing' and uncertified forms of expertise (Collins and Evans, 2002), as well as ontological arguments for acknowledging wider public meanings and non-scientific 'ways of being' with nature, whose dismissal by experts is often a source of public discontent with science (Wynne, 2005). A further participatory rationale, learning, can be identified as cutting across all three categories defined by Fiorino (Webler et al., 1995).

Such claims about the instrumental and substantive benefits of participation are rarely backed up with evidence of their realisation in practice. As a result some have characterised participation as an ideology or an 'act of faith' (Cleaver, 2001). The most obvious operational criticisms of (or 'barriers' to) participation, often aired in technocratic policy cultures, are that it is unpredictable, causes conflict and wastes time and resources (Petts, 2004). Participation is also plagued by a number of practical problems, including how it relates to expert and representative democratic systems, widespread 'consultation fatigue', and an increasing fragmentation of effort (Pellizzoni, 2003).

Beyond these practical problems, there lie three more deep seated critiques, relating to issues of representation, power, and consensus. First, critics often complain that the small number of participants typically involved in deliberative processes are 'unrepresentative' of wider affected populations (Munton, 2003). O'Neill (2001) sees concerns of statistical representativeness overshadowed by much thornier normative questions about the political and ethical legitimacy of deliberation that relies on appeals to the presence of different groups often without clear sources of authorisation from and accountability to those being represented. This not only relates to existing human actors, but also future generations and nonhuman actors who present particular 'problems of representing those who cannot speak and have in that sense no possibility of voice or presence in processes of environmental decisionmaking' (O'Neill, 2001, p. 483). Representation of nonhuman actors in particular exposes Habermas's 'discursive specialisation' which limits speech acts claiming to represent the material world to the sphere of science and technology and thereby 'privileg[es] science to speak for nature, and morality to speak for society' (Davies, 2006a, p. 426).

A second, perhaps even more trenchant, critique of participation is that its proponents have been 'naïve about the complexities of power and power relations' (Cooke and Kothari, 2001, p. 14). Instead of seeing power as something held in the hands of a few waiting to be redistributed (as Arnstein and Habermas do), most critics adopt a Foucauldian approach to understanding power as pervasive and circulating through networks of discourse, practices, and relationships (Foucault, 1980). The very discourse of participation, and not just the institutionalised practice, embodies the potential to disempower, exclude, conceal oppressions and allow political co-option (Cooke and Kothari, 2001). Within deliberative processes Tewdwr-Jones and Allmendinger (1998) argue that Habermasian ideals of communicative action cannot control strategic behaviour, nor guarantee participants will act in an open and honest manner all of the time, due to power inherent within individuals. Shifting to external power, Stirling (2005) expands on Collingridge's (1982) discussion of 'decision justification' to highlight how participation can be used by decision-making institutions as a form either of *weak justification*, avoiding any future blame through appeals to process effectiveness or *strong justification*, determining particular decision outcomes through manipulation of the framing of participatory processes to achieve those ends.

A third criticism centres on overly consensual deliberative ideals, found in Habermasian aspirations for agreement on the 'single best' solution to a problem. It is questionable whether such consensus is possible or desirable in a world of increasing uncertainty, pluralism, complexity and social inequality (Pellizzoni, 2003). Furthermore, there is considerable danger that the determination to forge consensus can hide intractable differences and reinforce hegemonic power relations by excluding certain voices, framings, and forms of expression (Tewdwr-Jones and Allmendinger, 1998; Davies, 2006a). In response to these problems, political theorist James Bohman (1996) has reworked Habermas, doing away with the assumption of a unity of rationality and redefining successful deliberation in terms of the continued co-operation of actors rather than a requirement for unanimous agreement. Chantal Mouffe (2000) goes further in arguing that the universal reason of consensual approaches needs to be replaced with 'agonistic pluralism'. This alternative theory of public deliberation highlights power relations and exclusions through confrontation, antagonism, and the exploration of difference.

So, while participatory enthusiasts hold up deliberative processes as a solution to the deficiencies of technocratic science-centred approaches, emerging critiques of participation expose similar problems with representation, exclusion, power, framing effects, and narrowing down debate. Both technical-analytic *and* deliberative-participatory approaches have the potential to 'close-down' *and* 'open-up' wider environmental policy discourses (Stirling, 2005). This includes the possible 'technocracy of participation' (Chilvers, 2008), as well as the potential for technical appraisals to enhance learning and reflexivity (Owens et al., 2004). Given the predominance of consensual participatory theories and practices in late modernity, there remains a pressing need to open them up to difference, otherness and indeterminacy. These deliberations over the theoretical and conceptual basis of environmental participation impinge in different ways on the three main streams of empirical research in participatory environmental geography, to which we now turn.

Developing Deliberative and Participatory Practices

Research developing participatory practices and innovative deliberative methodologies has received the most active and sustained attention from participatory environmental geographers. While critics see this stream of inquiry as little more than 'methodological revisionism' and the uncritical promotion of participation (Cooke and Kothari, 2001), it has brought considerable advances and remains a fundamental research frontier. Key questions include: what are the aims and purposes of participation? Who participates and on what basis are they selected? How should deliberation be designed, structured, and integrated with quantitative or analytic approaches? How can issues of representation and scale be addressed? Is it possible to fit methods to specific contexts?

Participatory practices are very much shaped by their purpose(s), whether that is to undertake research or formulate policy, explore environmental values or co-produce knowledge, build consensus or map out difference, or some combination thereof. The action-orientation of participatory work tends to blur conventional distinctions between science, policy and the public. Central to any deliberative or participatory approach is the *facilitator*, most often a human geographer or other social scientist, who designs the process, moderates discussion, and attempts to ensure that all voices are heard. Participants can be defined as *publics*, *stakeholders* or *specialists* based on their different epistemic (knowledge) and ethical (value) claims to participation (Pellizzoni, 2003; Burgess and Chilvers, 2006). These forms of representation serve as a basis in table 24.1 for differentiating between different deliberative approaches which involve participants in various forms of 'talk', and hybrid 'analytic-deliberative' approaches that fuse participatory deliberation with forms of scientific analysis and calculation (Stern and Fineberg, 1996).

Qualitative social science methods for bringing various *publics* together in deliberation have been extensively used to explore underlying environmental understandings and behaviours of participants (e.g., Harrison et al., 1996; Macnaghten and Jacobs, 1997). Here the most common approach is the focus group, a core method in human geography that involves 6–10 individuals in group discussion for one to two hours (Cameron, 2000). Such one-off events are hardly empowering, and some question whether the extractive nature of standard focus groups counts as 'participation'. Burgess et al., (1988) developed the method of in-depth groups to address these concerns by convening a series of meetings over time to build mutual understandings and foster collective group learning.

Emerging approaches for engaging publics in environmental analysis or data collection are located to the top right of table 24.1. An important advance here is the rise of participatory monitoring of environmental change, often where environmental science lacks capacity or coverage (see, e.g., Dougill et al., 2002; Ellis and

Table 24.1 A typology of deliberative and analytic-deliberative approaches in environmental participation

Representation	Deliberative	Analytic-deliberative
Publics	Focus groups In-depth groups	Participatory monitoring Participatory research Public participation GIS
Stakeholders	Community advisory committees Conflict resolution Consensus building Mediation	Joint fact-finding Stakeholder decision analysis Public participation GIS
Specialists	Expert advisory committees Expert workshops	Delphi exercise Collaborative GIS
Interactive	Citizens' juries / panels Consensus conferences	Cooperative discourse Deliberative mapping Integrated assessment focus groups Collaborative and PPGIS

Waterton, 2005). Participatory research (PR), now a core method in geography (Kesby et al., 2005), has been applied to many environmental issues, especially under the banner of participatory rural appraisal (PRA) in the global south (Chambers, 1997; see, e.g., Mosse, 2001). Taking its methodological foundations from the work of Freire (1972) and Whyte (1991), the origins of PR can also be traced to North America through classic examples of 'lay epidemiology' and the mobilisation of civic science (see Fischer, 2000). One of the major areas of analytic-deliberative methods development in geography is GIS. Originating in North America in the mid-1990s, public participation GIS (PPGIS) has been used variously to broaden public involvement in policymaking (Craig et al., 2002), to promote environmental justice (Sieber, 2006), and to engage stakeholders and specialists in forms of 'collaborative GIS' (Jankowski and Nyerges, 2001).

Stakeholder processes (see table 24.1) seek to involve actors who are interested in or affected by an issue, with an overall emphasis on consensual deliberation and reaching agreement. Spreading out from North American origins, conflict resolution, mediation, consensus building and community advisory committees are now a common way of attempting to resolve environmental disputes and agree on contentious environmental solutions, often through employing joint fact-finding techniques that seek to develop a shared and trustworthy knowledge-base (Renn et al., 1995; Susskind et al., 1999). Other approaches place more emphasis on opening things up, such as Stakeholder Decision Analysis (SDA), which integrates multi-criteria options appraisal with interactive deliberation (Burgess, 2000). *Specialists*, including geographers and other scientific experts, are often called upon to act as 'independent' expert witnesses to support other deliberative process participants, although it is important to recognise them as possible participants in their own right (Pellizzoni, 2003). Specialist deliberative processes (see table 24.1) can facilitate negotiations within and between interdisciplinary teams, the consideration of analytical uncertainties, and the exploration of social and ethical implications of scientific practice.

Given this range of approaches to engaging public, stakeholder, and specialist participants, a major research challenge is to build connections between all these actors in interactive learning processes (Webler et al., 1995) and move engagement 'upstream' to the earliest stages of policy processes and decisions that shape environmental, science and technology futures (Wilsdon and Willis, 2004). A number of citizen panel processes exist, including consensus conferences (Joss and Durrant, 1995) and citizens' juries (Crosby, 1999), where small groups of publics develop recommendations after questioning specialist expert witnesses. A key methodological challenge is the development of participatory processes that are both interactive and analytic-deliberative. A classic example remains Renn et al's (1993) cooperative discourse model, a three-step decision procedure where stakeholders first take a lead in articulating value-based criteria by which to assess decision options; specialists assess the impacts of options; and then publics take the lead in making final recommendations for policy decision making. Higher degrees of interaction have been afforded in participatory integrated assessments of climate change where publics directly engage with scientists and scientific models (e.g., Kasemir et al., 2003). A particularly innovative analytic-deliberative method is Deliberative Mapping (DM), as trialled in the context of medical biotechnologies and radioactive waste management, which builds highly symmetrical and interactive relations between citizens, specialists and stakeholders, who are given the same opportunities to define options, develop criteria, and make decision judgements (Burgess et al., 2007).

As with environmental research more broadly, geographers could do more to exploit the intra/interdisciplinary opportunity of analytic-deliberative learning processes: human geographers often act as facilitators but physical geographers are rarely involved as specialists in such exercises. Beyond this, a crucial area of future development relates to the eternal problem of 'scaling up' participation in 'functionally complex, socially differentiated, and spatially and numerically extensive societies' (Barnett and Lowe, 2004: 8). Claims about the representativeness and epistemic relevance of participatory approaches are harder to sustain in the face of transboundary or global environmental problems that involve multi-scalar governance (Davies, 2002; Swyngedouw, 2005). Meeting the challenge of representativeness over larger scales could be addressed by linking multiple deliberative processes across geographic regions or attempting to integrate intensive deliberation with extensive quantitative surveys (see Fishkin, 1991). Many see the Internet as an obvious means of overcoming problems of scale, which is emerging as a key research opportunity in PPGIS (Balram and Dragicevic, 2006).

Another crucial question about deliberative processes is whether environmental understandings and actions developed through them are context dependent (i.e., always different in different fora) or contain elements that are stable and generalisable across contexts (Owens, 2000). For example, focus group based research on environmental and scientific citizenship has reached similar conclusions in different contexts, such as about the importance of public trust in institutions (e.g., Harrison et al., 1996; Macnaghten and Jacobs, 1997; Marris et al., 2001). This raises questions about whether such convergence is a function of method, facilitation, context or a 'real' consistency in public understanding. Teasing out relative influences in this regard requires retrospective and ongoing analysis across deliberative research projects. Any stable elements might then be built on in future processes (with obvious checks to reflexivity) rather than being repeatedly rediscovered.

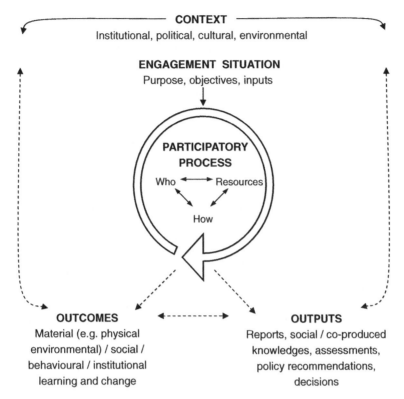

Figure 24.1 A contextual model of participatory process design and evaluation (adapted from Burgess and Chilvers, 2006).

These discussions emphasise the fundamental importance of context in participatory processes. As a rule of thumb, the more contentious and uncertain an issue is the greater the need for inclusive and interactive public deliberation (Funtowicz and Ravetz, 1993; Stern and Fineberg, 1996). In reality the range of deliberative approaches endlessly multiplies as established methods (in table 24.1) are tailored to specific decision situations, linked with other participatory methods, and integrated with forms of information provision on which they depend. This reflects an increasing realisation that participatory process design (the who, what, and how of participation) should be 'fit-for-purpose' (Burgess and Chilvers, 2006) vis-à-vis the immediate engagement situation and wider institutional, political, cultural, and environmental contexts, as illustrated in figure 24.1. A participatory process leads to a series of outputs and outcomes, which are influenced by, and in turn influence, aspects of context (as depicted by the two-way dashed arrows in figure 24.1).

Evaluating Deliberation and Participation

Questions about the effectiveness of these participatory processes and outcomes are the focus for a second stream of participatory inquiry in environmental geography. As participatory practices develop and become more widespread, systematic 'independent' evaluation becomes more important to deepen understanding of different approaches, promote 'good' practice, enhance transparency, and demonstrate their efficacy and legitimacy to sceptics (Renn et al., 1995; Rowe and Frewer, 2000). Key questions underpinning this stream of research include: what criteria define effective participatory processes and outcomes? Does participation make a difference to environmental governance? Are these impacts good or bad and for what/whom? How does context influence process effectiveness? What methods are appropriate to answer these questions?

In essence, such questions offer a way of testing whether rationales, claims, and assumptions about the virtues of participation are actually realised in practice. Evaluative work remains in its infancy, however. The current situation is defined by a lack of detailed empirical studies and evidence to answer these questions. These are not recent concerns by any means. Within geography, this second stream of work stretches back at least as far as Smith's (1984, p. 253) review of early public participation in Canada in which he called for 'more attention to the formal evaluation of participatory exercises' and outlined a similar, though some-what simpler, context-process-outcome model to that shown in figure 24.1. Excite-ment about doing participation (under stream one) means that evaluation has often been marginalised as little more than an informal or *ad hoc* afterthought. Neglect is also rooted in difficulties inherent in the subject-centred and multi-dimensional nature of participation where, 'whoever is doing the perceiving is crucial to any understanding of the effectiveness of citizen participation' (Rosener, 1978, p. 458).

Most evaluative attention to date has focused on process effectiveness. This partly reflects the emphasis on procedural justice in the deliberative democratic theories from which a number of evaluative criteria have been derived (e.g., Fiorino, 1990; Laird, 1993; Webler, 1995). Beyond this, other evaluative criteria have been derived from the views of individual authors (e.g., Rowe and Frewer, 2000), process partici-pants (e.g., Webler et al., 2001), decision makers and participatory researchers/ practitioners (e.g., Burgess and Clark, 2006). Across these studies there are at least seven broadly accepted effectiveness criteria. These state that participatory processes should:

1. be representative of all those interested in and affected by a decision and remove unnecessary barriers to participation (*representativeness and inclusivity*);
2. allow all those involved to put forward their views in interactive deliberation that develops mutual understanding (*fair deliberation*);
3. provide sufficient resources (information, expertise, time) for effective participa-tion (*access to resources*);
4. be transparent about objectives, boundaries, and the relationship of participa-tion to decision-making (*transparency and accountability*);
5. enhance social learning for all those involved, including participants, specialists, decision-makers and wider institutions (*learning*);
6. be conducted in an independent and unbiased way (*independence*); and
7. be cost-effective and timely (*efficiency*).

Process evaluations often apply criteria such as these in a 'check list' fashion to judge individual cases of participation based on documentary analysis, process observations, and interviews with participants. For example, Renn et al. (1995) evaluate eight participatory approaches employed in Europe and North America (including citizens' juries, community advisory committees, and mediation), while Rowe and Frewer (2000) include different methods (including consensus conferences and focus groups) in a similar generic evaluation based on criteria closely matching those listed above. Such generic analyses show that each approach has its own strengths and weaknesses, as illustrated by Petts' (2001) comparative analysis of community advisory committees and citizens' juries in UK waste management planning. The latter were more 'representative' of local publics while the former provided greater learning opportunities over longer time scales.

An emphasis on procedural evaluation assumes that better participatory processes lead to better environmental (and other) outcomes, yet there can, by definition, be no guarantee that this is the case (Munton, 2003). Very few existing studies consider outcomes, and fewer still the relationship between process and outcome. There are considerable methodological difficulties to be overcome in tracking emergent outcomes in the longer-term and detecting cause-effect relationships after an event. There are also conceptual challenges in defining 'outcomes' and then somehow measuring them, as well as institutional pressures for early evaluations to demonstrate process efficacy (Rowe and Frewer, 2004). Burgess and Chilvers (2006) differentiate between *outputs*, the immediate substantive products of participatory processes such as reports, assessments, and policy recommendations, and *outcomes*, the emergent impacts and resulting changes such as improvements in environmental quality, sustainability, social capital, individual/institutional learning, reflexivity and behaviour change (see figure 24.1). While Beierle and Konisky (2001) offer an optimistic assessment of environmental outcomes resulting from stakeholder engagement in the North American Great Lakes region, Bickerstaff and Walker (2005) provide a more pessimistic evaluation of two deliberative processes in UK local transport planning which had little substantive impact on policy outcomes due to institutional and national constraints (a finding shared by other retrospective evaluations, such as Goodwin, 1998; Davies, 2002).

Evaluating relations between participatory processes and outcomes poses significant methodological challenges. There is a pressing need for well designed longitudinal research involving retrospective and 'real time' studies that more effectively capture process dynamics and track emergent outcomes (Owens et al., 2004). The purpose of early and ongoing evaluation is not only summative; it should also play a formative role in shaping ongoing process design and enhancing reflexivity. Of particular importance are features such as learning that straddle the distinction between process and outcome and span individual through to institutional levels. Sophisticated research designs are required to establish whether and how participatory deliberation leads to transformations in participants' identities, knowledge, values, and competencies (Petts, 1997; Davies and Burgess, 2004) as well as changes in their environmental behaviour and action (Owens, 2000). If the claims of better environmental decisions are to be properly tested there is also a need to monitor and assess material environmental changes that result from participatory policymaking and appraisal processes. Here is another opportunity for the interdisciplinarity of environmental geography to make its mark. We must not forget, however, that evaluation comes with its own politics and sensitivities – between the evaluator, the evaluated, and wider audiences – which intensify as processes become larger in scale and more high profile (see Rowe et al., 2005). It is also important to recognise that evaluation can itself become wrapped up in a cycle of decision justification and be used by decision institutions for instrumental purposes.

Critical Studies of Participation

This highlights the crucial need to move beyond evaluation to undertake critical studies of the construction, performance, and discourse of participation. Whereas preceding streams often seek to promote and mainstream 'marginal' participatory practice, this third stream of research foregrounds the existence of participation 'as a legitimate object of study *in itself*' (Irwin, 2006, p. 310, original emphasis) through paying close attention to the processes and consequences of its construction and developing 'a more sophisticated and genuinely reflexive understanding of power' (Cooke and Kothari, 2001, pp. 14–15). Critical studies emerging at the interface between geography, STS and development studies are beginning to ask: how are participatory processes constructed and framed? How do various representations, boundaries, and inclusions/exclusions get made through the performance of (analytic-)deliberative practices? What about the manifestations and dynamics of internal/external power in participation? To what extent does participation express symmetry with respect to cultural and instrumental rationalities? Does participation make institutions more responsive, reflexive and responsible in the face of wider environmental concerns? Or does it represent a new form of technocracy that insulates neoliberal agendas and science-led progress from public challenge and dissent?

Such questions are being addressed in a variety of ways, ranging from situated ethnographies of 'participation in action' through to analyses of what happens beyond the formal 'invited' time-spaces of participation. An immediate, but often neglected, mode of inquiry is critical self-reflexivity. This has been demonstrated by Davies (2006b) in a highly reflective account of a DM experiment on medical biotechnology. She considers how framings of the process were partial and subject to 'overflows', how (multi-criteria) calculations kept framings and uncertainties open just as much as deliberation, and how such openness, while potentially more accountable, does not offer unequivocal justifications for policy decisions. In a similarly detailed account Elwood (2006) charts how participation and representation were negotiated through everyday knowledge practices in a PPGIS project with two Chicago community organizations, while Ferreyra (2006) has adopted a biographical approach to critically reflect on her experiences of participatory action research with a watershed partnership in Ontario, Canada.

Beyond self-reflection an increasing number of situated in-depth studies are beginning to expose critical issues of representation and power within and beyond formal spaces of participation. In a multi-sited ethnography of participatory monitoring in UK biodiversity action planning, Ellis and Waterton (2005) explore how the exclusion/inclusion of humans and non-humans was performed. They highlight how 'human investment as well as the significance of organisms in place may be made invisible' (p. 689) when recorded as a data point able to travel to distant centres of power for processing and decision-making. Hinchliffe's (2001) situated study of decision-making over BSE in Britain has also shown how the exclusion of indeterminacies, antagonisms and socio-natural diversity central to the crisis was performed through deliberative practices just as much as through scientific ones. There will also be self-exclusions linked to the willingness of individuals to participate, but, as Cleaver's (2001) work on water resource usage in Africa argues, rather than being an irresponsible act this is intimately tied up with one's own sense of identity and agency. Further exclusions can occur within the process, for example where 'front stage' performances of public deliberation mask 'back stage' power structures and concerns within local communities (Kothari, 2001). Back staging can also take the form of strategic behaviour where the more resourceful actors go 'round the back', as Hillier (2000) illustrates, or practice 'scale jumping' to effect change at higher orders of governance (Swyngedouw, 2005).

Critical studies are also focusing on (dis)connections between participants and decision-making institutions and on how framings and pre-commitments of the latter constrain deliberative spaces and construct notions of citizenship and expertise in powerful ways. Irwin's (2001) analysis of the UK Public Consultation on Developments in the Biosciences (PCDB) reveals how the process was instrumentally managed to allow the government's framing of the problem to define the decision situation and ensure that the organisation and outputs of the PCDB justified existing institutional arrangements. This resonates with Mosse's (2001) account of participatory farming system development in India which shows that, far from influencing development organisations, 'local knowledge' is a construct of the planning context itself through direct manipulation by external project agents (facilitators) and wider institutional contexts that require formal bureaucratic goals to be met.

It is increasingly realised that participation is becoming a policy 'bolt on' all too easily ignored by policymakers (Hajer and Kesselring, 1999). In this climate an uncritical focus on procedures is likely to raise expectations unrealistically, undermine institutional trust, and almost certainly fail in the long term (Owens, 2000). There is a pressing need for critical studies to focus more closely on institutional dimensions of participation, including the careful tracking of changes in their underlying epistemic and cultural pre-commitments in the longer term (Irwin, 2006).

Despite increasing recognition that participation is 'constructed by a cadre of . . . professionals' (Cooke and Kothari, 2001, p. 15) these actors remain understudied (Chilvers, 2005; 2008). Research in the UK environmental risk domain shows that public engagement experts are adopting increasingly powerful roles at the science-policy interface. However, intense fragmentation is limiting learning and reflexivity among them (Chilvers, 2007). Participation is becoming a lucrative industry with a wide variety of approaches competing in a market place of 'tools' – such as 'citizens' jury®', 'PRA', 'Stakeholder Dialogue', etc. – in which the resulting environmental knowledges are often commodified (see also Ellis and Waterton, 2005). This is a far cry from the idealistic origins of participation. Claims of 'democratisation' can begin to sound rather hollow given the irony that those committed to empowerment are contributing to the professionalisation of participation, potentially creating a new layer of technocracy (Chilvers, 2007). This raises critical questions about the politics of participatory process expertise and the need for more organic, spontaneous, citizen-led processes. In her historical review of international development Uma Kothari (2005) goes further in arguing that through professionalisation and 'technicalisation' participatory development has been captured by, and sustains, the neoliberal development agenda, thus depoliticising potentially critical discourses. She asks: what space is there for critical voices in the current climate?

While critical studies must resist participation where it becomes tyrannical, they should also be constructive. Kesby (2005, p. 2044) has criticised the valuable poststructural critiques in Cooke and Kothari's (2001) *Participation: The New Tyranny?* for being overly negative and upholding a binary logic of: 'power = bad / resistance = good'. As Foucault would have argued, participation as a field of power can be good as well as bad and there is a need to recognise its transformative potential (see also

Hickey and Mohan, 2004). Kesby (2005) calls for greater attention to the spatial dimensions of participation, arguing that closer exploration of participatory praxis can inform post-structural theorisation. Indeed, a focus on space and scale is an area where geographers can make a unique contribution to the environmental participation field. Other understudied aspects of participatory practice in need of critical scrutiny include: backstage negotiations, evaluative practices, official policy discourses, the role of social science/scientists, media representations, the publicity of participatory processes/outcomes and how they are viewed in society and so on. As participation becomes a thing, an object of study in itself, and deliberately more hybrid, there is a need to move beyond simplistic dichotomies such as technocratic/democratic, disempowerment/empowerment, consensual/agonistic and political/anti-political. Participation can exhibit a mixing of both, or foreground one or the other, in different time-spaces. Such complexity demands more nuanced, careful, situated studies of the openings and closings that occur through relations between actors, knowledge and power within and outside of participatory spaces.

Conclusions

In this chapter, I have mapped out three ongoing streams of research in environmental geography on the practice, evaluation and critical study of participation and deliberation. Their continued development requires a vibrant 'theory-praxis dialectic' (Webler, 1998) that forges innovative participatory practices in the light of critiques and develops empirically informed theories of participation grounded in the plural, complex, hybrid, uncertain and unequal realities of environmental research and policy. Such a constructive and cooperative project must overcome disconnects in certain quarters between naïvely optimistic practice and overly negative critiques. Fragmentation poses further impediments, not only in the wider epistemic community of researchers, practitioners, policymakers and activists shaping environmental participation, but also within geography. Such compartmentalisation can be seen, for example, between participatory work in environmental geography and in other sub-disciplinary areas such as GIS, social geography and development geography, which share common antecedents, principles, practices and challenges but do not engage with each other as much as they could or should.

The quest for science-policy legitimacy brought new waves of participation in the late 1960s and 1970s that did little to displace the dominance of rationalist natural science and economic approaches to environmental decision-making through the 1980s and early 1990s. Rather than posing an alternative to science, the recent upturn in environmental participation has levelled out and moved on, conceptually at least, towards hybrid forms of appraisal and policymaking where the question becomes: what is the desired nature, extent and interaction of science *and* participation, where and when? Only time will tell whether the current upturn is sustainable. Deliberative and participatory approaches have become a core method in environmental geography, and also, for some at least, a geographical way of being and acting. What happens in the wider world of environmental policy will ultimately depend on how institutions perform participation. To date there is little to suggest that participation will avoid being co-opted for managerialist and justificationlist ends. In this climate it is crucial that environmental geographers negotiate and take seriously their own participatory ethics, but, at the same time, do not downplay their own agency to effect change.

As repeatedly shown throughout this chapter, environmental participation is full of tensions and contradictions, to the extent that its qualities can be argued either way (and routinely are in policy and academic circles). Without the generation of detailed empirical evidence through systematic evaluation and critical study there is little to stop the possible onset of oppressive and technocratic forms of participation, nor the dogmatic resistance of participatory practices that have truly empowering and transformative potential. Both of these possibilities would hold dire consequences for environmental democracy, socio-environmental justice, and the pursuit of sustainability.

BIBLIOGRAPHY

Arnstein, S. (1969) A ladder of citizen participation. *Journal of the American Institute of Planners*, 35, 216–24.

Balram, S. and Dragicevic, S. (eds) (2006) *Collaborative Geographic Information Systems: Origins, Boundaries, and Structures.* London: Idea Group Publishing.

Barnett, C. and Lowe, M. (2004) Geography and democracy: an introduction. In C. Barnett (ed.), *Spaces of Democracy: Geographical Perspectives on Citizenship, Participation and Representation.* London: Sage, pp. 1–22.

Beierle, T. C. and Konisky, D. M. (2001) What are we gaining from stakeholder involvement? Observations from environmental planning in the Great Lakes. *Environment and Planning C*, 19(4), 515–27.

Bickerstaff, K. and Walker, G. (2005) Shared visions, unholy alliances: power, governance and deliberative processes in local transport planning. *Urban Studies*, 42(12), 2123–44.

Binns, T., Hill, T. and Nel, E. (1997) Learning from the people – participatory rural appraisal, geography and rural development in the 'new' South Africa. *Applied Geography*, 17(1), 1–9.

Bohman, J. (1996) *Public Deliberation: Pluralism, Complexity and Democracy.* Cambridge, MA: MIT Press.

Bohman, J. and Rehg, W. (eds) (1997) *Deliberative Democracy: Essays on Reason and Politics.* Cambridge MA: MIT Press.

Burgess, J. (2000) Situating knowledges, sharing values and reaching collective decisions: the cultural turn in environmental decision-making. In I. Cook, D. Crouch, S. Naylor and J. R. Ryan (eds), *Cultural Turns/Geographical Turns: Perspectives on Cultural Geography.* London: Prentice Hall, pp. 273–87.

Burgess, J. and Chilvers, J. (2006) Upping the *ante*: a conceptual framework for designing and evaluating participatory technology assessments. *Science and Public Policy*, 33(10), 713–28.

Burgess, J. and Clark J. (2006) Evaluating public and stakeholder engagement strategies in environmental governance. In A. Peirez, S. Vaz and S. Tognetti (eds), *Interfaces between Science and Society.* Sheffield: Greenleaf Press, pp. 225–52.

Burgess, J., Limb, M. and Harrison, C. M. (1988) Exploring environmental values through the medium of small groups: theory and practice. *Environment and Planning A*, 20(3), 309–26.

Burgess, J., Stirling, A., Clark, J., Davies, G., Eames, M., Staley, K. and Williamson, S. (2007) Deliberative mapping: a novel analytic-deliberative methodology to support contested science-policy decisions. *Public Understanding of Science*, 16, 299–322.

Cameron, J. (2000) Focussing on the Focus Group. In I. Hay (ed.), *Qualitative Methods in Geography.* Oxford: Oxford University Press, pp. 83–105.

Chambers, R. (1997) *Whose Reality Counts? Putting the First Last.* London: Intermediate Technology Publications.

Chilvers, J. (2005) Democratising science in the UK: the case of radioactive waste management. In M. Leach, I. Scoones and B. Wynne (eds), *Science and Citizens*. London: Zed Press, pp. 237–43.

Chilvers, J. (2007) Environmental risk, uncertainty and participation: mapping an emergent epistemic community. *Environment and Planning A*; advance online publication, doi:10.1068/a39279.

Chilvers, J. (2008) Deliberating competence: theoretical and practitioner perspectives on effective participatory appraisal practice. *Science, Technology and Human Values*, 33(2), 155–85.

Cleaver, F. (2001) Institutions, agency and the limitations of participatory approaches to development. In B. Cooke and U. Kothari (eds), *Participation: The New Tyranny?* London: Zed Books, pp. 36–55.

Coenen, F., Huitema, D. and O'Toole, L. J. (eds) (1998) *Participation and the Quality of Environmental Decision Making*. Dordrecht: Kluwer.

Collingridge, D. (1982) *Critical Decision Theory: A New Theory of Social Choice*. London: Frances Pinter.

Collins, H. M. and Evans, R. (2002) The third wave of science studies: studies of expertise and experience. *Social Studies of Science*, 32, 235–96.

Cooke, B. and Kothari, U. (eds) (2001) *Participation: The New Tyranny?* London: Zed Books.

Craig, W., Harris, T. and Weiner, D. (eds) (2002) *Community Participation and Geographic Information Systems*. London: Taylor and Francis.

Crosby, N. (1999) Using the citizens' jury® process for better environmental decision making. In K. Sexton, A. Marcus, K. W. Ester and T. Burkhardt (eds), *Better Environmental Decisions*. Washington, DC: Island Press, pp. 401–18.

Davies, A. R. (2002) Power, politics and networks: shaping partnerships for sustainable communities. *Area*, 34(2), 190–203.

Davies, G. (2006a) The sacred and the profane: biotechnology, rationality, and public debate. *Environment and Planning A*, 38, 423–43.

Davies, G. (2006b) Mapping deliberation: calculation, articulation and intervention in the politics of organ transplantation. *Economy and Society*, 35(2), 232–58.

Davies, G. and Burgess, J. (2004) Challenging the 'view from nowhere': citizen reflections on specialist expertise in a deliberative process. *Health and Place*, 10, 349–61.

Dougill, A. J., Twyman, C., Thomas, D. S. and Sporton, D. (2002) Soil degradation assessment in mixed farming systems of southern Africa: use of nutrient balance studies for participatory degradation monitoring. *Geographical Journal*, 168, 195–210.

Dryzek, J. (2002) *Deliberative Democracy and Beyond: Liberals, Critics, Contestations*. Oxford: Oxford University Press.

Eden, S. (1996) Public participation in environmental policy: considering scientific, counter-scientific and non-scientific contributions. *Public Understanding of Science*, 5, 183–204.

Ellis, R. and Waterton, C. (2005) Caught between the cartographic and the ethnographic imagination: the whereabouts of amateurs, professionals, and nature in knowing biodiversity. *Environment and Planning D*, 23(5), 673–93.

Elwood, S. (2006) Negotiating knowledge production: the everyday inclusions, exclusions, and contradictions of participatory GIS research. *Professional Geographer*, 58(2), 197–208.

Ferreyra, C. (2006) Practicality, positionality, and emancipation: reflections on participatory action research with a watershed partnership. *Systemic Practice and Action Research*, 19(6), 577–98.

Fiorino, D. (1990) Citizen participation and environmental risk: a survey of institutional mechanisms. *Science, Technology and Human Values*, 15, 226–43.

Fischer, F. (2000) *Citizens, Experts and the Environment.* Durham, NC: Duke University Press.

Fishkin, J. (1991) *Democracy and Deliberation: New Directions for Democratic Reform* London: Yale University Press.

Freire, P. (1972) *Pedagogy of the Oppressed.* Harmondsworth: Penguin.

Foucault, M. (1980) *Power/Knowledge: Selected Interviews and Other Writings.* Brighton: Harverster Wheatsheaf.

Funtowicz, S. O. and Ravetz, J. R. (1993) Science for the post-normal age. *Futures*, 25(7), 739–55.

Goodwin, P. (1998) 'Hired hands' or 'local voice': understandings and experience of local participation in conservation. *Transactions of the Institute of British Geographers* 23(4), 481–99.

Habermas, J. (1984) *Theory of Communicative Action – Volume 1: Reason and the Rationalisation of Society.* Boston: Beacon Press.

Habermas, J. (1987) *Theory of Communicative Action – Volume 2: System and Lifeworld.* Boston: Beacon Press.

Hajer, M. and Kesselring, S. (1999) Democracy in the risk society? Learning from the new politics of mobility in Munich. *Environmental Politics*, 8(3), 1–23.

Harrison, C. M., Burgess, J. and Filius, P. (1996) Rationalising environmental responsibilities: a comparison of lay public in the UK and the Netherlands. *Global Environmental Change*, 6(3), 215–34.

Healey, P. (1997) *Collaborative Planning: Shaping Places in Fragmented Societies.* London: Macmillan Press.

Hickey, S. and Mohan, G. (eds) (2004) *Participation: from Tyranny to Transformation?* London: Zed Books.

Hillier, J. (2000) Going round the back? Complex networks and informal action in local planning processes. *Environment and Planning A*, 32(1), 33–54.

Hinchliffe, S. (2001) Indeterminacy in-decisions – science, policy and politics in the BSE (Bovine Spongiform Encephalopathy) crisis. *Transactions of the Institute of British Geographers*, 26(2), 182–204.

Irwin, A. (1995) *Citizen Science.* London: Routledge.

Irwin, A. (2001) Constructing the scientific citizen: science and democracy in the biosciences. *Public Understanding of Science*, 10(1), 1–18.

Irwin, A. (2006) The politics of talk: coming to terms with the 'new' scientific governance. *Social Studies of Science*, 36(2), 299–320.

Jankowski, P. and Nyerges, T. (2001) *Geographic Information Systems for Group Decision Making: Towards a Participatory, Geographic Information Science.* London: Taylor & Francis.

Joss, S. and Durant, J. (eds) (1995) *Public Participation in Science: The Role of Consensus Conferences in Europe.* London: Science Museum.

Kasemir, B., Jager, J., Jaeger, C. C. and Gardner, M. T. (2003) *Public Participation in Sustainability Science.* Cambridge: Cambridge University Press.

Kesby, M. (2005) Retheorizing empowerment-through-participation as a performance in space: beyond tyranny to transformation. *Signs*, 30(4), 2037–65.

Kesby, M., Kindon, S. and Pain, R. (2005) 'Participatory' approaches and diagramming techniques. In R. Flowerdew and D. Martin (eds), *Methods in Human Geography.* Harlow: Pearson, pp. 144–66.

Kothari, U. (2001) Power, knowledge, and social control in participatory development. In B. Cooke and U. Kothari (eds), *Participation: The New Tyranny?* London: Zed Books, pp. 139–52.

Kothari, U. (2005) Authority and expertise: the professionalisation of international development and the ordering of dissent. *Antipode*, 37(3), 425–46.

Laird, F. (1993) Participatory analysis, democracy, and technological decision making. *Science, Technology and Human Values*, 18(3), 341–61.

Leach, M., Scoones, I. and Wynne B. (eds) (2005) *Science and Citizens: Globalization and the Challenge of Engagement*. London: Zed Press.

Macnaghten, P. and Jacobs, M. (1997) Public identification with sustainable development – investigating cultural barriers to participation. *Global Environmental Change*, 7(1), 5–24.

Marris, C., Wynne, B., Simmons, P. and Weldon, S. (2001) *Public Perceptions of Agricultural Biotechnologies in Europe*. Final report of the EU funded PABE research project FAIR CT98–3844 (DG12 – SSMI). Lancaster: Centre for the Study of Environmental Change, Lancaster University.

Mason, R. (1999) *Environmental Democracy*. London: Earthscan.

Mosse, D. (2001) 'People's knowledge', participation and patronage: operations and representations in rural development. In B. Cooke and U. Kothari (eds), *Participation: The New Tyranny?* London: Zed Books, pp. 16–35.

Mouffe, C. (2000) *The Democratic Paradox*. London: Verso.

Munton, R. (2003) Deliberative democracy and environmental decision-making. In F. Berkhout, M. Leach and I. Scoones (eds), *Negotiating Environmental Change: New Perspectives from Social Science*. Cheltenham: Edward Elgar, pp. 109–36.

O'Neill, J. (2001) Representing people, representing nature, representing the world. *Environment and Planning C*, 19, 483–500.

Owens, S. (2000) 'Engaging the public': information and deliberation in environmental policy. *Environment and Planning A*, 32(7), 1141–48.

Owens, S., Rayner, T. and Bina, O. (2004) New agendas for appraisal: reflections on theory, practice and research. *Environment and Planning A*, 36, 1943–59.

Pain, R. (2004) Social geography: participatory research. *Progress in Human Geography*, 28(5), 652–63.

Pellizzoni, L. (2003) Uncertainty and participatory democracy. *Environmental Values*, 12, 195–224.

Petts, J. (1997) The public-expert interface in local waste management decisions: expertise, credibility and process. *Public Understanding of Science*, 6(4), 359–81.

Petts, J. (2001) Evaluating the effectiveness of deliberative processes: waste management case studies. *Journal of Environmental Planning and Management*, 44, 207–26.

Petts, J. (2004) Barriers to participation and deliberation in risk decisions: evidence from waste management. *Journal of Risk Research*, 7(2), 115–33.

Renn, O., Webler, T., Rakel, H., Dienel, P. and Johnson, B. (1993) Public participation in decision making: a three-step procedure. *Policy Sciences*, 26, 189–214.

Renn, O., Webler, T. and Wiedemann, P. (eds) (1995) *Fairness and Competence in Citizen Participation: Evaluating Models for Environmental Discourse*. Dordrecht: Kluwer.

Rosener, J. (1978) Citizen participation: can we measure its effectiveness? *Public Administration Review*, 38, 457–63.

Rowe, G. and Frewer, L. (2000) Public participation methods: a framework for evaluation. *Science, Technology and Human Values*, 25(1), 3–29.

Rowe, G. and Frewer, L. (2004) Evaluating public participation exercises: a research agenda. *Science, Technology and Human Values*, 29(4), 512–57.

Rowe, G., Horlick-Jones, T., Walls, J. and Pidgeon N. (2005) Difficulties in evaluating public engagement initiatives: reflections on an evaluation of the UK GM Nation? Public debate about transgenic crops. *Public Understanding of Science*, 14(4), 331–52.

Sieber, R. (2006) Public participation geographic information systems: a literature review and framework. *Annals of the Association of American Geographers*, 96(3), 491–507.

Smith, L. G. (1984) Public participation in policy making. *Geoforum*, 15(2), 253–59.

Stern, P. and Fineberg, H. (eds) (1996) *Understanding Risk: Informing Decisions in a Democratic Society*. Washington, DC: National Research Council.

Stirling, A. (1998) Risk at a turning point? *Journal of Risk Research*, 1(2), 97–109.

Stirling, A. (2005) Opening up or closing down: analysis, participation and power in the social appraisal of technology. In M. Leach, I. Scoones and B. Wynne, (eds), *Science and Citizens*. London: Zed Press, pp. 218–31.

Susskind, L., McKearnan, S. and Thomas-Larmer, J. (eds) (1999) *The Consensus Building Handbook*. London: Sage.

Swyngedouw, E. (2005) Governance innovation and the citizen: the Janus face of governance-beyond-the-state. *Urban Studies*, 42(11), 1991–2006.

Tewdwr-Jones, M. and Allmendinger, P. (1998) Deconstructing communicative rationality: a critique of Habermasian collaborative planning. *Environment and Planning A*, 30, 1975–89.

Webler, T. (1995) Right discourse in citizen participation: an evaluative yardstick. In O. Renn, T. Webler and P. Wiedemann (eds) *Fairness and competence in citizen participation* (Dordrecht: Kluwer) pp. 35–86.

Webler, T. (1998) The craft and theory of public participation: a dialectical process. *Journal of Risk Research*, 2(1), 55–71.

Webler, T., Kastenholz, H. and Renn, O. (1995) Public participation in impact assessment: a social learning perspective. *Environmental Impact Assessment Review*, 15, 443–63.

Webler, T., Tuler, S. and Krueger, R. (2001) What is a good participation process? Five perspectives from the public. *Environmental Management*, 27(3), 435–50.

Whyte, W. F. (ed.) (1991) *Participatory Action Research*. Newbury Park: Sage.

Wilsdon, J. and Willis, R. (2004) *See-Through Science: Why Public Engagement Needs to Move Upstream*. London: Demos.

Wynne, B. (2005) Risk as globalising 'democratic' discourse: framing subjects and citizens. In M. Leach, I. Scoones and B. Wynne (eds), *Science and Citizens*. London: Zed Press, pp. 66–82.

Part IV   Topics

Chapter 25

Ecosystem Prediction and Management

Robert A. Francis

Introduction

Human survival depends on the resources and other so-called 'services' (e.g., nutrient cycling, climate regulation, soil formation) provided by ecosystems. Consequently, active management of biotic and abiotic ecosystem resources (e.g., production and consumption of plants and animals and their associated products, regulation and abstraction of water resources) has been central to the development of human civilisation since at least the early Holocene (e.g., Itzstein-Davey et al., 2007). All ecosystems are dynamic and exhibit notable complexity, variability, stochasticity and non-linearity (e.g., Pahl-Wostl, 1995; Arrow et al., 2000), although explicit recognition of these properties is a relatively recent development (e.g., May, 1987). Efficient and sustainable resource management depends on predictability of both the resources of interest and the wider ecosystems on which they depend. However, the variability and complexity inherent to ecosystems and their component parts means that prediction at any but the simplest levels is a substantial challenge, and one that we are ill-equipped to meet, given our current understanding of ecosystem dynamics.

The Development of the Ecosystem Concept

Scientific appreciation of the variability and complexity of ecosystems and their components has emerged in recent decades out of an interest in variability and complexity across a range of scientific disciplines, including mathematics, physics, biology and physical geography (e.g., Manson, 2001). Ecosystems (and indeed, many natural phenomena) were originally thought to operate in relatively simple, predictable ways (see discussions in Golley, 1993; Gaichas, 2008). Modern concepts of ecosystems and ecological processes developed out of long-standing philosophical traditions of natural history stretching back to classical antiquity. Aristotelian enquiries into nature began with simple investigations into the properties of biotic and abiotic ecosystem components, encapsulated within a teleological framework of idealised 'types' (Benson 2000). Subsequently, scientific thought and experimentation developed to focus more on interactions between these components, with the gradual realisation that the natural environment was more complex than at first supposed. Scientific enquiry generally focused on relatively fine-scale biological subjects (e.g., the human body, medicine) or much larger-scale physical subjects (e.g., hydrology, geology, astrophysics). Increasingly, these investigations found fundamental laws that could be applied to nature (e.g., the laws of thermodynamics, motion, relativity) but also that the *systems* under investigation were more variable, stochastic or chaotic than at first supposed (e.g., Gleick, 1987; Benson, 2000). This was also the case for the science of ecology and the study of ecosystems in the 19th and 20th centuries (see Golley, 1993; Benson, 2000).

Through much of ancient and pre-modern history, humans have generally managed ecosystems to obtain several resources at one time rather than being geared around the production of just one (e.g., Power and Campbell, 1992). For example, prior to the agricultural revolution in the mid-18th century European farming methods focused on using a single area of land to produce a range of food and other resources. Monoculture of a single crop was undesirable and uneconomic due to small population centres, limited ability to transport and preserve goods, and the possibility of disease or poor weather leading to excessive losses of a specific crop (Power and Campbell, 1992; Richardson, 2005). Until relatively recently, population pressure on ecosystems was also much lower, so that intensive exploitation of an ecosystem for a specific resource was unlikely to significantly influence the functional integrity of the ecosystem (e.g., Warner et al., 1996). It was only with the Agricultural and Industrial Revolutions of the 18th and 19th centuries when ecosystems were 'refined' for the intensive production of a specific resource that substantial modification of the structure and processes of ecosystems began. Consequently, understanding the dynamics of the desirable resources became a management priority. Other ecosystem components and processes (unless directly related) were typically seen as unimportant, irrelevant or tangential at best (e.g., Gaichas, 2008).

It was against this background of modified ecosystems and disrupted processes that modern ecological thought developed. Specific concepts of 'ecosystem' (an interactive system of biota and abiota), 'ecological community' (an assemblage of species with associated resource requirements) and 'ecosystem services' (functions provided by ecosystems which are of benefit to humans and the environment in general) developed in the 19th and 20th centuries as the discipline of ecology formed and ecologists became concerned with fundamental theoretical questions relating to how biotic and abiotic components of ecosystems interact and change (e.g., Pickett et al., 1994). This, in turn, led to increased experimentation into and quantification of ecosystem structure, processes, and variability (see, e.g., Gaichas, 2008).

The dynamical nature of ecosystems had been apparent for centuries, although this dynamism was poorly understood and its extent underrated (Pahl-Wostl, 1995). It was early theoretical ecologists such as Clements (1916) and Tansley (1935) who first began to consider the mechanisms of dynamism and to attempt to identify universal ecological laws that would enable ecosystem changes to be predicted (see Golley, 1993). Early ecological investigations were concerned mainly with changes in biodiversity, community composition, and the structure and arrangement of organisms within ecosystems, and focused primarily on plants (easily observable, sedentary organisms) as indicators of ecosystem change. Clements suggested that particular species would associate together to form a specific community. He conceptualised ecosystem seral dynamics as the simple replacement of particular communities over time until a climax stage was reached, with communities functioning in similar ways to discrete organisms or 'superorganisms'. This was an attractive model as it was simple and implied predictability of ecosystem dynamics. Tansley (1935), who coined the term 'ecosystem', refuted that communities acted as organisms but stated that ecosystems consisted of biotic and abiotic components that develop following disturbance to obtain a stable dynamic equilibrium. Both of these early viewpoints assumed that equilibrium states and predictability were inherent to ecosystems.

This meant that in theory it should be possible to create mathematical models that would explain and predict ecosystem dynamics and could therefore be used for ecosystem resource management. Early models, however, had poor predictive capacity, reinforcing the idea that ecosystem modelling and management should focus only on the very few ecosystem components directly relevant to the managing body (e.g., Gaichas, 2008). Despite increasing evidence of complexity and variability in ecosystem dynamics, particularly relating to seral changes in communities, the mechanistic approach (i.e., nature as machine) to ecosystem processes and concepts of equilibrium and predictability greatly influenced ecosystem modelling and resource management until the advent of chaos theory and investigations into complexity and complex systems (Pahl-Wostl, 1995). The legacy of this original focus on stability and equilibrium is still being felt today in resource management.

As the 20th century advanced, increasing evidence of individualistic and stochastic mechanisms operating within ecosystems made it clear that ecosystems were indeed more random, variable and unpredictable than originally supposed over all spatial and temporal scales (e.g., Pahl-Wostl, 1995). Recognition of stochasticity and ecosystemic variability across spatial and temporal scales raised important questions about the possibility of predicting maximum sustainable resource yields and consequently about the levels of uncertainty and acceptable risk in ecosystem resource management. This was further supported by an acknowledgement of the limitations and the poor predictive potential of resource models (e.g., Batchelor et al., 2002), which often ignored wider ecosystem variability and so compromised their accuracy. In part, this was due to the reductionist approach of monitoring and modelling components and processes in isolation, in attempts to understand their dynamics, rather than adopting a holistic approach for whole ecosystem understanding (Pahl-Wostl, 1995). By the 1990s, the concept of ecosystems as complex nonlinear systems was well established (e.g., Kay 2000).

Towards Whole Ecosystem Management for Resources and Ecosystem Services

At the same time that the inherent variability of ecosystem processes was being elucidated, the detrimental impacts of poor ecosystem management were also becoming apparent (e.g., Linton 1970; Darge and Kneese, 1980). In particular, changing land-use patterns and the intensification of single resource production (e.g., the clearing of Amazonian tropical forest for ranching) reduced biodiversity and affected the integrity, resilience and stability of ecosystems, potentially leading to local ecosystem collapse and, globally, the threat of a human-induced mass extinction event (e.g., Dale et al., 1994; cf van Loon, 2003). This, combined with the high-profile collapse of fish stocks, extinction or decline of charismatic species, and increased risks to human health or living standards meant that conservation and restoration became increasingly important scientifically and politically (e.g., Walters and Maguire, 1995; Petts, 2001; Dobson, 2005).

The concept of 'ecosystem services' was developed to acknowledge the range of functional processes within ecosystems that are indirectly necessary for human survival (e.g., Daily et al., 1997). These are mainly provided by biotic ecosystem components and include (as just a few examples) regulation of climate and atmosphere, cycling of water, sediments and nutrients, photosynthesis and production of biomass, pest control, pollination, and soil formation. Ecosystem services contribute to the functioning of ecosystems and ultimately human survival at a range of scales, from local (e.g., photosynthesis or nutrient cycling within an individual plant) to global (e.g., maintenance of the Earth's atmosphere and climate). The ecosystem services concept had implications for the prediction of ecosystem response to disturbance and management at different scales, and also raised questions regarding the scale at which management is appropriate. Ecosystem management for resource acquisition or production was almost always conducted at the local scale and with limited understanding of temporal variability. Consequently, there was very limited consideration of how alterations in functioning in local ecosystems would impact regional and global ecosystems.

The management of ecosystems for individual resources is increasingly being recognised as unsustainable and so whole ecosystem management is becoming more important, both for industry and conservation. This is reflected in trends towards land management at broad spatial scales, e.g., landscapes, regions or catchments, as well as increased interactions between geographers, ecologists, land managers and economists (e.g., Grumbine, 1994; Sparks, 1995). It is apparent that we need to preserve the complexity and dynamism of ecosystems, though we still do not know how best to achieve this. There are major gaps in our understanding of ecosystem variability and prediction that make ecosystem prediction and management less efficient and effective. Furthermore, the same considerations of ecosystem variability and complexity that apply to resource management in industry also apply to ecosystem conservation and restoration. Partly because ecosystems are poorly understood and partly because of the history of single resource management, conservation and restoration are often conducted with one resource or habitat in mind, and recent lessons relating to the complexity and variability of ecosystems have yet to be applied to this new form of ecosystem management, which is necessary if they are to be successful (e.g., Bond and Lake, 2003). This chapter goes on to discuss issues relating to the quantification and prediction of variability in ecosystems and the way in which this influences the management of ecosystem resources and services.

Understanding Ecosystem Interactions and Variability

Ecosystems as non-linear complex adaptive systems

Ecosystems on any scale involve extensive interconnections and interactions among their components that largely defy quantification and simplification. Despite the discovery of, for example, power laws linking some ecosystem patterns and processes over a range of scales and levels of organisation (Brown et al., 2002), ecosystems can display substantial non-linearity. They can be understood as complex adaptive systems (i.e., having the capacity to change). This means that they may have recognisable patterns and organisation, but are still difficult to understand and predict (e.g., Pahl-Wostl, 1995; Arrow et al., 2000; Gaichas, 2008). Such complexity also means that whole ecosystem investigations are problematic and resource intensive. Consequently, research and management usually focus on a small subset of interactions. Basing management on investigations of these subsets in isolation, without an understanding of broader ecosystem processes and linkages, can severely compromise management success. Chaotic, complex and/or non-linear systems commonly exhibit processes where effects are not proportional to causes, and where output is not equal to input, so that small changes in component structure and process can potentially have substantial effects on the wider ecosystem (Pahl-Wostl, 1995; Manson, this volume). Non-linearity in ecosystems has been shown in, for example, biogeochemical cycles (Qi et al., 2002), population dynamics (e.g., Turchin, 1993) hydrological cycles (Rodriguez-Iturbe et al., 1991), and climate (e.g., Ghil et al., 1991).

A theoretical example of non-linear response in an ecosystem process can be seen in figure 25.1. In this model, it is assumed that the structure and organisation of a system are related to its stability and resilience, and therefore its capacity to repair damage caused to biota (measured in biomass). In figure 25.1a, it is simply assumed that the various factors which together stress an ecosystem cause a reduction in growth rate of biota and a decline in biomass, but have little effect on the structure and organisation of the ecosystem (i.e., stability and resilience), creating a linear response to stress (figure 25.1c). In figure 25.1b, the negative effects of stress also change the structure and organisation of the ecosystem (reducing stability and

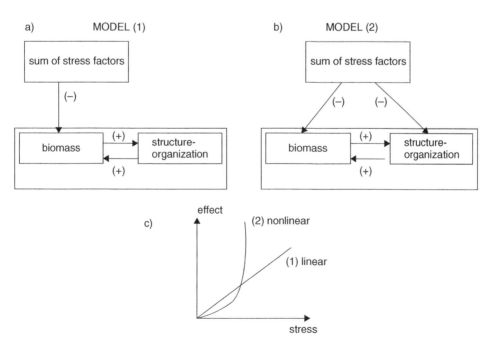

Figure 25.1 Examples of simple ecosystem interactions that can lead to either linear (Model 1) or non-linear (Model 2) responses (see text for explanation). Modified from Pahl-Wostl (1995).

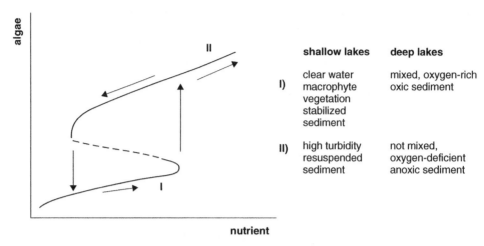

Figure 25.2 Example of a threshold effect and alternative states in eutrophied lacustrine ecosystems. Increasing nutrient content leads to expanding algal populations and increased oxygen consumption, decreasing biotic variability and species populations and creating a positive feedback in algal growth. When a critical threshold of nutrient supply and algal growth is exceeded, there is some discontinuity where the system changes from one state to another. These changes in state are accompanied by changes in both physical and biological characteristics, as indicated for both states in shallow and deep lakes. Arrows show a generalised path of recovery for decreasing nutrient loads, which is different to increasing loads. Taken from Pahl-Wostl (1995).

resilience) compromising the system's ability to compensate for this stress, so that the response is non-linear (figure 25.1c). This essentially means that a point is reached when the system cannot compensate for stress and the stress effect is exacerbated (i.e., a positive feedback occurs), so that the ecosystem 'crashes'. This is also termed a 'threshold effect', and results in a system rapidly changing from one state to another.

A classic example of a 'threshold effect' is eutrophication in lakes. With increasing nutrient inputs into such systems, algal populations increase steadily until the amount of algae reduces the level of oxygen in the water beyond a critical threshold, whereupon a series or 'cascade' of interactions occurs, such as a rapid decline in submerged plants, macrophytes, fish and invertebrates and further increases in algal populations (Pahl-Wostl, 1995). This then rapidly increases the rate of change in both biotic and abiotic structure and process in the ecosystem (figure 25.2), and is a useful example of a non-linear adaptive response to a simple change in ecosystem input, which may itself be a linear process. Petts and Gurnell (2005) discuss similar non-linear responses in river system morphology and ecology to dams. Dams reduce flows of energy, sediment and water into a river system, but the response to those changes depends on sediment type, species characteristics and the variability of the (regulated) hydrological regime.

Such non-linearity of system response inevitably causes problems for ecosystem and resource prediction. Qi et al. (2002) demonstrate how soil respiration (carbon emission) is sensitive to changes in temperature but has a non-linear response, being much more sensitive (and therefore variable) at lower temperatures compared to

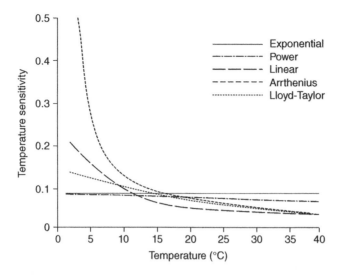

Figure 25.3 Variation in temperature sensitivity of soil respiration, with curves based on five different soil respiration models. Taken from Qi et al. (2002).

higher temperatures, and having a variable response along the temperature gradient (figure 25.3). This indicates that simple empirical models generally used to calculate soil respiration are ineffective, as they do not account for this non-linearity of response, and therefore, lead to substantial errors in predicting carbon budgets and flux within ecosystems at broad spatial scales. This is just a simple example of how a small non-linear variation in response at fine scales can potentially lead to inaccurate predictions at broader scales.

The recent suggested prevalence or dominance of chaotic or non-linear processes in ecosystems does not necessarily mean that all ecosystem processes are complex or non-linear; indeed ecosystems do have some characteristics of linearity, organisation, and predictability (e.g., Brown et al., 2002). Nevertheless, most ecosystems are complex and will have elements of non-linearity, which need to be accounted for in management.

Ecosystem individuality

For ease of interpretation and management, ecosystems have traditionally been categorised according to dominant physical or biological characteristics (e.g., Rodwell, 1991). Table 25.1 implies that ecosystems within the same category function in very similar ways, but in reality, each ecosystem and ecosystem component has notable individuality due to its historic context (e.g., sequence of past disturbances, management regimes, etc.), spatial variations in the characteristics and arrangements of biotic and abiotic components (from landscape configuration to genetics), and stochastic processes, among other reasons (e.g., Jørgensen et al., 2005). Consequently, although ecosystems may be categorized by type, and indeed show superficial similarities in functional processes, no two ecosystems will function identically. This context dependence frustrates generalisations and confounds management regimes that rely on predictive precision.

Table 25.1 Examples of classification of organisms at different scales of organisation. Despite these general classifications, variability is present at each level, from genetic variability between individuals to a wide range of climatic, geophysical, biotic and disturbance variability within each ecozone

Scale of organisation	Classification
Individual	Common oak (*Quercus robur* L.)
Community	W10 *Quercus robur – Pteridium aquilinum – Rubus fruticosus* woodland (NVC classification)
Ecosystem	Lowland oak woodland
Biome	Temperate broadleaved woodland
Ecozone	Palearctic

A useful case study of differences in ecosystem parameters between sites and natural variation within sites can be found in Osenberg et al. (1994), who investigated two undisturbed seabed communities off the coastline of California (depth ≈ 27 m), over a three-year period. They found that physico-chemical parameters (e.g., water temperature, sediment characteristics) and biological parameters relating to individuals (e.g., urchin body size and condition) were consistently different between sites, whereas population parameters (e.g., density) were more variable, increasing similarity and dissimilarity between sites over time in terms of species populations. Accordingly, a focus on the variability of only one or two parameters within ecosystems can potentially lead to incorrect conclusions regarding that variability and the 'prediction potential' of similar ecosystems.

Ecosystems as open, holarchically nested systems

Despite reductivist scientific and isolationist management approaches, no ecosystem is isolated from those around it. Instead every ecosystem is nested within a hierarchy of larger ecosystem assemblages (figure 25.4). This means that a specific hierarchical level (holon) is influenced by, and influences, the levels above and below it, and is maintained by a balance of allogenic and autogenic controls (see Kay, 2000). Each ecosystem therefore exchanges materials and energy with the systems nested above and below, and so all ecosystems may be considered open systems to some degree. Consequently, a holon cannot be truly understood in isolation: the nesting of holons within the overarching structure, or holarchy, and the nature of exchanges within such open systems must also be appreciated if system dynamics and future states are to be accurately quantified and predicted.

Equilibrium and Non-Equilibrium Ecosystem Paradigms

The concept of equilibrium states within ecosystems originates from Clementsian concepts of succession in ecological communities, wherein discrete assemblages of species develop through predictable stages to a climax ('mature') state whereupon the ecosystem is stabilised and communities and processes are considered to be in equilibrium (figure 25.5). In this sense, equilibrium essentially equates to a relatively 'steady state' system wherein flows of energy and materials are balanced within the

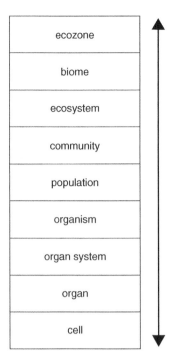

Figure 25.4 Holarchical organisation of biotic ecosystem components, from individual cells to ecozones.

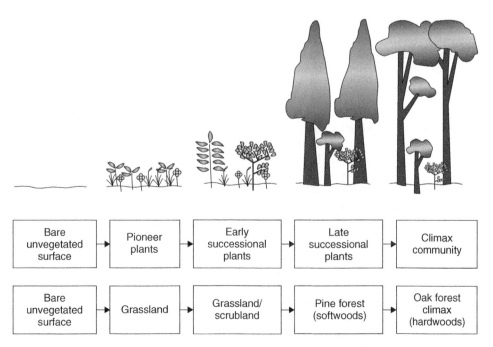

Figure 25.5 Seral stages in the classical model of succession, leading from bare ground to a mature 'climax' state which is considered to be in equilibrium.

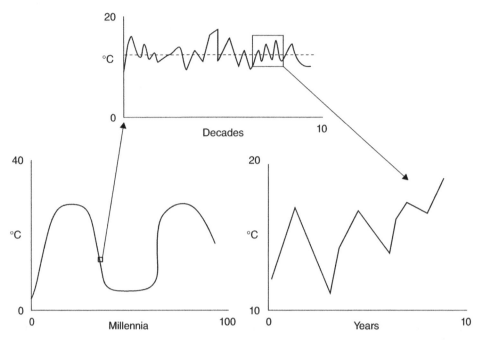

Figure 25.6 Theoretical example of how equilibrium of ecosystem processes is most commonly observed at medium temporal scales (e.g., decades). In this case, theoretical trends in average summer temperatures of an ecosystem are shown over years, decades and millennia. Non-equilibrium is observed over years, due to annual fluctuations in weather conditions and climatic variations. When these variations are recorded over decades, an approximate equilibrium (average temperature) can be observed. Over millennia however, broad changes in (for example) climate creates notable variation in summer temperatures, so that observations once again revert to non-equilibrium. Ecosystem management is often conducted on a short-term basis with equilibrium conditions assumed, despite these mainly being appropriate only at larger scales.

ecosystem, and community composition is relatively unchanging. Although the term 'steady state' is used here, it is apparent that some level of dynamism would still be exhibited by the ecosystem, but only at very small spatial and temporal scales.

In practice, the concept of 'steady state' equilibrium is linked directly to the spatial and temporal scale considered (see Sayre, this volume). At 'medium' spatial and temporal scales, which may be based around anthropocentric space and timescales (e.g., hectares of land, decades to centuries), the equilibrium concept remains valid; ecosystems do exhibit predictable changes and discrete stages observable over human lifetimes, leading up to a relatively 'stable state' ecosystem as given in figure 25.6. At finer spatial and temporal scales, equilibrium in biotic and abiotic patterns and processes is not always so apparent. Consider, for example, gap analysis of woodlands. If a tree dies, a new biological community and range of physical processes will occur within the relatively small space the tree previously occupied, which may operate at different rates to those in adjacent habitat. Likewise, at very broad spatial and temporal scales (centuries to millennia) ecosystems are likely to change according to broader changes in climate, geomorphological processes, and increased probability of disturbance. Consequently, steady-state equilibrium may only be observable or predictable at medium scales, and even at these scales non-linear responses or threshold effects may occur to create different states (figure 25.6).

An example of this can be seen in aquatic and riparian habitat turnover along an island-braided reach of the natural and dynamic River Tagliamento in Italy. Van der Nat et al. (2003) demonstrate that 75 percent of aquatic and 29 percent of vegetated island habitat was restructured over a 2.5-year period due to frequent channel movements, while Zanoni et al. (in press) examine habitat restructuring in the same reach, and conclude that although habitat rearrangement occurs frequently, over medium timescales (decades) the cover of different habitat types remains relatively constant. Over longer timescales (centuries), morphological changes along the river, along with variation in sedimentary processes and wood delivery, may result in these habitats changing notably or disappearing from a given reach completely (Spaliviero, 2003). Consequently, the degree of equilibrium apparent in a system depends on the spatial and temporal scale under consideration.

Unfortunately, much land and resource management is conducted with the concept of steady-state equilibrium in mind. Whereas such steady state equilibrium can only be said to exist at 'medium' scales (figure 25.6), management is typically conducted at relatively fine spatial and temporal scales (years to decades at best), where greater dynamism and variability are more typical. This mismatch creates problems for land management and prediction and can lead to unsustainable land-use practices. In effect, the concept of equilibrium states is subjective, difficult to quantify and may be misleading or inappropriate in a management context.

Measuring Variability in Ecosystems and Ecosystem Components

Measurement of variability in ecosystem pattern and process to inform decision making is an applied problem for ecologists, environmental geographers, and land managers. Comprehensive ecosystem measurement would require investigations of a multitude of processes by a large range of specialists in varying disciplines. Consequently, investigations into the variability of ecosystem patterns and processes are focused on a limited subset, and are necessarily limited in size and scope. Due both to the increasing acknowledgement that ecosystems are interconnected, open systems and to the consequent increase of management at the landscape or catchment level, there is now a move towards measuring variability at the landscape scale by geographers and landscape ecologsts using remote sensing techniques and landscape metrics analysis (e.g., Batchelor et al., 2002; Caseldine and Fyfe, 2006).

Measurements are usually conducted over a range of temporal scales using different techniques. Longer-term analysis of ecosystem variability is performed using landscape interpretation and reconstruction (e.g., Delcourt and Delcourt, 2005; Caseldine and Fyfe, 2006). Short-term measurements are more common for management and involve looking at spatio-temporal variability in ecosystem components, for example population and community dynamics, habitat turnover and heterogeneity, fluxes of carbon and water, and so on. Even within relatively simple ecosystems, such as agricultural systems, intensively managed for the production of a single resource over short temporal scales, a wide range of measurements need to be taken to quantify variability and allow even coarse predictions (table 25.2). The development and application of new geographical technologies are central to measurement

Table 25.2 Examples of key biotic and abiotic variables in arable agricultural ecosystems which would need to be quantified in terms of variability, in order to allow prediction of crop yields and to inform ecosystem management

Biotic	Abiotic
Crop genotype/variety	Soil texture
Crop density	Soil type
Crop yield (biomass)	Available soil moisture
Crop health	Soil organic content
Pathogen presence	Soil nutrients
Herbivore presence/abundance	Drainage (surface and subsurface water movement)
Weed presence/abundance	Topography
	Precipitation
	Temperature

of ecosystem variability (e.g., Kerr and Ostrovsky, 2003; Bocchi and Castignanò, 2007).

Regardless of the data collection methodology, common statistical metrics are used to define variation over set periods of time and space, such as mean, median, range, standard deviation, frequency, and size and shape of distributions (e.g., Landres et al., 1999). However, one of the main problems with measuring and quantifying variability in complex ecosystems is that different processes or components will require different descriptors of variability. For example, quantification of the variability of fire disturbance within a landscape will require measurement of fire frequency, intensity, size, and spatial arrangement. Measurements of flood disturbance will focus on flood frequency, discharge, stage variations, and spatial extent of riparian and floodplain inundation. Measurements of species population variability might include repeated counts or estimates of individuals, quantification of life history cycles, and metapopulation dynamics (Landres et al., 1999). All of these can be expected to vary notably over time, and so either long-term monitoring (which is often unfeasible) or reconstruction of past conditions is necessary, though this too has limitations. Time lags are a common feature of non-linear processes and also cause problems with obtaining and interpreting measurements.

For all of these reasons, quantifying the variability of ecosystem states and process is difficult (e.g., Adachi et al., 2005). Nevertheless, such measurements are essential if predictive models are to be utilised for ecosystem management.

Effectiveness of Prediction in Ecosystem Management

All of the issues discussed above create problems for ecosystem and resource prediction. But how unreliable are our current efforts at prediction? Modelling has traditionally been based on specific data and is closely related to the measured parameters of the ecosystem as it was during the measurement period, which limits its wider prediction potential (Tan et al., 2006). Empirical models use a broader range of data and parameters to make general assumptions about a system, and to predict system response where data are unknown. Often these assumptions greatly limit the applications of empirical models, and there is now more focus on non-linear models to predict ecosystem response and resource management (Tan et al., 2006;

Czerwinski et al., 2007). The complexity and range of data and parameters required for these models make them unattractive or unfeasible for most management situations however (Tan et al., 2006; Pitchford et al., 2007). The relative merits of such models are reviewed by Tan et al. (2006). Here, examples of predictive potential of models in two intensively managed ecosystem types (arable agriculture and marine fisheries) are given.

Predictability in intensively managed ecosystems: arable agriculture

Predictability of ecosystem processes can be limited even in ecosystems that have been greatly simplified or refined, and are intensively managed, to ensure that conditions remain relatively fixed. Agricultural ecosystems are artificial, but are designed and managed to ensure a fixed return on an ecological resource (e.g., biomass of a desired crop species) for a continued investment of resources (water, nutrients, labour). Agricultural ecosystems are simplified systems in which environmental conditions are made as homogeneous as possible (table 25.3). Despite these attempts to create ecosystems which are essentially biomass factories, spatial and temporal variation in crop yields is still common, due to variability in broader environmental factors which are hard to control or mitigate, e.g., changes in climate and weather conditions, and to residual variability in factors which are controlled, but cannot be controlled precisely enough, such as spatio-temporal variations in soil nutrients (e.g., Basso et al., 2001; Batchelor et al., 2002).

Because of the economic value of agricultural ecosystems, many attempts at yield prediction using crop models have been made, with varied success. Initial models were simple, and were based on regression techniques to compare yield to environmental parameters (e.g., Jones and Ritchie, 1996). These failed to adequately predict yields due to the many non-linear spatial and temporally variable factors which impact upon crop yield, such as intraspecific competition, crop population densities, weather, pest and pathogen dynamics and water and nutrient dynamics, all of which lead to complex patterns of plant stress (e.g., Batchelor et al., 2002). Process-oriented crop models examine the effects of this variability more explicitly and therefore predict single-crop yield under different environmental and management scenarios, but are still limited in their application to individual agricultural ecosystems and can display notable error margins (e.g., Basso et al., 2001). Examples include models such as CROPGRO (Boote et al., 1998) and CERES-Maize (Jones and Kiniry, 1986), which simulate daily growth of soybean and corn respectively in relation to differences in carbon, nitrogen, and water inputs. These models assumed spatial homogeneity and predicted yields based solely on temporal variation of environmental conditions.

More recently attempts have been made to refine these models to account for spatial variation in environmental parameters. Batchelor et al. (2002) evaluated the performance of these models using different methods of spatial validation. Environmental (field capacity, rainfall, temperature and solar radiation) and management (planting date, row spacing, genotypes) characteristics were recorded at a spatial resolution of 0.2 ha, with 100 points being measured in total (over a 20 ha field). These inputs were fed into the CERES-Maize model, and simulated yield was correlated to measured yield over three years, with an overall coefficient of 76 percent. Although the source of the remaining variability was unknown, measurements of

Table 25.3 Methods utilised to minimise variability in controlled arable agricultural ecosystems, including broad aims and unintended effects

Method	Aim	Unintended effects
Biotic variability		
1) Planting of a single species (often a single genotype) 2) Addition of herbicides and pesticides	1) Reduce species interactions such as competition and herbivory, which may reduce yield 2) Species/genotype chosen to ensure high yield and for other desirable characteristics (e.g. taste, appearance) to make a marketable product	1) Reduction of genetic variation within the planted species 2) Loss of ecosystem functioning and stability due to decreased interactions 3) Loss of biodiversity at a range of scales and taxa
Climate/hydrological variability		
1) Irrigation 2) Polytunnels	1) Provision of a constant or regular water supply to enable consistent growth and high yields 2) Mitigation of variations in climate and weather which may lead to e.g. dry periods, frost etc.	1) Degradation of water supplies outside the ecosystem 2) Increased soil erosion 3) Interruption of ecosystem functions which require such variability (e.g. life cycles of invertebrate species)
Soil variability		
1) Ploughing of fields 2) Addition of fertilisers	1) Break-up soil structure to reduce compaction, prevent distinct horizons forming in the upper layers, and allow surface organic matter to mix with lower layers where plant roots are common 2) To reduce spatial and temporal differences in the structure, texture and nutrient content of soils to ensure consistent yields	1) Reduction in soil biodiversity 2) Increased soil erosion

nitrogen, root depth, drainage and hydrological connectivity between measurement points were not made, and may help explain this residual variability. Furthermore, the model was unable to account for low plant populations and non-uniform plant spacings (Batchelor et al., 2002). Similarly, high-resolution measurements of spatial variability in environmental and management factors (including herbivory by soybean cyst nematodes and weed competition) were incorporated into the CROPGRO soybean model in the same fashion, explaining 80 percent of the variability in soybean yields (Batchelor et al., 2002). Although a large amount of variability is explained by these models, prediction errors are still substantial and are a source of uncertainty for resource managers. The models are also of limited use in off-site applications because of the level of detail of spatial variables needed for prediction. Recent attempts to use remotely sensed data have shown promise (Basso et al., 2001; Batchelor et al., 2002), and there is the possibility of linking models, for example water balance models based on Digital Elevation Models linked to Normalised Difference Vegetation Index, to highlight areas where water availability is a major limitation on yields (Batchelor et al., 2002). In general, predictive crop models have a reasonable level of yield prediction, but are either too vague and general, or are more accurate but require substantial data to validate for relatively small spatial areas.

Increasing evidence of the environmental impacts of intensive farming techniques have led to three farming scenarios with different attitudes to prediction and management of variability: (i) increased investment in precision agriculture to reduce spatial and temporal variability as much as possible; (ii) use of precision agriculture to reduce variability and improve yields in selected locations, while other areas (e.g., buffer zones) are left 'natural' and so allow some level of natural variability; and (iii) acceptance of natural variability and changing in farming practices to green/ organic farming, wherein crop yields are exchanged for variability (and the provision of ecosystem services, which society will pay increased prices for). This is an example of a move back towards variability and away from single-resource prediction and management.

Predictability in intensively managed ecosystems: marine fisheries

Marine fisheries are less intensively managed than arable fields, as it is not possible to exert the same level of control on the oceans that it is on land. Nevertheless, fish populations around the world are carefully managed to obtain maximum sustainable yield (MSY) and consequently accurate predictions of fish populations and their variability are crucial to the fishing industry (Gaichas, 2008). Fisheries management usually focuses on a single species, and fish stocks of several species have previously crashed in various locations due to over fishing (e.g., Roughgarden and Smith, 1996). Accurate predictions of population dynamics are essential because this will determine what the MSY is for a given fish population at a given time point (often calculated on an annual basis) to avoid population decline or collapse. Predictive models have generally been based on observations of variability in previous catches, sometimes supported by population monitoring. However, the focus on maximum sustainable yield poses risks, as such predictive models are not perfect, and taking perceived MSY does not allow for non-linearity and stochasticity in population dynamics (see Gaichas, 2008 for a detailed critique of MSY).

Despite long-standing management practice of ignoring non-linear dynamics and many ecosystem interactions (e.g., Hilborn and Walters, 1992), there is now a greater focus on developing non-linear models for fish population prediction. For example, the use of linear univariate time-series models (see, e.g., Czerwinski et al., 2007) were common predictive methods, but now non-linear models, for example artificial neural networks, are gaining in popularity. In order to evaluate the effectiveness of different predictive modelling approaches, Czerwinski et al. (2007)

compare predictive capabilities of both types of model ('linear' autoregressive moving average models versus 'non-linear' artificial neural network models) in relation to Pacific halibut (*Hippoglossus stenolepis*) catch per unit effort (CPUE), based on daily catches from May–September 1998–2003. The autoregressive moving average models produced population forecasts with explained variance levels (R^2) of between 9.0–39.8 percent and standard errors of prediction of around 50 percent. The artificial neural network models produced explained variance levels (R^2) of between 37.2–91.0 percent and standard errors of prediction of around 16 percent. In this case, the non-linear models were far more effective in predicting halibut CPUE, although there were notable differences between the explained variances of the different non-linear models. The difficulties faced by both types of model in explaining variance levels were probably due to limitations in the catch data used to calibrate and validate the models, which is a common problem in predictive modelling of resources. Nevertheless, investigations such as those conducted by Czerwinski et al. (2007) indicate that incorporating non-linearity in catch and population forecasting is essential, as fisheries may be particularly prone to rapid threshold transitions in population dynamics, and fishing to MSY can potentially lead to sudden stock collapses.

Pitchford et al. (2007) take this principle further and present a theoretical system where catch limits are set based on a simple deterministic model constructed using differential linear equations. Based on such a model, MSY is a fixed value and is sustainable, and so no further evaluation is needed. They then introduce elements of stochasticity and non-linearity to the recruitment and catch variability within the system (including human error), demonstrating that using a fixed MSY value would eventually lead to population collapse. They then consider two further management techniques: (i) harvest control, whereby if the population falls below a critical value (e.g., half the total carrying capacity), the yield is reduced to allow some recovery; and (ii) marine protected areas (MPA) wherein yield is taken from only half of the fish population (essentially leaving a protected area of ocean which is not harvested). This effectively guarantees some population survival, creates a buffer against variability, and allows continued resource harvest (figure 25.7). Testing the validity of these sorts of models using empirical data is important for the future of this industry, as is further research to investigate the significance of the many biotic and abiotic interactions in fishery ecosystems (e.g., Gertseva and Gertsev, 2006). Above all, management should allow for the maintenance of sustainable populations and resilient ecosystems.

Managing Variability

As noted above, variability is essential to ecosystem functioning and resilience and should not be ignored by resource managers. Indeed, it can now be regarded as an indicator of ecosystem health and can be a positive aspect of management. This is evidenced by the concept of 'natural variability', which can be best defined as 'the ecological conditions, and the spatial and temporal variation in these conditions, that are relatively unaffected by people, within a period of time and geographical area appropriate to an expressed goal' (Landres et al., 1999, p. 1180). This concept is an explicit acknowledgement that an understanding of ecosystem variability prior to anthropogenic disturbance is essential for sustainable resource management. The concept of natural variability has two main principles: (i) that past conditions and

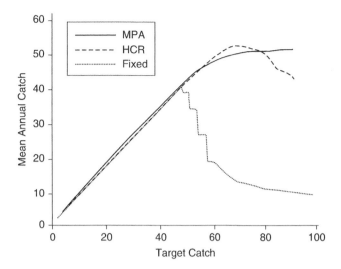

Figure 25.7 Simulation of mean annual catch in a marine fishery over a 100-year period, based on different target catches used in three management strategies: fixed (where the same catch is taken each year), harvest control (where catches are reduced if the population falls below a critical threshold) and marine protected areas (where yields are taken from only half of the population). Based on these models, only the marine protected areas management strategy would result in sustainable management. Taken from Pitchford et al. (2007).

processes provide context and guidance for current management of ecosystems; and (ii) spatio-temporal variability driven by disturbance is an intrinsic part of ecosystems. It therefore supports non-equilibrium paradigms over short temporal scales. As a management technique, the use of past, or undisturbed, conditions as reference templates for management and restoration was suggested by Leopold et al. (1963), in the form of natural vignettes which imply ecosystem integrity. Since then, greater appreciation, measurement, and quantification of variability has refined the simplistic idea that some natural, non-anthropogenically disturbed ecosystem state can be used as a benchmark for management. Nevertheless, understanding past ecosystem dynamics offers a useful way of predicting and reducing current management impacts, and the principles of natural variability certainly have a place in resource management.

It is important to acknowledge the chaotic, non-linear, and above all, unpredictable nature of ecosystems and their resources, and to avoid management practices that force ecosystems outside of their range of natural variability. This includes oversimplifying the system, restricting natural disturbances, or intensifying the frequency, severity, and duration of disturbances. There is increasing evidence that the principles of natural variability are being incorporated into management of some ecosystem resources, although there is still much progress to be made. A priority is to incorporate variability within biodiversity conservation and restoration, which is a relatively recent form of ecosystem resource management. The absence of objective scientific reasoning, pre- and post-project monitoring, and hard scientific training of practitioners, has resulted in many attempts at conservation and restoration around the globe being based on an idealised concept of steady-state ecosystems that have only been recorded within the last century (see, e.g., Pullin and Knight, 2001). For example, the aim of heathland conservation in the UK is to maintain heathland by preventing the encroachment of scrub. This approach mistakenly associates the halting of successional dynamics with the preservation of a 'rare' or 'endangered' habitat, and is both unnatural and doomed to failure. The same is true of species conservation targeting single species in isolation. Such strategies are unlikely to succeed if associated ecosystem structure, functioning and variability are not conserved or restored at the same time. The artificial preservation of habitats and species at a given, anthropogenically conceived state should not be the main target of conservation and restoration. Instead the return of ecosystem dynamism, functioning and integrity should be paramount (e.g., Murphy, 1989; van Kooten, 1998).

Future Challenges

Key future challenges in quantifying ecosystem resource variability and improving prediction include:

- integrating the principles of ecosystem variability, dynamism, and non-equilibrium states into ecosystem resource and service management, including an economic acceptance of the value of such variability;
- allowing for some level of natural variability in ecosystem management, including the acceptance of associated risks;
- relaxing management where appropriate to allow for the restoration of ecosystem variability and dynamism;
- developing and validating non-linear ecosystem and resource models to increase predictive potential;
- utilising new geographical technologies and landscape ecological metrics to obtain and analyse data regarding variability of ecosystem structure and process over a range of spatial and temporal scales (particularly broad scales), to inform modelling and management; and
- obtaining comprehensive measurements of biotic and abiotic interactions across whole ecosystems to increase confidence in our understanding, despite the cost and effort involved.

Conclusions

Our early understanding of ecosystems as simple, mechanistic, and predictable systems, conceptualised at the start of the 20th century, has recently developed to a more detailed understanding of the complex, chaotic, non-linear, and above all *unpredictable* nature of ecosystems and their resources. In particular, the variability of patterns and processes within ecosystems is problematic to quantify due to their complexity and our insufficient understanding of the many interactions between ecosystem components, their strengths, and their significance.

This lack of understanding creates problems for ecosystem and resource management, which requires predictability. Management is further complicated by uncertainty over which spatio-temporal scale is most relevant to both the specific resource and the wider ecosystem. With increasing evidence to support the importance of ecosystem dynamism and variability, these characteristics can now be regarded as resources with intrinsic value and which do ideally not need managing or restricting, but predicting. Consequently, although predictive models are useful for management, the inherent variability of ecosystems needs to acknowledged and incorporated, and the associated risk/uncertainty accounted for.

BIBLIOGRAPHY

Adachi, M., Bekku, Y. S., Konuma, A., Kadir, W. R., Okuda, T. and Koizumi, H. (2005) Required sample size for estimating soil respiration rates in large areas of two tropical forests and of two types of plantation in Malaysia. *Forest Ecology and Management*, 210, 455–59.

Arrow, K., Dailey, G., Dasgupta, P., Levin, S., Maler, K.-G., Maskin, E., Starrett, D., Sterner, T and Tietenberg, T. (2000) Managing ecosystem resources. *Environmental Science and Technology*, 34, 1401–6.

Basso, B., Ritchie, J. T., Pierce, F. J., Braga, R. P. and Jones, J. W. (2001) Spatial validation of crop models for precision agriculture. *Agricultural Systems*, 68, 97–112.

Batchelor, W. D., Basso, B. and Paz, J. O. (2002) Examples of strategies to analyze spatial and temporal yield variability using crop models. *European Journal of Agronomy*, 18, 141–158.

Benson, K. R. (2000) The emergence of ecology from natural history. *Endeavour*, 24(2), 59–62.

Bocchi, S. and Castignanò, A. (2007) Identification of different potential production areas for corn in Italy through multitemporal yield map analysis. *Field Crops Research*, 102, 185–97.

Bond, N. R. and Lake, P. S. (2003) Local habitat restoration in streams: constraints on the effectiveness of restoration for stream biota. *Ecological Management and Restoration*, 4(3), 193–98.

Boote, K. J., Jones, J. W., Hoogenboom, G. and Pickering, N. B. (1998) The CROPGRO for grain legumes. In G. Y. Tsuji, G. Hoogenboom and P. K. Thornton (eds), *Understanding Options for Agricultural Production*. Boston: Kluwer Academic Publishers, pp. 99–128.

Brown, J. H., Gupta, V. K., Li, B. L., Milne, B. T., Restrepo, C. and West, G. B. (2002) The fractal nature of nature: power laws, ecological complexity and biodiversity. *Philosophical Transactions of the Royal Society B – Biological Sciences*, 357, 619–26.

Caseldine, C. and Fyfe, R. (2006) A modelling approach to locating and characterising elm decline/landnam landscapes. *Quaternary Science Reviews*, 25, 632–44.

Clements, F. E. (1916) *Plant Succession: An Analysis of the Development of Vegetation*. Washington, DC: Carnegie Institution.

Czerwinski, I. A., Gutiérrez-Estrada, J. C. and Hernando-Casal, J. A. (2007) Short-term forecasting of halibut CPUE: linear and nonlinear univariate approaches. *Fisheries Research*, 86, 120–28.

Daily, G. C., Alexander, S., Ehrlich, P. R., Goulder, L., Lubchenco, J., Matson, P. A., Mooney, H. A., Postel, S., Schneider, S. H., Tilman, D. and Woodwell, G. M. (1997) Ecosystem services: benefits supplied to human societies by natural ecosystems. *Issues in Ecology*, 1(2), 1–18.

Dale, V. H., Pearson, S. M., Offerman, H. L. and O'Neill, R. V. (1994) Relating patterns of land-use change to faunal biodiversity in the central Amazon. *Conservation Biology*, 8(4), 1027–36.

Darge, R. C. and Kneese, A. V. (1980) State liability for international environmental degradation – an economic perspective. *Natural Resources Journal*, 20(3), 427–50.

Delcourt, H. R. and Delcourt, P. A. (2005) The legacy of landscape history: the role of paleoecological analysis. In J. Wiens and M. Moss (eds), *Issues and Perspectives in Landscape Ecology*. Cambridge: Cambridge University Press, pp. 159–66.

Dobson, A. (2005) Monitoring global rates of biodiversity change: challenges that arise in meeting the Convention on Biological Diversity (CBD) 2010 goals. *Philosophical Transactions of the Royal Society B – Biological Sciences*, 360(1454), 229–41.

Gaichas, S. K. (2008) A context for ecosystem-based fishery management: developing concepts of ecosystems and sustainability. *Marine Policy* 32(3), 393–401.

Gertseva, V. V. and Gertsev, V. I. (2006) A conceptual model of fish functional relationships in marine ecosystems and its application for fisheries stock assessment. *Fisheries Research*, 81, 9–14.

Ghil, M., Kimoto, M. and Neelin, J. D. (1991) Nonlinear dynamics and predictability in the atmospheric sciences. *Reviews of Geophysics*, 29, 46–55.

Gleick, J. (1987) *Chaos: Making a New Science*. New York: Penguin Books.

Golley, F. B. (1993) *A History of the Ecosystem Concept in Ecology*. London: Yale University Press.

Grumbine, R. E. (1994) What is ecosystem management? *Conservation Biology*, 8(1), 27–38.

Hilborn, R. and Walters, C. J. (1992) Quantitative fisheries stock assessment: choice, dynamics, and uncertainty. New York: Chapman and Hall.

Itzstein-Davey, F., Taylor, D., Dodson, J., Atahan, P. and Zheng, H. (2007) Wild and domesticated forms of rice (Oryza sp.) in early agriculture at Qingpu, lower Yangtze, China: evidence from phytoliths. *Journal of Archaeological Science*, 34, 2101–8.

Jones, C. A. and Kiniry, J. R. (1986) *CERES-Maize: A Simulation Model of Maize Growth and Development*. College Station: Texas A&M University Press.

Jones, J. W. and Ritchie, J. T. (1996) Crop growth models. In G. J. Hoffman, T. A. Howell and K. H. Soloman (eds), *Management of Farm Irrigation Systems*. St. Joseph, MI: ASAE.

Jørgensen, S. E., Costanza, R. and Xu, F.-L. (2005) *Handbook of Ecological Indicators for Assessment of Ecosystem Health*. Boca Raton: CRC Press.

Kay, J. J. (2000) Ecosystems as self-organizing holarchic open systems: narratives and the second law of thermodynamics. In S. E. Jørgensen and F. Müller (eds), *Handbook of Ecosystem Theories and Management*. Boca Raton: CRC Press, pp. 135–60.

Kerr, J. T. and Ostrovsky, M. (2003) From space to species: ecological applications for remote sensing. *Trends in Ecology and Evolution*, 18(6), 299–305.

Landres, P. B., Morgan, P. and Swanson, F. J. (1999) Overview of the use of natural variability concepts in managing ecological systems. *Ecological Applications*, 9(4), 1179–88.

Leopold, A. S., Cain, S. A., Cottom, C. M., Gabrielson, I. N. and Kimball, T. L. (1963) The wildlife management in the national parks. *Transactions of the North American Wildlife and Natural Resources Conference*, 28, 28–45.

Linton, R. M. (1970) *Terracide: America's Destruction of Her Living Environment*. Boston, Little Brown and Company.

Manson, S. M. (2001) Simplifying complexity: a review of complexity theory. *Geoforum*, 32, 405–14.

May, R. M. (1987) Chaos and the dynamics of biological populations. *Proceedings of the Royal Society of London A*, 413, 27–44.

Murphy, D. D. (1989) Conservation and confusion: wrong species, wrong scale, wrong conclusions. *Conservation Biology*, 3(1), 82–4.

Osenberg, C. W., Schmitt, R. J., Holbrook, S. J., Abu-Saba, K. E. and Flegal, A. R. (1994) Detection of environmental impacts: natural variability, effect size and power analysis. *Ecological Applications*, 4(1), 16–30.

Pahl-Wostl, C. (1995) *The Dynamic Nature of Ecosystems: Chaos and Order Entwined*. Chichester: John Wiley & Sons, Ltd.

Petts, G. (2001) Sustaining our rivers in crisis: setting the international agenda for action. *Water Science and Technology*, 43(9), 3–16.

Petts, G. E. and Gurnell, A. M. (2005) Dams and geomorphology: research progress and future directions. *Geomorphology*, 71, 27–47.

Pickett, S. T. A., Kolasa, J. and Jones, C. G. (1994) *Ecological Understanding: The Nature of Theory and the Theory of Nature*. San Diego: Academic Press, Inc.

Pitchford, J. W., Codling, E. A. and Psarra, D. (2007) Uncertainty and sustainability in fisheries and the benefit of marine protected areas. *Ecological Modelling*, 207, 286–292.

Power, J. P. and Campbell, B. M. S. (1992) Cluster-analysis and the classification of medieval demesne-farming systems. *Transactions of the Institute of British Geographers*, 17(2), 227–45.

Pullin, A. S. and Knight, T. M. (2001) Effectiveness in conservation practice: pointers from medicine and public health. *Conservation Biology*, 15, 50–54.

Richardson, G. (2005) The prudent village: risk pooling institutions in medieval English agriculture. *Journal of Economic History*, 65(2), 386–413.

Rodrigues-Iturbe, I., Entekhabi, D. and Bras, R. L. (1991) Nonlinear dynamics of soil moisture at climate scales: 1. stochastic analysis. *Water Resources Research*, 27(8), 1899–1906.

Rodwell, J. S. (1991) *British Plant Communities, Volume 1: Woodlands and Scrub*. Cambridge: Cambridge University Press.

Spaliviero, M. (2003) Historic fluvial development of the Alpine-foreland Tagliamento River, Italy, and consequences for floodplain management. *Geomorphology*, 52, 317–33.

Sparks, R. E. (1995) Need for ecosystem management of large rivers and their floodplains. *BioScience*, 45(3), 168–82.

Tan, C. O., Özesmi, U., Beklioglu, M., Per, E. and Kurt, B. (2006) Predictive models in ecology: comparison of performances and assessment of applicability. *Ecological Informatics*, 1, 195–211.

Tansley, A. G. (1935) The use and abuse of vegetational concepts and terms. *Ecology*, 16(3), 284–307.

Turchin, P. (1993) Chaos and stability in rodent population dynamics: evidence from nonlinear time-series analysis. *Oikos*, 68(1), 167–72.

Qi, Y., Xu, M. and Wu, J. (2002) Temperature sensitivity of soil respiration and its effects on ecosystem carbon budget: nonlinearity begets surprises. *Ecological Modelling*, 153, 131–42.

Van der Nat, D., Tockner, K., Edwards, P. J., Ward, J. V. and Gurnell, A. M. (2003) Habitat change in braided rivers (Tagliamento, NE-Italy). *Freshwater Biology*, 48, 1799–1812.

Van Kooten, G. C. (1998) Economics of conservation biology: a critical review. *Environmental Science and Policy*, 1, 13–25.

Van Loon, A. J. (2003) The dubious role of man in a questionable mass extinction. *Earth-Science Reviews*, 62, 177–86.

Walters, C. and Maguire, J.-J. (1995) Lessons for stock assessment from the northern cod collapse. *Reviews in Fish Biology and Fisheries*, 6(2), 125–37.

Warner, S., Feinstein, M., Coppinger, R. and Clemence, E. (1996) Global population growth and the demise of nature. *Environmental Values*, 5(4), 285–301.

Wigley, T. M. L. and Raper, S. C. B (1990) Natural variability of the climate system and detection of the greenhouse effect. *Nature*, 344, 324–27.

Zanoni, L., Gurnell, A. M., Drake, N. and Surian, N. (2008) Island dynamics in a braided river from analysis of historical maps and air photographs. *River Research and Applications*. DOI: 10.1002/rra.1086.

Chapter 26

Environment and Development

Tom Perreault

Introduction

'Environment and development' has long been a cornerstone of environmental geography. It is an inherently integrative field that incorporates a broad diversity of theoretical and methodological approaches, many of which are covered elsewhere in this volume, such as political ecology, environmental governance, sustainability, land use/land cover change, ecological modernisation, and environmental conservation. Moreover, geographers concerned with the relationship between environment and development draw heavily on economic and sociological theories of development as well as the fields of biogeography and ecology. As such, this chapter will examine environment and development not as a stand-alone subfield of geography, but rather as one that necessarily brings together diverse intellectual approaches trained on an array of social and environmental problems.

Since the 1970s, the twin themes of environment and development have become increasingly prominent in geography, and have continued to evolve in focus and scope. At its core, the field of environment and development geography is concerned with two fundamental realities: (i) social groups – households, rural communities, cities or nation states – are dependent upon nature and natural resources for their survival and welfare, and (ii) the practices and institutional arrangements social groups employ to ensure survival and welfare in turn impact environmental quality and the functioning of geo-ecological systems. This reciprocal relationship between resource use and environmental conditions, and the implications that these processes have for social welfare, have long been a focus of study for geographers. This vital area of research has undergone something of a sea change since the 1960s and 1970s, when dominant thinking in both development and conservation viewed 'local' and 'traditional' resource use practices – i.e., those strategies employed by the rural poor in the global South – as 'backward' and deleterious both to environments and to national development. In the 1970s and 1980s, critical scholarship began to challenge these views, as discussed below.

Following this introduction, the chapter traces the roots of the environment and development tradition in geography, emphasising its integrative nature. This is followed by a consideration of three themes that run through the literature in environment and development: conservation, livelihood, and sustainable development. The chapter does not propose new definitions of these concepts, nor can it provide a thorough review of the vast literatures on these topics. Rather, it attempts to illustrate, on the one hand, the importance of these concepts to the environment and development tradition in geography, and, on the other hand, the ways that these concepts relate to one another in the context of development and conservation initiatives.

Roots of the Environment-Development Tradition in Geography

The ambiguities of development and environment

If, as Raymond Williams (1976) suggests, 'nature' is the most complicated word in the English language, then 'development' cannot be far behind. Peet and Hartwick (1999, p. 1) call it a 'founding belief of the modern world' connoting progress, modernity and democratic values. It is a word that carries at once the aspirations of the poor and the designs of corporate elites. Indeed, it is this conceptual ambiguity – its capacity to be invested with distinct, even contradictory meanings – that makes development an 'arena for cultural contestation' (Escobar, 1995, p. 15). As Adams (2001, p. 6) notes, the word 'development' is used both descriptively, to explain economic, socio-cultural and environmental transformation, and normatively, as a prescription for how economies, societies and environments *should* be transformed. These distinct meanings are often conflated, contributing to the conceptual confusion and analytical complexity of the concept. In their insightful analysis of development discourses, Cowen and Shenton (1996) observe that discussions of development frequently conflate two related, but distinct, meaning of the term. On the one hand, development is often used to denote immanent macro-scale social and environmental transformations associated with capitalist expansion. As Bebbington (2002) notes, this is development understood as structural change: development at the scale of societies, nation states and regional economies. On the other hand, according to Cowen and Shenton, development can describe specific and intentional interventions – development projects and programmes – organised for particular (and usually limited) ends. This, then, is development as practice, at the scale of the local and the personal. Crucially, these two meanings of development are distinct, and their relationship is neither direct nor determinant. Specific development interventions may indeed promote broader capitalist expansion (for instance, projects aimed at market integration or commodity production). But projects that hold the language and practices of development in common may have quite different intentions and outcomes, and may even work to counter the negative effects of capitalism. For example, projects aimed at food self-sufficiency, rural healthcare, improved literacy or access to education, may serve to empower social groups opposed to particular objectives of state and capital (Perreault, 2003). It is worth noting that the very concept of 'development' is derived from biological understandings of growth, which serves to naturalise it, and give it an air of inevitability. Geographers must be attentive, then, to the divergent meanings ascribed to development, and the social and environmental implications of development discourses and practices.

It is crucial to recognise that development is always and necessarily an environmental project. In the most basic sense, processes of production (whether capitalist, socialist or subsistence-oriented) rely on nature to provide inputs of raw materials and ecological functions (e.g., metals, water, wood, fish, soil generation, photosynthesis) without which the productive activities we call development simply could not occur. Moreover, natural systems provide a sink for the by-products of these activities, as wastes are released into the atmosphere, waterways, or soils (Emel and Bridge, 1995). Apart from these most basic functions of source and sink, nature – both in its material forms and our varied understandings of it – figures importantly into manifold accumulation strategies, from shipping to ecotourism, hydroelectricity generation to housing development (Katz, 1998). Even for activities in which environmental management is neither explicit nor intentional, the unavoidable reality is that development carries important environmental implications. It is this inescapable reliance on the earth's natural systems, and the dialectical relationship between society and nature, that lay at the heart of geographical approaches to environment and development.

This recognition begs the question of what is meant by 'environment,' 'nature,' and 'natural resources.' An enormous body of literature addresses these questions, and a full review cannot be provided here. It bears recognising, however, that these terms are highly contested and cannot be taken as unproblematic. Scholars working in such disparate traditions as Marxism, post-structuralism and critical science studies all hold as axiomatic the social construction of nature. Though differing in their views of just how nature is constructed – whether this is primarily an epistemological or ontological construction, whether it is to be understood as discursive/ textual construction (Hacking, 1999), historical-materialist production (Smith, 1984), or as an artifact of the social processes of science (Demeritt, 2002) – these perspectives converge on the idea that any epistemological separation between nature and society is deeply problematic. As such, taken-for-granted categories such as 'wilderness' and 'natural resources,' that figure enormously in environment and development literature and practice require critical interrogation, and an unpacking of the sedimented human histories that are bound up with such understandings (Williams, 1973, see also Harvey, 1974; Cronon, 1996).

These epistemological complexities bear crucial implications for environment and development geography. If we take seriously the suggestion that such conceptual categories as 'development', 'nature', 'natural resources' and 'environment' cannot be taken for granted, and that their meanings are socially specific, temporally variable and always subject to contestation, then our assumptions as scholars and practitioners must similarly be subject to continual critical interrogation. This, then, points to the recognition that the relationship between environment and development is deeply political, and cannot be assessed in empirical, scientific terms alone.

Environment and development geography

We may reasonably inquire as to why, and to what effect, the concepts of 'environment' and 'development' are so commonly paired in geographical scholarship. On the one hand, the discipline of geography has long been concerned with practical application and 'relevant' research (Bebbington, 1994; Staheli and Mitchell, 2005). The subfield of human ecology, still influential in much geographical research in development, was founded on pragmatist ideals (Whyte 1986) – a pragmatism that at times has come under fire as naïve and under-theorised (Watts, 1983a). At its core, however, human ecology's pragmatic concerns for societal response to environmental change and risk, and the ways these shape human vulnerability to environmental conditions, remains a theme of fundamental importance to geographical approaches to environment and development. The similarly practical nature of geographical work in land-use/land-cover change, resource geography, and urban and regional planning is fundamentally concerned with the relationship between society and environment, and strongly influences geographical work in environment and development. Methodologically, GIS and remote sensing technologies are surely the geographical techniques most highly sought after by NGOs and state agencies working in development today. Thus, the enduring concern in geography for applied research and theoretical relevance lends itself well to the professional field and real-world concerns of environmental management and development policy.

On the other hand, it can convincingly be argued that the contemporary pairing of environment and development grew out of tensions that arose in the 1960s and 1970s between the emerging environmental movement in the global North and concerns about economic and social development (primarily, though not exclusively, in the global South). Poverty and underdevelopment were largely viewed as problems of the 'Third World' a view reinforced by the so-called Brundtland Report (WCED, 1987), which emphasised ecological degradation as largely a result of Third World poverty, as opposed to First World affluence. In the context of an expanding Cold War apparatus of international development (Escobar, 1995), North American and European intervention in the global South sought to replicate the model of industrial capitalism under the guise of 'modernisation', while dismissing pre-capitalist or communal forms of economic production and resource management as inefficient and 'backward' (Rostow, 1960). For instance, it was widely reported that smallholder farming practices in Nepal led to widespread deforestation and soil erosion, with potentially disastrous implications for downstream populations in India and Bangladesh (see Thompson et al., 1986). Similarly, as Turner (1993) reports, mal-adapted livestock management and associated overstocking were commonly thought to cause grassland degradation and desertification in sub-Sahelian Africa, in a classic 'tragedy of the commons' scenario (Hardin, 1968). In these cases and others, conventional wisdom held that, if traditional resource use practices were the problem, the logical solution was the modernisation of environmental management and development policy – an assumption that fit well with Rostowian notions of linear 'stages of growth' development and Green Revolution agricultural technologies supported by international financial institutions and the United Nations (Rostow, 1960; see also Escobar 1995; Peet and Hartwick 1999).

The industrial model of development promoted by US and European governments, and financed by the World Bank and other multilateral institutions (as well as private banks) was frequently premised on radical environmental transformation: the conversion of rainforest to cattle pasture, the damming of rivers, rapid urbanisation, the widespread use of chemical pesticides and fertilizers in agriculture. The often disastrous environmental implications of such practices were soon apparent, and by the 1970s and 1980s came under steady criticism from an increasingly influential and well-institutionalised environmental movement in the USA and Europe. These concerns emerged most visibly in the discourse of sustainable development, first proposed by the International Union for the Conservation of Nature

(IUCN) in its 1980 World Conservation Strategy report (IUCN 1980), and later enshrined in the World Commission on Environment and Development's *Our Common Future*, the so-called Brundtland Report, released in 1987 (WCED 1987). The concept of sustainable development, detailed elsewhere in this volume and briefly outlined below, proved particularly fertile ground for geographers concerned with the relationship between environment and development.

By the late 1970s, received wisdom regarding the salutary effects of industrial modernisation and scientific environmental management had to a large extent been turned on its head. Much geographical scholarship in the environment and development tradition of the 1980s – influenced by ascendant North Atlantic environmental movements, as well as 20 years of skepticism towards ideologies of scientific modernisation from environmentalist, Marxist and poststructuralist positions – critiqued prevailing ideas regarding both development and environment. This new wave of scholarship, much of it falling under the rather loose heading of political ecology, exposed the socially and environmentally negative effects of dominant development practices, while highlighting the many benefits of 'traditional' environmental management practices (Robbins, 2004). If foundational texts can be identified for political ecology, Watts' *Silent Violence* (1983b), Blaikie and Brookfield's *Land Degradation and Society* (1987), and Hecht and Cockburn's *Fate of the Forest* (1989) – each centrally concerned with questions of environment and development – are surely among them. Indeed, it is no mere coincidence that critical approaches to environment and development came to the fore at the same time that sustainable development emerged as a prominent discourse and policy objective among international institutions (if not in the halls of power in Washington and London). To a considerable extent, political ecologists writing in the 1980s were critiquing the same processes, institutions, and ideas as their contemporaries in the IUCN and the UN, though they typically did so from a more explicitly political, and politically radical, position and only rarely from within policy-making institutions. Insofar as these geographers critiqued conventional wisdom regarding, for instance, drought and famine, soil erosion, and deforestation, they raised vital questions about the functioning of international development programmes and the implications these have for environmental degradation, human welfare, and social justice (Adams, 2001; Robbins, 2004).

Key Themes: Conservation, Livelihood and Sustainability

Rather than attempting a comprehensive review of the environment and development literature, the chapter now turns to three themes that are central to environment and development geography: conservation, livelihoods, and sustainability. By directing analytical attention to questions of resource use, management and governance, these concepts help shed light on the complexities and contradictions inherent in the field. Though fully aware of the limitations of this brief discussion, I hope that it will help illuminate the ways in which geographers have approached these three interconnected core themes.

Nature conservation and protected areas

Questions of environmental conservation, and in particular the management of protected areas, lie at the heart of the environment and development tradition in geography. In the 1960s and 1970s concern over deteriorating environmental conditions grew, as oil spills, industrial accidents, and worsening air and water quality made plain the ecological effects of unchecked industry and resource extraction, and fueled a florescence of environmental activism and scholarship (Lean et al. 1990). The growing focus on environmental issues in turn led to a period of institutional reorganisation in Europe, the USA and Canada, as governments established agencies and legislation regulating, *inter alia*, air and water quality, pesticides application, industrial waste remediation, and resource-extraction activities.

During this period, neo-Malthusian alarmism about rapid population growth in the developing world and its assumed negative impacts on natural environments held sway among many ecologists and environmental activists (e.g., Ehrlich, 1972; Hardin, 1968; 1974). Deforestation, desertification, soil erosion and declines in wildlife populations were among the environmental problems blamed on 'overpopulation' and the mismanagement of natural resources which were commonly believed to characterise Third World societies. Such concerns became widespread among environmental NGOs and bilateral development agencies in Europe and North America, and by the mid-1970s were becoming increasingly linked to lending practices and development programmes. As the international lending boom of the 1970s turned into the debt crisis of the 1980s, the establishment of protected areas and other environmental programmes commonly became conditions for continued lending by the World Bank and other multilateral lending institutions. The 'greening' of the World Bank continued apace (Goldman, 2005). By 1992, when the World Bank focused its annual development report on the theme of development and environment to coincide with the UN's World Conference on Environment and Development, it could be said that environmental concerns had secured a place in mainstream development thinking (World Bank, 1992). By adopting the rhetoric of sustainable development and insisting that environmental conservation and economic growth could be made to complement one another, however, the World Bank and the UN only raised further questions about how these agencies conceived of development and nature, and their relationship to social justice.

In part as a result of the growing influence of the institutions of international development and Northern environmentalism, the 1970s and 1980s saw a dramatic upsurge in the number of national parks, wildlife reserves and other protected areas established in the developing world. Currently, some 6.9 percent of the earth's surface falls under protected area designation, a figure that jumps to over 10 percent if the IUCN's least-restrictive categories of protected areas are included (Wilhusen et al., 2003). While the oldest (and some of the largest) protected areas in the world are in the temperate countries of the 'First World' (the USA, Canada, New Zealand, Great Britain), international conservation efforts have, since the 1970s, focused especially on the biodiversity 'hotspots' found in the tropics regions of Latin America, Africa and Asia. Ecologist Norman Myers suggests that some 25 such hotspots may contain as much as 44 percent of vascular plant species and 35 percent of vertebrate species, but cover just 1.4 percent of earth's surface (Myers et al., 2000). The goals of protecting biodiversity hotspots has concentrated many protected areas and other conservation efforts in countries of the global South, many of which have relatively large resource-dependent rural populations, high levels of poverty and political-economic instability. These conditions pose a distinct set of conservation problems, which set them apart from most protected areas in the global North (Wilhusen et al., 2003).

The earliest national parks and protected areas – in both North and South – were established by elites, often colonial or white settler governments, with little concern for local peoples or their traditional resource use practices (Neumann, 2004). Local populations were in many cases strictly prohibited from residing in and using the resources of protected areas and were seen as the greatest threat to conservation objectives. Resource use activities were invariably coded according to the goals of conservation and recreation: wood gathering was redefined as theft, grazing as trespassing. Similarly, hunting was deemed 'poaching' when done for subsistence, but considered scientific management or economic benefit (or manly sport) when done by park managers or tourists (Adams, 2001). Inevitably, this view led to numerous clashes with local populations, many of which were greatly impacted by conservation interventions, and the concomitant reduction or outright loss of customary resource and territorial rights (Wilhusen et al., 2003).

Pathbreaking work by Peluso (1992; 1993) examined the coercive practices of an authoritarian Indonesian state, which saw in conservation programmes an opportunity to spatially segregate and politically control populations it considered unruly or undesirable. Relying on a mix of archival work, policy analysis and ethnographic approaches, Peluso's work revealed the dark side of seemingly benign conservation policies: that in some cases national parks were more effective at protecting vested political interests than they were at protecting endangered species and natural habitats. Similarly, Neumann (1998) has employed archival analysis to demonstrate that Tanzania's national parks system is rooted historically in the establishment of colonial-era game reserves and served the goals of the state's claims to sovereignty over territory. As he asserts elsewhere, Tanzanian conservation is part of the modern state's civilising mission: 'Containment and control of nature in conservation territories was inseparable from the colonising state's efforts to control its African subjects and ultimately create a new kind of person: civilised, productive, and observable' (Neumann, 2004, pp. 203–4). As with Yellowstone and Yosemite National Parks in the USA, the establishment of these protected areas was closely associated with spatialised racial hierarchies that excluded native peoples from the 'pleasuring grounds' of the dominant white settler population (Kosek, 2004). Like Zimmerer (2004), Kosek (2004) employs discourse analysis to disentangle the highly contested narratives of environmental conservation. The social and ecological implications of this 'fortress conservation' strategy came to be questioned by international agencies, which increasingly saw it as counter-productive to the goals of both conservation and development.

This shift came about in part as a result of the 1972 UN Conference on the Human Environment in Stockholm, which, in addition to establishing the United Nations Environment Programme, issued a declaration on human rights in relation to the environment. This was followed in 1975 by an IUCN resolution declaring that parks and protected areas should not be established without consulting local and indigenous populations (Fortwangler, 2003). The 1980 IUCN World Conservation Strategy (IUCN, 1980), a key document in sustainable development thinking, asserted that development could (and must) be reoriented to promote conservation and that conservation could in turn meet the needs of poor people (Adams, 2001). In 1982, the World Congress on Parks and Protected Areas encouraged greater local participation in the management of protected areas and other natural resources (McNeely and Miller, 1984). These efforts led to a diversity of approaches intended to incorporate local populations into protected areas management. Such approaches, known variously as 'community-based conservation', 'integrated conservation and development programmes', or 'conservation-with-development', sought to integrate the objectives of environmental conservation and local economic and social development. These efforts gave impetus to a variety of modified protected areas models, such as the biosphere reserve model, which allow for zones of sustainable resource use (Lean et al., 1990).

'Participation' thus became a watchword of international conservation policies in the 1980s and 1990s, as national and international NGOs – which flourished with the emergence of neoliberal economic policies in the 1980s – sought to implement conservation programmes that met the needs of nature protection and local livelihoods alike (Fortwangler, 2003). Many transnational environmental NGOs sought to incorporate local populations into the management of protected areas, wildlife, forests and other resources. Such efforts met with mixed results, however. Conservation, organisations and state environmental agencies were often frustrated by seemingly uncooperative local populations who failed to adopt 'appropriate' environmental practices. For their part, indigenous peoples and other rural populations frequently saw state agencies and environmental organisations as a threat to traditional resource rights and management practices (Sundberg, 1998). As Chapin (2004) has demonstrated, such frustrations have led many international environmental organisations to back away from the conservation-with-development approach, adopting once again a 'fortress conservation' approach that excludes resource uses by local populations, who are seen as a threat to the goals of biodiversity conservation. Retrenchment of this sort raises questions regarding the apparent contradictions between nature conservation and resource-dependent rural livelihoods, another key theme of environment and development geography, to which this chapter now turns.

Livelihoods

The concept of livelihood has long been a mainstay of environment and development geography, and has figured as a central analytical category for cultural ecology, political ecology and development geography. 'Livelihood' has been conceptualised variously as 'entitlements' (Leach et al., 1999), 'strategic assets' (Chambers, 1995) and 'capitals and capabilities' (Sen, 1997; Bebbington, 1999). The use and acceptance of the term far exceeds the level of agreement as to its exact meaning, though as Bebbington (2004) notes, there is considerable convergence among various approaches to livelihood, even if there is disagreement as to the precise language used to describe the term. Irrespective of exact definition and analytical metric, 'livelihood' implies the making and re-making of social life, maintaining some semblance of continuity and security. The notion of livelihood connotes relatively localised and immediate strategies of subsistence and use of resources (broadly conceptualised), aimed at ensuring the stability of life and lifeways. Indeed, despite its ambiguity, the enduring utility of livelihood as an analytical concept rests in large part on the way in which it highlights the everyday practices employed by individuals, households and communities to 'make themselves a living using their capabilities and their tangible and intangible assets' (de Haan, 2000, p. 343).

Geographers employ diverse methodological approaches in ways that shed light on the complex relations between livelihood and resource management, combining, for example, archival and ethnographic research with geographic information systems (GIS) and vegetation analysis (e.g., Robbins, 2001; McCusker and Weiner, 2003). Additionally, geographers have contributed in particularly fruitful ways to the literature on smallholder agricultural production and crop management, bringing empirical rigor and diverse methodological approaches to bear on questions of peasant livelihood strategies. For instance, in his large body of work, Zimmerer (see, *inter alia*, 1996; 1999) has combined detailed, household-level field studies of resource management practices and crop diversity with political and economic analysis of state development and conservation policies. Others have examined smallholder crop management, household income generation and regional economies in tropical forests (see, for instance, Coomes, 2004; McSweeny, 2005). The strength of this approach lies in its empirical rigor and focus on micro-scale strategies of resource management and income generation, placed within a context of translocal economic processes.

While this body of literature highlights the diverse practices involved in smallholder agricultural production and household-level livelihood strategies, only recently have geographers considered the diversity of non-agricultural strategies employed by peasant households to generate income and assure their reproduction. Recognising that rural livelihoods are seldom *only* rural or agricultural, some scholars have focused on the dynamics of livelihood diversification. For instance, Bebbington (2000) has demonstrated the diversification of rural livelihoods among indigenous peasant communities in the Andes, based on migration and remittance income, rural industry and temporary wage labour. Livelihood diversification of this sort does not mean that rural *places* are less important, however: indeed, many rural communities continue to be sites of cultural reproduction, investment and empowerment even as agriculture plays a diminishing role in livelihood strategies. As Jokisch (2002) has demonstrated, international migrants from the Ecuadorian Andes frequently use income earned overseas to build large homes and maintain farms in their home communities, with the hope of eventually retiring there. Such processes dramatically alter rural communities and the landscapes in which they are located.

These studies highlight the fact that the livelihoods of the rural poor, like the environments from which they are (partially) derived, are not static. Rather, livelihood strategies may change in response to perceived opportunities (for example migration, market integration, development interventions, expanded resource rights), as well as external stress (for instance drought, flood, pest infestation or loss of resource rights). Examination of rural livelihoods as dynamic and diverse rather than static and derived from a single economic activity (whether it be agriculture, pastoralism, fishing, or petty trading, to name a few obvious options) permits analysis of the diversity of resources and strategies employed in securing a livelihood. This in turns fosters a more geographically nuanced appreciation of livelihood strategies, as inquiry turns to the sources and combinations of assets that contribute to economic and social well-being: the sources of income, techniques of natural resource management, networks of social relations and forms of knowledge and education that people draw on to make a living (Bebbington, 1999; Sen, 1997).

If rural livelihoods are not static or unitary, neither are they socially homogeneous nor necessarily equitable. Geographical research into rural livelihoods has revealed tremendous gendered, ethnic and class-based differences in access to the natural resources necessary for livelihood security, as well as the benefits derived from resource rights, management strategies and development projects (Harcourt,

1994; Moser, 1993). Women and men often have vastly different opportunities available to them for income generation, resource management and food security. For instance, Prebisch et al. (2002) demonstrate that as Mexican men migrate out of their rural communities in search of waged employment, women are left to tend family farms, resulting in the feminisation of smallholder agriculture. Similar themes are taken up by Perreault (2005) who examines the gendered role of indigenous peoples' agrobiodiversity maintenance in swidden gardens in the Ecuadorian Amazon and the importance of this diversity for household food security. In this case, not only is there a highly gendered division of labour regarding the tending of gardens and the preparation of foods made from garden crops, but these same foods figure importantly in similarly gendered discursive and performative representations of indigenous identity. Gendered asymmetries in access to and control over natural resources also have important implications for the well-being of women and children. For instance, in her work in northern Pakistan, Halvorson (2002) examines the gendered ideologies of conservative Islam and the implications these have both for women's spatial mobility and differential outcomes on child health. As evident in these works, a critical approach to livelihood calls for examination of the micropolitics of intra- as well as inter-household power relations, resource access and strategies for production and reproduction.

Insofar as it focuses attention on the reproduction of social life and the management of resources (both natural and social) necessary for human well-being and the varied forms of environmental knowledge, the concept of livelihood has been, and will continue to be, central to geographical work on environment and development. It is important to acknowledge its limitations, however, and to adapt the concept to the changing political and economic conditions in which the world's poor live their lives. Paramount among these limitations is the tendency for livelihoods to be considered as primarily rural and local. Though there is nothing about the concept that precludes its consideration in urban contexts or at socio-spatial scales beyond the household or rural community, geographers tend to reserve 'livelihood' for studies of rural, often agrarian communities. This is likely a legacy of the cultural ecology and political ecology approaches, in which rural, agrarian livelihoods have long been a central concern. It is also surely a result of the complexity of researching urban and multi-scalar economic strategies. An illusion of isolation and simplicity can be sustained in rural settings in ways that urban contexts do not allow. But the rapid urbanisation of Latin America, Asia, the Middle East and much of Africa in recent decades necessitates a rethinking of how geographers study livelihood. If we accept that the concept of livelihood is inherently useful for environment and development geography, then we must find ways to employ it in geographical contexts relevant to the world's poor. Recent work by Potts (2006) and Potts and Bryceson (2006) on migration, urbanisation and livelihood in Africa points towards a much-needed extension of the livelihood concept into urban contexts. Following these authors' lead will necessitate a rethinking not only of the categories of livelihood strategies, but of our traditional conceptualisations of resources themselves.

Moreover, it is important that geographers conceptualise livelihoods as encompassing geographies at scales beyond the local (Bebbington, 2003). The concept of livelihood may be used to describe the strategies for household reproduction employed by a farmer in rural Africa. But can we use the same concepts – if not the same analytical metrics – to describe and analyze the transnational networks that articulate that farmer with financial institutions and development agencies in London or Washington, DC? Surely the livelihoods of farmers and those of development practitioners are imbricated in ways that demand a re-scaling of traditional development categories (Batterbury, 2001). For instance, smallholder agriculturalists may belong to peasant cooperatives through which they are inserted into domestic and foreign markets, which in turn may be connected to non-governmental organisations, state agencies, financial institutions, activists and academics. These complex, dynamic and often extensive networks channel development activities, acting both to facilitate and constrain individual agency. Whether as part of formal, self-acknowledged networks linking development organisations with common goals, or part of informal mesh-works that connect various actors across scales and at different sites, there is a growing focus among geographers on the role of networks in the flows of capital and knowledge (Bebbington, 2005). Crucially, the effects of such network relations depend to a considerable extent on the institutional context within which they operate (Bebbington and Kothari, 2006). As such, the best work on networks for environment and development focuses not only on the structure of the network itself, but also (and primarily) on the social relations and spatial forms that are produced and reproduced through the network; the ways that power relations are or are not reconfigured; and the influence that such networks have on environmental outcomes (Radcliffe, 2001; Pieck, 2006). The transnational, networked character of environment and development projects is particularly apparent in the discourses and practices of sustainable development, considered next.

Sustainability and sustainable development

Since its entry into both academic discourse and mainstream environmentalism in the 1980s, the concepts of sustainable development and sustainability have inspired much research, theorisation and critique. Famously characterised by the World Commission on Environment and Development (WCED, 1987, p. 43) as 'Development that meets the needs of the present without compromising the ability of future generations to meet their own needs,' the notion of sustainable development surely raises more questions than it answers (Redclift, 1987). One persistent problem lies in the very conceptualisation of what, precisely, sustainable development is to sustain. Here, the central tension between environment and development – between fostering human welfare and ensuring nature conservation – is most apparent. If we accept that human welfare unquestionably depends on some measure of environmental protection, it is equally true that development demands environmental transformation and, inevitably, some degree of degradation of biodiversity and the integrity of geo-ecological systems. The key issue then is not *whether* natural environments should be modified by development, but the rather more complex questions of quantity and quality: to what degree, and in what manner, should natural systems be modified to meet human needs. Fundamentally, then, sustainable development revolves around normative questions of how people should arrange their social relations with one another and with their natural environment. In spite of its technocratic roots, then, sustainable development remains at heart a fundamentally political concept, though its political nature is seldom acknowledged in the technical reports of development agency.

As outlined above, the foundational ideas of sustainable development emerged in the 1970s and 1980s, largely the result of international conservation and development agencies, led by the UN and the IUCN, but with the support and critique of scholars, non-governmental organisations and bi- and multilateral aid agencies in Europe and the USA. Sustainable development's place in mainstream development thought was secured by the 1987 publication of *Our Common Future* (the so-called Brundtland report), the UN Conference on Environment and Development in Rio de Janeiro in 1992, and the associated release of Agenda 21, which served as an agenda for sustainable development in the 21st century (Conca and Dabelko, 1998; WCED, 1987). These events led to a fluorescence of scholarship, activism and initiatives on the part of academics, non-governmental organisations, bilateral developmental agencies and multilateral lending institutions. These efforts also led to considerable critique of the notion as hollow rhetoric and oxymoronic, as discourse of sustainable development was adopted by development organisations, Third World activists, transnational corporations and even the US Army. A concept that appeals to such a broad range of seemingly irreconcilable interests, critics argued, in reality stands for very little indeed (Redclift, 2005).

As its many critics have pointed out, the concept of sustainable development is at once under-theorised and impossibly broad in scope, having run into something of an intellectual cul-de-sac from which it has yet to be extracted. For this reason, Sneddon (2000) argues that geographers could more usefully focus on the concept of *sustainability*, which, as he notes, has more specific applications to social and ecological systems. For Sneddon, a conceptual decoupling of sustainability from sustainable development would have the benefit of introducing specificity and potential for critical evaluation into muddled debates regarding development and economic models. As Sneddon (2000, p. 525) argues,

> The advantage of 'sustainability' lies in how researchers invoking it must reference it against specific geographic, temporal and socioecological contexts. This context-specificity forces the crucial questions: what exactly is being sustained, at what scale, by and for whom, and using what institutional mechanisms?

A focus on sustainability trains analytical attention on specific social and ecological processes: rates of resource extraction vis-à-vis resource regeneration (in the case of water, wood or fish, for instance), or the maintenance (or otherwise) of biodiversity. Inherent to this approach, moreover, is a focus on particular geographies of sustainability: the spaces and scales in which social and ecological processes occur and intersect, and at which the sustainability of these processes can and should be evaluated. Thus, a focus on sustainability, more than sustainable development, opens terrain for explicitly geographical analysis.

Such analysis has been undertaken in attempts to quantify the ecological impacts and sustainability of economic activities at the urban and national scales. Perhaps the best known of these indexes is the 'Ecological Footprint' (Wackernagel and Rees, 1996), an attempt to measure the area of land and water necessary to support a given population's patterns of resource consumption and waste emission (see www.footprintnetwork.org). Another major sustainability measure, the Environmental Sustainability Index (ESI), was developed jointly by the Yale University Center for Environmental Law and Policy and the Columbia University Center for International Earth Science Information Network (http://sedac.ciesin.columbia.edu/es/esi/). Like the Ecological Footprint, the ESI quantifies the sustainability of resource use according to country-level indicators, assigning scores and rankings to individual countries based on exhaustive quantitative analysis (Esty et al., 2005). Both indices represent rigorous attempts to measure the notoriously slippery concept of sustainability, and while each presents useful and thoughtful analysis of resource use patterns, they necessarily present partial views. The ESI, for instance, while accounting for social well-being as well as environmental indicators, focuses too rigidly on the scale of the nation-state. Because the ESI is concerned primarily with national-level policy frameworks for pollution remediation, resource consumption and conservation, it under-appreciates the impacts of consumption patterns *between* countries. For instance, strong environmental laws and low population density allow Canada to rank 6th among the 146 countries listed, in spite of the often rapacious environmental practices of Canadian extractive firms, at home and abroad. Employing methods of 'resource accounting,' the Ecological Footprint approach is better attuned to resource flows between countries, and provides measures of countries with 'ecological reserves' and those with 'ecological deficits.' But because the Ecological Footprint is focused exclusively on resource consumption and pollution generation, it fails to consider indicators of human welfare and social development. Moreover, it does not adequately take into account social adaptation to changing environmental conditions, such as technological innovation or transformations in patterns of resource use or pollution. Although valuable for the macro-level comparisons and easily digestible policy recommendations they provide, these sustainability indicators are rather too blunt an instrument for parsing the complexities of environment-development relations (cf. Warren et al., 2001).

More nuanced, micro-level analyses have been undertaken by numerous geographers in recent years, who have employed a diversity of methods to measure both social processes (e.g., income generation, health status, participation in resource governance) and biophysical processes (e.g., soil erosion, deforestation, contamination or depletion of water sources). For instance, geographers have examined the sustainability or otherwise of development efforts using demographic analysis (Lado, 1999), household income and crop surveys (Goldman, 1995), and case studies of particular development projects and policies (Klepeis and Laris, 2006). Micro-scale analyses such as these, relying largely on case studies of particular places and/or resource use practices, can reveal much about the intricate relationships between human welfare and natural resource flows, exposing complexities that broad-brush indexes cannot. However nuanced their findings may be, case study research is seldom generalisable in any straightforward manner, and such studies are hampered by difficulties in 'scaling up' their results. Bridging the gap between research and policy – an implicit, if not explicit goal of sustainability science – therefore requires methodological and theoretical approaches flexible enough to link scales of analysis. One such approach may be found in work on urban sustainability, which accounts for the complexities of multiform and dynamic places, social relations and resource -use patterns (Hanson and Lake 2000). Recent work on urban political ecology and urban metabolism provides one such model (Heynen et al., 2006).

Recognising the thorny issues involved in implementing policies for sustainable development, O'Riordan (2004) calls for a more critical and politically engaged sustainability science, and argues that geographers must involve themselves more fully in debates of environmental governance. As discussed elsewhere in this volume, environmental governance signals the decentering of environmental management away from the state apparatus, and towards new (and often unstable) institutional configurations involving state, market, and civil society. While calling for structures of environmental governance that are democratic, accountable, and sensitive to local livelihoods, social inequities and ecological limits, O'Riordan acknowledges that such governance structures also carry inherent political risks: 'Governance generally, and sustainability governance in particular, may well be replicating the existing order of economic power, military hegemony and local elitism' (p. 241). As O'Riordan emphasises, however, we may have little choice but to develop new modes of sustainability governance, as continued economic growth and human welfare surely depend on the maintenance of environmental systems and the resources and ecological services they provide. Linking critical analyses of sustainability and livelihoods across spatial scales, institutional contexts, and ecological systems, and bridging the gap between research and policy remain central tasks for geographers working in the environment and development tradition (Liverman, 2004).

Conclusion

Fundamentally, environment and development geography is focused on (i) the ways that social groups are dependent upon nature and natural resources for their survival and welfare; and (ii) the environmental transformations that result from resource use and economic activity. This dialectical relationship in turn draws attention to the apparent contradiction between environmental conservation and economic development, which lies at the heart of environment and development geography. In this chapter, I have examined this tension in the context of three themes central to the environment and development tradition in geography: conservation, livelihoods, and sustainability. Geographers have long been concerned with the social and ecological implications of nature conservation. Such efforts have been integrally linked to development policies insofar as they represent state efforts not only to protect wild nature, but also to exert control over resources and population. Moreover, nature conservation has figured prominently in international development programmes in recent decades, and has been a major focus for transnational environmental NGOs, bilateral development agencies, and multilateral lending institutions alike. Ecological and social questions are raised, however, by conflicts involving conservation projects and the livelihood strategies of local populations, and by the return of some conservation organisations to a 'fortress conservation' model of protected areas management. More geographical attention is needed on the dynamics of conservation efforts and their potential impacts – both positive and negative – for rural populations.

Insofar as it focuses attention on the quotidian practices by which individuals, households, and communities manage resources (both natural and social) necessary for life, the concept of livelihood has long been a central organising principle for environment and development geography. This work has highlighted the diversity in resource use systems and environmental knowledge, as well as the strengths and limitations of the concept of livelihood itself. Additional theoretical and empirical work is required in order to expand the livelihood concept to urban areas and scales beyond the local. To a considerable extent, the tensions between livelihood strategies and nature conservation gave rise in the 1980s to the concept of sustainable development, understood here as an institutional, discursive and practical attempt to overcome the contradictions between economic production and environmental conditions – what O'Connor (1996) calls the second contradiction of capitalism.

Greater critical attention is needed on sustainability indexes, in order to devise measures that are methodologically and theoretically capable of linking micro- and macro-analyses. Geographers in the environment and development tradition have drawn attention to the workings of power in relation to these themes: how the practices of conservation, livelihood and sustainable development each entail power asymmetries that must be negotiated through a constant reconfiguration of the relations between civil society actors, state institutions and transnational actors at a diversity of interconnected spatial scales. This ongoing negotiation of power in the context of environment and development will no doubt continue to be a central concern for geographers.

BIBLIOGRAPHY

Adams, W. M. (2007) *Green Development: Environment and Sustainability in the Third World*, 3rd edn. London: Routledge.

Batterbury, S. (2001) Landscapes of diversity: a local political ecology of livelihood diversification in south-western Niger. *Ecumene*, 8(4), 437–64.

Bebbington, A. (1994) Theory and relevance in indigenous agriculture: knowledge, agency and organization. In D. Booth (ed.), *Rethinking Social Development: Theory, Research and Practice*. Essex: Longman Scientific and Technical, pp. 202–25.

Bebbington, A. (1999) Capitals and capabilities: a framework for analyzing rural livelihoods and poverty alleviation. *World Development*, 27(12), 2021–44.

Bebbington, A. (2000) Reencountering development: livelihood transitions and place transformations in the Andes. *Annals of the Association of American Geographers*, 90(3), 495–520.

Bebbington, A. (2002) Geographies of development in Latin America? *Conference of Latin Americanist Geographers Yearbook*, 27, 105–48.

Bebbington, A. (2003) Global networks and local developments: agendas for development Geography. *Tijdschrift voor Economische en Sociale Geografie*, 94(3), 297–309.

Bebbington, A. (2004) NGOs and uneven development: geographies of development intervention. *Progress in Human Geography*, 28(6), 725–45.

Bebbington, A. (2005) Donor-NGO relations and representations of livelihood in nongovernmental aid chains. *World Development*, 33(6), 937–50.

Bebbington, A. and Kothari, U. (2006) Transnational development networks. *Environment and Planning A*, 38, 849–66.

Blaikie, P. and Brookfield, H. (1987) *Land Degradation and Society*. London: Methuen.

Chambers, R. (1995) *Poverty and Livelihoods: Whose Reality Counts?* Brighton: Institute for Development Studies.

Chapin, M. (2004) A challenge to conservationists. *WorldWatch*, November–December, 17–31.

Coomes, O. (2004) Rain forest 'conservation through use'? *Chambira* fibre extraction and handicraft production in a land-constrained community, Peruvian Amazon. *Biodiversity and Conservation*, 13(2), 351–60.

Cowen, M. P. and Shenton, R. W. (1996) *Doctrines of Development*. London: Routledge.

Cronon, W. (1996) The trouble with wilderness, or, getting back to the wrong nature. In W. Cronon (ed.), *Uncommon Ground: Rethinking the Human Place in Nature*. New York: Norton, pp. 69–90.

de Haan, L. J. (2000) Globalization, localization and sustainable livelihood. *Sociologia Ruralis*, 40(3), 339–65.

Demeritt, D. (2002) What is the 'social construction of nature'? A typology and sympathetic Critique. *Progress in Human Geography*, 26(6), 767–90.

Ehrlich, P. R. (1972) *The Population Bomb*. London: Ballantine.

Emel, J. and Bridge, G. (1995) The earth as input: resources. In R. J. Johnston, P. Taylor and M. J. Watts (eds), *Geographies of Global Change: Remapping the World in the Late Twentieth Century*. Oxford: Blackwell, pp. 318–32.

Escobar, A. (1995) *Encountering Development: The Making and Unmaking of the Third World*. Princeton, NJ: Princeton University Press.

Esty, D. C., Levy, M., Srebotnjak, T. and de Sherbinin, A. (2005) *2005 Environmental Sustainability Index: Benchmarking National Environmental Stewardship*. New Haven: Yale Center for Environmental Law and Policy.

Fortwangler, C. L. (2003) The winding road: incorporating social justice and human rights into protected area policies. In S. R. Brechin, P. R. Wilhusen, C. L. Fortwangler and P. C. West (eds), *Contested Nature: Promoting International Biodiversity with Social Justice in the Twenty-First Century*. Albany: State University of New York Press, pp. 25–40.

Goldman, A. (1995) Threats to sustainability in African agriculture: searching for appropriate Paradigms. *Human Ecology*, 23(3), 291–335.

Goldman, M. (2005) *Imperial Nature: The World Bank and Struggles for Social Justice in the Age of Globalization*. New Haven: Yale University Press.

Hacking, I. (1999) *The Social Construction of What?* Cambridge, MA: Harvard University Press.

Halvorson, S. J. (2002) Environmental health risks and gender in the Karakoram-Himalaya, Pakistan. *The Geographical Review*, 92(2), 257–81.

Hanson, S. and Lake, R. W. (2000) Needed: geographic research on urban sustainability. *Economic Geography*, 76, 1–3.

Harcourt, W. (1994) *Feminist Perspectives on Sustainable Development*. London: Zed.

Hardin, G. (1968) The tragedy of the commons. *Science*, 1628, 1243–48.

Hardin, G. (1974) Living on a lifeboat. *Bioscience*, 24, 561–68.

Harvey, D. (1974) Population, resources, and the ideology of science. *Economic Geography*, 50(3), 256–77.

Hecht, S. and Cockburn, A. (1989) *The Fate of the Forest: Developers, Destroyers and Defenders of the Amazon*. New York: HarperPerennial.

Heynen, N., Kaika, M. and Swyngedouw, E. (eds) (2006) *In the Nature of Cities: Urban Political Ecology and the Politics of Urban Metabolism*. London: Routledge.

IUCN (1980) *The World Conservation Strategy*. Geneva: International Union for the Conservation of Nature and Natural Resources, United Nations Environment Programme, World Wildlife Fund.

Jokisch, B. (2002) Migration and agricultural change: the case of smallholder agriculture in the highlands of south-central Ecuador. *Human Ecology*, 30(4), 523–50.

Katz, C. (1998) Whose nature, whose culture?: Private productions of space and the 'preservation' of nature. In B. Braun and N. Castree (eds), *Remaking Reality: Nature at the Millenium*. London, Routledge), pp. 46–63.

Klepeis, P. and Laris, P. (2006) Contesting sustainable development in Tierra del Fuego. *Geoforum*, 37, 505–18.

Kosek, J. (2004) Purity and pollution: racial degradation and environmental anxieties. In R. Peet and M. Watts (eds), *Liberation Ecologies: Environment, Development, Social Movements*, 2nd edn. London: Routledge, pp. 125–65.

Lado, C. (1999) Environmental resources, population and sustainability: evidence from Zimbabwe. *Singapore Journal of Tropical Geography*, 20(2), 148–68.

Leach, M., Mearns, R. and Scoones, I. (1999) Environment entitlements: dynamics and institutions in community-based natural resource management. *World Development*, 27(2), 225–47.

Lean, G., Hinrichsen, D. and Markham, A. (1990) *WWF Atlas of the Environment*. New York: Prentice Hall.

Lehmann, D. (1997) An opportunity lost: Escobar's deconstruction of development. *Journal of Development Studies*, 33(4), 568–78.

Liverman, D. (2004) Who governs, at what scale, and at what price? Geography, environmental governance, and the commodification of nature. *Annals of the Association of American Geographers*, 94(4), 734–38.

Liverman, D. M., Yarnal, B. and Turner, B. L., II (2004) The human dimensions of global environmental change. In G. Gaile and C. Wilmott (eds), *Geography in America at the Dawn of the 21st Century*. New York: Oxford University Press, pp. 267–82.

McCusker, B. and Weiner, D. (2003) GIS representations of nature, political ecology, and the study of land use and land cover change in South Africa. In K. S. Zimmerer and T. J. (eds), Bassett *Political Ecology: An Integrative Approach to Geography and Environment-Development Studies*. New York: Guilford Press, pp. 201–18.

McNeely, J. A. and Miller, K.R. (1984) *National Parks, Conservation and Development: The Role of Protected Areas in Sustaining Society*. Washington, DC: Smithsonian Institution.

McSweeney, K. (2005) Natural insurance, forest access and compounded misfortunte: forest resources in smallholder coping strategies before and after Hurricane Mitch, eastern Honduras. *World Development*, 33(9), 1453–71.

Moser, C. O. N. (1993) *Gender Planning and Development: Theory, Practice and Training*. London and New York: Routledge.

Myers, N., Mittermeirer, R. A., Mittermeirer, C. G., da Fonseca, G. A. B. and Kent, J. (2000) Biodiversity hotspots for conservation priorities. *Nature*, 403, 853–58.

Neumann, R. (1998) *Imposing Wilderness: Struggles Over Livelihood and Nature Preservation in Africa*. Berkeley: University of California Press.

Neumann, R. (2004) Nature-state-territory: Toward a critical theorization of conservation Enclosures. In R. Peet and M. Watts (eds), *Liberation Ecologies: Environment, Development, Social Movements*, 2nd edn. London: Routledge, pp. 195–217.

O'Connor, J. (1996) The second contradiction of capitalism. In T. Benton (ed.), *The Greening of Marxism*. New York: Guilford, pp. 197–221.

O'Riordan, T. (2004) Environmental science, sustainability and politics. *Transactions of the Institute of British Geographers (New Series)*, 29, 234–47.

Peet, R. and Hartwick, E. (1999) *Theories of Development*. New York: Guilford Press.

Peet, R. and Watts, M. (2004) Liberating political ecology. In R. Peet and M. Watts (eds), *Liberation Ecologies: Environment, Development, Social Movements*, 2nd edn. London: Routledge, pp. 3–47.

Peluso, N. L. (1992) *Rich Forests, Poor People: Forest Access and Control in Java*. Berkeley: University of California Press.

Peluso, N. L. (1993) Coercing conservation? The politics of state resource control. *Global Environmental Change*, 3(2), 199–218.

Perreault, T. (2003) 'A people with our own identity': toward a cultural politics of development in Ecuadorian Amazonia. *Environment and Planning D: Society and Space*, 21, 583–606.

Perreault, T. (2005) Why *chacras* (swidden gardens) persist: agrobiodiversity, food security, and cultural identity in the Ecuadorian Amazon. *Human Organization*, 64(4): 327–39.

Pieck, S. K. (2006) Opportunities for transnational indigenous eco-politics: the changing landscape in the new millennium. *Global Networks*, 6(3), 309–29.

Potts, D. (2006) 'All my hopes and dreams are shattered': urbanization and migrancy in an imploding African economy – the case of Zimbabwe. *Geoforum*, 37(4), 536–51.

Potts, D. and Bryceson, D. (eds) (2006) *African Urban Economies: Viability, Vitality or Vitiation*. Houndmills: Palgrave Macmillan.

Preibish, K. L., Rivera Herrejón, G. and Wigins, S. L. (2002) Defending food security in a free-market economy: the gendered dimensions of restructuring in rural Mexico. *Human Organization*, 61(1), 68–79.

Radcliffe, S. (2001) Development, the state, and transnational political connections: state and subject formations in Latin America. *Global Networks*, 1(1), 19–36.

Redclift, M. (1987) *Sustainable Development: Exploring the Contradictions*. London: Routledge.

Redclift, M. (2005) Sustainable development (1987–2005): an oxymoron comes of age. *Sustainable Development*, 13(4), 212–27.

Robbins, P. (2001) Fixed categories in a portable landscape: the causes and consequences of land-cover categorization. *Environment and Planning A*, 33, 161–79.

Robbins, P. (2004) *Political Ecology: A Critical Introduction*. Oxford: Blackwell.

Rostow, W. W. (1960) *The Stages of Economic Growth: A Non-Communist Manifesto*. Cambridge: Cambridge University Press).

Sen, A. (1997) Editorial: Human capital and human capability. *World Development*, 25(12), 1959–61.

Smith, N. (1984) *Uneven Development: Nature, Capital, and the Production of Space*. Oxford: Blackwell.

Sneddon, C. S. (2000) 'Sustainability' in ecological economics, ecology and livelihoods: a review. *Progress in Human Geography*, 24(4), 521–49.

Staheli, L. and Mitchell, D. (2005) The complex politics of relevance in geography. *Annals of the Association of American Geographers*, 95(2), 357–72.

Sundberg, J. (1998) Strategies for authenticity, space and place in the Maya Biosphere Reserve, Petén, Guatemala. *Yearbook, Conference of Latin Americanist Geographers*, 24, 85–96.

Thompson, M., Warbuton, M. and Hatley, T. (1986) *Uncertainty on a Himalayan Scale*. London: Ethnographica.

Turner, M. (1993) Overstocking the range: a critical analysis of the environmental science of Sahelian pastoralism. *Economic Geography*, 69(4), 402–21.

Turner, M. (1998) The interaction of grazing history with rainfall and its influence on annual rangeland dynamics in the Sahel. In K. S. Zimmerer and K. R. Young (eds), *Nature's Geography: New Lessons for Conservation in Developing Countries*. Madison: University of Wisconsin Press, pp. 237–61.

Wackernagel, M. and Rees, W. E. (1996) *Our Ecological Footprint: Reducing the Human Impact on the Earth*. Philadelphia: New Society Publishers.

Warren, A., Batterbury, S. and Osbahr, H. (2001) Sustainability and Sahelian soils: evidence from Niger. *The Geographical Journal*, 167(4), 324–41.

Watts, M. (1983a) On the poverty of theory: natural hazards research in context. In K. Hewitt (ed.), *Interpretations of Calamity: From the Viewpoint of Human Ecology*. Boston: Allen and Unwin, pp. 231–62.

Watts, M. (1983b) *Silent Violence: Food, Famine and Peasantry in Northern Nigeria*. Berkeley: University of California Press.

WCED (1987) *Our Common Future*. World Commission on Environment and Development. New York: Oxford University Press.

Whyte, A. V. T. (1986) From hazards perception to human ecology. In R. W. Kates and I. Burton (eds), *Geography, Resources, and Environment, Volume II: Themes from the Work of Gilbert F. White*. Chicago: University of Chicago Press, pp. 240–71.

Wilhusen, P., Brechin, S. R., Fortwangler, C. L. and West, P. C. (2003) Contested nature: conservation and development at the turn of the twenty-first century. In ed., S. R. Brechin, P. R. Wilhusen, C. L. Fortwangler and P. C. West (eds), *Contested Nature: Promoting International Biodiversity with Social Justice in the Twenty-first Century*. (Albany: State University of New York Press, pp. 1–22.

Williams, R. (1973) *The Country and the City*. New York: Oxford University Press.

Williams, R. (1976) *Keywords: A Vocabulary of Culture and Society*. New York: Oxford University Press.

World Bank (1992) *World Bank Development Report 1992: Development and Environment.* New York: Oxford University Press.

Zerner, C. (ed.) (2000) *People, Plants & Justice: The Politics of Nature Conservation.* New York: Columbia University Press.

Zimmerer, K. (1996) *Changing Fortunes: Biodiversity and Peasant Livelihood in the Peruvian Andes.* Berkeley: University of California Press.

Zimmerer, K. (1999) The overlapping patchworks of mountain agriculture in Peru and Bolivia: toward a regional-global landscape model. *Human Ecology,* 27(1), 135–65.

Zimmerer, K. (2004) Environmental discourses on soil degradation in Bolivia: sustainability and the search for the socioenvironmental 'middle ground'. In R. Peet and M. Watts (eds), *Liberation Ecologies: Environment, Development, Social Movements.* London: Routledge, pp. 107–24.

Zimmerer, K. and Young , K. R. (eds) (1998) *Nature's Geography: New Lessons for Conservation in Developing Countries.* Madison: University of Wisconsin Press.

Chapter 27

Natural Hazards

Daanish Mustafa

The Naturalness of Natural Hazards

The term 'natural hazards' refers to the potential, experience, and aftermath of environmental extremes such as earthquakes, volcanic activity, drought, storms and other extreme weather. As such, hazards are different from the term 'natural disasters', which refers only to the actual extreme event. By contrast the hazards perspective takes a longer-term view of adverse environmental extremes as integral to interactions between humans and their environments. Earthquakes, followed by floods and tropical cyclones, have been the most expensive and lethal hazards faced by humanity during the 20th century (Tobin and Montz, 1997). While there is clearly a geography to those particular hazards, the fact is that environmental extremes of one sort or another affect every square inch of planet Earth, though clearly what constitutes an extreme and how it comes to influence human affairs will vary enormously from one context to another. There were and continue to be very sound political, economic and cultural reasons for human habitation of various, apparently hazardous, environments (Burton et al., 1993). Flood plains, for example, may be more exposed to flooding but they also tend to have the most fertile soils. The hazards perspective developed by geographers and other scholars helps us appreciate how, why, and with what effects people adapt to the opportunities and perils offered by their environments.

Natural hazards are more appropriately called environmental hazards. Indeed, as we shall see many of the insights developed by geographers studying environmental hazards have also been successfully applied to understanding how societies deal with technological hazards, like nuclear accidents. Calling some hazards 'natural' implies that some uncontrollable Act of God or other external cause (aka Mother Nature) is responsible for the damage and suffering they cause. It ignores the role of social processes in influencing the frequency, degree of exposure, resulting consequences, and ability to resist or recover from a given environmental extreme. The reality is that there is nothing more typical than environmental variability – the issue for human societies is how we cope with it. One of the central paradoxes of contemporary hazards research is that the damage caused by hazards continues to rise despite massive expenditure in preventing losses from floods, earthquakes, and other environmental hazards to which 'less advanced' hunter gatherer and agricultural societies were apparently much better adapted (Burton et al., 1993; Tobin and Montz, 1997).

Therein lies the key conceptual concern of this chapter: How has research on environmental extremes accounted for the qualitatively different experience of environmental extremes under modernity? This chapter will explore this question by tracing the major conceptual developments in the field of hazards research.

Unpacking the Terminology of Environmental Hazards

Social scientific research on environmental hazards has understood them in terms of four variables – risk, exposure, vulnerability, and response – though as we will see precisely how those variables are defined and said to inter-relate is somewhat contested in the literature. Mitchell (1990), for example, conceptualises hazards as a multiplicative function:

$$\text{Hazards} = f(\text{risk} \times \text{exposure} \times \text{vulnerability} \times \text{response})$$

More commonly, however, risk is understood simply as the probability of particular extreme events occurring in particular place. This more prevalent conceptualisation of risk ignores the consequences of a given event captured by the more nuanced definition of risk as a multiplicative function of probability × consequences (Cutter, 1993). Nevertheless, the simple, and some would argue misleading, probabilistic formulation of hazard risk is particularly prevalent among engineering and actuary practitioners in hazards management (Cardona, 2004). For instance, flooding risk is typically discussed in terms of return periods, which represent the probability of flood event of a particular magnitude, such as the 100-year flood that is the default standard for many flood defence systems. Saying that the great Mississippi flood of 1993 was a 500-year flood is a way of stating that a flood of that magnitude has a 0.2 percent chance of occurring any particular year (Pitlick, 1997). Many physical scientists, engineers, and regrettably, policymakers tend to over-emphasise the importance of controlling the frequency and intensity of events. Many engineering interventions are geared towards controlling this narrowly defined aspect of hazard frequency, to the relative neglect of the remaining variables captured in Mitchell's formulation above. The building of levees along riverbanks to control floods is one such example of engineering-based risk mitigation. Although such physical interventions to reduce the frequency of extreme events can be a useful part of an overall disaster reduction and hazard mitigation strategy, they are not by themselves sufficient, and can at times be wasteful, if they encourage behaviour that increases exposure or vulnerability to a given hazard.

Exposure in the hazards context means the number of people and value of property or other social goods physically exposed to a given hazard. Both the northern coast of Australia and the state of Florida in the United States face a comparable risk of hurricanes making landfall any given year. Yet the potential damage suffered in the United States from such an event is much higher because the density of coastal development is much greater in Florida than in northern Australia. In other words the exposure is higher in the United States than the northern Australia. The relationship between the physical probability of an event and its consequences, whether economic, social or both, is captured by formulations of risk as a multiplicative function of both:

$$\text{Risk} = f(\text{probablility} \times \text{consequences})$$

Combining probability and consequences into a single mathematical function allows policymakers to use cost-benefit analysis and other decision support tools designed to optimise the aggregate level of expected annual financial loss. Cost-benefit analysis is now a standard feature of project appraisal for flood defense and other hazard mitigation schemes. Apart from the technical difficulties of actually quantifying and monetising levels of exposure to hazards whose probability and consequences are often highly uncertain, there is a fundamental ethical issue posed by this financial-based method for assessing hazard exposure. Measured in dollar terms the consequences of a given event will always be higher if it affects richer people. Financially based tools, like cost-benefit analysis, are blind to the distributional effects and to the consequences for the least worst off, which are central to Rawlsian conceptions of justice (Johnson et al., 2007).

One of the attractions of the concept of vulnerability is that it highlights questions about the uneven effects of a given hazard on differentiated societies. Different vulnerability profiles will determine socially differentiated impacts experienced by people otherwise similarly exposed to the same level of hazard intensity and frequency. In fact, many researchers argue that environmental hazards may be defined as situations where environmental extremes come into contact with vulnerable human populations (Mustafa, 1998). Vulnerability is defined as the susceptibility of an individual or group to suffer damage from environmental extremes and the relative inability to recover from that damage (Cutter et al., 1996, Adger, 2006). Levels of vulnerability depend upon long-term social, economic, political and physical factors. Poverty, age, gender, socio-economic status, ethnicity, state of the infrastructure, quality of governance or a combination of any of these factors are key determinants of whether certain populations will be vulnerable to hazards. The women and children living in the coast of Bangladesh without means of evacuation, for example, are much more vulnerable to coastal storms than affluent Florida residents able to get themselves out of harm's way. In addition to these huge differences in hazard vulnerability between developed and developing countries, there are also major differences within societies, as evidenced by the experience of Hurricane Katrina (Colten, 2006). Official statistics from the St. Gabriel Parish morgue in New Orleans show that the majority of the nearly 1,000 fatalities recorded there were poor, elderly and African American. Such vulnerable social groups were less able to evacuate the city as Hurricane Katrina approached (Louisiana Department of Health and Hospitals, 2006).

Hazards response may entail multiple steps including pre-disaster planning and preparation, vulnerability mitigation, disaster relief, long-term reconstruction and finally development directed towards building resistance and resilience in the face of existing and future hazards. Which particular step or configuration of steps is emphasised in hazards policy will bear upon the magnitude of damage suffered during specific disasters. Relatively recently, in the aftermath of the 2004 Asian Tsunami and the 2005 South Asian earthquake, national governments, international donors, and policymakers have turned their attention to operationalising the so-called disaster risk reduction (DRR) strategies aimed at addressing the sort of differential vulnerabilities that academic geographers have used a different terminology to describe. Despite their different intellectual genealogies, DRR and vulnerability reduction are really two ways of talking about the same set of issues, and so in this chapter they will be referred to interchangeably.

But to appreciate how these different terms have developed and the sorts of policies and practices they inform, it helps to know something about the history of hazards research.

From Divine Retribution to Scientism

Prior to the 20th century, natural calamities were often regarded as divine retribution for individual or collective sins. Many examples of pestilence or volcanic activity were reported in the scriptures of different religious traditions as signs of divine wrath. The Lisbon earthquake of 1755 was one such example in which the widespread devastation wrought by the earthquake was interpreted as a sign of divine displeasure resulting in temporary intensification of the inquisition (Alexander, 2002).

Increasingly through 18th and 19th centuries, religious and supernatural explanations for hazards gave way to scientific ones. Confident in their technological accomplishments, people in western societies in particular came to believe in the possibility and even desirability of subjugating nature to scientific control. Consequently, hazards came to be viewed as strictly technological problems, which superior application of engineering and science could solve. In the aftermath of the great Mississippi floods of 1927, for example, the US Army Corps of Engineers embarked upon massive basin-wide engineering works to prevent flooding (Platt, 1986). The flood control works in the Mississippi and earlier river engineering works in the Rhine River basin (Disse and Engel, 2001) were replicated throughout the 20th century in many other river basins of the world, e.g., the Indus (Mustafa and Wescoat, 1997), the Narmada (Gupta, 2001), the Huang He (Yellow River), and the Sacramento River basins (Golet et al., 2006) among others. Advances in environmental science and engineering made it possible to build safer, storm- and earthquake-resistant structures and provide advanced warnings of tornados, hurricanes and droughts to cushion the impact of environmental extremes (Tobin and Montz, 1997).

Physical geographers have made substantial contributions to understanding the physical processes involved in environmental hazards. For example, while the contributors to Cello and Malamud (2006) applied fractal mathematics to improve estimates of the frequency and magnitude of floods and other extreme events, others, such as Litschert (2004), have used GIS techniques to map and model landslide zones. In the sub-fields of geomorphology, hydrology, and climatology physical geographers have analyzed the physical processes effecting floods, droughts, landslides and other meteorological hazards (e.g., see Brooks, 2003; Todhunter and Rundquist, 2004; Dupigny-Giroux et al., 2006; Nicholls, 2007). Similarly, fire hazard has been a concern for biogeographers, particularly with an eye towards balancing the ecological benefits of fire cycles to forest ecosystems and societal needs (Bowman and Franklin, 2005; Lafon et al., 2005).

By highlighting the ecological effects of engineering interventions, research by physical geographers is also now informing explicit policy consideration of the trade-offs between environmental quality, ecosystem resilience and human needs. For example dams and levees were long considered a lynch pin of flood protection measures. But now with greater understanding of dynamic geomorphological and hydrological processes and their ecological significance (e.g., Graf, 2005), river engineering works have become less popular than they were. The United Kingdom's 'making space for water' flood risk management strategy is an important example of at least the realisation of the need to move beyond river engineering (DEFRA, 2005).

For the most part, however, this scientific capacity to quantify and model environmental hazards is still predominantly (mis)used to inform engineering and other physical interventions designed to prevent the occurrence or mitigate the physical effects of environmental hazards. For instance, levee systems and dams have been designed to prevent fluvial flooding, sea walls to prevent coastal flooding, and improved building design codes to enable buildings to withstand earthquakes as well as hazard maps and concomitant differential insurance rates to identify different hazards zones. Such efforts to prevent the incidence of environmental hazards can sometimes created new risks of their own. It was not the rain and wind from Hurricane Katrina that flooded New Orleans but a catastrophic failure of the levee system, which was designed to prevent localised flooding but encouraged rapid urban development in areas of the city below sea level.

Increasing recognition of the limits of science and of the impossibility of absolute technical control over the environment has helped foster new concerns about hazards posed by the economic and productive activities of industrialised societies. This new field of technological risk developed in parallel with environmental hazards research and has followed a similar intellectual trajectory. Most of the initial work on the risks posed by industrial hazards like nuclear power and chemical contamination was predicated upon a firm belief in the possibility of their scientific predictability and technical control. But as with research on natural and environmental hazards, work on technological hazards has moved beyond narrow scientism. There is now widespread recognition of the central importance of public perceptions and behaviour and increasing awareness of the intellectual resources available for hazards research from the political economic and post-structuralist traditions of analysis.

Hazards Perception and Behavioural Research

Despite the expenditure of billions of dollars in civil works to defend against environmental hazards, monetary losses, if not fatalities (at least in the industrialised North), have continued to increase (Burton et al., 1993, Tobin and Montz, 1997). The apparent failure of engineering and other technological approaches aimed at reducing hazard risk gave rise to new behaviourist approaches within hazards research and policy. The first behaviourists were mavericks. In contrast to the predominant scientific concern with the frequency and magnitude of environmental hazards, they instead emphasised the importance of public perceptions and behaviour in determining exposure and vulnerability to them.

The foundations of the behavioural approach were laid by Gilbert White's (1945) study of flood hazard in the United States. He argued that by focusing exclusively on engineering solutions to flooding, Americans were ignoring a range of other less capital intensive and potentially more effective measures for dealing with flood hazards, such as flood plain mapping, flood proofing, flood insurance and land-use planning. White's emphasis on the central role of perception in influencing individual and institutional responses to environment hazards inspired wealth of subsequent academic research on risk perception and behavioural response (White, 1974; Kasperson, et al., 1988; Mileti, 1999). It has also found increasing resonance in policy circles (Platt, 1986). Nevertheless, nearly half a century after his initial research, White was still decrying the over-reliance on levees in the Mississippi basin, which encouraged a false sense of security and the overdevelopment of hazardous flood plains (Myers and White 1993). Reviewing disaster policies across the United States, Mileti (1999) called for a move away from the arrogant attitude of controlling nature and towards living and adjusting to hazards. In the context of earthquakes in California for instance, he recommended microseismic hazard mapping, public education, and autonomous instead of integrated utility systems, so that in the event of an earthquake damage to one part of an integrated utility system did not have amplified effects on a much larger area.

In the 1960s, natural hazard researchers developed the first methods for measuring public perceptions of risk. Saarinen (1966), for example, used questionnaires to study farmers' perception of drought in the Great Plains of the United States and their role in informing decisions about behavioural responses such as varying cropping patterns or taking out crop insurance. At an international scale White (1974) undertook an international multi-country study on hazard perception to compare attitudes towards hazards and prospective adjustments across cultures. The study was pioneering in using a standard questionnaire across cultures, in addition to defining important parameters for perception studies in hazards research.

Another important area of research pioneered by this behavioural tradition is risk communication and its effects in changing behaviour. For example, Crozier et al. (2005) analyzed the impact of earthquake hazard zone information both on people's perception of likely damage from the hazard and on their possible adjustments to it. Controlling for demographic factors and social class, they found that residents in low risk zones deemed the earthquake risk manageable when given information about the hazard. In high-risk zones, however, information led to responses indicating a sense of resignation and fatalism in the face of the hazard. This practical concern with the potential for different kinds of information to change attitudes and behaviours to risk drew some of its inspiration from the American Pragmatist tradition of philosophy, which emphasised the importance of scientific research to inform practical action and democratic debate about hazards and their management (Wescoat, 1992).

The early work of hazards researchers on public perceptions of environmental hazards also informed research on technological hazards that blossomed through the 1970s in the wake of increasingly vocal public opposition to nuclear power in particular. Consistent with their pragmatic orientation, many behaviourists were concerned with the information needs and policies for mitigating technological risks. In this regard one of the central issues was the dissonance between expert views of technological risks and the public perception and reaction towards them (Slovic, 1987). Earlier on Starr (1969) pointed to the distinction between voluntary and involuntary risk, positing that people are much more likely to accept risks they take on voluntarily, such as from driving or smoking, than risks imposed on them involuntarily from outside, such as the location of a nuclear power facility in their community. Kasperson et al. (1988) proposed a framework of social amplification of risk where, risk messages are mediated by a series of lenses including the media, government, scientific debate, personal experience and culture that may either amplify or attenuate perceptions of risk. This work was based on communication theory and was directed towards understanding why certain risks cause more public anxiety than others. One of the critiques of this influential framework for understanding technological risk perceptions was its apparent focus on the hazards rather than the people who were experiencing the hazards (Baxter and Greenlaw, 2005). While attention to the social experience and construction of risk was one of the main contributions of the pragmatist tradition within technological hazards, this theme was to be developed further under the influence of the emerging political economic focus within hazards research.

From Political Economy to Social Nature

Since the 1980s, there has been increasing attention to the political economic factors contributing to social vulnerability to hazards. This newer, more radical hazards research tradition came to be closely aligned with the emerging political ecology research agenda within human geography. This was particularly so because many of the pioneers of the radical critique within hazards research in geography were also at the forefront of defining the political ecology research agenda in human-environment interactions and resource management (e.g., Watts, 1983; Blaikie and Brookfield, 1987; Blaikie et al., 1994). Because of their regional expertise, as well as the salience of drought in Sahel during the 1970s and 1980s, much of this early radical work was focused on Africa and on explaining the underlying social causes of famine (O'keefe et al., 1977; Blaike and Brookefield, 1987; Watts and Bohle, 1993). It also tended to be more oriented toward theoretical analysis than toward empirically based field study. Partly as a result it has been accused of practical and policy irrelevance (Proctor, 1998).

In keeping with its roots in Marxist political ecology, the radical tradition of hazards research focused on theorising the underlying structural features that make the economically and socially marginalised and disempowered also the most vulnerable to hazards. The pressure and release (PAR) model developed by Blaike et al. (1994) identified the structural root causes that translated into institutional level 'dynamic pressures' which in turn spawned the empirical 'unsafe conditions' to create vulnerability. Combining micro-level ethnographic research with analyses of macro level socio-economic trends, Mustafa (1998) showed how structural factors, such as economic marginalisation and political disempowerment within an authoritarian political system, translated into vulnerability to flooding for local communities. In later work, Mustafa (2002) devoted attention to conceptualising how differential social power levels translated into inequitable access to irrigation water and differential vulnerability to flood hazard for the same people. With reference to an ongoing hazard from volcanic activity in Ecuador, Tobin and Whitford (2002) used health indicators of evacuated communities as indicators of stress and vulnerability. Pelling (1998; 1999) documented the appropriation of the community participation and democratisation discourse and programming by the local elites in urban and peri-urban Guyana, thereby accentuating the vulnerability of the poor and the disenfranchised to flood hazard.

Gender specific configurations of vulnerability and disaster experience have also been documented. Using ethnographic data from Northern Pakistan, Halvorson

(2003) documents how changes in the local economy increased children's exposure and vulnerability to waterborne health hazards through their impacts on women's workloads. Fordham (1998) explored the gendered experience of flooding in Britain to make the case that socially produced gendered spaces make women differentially vulnerable to environmental hazards.

Beyond diagnosing the causality of vulnerability to environmental hazards, political ecological research has also been attentive to political economic drivers of post-disaster recovery issues. Empirical examples of post disaster policy changes and reconstruction have not been very encouraging in terms of addressing pre-existing vulnerability to hazards. Mustafa (2004) challenged the notion of restoring normalcy as an objective of flood relief and recovery when the pre-existing conditions were characterised by high levels of poverty, disempowerment and vulnerability. In the context of Hurricane Mitch, Brown (2000) traces the pattern of vulnerability that had been in place because of United States' strong support of neoliberal reforms in the Nicaraguan economy under the auspices of international financial institutions. Brown (2000) expresses skepticism that supposedly right words calling for transformation of Nicaraguan society through post-Mitch reconstruction will involve any profound shift in US policies that contributed to the country's vulnerability in the first place. Wisner (2001) in fact documents how post-Mitch recovery and reconstruction did not implement mitigation measures to prevent comparable losses from the 2001 earthquake in El Salvador. He attributes the failure to the El Salvador government's adherence to extreme form of free market ideology and the deep fissures in the society following the long and bloody civil war.

Running alongside these various radical but neo-realist approaches to understanding the social structuring of hazards and vulnerability to them, has been a relatively less influential but no less theoretically important strand of post-structuralist inspired research. Early work by Waddell (1977) and Hewitt (1983) pointed to the importance of human social systems and discursive constructs in framing hazards as 'unscheduled' or 'accidental' interruptions of 'normal' life. The initial invitation to engage with hazards from a post-structuralist perspective was later given an impetus by a turn towards post-structuralist thinking in wider human-environment relations research, as exemplified by the social-nature thesis advanced by Castree and Braun (2001). The 'social nature' or 'socionature' argument does not deny the materiality of non-human entities, but rather argues that we cannot separate their material existence from our knowledge of them. In other words there is no Olympian point from which we can gain value-free objective knowledge of non-human nature's existence. The socionature thesis is not intended mainly to stand judgement on the truth or falsity of claims about nature, but primarily points out how discourses on nature create their own truths (Castree, 2001, Demeritt, 2001). Drawing upon the socio-nature insights within geography, Mustafa (2005) outlined an approach to hazards he termed 'hazardscapes'. It posits that the material geographies of hazardousness and social responses are the outcome of various discursive constructs and interactions among hazard victims and managers with varying degrees of power/knowledge. In the case of urban flood plain in Pakistan, Mustafa (2005) contrasted the multifaceted understanding of flood plain residents of their environment as a hazardscape containing multiple interlinked social and environmental hazards with the dominant state vision of them as a series of administrative domains to be dealt with in isolation. The dominant Pakistani state's focus on the singular flood hazard, Mustafa (2005) argued, creates the material geographies of differential vulnerability and exposure in the urban hazardscape.

Such a post-structuralist perspective was used by Simpson and Corbridge (2006) to explore the politics of reconstruction and memorial practices that emerged after the 2001 earthquake in western India. Treating post-earthquake reconstruction as an exercise in place making, both discursively and materially, they illustrate how competing elite visions of Hindu nationalism and regionalism, manifest themselves spatially while silencing other memories of schedule castes and Muslims. The spatial production of exclusionary post-disaster geographies is likely to have profound and largely negative consequences for a multi-ethnic/religious country like India. In a similar vein but half a world away, Cupples (2007) cautions against strategic essentialism of gender sensitive disaster relief and reconstruction through a case study of post-Mitch reconstruction in Nicaragua. Beyond the negative material outcomes of disaster situations for women, which must be addressed through relief and aid, Cupples (2007) argues that attention must also be devoted to spatial shifts brought about by disasters so as to reveal new opportunities for women.

On the technological hazards side, Ulrich Beck takes the production of new, uncontrollable hazards from technological development to be the defining feature of our emergent 'risk society'. While science once delivered a sense of predictability and control over certain risks, e.g., traffic accidents, by making them probabilistically calculable and hence insurable, we are now entering a new age, according to Beck (1996), of widespread public recognition of the risks produced by technological progress itself:

> The hazards produced in the growth of industrial society become predominant. That both poses the question of the self-limitation of this development, and sets the task of redefining previously attained standards (of responsibility, safety, control, damage limitation and distribution of the consequences of loss) with reference to potential dangers. These, however, not only elude sensory perception and the powers of imagination, but also scientific determination. (pp. 28–29)

Public recognition of these new risks is helping to spark a new, more reflexive kind of modernisation associated with new forms of 'subpolitics' that operate outside and beyond the institutions of the nation-state.

Concerns about climate change illustrate many of Beck's claims about the risk society. Because of its anthropogenic origins and its global scale climate change is a mega technological hazard beyond the scope of traditional means of probabilistic calculations, insurance-based risk spreading, and damage limitation. Faced with deep scientific uncertainties about its likely extent and effects, conventional structures have struggled to cope. Into the breach have stepped new transnational forms of subpolitics, centred around environmental movements and new institutions of science, like the IPCC. To date, however, these efforts to address climate change have not resulted in much progressive socio-environmental change. Indeed, if anything, existing relations of power and international dominance have been reinforced (Bulkeley, 2001).

Though influential, Beck's risk society thesis is not without its critics (e.g., Bennett, 1999). The distinction between lay and expert knowledge is at the core of Beck's formulation of reflexive modernisation and risk society. While agreeing with the main thrust of Beck's arguments, Wynne (1996) uses the example of the sheep in Cumbria, England, contaminated by radioactivity from the Chernobyl nuclear accident, to question the distinction between lay and expert knowledge. He argues that expert scientific knowledge about radioactivity in the soils of Cumbria and possible courses of action was culturally problematic, based on inadequate models of human society, denigrated 'specialist lay' knowledge, and was known by the lay public to be problematic for these reasons. The public – Cumbrian farmers in this case, therefore treated expert knowledge with much more ambivalence than Beck is willing to give them credit.

In addition to these new approaches to understanding environmental hazards, the nascent post-structuralist trend has also generated work on non-traditional hazards, such as war (Hewitt, 1997) and terrorism (e.g., Mustafa, 2005 and Hewitt, 2001). In both cases, a particular concern has been discourses that legitimise practices of state and non-state terror and spatiality of that terror. Hewitt (2001) discusses the linkages between the spatiality of the Chilean state's self-image and its practice of terrorising certain spaces like plazas, shopping areas and streets. Mustafa (2005), in the same vein as Hewitt (2001), calls for empirically based research to find patterns in the spatiality of terrorist targets and the terrorists' (state and non-state) discursive construction of spaces as sacred and profane.

For all the insights provided by drawing on political economy and post-structural understandings, these radical strands of hazards research have had little impact on environmental and technological hazards policy. Part of the difficulty is that they are questioning the fundamental assumptions and praxis of contemporary society and its global capitalist economy. Therein lies their both greatest strength and the cause of their, hopefully limited, short-term impotence in the policy realm.

Conclusion: Key Geographical Contributions to Hazards Research

Hazards research is firmly embedded within the human-environment interactions research tradition of geography. Beyond the topical importance of hazard mitigation to human well-being and safety, hazards have been treated as special cases that illustrate wider problems within human societies and their interactions with, and understandings of, the non-human environment. Hazards research has helped broaden the policy agenda and move beyond simply controlling nature through scientific study of its underlying physical processes, to broader concerns with human exposure to hazards under the behaviourists, vulnerability under the political ecologist/radical geographers, and the nascent focus on hazardous socio-nature by the post-structuralists. Throughout, geographers have been attentive to both the conceptual and practical potentialities of hazards research. The attention to theory and social structures, with no immediately obvious practical or practicable guides to action, may seem indulgent and even frivolous when dealing with a subject of such profound importance for human life and well-being. But the review of different traditions above, illustrates the critical and long-term vision of the discipline of geography. Furthermore, the review also shows how geography's critical insights simultaneously confront the material basis and destabilise the legitimising discourses of dominant vulnerability producing social systems. The concern for immediate results in terms of safety and resilience does not distract geographical inquiry from seeking longer-term objectives of socially just, environmentally friendly, and materially sustainable human societies, because only such societies could be sustainably resilient in the face of environmental extremes.

The most recent consensus in the hazards research field includes the following propositions: (i) There is no trade-off between human well-being and environmental quality; in fact, there is a direct correlation between the two; (ii) The same social structures that cause inequality and poverty are also responsible for environmental degradation and inequitable resource distribution; (iii) Attempts at controlling and subduing nature by technically trained policy elites have in fact exchanged low intensity, high frequency environmental hazards, with high intensity but low frequency disasters; (iv) The damage and suffering from environmental hazards is caused by the dissonance between human social systems and the ecological systems in which they exist; (v) Allowing technically trained professionals the exclusive authority over technological or environmental hazards is a mistake. Participatory decision making informed, but not controlled, by scientific research is a must; (vi) The power/knowledge dynamics between hazards managers and disaster victims are important determinants of the geographies of vulnerability; and (vii) Policies directed towards ensuring social justice and environmental quality are the best guarantees against the damage and suffering caused by environmental hazards.

With the recent spate of mega disasters, there is an increasing demand for hazards' geographers skills and research outputs to educate the policymakers about vulnerability reduction and DRR so that the existing hazards/disasters paradigm moves from being reactive to proactive. Some of the major research themes being pursued to satisfy the demand for DRR strategies include development of quantifiable vulnerability indices (e.g., see Adger, 2006) and adaptation to climate change (O'Brien et al., 2006) and the accompanying need for greater communication between the epistemic communities of climate change, hazards research, and development (Schipper and Pelling, 2006). Vulnerability reduction and DRR are emerging areas of research within geographical hazards research with immediate policy implications. Geographical research on post-disaster relief and reconstruction from political ecological, post-structuralist and more practical orientations, promise to not only provide for more resilient communities, but also to point to the material and discursive root causes for unsafe conditions, which spawn the geographies of vulnerability in the first place.

Environmental hazards have been and will continue to be part of the human condition. The issue is whether humans will tackle the prospect of environmental extremes with arrogance and misplaced belief in technology, as they have in the recent past or will they approach their relationship with the environment and perhaps even technology with humility, reflexivity and respect. Greed and injustice render the weakest segments of the society differentially more vulnerable to hazards, just as the same lead to environmental degradation and hence more intense extreme events. Socially and ecologically sustainable development and not technical sophistication may yet be the best guarantee for a safer environment.

BIBLIOGRAPHY

Adger, W. N. (2006) Vulnerability. *Global Environmental Change*, 16, 268–81.
Alexander, D. (2002) Nature's impartiality, man's inhumanity; reflections on terrorism and world crisis in a context of historical disaster. *Disasters*, 26(1), 1–9.
Baxter, J. and Greenlaw, K. (2005) *Explaining perceptions of a technological environmental hazard using comparative analysis. The Canadian Geographer*, 49(1), 61–80.

Beck, U. (1992) *Risk Society: Towards a New Modernity*. London: Sage Publications.

Beck, U. (1996) *Risk Society and the Provident State*. In S. Lash, B. Szerszynski and B. Wynne (eds), *Risk, Environment and Modernity*. London: Sage Publications, 27–43.

Blaikie, P., Cannon, T., Davis, I. and Wisner, B. (1994) *At Risk: Natural Hazards, People's Vulnerability, and Disasters*. New York: Routledge.

Blaikie, P. M. and Brookfield, H. C. (1987) *Land Degradation and Society*. London: Methuen.

Bowman, D. M. J. S. and Franklin, D. C. (2005) *Fire Ecology*. *Progress in Physical Geography*, 29(2), 248–55.

Brooks, S. M. (2003) Slopes and slope processes: research over the past decade. *Progress in Physical Geography*, 27(1), 130–41.

Brown, E. (2000) Still their backyard? The US and post-Mitch development strategies in Nicaragua. *Political Geography*, 19, 543–72.

Bulkeley, H. (2001) Governing climate change: the politics of risk society? *Transactions of the Institute of British Geographers*, 26, 430–47.

Burton, I., Kates, R. W. and White, G. F. (1993) *The Environment as Hazard*, 2nd edn. New York: Guilford Press.

Cardona, O. D. (2004) The need for rethinking the concepts of vulnerability and risk from a holistic perspective: A necessary review and criticism for effective risk management. In G. Bankoff, G. Frerks and D. Hilhorst (eds), *Mapping Vulnerability: Disasters, Development & People*. London: Earthscan, pp. 37–51.

Castree, N. (2001) Socializing nature: theory, practice, and politics. In N. Castree and B. Braun (eds), *Social Nature: Theory Practice and Politics*. Malden: Routledge, pp. 1–21.

Castree, N. and Braun, B. (eds) (2001) *Social Nature: Theory Practice and Politics*. Malden: Routledge.

Cello G. and Malamud, B. D. (eds) *Fractal Analysis for Natural Hazards*. Geological Society of London, Special Publication, v. 261.

Colten, C. E. (2006) Vulnerability and place: flat land and uneven risk in New Orleans. *American Anthropologist*, 108, pp. 731–34.

Crozier, M., McClure, J., Vercoe, J. and Wilson, M. (2005) The effect of hazard zone information on judgments about earthquake damage. *Area*, 38(2), 143–52.

Cupples, J. (2007) Gender and Hurricane Mitch: reconstructing subjectivities after disaster. *Disasters*, 31(2), 155–75.

Cutter, S. L. (1993) *Living with Risk: The Geography of Technological Hazards*. New York: Edward Arnold.

Cutter, S. L. (1996) Vulnerability to environmental hazards. *Progress in Human Geography*, 20, 529–39.

DEFRA (2005) *Making Space for Water: Taking Forward a New Government Strategy for Flood and Coastal Erosion Risk Management in England. Delivery Plan*. London: DEFRA.

Demeritt, D. (2001) Being constructive about nature. In N. Castree and B. Braun (eds), *Social Nature: Theory Practice and Politics*. Malden: Routledge, pp. 22–40.

Disse, M. and Engel, H. (2001) Flood events in the Rhine basin: genesis, influences and mitigation. *Natural Hazards*, 23, 271–93.

Fordham, M. H. (1998) Making women visible in disasters: problematising the private domain. *Disasters*, 22(2), 126–43.

Golet, G. H., Roberts, M. D., Luster, R. A. and Werner, G. (2006) Assessing societal impacts when planning restoration of large alluvial rivers: a case study of the Sacramento river project, California. *Environmental Management*, 37(6), 862–79.

Graf, W. C. (2005) Geomorphology and American dams: the scientific, social, and economic context. *Geomorphology*, 71, 3–26.

Dupigny-Giroux, L., Hanning, J. R. and Engstrom, E. (2006) Orographic influence on frontally-produced flooding in Northern Vermont – The July 14–15, 1997, Event. *Physical Geography*, 27(1), 1–38.

Furley, P. A. (2007) Tropical savannas and associated forests: vegetation and plant ecology. *Progress in Physical Geography*, 31(2), 203–11.

Gupta, R. (2001) River basin management: a case study of Narmada Valley Development with special reference to the Sardar Sarovar Project in Gujarat, India. *International Journal of Water Resources Development*, 17(1), 55–78.

Hewitt, K. (2001) Between Pinochet and Kropotkin: state terror, human rights and the geographers. *The Canadian Geographer*, 45(3), 338–55.

Hewitt, K. (1997) *Regions of Risk: A Geographical Introduction to Disasters*. London: Longman.

Hewitt, K. (1983) *Interpretations of Calamity*. Boston: Allen & Unwin, Inc.

Johnson, C., Penning-Rowsell, E. and Parker, D. (2007) Natural and imposed injustices: the challenges in implementing 'fair' flood risk management policy in England. *The Geographical Journal*, 173(4), 374–90.

Kasperson, R. E., Renn, O., Slovic, P., Brown, H. S., Emel, J., Goble, R., Kasperson, J. X. and Ratick, S. (1988) The social amplification of risk: a conceptual framework. *Risk Analysis*, 8(2), 177–87.

Lafon, C. W., Hoss, J. A. and Grissino-Mayer, H. D. (2005) the contemporary fire regime of the central Appalachian mountains and its relation to climate. *Physical Geography*, 26(2), 126–46.

Litschert, S. (2004) Landslide hazard zoning using genetic programming. *Physical Geography*, 25(2), 130–51

Louisiana Department of Health and Hospitals (2006) *Deceased Katrina Victims Released 2-16-2006*. http://www.dhh.louisiana.gov/publications.asp?ID=192&Detail=1040&SearchFor=katrina (accessed 17 March 2008).

Mechler, R., Linnerooth-Bayer, J. and Peppiatt, D. (2006) *Disaster Insurance for the Poor? A Review of Microinsurance for Natural Disaster Risks in Developing Countries*. Geneva, Switzerland and Laxenburg, Austria: Provention Consortium & IIASA.

Mitchell, J. K. (1990) Human dimensions of environmental hazards, complexity, disparity and the search for guidance. In A. Kirby (ed.), *Nothing to Fear*: Tuscon: University of Arizona Press, pp. 131–75.

Mileti, D. S. (1999) *Disaster by Design: A Reassessment of Natural Hazards in the United States*. Washington, DC: Joseph Henry Press.

Mustafa, D. (1998) Structural causes of vulnerability to flood hazard in Pakistan. *Economic Geography*, 74(3), 289–305.

Mustafa, D. (2002) To each according to his power? Access to irrigation water and vulnerability to flood hazard in Pakistan, environment and planning. *Society and Space*, 20(6): 737–52.

Mustafa, D. (2004) Reinforcing vulnerability? The disaster relief, recovery and response to the 2001 flood Rawalpindi/Islamabad, Pakistan. *Environmental Hazards: Human and Policy Dimensions*, 5(3/4), 71–82.

Mustafa, D. (2005) The production of an urban hazardscape in Pakistan: modernity, vulnerability and the range of choice. *The Annals of the Association of American Geographers*, 95(3), 566–86.

Myers, M. F. and White, G. F. (1993) The challenge of the Mississippi flood. *Environment*, 35(10), 7–9, 25–35.

Nicholls, N. (2007) Has the climate become more variable or extreme: Progress 1992–2006. *Progress in Physical Geography*, 31(1), 77–87.

O'Brien, G., O' Keefe, P., Joanne, R. and Wisner, B. (2006) Climate change and disaster management. *Disasters*, 30(1), 64–80

Pelling, M. (1998) Participation, social capital and vulnerability to urban flooding in Guyana. *Journal of International Development*, 10, 469–86.

Pelling, M. (1999) The political ecology of flood hazard in urban Guyana. *Geoforum*, 30, 249–61.

Pitlick, J. (1997) A regional perspective of the hydrology of the 1993 Mississippi River Basin Floods. *Annals of the Association of American Geographers*, 87(1), 135–51.

Platt, R. (1986) Floods and man: a geographer's agenda. In R. Kates and I. Burton (eds), *Geography Resources and Environment: Themes from the Work of Gilbert F. White*, vol. 2. Chicago: University of Chicago Press, pp. 28–68.

Proctor, J. D. (1998) The social construction of nature: relativist accusations, pragmatist and critical realist responses. *Annals of the Association of American Geographers*, 88(3), 352–76.

Saarinen, T. F. (1966) *Perception of the Drought Hazard on the Great Plains*. Research paper no. 106. Chicago: Department of Geography, University of Chicago.

Schipper, L. and Pelling, M. (2006) Disaster risk, climate change and international development: scope for, and challenges to, integration. *Disasters*, 30(1), 19–38.

Simpson, E. and Corbridge, S. (2006) Geography of things that may become memories: the 2001 earthquake in Kachchh-Gujarat and the politics of rehabilitation in the prememorial era. *Annals of the Association of American Geographers*, 96(3), 566–85.

Slovic, P. (1987) Perception of risk. *Science*, 236(4799), 280–85.

Starr, C. (1969) Social benefit versus technological risk. *Science*, 165, 1232–38.

Tobin, G. A. and Montz, B. E. (1997) *Natural Hazards: Explanation and Integration*. New York: Guilford Press.

Tobin, G. A. and Whitford, L. M. (2002) Community resilience and volcano hazard: the eruption of Tungurahua and evacuation of the faldas in Ecuador. *Disasters*, 26(1), 28–48.

Todhunter, P. E. adn Rundquist, B. C. (2004) Terminal lake flooding and wetland expansion in Nelson County, North Dakota. *Physical Geography*, 25(1), 68–85.

Waddell, E. (1977) The hazards of scientism. *Human Ecology*, 5, 69–76.

Watts, M. J. and Bohle, H. J. (1993) The space of vulnerability: the causal structure of hunger. *Progress in Human Geography*, 17, 43–68.

Watts, M. J. (1983) On the poverty of theory: natural hazards research in context. In K. Hewitt (ed.), *Interpretations of Calamity*. London: Routledge, pp. 231–62.

Wescoat, J. (1992) Common themes in the work of Gilbert White and John Dewey: a pragmatic appraisal. *Annals of the Association of American Geographers*, 82, 587–607.

White, G. F. (1945) *Human Adjustment to Floods*. Research Paper 29. Chicago: University of Chicago, Department of Geography.

White, G. F. (1973) Natural hazards research. In R. J. Chorley (ed.), *Directions in Geography*. London: Methuen, pp. 193–216.

White, G. F. (1974) Natural hazard research: concepts, methods, and policy implications. In G. F. White (ed.), *Natural Hazards: Local, National, Global*. Oxford: Oxford University Press, pp. 3–15.

Wisner, B. (2001) Risk and the neoliberal state: why post-mitch lessons didn't reduce El-Salvador's earthquake losses. *Disasters*, 25(3), 251–68.

Wynne, B. (1996) May the sheep safely graze? A reflexive view of the expert-lay knowledge divide. In S. Lash, B. Szerszynski and B. Wynne (eds), *Risk, Enviornment and Modernity*. London: Sage Publications, pp. 44–83.

Chapter 28

Environmental Governance

Gavin Bridge and Tom Perreault

Introduction

The concept of 'environmental governance' has been in the ascendant since the mid-1990s. Drawing its intellectual credentials from a wave of social science research on non-environmental forms of 'governance,' environmental governance has gained rapid acceptance among geographers, sociologists, environmental managers, and development scholars to describe and analyse a qualitative shift in the manner, organisations, institutional arrangements and spatial scales by which formal and informal decisions are made regarding uses of nature. The term governance explicitly hinges the economic and the political, and its popularity within the social sciences reflects a broader institutional turn in which greater attention is paid to the relationships between institutional capacities, the coordination and coherence of economic processes, and social action. Our core claim in this chapter is that the popular appeal of the term may be proportional to its capacity to elide or conceal critical distinctions. Like 'sustainable development' and 'social capital' – with which environmental governance is occasionally allied – the vagueness and malleability of the term serve to obscure a broad range of interests and ideological positions.

Environmental governance, then, appears to have won widespread acceptance without the benefit of rigorous critique to review the range of circumstances in which it is deployed. Our modest hope is that this chapter advances such a critique. The chapter is divided into three main sections. Following this introduction, we distill six different meanings of governance in an effort to identify the epistemological and methodological tensions concealed within. We then focus on two distinctive lines of inquiry that have occupied the attention of environmental geographers: neoliberal modes of environmental governance, and eco-governmentality. Finally, we end the chapter by outlining an agenda for strengthening geographical research on environmental governance.

Unpacking Environmental Governance: Six Problematics

Governance refers to the fundamental question of how organisation, decisions, order and rule are achieved in heterogeneous and highly differentiated societies. At its core, governance addresses the problem of economic and political co-ordination in social life. Accounts of governance typically describe the form and geographical scale of socio-political institutions, identify key actors and organisations, and characterise how relations among these components may be changing (Wood and Valler, 2001; Jessop, 2002). The term's tap-roots trace to the 'new institutional economics' (North, 1990), economic sociology (Polanyi, 1944, Granovetter, 1985) and regime theory in the field of international relations (Rosenau, 1991). The academic emergence of governance marks an increasing skepticism towards traditional theories of economic and political action, such as the behavioural assumptions of neoclassical economics, Marxist analyses of the bourgeois state, or realist accounts of international relations premised on ideas of competitive relations among independent states.

The mainstreaming of 'governance' as an explanatory concept within the social sciences reflects the capacity of the term to carry several distinct meanings, according to disciplinary and ideological context. At its core, however, are three interrelated concerns. First, the concept explicitly problematises state-centric notions of regulation and administrative power and describes a putative shift in the institutional geometries of power. Governance thus recognises political authority as being multi-layered and often operating across several different spatial scales (Painter, 2000, p. 360, Lemos and Agrawal, 2006). Second, work on governance frequently highlights (or claims) the obsolescence of inherited analytical categories (e.g., private, public, state, sovereign, government) and the policy frameworks upon which these are based. Third, work on governance foregrounds the growing influence of non-traditional actors (such as NGOs, supra-national agencies, social movements, or sub-national administrative units) and qualitative shifts in the role of more familiar actors such as corporations and state agencies (McCarthy, 2005). Taken together, these shifting relations are understood as a restructuring of the social compact – the norms and expectations that differentiate and demarcate the arenas of 'private' and 'public'. Thus, in both analytical and policy approaches, the *discourse* of governance is strongly associated with social and economic change. It is not surprising, then, that environmental governance has assumed prominence as an explanatory concept at a time when the authority and legitimacy of the national state *viz.* environmental and resource issues is being increasingly called into question.

Packed into the slender frame of 'environmental governance' are a variety of different meanings, which we unravel below. Much of the social value, as opposed to analytical value, of the term environmental governance lies in its capacity to 'do political work' – that is, to suggest commonalities of purpose and interest that can obscure divergence and conflict. Indeed, to some observers the rise of environmental governance is symptomatic of a 'post-political condition' in which politics is reduced to the tactical practice of producing a consensus on the need for action in the face of an externalised threat (Swyngedouw, 2007). The popularity of environmental governance as an organising concept, then, is partly independent of intellectual currents within social science and stems from its capacity to articulate managerial concerns about 'environmental problems' (cf. Keil and Desfor, 2003). A definition of governance as 'attempts by governing bodies or combinations thereof to alleviate recognised environmental dilemmas' for example, reflects a managerial rather than analytical approach to governance that obscures the politics of definition (Davidson and Fickel, 2004). The *politics* of environmental governance, then, is a critical question to be brought to the fore. It is a question that may be highlighted by asking simply governance of what, by whom, and to what end?

In seeking a better grip on it, we unpack in the next section some of the competing claims that become loaded onto the concept of 'environmental governance'. Divergences in meaning originate from two sources: differences in the underlying 'problematic' or object of governance (i.e., the relations being governed); and different stances towards the function of knowledge-production (critical knowledge vs. instrumental knowledge). It is important to keep these two axes of difference separate, as the contested concepts at the heart of governance involve not only epistemological questions about the role of academic knowledge and practice but also more fundamental, ontological questions about the composition of society. For heuristic purposes we can identify a matrix bounded by a horizontal axis (along which we have arranged six different problematics) and a vertical axis (describing two different stances towards the function of knowledge). This produces the 6 × 2 matrix in figure 28.1. A key distinction, we suggest, is between approaches derived from political economy that understand governance as an *immanent* process rooted in the social relations of production (in its broadest sense); and those derived from realist approaches to international relations and development studies that approach governance as an *intentional* process – i.e., an active intervention to secure a particular outcome (cf. Hart, 2001, Cowen and Shenton, 1996). By borrowing this distinction from critical approaches to development, we can see not only how discourses of environmental governance describe significant shifts in the spatial, administrative and political relations of governing nature, but also how proposals for fixing various 'environmental' crises produce particular forms of social order. Environmental governance, then, is about *both* the social organisation of decision making with respect to the environment, *and* the production of social order via the administration of nature.

Governance as a problematic of scale

For many researchers the core problematic which environmental governance addresses is the de- and re-constitution of scalar relations. A range of environmental phenomena suggest the contemporary period is marked by a radical reworking of geographical scale: city and provincial governments seizing the initiative from national states in crafting 'global' initiatives around fair-trade and climate change (Bulkeley, 2001); the pervasiveness of transboundary material flows associated with pollution and the movement of hazardous and municipal wastes (Spaargaren et al., 2006); efforts to craft new regulatory structures for various 'global' natures (oceans, Antarctica, genetic diversity) that are outside interstate systems of regulation (Steinberg 2001). In considering how work on environmental governance addresses scale, however, we find a paradox. A key attribute of the concept of governance is that it is scale-free: like a number of other foundational geographical terms (e.g., 'ecosystem' or 'watershed'), governance may be used in reference to a rich range of spatial imaginaries from global climate and oceans to local species and neighborhoods. The concept of governance, then, is inherently agnostic when it comes to the question of an absolute scale at which governance is achieved. But at the same

PROBLEMATICS: Governance as......

	Re-scaling	Commodity chain co-ordination	Collective action for resource management	Political participation	State (re)-regulation	Rule and the production of (socio-natural) order
Analytical Stance	Swyngedouw (2000), McCarthy (2005), Perreault (2005)	Mutersbaugh et al. (2005), Taylor (2005)	Ostrom (1990)	Keil and Desfor (2003), Backstrand (2004), Swyngedouw (2007)	Bridge (2000), Bakker (2003a), Whitehead et al. (2007)	Darier (1996), Watts (2003), Dalby (2004)
Descriptive or Normative Stance	Uitto (1997), Borgese (1999)	Global Witness (1998), Bass et al. (2001)	Young (1994), Buck (1998)	Wapner (1995), Palmer (2006)	Mol and Sonnenfeld (2000), Fischhendler (2006)	Luke (1999)
Intellectual tradition and historical antecedents	Geographical political economy: uneven development	Political economy (dependency theory)	Anthropology, International relations	Political science, critical social theory	Institutional economics regulation theory, economic sociology	Governmentality, post-colonial studies, theories of empire

Figure 28.1 Six 'problematics of governance' and representative contributions to the literature.

time work on environmental governance is positively evangelical in its contention that governance is *all about scale*: recognition of the complex spatialities of environmental degradation and ecological interdependence, in other words, create the analytical and policy space that 'environmental governance' has come to fill. Long a core concern for geographers, the understanding of governance as a problematic of re-scaling is now found across a broad swathe of social science research on environmental governance.

Understood as a problematic of geographical scale, research on environmental governance expresses the broader 'spatial turn' within the social sciences. A distinction can be made, however, between research in which processes of re-scaling are the things to be explained (and where scalar outcomes are uncertain), and work in which different scales of governance are already assumed and governance jumps one or more notches in the scale hierarchy. The distinction here is between a processual understanding of scale as a produced outcome of political-economic activity, and Cartesian understandings of scale as a nested hierarchy describing different levels of territorial extensiveness (Bulkeley, 2005, p. 877). Much of the applied environmental management literature on governance, for example, adopts a processual view of scale as the outcome of deliberation and social decision, but looks to natural systems – watersheds, river catchments – for guidance on the geographical scale of governance regimes in order to avoid what it sees as 'the costs of dysfunctional environmental management caused by inappropriate jurisdictional boundaries' (Brunckhorst and Reeve, 2006, p. 147). This approach is also found in the normative literature on bioregionalism, with its conscious effort to re-imagine scales and spaces of governance in ways quite different to those of conventional political units.

Governance as commodity chain co-ordination

A core strand of research on governance concerns the coordination of exchanges within and between firms and the relative distribution of power among competing actors (e.g., between producers and consumers) along a production chain. A substantial literature within economic geography, for example, describes the organisation of production chains (also known as *filières*) and distinguishes between two ideal types of governance: hierarchy (where exchanges are internalised with a firm) and markets (where exchanges occur by contracts between firms) (see Lewis et al., 2002, Coe et al., 2004). Governance in this work describes 'patterns of authority and power relations which structure the parameters under which actors operate, including what is produced, how and when it is produced, how much is produced and at what price' (Humphrey and Schmitz, 2001; Taylor, 2005a, p. 130). Despite an extensive body of work on manufacturing and agro-food commodity chains that examines the implications of different forms of governance for regional development, industrial upgrading and labour practices (e.g., Gereffi et al., 2005, Ponte and Gibbon, 2005), the environmental implications of these relations of governance along commodity chains and across production networks have only recently begun to be investigated. Work on fair trade commodities, conflict diamonds, and forestry certification, for example, demonstrates the analytical possibilities of thinking about environmental governance along the structure of commodity chains (Cashore et al., 2004; Taylor, 2005a; 2005b). This work draws attention to the ways in which the structure of the production network influences how consumer demands translate into environmental (and social) conditions of production and, in turn, to how environmental regulations, certification and labelling can create new opportunities for value capture that restructure production networks (Stringer, 2006). While commodity networks are often deeply embedded in territorial structures, the focus in much of this work is on the emergence of new actors and spaces outside the territorial state and the way these can influence the environmental consequences of production and consumption in ways that exceed the reach of formal state regulation. Product boycotts, public campaigns and social activism are symptoms of the emergence of alternative or parallel 'regulatory' mechanisms that often articulate *with* the state (by, for example, pointing to the way corporate behaviour in one jurisdiction does not meet legal standards in another) but yet are not *of* the state. Product labeling and certification schemes – such as those associated with the Forestry Stewardship Council or Fair Trade – illustrate the increased importance of information and information provision in producing new modalities of governance (Mol 2006). Other work highlights a deliberate move towards the enrollment of third parties and market-based mechanisms – the outsourcing or 'privatisation of governance' – that complement the regulatory function of the state (O'Rourke, 2003, Cashore et al., 2004). Bennett (2000), for example, has shown how private insurers have become enrolled into a state regulatory framework to achieve regulatory goals of reducing environmental damage associated with pollution from ocean-going oil tankers. Research like this draws attention to modalities of regulation along commodity chains: in the case of oil tanker regulation, a concern for the effectiveness or quality of regulation positioned private insurers at the centre of the regulatory programme rather than merely a means to an end.

Governance as collective action for resource management

A third significant strand of research on environmental governance problematises governance as a social action problem. This work draws on both a long tradition of human ecology (which stresses social adaptation to environmental systems), and on the 'new institutional economics'. The latter highlights the role of social institutions – understood as formal and informal rules, conventions and codes of behaviour – in regulating human activities and rejects the individualism of neoclassical models of rational (economic) behaviour (Mehta et al., 2001). Geographers and anthropologists have examined the influence that different property regimes have on the governance of environments and resources. Much of this work challenges Hardin's (1968) highly influential metaphor of the 'tragedy of the commons', which asserts that collective resource management will inevitably fail and that, as a consequence, private ownership or state-control of resources is preferable. Research on a range of forestry, pasture and fisheries governance regimes around the world not only demonstrates that collective resource management regimes are widespread and can be resilient in the face of economic and environmental change; it also illustrates how many alternate forms of resource and environmental governance are possible beyond the stark choice posed by Hardin between 'state' or 'private'. The intellectual recognition of successful collective modes of environmental governance has provided a justification for decentralised, participatory and community-based natural resource management, and for hybrid management arrangements involving state and local groups (Mehta et al., 2001). Such co-management regimes are now ubiquitous, their uptake driven partly by evidence of successful self-management at small scales, the failures and exclusions of state management, and the problematic nature of local management when environmental systems and social drivers occupy translocal scales.

Recent research has thrown into question the equilibrium theories and models of ecological systems that underpinned work on the governance of common property resources. Highlighting the prevalence of disturbance events in ecological systems and the capacity of these systems for 'surprise', new research emphasises the importance of change, risk and uncertainty in the management of resources, and promotes adaptive resource management as an alternative framework for the governance of natural resources. Adaptive management explicitly recognises uncertainty and seeks to optimise decision making around natural resources via an iterative process of intervention, monitoring, evaluation and adaptation in the light of new information (Adger, 2001). Recent research on the governance of a range of resources – from fisheries to global climate – emphasises the possibilities for improving resilience and reducing vulnerability by designing responsive and flexible social mechanisms that allow adaptation as new information becomes available.

Governance as political participation

For some analysts, the primary problematic which governance addresses is the expansion of the political realm from the formal arena of representative democracy to include a range of other actors and political spaces, referred to by Beck (1992) as 'sub-politics'. While also the case in other areas of governance, it is in the environmental arena that extensive non-state activity has most forcefully expressed itself. The most visible symptom of this expansion of politics is the proliferation in number and variety of political actors on environmental issues – evidenced by the rapid and sustained growth in environmental non-governmental organisations over the last three decades – and the diversification of arenas in which politics is practiced. The net effect is a decentring of political authority on the environment, a shift of authority away from the state that is reflected in research on 'civic environmentalism' (John, 1993) and in the rise of new social movements that articulate a distinctive 'environmental' agenda (Howitt, 2001). The increasing prominence of non-governmental, non-elected bodies on matters of the environment raises questions about their implications – subversion, eclipse, augmentation – for formal politics (Bickerstaff and Walker, 2005). To what extent, for example, do contemporary modes of environmental governance unbundle environmental obligations once vested in the state, an eclipse of the public realm through the effective 'privatisation' of environmental decision making? It is apparent that notwithstanding the growth of non-governmental actors on environment, states retain substantial authority for environmental regulation. Not only have a variety of 'state natures' (from national parks to oil resources) proven remarkably durable over time, but also the 'environmental state' – which emerged in industrial economies from the late 1950s onwards to regulate and allocate the 'environmental bads' of pollution and resource degradation – cannot be written off as an historical artifact of late-Fordism (Whitehead et al., 2007).

More concretely, practices of contemporary environmental governance problematise the core political questions of whose voices get heard and who makes decisions. At the centre of environmental governance, then, are questions about the rights, obligations and responsibilities of political actors – issues that go to the heart of who – or, indeed, what – can be recognised as having political agency: indeed, as Bulkeley (2001, p. 442) observes, 'the scale, scope and nature of contemporary environmental risks poses novel challenges for "relations of definition"'. From this perspective a key part of governance, then, is the recasting of relationships via the language and models of partnership, participatory development, and stakeholder participation. These discursive tropes subvert old administrative, governmental hierarchies of ruler and ruled, state and citizen with their explicit indications of power and suggest an equality of agency among political actors. This occurs both in formal planning processes (e.g., via processes of consultation) and in the corporate social responsibility literature. Concern over the negative developmental effects of extractive industries, for example, has spurred a raft of initiatives to construct 'good governance' regimes around oil, minerals and forestry projects (primarily in the developing world). Institutionalised as new partnerships for development, these resource and environmental governance initiatives frame the politics of extraction as questions of inclusion and participation, rather than justice, rights and distribution. While these initiatives create new political spaces and facilitate the emergence of new actors, they can also serve to stabilise extractive regimes (Zalik, 2004).

The construction of consent has been the subject of several normative critiques of participatory forms of governance. Palmer (2006), for example, critiques the 'assimilative environmental governance arrangements' around national parks in Australia. Claiming that 'democratic projects that seek to create spaces of inclusion for indigenous peoples within existing environmental governance arrangements' are politically inadequate, she turns to alternative traditions of governance and sovereignty outside or preceding the nation state. The value of this critical perspective on governance as a process of negotiating the boundaries of the political is highlighted in Keil and Desfor's (2003, p. 28) assessment of environmental policy in Toronto and Los Angeles. They strive to recover local policymaking as a contested, political process 'about how we live our lives in cities for the generations to come . . . about sustainability, justice and redistribution' rather than 'a merely technical exercise, restricted to the arcane world of experts in the policy community.' As they point out, 'where policy making appears as a solely technical exercise, there are suspicions about what took the contentiousness out of the process'.

Governance as a problematic of state (re-)regulation

For researchers working from a neo-Marxian tradition, environmental governance is understood primarily as a matter of the social regulation of capitalist accumulation. Largely inspired by regulation theory and its focus on the institutional arrangements of state, market and civil society in relation to capitalism, this work focuses on the roles played by these 'extra-economic' institutions in stabilizing – politically and socially – particular regimes of accumulation. From this perspective, the rise of an administrative state apparatus for managing the environmental state appears analogous to the welfare state regimes put in place in industrial economies in the post-war period: both serve to mitigate some of the contradictory social/socio-ecological relations arising from the commodification of labour and nature (Hudson, 2001; Gandy, 2002). The 'environmentalisation' of the state, then, is an historical process that can be observed to have taken different trajectories around the world: from colonial administrative concerns over forestry and soil conservation to the introduction of regulations governing microtoxins, state-centric forms of environmental governance can be understood as efforts to offset various crises of underproduction (of nature, including the underproduction of human health) by shoring up the ecological conditions on which accumulation depends.

A range of critiques from the left, however, have challenged not only the efficacy of state-based environmental regulation, but also the assumptions of the relatively autonomous state that can be associated with accounts of the 'environmental Leviathan' (Paehlke and Torgerson, 1990). Accordingly, research on governance derived from regulation theory develops more socially embedded accounts of the state. In the Gramscian-inspired work of Jessop, in particular, the state becomes a social relation and a site of strategic action by different parts of civil society. This opens up a space for considering the way in which the 'regulation' of environment takes place not via the administrative and legal structures of government, but via the interactions and negotiated consent of many different actors. MacLeod and Goodwin (1999), for instance, highlight the institutional reconfiguration from state-centric govern*ment* to the multi-scalar ensemble of state and non-state actors involved in govern*ance*, which they discuss as a hallmark of urban administrative practice under neoliberalism (see also Painter, 2000; Wood and Valler, 2001). Central to this interpretation, then, is a focus on the 'hollowing out' of the state, and the role of non-state or quasi-state actors in carrying out functions that were previously the sole responsibility of the central state. This view has helped shape geographic analysis of urban and regional environmental governance and urban sustainability (Jonas and Gibbs, 2002). Gibbs and Jonas (2000; 2001), for instance, focus on the role of cities and localities in formulating environmental policy and enacting sustainable development plans, and highlight the rescaling and reinstitutionalisation of environmental policy at the urban scale. Rooted in regulation theory, this approach examines environmental decisions within the context of capitalist regimes of accumulation, arguing that urban environmental governance is a crucial spatial and scalar 'fix' to the after-Fordist crisis (Tickell and Peck, 1995). In much of this literature, then, 'environmental governance' captures the way actually-existing forms of neoliberal governance are multi-sited and multi-scaled, are products of social and political mobilisation, and can just as readily produce differentiation as convergence in norms relating to resource use and environmental quality (Prudham, 2004; Perreault, 2005).

A related approach is found in work on critical resource geography, sometimes referred to as 'First World political ecology' (Bakker, 2003a; Castree, 2006). Geographers working in this field seek to illuminate the ways that particular institutional configurations – for example resource rights, policies regarding resource extraction and conservation, or codified social norms and management practices – mediate the metabolic relationship between nature and society, and in so doing serve to stabilise environmental and social regulation within a given regime of accumulation. Such institutional arrangements are seen as responses to, and codifications of, the social and ecological contradictions of capitalism (e.g., Bridge, 2000; Bridge and Jonas, 2002; Bakker 2003b).

Governance as rule and the production of (socio-natural) order

A sixth problematic at the heart of some mobilisations of 'governance' concerns relations of power in the absence of a single, dominant authority. For researchers working in international relations, for example, the language of 'governance' is closely linked to the core concerns of regime theory, which emerged in the early 1990s as a way of thinking about relations between states in the absence of a clear hegemonic power. As Conca (2004, p. 15) explains, regime theory expresses the role of 'bargain-based co-operation to overcome barriers to collective action in response to . . . collective but differentiated responsibility.' There is a large literature (in international politics, public administration, and management studies, for example, and mainly outside of geography) which describes internationalist modes of governance for the 'global commons' and the management of transboundary pollution. Much of this work is premised on the assumption that addressing various international 'environmental crises' – climate change or species loss, for example – requires an unprecedented level of cooperation among states and the generation of new forums for interstate collaboration.

The emphasis on bargaining in this view of governance sits uneasily with another state-centric perspective on the problematic of rule which has emerged in relation to environment and resources in recent years: Empire (Harvey, 2003; Conca, 2004; Retort, 2005). Imperialism has a long and relatively complex intellectual history: its focus on the projection of state power via the control of extraterritorial land and resources has, however, made it a more muscular alternative to cooperation-based theories of governance for those seeking to understand the geographical expansion and extraterritorial control of contemporary resources and environments (Dalby, 2004).

Issues of Empire aside, a core tension within applications of governance-as-rule is whether one regards governance as a theory of *interstate* relations (an approach common to realist approaches to international relations), or whether it describes a 'non-statist reading of governmental rule' in which rule comes not via an authoritarian centre but through 'consensual-cum-socialized forms of political control' (Sparke, 2006, p. 358). The latter view actively decentres the state as the prime political authority and addresses instead the role of 'new regulatory coalitions, non-state-based regulatory mechanisms . . . and 'private authority' in world politics' (Conca, 2004, p. 17). Governance here explicitly problematises state-centric notions of regulation and administrative power: it alleges a transition from govern*ment* to govern*ance*, and is expressed in the proliferation of non-state actors with claims to represent particular environmental interests (NGOs in particular, but also TNCs). Thus the Commission on Global Governance, for example, defines governance as 'the sum of the many ways individuals and institutions, public and private, manage their common affairs. It is a continuing process through which conflicting or diverse interests may be accommodated and cooperative action may be taken' (cited in Dalby, 2002, p. 429). The focus on new modalities of rule has spurred policy proposals to change administrative scales of decision making, particularly via decentralisation and a limited devolution of authority to territorial units at scales below that of the nation state (Brannstrom et al., 2004).

Understood as a problematic of rule – and, in particular, as a problematic of the production of socio-natural order in the absence of a sovereign authority – the concept of environmental governance has a close affinity with the notion of (eco)-governmentality. Darier (1996), for example, neatly captures this problematic in his oxymoronic definition of governance as 'regulated autonomy'. Understood as a problem of rule, environmental governance draws attention to the normalisation of 'environmental' objectives and rationalities within a society, and the ways in which power – the capacity to get other people and things to align in particular ways – increasingly works through environmental rationalities. Geographers have made significant contributions to this perspective on governance-as-rule, as discussed in more detail below.

Assessing environmental governance

While we critique the widespread, uncritical use of 'environmental governance', our intention is not to discard the concept wholesale but to refine its application. Our view is that environmental governance retains some positive value for geographical analyses of nature-society relations for three broad reasons. First, insofar as environmental governance is centrally concerned with questions of spatial, ecological and administrative scale, it opens up a space for critical analyses of scale's production and contestation. A focus on the geographical scales of governance provides a way to think about scale as an inherently political and unstable spatial manifestation of socio-environmental relations. We have in mind here not only the way in which ecological processes and socio-political capacities can reside at different spatial scales, but also how particular scales become privileged as *the* appropriate sites for participation and decision making (Adger, 2001; Brannstrom et al., 2004). Geographers have readily accepted environmental governance's implicit invitation to think critically about scale, examining the scalar politics of environmental governance in the context of community forestry (McCarthy, 2006), fisheries (Mansfield, 2004b) and urban water systems (Swyngedouw, 2005).

Second, environmental governance focuses attention on the problem of *coherence* and the ways in which different peoples - and radically differentiated parts of the non-human world (such as atmospheric gases, tropical forests and fossil fuels) – may be brought into durable forms of alignment, despite problems of incommensurability and the many political, economic and social tensions that can exist around issues of the environment and resources. In contrast to the alternative concept of environmental management – which can imply a unitary 'manager' - environmental governance highlights the articulation of a range of actors. This perspective can enable relatively nuanced analyses of how power is produced and exercised over and through the non-human world, and the ends to which power is directed. Bakker (2002, 2007), for instance, has demonstrated how the commodification of water services requires the reconfiguration of not only market institutions but also those of the state and civil society, and how the whole process may be understood as an effort to stabilise a 'market environmental' or 'green neoliberal' regime of capital accumulation.

Third, by foregrounding decision making and the political process more generally, environmental governance helps us to think creatively about politics as the process of imagining, challenging and producing collective environmental futures. Environmental governance extends a broad embrace, encompassing relatively mundane issues like urban land use zoning or neighborhood recycling policy, as well as higher-profile concerns such as the patenting of life-forms or the regulation of access to the resources of Antarctica or the deep oceans. In each instance, however, an environmental governance perspective can focus attention on who participates in decisions large and small, and a sensitivity to the extent to which policies and proposals – i.e., the mechanisms through which environmental futures are enacted – express an elite vision or have 'social depth'. At its best, then, environmental governance can help to revitalise politics as a social practice – a struggle to define the future that is fundamentally part of, rather than apart from, the realm of normal life – and, consequently, to connect formal politics with everyday practice.

Geography's Contributions to Understanding Environmental Governance

Each of the six problematics outlined in the previous section finds an expression in geographical research on environmental governance (see figure 28.1). In the hands of geographers, environmental governance has become a broad analytical framework for addressing the institutional arrangements, spatial scales, organisational structures and social actors involved in decision making around different environments and resources. The term can also imply an attentiveness to network geographies and spatial assemblages produced via flows of materials, both commodities and uncommodified wastes. In short, geographical work on environmental governance focuses on the institutional (re-)alignments of state, capital and civil society actors in relation to the management of environments and resources, and the implications of these configurations for social and environmental outcomes. Although all six problematics are expressed in recent geographical scholarship on environmental governance, geographers have mobilised the language and concepts of environmental governance most extensively around two broad areas of inquiry: neoliberal modes of environmental governance, and eco-governmentality. These are examined in turn.

Neoliberal environmental governance

A central focus of work on environmental governance in geography has been the effects of neoliberal policies for environmental conditions and the management of environments and resources (Castree, 2008a,b; Himley, 2008). Neoliberalism is characterised by an institutional realignment away from state-centric (public-sphere) to market-based (private-sphere) forms of governance. As McCarthy and Prudham (2004, p. 279) point out, neoliberalism '. . . entails the construction of new scales ('the global market'), shifting relationships between scale ('glocalisation', the alleged hollowing out of the nation-state), and engagement with many scale-specific dynamics, all of which take shape and become tangible in the context of particular cultural, political and institutional settings'. In a broad sense, then, neoliberalism is an economic and political project that seeks to liberalise trade (particularly international trade), privatise state-controlled industries and services, and introduce market-oriented management practices to the reduced public sector (Jessop, 2002). Politically, neoliberalism seeks to 'roll back' selectively certain state functions, particularly the provision of social services and regulatory restraints on corporate practices. It comes as no surprise then that the governance of nature and resources would also be subject to neoliberal logic:

> [E]nvironmental governance itself is increasingly oriented toward market-based, rather than state-led, approaches: a prime example are emissions trading schemes as solutions to pollution, such as those proposed for reducing greenhouse gases that contribute to global warming. The rationale for this neoliberal turn in environmental governance is that market mechanisms will harness the profit motive to more innovative and efficient environmental solutions than those devised, implemented, and enforced by states. (Mansfield, 2004a, p. 313).

Neoliberal environmental governance, then, involves the reconfiguration of the institutional arrangements involved in managing nature and natural resources in such a way as to favor market-based actors and practices. This commonly involves the simultaneous rescaling of these institutions, as new actors and organisational forms are favored over others. As Bakker (2002; 2007) points out, the institutional realignment towards market principles also necessitates not just transformations in social relations and material practices, but also in the ways that natural resources themselves are conceptualised and discursively represented (see also Swyngedouw, 2005). Resources such as water are no longer conceived of as public goods that individuals have rights to as citizens, but rather as scarce commodities to which consumers have access via the allocative mechanism of the market.

The application of market principles to resource governance involves a fundamental shift towards private sector norms and institutions such as competition, markets, and efficiency indicators. The privatisation of (formerly) publicly controlled natural resources represents one such institutional reconfiguration, what Harvey (2003) terms 'accumulation by dispossession.' Harvey maintains that the mechanisms Marx (1967[1867]) described in his discussion of primitive accumulation have in recent years been refined through labour and social policy reform, trade agreements, resource privatisation, and economic and political restructuring, all of which have facilitated renewed rounds of accumulation. Rather than viewing such processes, with Marx, as the 'original sin of capitalism,' – a one-time, original enclosure of common property – Harvey views these processes as continual, and functional to – even necessary for – continued accumulation. While strategies for accumulation by dispossession have been a standard practice since the advent of capitalism, they have been facilitated and indeed encouraged by neoliberal restructuring, and attendant multi-scalar, diversified institutional frameworks for environmental regulation (McCarthy and Prudham, 2004).

For instance, Swyngedouw (2005) argues that recent efforts to privatise drinking water systems represent a common and particularly pernicious form of accumulation by dispossession. Geographical work on these processes has been valuable in explicating the institutional processes and multi-scalar politics involved in neoliberal environmental governance. Bakker (2002; 2007) illustrates how the neoliberalisation of water management occurs along one or both of two axes: privatisation and commercialisation. Privatisation involves an organisational transfer of ownership from public to private control, while commercialisation involves an institutional transformation, as efficiency measures, market mechanisms, and principles of competition gain primacy in resource management. The complexity and multiplicity of neoliberal institutional forms is similarly noted by Budds and McGranahan (2003), who argue that what is commonly referred to as (and decried) as water 'privatisation' in fact involves a variety of institutional and organisational arrangements, lying on a continuum from state control and non-market, to wholly private and market-based. These authors highlight the continued necessity of an activist state under neoliberalism, in establishing, regulating, and participating in markets, an irony that belies the neoliberal conceit of self-regulating markets.

At its core, environmental governance in a neoliberal era is concerned with the twin concepts of property and privatisation, and much recent geographical work has been focused on the contradictions and complexities involved in defining property and forming markets for resources and environments where none had previously existed (see, for instance, Mansfield, 2004a,b; Robertson, 2004). These are also central concerns of international trade agreements such as the North American Free Trade Agreement (NAFTA), which carry profound implications for environments and natural resources insofar as such agreements subsume environmental governance under the rubric of 'free market' capitalism. These processes have been investigated by McCarthy (2004; 2005), who has demonstrated the ways in which NAFTA has re-scaled and reinstitutionalised environmental governance. McCarthy's work also sheds light on new possibilities for agency opened up by neoliberal trade agreements. Indeed, resource-user organisations, environmental NGOs and other interest groups are important (at times the *most* important) implementers of resource governance decisions. McCarthy (2005) demonstrates that civil society actors such as environmental NGOs may have greater capacity to influence policy agendas under neoliberalism than under more centralised, Keynesian models of governance. Similar processes are at work in Bolivia, where indigenous and peasant social movements have played a major role in shaping water governance (Perreault, 2005).

From an administrative point of view, it is apparent that efforts to neoliberalise environmental governance are not far removed from the neoliberalisation of other economic sectors, such as the privatisation of pension funds or the opening of state-controlled telecommunications industries to market forces. As McCarthy (2006) notes, however, nature's irreducible materiality asserts itself in ways that set environmental governance apart from other regulatory domains. The careful attention paid by critical resource geographers and political ecologists to the biophysical qualities of natural resources – and therefore as factors of production distinct from one another and from human-made commodities – illuminates the particular problems each resource poses for its metabolism into capitalist relations of production (Bakker and Bridge, 2006). Drawing on Polanyi's (1944) discussion of nature as a 'fictitious commodity,' Bakker (2003b) exposes the 'uncooperative' nature of water as a commodity: although frequently subject to market-based modes of allocation, water's biophysicality – a product of climate, geology and ecology, an essential component of life - resists full commodification. Similarly, the biology and geography of the Douglas Fir – the preferred tree of timber firms in the North American northwest – militates against many practices that would optimise and rationalise the harvesting, processing and replanting of trees (Prudham, 2005). In this way nature presents barriers to, and opportunities for, accumulation that can necessitate a reconfiguration of the institutional form of capitalist processes. The fact that natural resources such as water, copper and wood are essential to the primary circuit of capital but are not themselves produced by capitalism sets them apart from other, human-produced commodities. The fact that the materiality of nature – the biophysical characteristics of particular natural resources – makes a difference to the way processes of accumulation 'work' in some sectors means that the governance of resource access, use and environmental impact in these sectors has become a vital area of research for environmental geographers.

Eco-governmentality

Governmentality is a concept allied with governance, yet also distinct from it. It shares with governance an interest in the process by which people, organisations and things come into alignment with political objectives. It focuses more explicitly, however, on the mechanisms of power and the specific question of how people and things come to be aligned in ways that enable their administration and rule. In short – and as the term itself indicates – governmentality is centrally concerned with the rationality of government. Where the starting point for work on governance is a putative shift in the actors and spaces of decision making (away from government to governance), governmentality explicitly returns to government as an analytical and historical problem: that is, to understand how '"the possible field of action of others" is structured' and the mechanisms through which governable subjects and governable objects are produced (Foucault, 1982, p. 221; Watts, 2003, p. 12). The origins of work on governmentality lie in Foucault's consideration of the character of modern power, an historically specific form of power emerging from the 17th century onwards marked by a shift in its 'point of application' away from territorial control to the governance and administration of 'things' (Scott 1995). Geographers (among others) have found this conception of power appealing as a way of thinking about the mechanisms and 'technologies of rule' through which states and other actors are able to secure certain forms of 'action at a distance' (Hannah, 2000; Mitchell, 2002). Governmentality highlights not an historical expansion in the capacity of the state (important as this may be in some contexts), but 'the emergence of a new field for producing the effects of power – the new, self-regulating field of the social' (Scott, 1995, pp. 202–3). Foucault's pithy definition of governmentality as 'the conduct of conduct' captures this attention to how people and things come to be brought together in such a way – neatly referred to by Scott (1995) as the 'right disposition' – that they are amenable to administration.

Eco-governmentality – or environmental governmentality (see Rutherford, 1994; Darier, 1996) – can be described as a concern with the way in which discourse and the apparatus of government (i.e., rule, more broadly) have come increasingly to centre on environmental phenomena. Accounts of eco-governmentality show how resources, ecosystems and bodies (both human and non-human) are subject to calculative procedures and practices of codification such that the administration of ecology and nature 'emerges as one more productive power formation' within modern society (Luke, 1999, p. 146). Eco-governmentality, then, is not so much an expanding application of governmentality *onto* environmental issues, as an exhumation and extrapolation of one of Foucault's initial observations about the centrality within modern government of calculative practices that pertain to *life* and, more explicitly, to the administration, optimisation and regulation of population. Foucault signaled this historical shift by labeling it 'biopower:' largely an anthropocentric concept to Foucault (Darier, 1999), it has subsequently been re-tooled by geographers, anthropologists, and sociologists to express the ways in which discourses about – and strategies towards – the management of biological, ecological, and biogeochemical processes are a key part of how social order is produced and maintained.

A substantial body of research on governmentality, environment and resources has developed since the mid-1990s. Broadly there are three different emphases within this work. The first relates to the generative *political* effects of 'environmental' knowledge associated with, for example, biodiversity conservation, sustainability, or climate change. Environmental science and the 'greening' of social science introduce calculative practices and administrative rationalities that create both new objects of rule – various novel 'spaces of nature' such as biodiversity hot spots, carbon sinks, or the 'interior geographies' of plants and animals – and new subject positions. Luke (1999), for example, describes the emergence of a new environmental power-knowledge nexus since the 1970s, a 'new environmental episteme' characterised, he argues, by the way power is exercised 'over, within and through Nature in the managerial structures of modern societies and economies'. This new, environmental modality of power finds its embodiment in the cadre of environmental managers, impact assessors, and environmental auditors graduating from higher-education institutions in Europe and North America. Goldman (2005) makes a more specific argument about the political effects of applied environmental science in his analysis of the 'greening' of the World Bank, a process centred on the epistemic transformation of a major hydro-electricity scheme in Laos into a show-case of sustainable development. The discourse of sustainability not only created new geographies of environmental degradation and resource commodification within Laos by introducing cultural and scientific logics that made the landscape legible in non-traditional ways: it also consolidated the position of the Bank as a global knowledge producer in the areas of environment and development.

A second emphasis is on the way individuals and communities internalise environmental objectives and rationalities, producing what can be called 'environmental subjects'. Among the most thorough working out of these ideas is the work of Agrawal (2005, p. 2) on forest conservation in northern India which examines the relationship between changes in the technologies of governing the environment and the emergence of an 'environmentally-oriented subject position'. At the core of Agrawal's analysis is an interest in the evolution of new 'technologies of government' to manage forests, such as the use of numbers, statistics, lists and rules, and the devolution of decision making to progressively smaller geographical scales. These technologies, he argues, not only materially changed the kinds of forests produced, but also the ways in which forest users in northern India came to under-stand their relationship with trees: technologies of government, in other words, not only produced the governable space of the 'forest reserve' but also the identity of individuals as 'environmental subjects'. A similar interest in the intersection of expert knowledge, identity, and the regulation of social practice can be seen in Robbins' (2007) work on the American lawn. Although his analysis largely eschews the language of governmentality, it exemplifies the shift within political ecology over the last decade towards a fuller engagement with environmental knowledge and the practices and techniques through which 'new natures' and social identities are co-produced.

A third emphasis addresses the modern techniques of power through which space and nature become incorporated into national projects (Peluso and Vandergeest, 2001; Mitchell, 2002). Research with this emphasis has been primarily historical, and hones in on the link between the production of specific forms of knowledge about the *qualities* of space and nature – via techniques of surveillance, calculation and abstraction that introduce new kinds of visibility and legibility – and the exten-sion of political and economic control over spatially-extensive socio-ecological systems. Key questions here centre on the techniques for establishing 'comprehensive epistemological access' to territory, enumerating the content and qualities of terri-tory, and the spatial organisation (i.e., centralisation/decentralisation) of knowledge management (Hannah 2002). Authors draw inspiration from Foucault's reference to governmentality as the governance of 'men and things' (Watts 2004). This explic-itly relational approach to the problem of modern government – to see 'men in their relations, their links, their imbrication with those things that are wealth, resources, means of subsistence . . .' – is also an invitation to examine the knowledge-power nexus that forms around socio-ecological relations. Peluso and Vandergeest (2001, p. 764), for example, identify in the colonial government's creation of 'political forests' in Indonesia a critical shift in the relationship between people and forest products. Defined by science as 'natural' land-cover and by law as 'state' territory, forests were a key site for the development of governmental institutions for security and disciplining. Geographers have been integral to this project of producing knowledge about the territorial qualities of the state, playing one of their longest-standing professional roles as an 'aid to statecraft' (Mackinder, 1904). Hannah's (2002) work on the US Census highlights the central role of data collection, mapping and the manipulation of 'spatial data' as a mechanism of social control. Mitchell (2002, p. 9) notes how colonial mapping projects in Egypt provided functions which far exceeded that of representing reality to administrators: national maps provided 'a means of recording complex statistical information in a centralized, miniaturized, and visual form . . . a mechanism for collecting, storing and manipulating multiple levels of information.' For Mitchell, the great national map – a 'technology of power' characterised by a combination of abstraction and the possibility of calculation – represented a prototype for the model of the 'national economy' which would emerge in the early 20th century. Like Braun's work on the role of earth sciences in the evolution of political rationality in Canada (Braun, 2000), Mitchell's study is testament to how – both historically and theoretically - the 'problem of government' and modern power is tied fundamentally to the problem of 'eco-governmentality' – that is, the governance of socio-ecological relations.

Conclusion: Strengthening Research on Governance

Environmental governance is a concept more popular than precise. It has been deployed in a variety of ways both critical and conservative, to describe and to occasionally critique the institutional arrangements of state, market and civil society through which decisions about environments and resources are made. It is worth asking, then, whether the concept of environmental governance is in danger of becoming – indeed whether it has already become – infinitely malleable, drained of analytical precision much like 'sustainable development' or 'social capital' before it. We hope that this brief review serves to stiffen the concept against the risk that popularity and widespread application render it overly malleable. We have argued that environmental governance specifically articulates the economic with the political, drawing attention to the relationships between institutional capacities and social action. In so doing, the term calls into question state-centric understandings of power and highlights the role of non-state actors – NGOs, supra-national agencies, social movements, or private firms – in allocating, administering and regulating environments and resources. Governance occurs at multiple sites and scales, which extend beyond those of formal institutions to include practices and norms through which key categories – nature, environment, citizens and resources – are contested, affirmed and reproduced. As an analytical framework, then, environmental governance provides a tool for examining the complex and multi-scalar institutional arrangements, social practices and actors engaged in environmental decision making.

But just as the language of governance highlights coherence and articulation in political and economic processes, so can it conceal dynamics of power, divergence and conflict that inhere in the process of managing resources and environments. Analyses of environmental governance can, therefore, lapse into a shallow institutionalism focused on describing changing organisational forms, or an environmental managerialism that is unreflexive about the relations of power enabled by talk of 'environmental crisis'. In short, environmental governance is often deployed in ways that flatten uneven relations of power, and which mask competing claims to, and about, the non-human world. Thus, we argue that a careful examination of differing epistemological, normative and rights claims to and about nature should be central to any treatment of environmental governance.

We identified six distinctive 'problematics' of environmental governance as a way to differentiate a large and expanding literature: spatial scale and its administrative reconfiguration; commodity chain coordination; management of common pool resources; popular participation and democratic action; institutional re-regulation under capitalism; and the production of social order. Although each of these themes appears in work by geographers on environmental governance, we identified two analytical areas in which geographers have mobilised concepts of governance critically to make significant contributions to a broader social science literature: neoliberal environmental governance, and eco-governmentality. Neoliberal environmental governance approaches are rooted in neo-Marxian theories of political-economic change, and attempt to understand and critique the relative coherence of meso- and macro-level political economic processes in the face of inherent contradictions arising from the socio-ecological organisation of production-consumption. By contrast, eco-governmentality draws explicitly on Foucauldian understandings of government to analyse the micropolitics of power, discipline and subject formation in relation to the administration of resources and environments. It is in these two areas, we suggest, that geography has made distinctive contributions to the literature on environmental governance.

Moreover, to the extent that geographers teach, research and write about environmental governance, we contribute to the ways these processes are understood and critiqued. It is essential, therefore, that we reflect upon the stakes for geography of adopting different perspectives towards environmental governance. In our view environmental governance describes an institutional arrangement that is not only a socio-spatial configuration: it is also, and fundamentally, an instantiation of – and resource for – political and economic power operating on and through the control of the non-human world. Because the institutions, organisations and relations of environmental governance are inherently power-laden, analyses of environmental governance should aim to lay bare these power geometries, and interrogate their origins and implications. It is our contention that managerial approaches to environmental governance can serve to mask the necessarily political-economic character of the 'environmental' objective on which management's sights are trained (such as improving air quality, stabilising carbon emissions or preserving biodiversity). Such pitfalls may be avoided, we suggest, by interpreting environmental governance not as the 'governance *of* nature' but as 'governance *through* nature' – that is, as the reflection and projection of economic and political power via decisions about the design, manipulation and control of socio-natural processes.

BIBLIOGRAPHY

Adams, W. M. (2007) *Green Development: Environment and Sustainability in the Third World*, 3rd edn. London: Routledge.

Adger, N. (2001) Scales of governance and environmental justice for adaptation and mitigation of climate change. *Journal of International Development*, 13(7), 921–31.

Agrawal, A. (2005) *Environmentality: technologies of government, and the making of subjects*. Durham and London: Duke University Press.

Bakker, K. (2002) From state to market? Water *mercantilización* in Spain. *Environment and Planning A*, 34, 767–90.

Bakker, K. (2003a) A political ecology of water privatization. *Studies in Political Economy*, 70, 35–58.

Bakker, K. (2003b) An uncooperative commodity: privatizing water in England and Wales. Oxford: Oxford University Press).

Bakker, K. (2007) The 'commons' versus the 'commodity': alter-globalization, anti-privatization and the human right to water in the Global South. *Antipode*, 39(3), 430–55.

Bakker, K. and Bridge, G. (2006) Material worlds? Resource geographies and the 'matter of nature'. *Progress in Human Geography*, 30(1), 1–23.

Bass, S., Thornber, K., Markopoulos, M., Roberts, S. and Grah, M. (2001) *Certification's Impacts on Forests, Stakeholders and Supply Chains*. London: International Institute for Environment and Development.

Beck, U. (1992) *Risk Society: Towards a New Modernity*. London: Sage.

Bennett, P. (2000) Environmental governance and private actors: enrolling insurers in international maritime regulation. *Political Geography*, 19, 875–99.

Bickerstaff, K. and Walker, G. (2005) Shared visions, unholy alliances: power, governance and deliberative processes in local transport planning. *Urban Studies*, 42, 2123–44.

Borgese, E. (1999) Global civil society: lessons from ocean governance. *Futures*, 31, 983–91.

Brannstrom, C., Clarke, J. and Newport, M. (2004) Civil society participation in the decentralisation of Brazil's water resources: assessing participation in three states. *Singapore Journal of Tropical Geography*, 25(3), 304–21.

Braun, B. (2000) Producing vertical territory: geology and governmentality in late Victorian Canada *Cultural Geographies*, 7, 7–46.

Bridge, G. (2000) The social regulation of resource access and environmental impact: production, nature and contradiction in the US copper industry. *Geoforum*, 31, 237–56.

Bridge, G. and Jonas, A. (2002) Governing nature: the reregulation of resource access, production and consumption. *Environment and Planning A*, 34, 759–66.

Brunckhorst, D. and Reeve, I. (2006) A geography of place: principles and application for defining 'eco-civic' resource governance regions. *Australian Geographer*, 37(2), 147–66.

Buck, S. (1998) *The Global Commons: An Introduction*. Washington, DC: Island Press.

Budds, J. and McGranahan, G. (2003) Are the debates on water privatization missing the point? Experiences from Africa, Asia and Latin America. *Environment and Urbanization*, 15(2), 87–113.

Bulkeley, H. (2001) Governing climate change: the politics of risk society? *Transactions of the Institute of British Geographers*, 26, 430–47.

Bulkeley, H. (2005). Reconfiguring environmental governance: Towards a politics of scales and networks. *Political Geography*, 24(8), 875–902.

Cashore, B., Auld, G. and Newsom, D. (eds) (2004) *Governing Through Markets: Regulating Forestry Through Non-State Environmental Governance*. New Haven: Yale University Press.

Castree, N. (2006) Commentary: from neoliberalism to neoliberalisation: consolations, confusions, and necessary illusions. *Environment and Planning A*, 38, 1–6.

Castree, N. (2008a) Neo-liberalising nature 1: the logics of de- and re-regulation. *Environment and Planning A*, 40, 131–52.

Castree, N. (2008b) Neo-liberalising nature 2: processes, outcomes and effects. *Environment and Planning A*, 40, 153–73.

Conca, K. (2004) Ecology in an age of empire: a reply to (and extension of) Dalby's imperial thesis. *Global Environmental Politics*, 4(2), 12–19.

Coe, N., Hess, M., Yeung, H., Dicken, P. and Henderson, J. (2004) Globalizing regional development: a global production networks perspective. *Transactions of the Institute of British Geographers*, 29(4), 468–85.

Cowen, M. and Shenton, R. (1996) *Doctrines of Development*. London and New York: Routledge.

Dalby, S. (2002) Environmental Governance. In R. Johnston, P. Taylor, P. and M. Watts (eds), *Geographies of Global Change: Remapping the World*. Oxford: Blackwell, pp. 427–40.

Dalby, S. (2004) Ecological politics, violence, and the theme of empire. *Global Environmental Politics*, 4(2), 1–11.

Darier, E. (1999) Foucault and the environment: an introduction. In E. Darier (ed.), *Discourses of the Environment*. Oxford: Blackwell, pp. 1–33.

Darier, E. (1996) Environemntal governmentality: the case of Canada's green plan. *Environmental Politics*, 5, 585–606.

Davidson, D. J. and Frickel, S. (2004) Understanding environmental governance. *Organization and Environment*, 17(4), 471–92.

Fischhendler, I. (2006) Governing climate risk: a study of international rivers. In G. Spaargaren, A. Mol and F. Buttel (eds), *Governing Environmental Flows: Global Challenges to Social Theory*. Cambridge: MIT Press, pp. 221–66.

Foucault, M. (1982) The subject of power. In H. Dreyfus and P. Rabinow (eds), *Michel Foucault: Beyond Structuralism and Hermeneutics*. Brighton: Harvester, pp. 208–26.

Gandy, M. (2002) *Concrete and Clay: Reworking Nature in New York City*. Cambridge, MA: MIT Press.

Gereffi, G., Humphrey, J., and Sturgeon, T. (2005) The governance of global value chains. *Review of International Political Economy*, 12(1), 78–104.

Gibbs, D. and Jonas, A. E. G. (2000) Governance and regulation in local environmental policy: the utility of a regime approach. *Geoforum*, 31, 299–313.

Gibbs, D. and Jonas, A. (2001) Rescaling and regional governance: the English Regional Development Agencies and the environment. *Environment and Planning C: Government and Policy*, 19, 269–88.

Global Witness (1998) *A Rough Trade*. http://www.globalwitness.org/media_library_detail. php/90/en/a_rough_trade (accessed 4 September 2008).

Goldman, M. (2005) *Imperial Nature: The World Bank and Struggles for Social Justice in the Age of Globalization*. New Haven and London: Yale University Press.

Granovetter, M. (1985) Economic action and social structure: the problem of embeddedness. *American Journal of Sociology*, 91, 481–510.

Hannah, M. (2000) *Governmentality and the Mastery of Territory in Nineteenth-Century America*. Cambridge: Cambridge University Press.

Hardin, G. (1968) The tragedy of the commons. *Science*, 162, 1243–48.

Hart, G. (2001) Development critiques in the 1990s: *culs de sac* and promising paths. *Progress in Human Geography*, 25, 649–58.

Harvey, D. (2003) *The New Imperialism*. Oxford: Oxford University Press.

Himley, M. (2008) Geographies of environmental governance: the nexus of nature and neoliberalism. *Geography Compass*, 2, 1–19.

Howitt, R. (2001) *Rethinking Resource Management: Justice, Sustainability and Indigenous Peoples*. London: Routledge.

Hudson, R. (2001) *Producing Places*. New York: Guilford Press.

Humphrey, J. and Schmitz, H. (2001) Governance in global value chains. *IDS Bulletin*, 32(3), 19–29.

Jessop, B. (1997) A Neo-Gramscian approach to the regulation of urban regimes: accumulation strategies, hegemonic projects and governance. In M. Lauria (ed.), *Reconstructing*

Urban Regime Theory: Regulating Urban Politics in a Global Economy. Thousand Oaks, CA: Sage, pp. 51–73.

Jessop, B. (2002) Liberalism, neoliberalism, and urban governance: a state-theoretical perspective. *Antipode*, 34, 452–72.

John, D. (1993) *Civic Environmentalism: Alternatives to Regulation in States and Communities.* Washington, DC: CQ Press.

Jonas, A and Gibbs, D. (2002) A tale of two areas: governance and regulation in local environmental policy making in the East of England. *Social Science Quarterly*, 84(4), 1018–37.

Keil, R. and Desfor, G. (2003) Ecological modernization in Los Angeles and Toronto. *Local Environment*, 8, 27–44.

Lemos, M. and Agrawal, A. (2006) Environmental governance. *Annual Review of Environment and Resources*, 31, 297–325.

Lewis, N., Moran, W., Perrier-Cornet, P. and Barker, J. (2002) Territoriality, enterprise and reglementation in industry governance. *Progress in Human Geography*, 26(4), 433–62.

Luke, T. (1999) Environmentality as green governmentality. In E. Darier (ed.), *Discourses of the Environment.* Oxford: Blackwell, pp. 121–51.

Mackinder, H. (1904) The geographical pivot of history. *The Geographical Journal*, 23, 421–37.

MacLeod G. and Goodwin, M. (1999) Space, scale and state strategy: rethinking urban and regional governance. *Progress in Human Geography*, 23(4), 503–27.

Mansfield, B. (2004a) Neoliberalism in the oceans: 'rationalization', property rights, and the commons question. *Geoforum*, 35, 313–26.

Mansfield, B. (2004b) Rules of privatization: contradictions in neoliberal regulation of North Pacific fisheries. *Annals of the Association of American Geographers*, 94(3), 565–84.

Marx K. (1967 [1867]) *Capital, Volume 1.* New York: International Publishers.

McCarthy, J. (2004) Privatizing conditions of production: trade agreements as neoliberal environmental governance. *Geoforum*, 35, 327–41.

McCarthy, J. (2005) Scale, sovereignty, and strategy in environmental governance. *Antipode*, 37(4), 731–53.

McCarthy, J. (2006) Neoliberalism and the Politics of Alternatives: community forestry in British Columbia and the United States. *Annals of the Association of American Geographers*, 96(1), 84–104.

McCarthy, J. and Prudham, S. (2004) Neoliberal nature and the nature of neoliberalism. *Geoforum*, 35, 275–83.

Mehta, L., Leach, M. and Scoones, I. (2001) Editorial: environmental governance in an uncertain world. *IDS Bulletin*, 32(4). http://www.ids.ac.uk/UserFiles/File/publications/classics/mehta_et_al_32_4.pdf (accessed 4 September 2008).

Mitchell, T. (2002) *Rule of Experts: Egypt, Techno-Politics, Modernity.* Berkeley: University of California Press.

Mol, A. (2006) The environmental state and informational governance. *Nature and Culture*, 1(1), 36–62.

Mol, A. and Sonnenfeld, D. (2000) *Ecological Modernization Around the World: Perspectives and Critical Debates.* London: Frank Cass.

Mutersbaugh, T., Klooster, D., Renard, M.-C. and Taylor, P. (2005) Certifying rural spaces: quality-certified products and rural governance. *Journal of Rural Studies*, 21, 381–88.

North, D. (1990) *Institutions, Institutional Change and Economic Performance.* Cambridge: Cambridge University Press.

O'Rourke, D. (2003) Outsourcing regulation: analyzing non-governmental systems of labour standards and monitoring. *The Policy Studies Journal*, 31(1), 1–29.

Ostrom, E. (1990) *Governing the Commons: The Evolution of Institutions for Collective Action.* Cambridge: Cambridge University Press.

Paehlke, R. and Torgerson, D. (eds) (1990) *Managing Leviathan: Environmental Politics and the Administrative State.* Peterborough, ON: Broadview Press.

Painter J. (2000) State and governance. In E. Sheppard and T. J. Barnes (eds), *A Companion to Economic Geography*. Oxford: Blackwell, pp. 359–76.

Palmer, L. (2006) 'Nature', place and indigenous polities. *Australian Geographer*, 37(1), 33–43.

Peluso, N. and Vandergeest, P. (2001) Genealogies of political forest and customary rights in Indonesia, Malaysia, and Thailand. *The Journal of Asian Studies*, 60, 761–812.

Perreault, T. (2005) State restructuring and the scale politics of rural water governance in Bolivia. *Environment and Planning A*, 37(2), 263–84.

Perreault, T. (2006) From the *Guerra del Agua* to the *Guerra del Gas*: Resource governance, neoliberalism, and popular protest in Bolivia. *Antipode*, 38(1), 150–72.

Polanyi K. (1944) *The Great Transformation*. New York: Farrar & Rinehart.

Ponte, S. and Gibbon, P. (2005) *Trading Down: Africa, Value Chains and the Global Economy*. Philadelphia: Temple University Press.

Prudham, S. (2004) Poisoning the well: neoliberalism and the containment of municipal water in Walkerton, Ontario. *Geoforum*, 35, 343–59.

Prudham, S. (2005) *Knock on Wood: Nature as Commodity in Douglas-Fir Country*. London: Routledge.

Retort. 2005. *Afflicted Powers: Capital and Spectacle in a New Age of War*. London: Verso.

Robbins, P. (2007) *Lawn People: How Grasses, Weeds and Chemicals Make Us Who We Are*. Philadelphia: Temple University Press.

Robertson, M. (2004) The neoliberalization of ecosystem services: wetland mitigation banking and problems in environmental governance. *Geoforum*, 35, 361–73.

Rosenau, J. (1991) Governance, order and change in world politics. In J. Rosenau and E. Czempiel (eds), *Governance Without Government: Order and Change in World Politics*. Cambridge: Cambridge University Press, pp. 1–29.

Rutherford, P. (1994) The Administration of life: ecological discourse as 'intellectual machinery of government. *Australian Journal of Communication*, 21(3), 40–55.

Scott, D. (1995) Colonial governmentality. *Social Text*, 43(Autumn), 191–220.

Spaargaren, G., Mol, A. and Buttel, F. (2006) *Governing Environmental Flows: Global Challenges to Social Theory*. Cambridge, MA: MIT Press.

Sparke, M. (2006) Political geography: political geographies of globalization (2) – governance. *Progress in Human Geography*, 30(2), 1–16.

Steinberg, P. (2001) *The Social Construction of the Ocean*. New York: Cambridge University Press.

Stringer, C. (2006) Forest certification and changing global commodity chains. *Journal of Economic Geography*, 6, 701–22.

Swyngedouw, E. (2000) Authoritarian governance, power, and the politics of rescaling. *Environment and Planning D: Society and Space*, 18, 63–76.

Swyngedouw, E. (2005) Dispossessing H_2O: the contested terrain of water privatization. *Capitalism, Nature, Socialism*, 16(1), 81–98.

Swyngedouw, E. (2007) Impossible/undesirable sustainability and the post-political condition. In R. Krueger and D. Gibbs (eds), *The Sustainable Development Paradox*. New York: Guilford Press, pp. 13–40.

Taylor, P. (2005a) In the market but not of it: fair trade coffee and forest stewardship council certification as market-based social change. *World Development*, 33(1), 129–47.

Taylor, P. (2005b) A fair trade approach to community forest certification? A framework for discussion. *Journal of Rural Studies*, 21, 433–47.

Tickell, A. and Peck, J. (1995) Social regulation *after* Fordism: regulation theory, neo- liberalism and the global-local nexus. *Economy and Society*, 24(3), 357–86.

Uitto, J. (1997) Environmental governance and the impending water crisis. *Global Environmental Change*, 7(2), 167–73.

Wapner, P. (1995) Politics beyond the state: environmental activism and world civic politics. *World Politics*, 47, 311–40.

Watts, M. (2004) Violent environments: petroleum conflict and the political ecology of rule in the Niger Delta, Nigeria. In R. Peet and M. Watts (eds), *Liberation Ecologies: Environment, Development, Social Movements*, 2nd edn. London: Routledge, pp. 272–98.

Watts, M. (2003) Development and governmentality. *Singapore Journal of Tropical Geography*, 24(1), 6–34.

Whitehead, M., Jones, R. and Jones, M. (2007) *The Nature of the State: Excavating the Political Ecologies of the Modern State*. Oxford: Oxford University Press.

Wood, A. and Valler, D. (2001) Turn again? rethinking institutions and the governance of local and regional economies. *Environment and Planning A*, 33, 1139–44.

Young, O. (1994) *International Governance: Protecting the Environment in a Stateless Society*. Ithaca, NY: Cornell University Press.

Zalik, A. (2004) The Niger Delta: 'petro-violence' and 'partnership development'. *Review of African Political Economy*, 31(101), 401–24.

Chapter 29

Commons

James McCarthy

Introduction

Commons are resources or other assets that members of a group of people have direct access to and some degree of control over by virtue of their membership in a community, without such relationships necessarily being mediated through the legal and economic structures of states or formal markets. Many resources have been considered to be commons of one sort or another: culture, open source software, the internet, the oceans and atmosphere, a dining hall in a building or piece of open land at the centre of a town that serve as community meeting places, and many more. As such diversity suggests, 'commons' is an evocative and broadly resonant word in the English language: it immediately brings to mind closely related and equally complex words and concepts such as commonwealth, commune, communal, and community, as well as 'common people', 'in common', 'common sense', and 'common law', to list just several well-established examples. It also conjures up important historical associations having to do with transitions from pre-capitalist feudal societies to capitalist liberal democracies. It thus approaches the status of what Raymond Williams (1976) called a 'keyword' – that is, a complex word with a wide range of active meanings, involving ideas and values, with which we attempt to understand, represent and influence the practices and relationships central to contemporary culture and society. Part of what Williams sought to emphasise through careful attention to such words was that we cannot resolve debates about their range of meanings by insisting upon a single correct definition, but that we should seek instead to understand how their multiple meanings in diverse contexts inform one another and change over time, and what is at stake when people emphasise one meaning over another. Thus, while 'commons' has come to have some quite specific and academically accepted definitions with respect to the environment and natural resources, it is vital to consider that those definitions exist within larger discourses in which the term is used to signify a wide range of meanings and political projects. This essay attempts to elucidate some of these complexities by examining the significance of the commons in three conversations: one focused on the historical significance of commons, a second centred on efforts to create abstract, analytical clarity around commons, and a third that articulates diverse contemporary political projects through invocations of commons. The subfields of historical, environmental, and radical geography have made important contributions to these three conversations, respectively, while geographic scholarship in the realm of political economy is of relevance to all three. Of course, such labels and distinctions are partly analytical devices to sharpen our focus on various active meanings of 'commons' in turn; in practice, the content of each of these conversations affects the others. Similarly, while some particular concerns and contributions of geographers are pointed out in each section, for the most part this essay treats commons as an inherently geographical subject, regardless of who is speaking about them: as Gregory (1994) and others have pointed out, professional geographers are far from the only actors producing geographical knowledges.

The Historical Significance of the Commons

Human history is full of commons, that is, of instances in which groups of people have used, controlled, and governed resources collectively and directly. As might be expected, the outcomes of such management have varied widely, but in many cases, natural resources have been managed as commons successfully and sustainably for generations, over centuries in some cases. The historical existence of such successful commons throughout the world has been critical in efforts to think about commons in analytical and institutional terms, as discussed in the following section.

For our purposes here, though, we will focus on the history of the commons in England, which remains in many respects the paradigmatic case and referent for most discussions of the commons in contemporary geography and many contemporary political conversations. Many discussions of historical commons centre on the common lands and associated rights that existed in parts of England prior to the development of capitalism, and whose elimination or curtailment played a central role in the transition to a capitalist economy and society.

A full history of these common lands and rights, and ongoing debates about their precise contours, is beyond the scope of this essay; see Thompson (1991), Williams (1973). For our purposes here, what matters is that up until the 19th century, residents of much of rural England (commoners) enjoyed a wide variety of use rights on nearby common lands, including rights to dwell, hunt, graze livestock, gather wood or other materials for fuel, building, and fencing, gather fruits, glean leftovers from fields after the harvest, collect stones and sand for building, and more. As this partial list suggests, these rights often played critical roles in livelihoods, enabling the survival of individuals and entire communities. Crucially, these rights were established and defended on the basis of custom and tradition, rather than through market exchanges; being born into a particular social group in a particular place sufficed to secure some of these rights, which played a major role in the feudal social order.

This sketch is far too singular and static, however: in actuality, common lands and rights were varied and dynamic in several dimensions (see Williams, 1973; Thompson, 1991). First, they varied considerably over space, with different regions, counties, and local communities differing with respect to the existence, extent, and content of common lands and rights. Moreover, different common rights applied to different types of land. For instance, some lands were 'waste' and nearly unowned, with use rights for adjacent residents being quite extensive; some cultivated lands were the property of individuals such as local landlords, with nearby commoners having quite limited rights to gather what they could from the fields after the harvest; still other lands were royal hunting preserves, where commoners could gather wood and other materials but not hunt. Second, common lands and rights changed considerably over time, with contestation over them constant, and enclosures severely curtailing or eliminating them beginning as early as the 12th century. Third, the strength and stability of commoners' rights varied, with some being codified into formal law, others being well-established and regularised in local custom, and still others being much more informal, fluid, and contested through daily practice. The variety of combinations of land tenures and use rights was thus highly complex and dynamic; such diversity is often glossed over in gestures towards 'the commons' as a unitary set of rights and relationships destroyed by capitalism (Williams, 1973).

It is certainly true, however, that attacks upon, and the eventual near-complete elimination of, the commons played a central role in the development of capitalism. Many commons were subject to enclosure, or privatization, in two major waves. From the late fifteenth through the mid-16th century in particular, many landowners, motivated by rapidly rising wool prices, enclosed their own properties in order to convert large areas into pastures for sheep, forcing off tenant farmers and entire communities with traditional rights of residence and use in the process. In the 18th and 19th centuries, a steady stream of parliamentary acts – thousands of them – authorised the consolidation and enclosure of many remaining commons, transforming them into private property by legal fiat and eliminating many commoners' rights in the process. The scale of these appropriations must be appreciated: by some accounts, the enclosures of the eighteenth and nineteenth centuries alone transformed some six million acres, including about a quarter of the cultivated land in the country, from common to private property (Williams, 1973, p. 96). The privatization of formerly common lands and the elimination of traditional use rights was a long, contested, and frequently violent process: hedges and walls were erected and broached; hunters and gatherers continued to take resources they felt they had rights to even after such actions were criminalised, and sometimes paid with their lives; and in the most extreme cases, soldiers evicted entire communities and burned their homes behind them (Marx, 1967; Thompson, 1975; 1991; Blomley, 2007).

Marx saw such enclosures as fundamental to the development of capitalism because they forcibly separated laborers from the land and created a legal framework in which labor power and nature were redefined as privately owned commodities that could only be brought together again through market exchanges. He described this as a process of 'primitive accumulation', meaning processes of accumulation that logically had to precede capitalist accumulation per se and that created the preconditions for it through extra-economic means (Marx, 1967; De Angelis, 2004; see Glassman, 2006). On the one hand, the mass of the population was cut off from any direct access to the land and from the possibility of legally supporting themselves through direct production for their own benefit, that is, producers were separated from the means of production. 'Freed' from their feudal and community networks of reciprocal obligation, their only option for subsistence was now to enter the market as individuals and sell their labor power to the capitalists who could afford to purchase both it and the means of production. At the same time, by converting many lands and associated natural resources from commons to private property, the enclosures dramatically increased the assets directly owned by nascent agrarian capitalists; freeing those assets from their network of feudal social obligations greatly facilitated their circulation as commodities and their more 'efficient' economic use by allowing them to be more divisible, alienable, calculable, usable as collateral, and so on (Castree, 2003). Thus, a fundamentally severing of one set of social and socio-natural relationships, that prevailing under and remaining from feudalism, was followed by a reconstitution of necessary relationships among people and their environments through capitalist relations of ownership, production and exchange.

Such wrenching changes were effected not only through everyday struggles between commoners and landowners, but through sweeping ideological transformations as well. The centuries that saw the most active enclosures also saw dramatic changes in how economic value, property and the relationships between individuals and society were theorised and prioritised, changes that helped to build support for and legitimate the enclosures. Physiocratic theories of value insisted on the centrality of agricultural production to national wealth, encouraging states to initiate national strategies and policies to maximise agricultural productivity. Physiocratic and labor theories of value agreed that value was created by the application of human labor to nature; therefore, society ought to be structured in ways that would maximise individuals' incentives to work, and particularly to engage in agricultural improvement designed to increase the productivity of both land and labor. John Locke, Jeremy Bentham and many others argued that strong, clear private property was the type of property most conducive to encouraging such work and the creation of value, since people would work hardest and take the greatest care with their resources if they would reap all of the resulting benefits, whereas the ambiguities, collective hazards, and safety net of common property would discourage work, investment, and stewardship. Finally, Thomas Malthus became the best known of many to make the argument that society had no place for or obligation to those who failed to support themselves through individual work within the context of a fully owned and privatised landscape: in a nutshell, those who would not or could not work for wages (and who were not fortunate enough to own property already) should be allowed to starve to death. Ironically, Malthus claimed to found his argument on humanitarian grounds: feeding the poor, of whom there were clearly too many for the limited agricultural base of the country to support, would simply lead to a larger number of insupportable poor people in future generations; not only would the latter starve in turn, but they would likely turn violent and attack those whom the resource base could support, leading to far more total human suffering, as well as waste and inefficiency. Thus, the strong individualism of liberalism and capitalism, in which relationships among people and between people and nature were mediated through commodity exchanges within a fully privatised landscape and no one was owed any living or support, was presented as more humane and more conducive to the careful stewardship of natural resources than the commons, in which people existed in complex webs of reciprocal obligations and overlapping rights to local environments.

As noted above, much historical and geographical detail about actual historical commons is glossed over in the increasingly generalised and abstracted narrative laid out in the previous few paragraphs. Reconstructing more precise and empirically substantiated understandings of the many varieties of pre-industrial commons through the use of both archival and field methods has been the work of many historians and historical geographers (it is a perennial topic in the *Journal of Historical*

Geography, for instance, as well as the major focus of the Department of Human Geography at Stockholm University, to name just one prominent example). Additionally, much subsequent scholarship has demonstrated that the stylised narrative above is excessively Anglocentric and Eurocentric in important respects: commons and struggles over their incorporation into evolving capitalist economies can be found throughout the world and up to the present, not only in the past of one country. Indeed, much of the history of colonialism is the history of colonial administrations appropriating collectively owned resources as state or private property, in part to create labor markets, a legacy that geographers have done much to document around the world (see, e.g., Neumann, 2005). Moreover, extra-economic forms of accumulation appear to have played perennial and ongoing roles in capitalist accumulation, rather than simply setting the stage for original accumulation and then disappearing – that is, forcible separations of producers from the means of production are a permanent ontological feature of capitalism, rather than only an historical precondition. Geographers working in the tradition of political economy have been among the major contributors to the latter line of analysis (e.g., Harvey 2003; Glassman 2006; Hart 2006; see also Perelman 2000; De Angelis 2004). It remains true, however, that enclosures of the commons in England and Scotland were ideologically and materially central to the development of capitalism. It is thus not surprising that ideological critiques of common property and assertions regarding the superiority of private property have remained central to ongoing capitalist development, nor that appeals to 'the commons' as an iconic alternative to capitalism remain prevalent and powerful. Both trends are evident in the sections that follow.

Analysing the Commons

A major new chapter in discussions of the commons can be traced to 1968 and the publication in *Science* by Garret Hardin, a population biologist, of an article entitled, 'The Tragedy of the Commons'. Hardin's article, a major text in a wave of neo-Malthusian environmentalism in the 1960s and 1970s, focused on the alleged environmental dangers posed by the rapid and seemingly exponential growth of the world's human population. The core of his argument was that natural resources around the globe were finite and that infinite population growth upon a finite resource base was impossible; therefore, continued growth of the total human population would necessarily result, sooner or later, in a population that exceeded the global environment's ability to support it: in the language of population biology, the species would exceed its environment's carrying capacity, and a dramatic plunge in its numbers would necessarily follow. From this basic logical proposition, Hardin argued that draconian controls on reproductive rights were necessary to save humanity from itself. It was essentially the same argument made by Malthus, but where Malthus focused narrowly on England and its ability to produce enough food to feed a growing population, Hardin generalised the argument to the entire globe and to a range of renewable and non-renewable environmental resources.

Hardin made a metaphorical model of a commons the centrepiece of his argument. He asked readers to imagine a pasture on which many commoners had unlimited rights to graze cows. The core problem, in his view, was that each user had strong economic incentives to overgraze, because they captured all of the economic gain from each additional cow that they brought to graze on the common – a right for which they paid nothing – but they bore only a fractional cost of the ecological damage and declining productivity caused by that additional cow. Since all users of the common had the same incentives, the result would be that each would bring more and more cows to graze on the common pasture until it was entirely destroyed, leaving both the resource and its users ruined. Hardin popularised this argument, but his version of it was hardly *sui generis*: the same argument had already been developed with respect to fisheries, about which it had often been observed that each party fishing had strong incentives to catch every fish possible, engaging in an arms race of constant investment in larger boats, bigger nets and other technological improvements in order to do so, because any fish any one of them left behind could just be caught and sold by someone else. Thus, participants in fisheries would often race towards their own collective ruin. For Hardin, this was the tragedy of the commons: that a resource freely available to all would inevitably and inexorably be overexploited to the point of ruination, leaving all of those who depended upon it devastated. He saw two viable solutions: privatisation or strong state control. Turning commons into private property (i.e., enclosing them) would leave their owners with strong economic incentives to manage resources sustainably, while removing the ruinous competition with others for the last fish or blade of grass that lay at the heart of the tragedy of the commons. Strong state control was also a viable option; Hardin described this, a version of social contract theory, as, 'mutual coercion, mutually agreed upon'; people would, in effect, empower a sovereign to enforce laws to protect their self-interest, even though the latter sometimes curtailed the maximisation of their opportunities for accumulation.

Hardin's article became extraordinarily popular and influential; it is among the most-cited academic articles ever, and its legacy remains strong among many environmentalists, resource economists, and international lending agencies and NGOs. Its basic logic can be and has been clearly and directly extrapolated to a range of resources, and to pollution sinks as well as material sources. For instance, it is easy to interpret atmospheric pollution as a tragedy of the commons (each user pollutes freely and captures all of the economic benefits from the associated production, but the costs of the pollution are distributed globally).

It also provoked immediate and powerful critiques, however, engendering a set of debates about the commons that continues up to the present. Critics attacked Hardin's argument on both empirical and theoretical grounds. Many geographers, anthropologists, and other scholars of human–environment relations quickly pointed out that Hardin's thesis regarding the inevitable degradation of common resources was easily falsifiable empirically: many societies around the world and throughout human history, up through the present, had managed common resources for generations without apparent degradation. In fact, some had even improved common resources over time. Examples ranged from fisheries to forests, from pastures in the Swiss Alps to irrigation systems in Nepal. So, Hardin had clearly missed something. Attempts to discern and explicate what he had missed – that is, what differentiated the many cases of successful commons from Hardin' scenario – eventually developed into a more abstract and systematic critique of the tragedy of the commons thesis, and complementary theories regarding what was necessary for commons to operate successfully.

First and most fundamental, perhaps, was the fact that successful commons were not actually freely available to everyone; they were typically controlled by a fairly small group of users who either faced no competition for the resource (e.g., because they were isolated), or were able to exclude any other potential users (e.g., from nearby villages). Thus, critics suggested that Hardin had made the mistake of confusing commons with *open-access regimes*, with the latter being truly available to anyone. For resources that fit the latter description – fisheries in international waters and the ability to emit pollutants into the atmosphere are frequently invoked as examples – the tragedy of the commons is far more likely to occur.

Second, Hardin made a number of assumptions about the users of the commons, both as individuals and as a group; these assumptions, although unstated, turned out to be both critical to his argument and quite debatable. He assumed that each user of the commons would seek to maximise their short-term income and their wealth, that they would engage in near-constant calculations (conscious or otherwise) about their daily activities in order to do so, and that they would continue to do so even once they realised that they were harming other users and the resource. In short, he assumed that they would behave as the 'economic man' of neoclassical economics. But this allegedly universal model of human behavior is in fact quite parochial and culturally situated; much evidence demonstrates that it does not describe the actual practices of people in many societies throughout history. Even more questionably, Hardin assumed that small numbers of people living in close proximity and depending upon the same resource for their livelihoods would not talk to each other, even as their shared environment degraded, and that they would not act to sanction the individuals most responsible. It is here that actual commons most directly contradict Hardin's vision: most functioning commons are in fact governed by complex and well-articulated sets of rules designed to maintain the viability of the resource over extended periods (recall, for instance, the precision and diversity of 'the commons' discussed in the previous section). Such rules typically define not only who has rights to that commons, but users' responsibilities as well, with provisions for varying levels of access, adjudicating disputes, punishing violators, resting the resource when needed and adjusting the rules over time. Thus, far from being asocial and ungoverned spaces, successful commons are highly structured relationships between human communities and biophysical environments.

Certainly, commons have not always succeeded in producing and maintaining sustainable human–environment relationships; a review of the literature in environmental geography and related fields reveals many instances of failures of commons alongside the successes. Yet the same is true of privatisation and centralised state control, which brings us to a third major criticism of Hardin's argument: it is not at all clear that his preferred approaches to environmental governance necessarily produce better outcomes. Many instances of environmental degradation and collapse can be traced to excesses of privatisation or state authority. Patterns of suburban sprawl in the United States, for instance, have been likened to a tragedy of the commons brought about by excessive deference to private property and individual self interest, with millions of households seeking to maximise their utility through actions in private property markets collectively destroying the very environmental qualities each is seeking, and producing landscapes that nearly all find undesirable (Donahue, 1999; McCarthy, 2005). A glance at the environmental records of the former Soviet Union or contemporary China, meanwhile, is typically enough to make the point that centralised state control is no guarantor of environmental quality either, a point that has also been well demonstrated through documentation of the failures of state-centred sustained yield forest management.

A strong tradition of scholarship around commons has developed out of the critiques above, one that has long since moved from criticism of the tragedy of the commons thesis towards a positive project of researching how commons work, articulating the key relevant concepts and relationships, and exploring potential new applications of these lessons. The political scientist economist Elinor Ostrom has been perhaps the most central figure in crafting a general conceptual framework in this area (e.g., Ostrom, 1990; 2005; Ostrom et al., 1994; Ostrom and Hess, 2007; see also McCay and Acheson, 1987; Ostrom et al., 1999; Dietz et al. 2003), but such efforts are built upon a foundation of many intensive empirical cases studies in heterogeneous locations around the globe. Geographers, including most centrally those in the traditions of cultural and political ecology, have been among the most important contributors to that corpus (reviews and bibliographies of this large body of literature can be found in Robbins, 2004 and Neumann, 2005). Most recently, forms of public-participation GIS have proved invaluable in documenting past and present patterns of resource use and associated territorial claims integral to historical and contemporary commons.

Commons scholars are now careful to distinguish between *resource characteristics* and *property and governance relationships* in discussions of commons. The former are referred to as *common pool resources*, the latter as *common property regimes*, and there are no necessary relationships between them: common pool resources can still become state or private property in some cases, while common property regimes can be designed to govern resources that do not have all the characteristics of common pool resources. The defining characteristics of common pool resources are that is it difficult to exclude users, and that exploitation by one user reduces the resource's availability to other users (Ostrom et al., 1999). Thus, a fishery is typically a common pool resource, whereas a mine is not: it is far easier to exclude potential users from the latter than the former. And while language can be thought of as a common asset of a sort, its use by one person does not reduce a finite material supply available to other users; in fact, each additional user adds to the resource, a dynamic also found in other 'inverse commons' such as culture or open source software. Therefore, language is not a common pool resource by this definition. With respect to biophysical systems and assets, some of the key characteristics examined by common property scholars are size, mobility (e.g., wildlife versus medicinal plants), carrying capacity, storage capacity (e.g., in an aquifer), rates of renewal, resilience, and the amount and quality of information about all of the above that are available or obtainable (see McCay and Acheson, 1987; Ostrom, 1990; Ostrom et al., 1994; Ostrom et al., 1999; Dietz et al., 2003). Each clearly affects what governance relationships users will deem feasible and desirable, reminding us that the social regulation of human–environment relationships must grapple seriously with a heterogeneous material world. Similar constraints and complexities are to be found on the institutional side: commons scholars have found that in order for common property regimes to be successful, members of the user group must be able to exclude others if necessary, communicate amongst themselves, develop and modify rules as needed, monitor the condition and use of the resource and enforce sanctions against users who violate the rules. Trust within the group is critical, not least because it lowers the costs of monitoring and enforcement, as is a relatively high degree of equality within the group, which increases legitimacy and reduces the incentives for violating rules and norms.

This body of theory has been used to help understand, advocate for, and in some cases design governance regimes for larger-scale resources that fit the description of common pool resources, and might be governable through common property regimes. Many have argued, for instance, that the atmosphere and the oceans would be best understood and managed as global-scale common pool resources, and that biological resources such as biodiversity or tropical rainforests ought to be reimagined as, respectively, the 'common heritage of mankind', or the 'lungs of the planet'. Other examples include Antarctica (Joyner, 1998), large fisheries in international waters, outerspace and minerals under the deep ocean floor (Buck, 1998). Many see such efforts as vital because the absence of a global sovereign or globally shared norms around private property mean that neither of Hardin's solutions are necessarily available for common pool resources that cross international borders; rather, equal users must be able to govern themselves if there is to be hope for global environments, now often referred to as 'global commons' (see Goldman, 1998; Eckersley, 2004). Yet the difficulties involved in attempts to 'scale up' common property lessons and approaches to larger biophysical and social systems are legion: even if nation states are imagined as the individual 'users' of global commons, as they are in many scenarios, it is not clear if the problems of incomplete information about environmental resources, inequality and mistrust among participants in the commons, and the difficulties of excluding or sanctioning violators (to name just a few) can be overcome. Without romanticising 'communities', it is far easier to imagine all of the conditions for successful commons being met in relatively small communities with a high degree of informal interaction among group members, governing a relatively small-scale nearby resource, than in larger-scale societies dependent upon more anonymous and bureaucratised structures attempting to govern truly global systems, such as climate (Dietz et al., 2003). Moreover, the discursive or institutional creation of 'global' commons necessarily entails an ironic willingness to overlook many local claims to resources for the sake of making a more general claim on behalf on a spuriously unified 'humanity' (Neumann, 2005). At best, there are tensions involved in conceiving of resources at such large scales as commons, inasmuch as many commons historically have been designed precisely to enforce quite locally specific rights and norms and exclude other claimants (cf. Thompson 1993: 184).

While common property scholars have heavily criticised Garret Hardin's thesis and gone on to research and imagine quite different scenarios around the commons, much work in this tradition still shares important assumptions with Hardin: it takes as given that the world is populated by rational, utility-maximising individuals, but then asks under what circumstances it is possible and advantageous for such individuals to construct and operate common property regimes. Much common property theoretical work has thus taken the form of game theories and role playing exercises (e.g., the prisoner's dilemma), and the refinement of institutional structures (see, e.g., Ostrom, 1990; 2005; Ostrom et al., 1994). Such methodologies have their own historical and political contexts: game theory exploring non-destructive outcomes of conflicts between rational adversaries became very popular during the cold war, for example, while the assumption that optimal governance institutions can be designed by academic experts and then handed to users is inseparable from the history of colonial and post-colonial western interventions under the flag of 'development'; in practice, interventions by 'experts' in institutional design have disrupted functioning commons far more often than they have helped to establish them (Goldman, 1998).

Some meta-critiques of the literature on common property have questioned these assumptions and silences, noting for instance that formal property theory typically has strong but implicit normative dimensions, in which powerful narratives and thought experiments actually help to create and mold particular subjectivities and orientations with respect to property and the environment (Rose, 1994; St. Martin, 2007). Similarly, discussions of actual or potential global commons that do not speak to the origins of existing inequalities, exploitations, and oppressions elide many of the central facts about contemporary patterns of resource use. Geographical scholarship related to environmental commons, most strongly in the tradition of political ecology in recent years, has thus tended to focus less on the construction of abstract models or rules of whatever sort, and more on context-specific investigations of the actual politics, power relations and forms of rationality and meaning relevant to particular cases. Most notably, perhaps, political ecology has demonstrated that in the modern era, instability in common property regimes has often been rooted less in internal dynamics than in the forcible integration of entire societies into a global capitalist economy – a problem that may not be fixable through modifications of local rules. In a similar vein, political ecology has sharpened understanding of property rights as dynamic political arrangements inseparable from production, and emphasised forms of differentiation and conflict (e.g., along axes of gender or class) within rights-holding 'communities' often treated as homogenous and egalitarian in some of the literature on common property regimes (see Robbins, 2004; Neumann, 2005).

The Contemporary Politics of the Commons

A third distinct conversation around commons has proliferated over just the past decade or so. It centres on the contention that the most recent global round of capital accumulation – post-Keynesian, post-Cold War, labelled as 'globalization' by some, 'neoliberalism' by others – has relied especially heavily upon the appropriation of assets and values by extra-economic means, i.e., the recent enclosure of many commons in a new round of primitive accumulation. 'Commons' are understood widely in this conversation, referring not only to the sorts of common-pool natural resources and land-based use rights discussed in the preceding sections, but also to a much broader set of public goods, trusts, spaces and interests, many of which have been variously appropriated, privaticised, marketised or simply eliminated during the neoliberal era. For instance, the widespread, often IMF-instigated privatisation of state assets and industries, the globalisation of ruthlessly self-serving western intellectual property regimes via the WTO, the patenting of gene sequences and entire organisms, the World Bank's insistence on private land titling at the expense of collective rights or redistributive land reform, the dramatic expansion of the doctrine of 'regulatory takings', and increased corporate control over research at public universities have all been invoked as examples of neoliberal enclosures of commons. For activists and social movements, these enclosures have prompted widespread resentment, resistance and calls for the creation or reconstitution of commons with respect to a wide variety of resources and spaces (e.g., *The Ecologist*, 1998; Klein, 2001; 2002; see McCarthy, 2005 for a review). For scholars of political economy, meanwhile, they have prompted renewed attention to primitive accumulation's role in capitalist accumulation (e.g., Harvey, 2003; De Angelis, 2004; Glassman, 2006; Hart, 2006;) and sparked new interest in the significance of neoliberalism for environmental governance and quality in particular (see Goldman, 1998; McCarthy and Prudham, 2004; Heynen et al., 2007).

As might be expected, activists and social movements around the world have interpreted and framed these developments in theoretically and politically disparate ways, with reference to a wide array of empirical situations. A single narrative thus does none of them justice. Still, a remarkable convergence is visible around the position that: (i) recent enclosures are the product of recent forms of corporate-led globalisation; (ii) states facilitate this globalization and also appropriate resources on their own behalfs, meaning that they are more likely to be enemies of public goods and commons than their defenders; (iii) the reclaiming or establishment of commons governed with a high degree of local autonomy is the most promising strategy to combat these trends; and (iv) inasmuch as many groups around the world may share this analysis and broad and underdetermined political project, they are natural allies in the 'anti-globalisation' movement. This basic analysis was articulated quite clearly in a 1998 collection by *The Ecologist*, for example, while in recent years it has become quite mainstream, appearing in summary form in mass-market environmental magazines such as *Sierra* (Rowe, 2005). Different elements of this discourse are prominent in different instances, of course, but the common threads are striking, as the following examples demonstrate.

Canadian journalist Naomi Klein, one of the most prominent voices advocating different sorts of commons at global scales, seems to articulate precisely the position above: 'Reclaiming the Commons' is the title of her 2001 article-cum-manifesto in the *New Left Review*, while her 2002 book, *Fences and Windows: Dispatches from the Front Lines of the Globalization Debate*, chronicles the struggles of countless others around the globe calling for a, 'radical reclaiming of the commons' in the contexts of particular struggles against privatization and other forms of enclosure. Her examples include struggles over water privatization in South Africa and Bolivia, over Napster and public sector jobs in the United States, and over community forests and fisheries in Canada. Klein explicitly views these proliferating calls for commons as reactions to neoliberal privatisations of formerly public domains, and uses 'the commons' to mean public goods, civic space and collective enterprises, as well as the common pool resources and property regimes discussed in the previous section. In a similar vein, Sitze (2004) constructs a brilliant and impassioned argument for life-saving drugs, particularly antiretroviral drugs effective against HIV, to be the common property of humanity. He criticises equally the corporations that profit from exclusive ownership of the drugs and the states that respect and enforce their patents, and he establishes a clear line of argument from the specifics of those struggles to a broader radical challenge to the global reproduction of capitalism and the spatial, political and environmental separations upon which it depends. Finally, Sumner (2004) argues that an array of 20th-century liberal democratic social and environmental protections can be theorised as a 'civil commons', one threatened by neoliberal globalization, and that the globalization of the civil commons will pave the way towards sustainability.

Some members of the environmental movement in the United States have become enthusiastic proselytizers of the four-part analysis above as well (e.g., Rowe, 2005). For instance, David Bollier's *Silent Theft: The Private Plunder of Our Common Wealth* has as its goal, '[D]eveloping a discourse of the commons,' as a first step towards, 'invent[ing] the commons we need for the 21st century.' The book documents myriad enclosures in domains that had been 'commons', including state-owned forests, the broadcast spectrum, disciplinary gift economies, the Internet, and more, and Bollier casts all efforts to reassert a public interest in these domains as fundamentally 'commons' inasmuch as they provide alternatives to privatization (see http://www.bollier.org/reclaim.htm). Peter Barnes' 2001 book, *Who Owns the Sky? Our Common Assets and the Future of Capitalism*, argues that the United States should set up a 'U.S. Skytrust' – a non-governmental public trust that would charge for rights to emit atmospheric pollutants in the United States under a cap-and-trade scheme and then distribute the proceeds equally among citizens. Barnes calls this 'stakeholder trust' model, 'The New Commons', and advocates extending it to other collective assets as well as a way to, 'save capitalism from itself' (p. 106); his is thus a broadly Polanyian project. Finally, Brian Donahue's 1999 book, *Reclaiming the Commons: Community Farms & Forests in a New England Town*, argues that American suburbanites ought to tax themselves to buy up at least half the land in their towns and set it aside as common forest and conservation lands to be used to reduce residents' ecological footprints and foster their environmental awareness. These lands would be true village commons, used as woodlots, fields for sustainable agriculture, and sources of berries and other non-timber forest products. Thus, while village commons have been metaphors and symbols for many in recent years, they are literal goals for Donahue, albeit ones adapted to contemporary circumstances as an antidote to excessive privatization (pp. 295, 297).

The metaphor of the commons has been taken up with respect to regional governance as well: in 2001, the Alliance for Regional Stewardship, a national network of regional leaders in the United States, published, 'The Triumph of the Commons', a manifesto for the creation of successful regions. The title was intended as a direct counterpoint to Hardin's tragedy of the commons, and the monograph argued (imaginatively, if rather loosely) that while the sheep of Hardin's village-scale example had now been replaced by sports utility vehicles at regional scales, the dynamics of the commons were still relevant. But where Hardin prophesied tragedy, the Alliance argued that commons were precisely the right way to think about and design cohesive, competitive regions, ones that leveraged the shared self-interests of regional citizens, public–private partnerships, and federal resources to govern and grow, without getting bogged down in either the bureaucracy of the national state or the pettiness of municipal-scale politics.

The handful of quick examples above in no way does justice to the range of the most recent invocations of the commons: they are all by North American authors, and mostly about North American cases. But they do illustrate a few key points about this conversation, including the many differences below the surface of some major commonalties. First, it is clear that in many, if not most, contemporary calls for 'commons', what is desired are not commons at all, in the senses discussed in the previous sections; rather, commons have become a powerful language for asserting the existence and legitimacy of collective rights and interests not limited to those that have to be paid for in markets or sanctioned by states. Many calls for commons are not calls for common property regimes strictly speaking, but calls for new public trusts, goods, spaces, entitlements, or property; increased state regulation; or even the creation of new markets in some cases (e.g., cap-and-trade schemes). All rest on a foundation of inherent collective rights to resources, though, rights tied to being a resident of a certain place, a citizen of a given country, or even a human being; in that sense, they are very much like archetypal commons and fundamentally

different from Malthusian perspectives. Second, contemporary commons are imagined at a variety of different scales, with conflicting and often ambiguous criteria
for membership, rule-making, and other fundamental aspects usefully explicated in
the institutional literature on common property regimes: participation in a regional
governance coalition is, after all, quite different from saving seeds in violation of
patents. Alternative political visions cannot and should not always be forced into
the Procrustean bed of rational choice models; St. Martin (2007) is right to resist
thinking of commons in only the latter terms. Still, it is both useful and important
to ask of any proposed commons the sorts of questions emphasised by Ostrom and
her colleagues: Who is excluded? Who makes the rules? What difference does it
make to imagine a given commons at one scale versus another? For instance, some
proposed global commons (e.g., those casting tropical rainforests as the lungs of the
planet), depend upon the bold assertion of some claims and the marginalization of
others, and are quite compatible with the further institutionalization of anti-democratic forms of globalisation (Neumann, 2005), while others (e.g., Sitze's), are far
more radical and inclusive in their politics, making all of humanity the relevant
commoners and demanding commons as a bulwark against precisely the excesses
of neoliberal globalization. Third, underlying different and more and less examined
criteria for membership, appropriate scales, and so on are different politics: proposed commons that share a vision of collective public rights and a skepticism
regarding unchecked markets can still differ sharply in terms of whether they are
radical or liberal; internationalist, nationalist, or resolutely local; fundamentally
critical of capitalist social relations or merely their of 'excesses'; and so on. In sum,
within these proposals for new commons are major differences in proposed property
arrangements and policy solutions, including whether the state owns resources
directly or not, whether it administers environmental protections directly or not
(versus indirectly, as through a semi-autonomous emissions market), whether the
state is seen as a viable trustee or guardian of public goods or not, and perhaps
most fundamentally, whether the state is seen as equivalent to the public.

A fourth key point illustrated in the examples above is that activists and intellectuals representing many different positions seem to concur that the state is not
an effective or trusted guardian of public goods and interests. Many recent calls for
commons can be seen as defensive reactions to the aggressive economic liberalism
of the past quarter century or so. Such defensive reactions are not at all new (see
Polanyi, 1944). But where earlier reactions to the failures of self-regulating markets
turned primarily to the state and to expansions of state property, control and regulation (as in the Progressive movement of the late 19th and early 20th centuries, the
Keynesianism, fascism and state-centred communism of the interwar period that
motivated Polanyi, or the modern environmental movement of the 1960s and
1970s), contemporary critics are skeptical of the state and turn instead to communities and to commons, understood in the myriad senses above, as remedies to both
market and state failures. With respect to the state, the authors above range from
mistrustful but willing to use state power in limited ways so long as it is overseen
by NGOs and markets, to unremittingly hostile, seeing states as oppositional to any
sort of genuine commons.

One explanation for this turn is that current calls for commons are in part reactions to the many failures of centralised state control throughout the history of
capitalist modernity (see Scott, 1998). In the realm of environmental management,
those failures include the overriding of legitimate local claims and knowledges,

miscalculations of sustained yield, industry captures of regulatory agencies, and the continuation of colonial and imperial oppressions and extractions under the guises of nationalism or development. Another, more pessimistic explanation is that the recent popularity of commons is deeply structured by the neoliberal consensus their advocates claim to reject. It is striking how often many of the examples above deploy as self-evident truths major planks of the neoliberal consensus, such as that states are inefficient and untrustworthy, markets have near-magical powers to which people must defer, and communities are the most reliable sources of social innovation and protection against market failures.

Recent debates within Marxist geography and related fields have also responded to many of the foregoing developments, but from a more coherent and radical theoretical perspective. The more comprehensive and structural perspective provided by Marxist analyses suggests that 'corporate globalization' is an inadequate explanation of the origins of contemporary enclosures, and that the valorization of local control and community governance characteristic of many activist programs is an insufficient response. Where many 'progressive' or 'liberal' activists view the 'corporate globalization' and market fundamentalism of recent years as an anomaly, an excess within a fundamentally just and sustainable capitalist economy, radical geographers have been far more prone to seeing the neoliberal era as representing a return to perennial capitalist dynamics of extreme inequality, exploitation, and appropriation that were somewhat anomalously mitigated in some parts of the world during the Cold War, by Keynesian policies in some countries and by state socialism in others. They have interpreted the manifold enclosures above, as well as countervailing calls for commons, largely in the context of enduring debates within Marxism over the role of primitive accumulation in capitalist development. For some, notably David Harvey (2003; 2005), neoliberalism is best understood as a more or less conscious effort to pull back or create anew class power and privilege that was reduced during the Keynesian era, a project motivated by a declining rate of profit; primitive accumulation, or 'accumulation by dispossession', has taken centre stage during this era precisely because the global capitalist economy faces a crisis of overaccumulation, making continued accumulation through productive circuits of capital increasingly difficult. The resurgence of extra-economic forms of accumulation as central strategies for some capitalists has been something of a surprise for some theorists, particularly those in the global North: echoing the narrative of enclosure rehearsed in the first section above, many had viewed primitive accumulation as well in the past of capitalist development. Radical geographers and others more familiar with the global South, though, where the forcible separation of producers from the means of production remains a highly visible fact of everyday life in many places, have long emphasised the empirical existence and ontological necessity of continual primitive accumulation to the expansion and reproduction of global capitalism (see Perelman, 2000; De Angelis, 2004; Glassman, 2006; Hart, 2006).

Even from the latter perspective, however, important debates remain about contemporary primitive accumulation and the politics of the commons. One is over the possible relationships between the enclosure of land and the proletarianization of its residents; contrary to classical formulations that sometimes viewed these as two facets of the same process, it has become clear that industrialization can occur without workers being expelled from the land (see Hart, 2006), while some have argued that contemporary enclosures are distinct inasmuch as land and other commons are being appropriated without the people thereby dispossessed subsequently being incorporated into the wage economy (George, 1998, p. x). A second, implicit in the literature but vital, is over how narrowly to construe primitive accumulation and enclosure: most examples focus on the enclosure of material assets, but De Angelis (2004), Moore (2004), McCarthy (2004), Heynen et al. (2007) and others collectively make a strong case that the broadly Polanyian social and environmental protections won in the postwar period can and should be understood as commons of a sort, one now subject to fierce enclosure. Third and most important are questions regarding agency and politics: should the manifold struggles against enclosures briefly chronicled above be understood as purely reactive, or can struggles outside of the workplace still be central to capitalist development, and, relatedly, are there sufficient commonalties of interest or politics among these disparate struggles to forge effective coalitions among their participants? Harvey (2003, p. 166) famously answered both of these questions in the negative, arguing that the manifold struggles against recent forms of primitive accumulation around the world are so varied in form, content and politics that there is little hope for effective resistance to global capitalism emerging from them. In response, however, many have argued that the substantive commonalties underlying the myriad different forms of contemporary enclosure and extra-economic accumulation do offer sufficient grounds for transnational solidarity and organising, and that indeed many such movements and coalitions are already in evidence, not simply reacting but making history (De Angelis, 2006; Glassman, 2006; Hart, 2006). Surely, this is what animates and holds together much of the 'anti-globalization' movement and forms of organisation such as the World Social Forum, whatever theoretical language their participants may use.

Conclusion: Looking Back, Out and Forward to the Commons

A single thread unites these three very different (albeit related) conversations regarding 'commons.' It is the belief that, in the language of the World Social Forum, 'another world is possible,' one in which our relations with other people and our environments are not limited to those dictated by the logics of capitalism. The ability to truly imagine such alternative relations is critical and rare, and all too readily dismissed as mere utopian fancy. Geographers have important roles to play in this vitally important task, whether we are demonstrating that functioning commons have existed and tracing how they have changed, analyzing how those lessons can be applied to new situations, or working with communities at multiple scales to craft alternatives that can help to realise the seemingly deep and enduring desire for commons (see, e.g., St. Martin 2007; De Angelis 2006). We are far from the sole contributors to such efforts, of course, but geography's focus on carefully theorised, empirically grounded research that pays attention to context, specificity, and the multiplicity of possible human-environment relationships surely gives geographers a head start in such endeavors.

BIBLIOGRAPHY

Barnes, P. (2001) *Who Owns the Sky? Our Common Assets and the Future of Capitalism.* Washington, DC: Island Press.

Blomley, N. (2007) Making private property: enclosure, common right and the work of hedges. *Rural History*, 18(1), 1–21.

Bollier, D. (2002) *Silent Theft: The Private Plunder of Our Common Wealth*. New York: Routledge.

Buck, S. (1998) *The Global Commons: An Introduction*. Washington, DC: Island Press.

Castree, N. (2003) Commodifying what nature? *Progress in Human Geography*, 27(3), 273–92.

De Angelis, M. (2004) Separating the doing and the deed: capital and the continuous character of enclosures. *Historical Materialism*, 12, 57–87.

De Angelis, M. (2006) *The Beginning of History: Value Struggles and Global Capital*. London: Pluto Press.

Dietz, T., Ostrom, E. and Stern, P. (2003) The struggle to govern the commons. *Science*, 302, 1907–12.

Donahue, B. (1999) *Reclaiming the Commons*. New Haven: Yale University Press.

Eckersley, R. (2004) *The Green State*. Cambridge, MA: The MIT Press.

Glassman, J. (2006) Primitive accumulation, accumulation by dispossession, accumulation by 'extra-economic' means. *Progress in Human Geography*, 30(5), 608–25.

George, S. (1998) Preface. In M. Goldman (ed.), *Privatizing Nature: Political Struggles for the Global Commons*. New Brunswick, NJ: Rutgers University Press, pp. ix–xiv.

Goldman, M. (1998) *Privatizing Nature: Political Struggles for the Global Commons*. New Brunswick, NJ: Rutgers University Press.

Gregory, D. (1994) *Geographical Imaginations*. Cambridge, MA and Oxford: Blackwell.

Hardin, G. (1968) The tragedy of the commons. *Science*, 162, 1243–48.

Hart, G. (2006) Denaturalizing dispossession: critical ethnography in the age of resurgent imperialism. *Antipode*, 38(5), 977–1004.

Harvey, D. (2003) *The New Imperialism*. Oxford and New York: Oxford University Press.

Harvey, D. (2005) *A Brief History of Neoliberalism*. Oxford and New York: Oxford University Press.

Heynen, N., McCarthy, J., Prudham, S. and Robbins, P. (eds) (2007) *Neoliberal Environments: False Promises and Unnatural Consequences*. New York: Routledge.

Joyner, C. (1998) *Governing the Frozen Commons: The Antartic Regime and Environmental Protection*. Columbia: University of South Carolina Press.

Klein, N. (2001) Reclaiming the commons. *New Left Review*, 9(May–June), 81–89.

Klein, N. (2002) *Fences and Windows: Dispatches from the Front Lines of the Globalization Debate*. New York: Picador.

Marx, K. (1967) *Capital, Volume I*. New York: International Publishers.

McCarthy, J. (2004) Privatizing conditions of production: trade agreements as environmental governance. *Geoforum*, 35(3), 327–41.

McCarthy, J. (2005) Commons as counterhegemonic projects. *Capitalism, Nature, Socialism*, 16(1), 9–24.

McCarthy, J. and Prudham, S. (2004) Neoliberal nature and the nature of neoliberalism. *Geoforum*, 35(3), 327–41.

McCay, B. and Acheson, J. (eds) (1987) *The Question of the Commons*. Tucson: University of Arizona Press.

Moore, D. (2004) The second age of the Third World: from primitive accumulation to global public goods? *Third World Quarterly*, 25(1), 87–109.

Neumann, R. (2005) *Making Political Ecology*. London: Hodder Arnold.

Ostrom, E. (1990) *Governing the Commons: The Evolution of Institutions for Collective Action*. Cambridge: Cambridge University Press.

Ostrom, E. (2005) *Understanding Institutional Diversity*. Princeton: Princeton University Press.

Ostrom, E. and Hess, C. (eds) (2007) *Understanding Knowledge as a Commons: From Theory to Practice*. Cambridge, MA: The MIT Press.

Ostrom, E., Gardner, R. and Walker, J. (eds) (1994) *Rules, Games, and Common Pool Resources*. Anne Arbor: University of Michigan Press.

Ostrom, E., Burger, J., Field, C., Norgaard, R. and Policansky, D. (1999) Revisiting the commons: local lessons, global challenges. *Science*, 284, 278–82.

Perelman, M. (2000) *The Invention of Capitalism: Classical Political Economy and the Secret History of Accumulation*. Durham: Duke University Press.

Polanyi, K. (1944) *The Great Transformation: The Political and Economic Origins of Our Time*. New York: Farrar & Rinehart.

Robbins, P. (2004) *Political Ecology*. Oxford: Blackwell.

Rose, C. (1994) *Property and Persuasion: Essays on the History, Theory, and Rhetoric of Ownership*. Boulder: West View Press.

Rowe, J. (2005) The common good. *Sierra*, 90(54), 54ff.

Scott, J. (1998) *Seeing Like a State: How Certain Schemes to Improve the Human Condition Have Failed*. New Haven: Yale University Press.

Sitze, A. (2004) Denialism. *South Atlantic Quarterly*, 103(4), 769–811.

St. Martin, K. (2007) Enclosure and economic identity in New England fisheries. In N. Heynen, J. McCarthy, S. Prudham and P. Robbins (eds), *Neoliberal Environments: False Promises and Unnatural Consequences*. New York: Routledge, pp. 255–66.

Sumner, J. (2004) *Sustainability and the Civil Commons: Rural Communities in the Age of Globalization*. Toronto: University of Toronto Press.

The Ecologist (1998) *Whose Common Future? Reclaiming the Commons*. Philadelphia: New Society Publishers.

Thompson, E. P. (1975) *Whigs and Hunters: The Origin of the Black Act*. London: Allen Lane.

Thompson, E. P. (1991) *Customs in Common: Studies in Traditional Popular Culture*. London: Merlin Press.

Williams, R. (1973) *The Country and the City*. London: Chatto and Windus.

Williams, R. (1976) *Keywords*. London: Collins.

Chapter 30

Water

Karen Bakker

Introduction

Water is an archetypal subject for environmental geographers. Essential for life and imbued with rich symbolism, water both shapes and is shaped by our cultures, economies, and landscapes.

In the contemporary period, hydrology is, of course, a well-established specialty within geography. Physical geographers also engage in the study of the role of water in shaping landscapes (geomorphology). This research is articulated with other water-related disciplines such as meteorology, climatology, ecology and hydrogeology. Yet the study of water by geographers is not limited to its purely biophysical aspects. Human geographers have also studied water's role in economic development, religious life, environmental politics, hazards and vulnerability, and urbanisation, to mention just a few topics.

This chapter explores themes in environmental geography pertaining to water, an area in which both human and physical geographers have made substantive contributions. It will focus on four research themes highlighted in the first volume of this series: landscapes; risk and hazards; sustainability; and scale. Much of this research is necessarily interdisciplinary, requiring an appreciation of the complex physical and social systems within which the hydrological (or, as some geographers prefer to term it, the 'hydrosocial') cycle is embedded. This interdisciplinarity is the central theme of this chapter: for each topic, the contributions of physical and human geographers will be explored together. The 'physical/human' split so often imposed on the discipline (mirroring the nature/society binary which underpins much of Western thought) is difficult to sustain when studying water; the study of topics like water scarcity, conflict over water resources or flood hazards invariably remind us of the mutual constitution of the 'social' and 'natural' aspects of water, which research by both human and physical geographers serves to elucidate.

Water and Landscapes

Water and land use are almost always treated separately in academic research, planning and government. Yet the two are inextricably interrelated. Environmental geographers have explored the interrelationship between land and water in detail. Their findings, as explored in this section, have significant implications for applied water management. In conceptual terms, this work's importance stems from its ability to demonstrate the mutual causality between human societies and landscapes: how human activities shape landscapes and waterscapes; and, simultaneously, how the water cycle shapes human societies.

One obvious way in which water and landscapes are interrelated is through the influence of land use and land cover on the hydrological cycle, which has long been the focus of study by geographers (see, e.g., Clifford, 2002; Brazier, 2004; Dollar, 2004; Holden et al., 2004, and the classic works by Chorley, 1969, Pereira, 1973, and White et al, 1972). Given the rapid rate of change in land use in many regions, this is one of the most pressing issues in water management. Indeed, the central problem of hydrology has frequently been characterised as an attempt to refine and solve the 'water balance' equation, which necessarily implies analysis of both land and water use. However, a central difficulty faced by physical geographers in generalising their results about the impact of land use is the fact that the majority of studies are field-based, hampering efforts to extrapolate from 'a series of mainly small-scale, short-term empirical studies of land-use effects . . . to a generalised body of scientific results operable in river basin management' (Newson, 1997: 96). Indeed, this problem characterises work in the environmental sciences more generally. Recent advances in hydrology have developed conceptual frameworks upon which such generalisations might soundly rest (see, e.g., Eagleson 2002, Eaton et al., 2004; Rodriguez-Iturbe and Porporato, 2004).

Given this field-based emphasis, it is not surprising that geographers have made important contributions to the study of the impacts of human land use on the hydrological cycle. For example, geomorphological and hydrological research on rivers has examined the role of riparian vegetation as a control on bank stability (Bennett and Simon, 2004) or as a buffer for material entering the channel from the hillslope (Burt and Pinay, 2005), and the effect of channel morphology and instream vegetation growth on river dynamics, such as sediment transport (Clarke, 2002; Nistor and Church, 2005). The empirical studies by geographers in this area complement theoretical work done in eco-hydrology on the relationship between soils, climate and vegetation (see, e.g., Budyko, 1974).

The implications of this work for the relationship between water and landscapes are particularly important for applied water management (e.g., Bonell and Bruijnzeel, 2005, Pereira 1989). For example, geographers have documented the impacts of dam construction, with its concomitant changes in fluvial flow regime, sediment transport, channel morphology and river ecosystems, as well as the implications of logging on stream temperature and water yields (Graf, 2001; Moore et al., 2005; Robinson and Dupeyrat, 2005). As the work of hydrologists has demonstrated, this implies that debates over restoring the physical integrity of rivers necessarily imply choices about how to intervene in landscapes which are already actively managed and shaped by human hands (see, for example, Graf, 2001; Downs and Gregory, 2004; McDonald et al., 2004; Hillman and Brierly,

2005). This research throws up, at times, counter-intuitive results: for example, improved understandings of the interactions between geomorphological, hydrological and ecosystem processes mean that the recent trend towards decommissioning dams in the United States may not always have positive impacts on all fish species because of the disruption of aspects of ecosystems dependent upon the post-dam fluvial regime, which may be favourable to certain fish species (Marks et al., 2006).

These findings lend urgency to the task of exploring how water management decisions might be influenced by broader social and political considerations, and in turn how the water cycle might shape and constrain human societies (Swyngedouw et al., 2002; Forsyth, 2005; Perrault, 2005). For example, Evenden's study of the failure to dam Canada's Fraser River (the largest salmon river in the world and the only large undammed river on North America's west coast) illustrates that conflicts over resources (in this case, hydroelectricity versus the salmon fisheries that would be negatively affected by large dams) rest centrally on questions of political identity and social power, rather than cost-benefit analyses or narrow technocratic considerations (Evenden, 2004). Other recent work in environmental geography has drawn on political economy and political ecology to explore the multifaceted (social, ecological, political and economic) interrelationship between water infrastructure such as dams and water supply networks, and processes of modernisation, urbanisation and industrialisation (e.g. Gandy, 1997; 2002; Desfor and Keil, 2000; Kaika, 2004; Swyngedouw, 2004a). The large-scale mobilisation of water – rendered difficult by technical challenges and political contestation – in turn has important implications for political governance. Kaika, for example, argues that nationalism in Greece in the 19th century came to be defined through a project of 'hydrological modernisation', whereby rural zones were sacrificed for water provision to a thirsty and rapidly growing Athens in the 19th century. Gandy argues that the public health implications of private-sector-run water supply systems in 19th-century New York were an important impetus for municipalisation and the rise of the welfare state at the urban scale (Gandy, 1997; 2002). Swyngedouw, in another example, argues that Spain's 20th-century 'hydraulic modernisation' was both a response to failed colonialism and a vehicle through which its mid-century dictatorship was entrenched (Swyngedouw, 2004b).

Yet the links between the water cycle and human societies are much broader than structures of political governance. Human geographers emphasise the mutual interrelationship between environment, material practices and symbolic culture, anticipating current debates in geography by several decades (e.g. White et al., 1972; Cosgrove et al., 1992; Matless 1992; Oliver, 2000; Howarth 2001). This work has been deliberately interdisciplinary, incorporating history, anthropology and cultural theory; for example, the University of Nottingham's Water, Culture and Society project, headed by geographer Stephen Daniels.[1] Urban geographers, on the other hand, have documented how urban form and urbanisation processes are predicated upon water availability and the ways in which water is incorporated into urban infrastructure (Kaika, 2004; 2006; Heynen et al., 2005; Keil 2005).

In so doing, of course, geographers must walk a fine line between incorporating materiality – in the varied senses of that term (Bakker and Bridge 2006) – and avoiding the spectre of environmental determinism, which has long haunted the discipline. Geographers seem to have done this more successfully than many other disciplines, largely through their careful assertion (and empirical verification) of mutual causality between environmental change, landscape change and human societies in a non-deterministic fashion. This stands in sharp contrast to the work of scholars from other disciplines on the relationship between human societies and their environments, in which the environment is seen to play a simplistically deterministic role, overly limiting human agency, but also, at times, insidiously displacing blame for inequality within human societies to 'natural' environmental factors (e.g. Wittfogel, 1957; Landes, 1998; Sachs, 2005).

Water and Risk: (De)Constructing Water Hazards

The mutual imbrication of nature and society has also been a core theme of geographical analyses of natural hazards, a focal point of which has been water-related hazards: floods, droughts, landslides and hurricanes. In the view of the 'hazards school' in environmental geography, research on physical processes must necessarily be integrated with research on the social construction of hazards if risk and vulnerability are to be comprehensively analysed and mitigated. For example, environmental geographers have evaluated flooding and landslide hazards by studying the siting of residential housing developments in floodplains or on landslide-prone slopes. Physical geographers have documented the relationship between geomorphological and hydrological processes and risk; for example, research on slope stability and landslide risk is used to support analyses of slope failures (see, e.g., Petley, 2004; Glade et al., 2005; Hufschmidt et al., 2005).

Hazards research integrates the work of physical geographers with assessments of vulnerability and risk. The work of geographer Gilbert White, for example, entailed the identification and classification of perceptions of, and responses to, flood hazards, enabling more accurate analyses of responses to floods in the United States (Hinshaw, 2006). White argued that an over-reliance on structural hydraulic works (such as levees, dams and barrages) in the United States had increased damages caused by flooding, rather than decreasing them, because the high degree of public confidence in structural works enticed development in flood-prone areas (White et al., 1958). White and his students demonstrated a lack of willingness of individuals living in flood-prone areas to adopt protective measures such as insurance or flood-proof doors and windows, reflecting a reduction in perception of risk due to the presence of engineered flood control structures (Kates 1962; Kreutzwiser et al., 1994; Shrubsole et al., 1997). White's work resulted in the development of the National Flood Insurance Program in the United States and was influential in the promotion of 'non-structural' solutions to flooding, such as regulatory restrictions on the use of floodplains (Kates and Burton, 1986; Tobin and Montz, 1997).

Hurricane Katrina serves as an illustration of the relevance of hazards research on floodplain development and flood control. Sited in a region in which hurricanes regularly make landfall, New Orleans' urban expansion into the surrounding floodplain has been enabled by the construction of levees and draining of wetlands (Lewis, 2003; Colten, 2005). As the city sank, the coastal wetlands protecting it from storm surges in the Gulf shrank. The extensive wetlands which used to absorb annual floods and storm surges in southern Louisiana have dramatically subsided, as upstream damming and channelisation of the river has reduced sediment loads and diverted them out to the Gulf; Louisiana has lost an estimated 1 million acres of coastal wetlands since the 1930s (Reed and Wilson, 2004). The combination of shipping channels and disappearing wetlands opened up what some term a 'hurricane highway' into the city, through which storm surges can be channelled and amplified (Hallowell, 2001). Despite numerous government reports foretelling the impact of a major hurricane (Travis, 2005), development continued on New Orleans' extensive floodplain, despite misgivings that containment measures, never foolproof, would increase the severity and catastrophic power of floodwaters when they inevitably overtopped the concrete and earthen barriers ringing the city (Bakker, 2005).

Despite numerous government reports foretelling the impact of a major hurricane (Travis, 2005), engineering hubris prevailed in Louisiana. Notwithstanding public celebrations of Cajun culture and despite Louisiana's rich seafood harvests (once the most productive coastal seafood fishery in the United States), its wetlands were sacrificed to commercial transport and oil extraction, underpinned by a deep-rooted cultural imaginary of wetlands as 'swamps' (Fritzell, 1978). These 'dark edens' (Miller, 1989), actively targeted for draining in the 19th and 20th centuries in a civilising mission directed at both the American landscape and at its inhabitants (Marx, 1964), were converted into dumping grounds for toxic by-products of the oil industry. Ironically, Katrina and other recent hurricanes were caused by unprecedented warming in surface temperatures in the Gulf of Mexico in recent years, which some scientists have attributed to climate change (Emanuel, 2005; Travis, 2005; Webster et al., 2005); if true, a tragically ironic example of Louisiana's oil economy coming full political-ecological circle.

Other work by environmental geographers on water-related hazards emphasises the differentiation of vulnerability by region, race, class and gender, documenting the disproportionate impacts of hazards on lower-income and minority communities and women (Blaikie et al., 1994; Cutter, 2001; Liverman, 2001; Mustafa, 2005). For example, as Susan Cutter notes in her analysis of the impacts of Hurricane Katrina,[2] the segregated past of the American South is still visible in the spatial and social geography of cities such as New Orleans, where housing for black, working-class communities is located in the least desirable areas, with limited social services and amenities and higher exposures to environmental risks, including floods. The result is well documented by environmental justice and political ecology research: wealthier, largely white individuals have secured relatively cleaner, safer environments in American urban centres (e.g. Pulido, 2000), leaving poor, largely black communities to locations with higher-pollution and higher-hazard probabilities and impacts (e.g. Cutter et al., 2003). As Neil Smith has observed, Katrina revealed that topographical gradients were proxies for race and class in New Orleans (Smith, 2005), with largely white neighbourhoods situated on higher, drier ground. Simply put, white privilege underlays the spatial location and racial composition of communities most vulnerable to flooding.

The work of geographers has explored how the severity of impacts caused by Hurricane Katrina was due to both social and natural causes; it was, in short, no 'natural' disaster. The impacts of Katrina thus serve as a tragic, emblematic example of the results of recent environmental geography research on hazards, in which socio-economic and psychological variables are as important as biophysical variables in understanding vulnerability and risk to water-related hazards. In turn, geographical research has demonstrated how these terms – risk, vulnerability, hazard and disaster – are deeply anthropocentric, insofar as they reflect concern about impacts on humans. As explored in the following section, rethinking this anthropo-centric attitude, at least in part, may hold the key to more sustainable relationships between water and humans.

Water and Sustainability: Integrated Water Resources Management

The example of Katrina is suggestive of a third key focus of water-related research by environmental geographers: sustainable water management. Here, it is important to note that water is a messy resource to manage. Water is a flow resource that constantly transgresses political boundaries. Unlike most of the resources central to our livelihoods and communities, water is constantly on the move. This means that water connects communities in ways that most other resources do not. Impacts of water use – both positive and negative – are felt far downstream, in other communi-ties and jurisdictions. Yet water is most often used close to the point of abstraction. Thus, water presents managers with three complex issues which are difficult to resolve: dealing with competition between multiple users of water resources (agri-culture, energy production (hydropower), industry, urban water supply, recreation, tourism and ecosystem services); balancing the multiple scales at which water is managed; and responding to the mismatch between geopolitical and administrative boundaries, on the one hand, and hydrological boundaries on the other. Research on water-related issues tends to mimic this fragmentation: knowledge on hydrologi-cal, ecological, biological, socio-political and economic processes tends to be pro-duced in separate fields, with little interaction between researchers. Recognition of the negative effects of fragmentation in the early 20th century led to attempts to develop organisational and institutional structures that could coordinate water resources management at a watershed scale, such as the Tennessee Valley Authority, through 'integrated water resources management' (IWRM).

According to its proponents, IWRM is intended to address some of the resource management flaws illustrated by Katrina, through the comprehensive, integrated assessment and governance of water in concert with other resources – particularly land – at the watershed scale. In particular, IWRM includes the implementation of governance mechanisms designed to reduce or eliminate conflict through, for example, integrating land use and water resource planning mechanisms. These gov-ernance mechanisms are informed by integrated scientific approaches that manage multiple resources (e.g. soil, vegetation, water) on a watershed scale, usually with explicit goals of constraining point and non-point source pollution, source protec-tion, and soil and water conservation (Mitchell 1995; 2005; Newson, 1997; White, 1997; Wescoat, 2000; Ducros and Watson, 2002; Wescoat and White, 2003; Watson, 2004; Shrubsole and Watson, 2005; Ivey et al., 2006; and for an early discussion see Weber, 1964).

As geographers have demonstrated, IWRM is particularly relevant in the global South, where critical challenges in water-related health and water security are prevalent and compounded by jurisdictional fragmentation and weak governance (see, e.g., Young et al., 1994; Lonergan et al., 2002; Jones and van de Walt, 2004). An integrated approach, many environmental geographers have argued, is better able to effectively address these water management challenges. A notable example is the work of geographer Gordon Young in developing the United Nations World Water Assessment Programme,[3] which coordinated 24 UN agencies to produce a comprehensive, integrated assessment of the world's key water problems, including the periodic World Water Development Report, which aims to give an overall picture of the state of the world's freshwater resources and to provide decision makers with the tools to implement sustainable water management (UN, 2003; 2006). More generally, a focus on the geographical dimensions of water, poverty and development has been characteristic of geographers' work on these issues, who pay particular attention to questions of scale and space, and the interrelationship between human and aquatic systems in their study of the links between water, poverty, geography and development (see, e.g., Wescoat et al., 2000; D'Souza 2002; Giordano et al., 2002; Sneddon et al., 2002; Halvorson, 2003, p. 2005).

Successful integrated water resources management is in turn dependent upon data and analyses produced by environmental geographers and other environmental scientists. For example, adequate source protection (the safeguarding of upland water sources critical for drinking water supply) is dependent upon a fairly comprehensive assessment of pollutant sources and pathways (atmospheric as well as hydrological), development activities and land use change, and meso- and macro-scale climatological and meteorological processes such as the changing frequency and intensity of extreme water-related events (such as droughts, floods and storms) (Durley and de Loë, 2005). Using groundwater hydrology, for example, using groundwater hydrology, geographers have assessed the history of groundwater recharge and climate change in various locales using isotopic compositions to infer the age of groundwater, and a chloride mass balance method to determine the long-term recharge rate (Ma et al., 2005). Focusing on the unsaturated profiles of specific aquifers (particularly those not receiving contemporary recharge) allows geographers to document wet and dry periods in the past, effectively using groundwater as an archive of past environmental and climatic change (Edmunds, 1996; 2005). These findings are of critical importance for understanding possible variations in precipitation within watersheds, and linking historical studies to contemporary analyses of variability (e.g. Mote et al., 2003). These micro-scale field-based findings (which are often characteristic of the work of environmental geographers, as opposed to other environmental scientists) are complemented by research linking macro- and meso-scale relationships between climate, weather and local hydrological conditions; for example, through analysing the relationship between ocean surface temperatures, regional climate and local weather patterns (see, e.g., Grundstein and Leathers 1998; McKendry et al., 2006).

Integrating these types of research at different scales has long been a goal of environmental geographers. A persistent drawback has been the difficulty of generalisation and replicability given the tendency of environmental geographers to focus on case-specific, fieldwork-intensive research.[4] However, remote sensing opens up the possibility of gathering large datasets relatively inexpensively, enabling geographers to more fully test hypotheses, and inductively generate falsifiable hypotheses and more general theories in a way that was not previously possible. Research on integrated water management is thus emblematic of the difficulties that have faced environmental geography, and illustrative of the ways in which new developments are opening up the possibility of a 'coming of age' of environmental geography within the environmental sciences. A critical aspect of this work will rest upon the integration of research by human and physical geographers. Indeed, in practice, water management is a field which draws simultaneously on the work of social and natural scientists (with no obvious boundary between the two); IWRM is thus an issue on which 'human' and 'physical' geographers have conducted a great deal of collaborative research (see, e.g., Watson et al., 1996) in pursuit of the elusive 'synthesis' which has long been a goal (or, some would argue, chimera) of geographical research.

This integration has, as with other areas of the natural and social sciences, recently been unsettled by the work of (largely social science) researchers interested in the social construction of scientific knowledge. Much of this work has focused on contentious issues in environmental debates, such as global climate change (Benton and Redclift, 1994; Demeritt, 2001). But environmental geographers have also deployed techniques such as deconstruction and discourse analysis to examine issues of water management. Geographer Jamie Linton, for example, examines academic and popular discourses of the global 'water crisis' which became prominent (particularly in the media) in the 1990s (Linton, 2004). Linton argues that rapid dissemination of the 'water crisis' narrative was not due to an equally rapid process of biophysical or environmental changes in actual water availability. While acknowledging the very real constraints that water scarcity poses in some areas, Linton argues that the method of discourse analysis demonstrates that the 'water scarcity' storyline gained prominence due to a conjunction of actors and interests that produced a range of artefacts enabling the 'water crisis' storyline to emerge. First, the development of this storyline was predicated upon the regional and eventually global hydrological models developed by Soviet hydrologists in response to the information demands of centralised planning and resource management in the former USSR. Initially suspicious of Soviet methods, Western hydrologists began (many reluctantly) in the 1980s to think in 'global terms' about core hydrological concepts such as runoff, renewable water resources and water balance. The subsequent development of indices of water stress and water scarcity married neatly with the desires of international aid agencies to secure additional funding for water-related development, effacing concerns voiced by hydrologists over uncertainty in global and even regional estimates of water availability. Finally, advocates of water commercialisation proved only too eager to mobilise a discourse of water scarcity and a 'water crisis' in order to argue in favour of full-cost water pricing. Advocates of privatization were equally supportive of the reframing of water as an economic good which required efficient management by the private sector, thereby attempting to naturalise the contentious processes of privatisation, marketisation and commercialisation of water supply and resources which became widespread in the 1990s (see, e.g., Shirley, 2002; Shiva, 2002).

Linton deconstructs and questions the 'water crisis' narrative, both querying the soundness of its empirical basis, and questioning the political and ideological agendas that it serves. This approach is a fairly common tactic adopted by human geographers working on water issues (and environmental issues more generally), and has led to some fruitful insights into the actual impacts of, for example, the privatisation and commercialisation of water supply systems (Bakker, 2004), as well as analyses of the ideological and economic goals embedded in seemingly 'apolitical' water supply systems (Swyngedouw, 2004a). Swyngedouw's work, for example, documents the relationship between political economic interests, ecological processes, and the complex and the highly inequitable water supply system in the city of Guayaquil, Ecuador – which, like most cities in the Third World, has a limited water supply network delivering subsidised water to the wealthy, leaving 600,000 mostly poor residents to rely on tankers or water vendors charging much higher prices per unit volume. In demonstrating the links between the water supply system, urbanisation, economic interests (such as the international trade in bananas) and political power, Swyngedouw's study was a considerable conceptual breakthrough. Moreover, it had significant policy influence – inspiring, for example, the theme and approach of the UNDP's 2006 Human Development Report on water, which sought to demonstrate (in contrast to much of the water-related research in the development literature) the links between power, inequitable political economies and households' access to water (UNDP 2006). Indeed, the UNDP's argument – that water scarcity is largely a social and economic construct that is the result of human actions rather than natural events – is directly opposed to arguments in favour of privatisation and commercialisation predicated upon an assumption of water scarcity, paralleling the divide between mainstream economists and critical geographers who have studied water privatisation in the global South (e.g. Bakker, 2003; Bond, 2004; McDonald and Ruiters, 2005).

Another example of this 'unsettling' of conventional concepts can be found in recent debates over the 'watershed' as a socially constructed scale. The desirability of organising environmental management at the watershed scale is rarely disputed within both academic and policy literatures. Similarly, the ecological relevance of the watershed scale is usually taken to be self-evident within the water management literature. Environmental geographers have critically examined these arguments, inspired by Neil Smith's (1984) arguments about the social construction of scales – whether evidently social (such as a parish, or a region) or putatively 'natural'. Geographers have argued, for example, that watersheds appear as meaningful scales to some environmental scientists, but are largely ignored by land users and managers (such as farmers and agricultural ministries) which operate within cadastral or local political boundaries such as the county, parish or nation-state (Fischhendler and Feitelsen, 2005; Ivey et al., 2006). Hydrologists and groundwater hydrogeologists have also argued that the watershed is an arbitrary scale upon which to base water management, insofar as groundwater (aquifer) boundaries and surface watershed boundaries almost never coincide, and insofar as topographical, meteorological and soil conditions can heavily influence the degree to which runoff is interconnected at a watershed scale (De Vito et al., 2005). And environmental scientists – particularly ecologists – argue that a watershed can be a relatively meaningless scale in ecological terms, and argue that spatial scales such as range or biome are far more relevant to integrated environmental assessment and management.

Water and Scale

The research discussed above opens up the question of the scales at which water is best managed. Water is a local resource par excellence: cheap to store but expensive to transport, with variations in water quality posing difficulties (both in ecological and public health terms) for long-distance transport. For these reasons, it would seem that water should best be governed at a local level. But the biophysically hierarchical nature of surface runoff organisation, the existence of transboundary waters, the multiple competing uses to which water is subject (exacerbated by inevitable conflicts between upstream and downstream users), frequent scalar mismatches between supply and demand, and the implications of poor water management for public and environmental health all imply the need for a higher order of governance – usually regional or national. On the one hand, distribution of governance to local levels of government makes sense, particularly where different regions have dramatically different hydrology, topography and political economy. Yet some water issues are best dealt with at regional or national scales, particularly where they affect public health. This governance-scale conundrum has long beset environmental geographers working on water resources, who have been influential in public policy debates on these issues.[5]

Research in hydrology and physical geography faces similar conceptual problems. Given that water is a flow resource, any study of water must deal simultaneously with multiple (and nested) spatial and temporal scales. Given a high degree of space-time variability, how can we make generalisable statements (or derive theoretical models) about the behaviour and evolution of hydrological systems? And given this high degree of variability and scale-interdependence, at what scale(s) is water best studied?

The problem of scaling (both upscaling and downscaling) is thus an important one in the study of water, both because of the interrelatedness of scales (the hydrological cycle, the global climate system and a local watershed) and because of a high spatial and temporal variability of key variables (soil moisture, precipitation, land cover and land use). Hydrologists need to make useful statements about variability in order to compare, generalise, and extrapolate results, but scalar change complicates this task, because variability changes as scale changes (Culling and Datko, 1987; Woods, 2005). This 'scaled variance' phenomenon is of critical importance to hydrology because 'virtually any quantitative approach to [the problem of variability] requires the selection of a limited set of spatial and temporal scales . . . [which] has a major influence on which aspects of this hydrological variability are perceived' (Bloschl, 2005).

Taking this 'scaled variance' phenomenon into account has, in fact, led environmental geographers to groundbreaking research on water-related topics, such as urban micro-meteorology and urban climatology. Urban climatology, urban hydrology and the urban water balance are typical of the sorts of problems studied by physical geographers, as they are characterised by finer spatial resolution than atmospheric scientists, tight feedbacks between human and natural systems, and important management and policy implications. The legitimacy for this research within geography stems in part from a broader tradition of studying 'human impacts on the environment' (Clark et al., 1990; Kasperson et al. 1995; Goudie 2000). Urban climatologists and hydrologists have demonstrated that the urban scale is markedly different from other scales and locales due to a high degree of alteration of the landscape, which significantly changes the water balance (Grimmond and Oke 1986; Grimmond et al., 1986; Aronica and Lanza 2005).

Specifically, urban runoff is greater than rural runoff and the peak in runoff following precipitation events happens much more quickly (due to a lower proportion of permeable plant and ground cover). A high degree of imperviousness results in lower evaporation and transpiration. And, since human water use has diurnal and seasonal peaks, urban water bodies from which water is removed and to which effluent experience (often significantly) altered water quality. These changes in local meteorological conditions may be further impacted by the 'urban heat island' effect, the study of which was first scientifically systematised by geographer Tim Oke (Oke 1982; Roth et al., 1989). Many, if not all, of these impacts would be obscured if analysis was conducted at a regional watershed scale. The choice of scale, in short, is of crucial importance to an assessment of environmental impacts.

The choice of scale is a similarly critical issue for water governance. In the European Union, for example, the allocation of water management responsibility is decided on the basis of 'subsidiarity' (Bermann and Pistor, 2004). This concept means, simply put, that decisions should be taken and policies implemented by the smallest (or lowest) competent authority. Subsidiarity is balanced, in the European approach to federalism, with 'harmonisation', in which legislation and policies are selectively standardised. In some cases, this means that member states make the decisions; in others, it is more appropriate and effective to make decisions at the European level. In general, the balance has tilted towards harmonisation; most national water policy in member states is now determined in Brussels and water legislation is one of the most harmonised components of European environmental legislation (Kaika and Page, 2002). The EU's Water Framework Directive and associated approach to water governance are now widely recognised as being amongst the most advanced in the world, at least potentially capable of redressing or attenuating many of the persistent problems that plague water governance globally.

The work of environmental geographers has illustrated that scale, in short, is both a social construct and a powerful lens through which to study and manage water. The need to choose one or more scales of analysis is inevitable, as are the constraints which a specific scale places upon research and management. However, through research which articulates scales and clarifies the bases upon which scales are constructed and chosen, environmental geography allows us to refine both our analyses and our stewardship of water resources.

Conclusions

Whereas engineering and hydrogeology (the other modern disciplines which can lay a claim to a sustained focus on water issues) have largely focused on questions of assessment and technique (associated with water supply and hydraulic technologies, and hydrogeological processes in the latter), geographers' studies of water have been concerned with interactions between humans and the environment in a much broader sense (although focusing, understandably but perhaps rather myopically, on the surface water cycle to the relative neglect of groundwater processes).

Through this research, geographers have made important contributions to broader debates in geography on 'socio-nature': a concept that refutes conventional nature/society binaries and asserts the mutual constitution of human and non-human worlds. This work has important implications for applied water management because it documents the effects of human actions on landscapes, allows us to assess and predict water-related hazards, and analyses the relative risks and vulnerability to those hazards across human societies. Moreover, work by geographers also documents how the hydrosocial cycle influences and impacts human societies, through, for example, the mutual constitution of waterscapes and cultural norms, the politics of water governance, and urban form. As a result, this work speaks to very general concerns in 21st-century academia: accurately describing the relationship between humans and environment, and shedding deterministic and anthropocentric assumptions about causality within that relationship.

This approach is in line with broader trends within academic and policy circles. Water managers, for example, are increasingly cognisant of the fact that water management must move beyond hydrology, biogeochemistry and engineering to include the dynamics of what Shiela Jasanoff terms the 'co-production' of socio-natural systems (Jasanoff, 2004; Bonnell, 2005;). With respect to water, geographers have developed the tools to accomplish this task. Perhaps surprisingly, the conceptual heterodoxy of approaches to the study of water (including the positivist, instrumentalist and Marxist approaches discussed in this chapter) has not proved to be a barrier to collaboration within and beyond the discipline (e.g. Bonell and Bruijnzeel, 2005), nor has it impeded the sustained impact of geographers on water policy. Indeed, this heterodoxy may partly explain why geography, more than any other modern discipline, best exemplifies an integrated, multidisciplinary approach to the study of water issues, which is increasingly recognised to be crucial for the improved stewardship and management of water resources – for the benefit of humans and non-humans alike. Geography's sometimes uneasy bridging of the binaries besetting the modern academy (physical versus social sciences, humans versus the environment) will continue to be the key to its unique contribution in this field.

NOTES

1. http://www.nottingham.ac.uk/hrc/water/about.php.
2. http://understandingkatrina.ssrc.org/Cutter/#5. An excellent exploration of the issues is provided on the Social Science Research Council's website 'Understanding Katrina: Perspectives from the Social Sciences' (http://understandingkatrina.ssrc.org/), with contributions from geographers Neil Smith, Susan Cutter, James K Mitchell, Stephen Graham and others.
3. www.unesco.org/water/wwap/wwdr/
4. These two issues are not necessarily so acute in areas of environmental science that are more lab-based or modelling-intensive. I am indebted to Professor Michael Church, University of British Columbia, for these insights.
5. For example, Bill Graf (past president of the Association of American Geographers) chaired the United States' National Resource Council's Committee on Watershed Management, charged with developing a new strategy for American watersheds (NRC 1998; Graf 2001).

BIBLIOGRAPHY

Aronica, G. and Lanza, L. (2005) Hydrology in the urban environment. *Hydrological Processes*, 19(5), 1005–6.
Bakker, K. (2003) From archipelago to network: urbanization and water privatization in the South. *The Geographical Journal*, 169(4), 328–41.
Bakker, K. (2004) *An Uncooperative Commodity: Privatising Water in England and Wales.* Oxford: Oxford University Press.
Bakker, K. (2005) Katrina: The public transcript of disaster. Guest Editorial in *Environment and Planning D: Society and Space*, 23, 795–809.
Bakker, K. and Bridge, G. 2006 Material worlds? Resource geographies and the 'matter of nature'. *Physical Geography*, 30(1), 1–23.
Bennett, S. and Simon, A. (eds) (2004) *Riparian Vegetation and Fluvial Geomorphology* Water Science and Application series. American Geophysical Union.
Bermann, G. and Pistor, K. (eds) (2004) *Law and Governance in an Enlarged Europe.* Portland: Hart.
Blaikie, P., Cannon, T., Davis, I., and Wisner, B. (1994) *At Risk: Natural Hazards, People's Vulnerability, and Disasters.* London: Routledge.

Benton, T. and Redclift, M. (1994) *Social Theory and the Global Environment*. London: Routledge.

Bloschl, G. (2005) On the fundamentals of hydrological sciences. In M. Anderson and J. McDonnell (eds), *Encyclopedia of Hydrological Sciences*. London: John Wiley & Sons, Ltd, pp. 1–12.

Bond, P. (2004) Water commodification and decommodification narratives: pricing and policy debates from Johannesburg to Kyoto to Cancun and back. *Capitalism Nature Socialism*, 15(1), 7–25.

Bonell, M. and Bruijnzeel, L. A. (eds) (2005) *Forests, Water and People in the Humid Tropics*. Cambridge: Cambridge: University Press.

Brazier, R. (2004) Quantifying soil erosion by water in the UK: a review of monitoring and modelling approaches. *Progress in Physical Geography*, 28(3), 340–65.

Budyko, M. I. (1974) *Climate and Life*. New York: Academic Press.

Burt T., and Pinay, G. (2005) Linking hydrology and biogeochemistry in complex landscapes. *Progress in Physical Geography*, 29, 297–316.

Chorley, R. (1969) *Water, Earth, and Man: A Synthesis of Hydrology, Geomorphology and Socio-economic Geography*. London: Methuen.

Clark, W. C., Turner, B. L., Kates, R. W., Richards, J., Mathews, J. T. and Meyer, W. (eds) (1990) *The Earth as Transformed by Human Action*. Cambridge: Cambridge University Press.

Clarke, S. (2002) Vegetation growth in rivers: influences upon sediment and nutrient dynamics. *Progress in Physical Geography*, 26(2), 159–72.

Clifford, N. (2002) Hydrology: the changing paradigm. *Progress in Physical Geography*, 26(2), 290–301.

Cosgrove, D., Roscoe, B. and Rycroft, S. (1992) Landscape and identity at Ladybower Reservoir and Rutland Water. *Transactions of the Institute of British Geographers*, 21, 534–51.

Colten, C. (2005) *An Unnatural Metropolis: Wrestling New Orleans from Nature*. Baton Rouge: Louisiana State University Press.

Culling, W. and Datko, M. (1987) The fractal geometry of the soil-covered landscape. *Earth Surface Processes and Landforms*, 12, 369–85.

Cutter, S. (ed) 2001 *American Hazardscapes; The Regionalization of Hazards and Disasters*. Washington, DC: Joseph Henry Press.

Cutter, S., Boruff, B. J. and Shirley, W. L. (2003) Social vulnerability to environmental hazards. *Social Science Quarterly*, 84(1), 242–61.

Demeritt, D. (2001) The construction of global warming and the politics of science. *Annals of the Association of American Geographers*, 91(2), 307–37.

Desfor, G. and Keil, R. (2000) Every river tells a story: the Don river (Toronto) & the Los Angeles river (Los Angeles) as articulating landscapes. *Journal of Environmental Policy and Planning*, 2(1), 5–23.

Devito, K., Creed, I. and Fraser, C. (2005) Controls on runoff from a partially harvested aspen-forested headwater catchment, Boreal Plain, Canada. *Hydrological Processes*, 19(1), 3–25.

Dollar, S. (2004) Fluvial geomorphology. *Progress in Physical Geography*, 28(3), 405–50.

Downs, P. and Gregory, K. (2004) *River Channel Management: Towards Sustainable Catchment Hydrosystems*. London: Arnold.

D'Souza, R. (2002) At the confluence of law and geography: contextualising inter-state water disputes in India. *Geoforum*, 33(2), 255–69.

Ducros, C. and Watson, N. M. (2002) Integrated land and water management in the United Kingdom: narrowing the implementation gap. *Journal of Environmental Planning and Management*, 45(3), 403–23.

Durley, J. L. and de Loë, R. C. (2005) Empowering communities to carry out drought contingency planning. *Water Policy*, 7(6), 197–210.

Eagleson, P. (2002) *Ecohydrology: Darwinian Expression of Vegetation Form and Function*. Cambridge: Cambridge University Press.

Eaton, B., Church, M. and Millar, R. (2004) Rational regime model of alluvial channel morphology and response. *Earth Surface Processes and Landforms*, 29(4), 511–29.

Edmunds, W. M. (1996) Indicators in the groundwater environment of rapid environmental change. In A. R. Berger and W. J. Iams *(eds), Geoindicators: Assessing Rapid Environmental Changes in Earth Systems*. Rotterdam: A.A. Balkema, pp. 121–36.

Edmunds, W. M. (2005) Groundwater as an archive of climatic and environmental change. In P. K. Aggarwal, J. Gat and K. Froehlich (eds), *Isotopes in the Water Cycle: Past, Present and Future of a Developing Science*. Dordrecht: Springer, pp. 341–52.

Emanuel, K. (2005) Increasing destructiveness of tropical cyclones over the past 30 years. *Nature*, 436(4), 686–88.

Evenden, M. (2004) *Fish versus Power: An Environmental History of the Fraser River*. Cambridge: Cambridge University Press.

Fischhendler, I. and Feitelson, E. (2005) The formation and viability of non-basin transboundary water management: the case of the U.S.–Canada Boundary Water. *Geoforum*, 36, 792–804.

Forsyth, T. (2005) Land use impacts on water resources: Science, social and political factors. In M. Anderson and J. McDonnell (eds), *Encyclopedia of Hydrological Sciences*. London: John Wiley & Sons, Ltd, pp. 2911–24.

Gandy, M. (1997) The making of a regulatory crisis: the restructuring of New York City's water supply. *Transactions of the Institute of British Geographers*, 22(3), 338–58.

Gandy, M. (2002) *Concrete and Clay: Reworking Nature in New York City*. Cambridge, MA: MIT Press.

Giordano, M., Giordano, M. and Wolf, A. (2002) The geography of water conflict and cooperation: internal pressures and international manifestations. *The Geographical Journal*, 168(4), 293–312.

Glade, T., Anderson, M. and Crozier, M. (eds) (2005) *Landslide Hazard and Risk*. London: John Wiley & Sons, Ltd.

Goudie, A. (2000) *The Human Impact on the Natural Environment*, 5th edn. Cambridge, MA: MIT Press.

Graf, W. L. (2001) Damage control: restoring the physical integrity of America's rivers. *Annals of the Association of American Geographers*, 91(1), 1–27.

Grimmond, C. and Oke, T. (1986) Urban water balance: results from a suburb of Vancouver, British Columbia. *Water Resources Research*, 22(10), 1404–12.

Grimmond, C., Oke, T. and Steyn, D. (1986) Urban water balance: a model for daily totals. *Water Resources Research*, 22(10), 1404–12.

Grundstein, A. J. and Leathers, D. A. (1998) A case study of the synoptic patterns influencing midwinter snowmelt across the northern great plains. *Hydrological Processes*, 12(15), 2293–305.

Hallowell, C. (2001) *Holding Back the Sea: The Struggle for America's Natural Legacy on the Gulf Coast*. New York: Harper Collins.

Halvorson, S. J. (2003) A geography of children's vulnerability: gender, household resources, and water-related disease hazard in northern Pakistan. *Professional Geographer*, 55(2), 120–33.

Heynen, N., Kaika, M. and Swyngedouw, E. (eds) (2005) *In the Nature of Cities: Urban Political Ecology and the Politics of Urban Metabolism*. London and New York: Routledge.

Hillman, M. and Brierley, G. (2005) A critical review of catchment-scale stream rehabilitation programmes. *Progress in Physical Geography*, 29(1), 50–76.

Hinshaw, R.E. (2006). *Living with Nature's Extremes: The Life of Gilbert Fowler White*. Boulder, CO: Johnson Books.

Holden, J., Chapman, P. J. and Labadz, J. C. (2004) Artificial drainage of peatlands: hydro-logical and hydrochemical process and wetland restoration. *Progress in Physical Geography*, 28(1), 95–123.

Howarth, W. (2001) Reading the wetlands. In P. C. Adams et al. *Textures of Place: Exploring Humanistic Geographies*. Minneapolis: University of Minnesota.

Hufschmidt G., Crozier M. and Glade T. (2005). Evolution of natural risk: research framework and perspectives. *Natural Hazards and Earth System Sciences*, 5, 375–87.

Ivey, J., de Loë, R. C., Kreutzwiser, R. D. and Ferreyra, C. (2006) An institutional perspective on local capacity for source water protection. *Geoforum*, 37(6), 944–57.

Jasanoff, S. (2004) The idiom of co-production. In S. Jasanoff (ed.), *States of Knowledge: The Co-Production of Science and the Social Order*. London: Routledge.

Jones, J. A. A. and van de Walt, I. J. (eds) (2004) Barriers and solutions to sustainable water resources in Africa. Special Issue, *Geojournal*, 61(2), 105–214.

Kaika, M. (2004) *City of Flows: Modernity, Nature and the City*. New York: Routledge.

Kaika, M. (2006) Dams as symbols of modernization: the urbanization of nature between geographical imagination and materiality. *Annals of the Association of American Geographers*, 96(2), 276–301.

Kalinin, G. P. (1971) *Global Hydrology*. Jerusalem: Kater Press.

Kasperson, J. X., Kasperson, R. E. and Turner, B. L. II, (eds) (1995) *Regions at Risk: Comparisons of Threatened Environments*. Tokyo: United Nations University.

Kates, R. (1962) *Hazard and Choice Perception in Flood Plain Management*. Department of Geography Research Paper no. 78, University of Chicago, Chicago.

Kates, R. and Burton, I. (1986) *Geography, Resources and the Environment*. Chicago: University of Chicago Press.

Keil, R. (2005) Urban political ecology. *Urban Geography*, 26(7), 640–51.

Kreutzwiser, R., Woodley, I. and Shrubsole, D. (1994) Perceptions of flood hazards and flood-plain management regulations in Glen Williams, Ontario. *Canadian Water Resources Journal*, 19, 115–24.

Landes, D. 1998: *The Wealth and Poverty of Nations: Why Some Are So Rich and Some So Poor*. New York: W. W. Norton.

Lanza, L. and Gallant, J. (2005) Fractals and similar approaches in hydrology. In M. Anderson and J. McDonnell (eds), *Encyclopedia of Hydrological Sciences*. London: John Wiley & Sons, Ltd, pp. 123–33.

Lewis, P. (2003) *New Orleans: The Making of an Urban Landscape*. Charlottesville: University of Virginia Press.

Linton, J. (2004) Global hydrology and the construction of a water crisis. *Great Lakes Geographer*, 11(2), 1–13.

Liverman, D. M. (2001) Vulnerability to drought and climate change in Mexico. In J. X. Kasperson and R. Kasperson (eds), *Global Environmental*. New York: UNU and Earthscan, pp. 201–216.

Lonergan, S. C., Wolf, A. and Cocklin, C. (eds) (2002) *Water and Human Security in Southeast Asia*. Tokyo: UNU Press.

Ma, J. Z., Wang, X. S. and Edmunds, W. M. (2005) The characteristics of groundwater resources and their changes under the impacts of human activity in the arid north-west China – a case study of the Shiyang river basin. *Journal of Arid Environments*, 61, 277–95.

Marks J. C., Parnell R., Carter C., Dinger E. C. and Haden, G. A. (2006) Interactions between geomorphology and ecosystem processes in travertine streams: Implications for decommissioning a dam on Fossil Creek, Arizona. *Geomorphology*, 77(3), 299–307.

Marx, L. (1964) *The Machine in the Garden: Technology and the Pastoral Ideal in America*. Oxford: Oxford University Press.

Matless, D. (1992) A modern stream: water, landscape, modernism and geography. *Environment and Planning D: Society and Space*, 10(5), 569–88.

McDonald, A., Lane, S. N., Chalk, E. and Haycock, N. (2004) Rivers of dreams: on the gulf between theoretical and practical aspects of river restoration. *Transactions of Institute of British Geographers*, 29, 257–81.

McDonald, D. and Ruiters, G. (2005) *The Age of Commodity: Water Privatization in Southern Africa*. London: Earthscan.

McKendry, I. G., Stahl, K. and Moore, R. D. (2006) Synoptic sea-level pressure patterns generated by a general circulation model: comparison with types derived from NCEP re-analysis and implications for downscaling. *International Journal of Climatology*, 26, 1727–36.

Miller, D. (1989) *Dark Eden: The Swamp in Nineteenth-Century American Culture*. Cambridge: Cambridge University Press.

Mitchell, B. (1995) *Integrated Water Management: International Experiences and Perspectives*. London: Belhaven.

Mitchell, B. (2005) Integrated water resource management, institutional arrangements, and land-use planning. *Environment and Planning A*, 37(8), 1335–52.

Moore, R., Sutherland, P., Gomi, T. and Dhakal, A. (2005) Thermal regime of a headwater stream within a clear-cut, coastal British Columbia, Canada. *Hydrological Processes*, 19, 2591–608.

Mote, T. L., Grundstein, A. J., Leathers, D. J. and Robinson, D. A. (2003) A comparison of modeled, remotely sensed, and measured snow water equivalent in the northern Great Plains. *Water Resources Research*, 39(8), 1209.

Mustafa, D. (2005) The production of an urban hazardscape in Pakistan: modernity, vulnerability, and the range of choice. *Annals of the Association of American Geographers*, 95(3), 566–86.

Newson, M. (1997) *Land, Water and Development: Sustainable Management of River Basin Systems*, New York: Routledge.

Nistor, C. and Church, M. (2005) Suspended sediment transport regime in a debris-flow gully on Vancouver Island, British Columbia. *Hydrological Processes*, 19(4), 861–85.

NRC (1998) *New Strategies for America's Watersheds* (Chair of Committee on Watershed Management is William Graf from Arizona State). Washington, DC: National Research Council.

Oke, T. (1982). The energetic basis of the urban heat island. *Quarterly Journal of the Royal Meteorological Society*, 108, 1–24.

Oliver, S. (2000) The Thames Embankment and the disciplining of nature in modernity. *The Geographical Journal*, 166(3), 227–38.

Page, B. (2005) Paying for water and the geography of commodities. *Transactions of the Institute of British Geographers*, 30(3), 293–306.

Pereira, H. C. (1973) *Land Use and Water Resources*. Cambridge: Cambridge University Press.

Pereira, H. C. (1989) *Policy and Practice in the Management of Tropical Watersheds*. London: Bellhaven Press.

Perrault, T. (2006) From the Guerra del Agua to the Guerra del Gas: resource governance, popular protest and social justice in Bolivia. *Antipode*, 38(1), 150–72.

Petley, D. (2004) The evolution of slope failures: mechanisms of rupture propagation. *Natural Hazards and Earth System Sciences*, 4(1), 147–52.

Pulido, L. (2000) Rethinking environmental racism: white privilege and urban development in Southern California, *Annals of the Association of American Geographers*, 90(1), 12–40.

Reed, D. and Wilson, J. (2004) Coast 2050: A new approach to restoration of Louisiana coastal wetlands. *Physical Geography*, 25(1), 4–21.

Robinson, M., Dupeyrat, A. et al. (2005) Effects of commercial timber harvesting on streamflow regimes in the Plynlimon catchments, mid-Wales. *Hydrological Processes*, 19(6), 1213–26.

Rodriguez-Iturbe, I. and Porporato, A. (2004) *Ecohydrology of Water Controlled Ecosystems: Soil Moisture and Plant Dynamics*. Cambridge: Cambridge University Press.

Roth, M., Oke, T. and Emery, W. J. (1989). Satellite-derived urban heat islands from three coastal cities and the utilization of such data in urban climatology. *International Journal of Remote Sensing*, 10, 1699–720.

Sachs, J. (2005) *The End of Poverty: Economic Possibilities for Our Time*. New York: Penguin Press.

Shirley, M. (2002) *Thirsting for Efficiency: The Economics and Politics of Urban Water System Reform*, Elsevier, Oxford.

Shiva, V. (2002) *Water Wars: Privatization, Pollution and Profit*. London: Pluto Press.

Shrubsole, D., Green, M. and Scherer, J. (1997) The actual and perceived effects of floodplain land use regulation son property values in London, Ontario. *Canadian Geographer*, 41, 166–78.

Shrubsole, D. and Watson, N. M. (eds) (2005), *Sustaining Our Futures: Perspectives on Environment, Economy and Society*, 14 Chapters, Geography Publication Series, University of Waterloo, Canada.

Smith, N. (1984) *Uneven Development: Nature, Capital and the Production of Space*, Blackwell, Oxford.

Smith, N. (2005) There's no such thing as a natural disaster. *Understanding Katrina: Perspectives from the Social Sciences*, http://understandingkatrina.ssrc.org/Smith/ (accessed 24-01-08).

Sneddon, C., Harris, L., Dimitrov, R. and Özesmi, U. (2002) Contested waters: conflict, scale, and sustainability in aquatic socioecological systems. *Society and Natural Resources*, 15(8), 663–75.

Swyngedouw, E. (2004a) *Social Power and the Urbanization of Water – Flows of Power*. Oxford: University Press, Oxford.

Swyngedouw, E. (2004b) Modernity and hibridity: nature, *Regeneracionismo*, and the production of the Spanish Waterscape, 1890–1930. *Annals of the Association of American Geographers*, 89(3), 443–65.

Swyngedouw, E., Kaïka, M. and Castro, J. (2002) Urban water: a political-ecology perspective, in *Built Environment* Special Issue on 'Sustainable water use in urban areas', 28(2), 124–37.

Tobin, G. and Montz, B. (1997). *Natural Hazards: Explanation and Integration*. New York: The Guilford Press.

Travis, J. (2005) 'Scientists' fears come true as hurricane floods New Orleans. *Science*, 309(9 September), 1656–59.

UN (2003) *Water for People, Water for Life*. Geneva: UNESCO.

UN (2006) *Water: A Shared Responsibility*. Geneva: UNESCO.

Watson, N. M. (2004), Integrated river basin management: a case for collaboration. *International Journal of River Basin Management*, 2(3), 1–15.

Watson, N., Mitchell, B. and Mulamoottil, G. (1996) Integrated resource management: Institutional arrangements regarding nitrate pollution in England. *Journal of Environmental Planning and Management*, 39, 45–64.

Weber, E. (1964) Comprehensive river basin: development of a concept. *Journal of Soil and Water Conservation*, 19(4), 133–38.

Webster, P., Holland, G., Curry, J. and Chang, H. (2005) Changes in tropical cyclone number, duration and intensity in a warming environment. *Science*, 309(5742), 1844–46.

Wescoat, J. (2000) 'Watersheds' in Regional Planning. In R. Fishman (ed.), *The American Planning Tradition*. Baltimore, MD: Johns Hopkins University Press, pp. 147–71.

Wescoat, J. L. and White, G. (2003) *Water for Life: Water Management and Environmental Policy*, (Cambridge Studies in Environmental Policy). Cambridge: Cambridge: University Press.

Wescoat, J., Halvorson, S. and Mustafa, D. (2000) Water management in the Indus basin of Pakistan: a half-century perspective. *International Journal of Water Resources Development*, 16(3), 391–406.

White, G. (1997) The river as a system: a geographer's view of promising approaches. *Water International*, 22(2), 79–81.

White, G. (1998). Reflections on the 50-year international search for integrated water management. *Water Policy*, 1(1), 21–27.

White, G., Calef, W., Hudson, H., Mayer, H., Sheaffer, J. and Volk, D. (1958). *Changes in Urban Occupance of Flood Plains in the United States*. University of Chicago Department of Geography Research Paper no. 57. Chicago: University of Chicago Press

White, G., Bradley, D. and White, A. (1972) *Drawers of Water: Domestic Water Use in East Africa*. Chicago: University of Chicago Press.

Wittfogel, K. (1957) *Oriental Despotism: A Study of Total Power*. New Haven: Yale University Press.

Woods, R. (2005) Hydrologic concepts of variability and scale. In M. Anderson and J. McDonnell (eds), *Encyclopedia of Hydrological Sciences*. London: John Wiley & Sons, Ltd, pp. 23–40.

Young, G., Dooge, J. and Rodda, J. 1994. *Global Water Resource Issues*. Cambridge: Cambridge University Press.

Chapter 31

Energy Transformations and Geographic Research

Scott Jiusto

Introduction

The challenge of fostering just and sustainable societies cannot be met without fundamentally transforming global energy systems. Socio-ecological contradictions are increasingly apparent in 'conventional' energy systems predicated on exponentially growing demand met principally through expanding supplies of fossil fuels and massive nuclear and hydropower projects. Energy underwrites developmental aspirations, yet the twin specters of climate change and oil wars are generating public support for alternative energy systems with potential also to reduce related problems of smog, respiratory disease, acid rain, strip mining, oil spills, forced rural resettlement, and inequities in access to energy services. Despite these problems, virtually all 'business as usual' forecasts expect continued rapid growth in energy consumption and production regimes, reflecting the socio-economic power imparted to the network of technologies, policies, institutions, and practices that constitute conventional energy systems. These networks of power are continuously contested and reproduced, however, and just as key social movements of the 19th and 20th centuries were built largely on challenging the labour, environmental, and financial practices of coal and oil industries, so the nascent 'sustainable energy' movement presents a potential vehicle for contesting global inequality and underdevelopment in the present century.

With so much at stake and with such strong implications for virtually every field of geographic scholarship, it is surprising how little geographic research focuses squarely on energy issues. This disinclination may reflect a mismatch between the heterogeneous, social-theoretically informed methods and concerns of geographic scholarship and the all-too-often technical and economistic nature of social science energy research. Climate change concerns, however, have opened up new space for contesting energy policy and investment decisions around the world in ways that will shape the meaning and prospects for sustainability in human-environment systems. While geographers productively study energy issues from many perspectives (see review by Solomon et al. 2004), this chapter emphasises research that treats energy system sustainability as a contestable process in which political-economic and cultural factors co-evolve with changes in the quality, location, and environmental impact of energy resources.

Contesting the Next Energy Revolution

The concept of 'energy system sustainability' refers broadly to policies and practices that promote the evolution of systems to provide desired energy services in a socially just and environmentally sensitive manner. The struggle for energy system sustainability is central to any larger vision of sustainable economic development due to the fundamental role energy systems play in economic activity and the evolution of human and environmental systems (Simmons, 1989; Smil, 1999; Hall et al., 2003). Global development over the past three centuries cannot be understood without appreciating the central role played by energy system transitions – roughly from wood to coal to oil and electricity (e.g., figure 31.1) – reinforced and embedded in complementary cycles of innovation in transportation, industry, agriculture, communication and war making (Podobnik 2000). The nations that best exploited the scientific, commercial and military potential of these changes achieved wealth, empire and a world order predicated on continuous social and environmental transformation. The British and US empires of the 19th and 20th centuries were based on the ability to access, control, and develop the economic and military potential of each era's cutting-edge energy resource. Residing within the current notion of sustainable energy transformation, therefore, is the possibility for social and ecological revolution as unimaginable as that stimulated by the first electrical power systems in 19th century New York, Chicago and London.

One challenge to energy system sustainability is that once transformative energy resources – oil, coal, natural gas, nuclear fission and large hydropower – are now so socially, technically, and economically embedded in industrialised societies that they are highly resistant to displacement (Hughes 1987). In 2006, five of the ten largest Fortune 500 corporations were oil companies, while another four produced

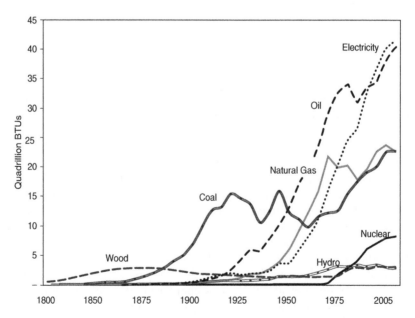

Figure 31.1 Energy transitions in the USA: primary fuel consumption. (Source: EIA 2007)

the automobiles that constituted their primary market (Fortune Magazine, 2006). 'Fossil fuel interests'– corporate and state alliances committed to those industries – are thus entrenched at the highest reaches of power around the world. As energy demand has grown, new energy resources have supplemented, rather than replaced, once dominant resources. Indeed, more coal is now consumed than ever before.

The great social and economic power vested in conventional energy systems makes challenges to their hegemony fiercely contested. For more than a century, key developments in social welfare have been won through struggle against prevailing energy interests, beginning with basic labour rights advanced through coal mine actions. The rise of the modern environmental movement in the 1970s brought with it national-level policies, such as the US Clean Air and Clean Water Acts, that significantly reduced pollution and other local- and regional-scale impacts of fossil fuel use, at least in the developed world. Thus, while air pollution remains a leading cause of premature death in the developed world, London, for example, is no longer in danger of having 4,000 people die in one week from an inversion of choking coal smog as happened in 1952 (Davis 2002). Instead, the most dangerous health and safety threats have migrated now to places like Mexico City, Beijing, and Delhi. This success in mitigating or spatially displacing some problems of conventional energy systems, combined with low energy prices beginning in the mid-1980s, vitiated much of the social movement pressing for sustainable energy transformation in the US and elsewhere. In an ironic twist, the movement has been reinvigorated in recent years through a growing international coalition concerned with the impacts of carbon dioxide, a gas that among fossil fuel by-products would be completely innocuous if it didn't constitute two-thirds of the greenhouse gases (GHG) threatening to destabilise the global climate system. Geographic energy research takes place within this context of an ongoing global, yet highly differentiated, struggle for sustainability against the hegemony of fossil fuels, and the following sections explore how issues of resource adequacy and location articulate with political economic dimensions of this struggle.

Uneven Geographies and the Geopolitics of Fossil Fuel Hegemony

'America is addicted to oil' declared George W. Bush in his 2006 State of the Union message, as a sharp rise in oil prices and war in Iraq increased the political saliency of problems stemming from an economy reliant on petroleum for 40% of its total energy needs and virtually all of its transportation. The administration's principal policy prescription for addressing this dependence – satisfy it with more oil produced domestically – was, however, at odds with research by Cleveland and Kaufmann (2003) demonstrating the steadily diminishing *energy return on investment* (EROI) of the US and global oil industries. When US oil production peaked in 1970, the US oil EROI was 50, meaning that for each unit of energy used to produce oil, an 'energy surplus' of 49 units subsidised other activities throughout the economy. Over time, as the largest, highest quality and easiest to extract oil reserves were drawn down, EROI fell to ~15, and production levels never recovered, despite improved drilling technology and enormous federal subsidies. Proposals for achieving national 'energy independence' through expanded domestic oil production on a dwindling resource base are thus likely to simply flush additional 'billions of dollars . . . down a dry hole' (Cleveland and Kaufmann 2003, p. 488).

Similar forces are expected to drive the global oil market to a production peak in the next decade or two (Kaufmann, 2006), just as consumption levels in China, India, and other developing countries rise towards those of industrialised nations. While long-standing Malthusian fears of fossil fuel depletion leading to economic collapse have not yet materialised (Jevons, 1965[1865]; Meadows and Meadows, 1972), the 1970s demonstrated the vulnerability of the global economy to tighter oil markets and political instability in oil-producing regions. With two-thirds of all reserves in a handful of Middle East countries (figure 31.2), the strategic importance and highly uneven geography of oil have vested enormous power in a handful of state and corporate actors. Watts (2005) traces the evolution of a global oil complex back to the 1930s and the establishment of Iraq as a British client state serving the interests of British, French and US oil companies and governmental allies. Following World War II, nationalist movements in oil-producing countries led eventually to the 'OPEC revolution [that] ushered cycles of conflict, militarisation and revolutionary upheaval – the so-called energy wars – in the major oil-producing regions' (Watts, 2005, p. 378). Capital and conflict continue to cycle as oil-producing elites reinvest large sums of petro-dollars into Western multinationals that sell weapons and manage massive construction projects in OPEC nations and elsewhere. This 'virtuous circle' of oil, money and weapons creates an industry in which business as usual is largely an undertaking of undemocratic multinational corporations and 'petro-states,' with the US engaged to assure the flow of oil for strategic, economic and corporate oil interests. The political economy of oil varies greatly by region, but oil figures heavily in redefining the role of state and capital in places as varied as Russia, Venezuela and Kuwait. Watts (2004) uses experience in Nigeria to argue that places of oil extraction have become enmeshed in, and reconfigured by, a distinctive 'petro-capitalism' that systematically undermines development, democracy and community.

The resource depletion and uneven locational characteristics of oil increasingly apply to the natural gas market (figure 31.3) that began growing rapidly in the 1980s for heating and electrical generation purposes. Both the USA and the UK relied heavily on natural gas to meet new electrical power requirements in recent years, but domestic supplies of this cleaner-burning and lower-carbon alternative to coal are expected to fall short of future demand (Brown et al., 2006). Meanwhile, the emerging geopolitical significance of natural gas was demonstrated on New Year's Day 2006, when Russia cut gas supplies to Ukraine, only to reverse course under intense pressure from European Union member states whose energy supplies were also pinched. Ostensibly a dispute over pricing and payment, the disruption also expressed Russia's opposition to growing ties between Ukraine and western Europe (Klare, 2006).

The latest Iraq war has prompted a number of geographic analyses of linkages between the US invasion and 'oil imperialism'. Iraq has the second largest pool of proven reserves in the world and likely vast undiscovered reserves, as well, since only one-fifth of its known, accessible fields have been developed (Jhaveri, 2004). The direct profit potential of these resources is substantial, as is the potential they confer upon Iraq to join Saudi Arabia as a 'swing producer' capable of influencing global oil prices by managing marginal production. Harvey (2005) sees the US invasion as an expression of these factors and of the pressure states and multinational energy corporations feel to constantly expand their territorial and market reach. The costly, destabilising effect of the Iraq war, however, is

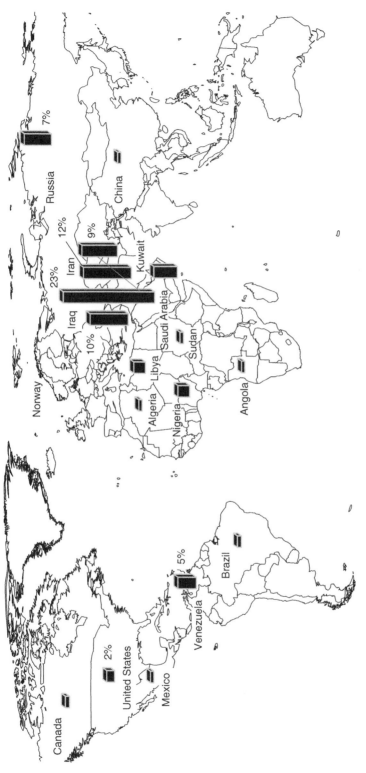

Figure 31.2 Oil reserves estimate 2006. (Source: EIA 2007)

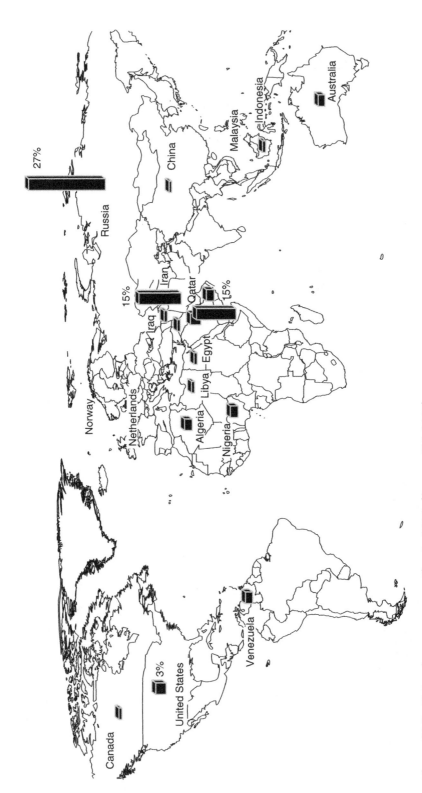

Figure 31.3 Natural gas reserves estimate 2006. (Source: EIA 2007)

certain to dampen enthusiasm for the 'primitive accumulation' strategy of securing oil through direct invasion. Instead, nations and multinational energy partners are likely to recommit to a diversified portfolio approach of direct investment, influence-buying, and military exchanges such as China is currently pursuing in Africa (Carmody and Owusu, 2007). Such strategies, however, are embedded in a global oil marketplace that, come a post-peak era, will be far harder to manage and navigate than before.

Whereas global oil reserves are localised and peaking, coal reserves are widespread and abundant (figure 31.4). If extracted at 1995 rates, global measured recoverable reserves would last 250 years (Smil 1999). Most coal (90 percent) is consumed in the country of origin, and so too the politics of coal have typically operated within national boundaries, but potently so – think of US coal miners contesting virtual indentured servitude in 19th and 20th century 'company store'-era Appalachia, or the signal importance of Thatcher's defeat of coal unions in restructuring the UK economy in the 1980s. Climate change, however, is bringing coal into the international arena, not as a matter of production access, as with oil, but instead as a matter of regulating consumption, because coal is dirtier and emits 20 percent more carbon than oil and 60 percent more than natural gas. The imminent threat from coal is that, unless curbed by carbon reduction policies, its use will continue accelerating because it is comparatively cheap and simple to use for electrical power generation.

Coal-fired power plants are often sited in rural areas near mining operations that increasingly use *mountain top removal* techniques that are every bit as ecologically subtle as the name suggests. Resulting power is than transmitted over power lines hundreds of miles to urban consumption centres. China's coal and power industry is quite inefficient (Xie and Kuby, 1997), yet a new coal-burning power plant, with the capacity to serve all the households in Dallas, opens in China every week to ten days producing not just GHGs, but also acid rain and choking smog responsible for an estimated 400,000 premature deaths annually (Bradsher and Barboza, 2006). In the USA, 150 new coal plants were in the proposal phase as of 2006 (Madsen and Sargent, 2006). The pace of future investment in such plants, each with a potential lifespan of decades, will impact both national politics in places like China, where smog is increasingly contested by a nascent environmental movement, and global climate-change progress. Prospects for capturing and sequestering carbon emissions underground, though much touted by coal and electrical power industries, appear limited, at best.

Climate Change and the Politics of Energy Sustainability

The hegemony of fossil fuels remains firmly anchored in strong, if crisis-prone, networks of capital, power, and sunk investment. Climate change concerns, however, accelerated by public concern over the Iraq war and higher energy prices, have engendered a sophisticated, multi-scalar sustainable energy advocacy network. It has succeeded in creating the rudimentary international and local institutions and policy frameworks with which to mount a serious challenge to the ever-upward spiral of oil, gas and coal consumption. The network dynamics and strategies of this movement are critical to reconfiguring how the roles of states, markets, and civil society are conceptualised and institutionalised in the pursuit of sustainable development.

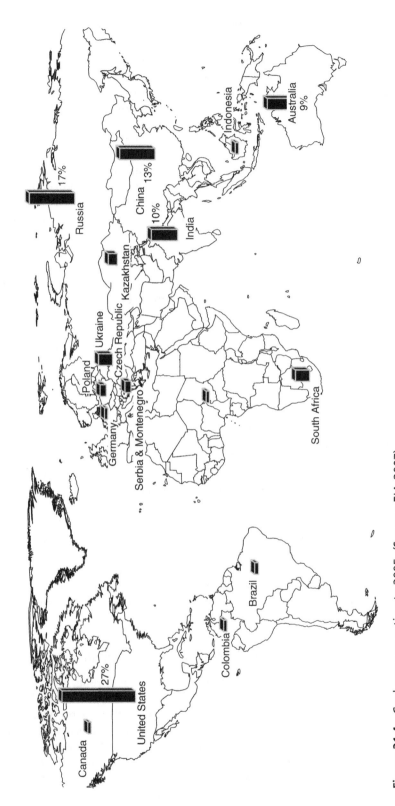

Figure 31.4 Coal reserves estimate 2005. (Source: EIA 2007)

At the international scale, 172 countries have ratified the Kyoto Protocol that commits industrialised nations to reducing their GHG emissions by an average of ~7 percent below 1990 levels by the year 2012. Few nations, however, are on target to meet this goal. Emissions also continue growing in the USA, which along with Australia has rejected the protocol, and in developing countries, which are not yet required to reduce emissions. While having little impact on emissions levels, the Kyoto process is a grand experiment in new institutional arrangements for pursuing international environmental and economic sustainability. The Protocol relies heavily on market-based 'new environmental policy instruments' (Bailey and Rupp, 2005), most notably emissions 'cap and trade' provisions that allocate tradable GHG emission permits so that those who are able to reduce emissions at relatively low costs might sell permits to high-cost emitters, thus reducing the total cost of compliance but introducing complex institutional and ethical considerations (Solomon and Lee 2000). The European Union Emission Trading Scheme and the UN Clean Development Mechanism are presently the two largest institutions facilitating the trading of emissions and emissions credits, and serious questions have been raised as to the effectiveness and transparency of each. Supporters argue problems to date are growing pains to be expected in the complex process of creating a functional GHG market, while others argue the concept of 'pollution trading' is flawed and inherently subject to political manipulation by industry (Davies, 2007). Despite the Kyoto Protocol's strong geographic and social justice implications, it has not been a direct subject of much geographic research.

In the USA, federal intransigence drove environmental activists to focus increasingly on states and localities as potential sites of policy innovation, particularly in the electricity sector that accounts for 40 percent of all US GHG emissions. This effort got caught up in the wave of neoliberal electrical sector privatisation and deregulation that reached America's shores in California following a decade of decidedly mixed results elsewhere (Bacon, 1995). Restructuring overthrew the model of state-regulated territorial utility monopolies that had guided electrical power infrastructure development for a century and replaced it with one based on competition and 'consumer choice' in buying electricity from newly deregulated independent power producers. The California model was a short-lived failure that cost the state and consumers billions of dollars due to a confluence of poor market design, fraud, transmission bottlenecks, constrained hydropower supplies, elevated natural gas price and rising electricity demand (see Solomon and Heiman, 2001). Nevertheless the political horse-trading that accompanied power sector restructuring in some two dozen states also ushered in new, more progressive climate-related energy policies, many based paradoxically on new institutions of state intervention. California, for example, recently embraced Kyoto-like mandatory emissions reduction targets, and a number of states support renewable energy and conservation through regulation and financial incentives. While there has been a strong process of policy diffusion and regionalisation embedded in state policy strategies in the USA (Peterson and Rose 2006) and elsewhere (Kent and Mercer, 2006), translating state 'leadership' into serious national progress remains enormously challenging (Heiman and Solomon, 2004; Heiman, 2006). With sustainable energy progress increasingly defined in terms of carbon and GHG trends, geographers have analysed the relationship of state emissions trends to demographic, economic and policy concerns (Rose et al., 2005). Although typically seen as a straightforward empirical exercise, Demeritt (2001) has deconstructed 'greenhouse gas' as a metric of climate change accountability, while Jiusto (2006) shows how seemingly technical choices in the design of indicators to account for emissions associated with interstate power flows produce dramatically different pictures of state emissions levels and trends and embody sharply different state policy incentives.

In contrast with these 'bottom-up' strategies for addressing climate change, the UK set ambitious national climate goals that required local and regional authorities to develop new energy management capabilities. McEvoy, Gibbs, and Longhurst (2000; 2001) looked at these new sub-national responsibilities and concluded that their potential to 'reduce energy costs, increase local employment, mitigate both local and global pollution, and achieve social goals through the relief of fuel poverty and improved living conditions' (2001, pp. 18–19) will likely go unrealised barring a radical commitment to sustainability in every aspect of local development and more effective intergovernmental collaboration.

In the developing world, where many lack basic energy services and investment capital is limited, priorities for energy sustainability differ and countries face difficult tensions in their energy investment decisions. Taylor (2005), for example, shows that although a major rural electrification program in Guatemala has brought grid access within spatial reach of 90 percent of the population, many rural people simply cannot afford to buy power or electrical appliances. The investment therefore offers little immediate benefit to rural people and leaves unaddressed a crisis in fuelwood supply that represents half the national energy balance and the essential cooking and heating fuel for almost all rural Guatemalans. The study is an interesting example of political ecology concerns and methods – village-level surveys exploring how changing land-use patterns and institutions increasingly limit rural people's access to fuelwood – combined with national energy policy analysis. The gendered, social impacts that come with land degradation and institutional restrictions on access to lands where biomass fuel harvesters compete with others have been well studied by geographers (e.g., Robbins 2001), but rarely in an energy policy framework. Such work will be increasingly important as the Kyoto Protocol fosters financial flows and regimes of accountability for 'clean development' carbon offset projects in the developing world that, while often of questionable value, reduce pressure for curbing fossil fuel use elsewhere.

Nowhere are the unevenly distributed spatial consequences and contradictions of low-carbon energy development more apparent than in the case of large hydropower dams, now pursued mainly in developing countries with large untapped hydropower potential. The scale of such projects can be staggering: China's Three Gorges project has inundated over 1,000 square kilometers and 'displace[d] the most people in a single project in human history' (Heming and Rees 2000, p. 440). The developmental discourses and politics legitimizing energy megaprojects are explored by Magee (2006) in China's Yunnan Province, where the physical and discursive construction of a 'powershed' of eight hydropower dams served also to reconfigure institutional relationships and decision-making processes among state and provincial agencies and power utility companies. Such projects illustrate tensions between the need for electricity, irrigation and flood control and the upheaval often experienced by rural communities and river ecosystems with few effective civil society institutions to represent their interests. Ironically, in the USA, problems of a massive, ageing hydropower infrastructure have made *eliminating* rather than erecting dams the principal policy focus (Kuby et al. 2005).

Consuming Passions

Given the momentum imparted to conventional energy technologies, the quickest, cheapest and most equitable way to reduce global GHG emissions is to reduce the high rates of energy consumption in developed countries (figure 31.5). In one of the most comprehensive assessments of US national energy policy and technology potential, Brown and colleagues (2001, p. 1179) 'conclude[d] that policies exist that can significantly reduce oil dependence, air pollution, carbon emissions, and inefficiencies in energy production and end-use systems at essentially no net cost to the US economy,' largely by eliminating a large national 'efficiency gap' between actual and optimal investment in energy efficiency due to well-understood market failures and barriers (Brown, 2001; Banerjee and Solomon, 2003). The problem is not that the potential for significant efficiency gains is uncertain – the 'energy intensity' (energy use per unit of economic output) of developed countries has declined for decades as technologies became more efficient, economies shed certain high-energy industries, and government environmental policies encouraged energy efficiency. Indeed, from the mid-1970s to mid-1980s, when these factors were accelerated by energy price increases and a sense of national urgency, US GDP grew by one-third with no net increase in energy use and emissions. Rather, curbing global energy consumption requires dealing with a structural economic problem that is compounded by political and cultural forces.

The structural problem is that economies predicated on continuous growth create constant pressure for increased energy use to produce more goods and services. Although there is an increasingly influential discourse of ecological modernisation based on ideas of the triple bottom line and technical improvements to 'do more with less' (see Mol's chapter), the political constituency seeking to support economic growth by expanding conventional energy systems is typically better funded, organised and politically connected than are the ecological modernizing advocates of conservation and efficiency improvements. For example, until 2007, automakers and auto labour unions had for two decades successfully defeated legislation that would have improved US automobile fleet efficiency. Beyond politics, Hinchliffe (1997) finds cultural reasons, such as the distance many people feel between their own actions and the causes and consequences of socio-economic problems, for why individuals and communities might act neither as economic rationalists nor as environmentally conscious consumers in their energy behaviours. Lovell (2005) uses science and technology studies (STS) concepts to explain why public investment in low energy social housing may not be a sound strategy for diffusing efficiency innovation across multiple housing sectors. Both political and cultural insights help explain why, even in Europe, where support for progressive climate action is strong, efficiency 'policy is progressing too slowly and . . . (t)he most effective policy – minimum standards – is being replaced with the much weaker industry-promoted voluntary agreements' (Boardman, 2004, p. 1932).

Mammals in the Land of Dinosaurs: Prospects for Renewable Energy Resources

Despite the high value of energy conservation and efficiency, they do not eliminate the need to develop low-carbon alternatives to fossil fuels. The best positioned, but most problematic, of these alternatives is nuclear power, for which climate change

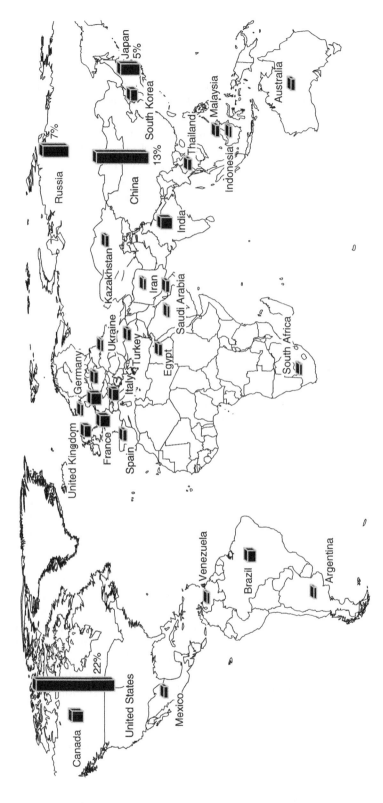

Figure 31.5 Primary energy consumption 2004. (Source: EIA 2007)

has been like a fresh spring rain promising renewal through state subsidies and carbon trading credits. Following accidents at Three Mile Island and Chernobyl, nuclear power became for many a paradigmatic symbol of technological hubris ('cutting butter with a chainsaw') posing a range of hazards – in uranium mining, plant operations, radioactive waste disposal, plant decommissioning and nuclear proliferation – that remain largely unsolved (e.g., Pasqualetti and Pijawka, 1996). Although a new plant has not been started in the USA since the 1980s, the industry remains strong in France, and elsewhere, nuclear politics and perceptions are shifting. Sweden, for example, recently put two plants into 'early retirement' before halting its program of nuclear phase-out (Lofstedt, 2001). In Iran, President Ahmadinejad's claim to an 'inalienable right' to develop nuclear power – building upon a program begun with US support prior to the 1979 revolution – has been viewed by many as a pretense for pursuing nuclear weapons development. The regimes of control needed to manage the contradictions of nuclear power make it an inherently anti-democratic technology (Lovins, 1977; Winner, 1986), and the enormous associated costs and risks have private capital refusing to invest absent massive state subsidies in everything from R&D, facility development and radioactive waste disposal to limiting liability in the event of catastrophe. These subsidies are antithetical to neoliberal principals of market competition underpinning power sector restructuring and threaten to divert investment capital away from more sustainable alternatives. Nevertheless, they are gaining serious traction, even among longtime nuclear critics.

Whereas nuclear power reinforces the conventional 'hub and spoke' geography of electrical power grids, renewable energy resources such as solar, wind, small hydro, geothermal and biomass can often be exploited with small, mass-producible technologies distributed throughout the grid. Although accounting for little more than 2 percent of the global commercial energy mix, the resource base for renewables is immense. The wind power potential in just three states (North Dakota, Texas, and Kansas) could meet current US electrical power demand (Pasqualetti, 2004), and solar potential is similarly great. While technical challenges are not insignificant – wind sites are often far from transmission lines capable of handling large, variable generation sources – each type of renewable resource also presents different political potentialities and liabilities.

Wind power, for example, is the fastest growing source of electricity in the USA because the technology has advanced rapidly to become economically competitive with conventional power resources and has been encouraged through state policy (Pasqualetti, 2004). However, as wind turbines become larger and more visible, they become increasingly controversial. Large wind farm proposals, for example, often generate conflictual discourses pitting the benefits of clean energy and rural economic development against those of landscape preservation, tourism, and other land-use options. Pasqualetti (2000) explores how competing social and cultural interests intersect with technological and ecological constraints to produce 'landscapes of power' in the American West and elsewhere. With his collaborators Pasqualetti (2002) offers guidance for reducing conflict that can impede windpower development even in supportive areas like Germany and California. By contrast, Mercer (2003, p. 91) fears geographers and others are giving inadequate weight to 'the place of landscape values within the ecologically sustainable development paradigm.'

Unlike windpower, solar photovoltaic (PV) technology is rarely controversial as economies of scale in siting are comparatively few, so that PV arrays need not be overly large or agglomerated. Indeed, their most promising applications are as highly distributed micropower generators located very close to sources of consumption, such that they *reduce* rather amplify burdens on transmission and distribution systems. PV is reliable, long-lived and technically feasible even in locations of moderate sunshine, but comparatively expensive. Thus, PV growth forecasts are meager (Energy Information Administration, 2005), despite increased state and federal policy support that only marginally offset historical subsidies for conventional energy sources. New thin film and other technologies for integrating PV into roofing shingles, exterior siding and other building components will likely open up other applications. Furthermore, PV production is based on techniques and principles of the semi-conductor industry, suggesting far greater potential for continuous, shorter cycles of innovation and deployment than conventional power technologies like coal and nuclear. Other solar energy applications, such as daylighting and passive space and water heating, often pay back quickly, but remain underutilised due to design ignorance. Geographers have taken surprisingly little interest in solar energy issues, such as contextual analysis of solar power in the Third World, where the developmental vision of green, independently generated power can conflict with local perceptions of solar as an inferior and theft-prone 'poor people's' substitute for 'modern' grid power. Cultural perceptions in the West can likewise inhibit solar development. California is proposing renewable energy requirements in new construction that will make sound energy investments as routine and 'sensible' as buying ever-larger homes packed with hot tubs, home cinemas and other energy-hungry amenities.

The contradictory politics and analytics of energy are also exemplified in the case of biomass fuels, which include everything from woodland forage to 'energy crops' to sewerage. While residues from agricultural, forestry and mill operations comprise 70 percent of the biomass energy potential in the USA (Milbrandt, 2005), it is ethanol for transportation, currently just 3 percent of the US renewable energy total, that has received the lion's share of attention, driven largely by a discourse of reducing dependence on imported oil. In current practice, ethanol does little to achieve this goal because it is mostly based on fossil fuel-intensive corn monocultures and federal and state incentives that reflect the interests of agribusiness more than sustainability. The estimated EROI of corn ethanol is less than 2:1, versus 15:1 for oil, meaning that most ethanol investment (in crops, money, labour and fossil fuels used as fertilizers) goes simply to reproduce the ethanol industry, rather than for other purposes (Cleveland et al., 2006). Potentially much more promising are cellulosic ethanol systems using switchgrass or other plants grown on marginal cropland with few inputs, and venture capital is pouring into such schemes. A geopolitics of ethanol is beginning to emerge as the USA and Brazil, the world leader in ethanol production based on comparatively high EROI sugar cane, seek a strategic ethanol alliance that will open up new opportunities for capital investment in Brazil and Latin American following defeat of the Free Trade Agreement of the Americas and counter-leftist Venezuelan President Hugo Chavez's petro-financed regional ambitions (Zibechi, 2007). A prominent promoter of the alliance is the Inter-American Ethanol Commission under the direction of Jeb Bush, brother of oilman and US President George W. Bush, underscoring in a small way the kind of shifting and contradictory allegiances that maintain, and may yet undermine, a century of oil hegemony.

Energy and Geographic Thought

As we have seen, the paradox of conventional energy systems is that they are, at present, essential to economic productivity and social well-being and yet enormously destructive, crisis-prone and unsustainable. While the thermodynamic potential of renewable energy resources far exceeds demand for energy services worldwide, it remains that in most places and in global aggregate, far more social investment goes to maintaining dependence on fossil fuels and other conventional resources than to developing clean and efficient alternative energy systems. The question is why?

One answer is that the conceptual underpinnings of the dominant economic theory informing conventional energy policy are fundamentally flawed. Geographers active in the field of ecological economics argue that energy insights are essential to conceiving alternatives to neo-classical economics that recognise human economies as embedded in environmental systems which provide essential services that must be adequately accounted for in economic decision making (Hall et al., 1992). Critical among these services are the supply of low entropy, high-quality energy resources needed for productive activity and the reprocessing of high-entropy waste from human activity back into sustainable cycles of ecological renewal. This line of research challenges the bases of mainstream economic estimation of climate change mitigation costs, and certainly those studies used by the Bush administration and others to claim that carbon reduction policies pose grave threats to economic well-being. It also challenges the wider assertion that 'economic growth is the best environmental policy' espoused by some opposing GHG limits, and echoed in academic research suggesting that industrialisation brings with it processes of ecological modernisation that reliably engender the capital, knowledge and politically empowered citizenry necessary to move countries along an 'environmental Kuznets curve' of decreasing pollution and decarbonisation (Selden and Song, 1994). The empirical basis for such claims is weak (Cleveland and Ruth, 1999), and offers little hope that 'autonomous' economic processes will resolve the problems posed by conventional energy systems (Richmond and Kaufmann, 2006). Similarly, the idea that neoliberal privatisation and deregulatory restructuring of energy systems will improve economic and environmental performance rests far more comfortably in the realm of theory than experience (Solomon and Heiman, 2001; Heiman and Solomon, 2004; Perkins, 2005). Clearly, better conceptual understanding of systems of energy, ecology and economics is needed to guide decision making.

Such work often lacks a critical, political economic perspective that can help explain the entrenched, though hardly static, power of fossil fuel and nuclear industries, and illuminate emergent strategies that might reshape these configurations of power (though see Kaufmann, 1987). Presently, geographic research, like most energy research, all too often seems to assume that the route to sustainability lies largely in sound technical analysis and sober planning. It tends to overlook the often brutal way in which 'policy' gets executed on the ground around the world. Beyond the perils of imperial oil lie largely unexamined, geographically contextual struggles over the future of energy industries playing out globally in highly varied ways. Similarly, the complex top-down and bottom-up scalar strategies of sustainable energy advocacy operating through cities, states and nations are crucial to understanding and enhancing transformational potentialities. One avenue to realising these potentialities lies in overcoming the discursive constraints that too often limit 'serious' policy analysis to narrowly technical and economistic discourses. Another need is to better understand and counter the processes of commodification that obscure the full social and environmental impacts of conventional energy systems and the power interests they serve. Climate change and the US engagement in Iraq provide unwelcome, but nonetheless important, new opportunities to make visible more of the trauma embedded in 'business as usual' energy scenarios.

Ultimately, the reality of energy transformation plays itself out on the ground and in the air, as social and technological networks become manifest in energy capital stock. 'Landscapes of power' are produced and reproduced at various scales through conflict over dams, wind turbines, coal and nuclear plants, and access to oil, forests, and fields. Presently, the gigantism of conventional energy systems produces, on the supply side, unhealthy concentrations of social power and ecological impacts, along with, on the demand side, profligate, disconnected, commodified consumers. Because investments planned for coal, oil and nuclear are simultaneously investments not made in wind, solar, or efficiency, virtually all business-as-usual forecasts suggest the age of massive, centralised energy systems and their problems is far from over. There are, however, indications that emerging within the interstices of conventional energy systems are possibilities for more highly distributed energy networks, ones that are composed of far smaller, more numerous, and 'smarter' technologies that could be aligned more closely with the ecological and social conditions of particular places (Lovins and Rocky Mountain Institute, 2002). This vision of the next energy revolution – one that emerges from a deep understanding of the transformative power of energy systems and a commitment to local empowerment – can surely arise only as part of a movement redirecting political and corporate incentives towards sustainable communities, small and large. A tall order, but one full of possibility for geographers.

ACKNOWLEDGEMENTS

Thanks to Annie Fine, Matt Huber and the editors of this volume for their helpful suggestions.

BIBLIOGRAPHY

Bacon, R. W. (1995) Privatization and reform in the global electrical supply industry. *Annual Review of Energy and the Environment*, 20, 119–43.

Bailey, I. and Rupp, S. (2005) Geography and climate policy: a comparative assessment of new environmental policy instruments in the UK and Germany. *Geoforum*, 36(3), 387–401.

Banerjee, A. and Solomon, B. D. (2003) Eco-labeling for energy efficiency and sustainability: A meta-evaluation of US programs. *Energy Policy*, 31(2), 109–23.

Boardman, B. (2004) New directions for household energy efficiency: evidence from the UK. *Energy Policy*, 32, 1921–33.

Bradsher, K. and Barboza, D. (2006) Pollution from Chinese coal casts shadow around globe. *The New York Times*, 11 June, A1(L).

Brown, M. A. (2001) Market failures and barriers as a basis for clean energy policies. *Energy Policy*, 29(14), 1197–207.

Brown, M. A., Levine, M. D., Short, W. and Koomey, J. G. (2001) Scenarios for a clean energy future. *Energy Policy*, 29(14), 1179–96.

Brown, M. A., Sovacool, B. K. and Hirsh, R. F. (2006) Assessing U.S. energy policy. *Daedalus*, 135(3), 5–11.

Carmody, P. R. and Owusu, F. Y. (2007) Competing hegemons? Chinese versus American geo-economic strategies in Africa. *Political Geography*, 26(5), 504–24.

Cleveland, C. J. and Kaufmann, R. K. (2003) Oil supply and oil politics: déjà vu all over again. *Energy Policy*, 31(6), 485–89.

Cleveland, C. J., Hall, C. A. S. and Herendeen, R. A. (2006) Energy returns on ethanol production. *Science*, 312(5781), 1746.

Cleveland, C. J. and Ruth, M. (1999) Indicators of dematerialization and the materials intensity of use. *Journal of Industrial Ecology*, 2(3), 15–50.

Davies, N. (2007) Truth about Kyoto: huge profits, little carbon saved. *The Guardian*, 2 June.

Davis, D. L. (2002) *When Smoke Ran Like Water*. New York: Basic Books.

Energy Information Administration (2005) Table 16. Renewable energy generating capacity and generation. *Annual Energy Outlook 2006 with Projections to 2030*. Washington, DC: EIA.

Energy Information Administration (2007) International data. Web page. http://www.eia.doe.gov/emeu/international/contents.html (accessed 10 June 2007).

Fortune Magazine (2006) Web page. http://money.cnn.com/magazines/fortune/global500/2006/full_list/ (accessed 21 June 2007).

Hall, C. A. S., Cleveland, C. J. and Kaufmann, R. (1992) *Energy and Resource Quality: The Ecology of the Economic Process*. Niwot: University Press of Colorado.

Hall, C., Tharakan, P., Hallock, J., Cleveland, C. and Jefferson, M. (2003) Hydrocarbons and the evolution of human culture. *Nature*, 426(6964), 318–22.

Harvey, D. 2005. *The New Imperialism*. Oxford: Oxford University Press.

Heiman, M. K. (2006) Expectations for renewable energy under market restructuring: the U.S. experience. *Energy*, 31(6–7), 1052–66.

Heiman, M. K. and Solomon, B. D. (2004) Power to the people: electric utility restructuring and the commitment to renewable energy. *Annals of the Association of American Geographers*, 94(1), 94–116.

Heming, L. and Rees, P. (2000) Population displacement in the Three Gorges Reservoir area of the Yangtze River, Central China: relocation policies and migrant views. *International Journal of Population Geography*, 6(6), 439–62.

Hinchliffe, S. (1997) Locating risk: energy use, the 'ideal' home and the non-ideal world. *Transactions of the Institute of British Geographers*, 22(2), 197–209.

Hughes, T. P. (1987) The evolution of large technological systems. In W. E. Bijker, T. P. Hughes, and T. Pinch (eds), *The Social Construction of Technological Systems*. Cambridge, MA and London: MIT Press, pp. 51–82.

Jevons, W.S. (1965 [1865]). *The Coal Question*, 3rd edn. Edited by A. W. Flux. New York: Augustus M. Kelley.

Jhaveri, N. (2004) Petroimperialism: US oil interests and the Iraq War. *Antipode*, 36(1), 2–11.

Jiusto, S. (2006) The differences that methods make: cross-border power flows and accounting for carbon emissions from electricity use. *Energy Policy*, 34(17), 2915–28.

Kaufmann, R. K. (2006) Planning for the peak in world oil production. *World Watch*, 19(1), 19–21.

Kaufmann, R. K. (1987) Biophysical and Marxist economics: learning from each other. *Ecological Modeling*, 38, 91–105.

Kent, A. and Mercer, D. (2006) Australia's mandatory renewable energy target (MRET): an assessment. *Energy Policy*, 34(9), 1046–62.

Klare, M. T. (2006) The geopolitics of natural gas. *The Nation*, 282, 19.

Kuby, M., Fagan, W., ReVelle, C. and Graf, W. (2005) A multiobjective optimization model for dam removal: an example trading off salmon passage with hydropower and water storage in the Willamette Basin. *Advances in Water Resources*, 28(8), 845–55.

Lofstedt, R. E. (2001) Playing politics with energy policy: the phase-out of nuclear power in Sweden. *Environment*, 43(4), 20–33.

Lovell, H. (2005) Supply and demand for low energy housing in the UK: insights from a Science and Technology Studies approach. *Housing Studies*, 20(5), 815–29.

Lovins, A. B. (1977) *Soft Energy Paths: Toward a Durable Peace*. Cambridge: Ballinger.

Lovins, A. B. and Rocky Mountain Institute (2002) *Small Is Profitable*, 1st edn. Snowmass, CO: Rocky Mountain Institute.

Madsen, T. and Sargent, R. (2006) Making sense of the 'coal rush': the consequences of expanding America's dependence on coal. U.S. PIRG Education Fund; National Association of State PIRGs.

Magee, D. (2006) Powershed politics: Yunnan hydropower under Great Western Development. *The China Quarterly*, 185, 23–41.

McEvoy, D., Gibbs, D. C. and Longhurst, J. W. S. (2000) City-regions and the development of sustainable energy-supply systems. *International Journal of Energy Research*, 24(3), 215–37.

McEvoy, D., Gibbs, D. C. and Longhurst, J. W. S. (2001) Reducing residential carbon intensity:the new role for English local authorities. *Urban Studies*, 38(1), 7–21.

Meadows, D. H. and Meadows, D. L. (1972) *The Limits to Growth*. New York: Universe Books.

Mercer, D. (2003) The great Australian wind rush and the devaluation of landscape amenity. *Australian Geographer*, 34(1), 91–121.

Milbrandt, A. (2005) *A Geographic Perspective on the Current Biomass Resource Availability in the United States*. NREL/TP-560-39181. Golden, CO: National Renewable Energy Laboratory.

Pasqualetti, M. J. (2000) Morality, space, and the power of wind-energy landscapes. *Geographical Review*, 90(3), 381–94.

Pasqualetti, M. J. (2004) Wind power: Obstacles and opportunities. *Environment* 46, no. 7: 22–38.

Pasqualetti, M. J. and Pijawka, K. D. (1996) Unsiting nuclear power plants: decommissioning risks and their land use context. *Professional Geographer*, 48(1), 57–69.

Pasqualetti, M. J., Gipe, P. and Righter, R. W. (2002) *Wind Power in View: Energy Landscapes in a Crowded World*. Sustainable World Series. San Diego: Academic Press.

Perkins, R. (2005) Electricity sector restructuring in India: an environmentally beneficial policy? *Energy Policy*, 33(4), 439–49.

Peterson, T. D. and Rose, A. Z. (2006) Reducing conflicts between climate policy and energy policy in the US: the important role of the states. *Energy Policy*, 34(5), 619–31.

Podobnik, B. (2006) *Global Energy Shifts: Fostering Stability in a Turbulent Age*. Philadelphia: Temple University Press.

Richmond, A. K. and Kaufmann, R. K. (2006) Is there a turning point in the relationship between income and energy use and/or carbon emissions? *Ecological Economics*, 56(2), 176–89.

Robbins, P. (2001) Fixed categories in a portable landscape: the causes and consequences of land-cover categorization. *Environment and Planning A*, 33(1), 161–79.

Rose, A., Neff, R., Yarnal, B. and Greenberg, H. (2005) A greenhouse gas emissions inventory for Pennsylvania. *Journal of the Air and Waste Management Association*, 55(8), 1122–33.

Selden, T. M. and Song, D. (1994) Environmental quality and development: is there a Kuznets curve for air pollution emissions? *Journal of Environmental Economics and Management*, 27(2), 147–52.

Simmons, I. G. (1989) *Changing the Face of the Earth*. Oxford: Basil Blackwell.

Smil, V. (1999) *Energies: An Illustrated Guide to the Biosphere and Civilization*. Cambridge: MIT Press.

Solomon, B. D. and Heiman, M. K. (2001) The California electric power crisis: lessons for other states. *Professional Geographer*, 53(4), 463–68.

Solomon, B. D. and Lee, R. (2000) Emissions trading systems and environmental justice. *Environment*, 42(8), 32–45.

Solomon, B. D., Pasqualetti, M. J. and Luchsinger, D. A. (2004) Energy geography. In G. Gaile and C. Willmott (eds), *Geography in America at the Dawn of the 21st Century*. Oxford: Oxford University Press.

Taylor, M. J. (2005) Electrifying rural Guatemala: central policy and rural reality. *Environment and Planning C: Government and Policy*, 23, 173–89.

Watts, M. J. (2004) Antinomies of community: some thoughts on geography, resources and empire. *Transactions of the Institute of British Geographers*, 29(2), 195–216.

Watts, M. J. (2005) Righteous oil? Human rights, the oil complex, and corporate social responsibility. *Annual Review of Environment and Resources*, 30, 373–407.

Winner, L. (1986) *The Whale and the Reactor: A Search for Limits in an Age of High Technology*. Chicago: University of Chicago Press.

Xie, Z. J. and Kuby, M. (1997). Supply-side, demand-side optimization and cost-environment tradeoffs for China's coal and electricity system. *Energy Policy*, 25(3), 313–26.

Zibechi, R. (2007) United States and Brazil: the new ethanol alliance. Web page. http://americas.irc-online.org/am/4051 (accessed 21 June 2007).

Chapter 32

Food and Agriculture in a Globalising World

Richard Le Heron

Introduction

In the 21st century, food and agriculture in many countries is increasingly understood as both shaped by and constitutive of two broad influences or regulatory trends, namely, material developments in the neoliberalising global economy and the rise of new moral economies around the rights and responsibilities of individuals and institutions with respect to food. Human geographers have been prominent contributors to understanding both trends over the past 25 or so years, making major and sustained contributions to the bourgeoning literature.

The chapter has two aims. First, it examines the big issues germane to 21st century food production, supply and consumption. The literature is, of course, made up of diverse threads, some disciplinary, others interdisciplinary, still others more popular in nature. Recognising these threads is important because over recent decades many researchers have changed or extended their interests, usually in response to the blurring of boundaries of once discrete specialisations. Second, the chapter situates the diverse and frequently programmatic attempts of social researchers, mostly from developing nations, to make sense of the changing set of issues covered in the field of food and agriculture. The chapter highlights human geographer's contributions, against a background of key sources from the wider literature.

The number of human geographers working in the field of food and agriculture studies at any time has never been very large. Neither agricultural nor food geography stand out as particularly visible subfields. Many geographers working on food and agriculture often identify themselves by other subfields, such as development studies, rural geography, cultural geography or economic geography. And no Geography journal on agricultural or food geography exists today. Any suggestion that human geographers can somehow cover the complete foodscape is unrealistic. Nevertheless, geographers with food and agriculture interests have published widely in journals, some in geography (e.g., *Environment and Planning A and D*, *Geoforum*), some notionally less geographic (e.g., *Journal of Rural Studies*), others with obvious disciplinary bases (e.g., *Sociologia Ruralis*, *Rural Sociology*, *European Journal of*

Planning Studies) or interdisciplinary heritages (e.g., *International Journal of the Sociology of Agriculture and Food, Agriculture and Human Values*). Human geographers have succeeded in obtaining research monies from national funding sources. A steady stream of empirically grounded theorisations from different research groupings has appeared in the international literature. All this said, a Google search of 'food geography' conducted in early 2008 reveals just seven university courses (only one was outside the USA) and the surprising paucity of geographical writings in the course reading lists. Yet, considerable contemporary interest in food and agriculture – within geography – can be found. In 2006, for instance, a theme of the International Geographical Union 'The Dynamics of Economic Spaces' Commission was 'Agri-food commodity chains and globalizing networks'. A year later, both the Association of American Geographers and the Institute of British Geographers conferences had multiple, well-attended and vibrant sessions dealing with 'food'.

The chapter opens by looking at pressing issues around global agriculture and food identified in an international literature that is biased towards what is happening in developing nations. The literature deals with food for affluent consumers as comprehended from Europe and North America, rather than livelihoods for the many and poor in other places. The chapter then turns to understandings of the origins and emergence of globalising food and agriculture developed since the late 1980s. This work is heavily influenced by human geographers using political economy approaches, and with a production emphasis. This is followed by discussion of food consumption, focusing on aspects of cultural economy, well-being and moral evaluations. In the 2000s, after cultural and social geographers had enriched the field in the previous decade, a significant convergence of ideas began. Increasingly, work centres on the nexus of political economy and moral economy, although there is still hesitation about acknowledging and using jointly what each tradition has to offer.

Situating Agriculture and Food

Food, after water, is the second concern of daily life for humans. Food is implicated in who dies, starves or goes hungry, where and why, on the planet today (Grigg, 1981; Watts and Bohle, 1993). The Food and Agriculture Organisation's (FAO) State of Food Insecurity Report (2006) estimated that over 850 million people worldwide suffer from hunger and malnutrition, including 820 million in developing countries. Those most affected live in countries dependent on food imports. Some 37 countries, 20 in Africa, 9 in Asia, 6 in Latin America and 2 in Eastern Europe currently face exceptional food shortages in food production and supply. Over 40, mostly developing countries, depend on a single agricultural commodity for more than 20 percent of their total export income (International Fund for Agricultural Development, 2004). Issues such as household food insecurity, the physical and economic access to adequate food for all members of the household, without undue risk of losing that access, have been considered by the International Geographical Union Commissions on the Geography of Famine and Vulnerable Food Systems. Plenty and poverty are different starting points when examining agriculture and food.

Today a wide spectrum of actors is trying to redirect the food agenda in developing and developed countries. Agriculture and food are contested arenas. That structural features – power relationships and moral concerns, and a mixture of vulnerabilities to human and biophysical processes – lie behind the unevenness of food production and consumption is hardly a novel claim. But in the 21st century there are new dimensions. These include agricultural trade regimes built-up around non-trade subsidies, food production for export in developing countries (via structural adjustment programmes) with diversion from immediate consumption, a resurgence of peasants and small market producers in many countries being driven off their land, the proliferation of industrial agriculture featuring a high dependence on agricultural inputs such as seeds, fertilizers, pesticides from multinational agri-food companies, the emulation of Western diets by consumers in developing countries and much more besides. As a result, 'my hand to my mouth' subsistence economies are rapidly making way for 'my hands feed other mouths' production and concentrations of consumers, increasingly in cities, who are very dependent on those 'other hands who supply their food'. This is not a new story but one with new plot lines, resources and a changing cast of characters.

The international literature on food and agriculture mostly deals with food for the few and affluent as comprehended from Europe and North America, rather than livelihood for the many and poor in other places. This literature nevertheless identifies worrying food dynamics and features that are now embracing both developed and developing nations. Much of the recent re-politicisation of agriculture and food springs from breakdowns in the chains of trusted hands that are integral to the contemporary food scene at every level and every place.

Globalising Agriculture and Food

This section examines material changes in agriculture and food that are interlinked with commodification and industrial developments over several centuries. These processes continue to be implicated in production and consumption dimensions of agriculture and food.

The making of markets, especially for land and labour, and the incorporation of agriculture and food into capitalist commodity relations surged in the 18th century when people in UK and Europe were denied traditional access to land for livelihood, through enclosure and dispossession. The expanding populations of towns and cities provided cheap labour for industrial activities. Agriculture and food quickly became a market for industrial products, equipment and technologies. Industrialisation has added extra steps in agricultural and food production, and created industrial substitutes for products from traditional agricultural systems. The then prescient phrasing of the Goodman et al.'s *From Farming to Biotechnology* (1987) captures this idea, identifying a number of phases associated with these interlocking processes. Agricultural production processes have been altered by changes in labour processes (e.g., mechanisation of handling in the form of tractors and farm implements that raised labour productivity), changes in natural production processes (e.g., fertilizers), the addition of science (e.g., hybrid seeds, high yielding varieties, feedlots) and adding properties to food by processing (e.g., preserving, canning, refrigeration, powdered products, freeze drying, irradiation). Industrial substitutes include margarine for butter, fructose for sugar and soy for meat. More recently, developments in the life sciences have widened and deepened the impact of 'science-industry' to include modifications to the genetic make-up of plants and animals (e.g., hormone-dosed milk cows) and attention to altered bodily performance through, for instance, functional foods, which comprise products that have proven health benefits, beyond basic nutrition, from targeting functions in the body (e.g., probiotic bacteria in yoghurt, Omega 3 milk). These interwoven developments have significantly remade and hybridised agriculture and food. Strong preferences have become deeply ingrained in consumption cultures – sweeteners over sugar, thickeners over flour or cornstarch, fats over palm oil or butter or margarine, and proteins over beef or cod. A key point is that food constituents (what a food item consists of) and their valuation (attributes deemed especially important) are actively created and are therefore always open to change. Significantly, the technological and scientific advances that characterise the several century trajectory of the industrialisation and commodification of agriculture and food means new production possibilities keep coming on the scene. A noticeable overall trend by the late 20th century was the shift by food companies from the mercantile (trade-oriented) strategies of diversifying sources of supply of specific crops to increasing reliance on interchangeable natural or chemically synthesised inputs. This strategy allows a higher degree of control by corporate agriculture by switching components and bypassing products and regions in sourcing industrial requirements.

The geographical trace of agricultural industrialisation and the related internationalisation of traded agricultural and industrial inputs into food production have been considered at several levels. The 20th century saw the rise of specialised agricultural areas and specialist agricultural producers – with both being increasingly tied into national food systems and globalising trade and production networks. Agricultural specialisation emerged out of more diversified farming approaches and the appearance of farmers who specialised in particular land uses and outputs. FitzSimmons (1986) and Le Heron and Roche (1996), for instance, document what happened in two agricultural areas, the Salinas Valley, California, and the Heretaunga Plains, New Zealand. In simple terms, the processes of regional transformation involved intensification and local integration. For the Salinas Valley, this meant the rise of truck crops, such as lettuces and grapes, while the Heretaunga Plains moved out of pastoral farming and process cropping into apples and then increasingly into grapes for wine. Local restructuring led to fewer and fewer growers/farmers and the consolidation of agribusiness processors as the volume and quality of specialised production grew. Such transformations shifted the balance of power between farmers and processors and paradoxically decontextualised industrial food while linking places of food production and consumption in more complex ways.

Agri-food researchers have tried at different times to reveal something of the geopolitical, geo-economic and geo-cultural dimensions to contemporary food. Gray et al. (2007), for instance, map how the dairy exports from New Zealand to the world have changed dramatically over three decades, bringing the key New Zealand dairy industry actor into complex local, national and international encounters with diverse moral and political orders. Lang (1999, p. 124) operationalises the idea of 'ghost acres' to map from where in the world animal feed imports for Europe's industrial-livestock complex derived in 1993. In their introduction to *The Atlas of Food* Millstone and Lang (2002, p. 7) write 'What we eat, where we eat and how we eat reveals a world of food and drink culture. How our daily bread – or rice – reaches our plates and palates is sometimes so complex that we cannot unravel its route in one bite'. Their atlas contains sections that disclose many dimensions about the internationalisation of contemporary food – trade flows, animal transport worldwide, transporting animals in Europe, food miles, subsidies and tariffs, trade disputes, developing trade and fair trade. Cartograms on world food and agriculture patterns can be found at www.worldmapper.org

So far this section has emphasised the incorporation of agricultural land and labour into the capitalist system and changes in the physical and production dimensions of agriculture and food. We turn now to the changing political economy of agriculture and food in which structural forces and large food sector actors are given prominence. By the beginning of the 1980s, both UK and US food and agriculture researchers were exploring political economy approaches (Newby and Buttel, 1978). This fuelled research into the impact of capitalist processes on farming and the impacts of agribusiness (Marsden et al., 1986). From this foundation new conceptualisations allowed agri-food research to explore new frontiers. These included the study of commodity systems (Friedland et al., 1981), the new political economy of agriculture and food (Friedland et al., 1991), global commodity chains (Gereffi and Korzeniewicz, 1994) and food regimes (Friedmann and McMichael, 1989). While I isolate these threads, it should be remembered that many in the relatively small community of agri-food researchers attended common conferences (in geography, sociology, rural sociology, agricultural economics) and that the interdisciplinary exchanges at these conferences encouraged both disciplinary and interdisciplinary advances. Throughout the 1990s the focus was more on agriculture and the production of food than on the consumption of food.

Agri-food chains

While often used interchangeably the concepts of agri-food commodity, value and supply chains originate from very different research traditions. Each approach asks different questions. The commodity chain approach draws upon Marxist political economy to consider capitalist commodification processes, power asymmetries and unjust and unfair outcomes that characterise contexts. The long-term vision is critical, seeking some alternative other than organising production and consumption around profits. The value chain tradition, coming out of business economics and marketing, focuses on individual actors improving their situation by repositioning within value chains, so as to increase margins and profits. The idea of the supply chain, in contrast, is more functional and utilitarian, looking at how the job of providing food gets done through synchronising within and across supply chains, usually ignoring injustices and maldistribution. Googling the terms in 2008 confirms that supply chain management thinking predominates.

Geographers have been among the most adventurous and ambitious in terms of expanding the agenda of agri-food commodity chain research. First, the simplistic physical conception from paddock to plate has been advanced by conceptual and empirical studies revealing the constitutive complexities of the agriculture–food relation (Bowler et al., 2000; Hughes and Reimer, 2004; Fold and Pritchard, 2005). This has broken the deterministic and economistic mould of much early work. These include the delineation of agri-food chains in network terms, so indicating the geography and temporality of power relations of actor connections (Whatmore and Thornes, 1997; Freidberg, 2004; Stringer and Le Heron, 2008). Second, investigations of alternative food networks such as for coffee and a variety of organic products document the distinctive ethical and moral foundations of many food networks and the political fights that go into their development, operation and maintenance (Maye et al., 2007). Strong warnings, however, are being made about the geographic and social exclusivity of alternative food networks originating from the European and North American contexts (Abrahams, 2007). Such analyses stress the diversity of economic relations (Gibson-Graham, 2006). Third, a strand of research involving 'following the commodity' and representing the encounters that emanate from such a research methodology (e.g., bananas, beans, chicken, coffee beans, cut flowers, tomatoes, papaya) has thrown new light on mobilising political possibilities around new visions of agri-food relations (Cook et al., 2006). Fourth, conventions theory emphasising the multiplicity of motivations and evaluations in economic relations (Rosin, 2008) has given additional understandings of the coordination of particular agri-food chains. Fifth, new research on fisheries management has extended the frontiers into ecological issues relating to food (Mansfield, 2003; Le Heron et al., 2008). Sixth, a post-production thread is addressing multifunctionality of land use in the context of reduced production subsidies (Wilson, 2005). Finally, a number of studies have explored how the research agenda might be shifted, by recognising the social construction of international food (Arce and Marsden, 1993), what happens to differently structured commodity chains under competition (Morgan and Murdoch, 2000), the need to resist abandoning political economy and instead utilising post-structural insights (Marsden 2000), leading to the conclusion that only new styles of politics will produce enduring alternative outcomes.

Given the wide acceptance of the supply chain idea, what do we know about changes in food supply chain drivers? In the 1960s–1970s, for instance, companies strove to develop new products and processes and cut prices, as national competition intensified from merger waves that led to bigger units and spawned the emergence of supermarket power. More recently, other factors have affected supply chain actors – privatisation, information technology, internationalisation, retailer-driven choice, risk containment, brands and concentration of supermarket power, full cost accounting, water shortages, fuel price rises, low-cost versus ethical and healthy products, ethical and fair trade sourcing, the rise of corporate social responsibility and so on. Supply chain pressure come in part from the falling proportion of household expenditure going on food. In the UK, this went from 24 percent in 1970 to 8 percent in 2006 (Bowyer and Lang, 2006).

Food regimes

The food regimes concept captures the patterning and dynamics of investment trajectories and the behaviours of a multiplicity of actors that underlie such arrangements. Friedmann (2005, p. 228) defines a food regime as 'a specific constellation of governments, corporations, collective organisations and individuals that allow renewed accumulation of capital based on shared definitions of social purpose by key actors while marginalising others'. Early food regime writing distinguished two regimes. The latest work argues that a third food regime is perhaps cohering, though its dimensions are by no means clear or its existence certain.

The first food regime, from the 1870s to World War I saw the rise of food and fibre flows, under colonial relationships from peripheral resource areas to the expanding metropolitan core of Europe and North America. This was in response to working-class movements in Europe and created a historically unprecedented class of commercial family farmers. These 'family farms which had never existed in history, could only exist through international trade and would suffer the most from the collapse of the regime' (Friedmann, 2005, p. 236). In product terms, this involved the addition of two new wage foods – red meat and wheat. When world markets collapsed in the 1920s and 1930s, those farmers entered into new alliances, including one settlement that led to the mercantile-industrial or second food regime, from the late 1940s into the 1970s. This regime was an aid-based order that paradoxically fuelled the emergence of the livestock complex – a mix of feed producers, feedlot technology and intensive livestock producers. In an inversion of the first food regime where grain came from around the world to the core, the mid-20th century saw wheat being sent from the core to the developing world as food aid. Within geography Le Heron and Roche (1995) and Roche et al. (1999) explored whether a 'fresh' food regime around fruit and vegetables existed, concluding that while theoretically appealing, empirical evidence was insufficient.

Friedmann (2005) and McMichael (2005) now hold that food regimes should be regarded as emerging from the politics around competing ideas (e.g., social movements), contending with powerful institutions of rule and wealth. The period between the first and second food regimes thus deserves as much attention as the food regimes themselves. This revised view exposes the struggles over framing issues and understandings (Fagan, 2005). This emphasises change rather than stability and reminds us that regimes are provisional compromises among some of the contending social actors who manage to create a new interpretive framework in common. In keeping with this view two moments in the 20th century were especially significant – for what was lost as well as what was gained. The first was 1947 when the internationalist World Food Organisation, planned during World War II as a way forward given the protectionist trade policies of the 1930s, failed to secure support from the USA. Instead, a US-dominated and US-advantageous framework was adopted. The second was the advent of the World Trade Organisation in 1995 and the signing of the Agreement on Agriculture. This moment is considered critical in destabilising the second food regime because new developments in commodity circuits are possible.

Globalised (Le Heron, 1993) and globalising (Goodman and Watts, 1997) agriculture and food refers not to 'the entirety of agriculture across the world but a transnational space of corporate agriculture and food relations integrated by commodity circuits' (McMichael, 2005, p. 284). Developments in this space include: agriculture reframed as production for trade, food security responsibility delegated to households and villages, decimation of peasant agriculture through appropriation of land and switching to export production, the targeting of the Third World (urban) consumers, loss of local biodiversity of food sources, minimal intergovernmental intervention, selective adoption of economic standards (national standards lower than private standards) and so on. The top 30 supermarket grocery chains in the world control an estimated 33 percent of all global sales (Burch and Lawrence, 2007, 21). The authority vested in these supermarkets has arisen from strategies involving trust building through associations with 'valued institutions, practices, people and portfolios of products and services . . . supply chain control is pivotal to the supermarkets ability to perform its role as guardian of household and family life' (Dixon, 2007, p. 48). The goal of international retailers is to be able to trade freely across borders and set up their preferred retail formats without restrictive national legislation. Reardon and Swinnen (2004) document this trend internationally. Campbell and Le Heron (2007, p. 149) qualify this, concluding that 'while a blanket claim of a shift in power from food producers to food retailers may be appealing, it actually misses a range of diverse power gains (and shifts) within (the) agri-food' sphere. Indeed, as Aksoy and Beghin (2005) show for agricultural trade patterns, contradictory international trade patterns are emerging, at the country and commodity levels. In their review of UK consumer attitudes, Bowyer and Lang (2006) identify several especially salient points: different government agencies are content to be boxed in by their remits, consumers want both value-for-money and values-for-money (Lang, 2007), attitudes may not equate to actions, government responsibility for meeting consumer attitudes is split and a long-term view needs to be taken on consumer attitudes.

If we accept the view that 'everything to do with food has experienced an unprecedented period of flux: on the farm, in the factory, on retail shelves, in transit, in marketing and the home' (Millstone and Lang, 2002, p. 7), then we need more than political economy insights. The agri-food chain and food regimes approaches with their focus on structural and contextual processes and actors, investment patterns and pressures and behaviours of actors, must be supplemented with a discussion of meanings, moral and cultural politics of food.

The Cultural and Moral Economy of Food

At the heart of any cultural and moral economy of food are two components: the complexity of valuation processes and the power that different actors derive from, and exercise around, valuations by virtue of their positions in these processes. Interestingly, cultural and moral economy thinkers arrive at similar conclusions to those taking the food regime and agri-food commodity chain paths.

But where do food values come from and how are they 'fixed' (however fleetingly)? By way of illustration, Dixon (2002, p. 157) contends that the 'chicken delivers a melange of values with less effort than other meats'. These values have not been static; moving from the chicken for festive occasions in the 1960s, to an emphasis on freshness in the 1970s, nutritional content in relation to red meat in the 1980s, and the ultimate convenience food in the 1990s. Much of the struggle over the valuing of chicken has been around influencing the practices at a number of key sites – household kitchens, the community-like kitchens of fast-food outlets and industrial kitchens supplying supermarkets and institutions such as hospitals (see also Watts [2005] for a provocative discussion of the chicken commodity in political and moral economy terms).

Using Appadurai's (1986, p. 57) concept of a 'regime of value', namely a broad set of agreements over what is desirable, what a reasonable 'exchange of sacrifices' entails and who is permitted to exercise what kind of effective demand in what circumstances. Dixon (2002; 2007) outlines the substitution nature of the chicken and food retailing industries more generally, where words are exchanged rather than goods, the primary role of professionals, increasingly in supermarkets, being to mobilise reputations and bias in order to shape regimes of value. Jackson et al. (2007, p. 329) express similar sentiments. They argue that chicken is 'emblematic of a wider process occurring within the food industry whereby mainstream retailers such as Marks and Spencer are appropriating the language that was formerly associated with "alternative" producers'. Friedmann (2005, p. 229) is less charitable, suggesting that the supermarkets seek to 'share perceptual frames, choosing demands that best fit with expanding market opportunities and profits'. Commodity status is thus bound up in discursive judgements made by people, firms and institutions not by markets. A commodity can move in and out of favour, as meanings are manufactured and values appropriated.

Bell and Valentine (1997) redirected food research by focusing on how our relationships to eating have altered, especially through the technologies of food. They explored how circuits of culinary culture could be mapped across space, moving from the body, home, community, city, region, nation, to the global, to conclude that all geographical scales could end up on one plate. By centring and beginning with the body, they probe tensions around what has been called the 'omnivore's paradox' – humans have sought a varied diet in order to survive and so are inclined towards innovation and experimentation, yet humans have to be wary about what they consume because unknown food is a potential source of danger (e.g., food poisoning, allergies, unhealthy diets, cancer).

Le Heron and Hayward's (2002) study of the Australasian breakfast cereals industry illustrates the importance of deep and long-run cultural economy processes. They found that even a 'simple' analysis of production organisation and change in breakfast cereals (an industry founded in the 19th century on religious principles and notions of improved diets and food) could not be separated from powerful cultural and social traditions particular to the industry. Competing definitions of cereal value come from the socially constructed symbolic content of breakfast cereals. Much 'gaming' in cereals ingredients occurs – higher fats mean less sugar, higher sugar means lower fats, but declines in both sugar and fat are rarely seen. Instability and variety in breakfast cereals as a category are influenced by pressures from a wide field of NGOs. The narratives of the contemporary industry still align with the industry's foundations, so adding further episodes in the 'cereal' of the moral commodity.

The alternative food literature (for a critical review see Maye et al. [2007]) in which food becomes the basis for a localised life commonly romanticises and idealises place in food (e.g., slow food, short supply chains). In some versions of the rural idyll, the range of place-bound connections with the corporate food regime are ignored. The comfortable conservatism of retreating to the local restricts the geographical imagination, suppresses debate over wider issues and reduces the need to articulate the geography of power relationships. Conversely, work on alternative food networks literature has inserted the marginalised south into the food literature (Murtesbaugh 2002) and re-enlivened development issues.

At the turn of the 21st century, the contemporary foodscape suddenly altered. Two principal influences were a growing consensus on diet and cancer spearheaded by the 1997 World Cancer Research Fund (WCRF) report on Food, Nutrition and Prevention of Cancer, and the global obesity epidemic, highlighted by the World Health Organisation (WHO) in 1998. Obesity activated the political imagination of governments (Morgan et al. 2006, pp. 168–72). That the conventional food industry failed to foresee obesity as perhaps its biggest challenge is ironic when set against the trajectories outlined in the chapter.

The advent, diffusion and aggressive marketing of low-cost, processed food, high in fat, salt and sugar, is widely believed to be one of the main causes of the epidemic. Convenience foods are an efficient way to deliver calories, and this happens in numerous ways: calorie density, super-sized portions, speed of eating, the frequency of their consumption through grazing or snacking, hand-sized packaging for eating on the move and so on. Statistics compiled from Lang and Heasman (2004, p. 206) give some measure of the character and depth of problems in food and agriculture. Vast advertising budgets are undermining staple diets in the North and South – the food industry's global advertising budget has been put at $40 billion, a figure greater than the Gross Domestic Product of 70 percent of the world's nations. For every dollar spent by the WHO on preventing diseases from western-style diets, more than $500 is spent by the food industry promoting diets. In industrialised countries, food advertising accounts for approximately half of all advertising during children's TV viewing times. For countries with transitional economies (e.g., Eastern Europe) more than half of foreign direct investment in food production is for sugar, confectionary and soft drinks.

It is no wonder that Lang (2005, p. 123) insists that 'Health should be at the heart of social scientific thinking about food and farming . . . the case for a more integrated approach to food and farming, linking health, environment and society is strong'. The new food and dietary 'models' emphasise plant-based diets rich in a variety of vegetables and fruits, pulses (legumes) and minimally processed starchy food. The WCRF model is emphatic – limit sugar, alcohol, red meat (and fish and chicken), fat and salt and avoid charred food, additives and dietary supplements. Such a vision strikes at the organisational core of conventional food and food supply chains.

It remains unclear, nonetheless, whether cultural attributes are more important than price even in a food supply characterised by relative abundance and cheap foods. In the UK, where low food prices have been privileged for more than 160 years, there is growing concern that some turning point may have been reached, foreshadowing a time of peak food. For food-wealthy-importing nations, local provenance, which is thought to imply safer and healthier food because of relations of proximity, has attracted attention. However, the UK Research Council's Rural Economy and Land Use programme (Trail, 2006) reports that the UK agricultural sector could not support everyone in the UK, from locally produced food and meet government healthy eating guidelines.

Conclusion: Making Food Futures for All

The chapter pictures the contemporary agri-foodscape in terms of the political economy of food supply and the moral economy of food valuing and evaluation in a globalising context. Figure 32.1 details influences upon the food supply chain, as identified by two food scene thinkers and activists, Friedmann (working very much within the heritage of food regime thinking and food movement activism) and Lang (a food policy specialist who has prioritised the supply chain in efforts to redirect food policy). The special value of figure 32.1 is that it portrays the potential range and interconnectedness of issues 'outside' and 'inside' the food domain. The range of political and ethical projects that have arrived from 'outside food' is lengthy, e.g., ecological footprints, climate change, ethical sourcing, biodiversity, race, gender, pollution, safety, product and process claims, fair trade, labelling and welfare. These all have their food counterparts. Moreover, the dynamics 'inside food' as covered in the chapter have generated other tensions. The figure suggests that food is not reducible to any single issue (e.g., food miles, food safety, dietary intake) or any single interest group (e.g., farmers, consumers, supermarkets, regulators). What is now widely recognised is that the nature and mix of agri-food actors is historically unprecedented. Food and agriculture are increasingly dominated by globalising entities: big supermarkets, big producers, processors and traders, big science, big public

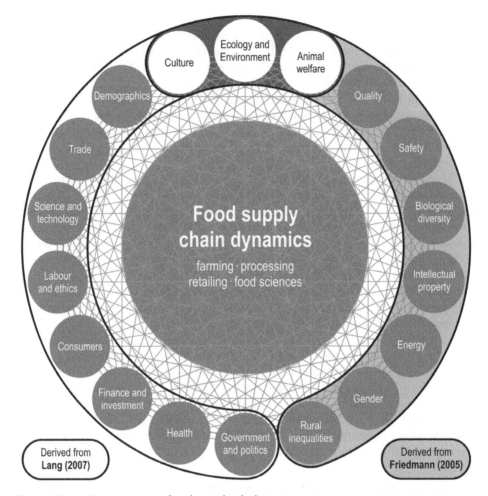

Figure 32.1 Contemporary food supply chain pressures.

regulators and private auditors, big citizen and consumer NGOs and big social movement voices. How attuned are we to this new reality when attempting to produce knowledge about food and agriculture? Overall, little is known about the field of actors. Less is known about the interrelations among actual actors, understood as a complex and dynamic whole.

Human geographers working within the field of food and agriculture are especially well positioned to use the rich insights from political and moral economy – which emphasise connectivity and responsibility – to address the ethical and political dilemmas that food and agriculture evoke. Recent books on food and agriculture edited by human geographers reveal a vibrancy to geographic research and scholarship (Fold and Pritchard, 2005; Maye et al., 2007). Positive though this conclusion might be, some serious concerns plague both the wider literature and geographic endeavour. First, ecologically centred accounts linked into political and moral economy are just appearing, at least in the context of developed-nation food, agriculture, aquaculture and fisheries. Second, urban food provisioning has received insufficient attention, despite over half the world's population living in large cities.

Third, the geopolitics of food security (globally and in particular countries) needs to be brought back onto the research agenda. Fourth, given the confluence of political and moral economy approaches in contemporary food and agriculture, is the nature of knowledge production itself. Cook et al. (2007, p. 1113) assert that 'commodity geographies are politically weak'. This challenge has profound implications for interventions. What advice and warnings are offered in the food and agriculture literature?

One mainstream answer is: prepare a food strategy. But as Maxwell and Slater (2004) note, this has problems – overloading policy with analysis, designing excessively complex organisational structures and planning in such detail as to make implementation impossible. Vorley et al. (2007, pp. 210–11) contend that the capacity for public policy response is limited, because changes are invisible to most policymakers. They say 'the debate around pro-poor growth and rural livelihoods is . . . proceeding as if national public policy is still the key determinant of rural livelihoods. . . . The focus should be on dynamic restructured national and regional markets that are displacing existing chains and their interactions with small-scale farmers and local rural economies'. In a sobering reflection Lang (2006, p. xv) writes of the complexity of mobilising actors,

> When it (food) enters our bodies, our identities are shaped through it and we gain an entire and sometimes poorly understood appreciation of the supply chain. Some contemporary industry analysts argue that consumers neither want to nor need to understand the complexities behind the check-out till. The brand is the seal of trust. But when we eat, we partake in an increasingly long chain of reactions. If brand is the sole mediator of trust, this is a fragile relationship.

According to Morgan et al. (2006, pp. 192–97) three major battle grounds are obvious in a globalising world. These are (i) the international, where concessions from the WTO are called for (around dumping and non-trade concerns to allow other models of agricultural and rural development); (ii) the national, where the emphasis should be on getting and maintaining state commitment; and (iii) the subnational level, where – in the context of a supportive multilevel polity – food chain revisioning can be promoted.

The contemporary moment of food and agriculture is distinguished by a proliferation of big and small actors fighting over new material and moral issues. This is a turbulent context calling for new research directions. Four are especially promising. The first is 'following' (Cook et al., 2006). This idea can be extended beyond the commodity, to include following the organisation – public, private and civil – and tracing its ramified connections in time and space. The second, 'entangling', is a strategy aimed at tying supply chains up in knots with monitoring, revealing the origins of food and detailing passage points, so stifling the free flow of power along chains. A third is pedagogy. This leads on from developing strategies of engagement in following and entangling that open up opportunities to get among agri-food actors. It must involve pro-activity in changing pedagogies and methodologies to confront the need to identity and get into decision spaces. The fourth strategy is 'being present'. This involves both resisting and engaging with food systems in multiple ways, and zeroes-in on key moments of decision making. Decision making involves creativity when power to remake the material and moral fibre of food is exercised. The inescapable knowledge-production challenge is seeing critique and resistance as necessary but insufficient for changing the outcomes and patterns of investment in food and agriculture. It is hoped that this chapter has established a more reflexive basis for making food futures for us all.

BIBLIOGRAPHY

Abrahams, C. (2007) Globally useful conceptions of alternative food networks in the developing South: the case of Johannesburg's urban food supply system. In D. Maye, L. Holloway, M. Kneafsey (eds), *Alternative Food Geographies: Representation and Practice*. Amsterdam: Elsevier Science, pp. 95–114.

Aksoy, M. and Beghin, T. (eds) (2005) *Global Agricultural Trade and Developing Countries*. Washington, DC: World Bank.

Appadurai, A. (ed.) (1986) *The Social Life of Things: Commodities in Cultural Perspective*. Cambridge: Cambridge University Press.

Arce, A. and Marsden, T. (1993) The social construction of international food: a new research agenda. *Economic Geography*, 69(3), 293–311.

Bell, D. and Valentine, G. (1997) *Consuming Geographies: We Are Where We Eat*. Oxford: Routledge.

Bowyer, S. and Lang, T. (2006) UK consumers and food security: first findings. A briefing report to the Royal Institute for International Affairs (Chatham House), Department of Health Management and Food Policy, City University, London, Working Party on Food Supply in the 21st Century, Centre for Food Policy.

Bowler, I., Bryant, C., and Cocklin, C. (eds) (2000) *Sustainability of Rural Systems Geographical Interpretations*. Dordrecht: Kluwer Academic.

Burch, D. and Lawrence, G. (eds) (2007) *Supermarkets and Agri-Food Supply Chains: Transformation in the Production and Consumption of Foods*. Cheltenham: Edward Elgar.

Campbell, H. and Le Heron, R. (2007) Big supermarkets, big producers and audit technologies: the constitutive micro-politics of food legitimacy and food system governance. In D. Burch and G. Lawrence (eds), *Supermarkets and Agri-Food Supply Chains*. Cheltenham: Edward Elgar, pp. 131–53.

Cook, I. (2006) Geographies of food: following. *Progress in Human Geography*, 30(5), 655–66.

Cook, I., Evans, J., Griffith, H., Morris, M., and Wrathmell, S. (2007) 'It's more than just what it is': defetishising commodities, expanding fields, mobilising change. . . . *Geoforum*, 38, 1113–26.

Dixon, J. (2002) *The Changing Chicken: Chooks, Cooks and Culinary Culture*. Sydney: University of New South Wales Press.

Dixon, J. (2007) Supermarkets as new food authorities. In D. Burch and G. Lawrence (eds), *Supermarkets and Agri-Food Supply Chains*. Cheltenham: Edward Elgar, pp. 29–50.

Fagan, B. (2005) Globalisation, the WTO and the Australian-Philippines 'banana war. In N. Fold and B. Pritchard (eds), *Cross-Continental Agro-Food Chains Structures, Actors and Dynamics of the Global Food System*. Oxford: Routledge, pp. 207–22.

FitzSimmons, M. (1986) The new industrial agriculture: the regional integration of specialty crop production. *Economic Geography*, 62, 334–53.

Fold, N. and Pritchard, B. (2005) (eds) *Cross-Continental Agro-Food Chains Structures, Actors and Dynamics of the Global Food System*. Oxford: Routledge.

Food and Agriculture Organisation (2006) *State of Food Insecurity*. Rome: Food and Agriculture Organisation.

Freidberg, S. (2004) *French Beans and Food Scares: Culture and Commerce in an Anxious Age*. Oxford: Oxford University Press.

Friedland, W., Barton, A. and Thomas, R. (1981) *Manufacturing Green Gold*. Cambridge: Cambridge University Press.

Friedland, W., Busch, L., Buttel, F. and Rudy, A. (eds) (1991) *Towards a New Political Economy of Agriculture*. Boulder, CO: Westview Press.

Friedmann, H. (2005) From colonialism to green capitalism: social movements and the emergence of food regimes. In F. Buttel and P. McMichael (eds), *New Directions in the Sociology of Global Development*. Research in Rural Sociology and Development, Volume 11. Amsterdam: Elsevier, pp. 227–64.

Friedmann, H. and McMichael, P. (1989) Agriculture and the state system: the rise and decline of national agriculture. *Sociologia Ruralis*, 19(2), 93–117.

Gereffi, G. and Korzeniewicz, M. (eds) *Commodity chains and global capitalism*. Westport, CT: Greenwood Press.

Gibson-Graham, J.-K. (2005) *A Postcapitalist Politics*. Minneapolis: University of Minnesota Press.

Goodman, D. and Watts, M. (eds) *Globalizing Food*. New York: Routledge.

Goodman, D. Sorj, B. and Wilkinson, J. (1987) *From Farming to Biotechnology*. Oxford: Blackwell.

Gray, S., Le Heron, R., Stringer, C. and Tamasy, T. (2007) Competing from the edge of the global economy: the globalising world dairy industry and the emergence of Fonterra's strategic networks. *Die Erde*, 138(2), 1–21.

Grigg, D. 1981 'The historiography of hunger: changing views on the world food problem 1945–1980', *Transaction of the Institute of British Geographers*, NS, 6, 3, 279–92.

Heasman, M. and Melletin, J. (2001) *The Functional Foods Revolution: Healthy People, Healthy Profits?* London: Earthscan.

Hughes, A. and Reimer, A. (eds) (2004) *Geographies of Commodity Chains*. Oxford: Routledge.

Jackson, P., Ward, N. and Russell, P. (2006) Mobilising the commodity chain concept in the politics of food and farming. *Journal of Rural Studies*, 22(2), 129–41.

International Fund for Agricultural Development (2004) Statement by C. Enweze to the High-Level Forum on Trade and Investment of the Group of 77, Doha, 5 December.

Lang, T. (1999) Towards a Sustainable Food Policy. In G. Tansey and J. d'Silva (eds), *The Meat Business*. London: Earthscan, pp. 120–35.

Lang, T. (2005) What is food and farming for? – The (re) emergence of health as a key policy driver. In F. Buttel and P. McMichael (eds), *New Directions in the Sociology of Global Development*. Research in Rural Sociology and Development, Volume 11. Amsterdam: Elsevier, 11, 123–44.

Lang, T. (2006) Preface. In S. Barrinetos and C. Dolan (eds), *Ethical Sourcing in the Global Food System*. London: Earthscan, pp. xiv–xv.

Lang, T. (2007) A transition from 'value-for-money' to 'values-for-money' in food? Food policy and ethical dilemmas. PowerPoint presentation to the Institute of British Geographers/Royal Geographical Society conference, Session on 'Ethical Foodscapes', 31 August.

Lang, T. and Heasman, M. (2004) *Food Wars: The Global Battle for Mouths, Minds and Markets*. London: Earthscan.

Le Heron, R. (1993) *Globalized Agriculture: Political Choice*. Oxford: Pergamon.

Le Heron, R. and Roche, M. (1995) A 'fresh' place in food's space. *Area*, 27(1), 22–32.

Le Heron, R. and Roche, M. (1996) Globalisation, sustainability and apple orcharding, Hawkes Bay, New Zealand. *Economic Geography*, 72(4), 416–32.

Le Heron, K. and Hayward, D. (2002) The moral commodity: production, consumption, and governance in the Australasian breakfast cereal industry. *Environment and Planning A*, 34(12), 2231–51.

Le Heron, R. Rees, E., Massey, E., Bruges, M. and Thrush, S. (2008) Progressing sustainable management of marine fisheries: developing a dialogue between fisheries science and ecosystem science', *Geoforum*, 39, 48–61.

McMichael, P. (2005) Corporate development and the corporate food regime. In F. Buttel and P. McMichael (eds), *New Directions in the Sociology of Global Development. Research in Rural Sociology and Development*, Volume 11. Amsterdam: Elsevier, pp. 265–99.

Mansfield, B. (2003) Spatializing globalization: a 'geography of quality' in the seafood industry. *Economic Geography*, 79(1), 1–16.

Marsden, T. (2000) Food matters and the matter of food: towards a new food governance. *Sociologia Ruralis*, 40(1), 20–29.

Maxwell, S. and Slater, R. (eds) (2004) *Food Policy Old and New*. Oxford: Blackwell.

Maye, D., Holloway, L. and Kneafsey, M. (eds) *Alternative Food Geographies: Representation and Practice*. Amsterdam: Elsevier Science.

Millstone, E. and Lang, T. (2002) *The Atlas of Food: Who Eats What, Where and Why*. London: Earthscan.

Morgan, K. and Murdoch, J. (2000) Organic vs conventional agriculture: knowledge, power and innovation in the food chain. *Geoforum*, 31, 159–73.

Morgan, K., Marsden, T. and Murdoch, J. (2006) *Worlds of Food: Place, Power, and Provenance in the Food Chain*. Oxford: Oxford University Press.

Mutersbaugh, T. (2002) The number is the beast: a political economy of organic-coffee certification and producer unionism. *Environment and Planning A*, 34, 1165–84.

Newby, H. and Buttel, F. (eds) (1978) *The Rural Sociology of the Advanced Societies*. Totowa, NJ: Allanheld Osmun.

Reardon, T. and Swinnen, J. (2004) Agrifood sector liberalization and the rise of supermarkets in former state-controlled economies: a comparative review. *Development Policy Review*, 22(5), 515–23.

Roche, M., McKenna, M. and Le Heron, R. (1999) Making fruitful comparisons: Southern hemisphere producers and the global apple industry. *Tijdschrift voor Economische en Sociale Geografie*, 90(4), 410–26.

Rosin, C. (2008) Contesting the rules of exchange: changing conventions of procurement in the MERCOSUR yerba mate commodity chain. In C. Stringer and R. Le Heron (eds), *Agri-Food Commodity Chains and Globalising Networks*. Aldershot: Ashgate, pp. 49–60.

Steinfeld, H., Gerber, P., Wassenaar, V., Castel, M., Rosales, M. and de Haan, C. (2006) *Livestock's Long Shadow: Environmental Issues and Options*. Rome: Food and Agriculture Organisation.

Stringer, C. and Le Heron, R. (eds) (2008) *Agri-Food Commodity Chains and Globalising Networks*. Aldershot: Ashgate (in press).

Trail, B. (2006) Suppose we all ate a healthy diet. . . .' University of Reading, Rural Economy and Land Use project. http://www.extra.rdg.ac.uk/news/details (accessed 9 September 2007).

Vorley, B., Fierne, A. and Ray, D. (eds) (2007) *Regoverning Markets: A Place for Small-Scale Producers in Modern Food Chains?* Aldershot: Gower.

Watts, M. (2005) Commodities. In P. Cloke, P. Crang, and M. Goodwin (eds), *Introducing Human Geographies*. Oxford: Hodder Arnold, pp. 527–43.

Watts, M. and Bohle, H. (1993) The space of vulnerability: the causal structure of hunger and famine. *Progress in Human Geography*, 17(1), 43–67.

Whatmore, S. and Thornes, L. (1997) Nourishing networks: alternative geographies of food. In D. Goodman and M. Watts (eds), *Globalizing Food*. New York: Routledge, pp. 287–304.

Wilson, G. (2001) From productivism to post-productivism . . . and back again? Exploring the (un)changed natural and mental landscapes of European agriculture. *Transactions of the Institute of British Geographers*, 26(1), 77–102.

Chapter 33

Environment and Health

Hilda E. Kurtz and Karen E. Smoyer-Tomic

Introduction

Environmental health can be viewed in terms of human impacts affecting the health of the environment, as well as the impacts of the environment on human health. Geographers theorise and examine both issues from a variety of sub-areas, including environmental justice, political ecology, hazards geography and health geography. Geography has long been concerned with understanding the complex interplay of human-environment relations both as a central theme as well as through subtexts arising in its systematic branches. In the face of rapid and adverse environmental changes, human and ecological health have become key focal points for evaluating relationships between human use of the environment, the ecosystemic disruptions that ensue, and their implications for human well-being. The discourse on sustainability has been influential in understanding the interplay of human–environment relations, as has a growing acceptance that we live in a 'risk society' (Beck, 1986). We refer to these complex relations as the *human-environment-health nexus*. While geographers and those in other disciplines have long recognised the environmental consequences of human activity, studies of disparities in health and disparities in environmental quality have been conducted in largely separate realms both within and outside of geography (Brulle and Pellow, 2006).

This chapter considers how geographers and those in related disciplines understand the human-environment-health nexus at a range of geographic scales. We review developments in four areas of geography: environmental justice (EJ), political ecology of disease, vulnerability analyses and health geography. In each, we note the increasing importance ascribed to environmental and health disparities, a growing emphasis on ecological analyses, and the ways in which particular operational definitions of health and environment shape research trajectories. Geographic scholarship on EJ problematises the human-environment-health nexus, politicising and interrogating the uneven impact of (primarily) industrial activity on both human health and the health of the environment, as well as the complex factors which produce such uneven effects. Geographic scholarship on the political ecology of disease and on the nexus of health and place is developing increasingly complex understandings of how environmental factors at a range of scales shape health outcomes. Elsewhere in geography, vulnerability analyses evolving from hazards research seek to understand how disruptions to the dynamic natural-human equilibrium impact ecosystems and societies, as well as how human actions and reactions to the environment foster new forms of vulnerability. Following a discussion of each of these approaches to the human-environment-health nexus, we consider their similarities and reflect on possibilities for dialogue between them.

Environmental Justice

The environmental justice (EJ) movement and its related scholarship have drawn attention to uneven patterns of health and environmental quality. Originating with concerns about the adverse health impacts on vulnerable populations of exposure to environmental toxins, EJ discourse developed initially within the US legal system as public interest lawyers sought civil rights and environmental law remedies for environmental inequities. The legal arena encouraged an emphasis on 'naming, blaming and claiming', and produced a rather narrow 'perpetrator-victim' conceptualisation of environmental inequality (Pellow, 2000), focused on a local, regional or national (USA) scale. As EJ research moved beyond the constraints of legal scholarship, geographers were among the first to investigate the inherently geographical problem of an uneven distribution of environmental hazards and potentially related health outcomes (Bowen et al., 1996; Cutter and Solecki, 1996). Whereas geographers' early engagement with the uneven quality of environmental conditions focused on measuring environmental inequity, and remained focused at local and regional scales, more recent geographic EJ work investigates a broader 'range of structural, institutional and social forces which [contribute] to a landscape of inequality' (Pulido, 1996, p. 142) at a range of geographic scales, from the local to the global.

EJ activism and scholarship reconceptualise the environment as those places where we live, work and play. Conceptually, such a stance reinserts human beings into complex (and humanly mediated) ecosystems. Politically, such a stance introduces an anthropocentric emphasis to mainstream (American) environmentalism, placing human health and well-being at the centre of concern about ecosystem health. From an EJ perspective, then, humans have been explicitly included in ecosystems and yet politically take precedence over other entities within a given ecosystem.

Geography's tradition of Marxian analysis contributes much to this view, in which the environment in the human-environment-health nexus is not protected wilderness, but rather those elements of ecosystems that have been brought into circuits of capital, and hence, made directly or indirectly responsive to political economic interests in the context of industrial capitalism (Lake and Disch, 1992; Harvey, 2006). A central premise of much of this research is that 'exploitation of the environment and exploitation of human populations are linked' (Brulle and Pellow, 2006, p. 108) and that devaluing one can imperil the other. The task of much recent EJ scholarship is to explore the nature of these linkages, conceptualising environmental injustice as a socio-historical process rather than a discrete event (Pellow, 2000). Indeed, geographers' interest in the political economy of environmental hazards has influenced partially by Beck's (1986) argument in *Risk Society* that '[e]nvironmental problems are fundamentally based in how a society is organized' (p. 81).

Myriad environmental problems are incurred by society's organisation around industrial capitalism. In the process of transforming first nature into second nature, materials which may have been harmless in an ecosystem become potential hazards to health and environment. Sodium chloride, for instance, is mined from naturally occurring salt domes and used to manufacture chlorine, a hazardous material which is in turn used to produce polyvinyl chloride (PVC), a feedstock for plastic production that ranks among the top five hazardous materials monitored by the US Environmental Protection Agency. The demand for PVC, then, and its production as enabled by science in the service of capitalism, are central to the problem of environmental inequality (Pellow, 2000).

EJ research is largely motivated by concerns about the consequences of industrial-environmental hazards for human health, yet most EJ researchers are not health geographers. While a holistic concept of environment shaped EJ research early on, operational concepts of health remained focused on specific health concerns raised by exposure to industrial toxins – such as respiratory illnesses in the case of airborne contaminants, and cancers and reproductive problems in the case of water-borne toxins. As the EJ movement matured and vulnerable communities voiced their concerns about environmental justice, a more holistic and integrative view of health emerged. It became recognised that:

> [m]ore often than not, issues of environmental justice comprise a complex web of public health, environmental, economic and social concerns. Given the multiple stressors that impact low-income, people of color, and tribal communities, such groups do not have the luxury of addressing one issue at a time. They require holistic, integrative and unifying strategies that address social, economic and health improvement simultaneously (Lee 2001: 141).

Currently, as EJ concerns percolate widely around the globe and through many academic disciplines, more expertise in health research (often outside of geography) is being brought to bear on EJ issues. Cross-cutting research examines how one type of environmental stressor can influence susceptibility to another type. For example, Gee and Payne-Sturges (2004) make the case that psychosocial stresses, such as experienced by marginalised groups, can weaken immune systems, and thus, make people more susceptible to environmental toxins. Although often not explicitly stated, the inherently geographical concepts of space and place, site and situation, underlie the newest EJ research, which suggests that residential segregation and isolation result in combined stressors from built, social and natural environments that are manifested in limited services, fewer community resources and greater likelihood of exposure to pollutants.

Significantly, EJ scholars have not focused much attention on how ecological processes shape environmental inequalities, yet EJ scholarship increasingly informs geographic scholarship on urban political ecology (Swyngedouw and Heynen, 2003; Njeru, 2005; Heynen et al., 2006). Indeed, the agendas of urban political ecology and EJ research may be converging, aided in part by the political ecology of disease framework within medical geography.

Political Ecology and Health

Within geography and related disciplines, political ecology has influenced work on the human-environment-health nexus both directly as it has taken on disease etiology as an object of study (e.g., Mayer, 1996; 2000) and indirectly, as urban political ecology makes its mark on the new urban environmental scholarship. Political ecology's traditional strengths lie in detailed studies of how local actors manage environmental resources within particular social, economic and political conditions, in traditionally agrarian societies in peripheral regions of the world economy. Political ecologists explore the impacts of state practices, social norms, and economic and political marginalisation on land degradation. Although the human health consequences of environmental degradation have been noted, they have not been of central concern in traditional political ecology.

More recently, geographers (McCarthy, 2002; Robbins, 2002) have called for using the conceptual and methodological repertoire of political ecology to explicate 'First World' environmental problems. Converging with this agenda has been the development of an increasingly robust framework for thinking through the political ecologies of cities, where marginalisation of vulnerable groups, power wielded by local and state institutions, and environmental management fraught with corporate and state agendas are as relevant as they are in peripheral regions of the world system, but in different ways.

Braun (2005) traces a history of intermittent, yet intellectually powerful, attention to ecological processes in urban scholarship. Such work has been limited by a conceptual separation of the urban from the rural landscape that is deeply engrained in Western culture (Gottlieb, 1993). Consequently, urban scholarship has had more to say about the effects of urbanisation on human health than on the ecological processes, which sustain cities.

Recent work in urban political ecology, however, directly challenges the conceptual separation of city and country that has informed so much geography's urban and EJ scholarship. Much of this work owes a debt to William Cronon's magnificent environmental history of Chicago, *Nature's Metropolis*. That book, and others to follow, by Davis (1999), Wilson (1992), Ross (1994) and Gandy (2002), explicates in insightful detail Raymond Williams' (1973) argument that city and country are inextricable parts of larger economic and ecological systems. In this view, properties of an 'urban environment' cannot be fully understood at the scale of the city alone; rather, environmental conditions in the city (including those which impinge on human health) must be viewed ecologically in relation to their consequences for and dependence upon processes operating at lower (e.g., the body) and higher (e.g., the bioregion or globe) scales of analysis. As Haughton and McGranahan (2006, p. 3) note:

> The very notion of urban ecology has become multi-scalar, extending from individual urban systems to systems of cities and towns, and from ecosystems within urban settlements to urban settlements as ecosystems, to the ways in which cities and towns shape ecosystems beyond as well as within urban boundaries.

The multi-scalar dimension of urban ecology takes on particular salience in relation to the goal of sustainability, which is an indicator of environmental health, as well as a keystone to population health. As Haughton (1999) argues, a city (or any other discrete areal unit) cannot be an island of sustainability; it must relate and contribute to the sustainability of larger, more extended and quite distant ecosystems as well.

Blurring the multi-scalar boundaries between city and country opens for analysis a vast and complex ecological web with the potential for both good and ill human and environmental health. Braun (2005) notes the analytical focus within urban political ecology on water, energy, food and waste explodes traditional understandings of city and country as discrete places. Water, for instance, both incubates disease and disease transmitting insects and cleanses wounds that might otherwise lead to ill-health and even death. Too much water in the wrong places and times erodes vulnerable soil, compromising food security, but without water, food production is impossible. Vital to human survival, water flows are increasingly constrained by human technologies, and yet still escape human control in both predictable and unpredictable ways.

Yet, even while geographers working in urban political ecology consider the sustainability of urban ecological systems, there is little attention to the human health implications of unsustainability. For instance, in an otherwise sophisticated and insightful analysis of New York City's 'metropolitan natures' – ideologies and incumbent transformations of nature within and beyond the city to effect the city – Gandy (2003) pays relatively little attention to the (human) body. As Braun notes, Gandy is thus

> continuing a tradition in geography to imagine nature to be 'outside' the body. Yet, if bodies are truly 'composites' and urban natures – water, air, food – crucial to their capacity for life, this is an odd elision. It points also to the vast gulf that remains between the new urban environmental geography and medical [i.e., health] geography, despite common knowledge of the historical links between public health, the state and urban form. (Braun, 2005, p. 647, brackets added).

Despite the general absence of human health in urban political ecology research, pioneering health geographers have begun to address health outcomes in wider political ecological contexts. This approach derives directly from a long tradition of disease ecology within health geography, which examines 'the ways human behavior, in its cultural and socioeconomic context, interacts with environmental conditions to produce or prevent disease' (Meadeet al., 1988, p. 29). In this view, the intersection of habitat (physical and built), population, and behaviour is seen to underlie human health. The task for disease ecologists, then, is to understand how social, cultural, political and behavioural factors, on the one hand, and environmental characteristics, on the other, shape human exposure to disease vectors (May, 1954). The main limitation of this approach is its tendency to focus on a local scale, with limited attention to structural forces operating at regional, national and global scales to generate those mediating social, cultural and political factors in the first place. Mayer (1996; 2000) expanded the concept of disease ecology and reframed it as the *political ecology of disease* that takes into account the political economy as well as the cultural ecology within which humans interact with the environment. Mayer (1996) argued that political ecological analyses of disease situate disease in its local socio-economic and biophysical context to processes and circumstances stretching across broader geographic scales. In ecological terms, the political ecology of disease perspective views population, society and environment in dynamic equilibrium, which can be thrown off balance by events and actions occurring at diverse scales.

The political ecology of disease perspective has been used to link health outcomes with social, political and environmental factors at multiple scales, in both developed and developing world contexts. Salehi and Ali (2006), for example, use a political ecology framework to examine in detail the human-environment interactions that enabled the transmission of SARS in Toronto. Linking interactions within the city to a globally extended network of actors, they demonstrate the inadequacies of global health governance in relation to the spread of SARS. Collins (2001) examines the traditional concerns of political ecologists – land degradation – in relation to (adverse) health outcomes. He suggests that policy responses to the sustainability problem should take both human and environmental health into account simultaneously.

While Mayer (1996) restricts his focus to the implications of disequilibrium for new and remerging infectious diseases, the framework can be applied to health, broadly defined. Richmond et al. (2005) develop this potential with a fine-grained political ecological study of linkages between the economic marginalisation of members of the 'Nangis First Nation in British Columbia as a result of commercial aquaculture, and consequent constraints on cultural and social activities related to health and well-being. Elsewhere in geography, the political ecology framework is being robustly applied to understand a range of health outcomes within the vulnerability approach to natural hazards.

Vulnerability Analyses

Hazards geographers investigate the adverse outcomes (including health) of the intersection between environment (natural and technological) and humans. Early work in this vein tended to focus on describing the spatial distribution and historical frequency of hazard occurrence, and then quantifying hazard impacts in terms of population sensitivity (Burton et al., 1978; Kates and Kasperson, 1983; Palm, 1990). Over the past two decades, the hazards literature has been moving towards a vulnerability approach that seeks to understand environmental and human factors that underlie differential exposure to hazards, population sensitivity to these exposures and resilience after a hazard event has occurred (Hewitt, 1983; Blaikie et al., 1994, Bohle et al., 1994; Cutter, 1996; Cutter, 2003). Thus, vulnerability arises from within the linked human-environment system.

The concept of vulnerability has ties to political ecology in its attention to how disruptions of the dynamic equilibrium of ecosystems (including the human component) intersect with societal constructions of risk (Oliver-Smith, 1996). Vulnerability approaches to hazards research seek to understand how disparities in resource endowments and entitlements underlie the differential exposure of people and places to hazards and their consequences, as well as their predisposition to future hazard risk (Sen, 1981; Blaikie, et al., 1994; Cutter et al. 2000; Turner et al., 2003). Within the vulnerability framework, place is where population health is compromised and where responses occur in terms of disaster relief and healthcare. At the same time, vulnerability analyses are also highly cognizant of scale, with attention to uneven local impacts of events occurring at diverse spatial and temporal scales.

Useful insights stemming from vulnerability analysis include problematising exposure to environmental stressors, population sensitivity to these stressors and population resilience in recovering from harmful exposures as well as sensitivity to future environmental stressors. Exposure, sensitivity and resilience differ over space and time, with local conditions affected by agents operating outside the local system and at larger scales. Environmental exposures may be acute or cumulative, ongoing or intermittent. Subtle shifts in the human-environment equilibrium (including reactive disaster mitigation responses) can lead to large-scale impacts later in time, or outside of the locale where major environmental disruption is occurring (Turner et al., 2003, Ingram et al., 2006).

The vulnerability approach 'requires a new way of viewing the world, one that integrates perspectives from the sciences, social sciences and humanities' Cutter (2003, p. 6). Turner et al., (2003) situate vulnerability firmly within what is known as sustainability science. They attempt to redress the conceptual separation of environment from population, with the implementation of a coupled human-environmental systems framework that incorporates feedbacks between human actions (and reactions) in the environment, environmental change and subsequent shifting of multiple equilibria between human and environmental systems in a continuing cycle.

Health is embedded in the subtext of vulnerability scholarship in two quite distinct ways. First, the concept of *environmental* health emerges in terms of environmental hazard likelihood based on an area's biophysical properties, which can be traced back to conditions resulting both from natural features (e.g., landforms, climate, location) and from how humans have modified the environment to create a new steady state. Second, *human* health arises in the treatment of the exposure, sensitivity, and resilience of an area's population. Researchers recognise that people in certain demographic categories (i.e., low-income or socially marginalised groups) may be more exposed to some hazards by living in higher-risk areas, close to, for example, polluting factories. Likewise, some groups (i.e., children or those in ill-health) may be more susceptible to a given level of hazard exposure, and less resilient, and thus, slower to recover from hazard impacts (Blaikie et al., 1994). Recent work has made health more central by discussing commonalities between the social, physical and economic factors underlying population health (referred to as the 'determinants of health') and population vulnerability to disasters (Lindsay, 2003). Continued contact between hazards and health geographers will no doubt build on commonalities in the fields, and serve to strengthen conceptual as well as practical understanding of the human-environment-health nexus.

Health and Place

As medical geography 'reinvente[d] itself as health geography' Dyck (1999, p. 243), it incorporated more holistic definitions of health and increasing attention to socio-environmental factors, moving beyond a focus on biomedical outcomes and health care provision. As Elliott (1999) notes (citing Wilkinson, 1996), with this shift health [geography] emerged as a social science. Newly engaged with social theory, health geography has become positioned to draw from and contribute to the commitments of EJ scholarship, thus linking environmental and health disparities to political economy.

Within health geography, and extending into public health and health promotion, is a growing body of scholarship on 'health and place'. Health and place research incorporates a broad, integrated view of the environment and explores the diverse impacts of environment on health (Macintyre et al, 1993, Smoyer, 1998, Frumkin 2003). Attention is given to uneven access to resources and exposure to harmful substances arising from economic, political and social structures, and to disparate health outcomes to which those inequalities give rise. Environment is conceptualised and articulated as the *places* where people spend their time, which in urbanised societies are more likely than not built (and often indoor) environments of cities. The lived environment is important in health and place research as it is the source of acute and chronic exposures, both harmful and beneficial, that are experienced over the life course. The scale of health and place research tends to be highly local-ised around spheres of human activity, often the home.

Places have associated with them material infrastructure (housing, investment, institutions, connectivity to other places) and collective social functioning and prac-tices (empowerment, norms, values, social capital and capacity) that affect health (Macintyre et al., 2002; Bolam et al., 2006). Macintyre (1997) and colleagues (Macintyre et al., 2002) have theorised three pathways linking place to human health. First, *compositional* effects relate to health behaviours or outcomes arising from the aggregate characteristics of people who populate a place, such as age, income, or social class. Second, *contextual* effects arise from the characteristics of the places themselves, such as dilapidated housing, crime, or limited access to ser-vices. Finally, *collective* effects stem from the historical or socio-cultural aspects of communities including norms and values, such as a lack of economic investment due to negative perceptions about a place. These pathways can be understood in terms of five types of features: physical features of the environment shared by all residents; availability of healthy environments at home, work, school, and play; availability of services needed to support people in their daily lives; socio-cultural features of the area; and the area's reputation (Macintyre et al., 2002).

Early work in health and place theorised the differences between compositional effects and contextual effects (Macintyre et al., 1993), while more recent research challenges include conceptualising, operationalising and measuring neighbourhood effects, understood as both contextual and collective effects of and on place (Diez-Roux, 2000; Macintyre et al., 2002; Oakes 2004). Multi-levelling modelling, which arose from educational research, has been valuable to geographers in separating out individual or household effects from neighbourhood effects, with relatively modest variation in a range of health impacts attributed to neighbourhood-level variables (Duncan et al., 1993; Pickett and Pearl, 2001).

Environment conditions, often byproducts of industrialisation and capitalism, as well as those emanating from socio-political processes, are problematised in terms of how they affect health and well-being. Factors studied are diverse: rat bites in children (Bunge and Bordessa, 1975); accessibility to supermarkets (Morland et al., 2002; Smoyer-Tomic et al., 2006); and harmful social environments (Sampson, 2001). The physical features, availability of health-promoting environments, service provision, socio-cultural features and reputation of a place create a suite of inter-related, dynamic processes that influence health behaviours and health outcomes (Macintyre et al., 2002).

Significantly, health and place research tends to frame humans separately from the environment, with people experiencing adverse health effects caused by their environments. To more fully engage with the human-environment-health nexus, health and place researchers can benefit from EJ, political ecology and vulnerability approaches that pay more attention to how people constitute their environments. In this way, geographers can continue to refine theoretical understandings of the duality of human-environment relations, with people as a constitutive part of, rather than outside of, their environments.

Shared Assumptions and Concerns

Both *health* and *environment* are subject to definition from differing perspectives that vary in emphasis, scope and complexity. Health can be defined narrowly in terms of absence, or effective management, of illness and disability (US NIEHS, *n.d.*). Health can also be defined holistically in terms of well-being beyond absence of disease or infirmity. Embraced by the World Health Organization, this is a definition that acknowledges social as well as biophysical factors (WHO, 1957; 1986). Environment can be understood in terms of its raw physical, atmospheric, terrestrial, and biotic components along with 'chemicals, radiation, and some biological agents' (WHO, Regional Office for Europe, *n.d.*), or broadly as 'where we live, work and play' within the EJ and health geography literature. In each of the four research areas addressed here, the concept of environment has been broadened from a focus on the 'natural' environment, to encompass the built environment as well as certain social, cultural, political and economic characteristics of human spheres of activity. Different approaches to environment and health pay varying amounts of attention to the ecological complexity of the environment. In some cases, environment is viewed as the location for and partial cause of the health outcomes under study. Other approaches view environment as the 'patient' whose disease comes from human activity. Still other research places humans centrally into the ecological web of life.

Given the multidisciplinary traditions of research on the human-environment-health nexus, both within and beyond geography, different deployments of the individual terms 'health' and 'environment' have implications for how the relationship between the two can be conceptualised and investigated. Differing frameworks for considering that relationship (in all its complexity) tend to be used in diverse research settings; for instance, vulnerability analyses have evolved from studies of food insecurity in rural developing world locales (e.g., Sen, 1981; Blaikie et al., 1994), while health and place research emerged from studies of health disparities among urban neighbourhoods in developed countries (Greenberg and Schneider, 1993; Macintyre et al., 1993). Such divergent settings bring to the fore quite different components of both health and environment. Vulnerability analyses focus on issues such as malnutrition as an effect of desertification and various political economic constraints, while health and place research has investigated issues such as high rates of respiratory illness in congested urban environments. Despite an underlying geographical approach, research with such diverse empirical foci can be expected to develop different models of how interactions between health and environment play out at multiple spatial and temporal scales.

Research on the human-environment-health nexus is converging on a common set of problems from diverse disciplinary directions and understandings of environment and health, and several key commonalities emerge from our brief overview. First, each of the areas of research noted above ascribes increasing significance to geographically uneven health and environmental conditions and outcomes, as both pressing socio-political problems and as rather knotty analytical problems as well. In both vulnerability analyses and political ecology approaches, uneven environmental quality serves as a sensitising device for research, rather than as a central analytical problem. Geographers undertaking EJ and health and place research address this problem most directly. In EJ research, not only do patterns of environmental inequality serve as a persistent backdrop for case studies of EJ issues and activism (Harwood, 2005; Saha and Mohai, 2005); but environmental inequality is an object of analysis in its own right (Pellow, 2000). The focus in health and place research on factors producing uneven environments and health outcomes at a micro-scale may eventually inform EJ activism and scholarship more productively.

Second, in each of the geographical sub-areas of research discussed here, an increasingly holistic, integrative and ecological approach to health and environment is becoming a dominant paradigm. All but the health and place agenda incorporate an ecological framework for studying the human-environment-health nexus that embeds human beings within ecosystems, and recognises the central importance of many human-induced ecological perturbations for the health both of the system as a whole and of the entities within it (vulnerable populations, human and otherwise). An ecological lens also offers a way for geographers to refresh their thinking on how different places and regions are linked. It blurs city and country, emphasising that both urban and rural areas are parts of a greater whole, and that they have functionality in relation to one another. An ecological perspective further under-scores the limitations of focusing exclusively on any one site or at any one geo-graphic scale in the effort to tease out relations within the human-environment-health nexus. It reminds us that even while research on human-centred factors in uneven environment and health conditions becomes more complex, we must not lose sight of ecological processes. The natural environment, too, has agency, or in the language of critical realism, has causal properties.

Third, each of the areas of research takes on the problem of geographic scale to a greater or lesser degree. Introducing a goal such as 'sustainability' or a measure such as 'health' raises the question of a geographic scale at which the phenomenon is to be measured and evaluated. Increasing recognition of the multi-scalar nature of ecological processes and their direct and indirect implications for human health has led to multi-scalar research designs that can investigate processes and dynamics between and across geographic scales. In each of the research areas discussed, it is clear that local actions have broad-reaching, even global impacts, and also that local, national, and international policies have impacts on environmental quality. The geographic scale at which a problem of environmental injustice can be demonstrated is a central problem in EJ research and activism, and EJ scholarship explores both how structural actors operating at a range of scales foster environmental inequalities (Harvey, 1996; Pellow, 2000) as well as how EJ activists create possibilities for solutions to environmental injustice at various geographic scales (Towers, 2000; Kurtz, 2003). Health and place research focuses on environmental factors impinging on human health and well-being at the relatively micro-scale of the (urban) neigh-bourhood and tends to focus on built and social environments rather than natural environment. In the political ecology of disease, geographic scale is extended from a localised set of circumstances, with the researcher working out in concentric circles to identify increasingly macro-factors contributing to poor land management. Simi-larly, nested scales are central to vulnerability analyses in terms of hazard risk, exposure, sensitivity and resilience.

The Future: Possibilities for Dialogue

The human-environment-health nexus comprises a complex, challenging and cru-cially important set of issues for geographic and interdisciplinary research. Geogra-phy offers sophistication, nuance, and analytical power in three key areas of interest in human-environment-health scholarship. First, geography as a discipline offers a long history of developing different ideas, definitions and operationalisations of 'where we live, work and play' – as place, environment, locale, location, to name just a few terms in use. As environment and health researchers grapple with the complexity of human environments, dialogue about different conceptualisations of what is meant by environment can be a powerful tool by which to open new avenues of inquiry. Second, a robust engagement by geographers with the complexities of scale fosters opportunities for investigating how multi-scalar interactions between socially constructed scales of human activity and the scales of biotic processes affect the prospects for human and environmental health. The current trend towards multi-scalar research designs such as those now common in vulnerability analyses is promising in this regard. Third, the increasing influence of ecology and ecological perspectives on both human and physical geographic research provides rich ground from which to make the linkages just noted. Geography, then, is poised to make significant and deeply relevant contributions to understanding the complexity of environmental conditions within and among places, and their implications for human health.

BIBLIOGRAPHY

Beck, U. (1986) *Risk Society: Towards a New Modernity*. London: SAGE.

Blaikie, P., Cannon, T., Davies, I. and Wisner, B. (1994) *At Risk: Natural Hazards, People's Vulnerability, and Disasters*. London: Routledge.

Bohle, H., Downing, T. and Watts, M. (1994) Climate change and social vulnerability-toward a sociology and geography of food security. *Global Environmental Change*, 4, 37–48.

Bolam, B., Murphy, S. and Gleeson, K. (2006) Place-identity and geographical inequalities in health: A qualitative study. *Psychology and Health*, 21(3), 399–420.

Bowen, M., Salling, M., Haynes, K. and Cyran, E. (1995) Toward environmental justice: spatial equity in Ohio and Cleveland. *Annals of the Association of American Geographers*, 85(4), 641–63.

Braun, B. (2005) Environmental issues: writing a more-than-human urban geography. *Progress in Human Geography*, 29(5), 635–50.

Brulle, R. and Pellow, D. (2006) Environmental justice: human health and environmental inequalities. *Annual Review of Public Health*, 27, 103–24.

Bunge, W. W. and Bordessa, R. (1975) *The Canadian Alternative: Survival, Expeditions, and Urban Change*. Geographical Monographs No. 2, Department of Geography, Atkinson College, York College. Toronto, Ontario.

Burton, I., Kates, R. W. and White, G. F. (1978) *The Environment as Hazard*. Oxford: Oxford Univ. Press.

Collins, A. (2002) Health ecology and environment management in Mozambique. *Health and Place*, 8, 263–72.

Cronon, W. (1991) *Nature's Metropolis: Chicago and the Great West*. New York: W.W. Norton.

Cutter, S. (1996) Vulnerability to environmental hazards. *Progress in Human Geography*, 20(40), 529–39.

Cutter, S. (2003) The vulnerability of science and the science of vulnerability. *Annals of the Association of American Geographers*, 93(1), 1–12.

Cutter, S. L., Mitchell, J. T. and Scott, M. S. (2000) Revealing the vulnerability of people and places: a case study of Georgetown County, South Carolina. *Annals of the Association of American Geographers*, 90(4), 713–37.

Cutter, S. and Solecki, W. (1996) Setting environmental justice in space and place: Acute and chronic airborne toxic releases in the southeastern United States. *Urban Geography*, 17(5), 380–99.

Davis, M. (1998) *Ecology of Fear: Los Angeles and the Imagination of Disaster*. New York: Metropolitan Books.

Diez-Roux, A. V. (2000) Multilevel analysis in public health research. *Annual Review of Public Health*, 21, 171–92.

Duncan, C., Jones, K. and Moon, G. (1993) Do places matter? A multi-level analysis of regional variations in health-related behaviour in Britain. *Social Science and Medicine*, 37, 725–33.

Dyck, I. (1999) Using qualitative methods in medical geography: a deconstructive moment in a subdiscipline? *The Professional Geographer*, 51(2), 243–53.

Elliott, S. (1999) And the question shall determine the method. *Professional Geographer*, 51(2), 240–43.

Frumkin, H. (2003) Health places: exploring the evidence. *American Journal of Public Health*, 93(9), 1451–56.

Gandy, M. (2002) *Concrete and clay: reworking nature in New York City*. Cambridge, MA: MIT Press.

Gee, G. and Payne-Sturges, D. C. (2004) Environmental health disparities: a framework integrating psychosocial and environmental concepts. *Environmental Health Perspectives*, 112, 1645–53.

Gottlieb, R. (1993) *Forcing the Spring: The Transformation of the American Environmental Movement*. Washington, DC: Island Press.

Greenberg, M. and Schneider, D. (1993) Violence in American cities: young black males is the answer, but what was the question? *Social Science and Medicine*, 39, 179–87.

Harlan, S. L., Brazel, A., Prashad, L., Stefanov, W. L. and Larsen, L. (2006) Neighborhood microclimates and vulnerability to heat stress. *Social Science and Medicine*, 63, 2847–63.

Harvey, D. (1985) *Justice, Nature and the Geography of Difference*. Oxford: Blackwell.

Harwood, S. (2005) Environmental justice on the streets: advocacy planning as a tool to contest environmental racism. *Journal of Planning Education and Research*, 23(1), 24–38.

Haughton, G. (1999) Environmental justice and the sustainable city. *Journal of Planning Education and Research*, 18, 233–43.

Haughton, G. and McGranahan, G. (2006) Editorial: urban ecologies. *Environment and Urbanization*, 18(1), 3–8.

Hewitt, K. (1983) *Interpretations of Calamity from the Viewpoint of Human Ecology*. Boston: Allen & Unwin.

Heynen, N., Perkins, H. A. and Roy P. (2006) The political ecology of uneven urban green space – the impact of political economy on race and ethnicity in producing environmental inequality in Milwaukee. *Urban Affairs Review*, 42(1), 3–25.

Ingram, J. C., Franco, G., Rumbaitis-del Rio, C. and Khazai, B. (2006) Post-disaster recovery dilemmas: challenges in balancing short-term and long-term needs for vulnerability reduction. *Environmental Science and Policy*, 9, 607–13.

Kates, R. W. and Kasperson, J. X. (1983) Comparative risk analysis of technological hazards: a review. *Proceedings of the National Academy of Science USA*, 80, 7027–38.

Lee, C. (2002) Environmental justice: building a unified vision of health and the environment. *Environmental Health Perspectives*, 110(2), 141–44.

Macintyre, S. (1997) What are spatial effects and how can we measure them? In A. Dale (ed.), *Exploiting National Survey Data: The Role of Locality and Spatial Effects*. Manchester: Faculty of Economic and Social Studies, University of Manchester, pp. 1–17.

Macintyre, S., Ellaway, A. and Cummins, S. (2002) Place effects on health: how can we conceptualise, operationalise and measure them? *Social Science and Medicine*, 55, 125–39.

Macintyre, S., Maciver, S. and Sooman, A. (1993) Area, class and health: should we be focusing on places or people? *Journal of Social Policy*, 22, 213–34.

Mayer J. D. (1996) The political ecology of disease as one new focus for medical geography. *Progress in Human Geography*, 20(4), 441–56.

Mayer, J. D. (2000) Geography, ecology and emerging infectious diseases. *Social Science and Medicine*, 50, 937–52.

McCarthy, J. (2002) First world political ecology: lessons from the Wise Use Movement. *Environment and Planning A*, 34, 1281–1302.

McCarthy, J. (2005) Guest editorial: first world political ecology: directions and challenges. *Environment and Planning A*, 37, 953–58.

Meade, M., Florin, J. and Gesler, W. (1988) *Medical Geography*. New York: Guilford Press, pp. 103–6.

Morland, K., Wing, S. and Diez Roux, A. (2002) The contextual effects of the local food environment on residents' diets: the atherosclerosis risk in communities study. *American Journal of Public Health*, 92(11), 1761–67.

Njeru, J. (2006) The urban political ecology of plastic bag waste problem in Nairobi, Kenya. *Geoforum*, 37(6), 1046–58.

Oakes, M. S. (2004) The (mis)estimation of neighborhood effects: causal inference for a practical social epidemiology. *Social Science and Medicine*, 58, 1929–52.

Palm, R. I. (1990) *Natural Hazards: An Integrative Framework for Research and Planning*. Baltimore: Johns Hopkins University Press.

Pellow, D. (2000) Environmental inequality formation: toward a theory of environmental injustice. *American Behavioral Scientist*, 43, 581–601.

Pickett, K. E. and Pearl, M. (2001) Multilevel analyses of neighborhood socioeconomic context and health outcomes: a critical review. *Journal of Epidemiology and Community Health*, 55, 111–22.

Pulido L. (1996) A critical review of the methodology of environmental racism research. *Antipode*, 28, 142–59.

Robbins, P. (2002) Obstacles to a first world political ecology: looking near without looking up. *Environment and Planning A*, 34, 1509–13.

Saha, R. and Mohai, P. (2005) Historical context and hazardous waste facility siting: understanding temporal patterns in Michigan. *Social Problems*, 52(4), 618–48.

Salehi, R. and Ali, S. (2006) The social and political context of disease outbreaks: the case of SARS in Toronto. *Canadian Public Policy – Analyse de Politiques*, 32(4), 373–85.

Sampson, R. J. (2001) How do communities undergird or undermine human development? Relevant contexts and social mechanisms. In A. Booth and N. Crouter (eds), *Does It Take a Village? Community Effects on Children, Adolescents, and Families*. Mahwah, NJ: L Erlbaum, pp. 3–30.

Sen, A. (1981) *Poverty and Famines*. Oxford: Oxford University Press.

Smoyer, K. E. (1998) Putting risk in its place: methodological considerations for investigating extreme event health risk. *Social Science and Medicine*, 47(11), 1809–24.

Smoyer-Tomic, K. E., Spence, J. C. and Amrhein, C. (2006) Food deserts in the prairies? Supermarket accessibility and neighborhood need in Edmonton, Canada. *The Professional Geographer*, 58(3), 307–26.

Swyngedouw, E. and Heynen, N. (2003) Urban political ecology, justice and the politics of scale. *Antipode*, 35(5), 898–918.

Towers, G. (2000) Applying the political geography of scale: grassroots strategies and environmental justice. *Professional Geographer*, 52(1), 23–36.

Turner, B. L., II, Kasperson, R. E., Matson, P. A., McCarthy, J. J., Corell, R. W., Christensen, L., Eckley, N., Kasperson, J. X., Luers, A., Martello, M. L., Polsky, C., Pulsipher, A. and Schiller, A. (2003) A framework for vulnerability analysis in sustainability science. *PNAS*, 100(14), 8074–79.

US NIEHS, (*n.d.*) U.S. National Institutes of Health, National Institute of Environmental Health Sciences Mission Statement. http://www.niehs.nih.gov/external/intro.htm (accessed 17-01-07).

Watts, M. J. and Bohle, H. G. (1993) The space of vulnerability: the causal structure of hunger and famine. *Progress in Human Geography*, 17(1), 43–67.

WHO, Regional Office for Europe (*n.d.*) *What Is Environmental Health?* http://www.euro.who.int/envhealth/20060609_1 (accessed 17-01-07).

Wilkinson, R. (1996) *Unhealthy Societies*. London: Routledge.

Williams, R. (1973) *The Country and the City*. New York: Oxford University Press.

Index